MW01717346

Intermediate Algebra

Washtenaw Community College Edition
McKeague

THOMSON

Australia • Canada • Mexico • Singapore • Spain • United Kingdom • United States

THOMSON

McKeague
Intermediate Algebra

Executive Editors:
Michele Baird, Maureen Staudt &
Michael Stranz

Project Development Manager:
Linda deStefano

Sr. Marketing Coordinators:
Lindsay Annett and Sara Mercurio

Production/Manufacturing Manager:
Donna M. Brown

Production Editorial Manager:
Dan Plofchan

Pre-Media Services Supervisor:
Becki Walker

Rights and Permissions Specialist:
Kalina Ingham Hintz

Cover Image
Getty Images*

Printed in the
United States of America
1 2 3 4 5 6 7 10 09 08 07

For more information, please contact Thomson Custom Solutions, 5191 Natorp Boulevard, Mason, OH 45040. Or you can visit our Internet site at www.thomsoncustom.com

The Adaptable Courseware Program consists of products and additions to existing Thomson products that are produced from camera-ready copy. Peer review, class testing, and accuracy are primarily the responsibility of the author(s).

Intermediate Algebra
McKeague

ISBN-13: 978-0-495-45595-0
ISBN-10: 0-495-45595-4

International Divisions List

Asia (Including India):
Thomson Learning
(a division of Thomson Asia Pte Ltd)
5 Shenton Way #01-01
UIC Building
Singapore 068808
Tel: (65) 6410-1200
Fax: (65) 6410-1208

Australia/New Zealand:
Thomson Learning Australia
102 Dodds Street
Southbank, Victoria 3006
Australia

Latin America:
Thomson Learning
Seneca 53
Colonia Polano
11560 Mexico, D.F., Mexico
Tel (525) 281-2906
Fax (525) 281-2656

Canada:
Thomson Nelson
1120 Birchmount Road
Toronto, Ontario
Canada M1K 5G4
Tel (416) 752-9100
Fax (416) 752-8102

UK/Europe/Middle East/Africa:
Thomson Learning
High Holborn House
50-51 Bedford Row
London, WC1R 4LS
United Kingdom
Tel 44 (020) 7067-2500
Fax 44 (020) 7067-2600

Spain (Includes Portugal):
Thomson Paraninfo
Calle Magallanes 25
28015 Madrid
España
Tel 34 (0)91 446-3350
Fax 34 (0)91 445-6218

*Unless otherwise noted, all cover images used by Thomson Custom Solutions have been supplied courtesy of Getty Images with the exception of the *Earthview* cover image, which has been supplied by the National Aeronautics and Space Administration (NASA).

Table of Contents

Preface to the Instructor

This seventh edition of *Intermediate Algebra: A Text/Workbook* retains the same basic format and style as the previous edition. The book can be used in either a traditional lecture class or in a self-paced situation. In a lecture-format class, each section of the book can be discussed in a 45- to 50-minute class session. In a self-paced situation, the Practice Problems in the margins allow the student to become actively involved with the material before working the problems in the Problem Set.

Features of the Book/New to This Edition

Practice Problems The Practice Problems, with their answers and solutions, are the key to moving students through the material in this book. We call it the EPAS system for Example, Practice Problem, Answer, Solution. Here is how it works: In the margin, next to each example, is a Practice Problem. Students begin by reading through the text, stopping after they have read through an example. Then they work the Practice Problem in the margin next to the example. When they are finished, they check the answers to the Practice Problem at the bottom of the page. If they have made a mistake, they try the problem again. If they still cannot get it right, they look at the Solutions to the Practice Problems in the back of the book. After working their way through the section in this manner, they are ready to start on the problems in the Problem Set. As an added bonus, there is a video lesson to accompany each section of the text, on videotape or CD-ROM. On the video I work some of the odd-numbered problems in the Problem Set.

Objectives **New to this edition** are learning objectives for each section of the text. They are listed in the margin at the beginning of each section. They are referred to again at the end of each section with a table that gives students the opportunity to match each example in the section with one of the learning objectives. Having students match examples with learning objectives requires them to think about how the specific problems they are working correlate with the more general learning objectives.

Expanded Review Problems **New to this edition** is an expanded treatment of the review problems found toward the end of the problem sets. The review problems begin in Chapter 2 and are divided into two categories: *Getting Ready for the Next Section* and *Maintaining Your Skills.* Under the *Getting Ready for the Next Section* heading are problems that students will see as they read through the examples and explanations in the next section. These problems are an easy way for students to see if they are prepared for the material they will see next. Students who consistently work these problems will be much better prepared for class than students who do not work these problems. The second category of review problems, the *Maintaining Your Skills* problems, review the most important concepts from the previous chapter or they review concepts that form the foundation of the course. These are the problems we expect students to know when they finish the course. In most cases we review these problems more than once.

Early Coverage of Graphing The material on graphing equations in two variables begins in Chapter 3, where it is combined with an introduction to functions and function notation.

Early Coverage of Functions The concept of a function is used more throughout this edition than the previous edition. It is introduced in Chapter 3 and then used throughout the remainder of the book, particularly in the chapters on rational functions, quadratic functions, and exponential and logarithmic functions. As a corollary to this, a concerted effort was made to use function notation whenever possible and appropriate in this new edition.

Using Technology Scattered throughout the book is new material that shows how graphing calculators, spreadsheet programs, and computer graphing programs can be used to enhance the topics being covered. This material is easy to find because it appears under the heading "Using Technology."

Blueprint for Problem Solving The Blueprint for Problem Solving is a detailed outline of the steps needed to successfully attempt application problems. Intended as a guide to problem solving in general, the Blueprint overlays the solution process to the application problems in the book.

Facts From Geometry Many of the important facts from geometry are listed under this heading. In most cases, an example or two accompanies each of the facts to give students a chance to see how topics from geometry are related to the algebra they are learning.

Number Sequences and Inductive Reasoning An introductory coverage of number sequences and inductive reasoning is contained in Section 1.5. This material is expanded upon in Chapter 2, and then carried through to other parts of the book. I find that there are many interesting topics I can cover if students have some experience with number sequences. It is also the easiest way to demonstrate inductive reasoning.

Unit Analysis Chapter 6 contains problems requiring students to convert from one unit of measure to another. The method used to accomplish the conversions is the method they will use if they take a chemistry class. Since this method is similar to the method we use to multiply rational expressions, unit analysis is covered in Section 6.7 as an application of rational expressions.

Chapter Openings Each chapter opens with an introduction in which a real-world application is used to stimulate interest in the chapter. These opening applications are expanded on later in the chapter. Following the opening application are the following items:

> **Study Skills** These appear in the first six chapters to help students become organized and efficient with their time in order to point them in the direction of success with the course.
>
> **Getting Ready for Chapter X** is a set of problems from previous chapters that students need in order to be successful in the current chapter. These are review problems intended to reinforce the idea that all topics in the course are built on previous topics.

Getting Ready for Class Just before each problem set is a list of four questions under the heading *Getting Ready for Class.* These problems require written responses from students and are to be done before students come to class. The answers can be found by reading the preceding section. These questions reinforce the importance of reading the section before coming to class.

Quantitative Literacy Problems **New to this edition** are problems that address quantitative literacy. They appear at the end of a number of problem sets, under the heading *Improving Your Quantitative Literacy.*

Organization of the Problem Sets The following features are integrated into the problem sets.

1. ***Drill*** There are enough problems in each set to ensure student proficiency in the material.
2. ***Progressive difficulty*** The problems increase in difficulty as the problem set progresses.
3. ***Odd-even similarities*** Each pair of consecutive problems is similar. Since the answers to the odd-numbered problems are listed in the back of the book, the similarity of the odd-even pairs of problems allows your students to check their work on an odd-numbered problem and then to try the similar even-numbered problem.
4. ***Applying the Concepts problems*** Students are always curious about how the algebra they are learning can be applied, but at the same time many of them are apprehensive about attempting application problems. I have found that they are more likely to put some time and effort into trying application problems if they do not have to work an overwhelming number of them at one time and if they work on them every day. For these reasons, I have placed a few application problems in almost every problem set in the book. Many of these application problem are new to this edition, and we have made a considerable effort to use real data whenever possible.
5. ***Review problems*** As we mentioned earlier, the review problems in this edition have been expanded. See the earlier description in this preface for details.
6. ***Extending the Concepts problems*** Many of the problem sets end with a few problems under this heading. These problems are more challenging than those in the problem set are, or they are problems that extend some of the topics covered in the section.
7. ***Improving Your Quantitative Literacy*** These are problems that address specific issues in quantitative literacy, along the guidelines given by AMATYC and MAA.

End-of-Chapter Retrospective Each chapter ends with the following items that together give a comprehensive reexamination of the chapter.

1. ***Chapter Summary*** Lists all main points from the chapter. In the margin, next to each topic, is an example that illustrates the type of problem associated with the topics being reviewed.
2. ***Chapter Review/Test*** Following the Chapter Summary in each chapter is the Chapter Review/Test. It contains an extensive set of problems that review all the main topics in the chapter. The odd-numbered problems in the Chapter Review/Test can be used as a chapter test. The same is true of the even-numbered problems.
3. ***Chapter Projects*** One group project, for students to work on in class, and one research project for students to do outside of class.

Supplements to the Textbook

Available to qualified adopters. Please consult your local sales representative for details.

For the Instructor

Annotated Instructor's Edition This special version of the complete student text contains a Resource Integration Guide that correlates the supplementary items for this text. With this easy-to-use tool, you can quickly compile a teaching and learning program that complements both the text and your personal teaching style. Answers are also printed next to all respective exercises. Graphs, tables, and other answers appear in a special answer section in the back of the text.

Test Bank The test bank includes eight tests per chapter as well as three final exams. The tests are made up of a combination of multiple-choice, free-response, true/false, and fill-in-the-blank questions.

Complete Solutions Manual The complete solutions manual provides worked-out solutions to all of the problems in the text.

iLrn™ Instructor Version Efficient and versatile, **iLrn** gives you the power to transform the teaching and learning experience. **iLrn Instructor Version** is made up of two components, **iLrn Testing** and **iLrn Tutorial. iLrn Testing** is an Internet-ready, text-specific testing suite that allows instructors to customize exams and track student progress in an accessible, browser-based format. **iLrn** offers full algorithmic generation of problems and free-response mathematics. **iLrn Tutorial** is a text-specific, interactive tutorial software program that is delivered via the web (at http://www.iLrn.com) and is offered in both student and instructor versions. Like **iLrn Testing,** it is browser-based, making it an intuitive mathematical guide even for students with little technological proficiency. **iLrn Tutorial** allows students to work with real math notation in real time, providing instant analysis and feedback. The tracking program built into the instructor version of the software enables instructors to carefully monitor student progress. The complete integration of the testing, tutorial, and course management components simplifies your routine tasks. Results flow automatically to your grade book, and you can easily communicate with individuals, sections, or entire courses.

For the Student

A Digital Video Companion (DVC) containing MathCue This single CD-ROM containing over 6 hours of video instruction is packaged with each book. The foundation of the DVC is the video lessons. There is one video lesson for each section of the book. Each video lesson is 8 to 12 minutes long. The problems worked during each video lesson are listed next to the viewing screen so that students can work them ahead of time, if they choose.

Integrated on the DVC is MathCue tutorial and testing software. MathCue is keyed to the text and includes:

MathCue Skill Builder Presents problems to solve, evaluates answers, and tutors students by displaying complete solutions with step-by-step explanations. Students may view partial solutions to get started on a problem, see a continuous record of progress, and back up to review missed problems.

MathCue Quiz Allows students to generate large numbers of quiz problems keyed to problem types from each section of the book. Students work default section quizzes or customize sessions to include problems from one or more sections and/or chapters as desired. Students receive immediate feedback after entering their answers.

MathCue Chapter Test Also provides large numbers of problems keyed to problem types from each chapter. Students work default chapter tests or customize sessions to include problems from one or more chapters. Feedback is given only after the test is completed.

Score reports for any MathCue session can be printed and handed in for credit or extra credit.

The DVC also includes **MathCue Solution Finder.** This unique tool allows students to enter their own basic problems and receive step-by-step help as if they were working with a tutor.

Text-Specific Videos These text-specific videotape sets, available at no charge to qualified adopters of the text, feature 10- to 20-minute problem-solving lessons that cover each section of every chapter. On the videotapes I work selected problems from the text.

iLrn™ Tutorial Student Version Automatic access to this text-specific, interactive, web-based tutorial system is included with the text. **iLrn Tutorial Student Version** is browser-based, making it an intuitive mathematical guide even for students with little technological proficiency. Simple to use, **iLrn Tutorial** allows students to work with real math notation in real time, providing instant analysis and feedback. The entire textbook is available in PDF format through **iLrn Tutorial,** as are section-specific video tutorials, unlimited practice problems, and additional student resources such as a glossary, web links, and more. And, when students get stuck on a particular problem or concept, they need only log on to **vMentor™,** accessed through **iLrn Tutorial,** where they can talk (using their own computer microphones) to **vMentor** tutors who will skillfully guide them through the problem using an interactive whiteboard for illustration.

Student's Solutions Manual This manual contains complete annotated solutions to all odd problems in the problem sets (except answers to "Improving Your Quantitative Literacy" problems where correct answers may vary) and all chapter review and chapter test exercises.

Thomson Brooks/Cole Mathematics Website

http://www.thomsonedu.com/math. When you adopt a Thomson Brooks/Cole mathematics text, you and your students will have access to everything from book-specific resources to newsgroups. It's a great way to make teaching and learning an interactive and intriguing experience.

Acknowledgments

A project of this size cannot be handled by one person; I had lots of help. Jennifer Laugier and John-Paul Ramin, my editors at Brooks/Cole, were involved in all parts of the revision process. Cheryll Linthicum of Brooks/Cole and Carol O'Connell of Graphic World Publishing Services turned the manuscript into a book. Tammy Fisher-Vasta and Patrick McKeague assisted me with all parts of this revision, from manuscript preparation to proofreading page proofs, and finally preparing the index. Ann Ostberg and Sudhir Goel checked all the problems in all the problem sets. Shane McKeague, Devin Christ, Erin Williams, and Janai Wallace assisted with error checking and proofreading. Liz West, Cuesta College, along with Mary Gentilucci and Annie Stephens checked the manuscript for errors. Edie Carter, Amarillo College, helped create many of the Quantitative Literacy problems that are new in this edition. In the office at Brooks/Cole, Rebecca Subity and Sarah Woicicki worked very hard to keep everything on schedule. Kathleen Olson spent many hours on photo research. And finally, Ross Rueger produced the solutions manuals that accompany the book. My thanks to all these people; this book would not have been possible without them.

Thanks also to Diane McKeague and Amy Jacobs for their encouragement with all my writing endeavors.

Finally, I am grateful to the following instructors for their suggestions and comments on this revision. Some reviewed the entire manuscript, while others were asked to evaluate the development of specific topics, or the overall sequence of topics. My thanks go to the following people:

Joye Elaine Gowan, Roane State Community College
Kevin Ratliff, Blue Ridge Community College
Kathryn Gundersen, Three Rivers Community College
Angelito Garcia, Moraine Valley Community College
Judy Carney, West Liberty State College
Annette Squires, Palomar College
Anne Franklin, Pima Community College
Harry McLaughlin, Montclair State University
Ali A. Jafarian, University of New Haven
Joseph Jordan, John Tyler Community College
James Namekata, Riverside Community College
Donna Bernardy, Lane Community College
David L. Schultz, Virginia Western Community College
Lynette King, Gadsden State Community College
Karin Chess, Owensboro Community College
Trudy Ann Bostick, Shawnee State University
Patricia Roux, Delgado Community College
Mustafa Hamdan, Wichita State University
Janna Liberant, SUNY/Rockland Community College
John F. Close, Salt Lake Community College
Maryanne Kirkpatrick, Laramie County Community College-Albany County Campus
Christopher Cox, Illinois Central College

Charles P. McKeague

Preface to the Student

I often find my students asking themselves the question "Why can't I understand this stuff the first time?" The answer is "You're not expected to." Learning a topic in mathematics isn't always accomplished the first time around. There are many instances when you will find yourself reading over new material a number of times before you can begin to work problems. That's just the way things are in mathematics. If you don't understand a topic the first time you see it, that doesn't mean there is something wrong with you. Understanding mathematics takes time. The process of understanding requires reading the book, studying the examples, working problems, and getting your questions answered.

How to Be Successful in Mathematics

1. **If you are in a lecture class, be sure to attend all class sessions on time.** You cannot know exactly what goes on in class unless you are there. Missing class and then expecting to find out what went on from someone else is not the same as being there yourself.
2. **Read the book and work the Practice Problems.** It is best to read the section that will be covered in class beforehand. Reading in advance, even if you do not understand everything you read, is still better than going to class with no idea of what will be discussed. As you read through each section, be sure to work the Practice Problems in the margin of the text. Each Practice Problem is similar to the example with the same number. Look over the example and then try the corresponding Practice Problem. The answers to the Practice Problems are at the bottom of the page. If you don't get the correct answer, see if you can rework the problem correctly. If you miss it a second time, check your solution with the solution to the Practice Problem in the back of the book.
3. **Work problems every day and check your answers.** The key to success in mathematics is working problems. The more problems you work, the better you will become at working them. The answers to the odd-numbered problems are given in the back of the book. When you have finished an assignment, be sure to compare your answers with those in the book. If you have made a mistake, find out what it is, and correct it.
4. **Do it on your own.** Don't be misled into thinking someone else's work is your own. Having someone else show you how to work a problem is not the same as working the same problem yourself. It is okay to get help when you are stuck. As a matter of fact, it is a good idea. Just be sure you do the work yourself.
5. **Review every day.** After you have finished the problems your instructor has assigned, take another 15 minutes and review a section you have already completed. The more you review, the longer you will retain the material you have learned.

6. **Don't expect to understand every new topic the first time you see it.** Sometimes you will understand everything you are doing, and sometimes you won't. That's just the way things are in mathematics. Expecting to understand each new topic the first time you see it can lead to disappointment and frustration. The process of understanding takes time. It requires that you read the book, work problems, and get your questions answered.
7. **Spend as much time as it takes for you to master the material.** No set formula exists for the exact amount of time you need to spend on mathematics to master it. You will find out as you go along what is or isn't enough time for you. If you end up spending 2 or more hours on each section in order to master the material there, then that's how much time it takes; trying to get by with less will not work.
8. **Relax.** It's probably not as difficult as you think.

iLrn Tutorial Quick Start Guide

iLrn Can Help You Succeed in Math

iLrn is an online program that facilitates math learning by providing resources and practice to help you succeed in your math course. Your instructor chose to use *iLrn* because it provides online opportunities for learning (Explanations found by clicking **Read Book**), practice (Exercises), evaluating (Quizzes), Personalized Learning Plans **(MathNOW),** and complete online lessons **(TLE Labs).** It also gives you a way to keep track of your own progress and manage your assignments.

The mathematical notation in *iLrn* is the same as that you see in your textbooks, in class, and when using other math tools like a graphing calculator. *iLrn* can also help you run calculations, plot graphs, enter expressions, and grasp difficult concepts. You will encounter various problem types as you work through *iLrn,* all of which are designed to strengthen your skills and engage you in learning in different ways.

Logging in to iLrn

Registering with the Access Code on the 1pass Card *Situation:* Your instructor has not given you an access code for an online course, but you have a textbook with a 1pass access code.

Initial Log-in

1. Go to **http://www.thomsonedu.com/math.**
2. In the menu at the right, click on **Create My Account.**
3. Enter your information in the **Account Information** fields on the left-hand side of the screen. Locate the 1pass card that came with your textbook and note the 1pass Access Code printed on the card. Enter this code into the **1pass Access Code** field.
4. Select McKeague's *Intermediate Algebra: A Text/Workbook, 7th Edition* on your **Bookshelf.**
5. Click on the MathNOW product on the right-hand side of the screen to enter iLrn, and begin using the personalized learning plan as well as the tutorials.

You will be asked to select a user name and password. Save your user name and password in a safe place. You will need them to log in the next time you use 1pass or iLrn. Only your user name and password will allow you to reenter 1pass or iLrn.

Subsequent Log-in

1. Go to **http://www.thomsonedu.com/math.**
2. Click on **Sign In** on the right-hand side of the page.
3. Type your user name and password (see boxed information above); then click on **Submit.** Click on the MathNOW product on the right-hand side of the screen to enter iLrn and begin using the personalized learning plan as well as the tutorials.

Navigating through iLrn

To navigate between chapters and sections, use the drop-down menu below the top navigation bar. This will give you access to the study activities available for each section.

The view of a tutorial in iLrn looks like this.

Math Toolbar

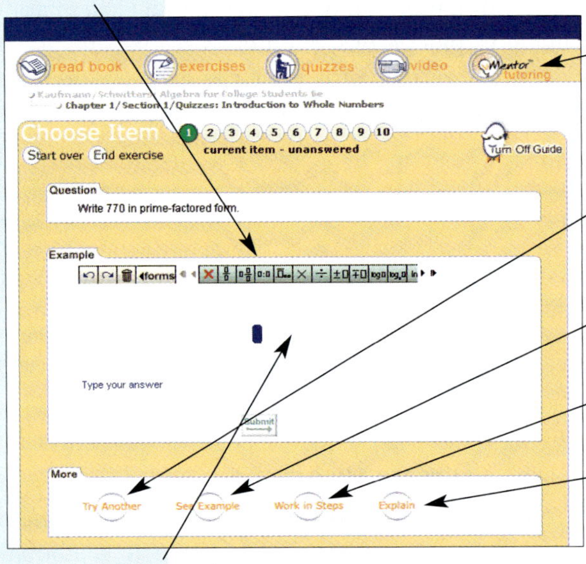

vMentor: Live online tutoring is only a click away. Tutors can take screen shots of your book and lead you through a problem with voice-over and visual aids.

Try Another: Click here to have iLrn create a new question or a new set of problems.

See Examples: Preworked examples provide you with additional help.

Work in Steps: iLrn can guide you through a problem step-by-step.

Explain: Additional explanation from your book can help you with a problem.

Type your answer here.

Online Tutoring with vMentor™

Access to iLrn also means access to online tutors and support through vMentor, which provides live homework help and tutorials. To access vMentor while you are working in the Exercises or "Tutorial" areas in iLrn, click on the **vMentor Tutoring** button at the top right of the navigation bar above the problem or exercise.

Next, click on the **vMentor** button; you will be taken to a web page that lists the steps for entering a vMentor classroom. If you are a first-time user of vMentor, you might need to download Java software before entering the class for the first class. You can either take an Orientation Session or log in to a vClass from the links at the bottom of the opening screen.

All vMentor Tutoring is done through a vClass, an Internet-based virtual classroom that features two-way audio, a shared whiteboard, chat, messaging, and experienced tutors.

You can access vMentor Monday through Thursday, as follows:

5 p.m. to 12 midnight Pacific Time
6 p.m. to 1 a.m. Mountain Time
7 p.m. to 2 a.m. Central Time
8 p.m. to 3 a.m. Eastern Time

You can access vMentor Sunday, as follows:

12 noon to 12 midnight Pacific Time
1 p.m. to 1 a.m. Mountain Time
2 p.m. to 2 a.m. Central Time
3 p.m. to 3 a.m. Eastern Time

If you need additional help using vMentor, you can access Technical Support at this website: **http://www.elluminate.com/support/index.jsp.**

Interact with TLE Online Labs

Found within the personalized learning plan **(MathNOW),** the **TLE Labs** can be used to explore and reinforce key concepts introduced in this text. These electronic labs give you access to additional instruction and practice problems, so you can explore each concept interactively, at your own pace. Not only will you be better prepared, but you will perform better in the class overall. You may access these labs through **MathNOW,** using the **1pass** login instructions above.

Intermediate Algebra

1

David Young-Wolff/PhotoEdit

CHAPTER OUTLINE

INTERMEDIATE Algebra f(x) Now™

Basic Properties and Definitions

The following table and diagram show how the concentration of a popular antidepressant changes over time once the patient stops taking it. In this particular case, the concentration in the patient's system is 80 ng/mL (nanograms per milliliter) when the patient stops taking the antidepressant, and the half-life of the antidepressant is 5 days.

CONCENTRATION OF AN ANTIDEPRESSANT

DAYS SINCE DISCONTINUING	CONCENTRATION (ng/mL)
0	80
5	40
10	20
15	10
20	5

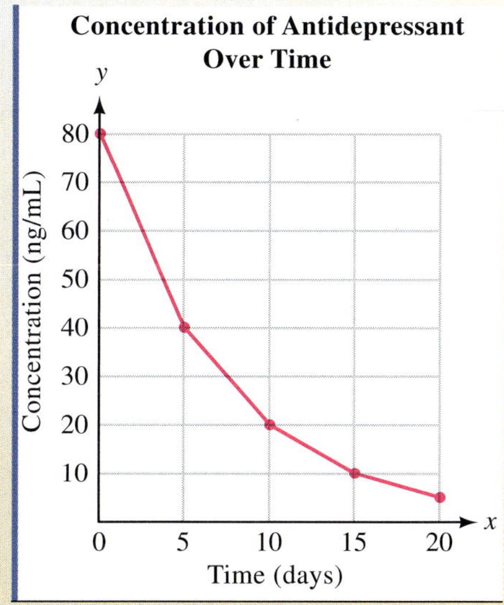

FIGURE 1

The half-life of a medication tells how quickly the medication is eliminated from a person's system: Medications with a long half-life are eliminated slowly, whereas those with a short half-life are more quickly eliminated. Half-life is the key to constructing the preceding table and graph. When you are finished with this chapter, you will be able to use the half-life of a medication to construct the table and graph.

Visit our web site at http://www.brookscole.com

Study Skills

Some of the students enrolled in my beginning algebra classes develop difficulties early in the course. Their difficulties are not associated with their ability to learn mathematics; they all have the potential to pass the course. Students who get off to a poor start do so because they have not developed the study skills necessary to be successful in algebra.

Here is a list of things you can do to begin to develop effective study skills.

1 Put Yourself on a Schedule

The general rule is that you spend 2 hours on homework for every hour you are in class. Make a schedule for yourself in which you set aside 2 hours each day to work on algebra. Once you make the schedule, stick to it. Don't just complete your assignments and stop. Use all the time you have set aside. If you complete an assignment and have time left over, read the next section in the book and work more problems.

2 Find Your Mistakes and Correct Them

There is more to studying algebra than just working problems. You must always check your answers with the answers in the back of the book. When you make a mistake, find out what it is and correct it. Making mistakes is part of the process of learning mathematics.

In the Prologue to *The Book of Squares,* Leonardo Fibonacci (ca. 1170–ca. 1250) has this to say about the content of his book:

> I have come to request indulgence if in any place it contains something more or less than right or necessary; for to remember everything and be mistaken in nothing is divine rather than human . . .

Fibonacci knew, as you know, that human beings make mistakes. You cannot learn algebra without making mistakes.

3 Gather Information on Available Resources

You need to anticipate that you will need extra help sometime during the course. There is a form to fill out in Appendix A to help you gather information on resources available to you. One resource is your instructor; you need to know your instructor's office hours and where the office is located. Another resource is the math lab or study center, if they are available at your school. It also helps to have the phone numbers of other students in the class, in case you miss class. You want to anticipate that you will need these resources, so now is the time to gather them together.

Getting Ready for Chapter 1

To get started in this book, we assume that you can do simple addition and multiplication problems with whole numbers and decimals. To check to see that you are ready for this chapter, work each of the problems below without using a calculator.

1. $5.7 - 1.9$ **2.** $10 - 9.5$ **3.** $0.3(6)$ **4.** $39 \div 13$ **5.** $28 \div 7$

6. $125 - 81$ **7.** $630 \div 63$ **8.** $210 \div 10$ **9.** $2 \cdot 3 \cdot 5 \cdot 7$ **10.** $3 \cdot 7 \cdot 11$

11. $2 \cdot 3 \cdot 3 \cdot 5 \cdot 7$ **12.** $18{,}000 - 4{,}500$ **13.** $1.6 + 2.0$ **14.** $7(0.05)$ **15.** $7{,}546 \div 35$

16. $1.2052 - 1$ **17.** $250(3.14)$ **18.** $109 - 36 + 14$ **19.** $0.08(4{,}000)$ **20.** $3.9 \div 1.3$

Fundamental Definitions and Notation 1.1

The diagram below is called a *bar chart.* This one shows the net price of a popular intermediate algebra textbook. (The net price is the price the bookstore pays for the book.)

OBJECTIVES	
A	Translate expressions written in English into algebraic expressions.
B	Expand and multiply numbers raised to positive integer exponents.
C	Simplify expressions using the rule for order of operations.
D	Find the value of an expression.
E	Find the union and intersection of two sets.

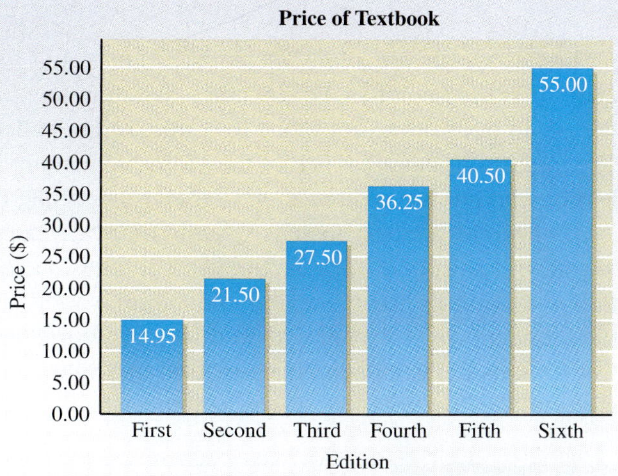

From the chart, we can find many relationships between numbers. We may notice that the price of the third edition was less than the price of the fourth edition. In mathematics we use symbols to represent relationships between quantities. If we let P represent the price of the book, then the relationship just mentioned, between the price of the third edition and the price of the fourth edition, can be written this way:

$$P(3) < P(4)$$

This section is, for the most part, simply a list of many of the basic symbols and definitions we will be using throughout the book.

Comparison Symbols

In Symbols	*In Words*
$a = b$	*a* is equal to *b*
$a \neq b$	*a* is not equal to *b*
$a < b$	*a* is less than *b*
$a \leq b$	*a* is less than or equal to b
$a \nless b$	*a* is not less than *b*
$a > b$	*a* is greater than *b*
$a \geq b$	*a* is greater than or equal to *b*
$a \ngtr b$	*a* is not greater than *b*
$a \Leftrightarrow b$	*a* is equivalent to *b*

Operation Symbols

Operation	*In Symbols*	*In Words*
Addition	$a + b$	The sum of a and b
Subtraction	$a - b$	The difference of a and b
Multiplication	ab, $a \cdot b$, $a(b)$, $(a)(b)$, or $(a)b$	The product of a and b
Division	$a \div b$, a/b, or $\frac{a}{b}$	The quotient of a and b

NOTE

For each example in the text there is a corresponding practice problem in the margin. After you read through an example in the text, work the practice problem with the same number in the margin.

The key words are *sum, difference, product,* and *quotient.* They are used frequently in mathematics. For instance, we may say the product of 3 and 4 is 12. We mean both the expressions $3 \cdot 4$ and 12 are called the products of 3 and 4. The important idea here is that the word *product* implies multiplication, regardless of whether it is written $3 \cdot 4$, 12, 3(4), (3)4, or (3)(4).

When we let a letter, such as x, stand for a number or group of numbers, then we say x is a *variable* because the value it takes on can vary. In the lists of notation and symbols just presented, the letters a and b are variables that represent the numbers we will work with in this book. In the next example we show some translations between sentences written in English and their equivalent expressions in algebra. Note that all the sentences contain at least one variable.

PRACTICE PROBLEMS

1. Write a statement, using symbols, that is equivalent to each English statement.
   - **a.** The sum of x and 2 is greater than 5.
   - **b.** The product of 2 and y is 18.
   - **c.** The quotient of a and 3 is 7.
   - **d.** Twice the product of y and z is less than 2.
   - **e.** The product of twice y and z is less than 2.

EXAMPLE 1

In English	*In Symbols*
The sum of x and 5 is less than 2.	$x + 5 < 2$
The product of 3 and x is 21.	$3x = 21$
The quotient of y and 6 is 4.	$\frac{y}{6} = 4$
Twice the difference of b and 7 is greater than 5.	$2(b - 7) > 5$
The difference of twice b and 7 is greater than 5.	$2b - 7 > 5$

Exponents

Consider the expression 3^4. The 3 is called the *base* and the 4 is called the *exponent.* The exponent 4 tells us the number of times the base appears in the product. That is:

$$3^4 = 3 \cdot 3 \cdot 3 \cdot 3 = 81$$

The expression 3^4 is said to be in *exponential form*, whereas $3 \cdot 3 \cdot 3 \cdot 3$ is said to be in *expanded form*.

Expand and multiply.

2. 4^2
3. 2^4
4. 3^3

EXAMPLES Expand and multiply.

2. $5^2 = 5 \cdot 5 = 25$ Base 5, exponent 2
3. $2^5 = 2 \cdot 2 \cdot 2 \cdot 2 \cdot 2 = 32$ Base 2, exponent 5
4. $4^3 = 4 \cdot 4 \cdot 4 = 64$ Base 4, exponent 3

Order of Operations

It is important when evaluating arithmetic expressions in mathematics that each expression have only one answer in reduced form. Consider the expression

$$3 \cdot 7 + 2$$

Answers

1. **a.** $x + 2 > 5$ **b.** $2y = 18$ **c.** $\frac{a}{3} = 7$ **d.** $2(yz) < 2$ **e.** $(2y)z < 2$
2. 16 **3.** 16 **4.** 27

If we find the product of 3 and 7 first, then add 2, the answer is 23. On the other hand, if we first combine the 7 and 2, then multiply by 3, we have 27. The problem seems to have two distinct answers depending on whether we multiply first or add first. To avoid this situation, we will decide that multiplication in a situation like this will always be done before addition. In this case, only the first answer, 23, is correct.

The complete set of rules for evaluating expressions follows.

Rule (Order of Operations)

When evaluating a mathematical expression, we will perform the operations in the following order, beginning with the expression in the innermost parentheses or brackets and working our way out.

1. Simplify all numbers with exponents, working from left to right if more than one of these expressions is present.
2. Then, do all multiplications and divisions left to right.
3. Perform all additions and subtractions left to right.

NOTE

This rule is very important. We will use it many times throughout the book. It is a simple rule to follow. First we evaluate any numbers with exponents; then we multiply and divide; and finally we add and subtract, always working from left to right when more than one of the same operation symbol occurs in a problem.

EXAMPLES Simplify each expression using the rule for order of operations.

5. $5 + 3(2 + 4) = 5 + 3(6)$ **Simplify inside parentheses**
$= 5 + 18$ **Then, multiply**
$= 23$ **Add**

6. $5 \cdot 2^3 - 4 \cdot 3^2 = 5 \cdot 8 - 4 \cdot 9$ **Simplify exponents left to right**
$= 40 - 36$ **Multiply left to right**
$= 4$ **Subtract**

7. $20 - (2 \cdot 5^2 - 30) = 20 - (2 \cdot 25 - 30)$
$= 20 - (50 - 30)$
$= 20 - (20)$
$= 0$
Simplify inside parentheses, evaluating exponents first, then multiplying, and finally subtracting

8. $40 - 20 \div 5 + 8 = 40 - 4 + 8$ **Divide first**
$= 36 + 8$
$= 44$
Then, add and subtract left to right

9. $2 + 4[5 + (3 \cdot 2 - 2)] = 2 + 4[5 + (6 - 2)]$
$= 2 + 4(5 + 4)$
$= 2 + 4(9)$
Simplify inside inner-most parentheses
$= 2 + 36$ **Then, multiply**
$= 38$ **Add**

Simplify each expression using the rule for order of operations.

5. $6 + 2(3 + 4)$

6. $5 \cdot 3^2 - 2 \cdot 4^2$

7. $30 - (2 \cdot 3^2 - 8)$

8. $60 + 20 \div 2 - 40$

9. $3 + 5[2 + (7 \cdot 2 - 10)]$

Finding the Value of an Algebraic Expression

An algebraic *expression* is a combination of numbers, variables, and operation symbols. For example, each of the following is an algebraic expression

$$5x \qquad x^2 + 5 \qquad 4a^2b^3 \qquad 5t^2 - 6t + 3$$

An expression such as $x^2 + 5$ will take on different values depending on what number we substitute for x. For example:

When $x = 3$ the expression $x^2 + 5$ becomes $3^2 + 5 = 9 + 5 = 14$

and When $x = 7$ the expression $x^2 + 5$ becomes $7^2 + 5 = 49 + 5 = 54$

Answers
5. 20 **6.** 13 **7.** 20 **8.** 30 **9.** 33

As you can see, our expression is 14 when x is 3, and when x is 7, our expression is 54. In each case the value of the expression is found by replacing the variable with a number.

10. Evaluate $(a + 4)^2$ when $a = 0, 1, 2, 3, 4$

EXAMPLE 10 Evaluate the expression $a^2 + 8a + 16$ when a is 0, 1, 2, 3, and 4.

Solution We can organize our work efficiently by using a table.

WHEN a IS	THE EXPRESSION $a^2 + 8a + 16$ BECOMES
0	$0^2 + 8 \cdot 0 + 16 = 0 + 0 + 16 = 16$
1	$1^2 + 8 \cdot 1 + 16 = 1 + 8 + 16 = 25$
2	$2^2 + 8 \cdot 2 + 16 = 4 + 16 + 16 = 36$
3	$3^2 + 8 \cdot 3 + 16 = 9 + 24 + 16 = 49$
4	$4^2 + 8 \cdot 4 + 16 = 16 + 32 + 16 = 64$

11. Evaluate $2x + 7y - 5$ when x is 4 and y is 1.

EXAMPLE 11 Find the value of the expression $3x + 4y + 5$ when x is 6 and y is 7.

Solution We substitute the given values of x and y into the expression and simplify the result:

When $x = 6$ and $y = 7$
the expression $3x + 4y + 5$
becomes $3 \cdot 6 + 4 \cdot 7 + 5 = 18 + 28 + 5$
$= 51$

The next concept we will cover, that of a set, can be considered the starting point for all the branches of mathematics.

Sets

DEFINITION

A **set** is a collection of objects or things. The objects in the set are called **elements,** or **members,** of the set.

Sets are usually denoted by capital letters, and elements of sets are denoted by lowercase letters. We use braces, { }, to enclose the elements of a set.

To show that an element is contained in a set we use the symbol $\in$. That is,

$x \in A$ is read "x is an element (member) of set A"

For example, if A is the set $\{1, 2, 3\}$, then $2 \in A$. However, $5 \notin A$ means 5 is not an element of set A.

DEFINITION

Set A is a **subset** of set B, written $A \subset B$, if every element in A is also an element of B. That is:

$A \subset B$ if and only if A is contained in B

Answers
10. 16, 25, 36, 49, 64
11. 10

EXAMPLES

12. The set of numbers used to count things is $\{1, 2, 3, \ldots\}$. The dots mean the set continues indefinitely in the same manner. This is an example of an *infinite set*.

13. The set of all numbers represented by the dots on the faces of a regular die is $\{1, 2, 3, 4, 5, 6\}$. This set is a subset of the set in Example 12. It is an example of a *finite set* because it has a limited number of elements. ■

Identify each statement as either true or false.

12. The set $\{2, 4, 6, \ldots\}$ is an example of an infinite set.

13. The set $\{2, 4, 8\}$ is a subset of $\{2, 4, 6, \ldots\}$.

DEFINITION

The set with no members is called the **empty, or null, set.** It is denoted by the symbol $\varnothing$. The empty set is considered a subset of every set.

Operations with Sets

Two basic operations are used to combine sets: union and intersection.

The diagrams shown here are called *Venn diagrams* after John Venn (1834–1923). They can be used to visualize operations with sets. The region inside the circle labeled A is set A; the region inside the circle labeled B is set B.

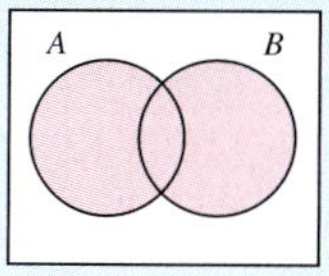

$A \cup B$

FIGURE 1 **The union of two sets**

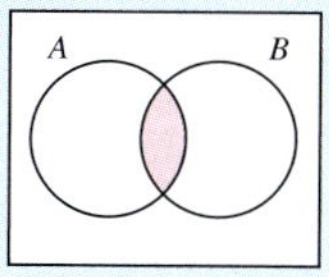

$A \cap B$

FIGURE 2 **The intersection of two sets**

DEFINITION

The **union** of two sets A and B, written $A \cup B$, is the set of all elements that are either in A or in B, or in both A and B. The key word here is *or.* For an element to be in $A \cup B$ it must be in A or B. In symbols, the definition looks like this:

$$x \in A \cup B \quad \text{if and only if} \quad x \in A \text{ or } x \in B$$

DEFINITION

The **intersection** of two sets A and B, written $A \cap B$, is the set of elements in both A and B. The key word in this definition is the word *and.* For an element to be in $A \cap B$ it must be in both A and B, or

$$x \in A \cap B \quad \text{if and only if} \quad x \in A \text{ and } x \in B$$

EXAMPLES

Let $A = \{1, 3, 5\}$, $B = \{0, 2, 4\}$, and $C = \{1, 2, 3, \ldots\}$. Then

14. $A \cup B = \{0, 1, 2, 3, 4, 5\}$

15. $A \cap B = \varnothing$ (A and B have no elements in common.)

16. $A \cap C = \{1, 3, 5\} = A$

17. $B \cup C = \{0, 1, 2, 3, \ldots\}$ ■

Let $A = \{1, 2, 3\}$, $B = \{0, 3, 7\}$, and $C = \{0, 1, 2, 3, \ldots\}$. Find the following.

14. $A \cup B$

15. $A \cap B$

16. $A \cap C$

17. $B \cap C$

Another notation we can use to describe sets is called *set-builder* notation. Here is how we write our definition for the union of two sets A and B using set-builder notation:

$$A \cup B = \{x \mid x \in A \text{ or } x \in B\}$$

The right side of this statement is read "the set of all x such that x is a member of A or x is a member of B." As you can see, the vertical line after the first x is read "such that."

EXAMPLE 18

If $A = \{1, 2, 3, 4, 5, 6\}$, find $C = \{x \mid x \in A \text{ and } x \geq 4\}$.

Solution We are looking for all the elements of A that are also greater than or equal to 4. They are 4, 5, and 6. Using set notation, we have

$$C = \{4, 5, 6\}$$ ■

18. Let $A = \{2, 4, 6, 8, 10\}$ and find $B = \{x \mid x \in A \text{ and } x < 8\}$.

Answers

12. True **13.** True **14.** $\{0, 1, 2, 3, 7\}$ **15.** $\{3\}$ **16.** $\{1, 2, 3\}$ **17.** $\{0, 3, 7\}$ **18.** $\{2, 4, 6\}$

Linking Objectives and Examples

Next to each **objective**, list the examples that are best described by that objective.

A	
B	
C	
D	
E	

Getting Ready for Class

Each section of the book will end with some problems and questions like the ones that follow. They are for you to answer after you have read through the section but before you go to class. All of them require that you give written responses in complete sentences. Writing about mathematics is a valuable exercise. If you write with the intention of explaining and communicating what you know to someone else, you will find that you understand the topic you are writing about even better than you did before you started writing. As with all problems in this course, you want to approach these writing exercises with a positive point of view. You will get better at giving written responses to questions as you progress through the course. Even if you are never comfortable writing about mathematics, just the process of attempting to do so will increase your understanding and ability in mathematics.

After reading through the preceding section, respond in your own words and in complete sentences.

1. What is the meaning of the expression 2^3?
2. Why do we have a rule for the order of operations?
3. What is the intersection of two sets?
4. Explain the operations that are associated with the words *sum, difference, product,* and *quotient.*

Problem Set 1.1

INTERMEDIATE Algebra f(x) Now™

Translate each of the following statements into symbols.

1. The sum of x and 5 is 2.

2. The sum of y and -3 is 9.

3. The difference of 6 and x is y.

4. The difference of x and 6 is $-y$.

5. The product of t and 2 is less than y.

6. The product of $5x$ and y is equal to z.

7. The sum of x and y is less than the difference of x and y.

8. Twice the sum of a and b is 15.

9. Three times the difference of x and 5 is more than y.

10. The product of x and y is greater than or equal to the quotient of x and y.

Expand and multiply.

11. 6^2

12. 8^2

13. 10^2

14. 10^3

15. 2^3

16. 5^3

17. 2^4

18. 1^4

19. 10^4

20. 4^3

21. 11^2

22. 10^5

The problems that follow are intended to give you practice using the rule for order of operations. Some of them are arranged so they model some of the properties of real numbers.

Simplify each expression.

23. **a.** $3 \cdot 5 + 4$
b. $3(5 + 4)$
c. $3 \cdot 5 + 3 \cdot 4$

24. **a.** $3 \cdot 7 - 6$
b. $3(7 - 6)$
c. $3 \cdot 7 - 3 \cdot 6$

25. **a.** $6 + 3 \cdot 4 - 2$
b. $6 + 3(4 - 2)$
c. $(6 + 3)(4 - 2)$

26. **a.** $8 + 2 \cdot 7 - 3$
b. $8 + 2(7 - 3)$
c. $(8 + 2)(7 - 3)$

27. **a.** $(7 - 4)(7 + 4)$
b. $7^2 - 4^2$

28. **a.** $(8 - 5)(8 + 5)$
b. $8^2 - 5^2$

29. **a.** $(5 + 7)^2$
b. $5^2 + 7^2$
c. $5^2 + 2 \cdot 5 \cdot 7 + 7^2$

30. **a.** $(8 - 3)^2$
b. $8^2 - 3^2$
c. $8^2 - 2 \cdot 8 \cdot 3 + 3^2$

31. **a.** $2 + 3 \cdot 2^2 + 3^2$
b. $2 + 3(2^2 + 3^2)$
c. $(2 + 3)(2^2 + 3^2)$

32. **a.** $3 + 4 \cdot 4^2 + 5^2$
b. $3 + 4(4^2 + 5^2)$
c. $(3 + 4)(4^2 + 5^2)$

33. **a.** $40 - 10 \div 5 + 1$
b. $(40 - 10) \div 5 + 1$
c. $(40 - 10) \div (5 + 1)$

34. **a.** $20 - 10 \div 2 + 3$
b. $(20 - 10) \div 2 + 3$
c. $(20 - 10) \div (2 + 3)$

35. **a.** $40 + [10 - (4 - 2)]$
b. $40 - 10 - 4 - 2$

36. **a.** $50 - [17 - (8 - 3)]$
b. $50 - 17 - 8 - 3$

37. **a.** $3 + 2(2 \cdot 3^2 + 1)$
b. $(3 + 2)(2 \cdot 3^2 + 1)$

38. **a.** $4 + 5(3 \cdot 2^2 + 5)$
b. $(4 + 5)(3 \cdot 2^2 + 5)$

The problems below will make certain you are using the rule for order of operations correctly.

39. $5 \cdot 10^3 + 4 \cdot 10^2 + 3 \cdot 10 + 1$

40. $6 \cdot 10^3 + 5 \cdot 10^2 + 4 \cdot 10 + 3$

41. $3[2 + 4(5 + 2 \cdot 3)]$

42. $2[4 + 2(6 + 3 \cdot 5)]$

43. $6[3 + 2(5 \cdot 3 - 10)]$

44. $8[7 + 2(6 \cdot 9 - 14)]$

45. $5(7 \cdot 4 - 3 \cdot 4) + 8(5 \cdot 9 - 4 \cdot 9)$

46. $4(3 \cdot 9 - 2 \cdot 9) + 5(6 \cdot 8 - 5 \cdot 8)$

47. $25 - 17 + 3$

48. $38 - 19 + 1$

49. $109 - 36 + 14$

50. $200 - 150 + 20$

51. $20 - 13 - 3$

52. $57 - 18 - 8$

53. $63 - 37 - 4$

54. $71 - 11 - 1$

55. $36 \div 9 \cdot 4$

56. $28 \div 7 \cdot 2$

57. $75 \div 3 \cdot 25$

58. $48 \div 3 \cdot 2$

59. $64 \div 16 \div 4$

60. $125 \div 25 \div 5$

61. $75 \div 25 \div 5$

62. $36 \div 12 \div 4$

The problems below are problems you will see later in the book. Work these problems without using a calculator.

63. $18{,}000 - 9{,}300$

64. $18{,}000 - 4{,}500$

65. $3.45 + 2.6 - 1.004$

66. $24.3 + 6(8.1)$

67. $275 \div 55$

68. $4.8 \div 2.4$

69. $4(2)(4)^2$

70. $230(5) - 20(5)^2$

71. $250(5) - 25(5)^2$

72. $3(3)^2 + 2(3) - 1$

73. $5 \cdot 2^3 - 3 \cdot 2^2 + 4 \cdot 2 - 5$

74. $125 \cdot 2^2$

75. $125 \cdot 2^3$

76. $7.5(10)$

77. $500(1.5)$

78. $39.3 \cdot 60$

79. $5(0.10)$

80. $2(0.25)$

81. $0.20(8)$

82. $0.30(12)$

83. $0.08(4{,}000)$

84. $0.09(6{,}000)$

We are assuming that you know how to use a calculator to do simple arithmetic problems. Use a calculator to simplify each expression. If rounding is necessary, round your answers to the nearest ten thousandth (4 places past the decimal point). You will see these problems again later in the book.

Simplify.

85. $0.6931 + 1.0986$

86. $1.6094 + 1.9459$

87. $3(0.6931)$

88. $2(1.9459)$

89. $250(3.14)$

90. $165(3.14)$

91. $4{,}628 \div 25$

92. $7{,}546 \div 35$

93. $65{,}000 \div 5{,}280$

94. $2{,}358 \div 5{,}280$

95. $0.8413 - 1$

96. $1.2052 - 1$

97. $16(3.5)^2$

98. $4(3.14)3^2$

99. $11.5(130) - 0.05(130)^2$

100. $10(130) - 0.04(130)^2$

101. Find the value of each expression when x is 5.

a. $x + 2$

b. $2x$

c. x^2

d. 2^x

102. Find the value of each expression when x is 3.

a. $x + 5$

b. $5x$

c. x^5

d. 5^x

103. Find the value of each expression when x is 10.

a. $x^2 + 2x + 1$

b. $(x + 1)^2$

c. $x^2 + 1$

d. $(x - 1)^2$

104. Find the value of each expression when x is 8.

a. $x^2 - 6x + 9$

b. $(x - 3)^2$

c. $x^2 - 3$

d. $x^2 - 9$

105. Find the value of $b^2 - 4ac$ if

a. $a = 2$, $b = 5$, and $c = 3$

b. $a = 10$, $b = 60$, $c = 30$

c. $a = 0.4$, $b = 1$, and $c = 0.3$

106. Find the value of $6x + 5y + 4$ if

a. $x = 3$ and $y = 2$

b. $x = 2$ and $y = 3$

c. $x = 0$ and $y = 0$

For the following problems, let $A = \{0, 2, 4, 6\}$, $B = \{1, 2, 3, 4, 5\}$, and $C = \{1, 3, 5, 7\}$.

107. $A \cup B$

108. $A \cup C$

109. $A \cap B$

110. $A \cap C$

111. $B \cap C$

112. $B \cup C$

113. $A \cup (B \cap C)$

114. $C \cup (A \cap B)$

115. $\{x \mid x \in A \text{ and } x < 4\}$

116. $\{x \mid x \in B \text{ and } x > 3\}$

117. $\{x \mid x \in A \text{ and } x \notin B\}$

118. $\{x \mid x \in B \text{ and } x \notin C\}$

119. $\{x \mid x \in A \text{ or } x \in C\}$

120. $\{x \mid x \in A \text{ or } x \in B\}$

121. $\{x \mid x \in B \text{ and } x \neq 3\}$

122. $\{x \mid x \in C \text{ and } x \neq 5\}$

Applying the Concepts

123. Minutes and Seconds The chart shows the five fastest winning times for the Fifth Avenue Mile run in New York City. The times are given in minutes and seconds, to the nearest hundredth of a second.

a. How much faster was Isaac Viciosa's time than Peter Elliott's time?

b. How much faster was Peter Elliott's time than Stephen Kipkorir's time?

From *USA Today.* Copyright 2005. Reprinted with permission.

124. Mobile Phone Sales Use the chart to answer the following questions.

a. For which year are estimated phone sales closest to 900,000,000 phones?

b. Estimate the increase in mobile phone sales from 2005 to 2006.

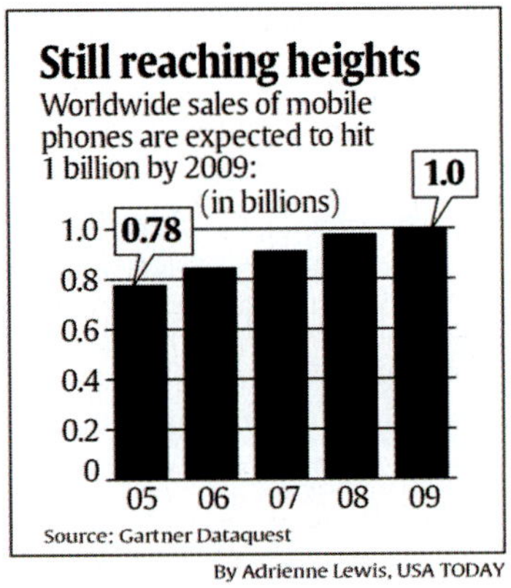

From *USA Today.* Copyright 2005. Reprinted with permission.

Extending the Concepts

Many of the problem sets in this book end with a few problems like the ones that follow. These problems challenge you to extend your knowledge of the material in the problem set. In most cases, there are no examples in the text similar to these problems. You should approach them with a positive point of view because even though you may not complete them correctly, just the process of attempting them will increase your knowledge and ability in algebra.

125. Fermat's Last Theorem The postage stamp shown here was issued by France in 2001 to commemorate the 400th anniversary of the birth of the French mathematician Pierre de Fermat. The stamp shows Fermat's last theorem, which states that if n is an integer greater than 2, then there are no positive integers x, y, and z that will make the formula $x^n + y^n = z^n$ true.

However, there are many ways to make the formula $x^n + y^n = z^n$ true when n is 1 or 2. Show that this formula is true for each case below.

a. $n = 1, x = 5, y = 8$, and $z = 13$

b. $n = 1, x = 2, y = 3$, and $z = 5$

c. $n = 2, x = 3, y = 4$, and $z = 5$

d. $n = 2, x = 7, y = 24$, and $z = 25$

The Real Numbers

The Real Number Line

The real number line is constructed by drawing a straight line and labeling a convenient point with the number 0. Positive numbers are in increasing order to the right of 0; negative numbers are in decreasing order to the left of 0. The point on the line corresponding to 0 is called the *origin*.

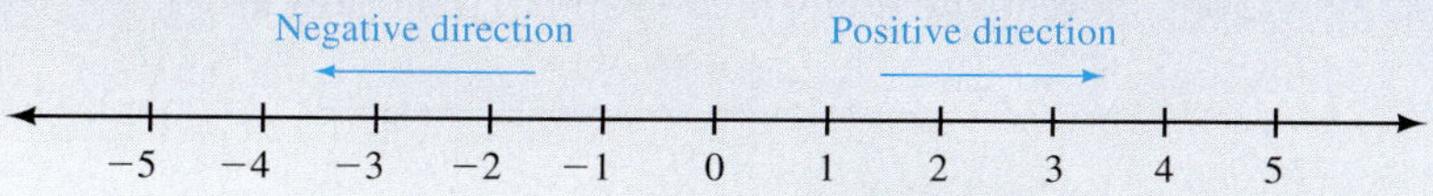

The numbers associated with the points on the line are called *coordinates* of those points. Every point on the line has a number associated with it. The set of all these numbers makes up the set of real numbers.

DEFINITION

A **real number** is any number that is the coordinate of a point on the real number line.

EXAMPLE 1 Locate the numbers -4.5, -0.75, $\frac{1}{2}$, $\sqrt{2}$, π, and 4.1 on the real number line.

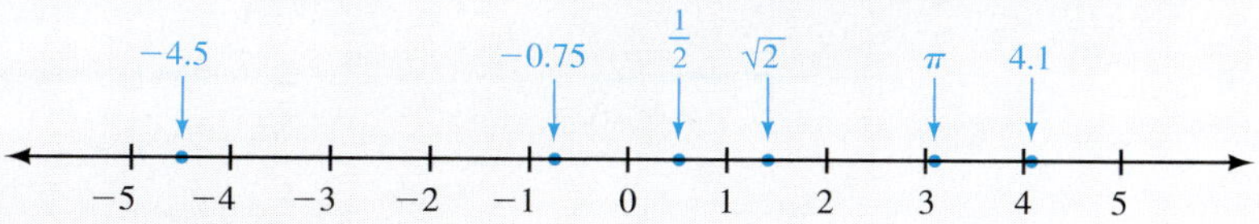

We can use the real number line to give a visual representation to inequality statements.

EXAMPLE 2 Graph $\{x \mid x \leq 3\}$.

Solution We want to graph all the real numbers less than or equal to 3—that is, all the real numbers below 3 and including 3. We label 0 on the number line for reference as well as 3 since the latter is what we call the endpoint. The graph is as follows:

We use a right bracket at 3 to show that 3 is part of the solution set.

Note: You may have come from an algebra class in which open and closed circles were used at the endpoints of number line graphs. If so, you would show the graph for Example 1 this way, using a closed circle at 3 to show that 3 is part of the graph.

The two number line graphs above are equivalent; they both show all the real numbers that are less than or equal to 3.

OBJECTIVES

A	Graph simple and compound inequalities.
B	Translate sentences and phrases written in English into inequalities.
C	List the elements in subsets of the real numbers.
D	Factor positive integers into the product of primes.
E	Reduce fractions to lowest terms.

NOTE

The numbers on the number line increase in size as we move to the right. When we compare the size of two numbers on the number line, the one on the left is always the smaller number.

PRACTICE PROBLEMS

1. Locate the numbers -3, -1.5, $-\frac{1}{4}$, 1, and 2.75 on the number line.

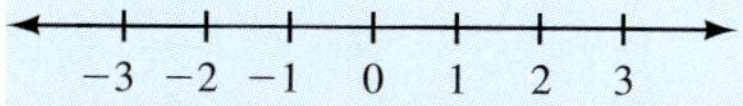

NOTE

In this book we will refer to real numbers as being on the real number line. Actually, real numbers are *not* on the line; only the points representing them are on the line. We can save some writing, however, if we simply refer to real numbers as being on the number line.

2. Graph $\{x \mid x \leq 2\}$.

Answers

1–2. For answers to practice problems that include graphs, see the section titled Solutions to Selected Practice Problems at the back of the book.

3. Graph $\{x \mid x < 2\}$.

EXAMPLE 3

Graph $\{x \mid x < 3\}$.

Solution The graph is identical to the graph in Example 2 except at the endpoint 3.

In this case we use a parenthesis that opens to the left to show that 3 is not part of the graph.

Note For those of you who are used to open/closed circles at the endpoints of number line graphs, here is the equivalent graph in that format.

The table below further clarifies the relationship between number line graphs that use parentheses and brackets at the endpoints and those that use open and closed circles at the endpoints.

Inequality notation	Graph using parentheses/brackets	Graph using open and closed circles
$x < 2$		
$x \le 2$		
$x \ge -3$		
$x > -3$		

In this book, we will use the parentheses/brackets method of graphing inequalities because that method is better suited to the type of problems we will work in intermediate algebra.

In the previous section we defined the *union* of two sets A and B to be the set of all elements that are in either A or B. The word *or* is the key word in the definition. The *intersection* of two sets A and B is the set of all elements contained in both A and B, the key word here being *and.* We can put the words *and* and *or* together with our methods of graphing inequalities to graph some *compound inequalities.*

4. Graph $\{x \mid x \le -1 \text{ or } x > 2\}$.

NOTE

It is not absolutely necessary to show these first two graphs. It is simply helpful to do so. As you get more practice at this type of graphing you can easily omit them.

EXAMPLE 4

Graph $\{x \mid x \le -2 \text{ or } x > 3\}$.

Solution The two inequalities connected by the word *or* are referred to as a **compound inequality.** We begin by graphing each inequality separately.

$x \le -2$ 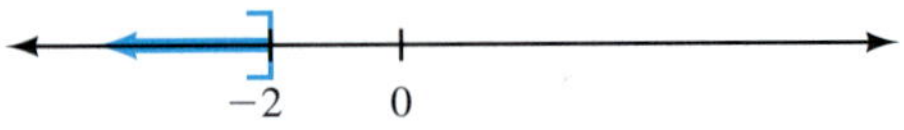

$x > 3$ 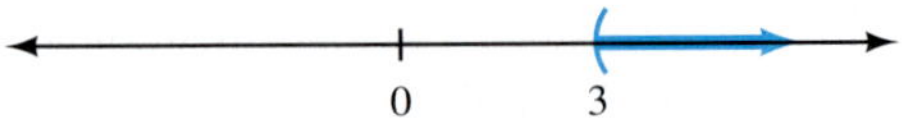

Answers

3–4. For answers to practice problems in this section that include graphs, see Solutions to Selected Practice Problems at the back of the book.

Because the two inequalities are connected by the word *or,* we graph their union; that is, we graph all points on either graph.

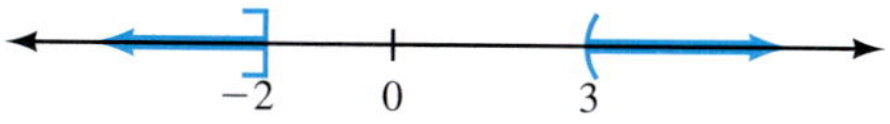

EXAMPLE 5

Graph $\{x \mid x > -1 \text{ and } x < 2\}$.

Solution We first graph each inequality separately.

$x > -1$

$x < 2$

Because the two inequalities are connected by the word *and,* we graph their intersection—the part they have in common.

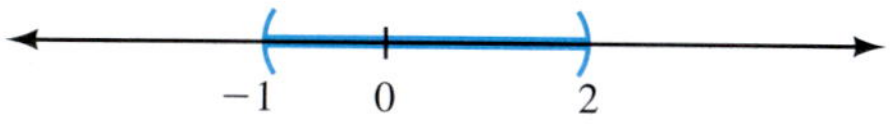

5. Graph $\{x \mid x > -2 \text{ and } x < 3\}$.

Note: Sometimes compound inequalities that use the word *and* as the connecting word can be written in a shorter form. For example, the compound inequality $-3 \le x$ and $x \le 4$ can be written $-3 \le x \le 4$. The word *and* does not appear when an inequality is written in this form. It is implied. Inequalities of the form $-3 \le x \le 4$ are called *continued inequalities.* This new notation is useful because it takes fewer symbols to write it. The graph of $-3 \le x \le 4$ is

EXAMPLE 6

Graph $\{x \mid 1 \le x < 2\}$.

Solution The word *and* is implied in the continued inequality $1 \le x < 2$; that is, the continued inequality $1 \le x < 2$ is equivalent to $1 \le x$ and $x < 2$. Therefore, we graph all the numbers between 1 and 2 on the number line, including 1 but not including 2.

6. Graph $\{x \mid -2 < x \le 1\}$.

The table that follows shows the connection between number line graphs for a variety of continued inequalities. Again, we have included the graphs with open and closed circles for those of you who have used this type of graph previously. Remember, however, that in this book we will be using the parentheses/brackets method of graphing.

Answers

5–6. For answers to practice problems in this section that include graphs, see Solutions to Selected Practice Problems at the back of the book.

Inequality notation	Graph using parentheses/brackets	Graph using open and closed circles
$-3 < x < 2$		
$-3 \le x \le 2$		
$-3 \le x < 2$		
$-3 < x \le 2$		

7. Graph $\{x \mid x \le -1 \text{ or } 2 \le x \le 4\}$.

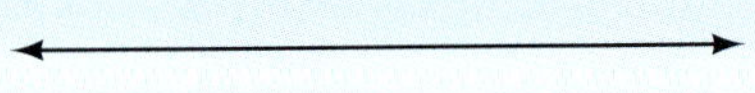

EXAMPLE 7

Graph $\{x \mid x < -2 \text{ or } 2 < x < 6\}$.

Solution Here we have a combination of compound and continued inequalities. We want to graph all real numbers that are either less than -2 or between 2 and 6.

In addition to the phrases that translate directly into inequality statements, we have the following translations:

IN WORDS	IN SYMBOLS
x is at least 40	$x \ge 40$
x is at most 30	$x \le 30$
x is no more than 20	$x \le 20$
x is no less than 10	$x \ge 10$
x is between 4 and 5	$4 < x < 5$

312 Help Wanted

ESPRESSO BAR OPERATOR
Must be dependable, honest, service-oriented. Coffee exp desired. 15-30 hrs per wk. $6.25/hr. Start 5/31. Apply in person: Espresso Yourself, Central Coast Mall. Deadline 5/23.

In the last case, we can include the endpoints 4 and 5 by using "x is between 4 and 5, inclusive," which translates to $4 \le x \le 5$.

8. Repeat Example 8 if you work more than 10 hours but less than 20 hours each week.

EXAMPLE 8

Suppose you have a part-time job that requires that you work at least 10 hours, but no more than 20 hours, each week. Use the letter t to write an inequality that shows the number of hours you work per week.

Solution If t is at least 10 but no more than 20, then $10 \le t$ and $t \le 20$, or equivalently, $10 \le t \le 20$. Note that the word *but,* as used here, has the same meaning as the word *and.*

9. Suppose your weight ranges from a low of 125 pounds to a high of 131 pounds over a 6-month period. Write an inequality using W that gives this range of weight.

EXAMPLE 9

If the highest temperature on Tuesday was 76°F and the lowest temperature was 55°F, write an inequality using the letter x that gives the range of temperatures on Tuesday.

Solution Since the smallest value of x is 55 and the largest value of x is 76, then $55 \le x \le 76$. We could say that the temperature on Tuesday was between 55°F and 76°F, inclusive.

Answers

7. See Solutions to Selected Practice Problems.
8. $10 < t < 20$
9. $125 \le W \le 131$

Subsets of the Real Numbers

Next, we consider some of the more important subsets of the real numbers. Each set listed here is a subset of the real numbers:

Counting (or natural) numbers $= \{1, 2, 3, \ldots\}$
Whole numbers $= \{0, 1, 2, 3, \ldots\}$
Integers $= \{\ldots, -3, -2, -1, 0, 1, 2, 3, \ldots\}$
Rational numbers $= \left\{\frac{a}{b} \,\middle|\, a \text{ and } b \text{ are integers, } b \neq 0\right\}$

Remember, the notation used to write the rational numbers is read "the set of numbers a/b, such that a and b are integers and b is not equal to 0." Any number that can be written in the form

$$\frac{\text{Integer}}{\text{Integer}}$$

is a rational number. Rational numbers are numbers that can be written as the ratio of two integers. Each of the following is a rational number:

$\frac{3}{4}$	Because it is the ratio of the integers 3 and 4
-8	Because it can be written as the ratio of -8 to 1
0.75	Because it is the ratio of 75 to 100 (or 3 to 4 if you reduce to lowest terms)
0.333 . . .	Because it can be written as the ratio of 1 to 3

Still other numbers on the number line are not members of the subsets we have listed so far. They are real numbers, but they cannot be written as the ratio of two integers; that is, they are not rational numbers. For that reason, we call them irrational numbers.

$$\text{Irrational numbers} = \{x \mid x \text{ is real, but not rational}\}$$

The following are irrational numbers:

$$\sqrt{2}, \quad -\sqrt{3}, \quad 4 + 2\sqrt{3}, \quad \pi, \quad \pi + 5\sqrt{6}$$

EXAMPLE 10 For the set $\{-5, -3.5, 0, \frac{3}{4}, \sqrt{3}, \sqrt{5}, 9\}$, list the numbers that are (a) whole numbers, (b) integers, (c) rational numbers, (d) irrational numbers, and (e) real numbers.

Solution

a. Whole numbers: 0, 9
b. Integers: -5, 0, 9
c. Rational numbers: $-5, -3.5, 0, \frac{3}{4}, 9$
d. Irrational numbers: $\sqrt{3}, \sqrt{5}$
e. They are all real numbers.

The following diagram gives a visual representation of the relationships among subsets of the real numbers.

NOTE

As you can see, the whole numbers are the counting numbers and 0 together, whereas the integers are the whole numbers along with the opposites of all the counting numbers. We can say that the counting numbers are a subset of the whole numbers, and the whole numbers are a subset of the integers.

NOTE

We can find decimal approximations to some irrational numbers by using a calculator or a table. For example, on an eight-digit calculator

$$\sqrt{2} = 1.4142136$$

This is not exactly $\sqrt{2}$ but simply an approximation to it. There is no decimal that gives $\sqrt{2}$ exactly.

10. For the set $\{-8, -2.1, -\sqrt{2}, 0, \frac{5}{8}, 3\}$ choose the following:
a. Whole numbers
b. Integers
c. Rational numbers
d. Irrational numbers
e. Real numbers

Answer

10. a. 0, 3 **b.** -8, 0, 3 **c.** $-8, -2.1, 0, \frac{5}{8}, 3$ **d.** $-\sqrt{2}$ **e.** All of them

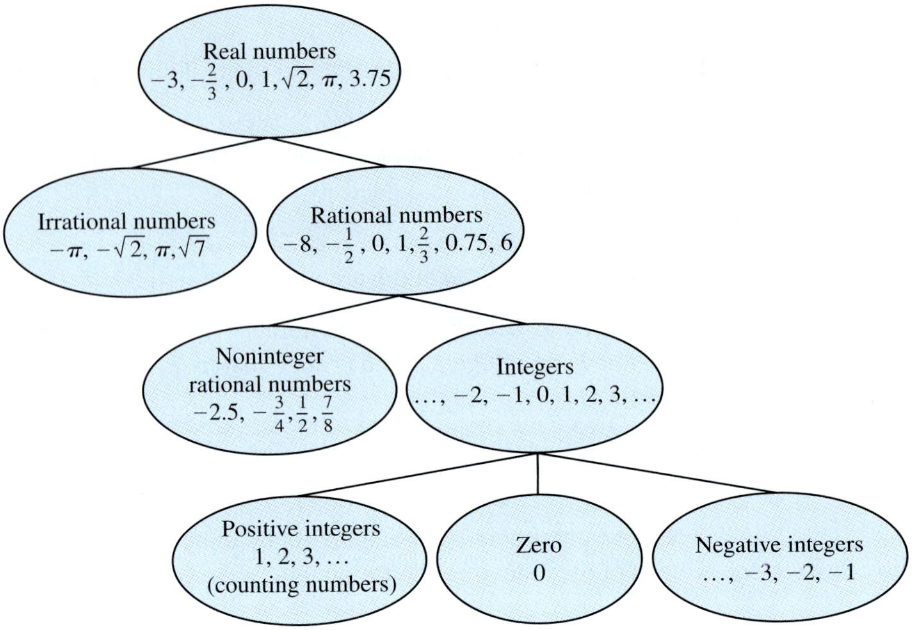

Prime Numbers and Factoring

The following diagram shows the relationship between multiplication and factoring:

Multiplication

Factors $\rightarrow 3 \cdot 4 = 12 \leftarrow$ Product

Factoring

When we read the problem from left to right, we say the *product* of 3 and 4 is 12, or we multiply 3 and 4 to get 12. When we read the problem in the other direction, from right to left, we say we have *factored* 12 into 3 times 4, or 3 and 4 are *factors* of 12.

The number 12 can be factored still further:

$$\begin{aligned} 12 &= 4 \cdot 3 \\ &= 2 \cdot 2 \cdot 3 \\ &= 2^2 \cdot 3 \end{aligned}$$

The numbers 2 and 3 are called *prime* factors of 12 because neither can be factored any further.

DEFINITION

If a and b represent integers, then a is said to be a **factor** (or divisor) of b if a divides b evenly—that is, if a divides b with no remainder.

DEFINITION

A **prime** number is any positive integer larger than 1 whose only positive factors (divisors) are itself and 1. An integer greater than 1 that is not prime is said to be **composite.**

Here is a list of the first few prime numbers:

$$\text{Prime numbers} = \{2, 3, 5, 7, 11, 13, 17, 19, 23, 29, 31, 37, 41, \ldots\}$$

When a number is not prime, we can factor it into the product of prime numbers. To factor a number into the product of primes, we simply factor it until it cannot be factored further.

NOTE

The number 15 is not a prime number because it has factors of 3 and 5; that is, $15 = 3 \cdot 5$.

EXAMPLE 11 Factor 525 into the product of primes.

Solution Because 525 ends in 5, it is divisible by 5:

$$\begin{aligned} 525 &= 5 \cdot 105 \\ &= 5 \cdot 5 \cdot 21 \\ &= 5 \cdot 5 \cdot 3 \cdot 7 \\ &= 3 \cdot 5^2 \cdot 7 \end{aligned}$$

11. Factor 420 into the product of primes.

NOTE

It is customary to write the prime factors in order from smallest to largest.

EXAMPLE 12 Reduce $\frac{210}{231}$ to lowest terms.

Solution First we factor 210 and 231 into the product of prime factors. Then we reduce to lowest terms by dividing the numerator and denominator by any factors they have in common.

$$\frac{210}{231} = \frac{2 \cdot 3 \cdot 5 \cdot 7}{3 \cdot 7 \cdot 11} \quad \textbf{Factor the numerator and denominator completely}$$

$$= \frac{2 \cdot \cancel{3} \cdot 5 \cdot \cancel{7}}{\cancel{3} \cdot \cancel{7} \cdot 11} \quad \textbf{Divide the numerator and denominator by } 3 \cdot 7$$

$$= \frac{2 \cdot 5}{11}$$

$$= \frac{10}{11}$$

The small lines we have drawn through the factors that are common to the numerator and denominator are used to indicate that we have divided the numerator and denominator by those factors.

12. Reduce $\frac{231}{770}$ to lowest terms.

Getting Ready for Class

After reading through the preceding section, respond in your own words and in complete sentences.

1. Explain why some, but not all, rational numbers are also integers.
2. Explain the difference between a prime number and a composite number.
3. Give a written description of the set $\{x \mid -2 \le x < 3\}$.
4. What is an irrational number?

Linking Objectives and Examples

Next to each **objective**, list the examples that are best described by that objective.

A	
B	
C	
D	
E	

Answers

11. $2^2 \cdot 3 \cdot 5 \cdot 7$
12. $\frac{3}{10}$

Problem Set 1.2

INTERMEDIATE Algebra $f(x)$ Now™

Graph the following on a real number line.

1. $\{x \mid x < 1\}$

2. $\{x \mid x > -2\}$

3. $\{x \mid x \leq 1\}$

4. $\{x \mid x \geq -2\}$

5. $\{x \mid x \geq 4\}$

6. $\{x \mid x \leq -3\}$

7. $\{x \mid x > 4\}$

8. $\{x \mid x < -3\}$

9. $\{x \mid x > 0\}$

10. $\{x \mid x < 0\}$

11. $\{x \mid 3 \geq x\}$

12. $\{x \mid 4 \leq x\}$

13. $\{x \mid x \leq -3 \text{ or } x \geq 1\}$

14. $\{x \mid x < 1 \text{ or } x > 4\}$

15. $\{x \mid -3 \leq x \text{ and } x \leq 1\}$

16. $\{x \mid 1 < x \text{ and } x < 4\}$

17. $\{x \mid -3 < x \text{ and } x < 1\}$

18. $\{x \mid 1 \leq x \text{ and } x \leq 4\}$

19. $\{x \mid x < -1 \text{ or } x \geq 3\}$

20. $\{x \mid x < 0 \text{ or } x \geq 3\}$

21. $\{x \mid x \leq -1 \text{ and } x \geq 3\}$

22. $\{x \mid x \leq 0 \text{ and } x \geq 3\}$

23. $\{x \mid x > -4 \text{ and } x < 2\}$

24. $\{x \mid x > -3 \text{ and } x < 0\}$

Graph the following on a real number line.

25. $\{x \mid -1 \le x \le 2\}$

26. $\{x \mid -2 \le x \le 1\}$

27. $\{x \mid -1 < x < 2\}$

28. $\{x \mid -2 < x < 1\}$

29. $\{x \mid -4 < x \le 1\}$

30. $\{x \mid -1 < x \le 5\}$

Graph each of the following.

31. $\{x \mid x < -3 \text{ or } 2 < x < 4\}$

32. $\{x \mid -4 \le x \le -2 \text{ or } x \ge 3\}$

33. $\{x \mid x \le -5 \text{ or } 0 \le x \le 3\}$

34. $\{x \mid -3 < x < 0 \text{ or } x > 5\}$

35. $\{x \mid -5 < x < -2 \text{ or } 2 < x < 5\}$

36. $\{x \mid -3 \le x \le -1 \text{ or } 1 \le x \le 3\}$

Translate each of the following phrases into an equivalent inequality.

37. x is at least 5

38. x is at least -2

39. x is no more than -3

40. x is no more than 8

41. x is at most 4

42. x is at most -5

43. x is between -4 and 4

44. x is between -3 and 3

45. x is between -4 and 4, inclusive

46. x is between -3 and 3, inclusive

For $\{-6, -5.2, -\sqrt{7}, -\pi, 0, 1, 2, 2.3, \frac{9}{2}, \sqrt{17}\}$, list all the elements of the set that are named in each of the following problems.

47. Counting numbers

48. Whole numbers

49. Rational numbers

50. Integers

51. Irrational numbers

52. Real numbers

53. Nonnegative integers

54. Positive integers

Factor each number into the product of prime factors.

55. 60

56. 154

57. 266

58. 385

59. 111

60. 735

61. 369

62. 1,155

Reduce each fraction to lowest terms.

63. $\dfrac{165}{385}$

64. $\dfrac{550}{735}$

65. $\dfrac{385}{735}$

66. $\dfrac{266}{285}$

67. $\dfrac{111}{185}$

68. $\dfrac{279}{310}$

69. $\dfrac{525}{630}$

70. $\dfrac{205}{369}$

71. $\dfrac{75}{135}$

72. $\dfrac{38}{30}$

73. $\dfrac{6}{8}$

74. $\dfrac{10}{25}$

75. $\dfrac{200}{5}$

76. $\dfrac{240}{6}$

77. $\dfrac{10}{22}$

78. $\dfrac{39}{13}$

79. Name two numbers that are 5 units from 2 on the number line.

80. Name two numbers that are 6 units from -3 on the number line.

81. Write an inequality that gives all numbers that are less than 5 units from 8 on the number line.

82. Write an inequality that gives all numbers that are less than 3 units from 5 on the number line.

83. Write an inequality that gives all numbers that are more than 5 units from 8 on the number line.

84. Write an inequality that gives all numbers that are more than 3 units from 5 on the number line.

Applying the Concepts

85. Goldbach's Conjecture The letter shown below was written by Christian Goldbach in 1742. In the letter Goldbach indicates that every even number greater than two can be written as the sum of two prime numbers. For example $4 = 2 + 2$, $6 = 3 + 3$, and $8 = 3 + 5$. His assertion has never been proven, although most mathematicians believe it is true. Because of this, it has come to be known as Goldbach's Conjecture. In March 2000, the publishing firm of Faber and Faber offered a $1 million prize to anyone who could prove Goldbach's Conjecture by March 20, 2002. Show that Goldbach's Conjecture holds for each of the following even numbers.

a. 10

b. 16

c. 24

d. 36

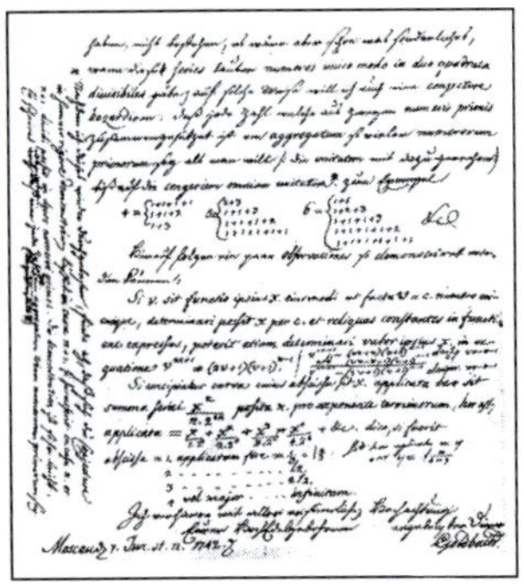

86. Perfect Numbers More than 2,200 years ago, the Greek mathematician Euclid (depicted on the stamp shown here) wrote: "A perfect number is a whole number that is equal to the sum of all its divisors, except itself." The first perfect number is 6 because $6 = 1 + 2 + 3$. The next perfect number is between 20 and 30. Find it.

87. Engine Temperature The temperature gauge on a car keeps track of the temperature of the water that cools the engine. The temperature gauge shown here registers temperatures from 50°F (read 50 degrees Fahrenheit) to 270° F. If F is the temperature of the water in the car engine, write an inequality that shows all the temperatures that can register on the temperature gauge below.

88. Triangle Inequality The *triangle inequality* is a property that is true for all triangles. It states that in any triangle, the sum of the lengths of any two sides is always greater than the length of the third side. Below is a triangle with sides of length x, y, and z. Use the triangle inequality to write three inequalities using x, y, and z.

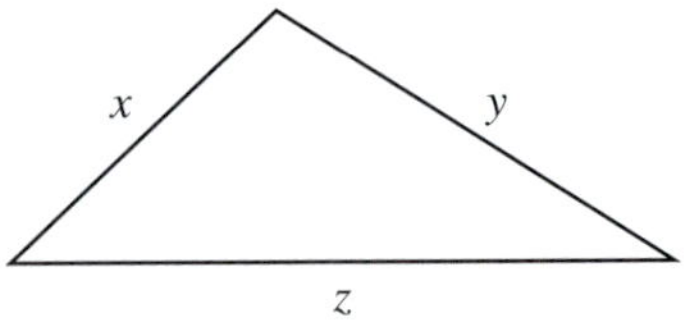

89. Televisions Use the chart to answer the questions below.

a. On average, what percentage of American homes have at least 1 television set?

b. On average, what percentage of American homes have no more than 3 television sets?

c. On average, what percentage of American homes have 3 or more television sets?

90. Hours of Work Suppose you have a job that requires that you work at least 20 hours but less than 40 hours per week. Write an inequality, using t, that gives the number of hours you work each week.

Extending the Concepts

The expression $n!$ is read "n factorial" and is the product of all the consecutive integers from n down to 1. For example,

$$1! = 1$$
$$2! = 2 \cdot 1 = 2$$
$$3! = 3 \cdot 2 \cdot 1 = 6$$
$$4! = 4 \cdot 3 \cdot 2 \cdot 1 = 24$$

91. Calculate $5!$

92. Calculate $6!$

93. Show that this statement is true: $6! = 6 \cdot 5!$

94. Show that the following statement is false: $(2 + 3)! = 2! + 3!$

Improving Your Quantitative Literacy

95. Quantitative literacy is a subject discussed by many people involved in teaching mathematics. The person they are concerned with when they discuss it is you. We are going to work at improving your quantitative literacy, but before we do that, we should answer the question: What is quantitative literacy? Lynn Arthur Steen, a noted mathematics educator, has stated that quantitative literacy is "the capacity to deal effectively with the quantitative aspects of life."

a. Give a definition for the word *quantitative*.

b. Give a definition for the word *literacy*.

c. Are there situations that occur in your life that you find distasteful, or that you try to avoid, because they involve numbers and mathematics? If so, list some of them here. (For example, some people find buying a car particularly difficult because they feel the details of financing it are beyond them.)

96. Drinking and Driving The chart shows that if the risk of getting in an accident with no alcohol in your system is 1, then the risk of an accident with a blood-alcohol level of 0.24 is 147 times as high.

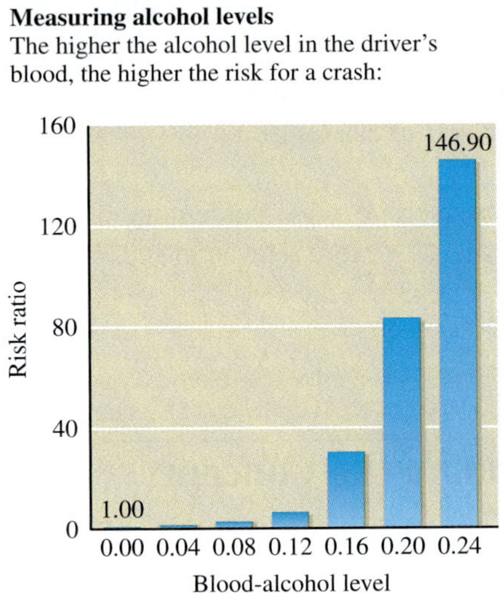

Data from *USA Today*. Copyright 2005.

a. A person driving with a blood-alcohol level of 0.20 is how many times more likely to get in an accident than if she was driving with a blood-alcohol level of 0?

b. If the probability of getting in an accident while driving for an hour on surface streets in a certain city in 0.02%, what is the probability of getting in an accident in the same circumstances with a blood-alcohol level of 0.20?

1.3 Properties of Real Numbers

The area of the large rectangle shown here can be found in two ways: We can multiply its length a by its width $b + c$, or we can find the areas of the two smaller rectangles and add those areas to find the total area.

Area of large rectangle: $a(b + c)$

Sum of the areas of two smaller rectangles: $ab + ac$

Because the area of the large rectangle is the sum of the areas of the two smaller rectangles, we can write:

$$a(b + c) = ab + ac$$

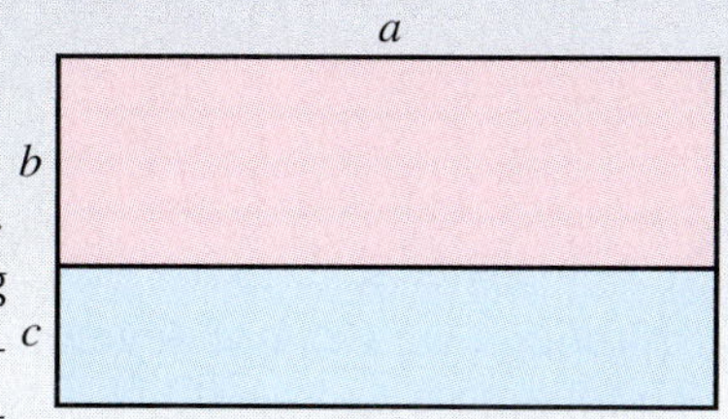

This expression is called the *distributive property.* It is one of the properties we will be discussing in this section. Before we arrive at the distributive property, we need to review some basic definitions and vocabulary.

Opposites and Reciprocals

DEFINITION

Any two real numbers the same distance from 0, but in opposite directions from 0 on the number line, are called **opposites,** or **additive inverses.**

EXAMPLE 1 The numbers -3 and 3 are opposites. So are π and $-\pi$, $\frac{3}{4}$ and $-\frac{3}{4}$, and $\sqrt{2}$ and $-\sqrt{2}$.

The negative sign in front of a number can be read in a number of different ways. It can be read as "negative" or "the opposite of." We say -4 is the opposite of 4, or negative 4. The one we use will depend on the situation. For instance, the expression $-(-3)$ is best read "the opposite of negative 3." Because the opposite of -3 is 3, we have $-(-3) = 3$. In general, if a is any positive real number, then

$-(-a) = a$ **(The opposite of a negative is a positive)**

Review of Multiplication with Fractions

Before we go further with our study of the number line, we need to review multiplication with fractions. Recall that for the fraction $\frac{a}{b}$, a is called the numerator and b is called the denominator. To multiply two fractions we simply multiply numerators and multiply denominators.

EXAMPLES Multiply.

2. $\frac{3}{5} \cdot \frac{7}{8} = \frac{3 \cdot 7}{5 \cdot 8} = \frac{21}{40}$

3. $8 \cdot \frac{1}{5} = \frac{8}{1} \cdot \frac{1}{5} = \frac{8 \cdot 1}{1 \cdot 5} = \frac{8}{5}$

4. $\left(\frac{2}{3}\right)^4 = \frac{2}{3} \cdot \frac{2}{3} \cdot \frac{2}{3} \cdot \frac{2}{3} = \frac{16}{81}$

OBJECTIVES	
A	Find the opposite of a real number.
B	Simplify expressions involving absolute value.
C	Multiply fractions.
D	Recognize and apply the properties of real numbers.
E	Recognize and apply the distributive property.
F	Add and subtract fractions.
G	Simplify algebraic expressions.

PRACTICE PROBLEMS

1. Give the opposite of each number.
   a. 5
   b. $\frac{1}{4}$
   c. -3
   d. $-\sqrt{5}$

Multiply.

2. $\frac{3}{7} \cdot \frac{2}{5}$

3. $9 \cdot \frac{1}{4}$

4. $\left(\frac{3}{4}\right)^3$

Answers

1. a. -5 **b.** $-\frac{1}{4}$ **c.** 3 **d.** $\sqrt{5}$

2. $\frac{6}{35}$ **3.** $\frac{9}{4}$ **4.** $\frac{27}{64}$

NOTE

In past math classes you may have written fractions like $\frac{8}{5}$ (improper fractions) as mixed numbers, such as $1\frac{3}{5}$. In algebra, it is usually better to leave them as improper fractions.

The idea of multiplication of fractions is useful in understanding the concept of the reciprocal of a number. Here is the definition.

DEFINITION

Any two real numbers whose product is 1 are called **reciprocals,** or **multiplicative inverses.**

Give the reciprocal of each number.

5. 5
6. $\frac{1}{4}$
7. $\frac{3}{4}$
8. x if $x \neq 0$

EXAMPLES Give the reciprocal of each number.

	Number	*Reciprocal*	
5.	3	$\frac{1}{3}$	Because $3 \cdot \frac{1}{3} = \frac{3}{1} \cdot \frac{1}{3} = \frac{3}{3} = 1$
6.	$\frac{1}{6}$	6	Because $\frac{1}{6} \cdot 6 = \frac{1}{6} \cdot \frac{6}{1} = \frac{6}{6} = 1$
7.	$\frac{4}{5}$	$\frac{5}{4}$	Because $\frac{4}{5} \cdot \frac{5}{4} = \frac{20}{20} = 1$
8.	a	$\frac{1}{a}$	Because $a \cdot \frac{1}{a} = \frac{a}{1} \cdot \frac{1}{a} = \frac{a}{a} = 1$ **($a \neq 0$)**

Although we will not develop multiplication with negative numbers until later in this chapter, you should know that the reciprocal of a negative number is also a negative number. For example, the reciprocal of -5 is $-\frac{1}{5}$.

The Absolute Value of a Real Number

DEFINITION

The **absolute value** of a number (also called its **magnitude**) is the distance the number is from 0 on the number line. If x represents a real number, then the absolute value of x is written $|x|$.

NOTE

It is important to recognize that if x is a real number, $-x$ is not necessarily negative. For example, if x is 5, then $-x$ is -5. However, if x were -5, then $-x$ would be $-(-5)$, which is 5.

This definition of absolute value is geometric in form since it defines absolute value in terms of the number line. Here is an alternative definition of absolute value that is algebraic in form since it involves only symbols.

ALTERNATIVE DEFINITION

If x represents a real number, then the **absolute value** of x is written $|x|$, and is given by

$$|x| = \begin{cases} x & \text{if } x \geq 0 \\ -x & \text{if } x < 0 \end{cases}$$

If the original number is positive or 0, then its absolute value is the number itself. If the number is negative, its absolute value is its opposite (which must be positive).

Write the following expressions as equivalent statements without absolute value symbols.

9. $|7|$
10. $|-4|$
11. $|-\frac{2}{3}|$
12. $-|-6|$
13. $-|7|$
14. $-|-\sqrt{3}|$

EXAMPLES Write each expression without absolute value symbols.

9. $|5| = 5$

10. $|-2| = 2$

11. $\left|-\frac{1}{2}\right| = \frac{1}{2}$

12. $-|-3| = -3$

13. $-|5| = -5$

14. $-|-\sqrt{2}| = -\sqrt{2}$

Properties of Real Numbers

We know that adding 3 and 7 gives the same answer as adding 7 and 3. The order of two numbers in an addition problem can be changed without changing the result. This fact about numbers and addition is called the *commutative property of addition.*

Answers

5. $\frac{1}{5}$ 6. 4 7. $\frac{4}{3}$ 8. $\frac{1}{x}$
9. 7 10. 4 11. $\frac{2}{3}$ 12. -6
13. -7 14. $-\sqrt{3}$

For all the properties listed in this section, a, b, and c represent real numbers.

Commutative Property of Addition

In symbols: $a + b = b + a$

In words: The *order* of the numbers in a sum does not affect the result.

Commutative Property of Multiplication

In symbols: $a \cdot b = b \cdot a$

In words: The *order* of the numbers in a product does not affect the result.

EXAMPLES

15. The statement $3 + 7 = 7 + 3$ is an example of the commutative property of addition.

16. The statement $3 \cdot x = x \cdot 3$ is an example of the commutative property of multiplication.

Complete each statement so it is an example of the commutative property.

15. $5 + x = x +$ __

16. $7 \cdot 3 = 3 \cdot$ __

Another property of numbers you have used many times has to do with grouping. When adding $3 + 5 + 7$, we can add the 3 and 5 first and then the 7, or we can add the 5 and 7 first and then the 3. Mathematically, it looks like this: $(3 + 5) + 7 = 3 + (5 + 7)$. Operations that behave in this manner are called *associative* operations.

NOTE

The other two basic operations (subtraction and division) are not commutative. If we change the order in which we are subtracting or dividing two numbers, we change the result.

Associative Property of Addition

In symbols: $a + (b + c) = (a + b) + c$

In words: The *grouping* of the numbers in a sum does not affect the result.

Associative Property of Multiplication

In symbols: $a(bc) = (ab)c$

In words: The *grouping* of the numbers in a product does not affect the result.

The following examples illustrate how the associative properties can be used to simplify expressions that involve both numbers and variables.

EXAMPLES

Simplify by using the associative property.

17. $2 + (3 + y) = (2 + 3) + y$ **Associative property**

$= 5 + y$ **Addition**

18. $5(4x) = (5 \cdot 4)x$ **Associative property**

$= 20x$ **Multiplication**

19. $\frac{1}{4}(4a) = \left(\frac{1}{4} \cdot 4\right)a$ **Associative property**

$= 1a$ **Multiplication**

$= a$

20. $2\left(\frac{1}{2}x\right) = \left(2 \cdot \frac{1}{2}\right)x$ **Associative property**

$= 1x$ **Multiplication**

$= x$

21. $6\left(\frac{1}{3}x\right) = \left(6 \cdot \frac{1}{3}\right)x$ **Associative property**

$= 2x$ **Multiplication**

Simplify.

17. $5 + (7 + y)$

18. $3(2x)$

19. $\frac{1}{3}(3a)$

20. $5\left(\frac{1}{5}x\right)$

21. $9\left(\frac{2}{3}x\right)$

Answers

15. $5 + x = x + 5$ **16.** $7 \cdot 3 = 3 \cdot 7$ **17.** $12 + y$ **18.** $6x$ **19.** a **20.** x **21.** $6x$

Our next property involves both addition and multiplication. It is called the *distributive property* and is stated as follows.

Distributive Property

In symbols: $a(b + c) = ab + ac$

In words: Multiplication *distributes* over addition.

NOTE

Although the properties we are listing are stated for only two or three real numbers, they hold for as many numbers as needed. For example, the distributive property holds for expressions like $3(x + y + z + 2)$. That is:

$3(x + y + z + 2) = 3x + 3y + 3z + 6$

You will see as we progress through the book that the distributive property is used very frequently in algebra. To see that the distributive property works, compare the following:

3(4 + 5)	3(4) + 3(5)
3(9)	12 + 15
27	27

In both cases the result is 27. Because the results are the same, the original two expressions must be equal, or $3(4 + 5) = 3(4) + 3(5)$.

EXAMPLES Apply the distributive property to each expression and then simplify the result.

22. $5(4x + 3) = 5(4x) + 5(3)$ **Distributive property**
$= 20x + 15$ **Multiplication**

23. $6(3x + 2y) = 6(3x) + 6(2y)$ **Distributive property**
$= 18x + 12y$ **Multiplication**

24. $\frac{1}{2}(3x + 6) = \frac{1}{2}(3x) + \frac{1}{2}(6)$ **Distributive property**
$= \frac{3}{2}x + 3$ **Multiplication**

25. $2(3y + 4) + 2 = 2(3y) + 2(4) + 2$ **Distributive property**
$= 6y + 8 + 2$ **Multiplication**
$= 6y + 10$ **Addition**

Apply the distributive property and then simplify.

22. $7(6x + 8)$

23. $9(7x + 11y)$

24. $\frac{1}{3}(3x + 6)$

25. $4(5y + 2) + 8$

We can combine our knowledge of the distributive property with multiplication of fractions to manipulate expressions involving fractions. Here are some examples that show how we do this.

EXAMPLES Apply the distributive property, then simplify if possible.

26. $a\left(1 + \frac{1}{a}\right) = a \cdot 1 + a \cdot \frac{1}{a} = a + 1$

27. $3\left(\frac{1}{3}x + 5\right) = 3 \cdot \frac{1}{3}x + 3 \cdot 5 = x + 15$

28. $6\left(\frac{1}{3}x + \frac{1}{2}y\right) = 6 \cdot \frac{1}{3}x + 6 \cdot \frac{1}{2}y = 2x + 3y$

Apply the distributive property, then simplify if possible.

26. $a\left(2 - \frac{1}{a}\right)$

27. $5\left(\frac{1}{5}x + 8\right)$

28. $12\left(\frac{2}{3}x + \frac{3}{4}y\right)$

Combining Similar Terms

The distributive property can also be used to combine similar terms. (For now, a *term* is a number, or the product of a number with one or more variables.) Similar terms are terms with the same variable part. The terms $3x$ and $5x$ are similar, as are $2y$, $7y$, and $-3y$, because the variable parts are the same.

EXAMPLES Use the distributive property to combine similar terms.

29. $3x + 5x = (3 + 5)x$ **Distributive property**
$= 8x$ **Addition**

30. $3y + y = (3 + 1)y$ **Distributive property**
$= 4y$ **Addition**

Combine similar terms.

29. $4x + 9x$

30. $y + 5y$

Answers

22. $42x + 56$ **23.** $63x + 99y$ **24.** $x + 2$ **25.** $20y + 16$ **26.** $2a - 1$ **27.** $x + 40$ **28.** $8x + 9y$ **29.** $13x$ **30.** $6y$

Review of Addition with Fractions

To add fractions, each fraction must have the same denominator.

DEFINITION

The **least common denominator** (LCD) for a set of denominators is the smallest number divisible by *all* the denominators.

The first step in adding fractions is to find a common denominator for all the denominators. We then rewrite each fraction (if necessary) as an equivalent fraction with the common denominator. Finally, we add the numerators and reduce to lowest terms if necessary.

EXAMPLE 31 Add $\frac{5}{12} + \frac{7}{18}$.

Solution The least common denominator for the denominators 12 and 18 must be the smallest number divisible by both 12 and 18. We can factor 12 and 18 completely and then build the LCD from these factors.

12 divides the LCD

$$\left.\begin{aligned} 12 &= 2 \cdot 2 \cdot 3 \\ 18 &= 2 \cdot 3 \cdot 3 \end{aligned}\right\} \quad \text{LCD} = 2 \cdot 2 \cdot 3 \cdot 3 = 36$$

18 divides the LCD

Next, we rewrite our original fractions as equivalent fractions with denominators of 36. To do so, we multiply each original fraction by an appropriate form of the number 1:

$$\frac{5}{12} + \frac{7}{18} = \frac{5}{12} \cdot \frac{\mathbf{3}}{\mathbf{3}} + \frac{7}{18} \cdot \frac{\mathbf{2}}{\mathbf{2}} = \frac{15}{36} + \frac{14}{36}$$

Finally, we add numerators and place the result over the common denominator, 36.

$$\frac{15}{36} + \frac{14}{36} = \frac{15 + 14}{36} = \frac{29}{36}$$

31. Add $\frac{3}{14} + \frac{7}{30}$.

Simplifying Expressions

We can use the commutative, associative, and distributive properties together to simplify expressions.

EXAMPLE 32 Simplify $7x + 4 + 6x + 3$.

Solution We begin by applying the commutative and associative properties to group similar terms:

$7x + 4 + 6x + 3 = (7x + 6x) + (4 + 3)$	**Commutative and associative properties**
$= (7 + 6)x + (4 + 3)$	**Distributive property**
$= 13x + 7$	**Addition**

32. Simplify $10x + 3 + 7x + 12$.

Answers
31. $\frac{47}{105}$ **32.** $17x + 15$

33. Simplify $9 + 5(4y + 8) + 10y$.

EXAMPLE 33

Simplify $4 + 3(2y + 5) + 8y$.

Solution Because our rule for order of operations indicates that we are to multiply before adding, we must distribute the 3 across $2y + 5$ first:

$4 + 3(2y + 5) + 8y = 4 + 6y + 15 + 8y$	**Distributive property**
$= (6y + 8y) + (4 + 15)$	**Commutative and associative properties**
$= (6 + 8)y + (4 + 15)$	**Distributive property**
$= 14y + 19$	**Addition**

The remaining properties of real numbers have to do with the numbers 0 and 1.

NOTE

0 and 1 are called the *additive identity* and *multiplicative identity*, respectively. Combining 0 with a number, under addition, does not change the identity of the number. Likewise, combining 1 with a number, under multiplication, does not alter the identity of the number. We see that 0 is to addition what 1 is to multiplication.

Additive Identity Property

There exists a unique number 0 such that

In symbols: $a + 0 = a$ and $0 + a = a$

Multiplicative Identity Property

There exists a unique number 1 such that

In symbols: $a(1) = a$ and $1(a) = a$

Additive Inverse Property

For each real number a, there exists a unique real number $-a$ such that

In symbols: $a + (-a) = 0$

In words: Opposites add to 0.

Multiplicative Inverse Property

For every real number a, except 0, there exists a unique real number $\frac{1}{a}$ such that

In symbols: $a\left(\frac{1}{a}\right) = 1$

In words: Reciprocals multiply to 1.

Identify the property that justifies each of the following.

34. $2 + 0 = 2$

35. $3(1) = 3$

36. $-5 + 5 = 0$

37. $\frac{3}{4} \cdot \frac{4}{3} = 1$

EXAMPLES

34. $7(1) = 7$ **Multiplicative identity property**

35. $4 + (-4) = 0$ **Additive inverse property**

36. $6\left(\frac{1}{6}\right) = 1$ **Multiplicative inverse property**

37. $(5 + 0) + 2 = 5 + 2$ **Additive identity property**

Linking Objectives and Examples

Next to each **objective**, list the examples that are best described by that objective.

A	
B	
C	
D	
E	
F	
G	

Getting Ready for Class

After reading through the preceding section, respond in your own words and in complete sentences.

1. Describe the commutative property of multiplication.
2. Give definitions for each of the following:
   a. The opposite of a number
   b. The absolute value of a number
3. Explain why subtraction and division are not commutative operations.
4. Explain why zero does not have a multiplicative inverse.

Answers

33. $30y + 49$
34. Additive identity
35. Multiplicative identity
36. Additive inverse
37. Multiplicative inverse

Problem Set 1.3

INTERMEDIATE Algebra $f(x)$ Now™

Complete the following table.

	NUMBER	OPPOSITE	RECIPROCAL
1.	4		
2.	-3		
3.	$-\frac{1}{2}$		
4.	$\frac{5}{6}$		
5.		-5	
6.		7	
7.		$-\frac{3}{8}$	
8.		$\frac{1}{2}$	
9.			-6
10.			-3
11.			$\frac{1}{3}$
12.			$-\frac{1}{4}$

13. Name two numbers that are their own reciprocals.

14. Give the number that has no reciprocal.

15. Name the number that is its own opposite.

16. The reciprocal of a negative number is negative—true or false?

Write each of the following without absolute value symbols.

17. $|-2|$

18. $|-7|$

19. $\left|-\frac{3}{4}\right|$

20. $\left|\frac{5}{6}\right|$

21. $|\pi|$

22. $|-\sqrt{2}|$

23. $-|4|$

24. $-|5|$

25. $-|-2|$

26. $-|-10|$

27. $-\left|-\frac{3}{4}\right|$

28. $-\left|\frac{7}{8}\right|$

Multiply the following.

29. $\frac{3}{5} \cdot \frac{7}{8}$

30. $\frac{6}{7} \cdot \frac{9}{5}$

31. $\frac{1}{3} \cdot 6$

32. $\frac{1}{4} \cdot 8$

33. $\left(\frac{2}{3}\right)^3$

34. $\left(\frac{4}{5}\right)^2$

35. $\left(\frac{1}{10}\right)^4$

36. $\left(\frac{1}{2}\right)^5$

37. $\frac{3}{5} \cdot \frac{4}{7} \cdot \frac{6}{11}$

38. $\frac{4}{5} \cdot \frac{6}{7} \cdot \frac{3}{11}$

39. $\frac{4}{3} \cdot \frac{3}{4}$

40. $\frac{5}{8} \cdot \frac{8}{5}$

Use the associative property to rewrite each of the following expressions and then simplify the result.

41. $4 + (2 + x)$

42. $6 + (5 + 3x)$

43. $(a + 3) + 5$

44. $(4a + 5) + 7$

45. $5(3y)$

46. $7(4y)$

47. $\frac{1}{3}(3x)$

48. $\frac{1}{5}(5x)$

49. $4\left(\frac{1}{4}a\right)$

50. $7\left(\frac{1}{7}a\right)$

51. $\frac{2}{3}\left(\frac{3}{2}x\right)$

52. $\frac{4}{3}\left(\frac{3}{4}x\right)$

Apply the distributive property to each expression. Simplify when possible.

53. $3(x + 6)$

54. $5(x + 9)$

55. $2(6x + 4)$

56. $3(7x + 8)$

57. $5(3a + 2b)$

58. $7(2a + 3b)$

59. $\frac{1}{3}(4x + 6)$

60. $\frac{1}{2}(3x + 8)$

61. $\frac{1}{5}(10 + 5y)$

62. $\frac{1}{6}(12 + 6y)$

63. $(5t + 1)8$

64. $(3t + 2)5$

The problems below are problems you will see later in the book. Apply the distributive property, then simplify if possible.

65. $3(3x + y - 2z)$

66. $2(2x - y + z)$

67. $10(0.3x + 0.7y)$

68. $10(0.2x + 0.5y)$

69. $100(0.06x + 0.07y)$

70. $100(0.09x + 0.08y)$

71. $3\left(x + \frac{1}{3}\right)$

72. $5\left(x - \frac{1}{5}\right)$

73. $2\left(x - \frac{1}{2}\right)$

74. $7\left(x + \frac{1}{7}\right)$

75. $x\left(1 + \frac{2}{x}\right)$

76. $x\left(1 - \frac{1}{x}\right)$

77. $a\left(1 - \frac{3}{a}\right)$

78. $a\left(1 + \frac{1}{a}\right)$

79. $8\left(\frac{1}{8}x + 3\right)$

80. $4\left(\frac{1}{4}x - 9\right)$

81. $6\left(\frac{1}{2}x - \frac{1}{3}y\right)$

82. $12\left(\frac{1}{4}x - \frac{1}{6}y\right)$

83. $12\left(\frac{1}{4}x + \frac{2}{3}y\right)$

84. $12\left(\frac{2}{3}x - \frac{1}{4}y\right)$

85. $20\left(\frac{2}{5}x + \frac{1}{4}y\right)$

86. $15\left(\frac{2}{3}x + \frac{2}{5}y\right)$

Apply the distributive property to each expression. Simplify when possible.

87. $3(5x + 2) + 4$

88. $4(3x + 2) + 5$

89. $4(2y + 6) + 8$

90. $6(2y + 3) + 2$

91. $5(1 + 3t) + 4$

92. $2(1 + 5t) + 6$

93. $3 + (2 + 7x)4$

94. $4 + (1 + 3x)5$

Add the following fractions.

95. $\frac{2}{5} + \frac{1}{15}$

96. $\frac{5}{8} + \frac{1}{4}$

97. $\frac{17}{30} + \frac{11}{42}$

98. $\frac{19}{42} + \frac{13}{70}$

99. $\frac{9}{48} + \frac{3}{54}$

100. $\frac{6}{28} + \frac{5}{42}$

101. $\frac{25}{84} + \frac{41}{90}$

102. $\frac{23}{70} + \frac{29}{84}$

The problems below are problems you will see later in the book. Simplify each expression.

103. $\frac{3}{14} + \frac{7}{30}$

104. $\frac{3}{10} + \frac{11}{42}$

105. $32\left(\frac{3}{4}\right) - 16\left(\frac{3}{4}\right)^2$

106. $32\left(\frac{3}{2}\right) - 16\left(\frac{3}{2}\right)^2$

Use the commutative, associative, and distributive properties to simplify the following.

107. $5a + 7 + 8a + a$

108. $6a + 4 + a + 4a$

109. $3y + y + 5 + 2y + 1$

110. $4y + 2y + 3 + y + 7$

111. $2(5x + 1) + 2x$

112. $3(4x + 1) + 9x$

113. $7 + 2(4y + 2)$

114. $6 + 3(5y + 2)$

115. $3 + 4(5a + 3) + 4a$

116. $8 + 2(4a + 2) + 5a$

117. $5x + 2(3x + 8) + 4$

118. $7x + 3(4x + 1) + 7$

Identify the property of real numbers that justifies each of the following.

119. $3 + 2 = 2 + 3$

120. $3(ab) = (3a)b$

121. $5 \cdot x = x \cdot 5$

122. $2 + 0 = 2$

123. $4 + (-4) = 0$

124. $1(6) = 6$

125. $x + (y + 2) = (y + 2) + x$

126. $(a + 3) + 4 = a + (3 + 4)$

127. $4(5 \cdot 7) = 5(4 \cdot 7)$

128. $6(xy) = (xy)6$

129. $4 + (x + y) = (4 + y) + x$

130. $(r + 7) + s = (r + s) + 7$

131. $3(4x + 2) = 12x + 6$

132. $5\left(\frac{1}{5}\right) = 1$

Applying the Concepts

Area and the Distributive Property Find the area of each of the following rectangles in two ways: first by multiplying length and width, and then by adding the areas of the two smaller rectangles together.

133.

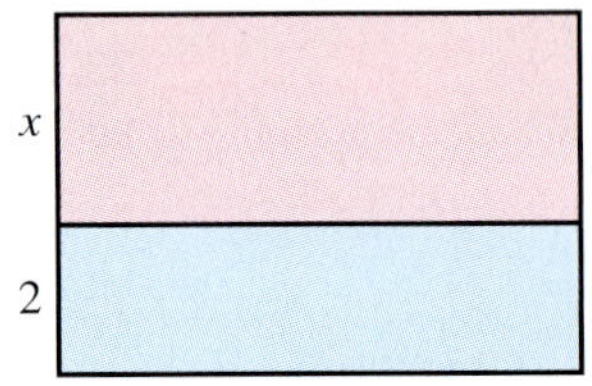

134.

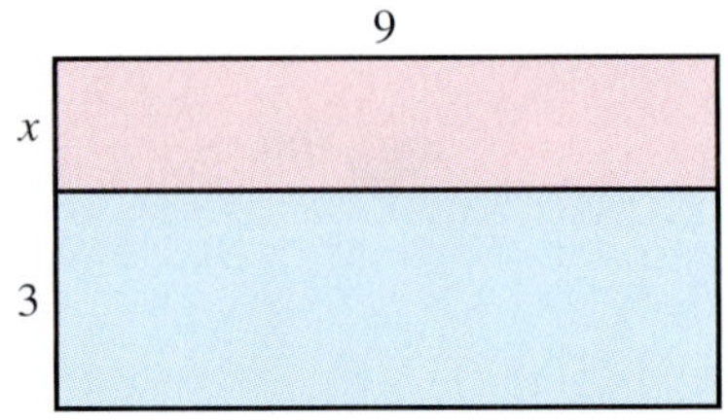

135.

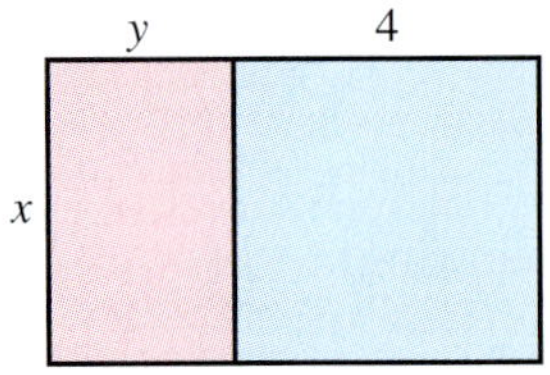

136.

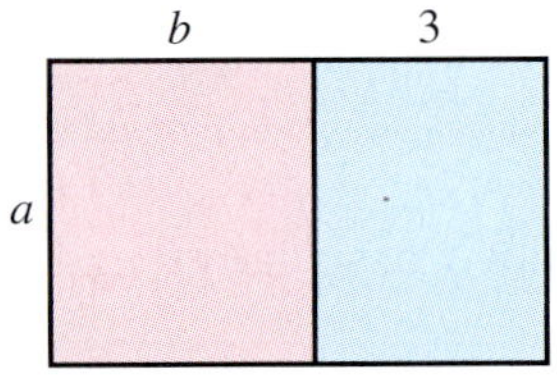

137. Rhind Papyrus In approximately 1650 B.C., a mathematical document was written in ancient Egypt called the *Rhind Papyrus* (part of which is shown here). An "exercise" in this document asked the reader to find "a quantity such that when it is added to one fourth of itself results in 15." Verify this quantity must be 12.

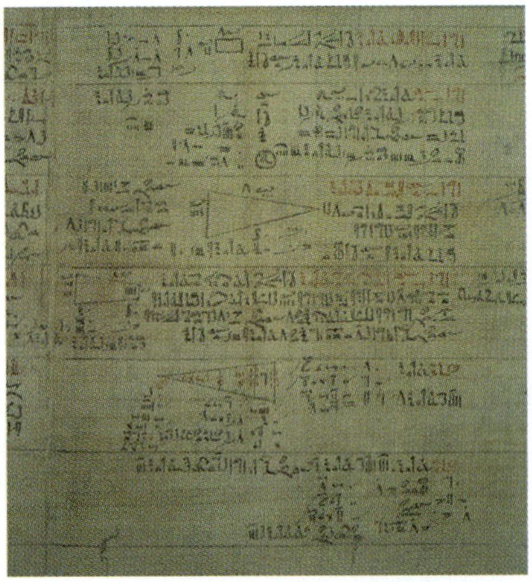

British Museum/Bridgeman Art Library

138. Clock Arithmetic In a normal clock with 12 hours on its face, 12 is the additive identity because adding 12 hours to any time on the clock will not change the hands of the clock. Also, if we think of the hour hand of a clock, the problem $10 + 4$ can be taken to mean: The hour hand is pointing at 10; if we add 4 more hours, it will be pointing at what number? Reasoning this way, we see that in clock arithmetic $10 + 4 = 2$ and $9 + 6 = 3$. Find the following in clock arithmetic:

10(o'clock) + 4(hours) = 2(o'clock)

a. $10 + 5$
b. $10 + 6$
c. $10 + 1$
d. $10 + 12$
e. $x + 12$

Improving Your Quantitative Literacy

139. Incarceration Rate Use the chart to answer the questions below.

a. For the most part, the chart shows the number of people incarcerated increase as the years increase. How do you explain the title *Incarceration rate slows*?

b. Which year showed a decline in the numbers of people incarcerated?

c. Was the rate of increase in incarcerations from 2000 to 2001 a positive number or a negative number?

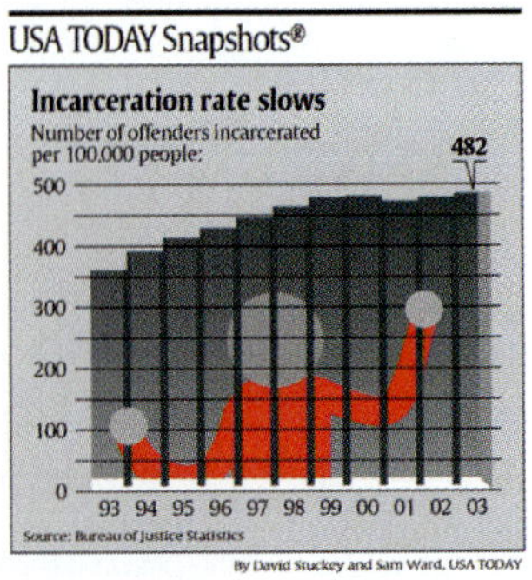

From *USA Today*. Copyright 2005. Reprinted with permission.

Arithmetic with Real Numbers

1.4

The temperature at the airport is 70°F. A plane takes off and reaches its cruising altitude of 28,000 feet, where the temperature is -40°F. Find the difference in the temperatures at takeoff and at cruising altitude.

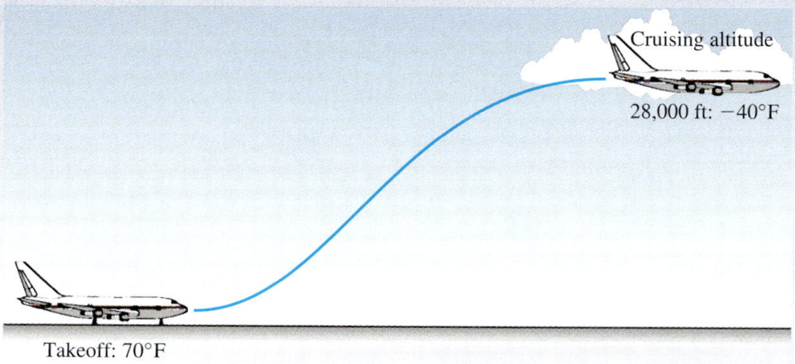

We know intuitively that the difference in temperature is 110°F. If we write this problem using symbols, we have

$$70 - (-40) = 110$$

In this section we review the rules for arithmetic with real numbers, which will include problems such as this one.

OBJECTIVES	
A	Add and subtract real numbers.
B	Multiply and divide real numbers.
C	Apply the rule for order of operations.
D	Simplify algebraic expressions.
E	Divide fractions.
F	Find the value of an expression.

Adding Real Numbers

The purpose of this section is to review the rules for arithmetic with real numbers and the justification for those rules. We can justify the rules for addition of real numbers geometrically by use of the real number line. Consider the sum of -5 and 3:

$$-5 + 3$$

We can interpret this expression as meaning "start at the origin and move 5 units in the negative direction and then 3 units in the positive direction." With the aid of a number line we can visualize the process.

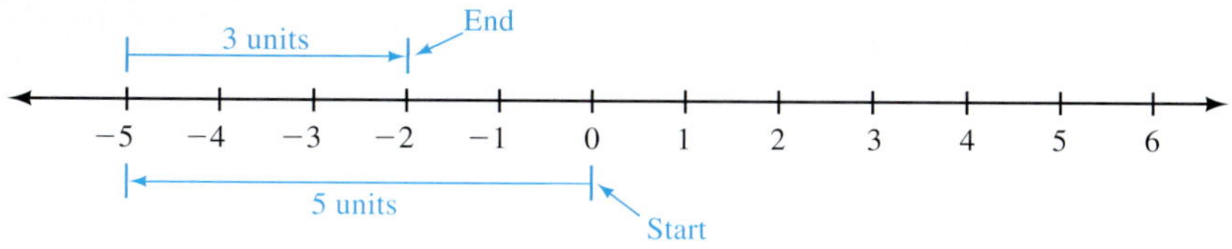

Because the process ends at -2, we say the sum of -5 and 3 is -2:

$$-5 + 3 = -2$$

We can use the real number line in this way to add any combination of positive and negative numbers.

The sum of -4 and -2, $-4 + (-2)$, can be interpreted as starting at the origin, moving 4 units in the negative direction, and then 2 more units in the negative direction:

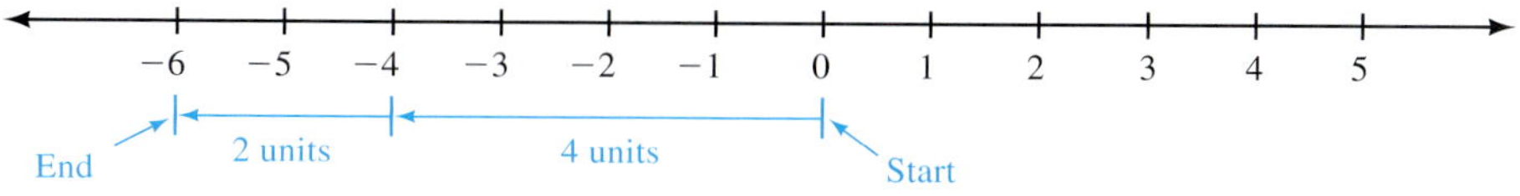

NOTE

We are showing addition of real numbers on the number line to justify the rule we will write for addition of positive and negative numbers. You may want to skip ahead and read the rule on the next page first, and then come back and read through this discussion again. The discussion here is the "why" behind the rule.

Because the process ends at -6, we say the sum of -4 and -2 is -6:

$$-4 + (-2) = -6$$

We can eliminate actually drawing a number line by simply visualizing it mentally. The following example gives the results of all possible sums of positive and negative 5 and 7.

PRACTICE PROBLEMS

1. Add.
   $3 + 5 =$
   $-3 + 5 =$
   $3 + (-5) =$
   $-3 + (-5) =$

EXAMPLE 1 Add all combinations of positive and negative 5 and 7.

Solution

$$5 + 7 = 12$$
$$-5 + 7 = 2$$
$$5 + (-7) = -2$$
$$-5 + (-7) = -12$$

Looking closely at the relationships in Example 1 (and trying other similar examples if necessary), we can arrive at the following rule for adding two real numbers.

NOTE

This rule is the most important rule we have had so far. It is very important that you use it exactly the way it is written. Your goal is to become fast and accurate at adding positive and negative numbers. When you have finished reading this section and working the problems in the problem set, you should have attained that goal.

To Add Two Real Numbers

With the *same* sign:

Step 1: Add their absolute values.

Step 2: Attach their common sign. If both numbers are positive, their sum is positive; if both numbers are negative, their sum is negative.

With *opposite* signs:

Step 1: Subtract the smaller absolute value from the larger.

Step 2: Attach the sign of the number whose absolute value is larger.

Subtracting Real Numbers

In order to have as few rules as possible, we will not attempt to list new rules for the *difference* of two real numbers. We will instead define it in terms of addition and apply the rule for addition.

DEFINITION

If a and b are any two real numbers, then the **difference** of a and b, written $a - b$, is given by

$$\underbrace{a - b} = \underbrace{a + (-b)}$$

To subtract b, add the opposite of b.

We define the process of subtracting b from a as being equivalent to adding the opposite of b to a. In short, we say, "subtraction is addition of the opposite."

Subtract.

2. $7 - 4$
3. $-6 - 3$
4. $8 - (-2)$
5. $-4 - (-6)$

EXAMPLES Subtract.

2. $5 - 3 = 5 + (-3)$ **Subtracting 3 is equivalent to adding -3**
   $= 2$
3. $-7 - 6 = -7 + (-6)$ **Subtracting 6 is equivalent to adding -6**
   $= -13$
4. $9 - (-2) = 9 + 2$ **Subtracting -2 is equivalent to adding 2**
   $= 11$
5. $-6 - (-5) = -6 + 5$ **Subtracting -5 is equivalent to adding 5**
   $= -1$

Answers

1. 8, 2, −2, −8 2. 3 3. −9
4. 10 5. 2

EXAMPLE 6 Subtract −3 from −9.

Solution Because subtraction is not commutative, we must be sure to write the numbers in the correct order. Because we are subtracting −3, the problem looks like this when translated into symbols:

$$-9 - (-3) = -9 + 3 \quad \text{Change to addition of the opposite}$$
$$= -6 \quad \text{Add}$$

6. Subtract −5 from −3.

EXAMPLE 7 Add −4 to the difference of −2 and 5.

Solution The difference of −2 and 5 is written −2 − 5. Adding −4 to that difference gives us

$$(-2 - 5) + (-4) = -7 + (-4) \quad \text{Simplify inside parentheses}$$
$$= -11 \quad \text{Add}$$

7. Add −8 to the difference of −7 and 4.

Multiplying Real Numbers

Multiplication with whole numbers is simply a shorthand way of writing repeated addition.

For example, 3(−2) can be evaluated as follows:

$$3(-2) = -2 + (-2) + (-2) = -6$$

We can evaluate the product −3(2) in a similar manner if we first apply the commutative property of multiplication:

$$-3(2) = 2(-3) = -3 + (-3) = -6$$

From these results it seems reasonable so say that the product of a positive and a negative is a negative number.

The last case we must consider is the product of two negative numbers, such as −3(−2). To evaluate this product we will look at the expression −3[2 + (−2)] in two different ways. First, since 2 + (−2) = 0, we know the expression −3[2 + (−2)] is equal to 0. On the other hand, we can apply the distributive property to get

$$-3[2 + (-2)] = -3(2) + (-3)(-2) = -6 + ?$$

Because we know the expression is equal to 0, it must be true that our ? is 6 because 6 is the only number we can add to −6 to get 0. Therefore, we have

$$-3(-2) = 6$$

Here is a summary of what we have so far:

ORIGINAL NUMBERS HAVE		THE ANSWER IS
The same sign	3(2) = 6	Positive
Different signs	3(−2) = −6	Negative
Different signs	−3(2) = −6	Negative
The same sign	−3(−2) = 6	Positive

NOTE

This discussion is to show why the rule for multiplication of real numbers is written the way it is. Even if you already know how to multiply positive and negative numbers, it is a good idea to review the "why" that is behind it all.

To Multiply Two Real Numbers

Step 1: Multiply their absolute values.

Step 2: If the two numbers have the *same* sign, the product is positive. If the two numbers have *opposite* signs, the product is negative.

Answers

6. 2 **7.** −19

8. Multiply.

$4(5)$

$4(-5)$

$-4(5)$

$-4(-5)$

EXAMPLE 8 Multiply all combinations of positive and negative 7 and 3.

Solution

$$7(3) = 21$$
$$7(-3) = -21$$
$$-7(3) = -21$$
$$-7(-3) = 21$$

Dividing Real Numbers

DEFINITION

If a and b are any two real numbers, where $b \neq 0$, then the **quotient** of a and b, written $\frac{a}{b}$, is given by

$$\frac{a}{b} = a \cdot \left(\frac{1}{b}\right)$$

Dividing a by b is equivalent to multiplying a by the reciprocal of b. In short, we say, "division is multiplication by the reciprocal."

Because division is defined in terms of multiplication, the same rules hold for assigning the correct sign to a quotient as held for assigning the correct sign to a product; that is, *the quotient of two numbers with like signs is positive, while the quotient of two numbers with unlike signs is negative.*

Divide.

9. $\frac{12}{4}$

10. $\frac{12}{-4}$

11. $\frac{-12}{4}$

12. $\frac{-12}{-4}$

EXAMPLES Divide.

9. $\frac{6}{3} = 6 \cdot \left(\frac{1}{3}\right) = 2$

10. $\frac{6}{-3} = 6 \cdot \left(-\frac{1}{3}\right) = -2$

11. $\frac{-6}{3} = -6 \cdot \left(\frac{1}{3}\right) = -2$

12. $\frac{-6}{-3} = -6 \cdot \left(-\frac{1}{3}\right) = 2$

Notice these examples indicate that if a and b are positive real numbers then

$$\frac{-a}{b} = \frac{a}{-b} = -\frac{a}{b}$$

and

$$\frac{-a}{-b} = \frac{a}{b}$$

The second step in the preceding examples is written only to show that each quotient can be written as a product. It is not actually necessary to show this step when working problems.

In the examples that follow, we find a combination of operations. In each case we use the rule for order of operations.

Answers

8. 20, −20, −20, 20 **9.** 3 **10.** −3 **11.** −3 **12.** 3

EXAMPLES Simplify each expression as much as possible.

13. $(-2-3)(5-9) = (-5)(-4)$ **Simplify inside parentheses**
$= 20$ **Multiply**

14. $2 - 5(7-4) - 6 = 2 - 5(3) - 6$ **Simplify inside parentheses**
$= 2 - 15 - 6$ **Then, multiply**
$= -19$ **Finally, subtract, left to right**

15. $2(4-7)^3 + 3(-2-3)^2 = 2(-3)^3 + 3(-5)^2$ **Simplify inside parentheses**
$= 2(-27) + 3(25)$ **Evaluate numbers with exponents**
$= -54 + 75$ **Multiply**
$= 21$ **Add**

Simplify.
13. $(-6-4)(8-12)$
14. $3 - 7(9-5) - 2$
15. $3(5-8)^3 - 2(-1-1)^2$

We can combine our knowledge of the properties of multiplication with our definition of division to simplify more expressions involving fractions. Here are two examples:

EXAMPLES

16. $6\left(\frac{t}{3}\right) = 6\left(\frac{1}{3}t\right)$ **Dividing by 3 is the same as multiplying by $\frac{1}{3}$**
$= \left(6 \cdot \frac{1}{3}\right)t$ **Associative property**
$= 2t$ **Multiplication**

17. $3\left(\frac{t}{3} - 2\right) = 3 \cdot \frac{t}{3} - 3 \cdot 2$ **Distributive property**
$= t - 6$ **Multiplication**

16. $2\left(\frac{x}{2}\right)$
17. $2\left(\frac{x}{2} - 3\right)$

Our next examples involve more complicated fractions. The fraction bar works like parentheses to separate the numerator from the denominator. Although we don't write expressions this way, here is one way to think of the fraction bar:

$$\frac{-8-8}{-5-3} = (-8-8) \div (-5-3)$$

As you can see, if we apply the rule for order of operations to the expression on the right, we would work inside each set of parentheses first, then divide. Applying this to the expression on the left, we work on the numerator and denominator separately, then we divide or reduce the resulting fraction to lowest terms.

EXAMPLES Simplify as much as possible.

18. $\frac{-8-8}{-5-3} = \frac{-16}{-8}$ **Simplify numerator and denominator separately**
$= 2$ **Divide**

19. $\frac{-5(-4)+2(-3)}{2(-1)-5} = \frac{20-6}{-2-5}$
$= \frac{14}{-7}$
$= -2$

20. $\frac{2^3+3^3}{2^2-3^2} = \frac{8+27}{4-9}$
$= \frac{35}{-5}$
$= -7$

Simplify.
18. $\frac{-6-6}{-5-3}$
19. $\frac{5(-6)+3(-2)}{4(-3)+3}$
20. $\frac{3^3-4^3}{3^2+4^2}$

Remember, since subtraction is defined in terms of addition, we can restate the distributive property in terms of subtraction; that is, if a, b, and c are real numbers, then $a(b-c) = ab - ac$.

Answers
13. 40 **14.** -27 **15.** -89 **16.** x **17.** $x-6$ **18.** $\frac{3}{2}$ **19.** 4 **20.** $-\frac{37}{25}$

21. Simplify $2(5y - 1) - y$.

EXAMPLE 21 Simplify $3(2y - 1) + y$.

Solution We begin by multiplying the 3 and $2y - 1$. Then, we combine similar terms:

$$3(2y - 1) + y = 6y - 3 + y \quad \textbf{Distributive property}$$
$$= 7y - 3 \quad \textbf{Combine similar terms}$$

22. Simplify $6 - 2(5x + 1) + 4x$.

EXAMPLE 22 Simplify $8 - 3(4x - 2) + 5x$.

Solution First we distribute the -3 across the $4x - 2$.

$$8 - 3(4x - 2) + 5x = 8 - 12x + 6 + 5x$$
$$= -7x + 14$$

23. Simplify $4(3a + 1) - (7a - 6)$.

EXAMPLE 23 Simplify $5(2a + 3) - (6a - 4)$.

Solution We begin by applying the distributive property to remove the parentheses. The expression $-(6a - 4)$ can be thought of as $-1(6a - 4)$. Thinking of it in this way allows us to apply the distributive property.

$$-1(6a - 4) = -1(6a) - (-1)(4) = -6a + 4$$

Here is the complete problem:

$$5(2a + 3) - (6a - 4) = 10a + 15 - 6a + 4 \quad \textbf{Distributive property}$$
$$= 4a + 19 \quad \textbf{Combine similar terms}$$

Dividing Fractions

Next we review division with fractions and division with the number 0.

Divide and reduce to lowest terms.

24. $\frac{3}{5} \div \frac{6}{7}$

25. $12 \div \frac{3}{4}$

26. $-\frac{5}{6} \div 10$

EXAMPLES Divide and reduce to lowest terms.

24. $\frac{3}{4} \div \frac{6}{11} = \frac{3}{4} \cdot \frac{11}{6}$ **Definition of division**

$= \frac{33}{24}$ **Multiply numerators, multiply denominators**

$= \frac{11}{8}$ **Divide numerator and denominator by 3**

25. $10 \div \frac{5}{6} = \frac{10}{1} \cdot \frac{6}{5}$ **Definition of division**

$= \frac{60}{5}$ **Multiply numerators, multiply denominators**

$= 12$ **Divide**

26. $-\frac{3}{8} \div 6 = -\frac{3}{8} \cdot \frac{1}{6}$ **Definition of division**

$= -\frac{3}{48}$ **Multiply numerators, multiply denominators**

$= -\frac{1}{16}$ **Divide numerator and denominator by 3**

Answers
21. $9y - 2$ **22.** $-6x + 4$
23. $5a + 10$ **24.** $\frac{7}{10}$ **25.** 16
26. $-\frac{1}{12}$

Division with the Number 0

For every division problem an associated multiplication problem involving the same numbers exists. For example, the following two problems say the same thing about the numbers 2, 3, and 6:

Division	*Multiplication*
$\frac{6}{3} = 2$	$6 = 2(3)$

We can use this relationship between division and multiplication to clarify division involving the number 0.

First of all, dividing 0 by a number other than 0 is allowed and always results in 0. To see this, consider dividing 0 by 5. We know the answer is 0 because of the relationship between multiplication and division. This is how we write it:

$$\frac{0}{5} = 0 \qquad \text{because} \qquad 0 = 0(5)$$

On the other hand, dividing a nonzero number by 0 is not allowed in the real numbers. Suppose we were attempting to divide 5 by 0. We don't know whether there is an answer to this problem, but if there is, let's say the answer is a number that we can represent with the letter n. If 5 divided by 0 is a number n, then

$$\frac{5}{0} = n \qquad \text{and} \qquad 5 = n(0)$$

But this is impossible because no matter what number n is, when we multiply it by 0 the answer must be 0. It can never be 5. In algebra, we say expressions like $\frac{5}{0}$ are undefined because there is no answer to them; that is, division by 0 is not allowed in the real numbers.

Finding the Value of an Algebraic Expression

As we mentioned earlier in this chapter, an algebraic expression is a combination of numbers, variables, and operation symbols. Each of the following is an algebraic expression

$$7a \qquad x^2 - y^2 \qquad 2(3t - 4) \qquad \frac{2x - 5}{6}$$

An expression such as $2(3t - 4)$ will take on different values depending on what number we substitute for t. For example, if we substitute -8 for t then the expression $2(3t - 4)$ becomes $2[3(-8) - 4)]$ which simplifies to -56. If we apply the distributive property to $2(3t - 4)$ we have

$$2(3t - 4) = 6t - 8$$

Substituting -8 for t in the simplified expression gives us $6(-8) - 8 = -56$, which is the same result we obtained previously. As you would expect, substituting the same number into an expression, and any simplified form of that expression, will yield the same result.

27. Evaluate each expression when x is -1.

a. $7(4x - 6)$
b. $28x - 42$
c. $28x - 6$

EXAMPLE 27 Evaluate the expressions $(a + 4)^2$, $a^2 + 16$, and $a^2 + 8a + 16$ when a is -2, 0, and 3.

Solution Organizing our work with a table, we have

a	$(a + 4)^2$	$a^2 + 16$	$a^2 + 8a + 16$
-2	$(-2 + 4)^2 = 4$	$(-2)^2 + 16 = 20$	$(-2)^2 + 8(-2) + 16 = 4$
0	$(0 + 4)^2 = 16$	$0^2 + 16 = 16$	$0^2 + 8(0) + 16 = 16$
3	$(3 + 4)^2 = 49$	$3^2 + 16 = 25$	$3^2 + 8(3) + 16 = 49$

When we study polynomials later in the book, you will see that the expressions $(a + 4)^2$ and $a^2 + 8a + 16$ are equivalent, and that neither one is equivalent to $a^2 + 16$.

Linking Objectives and Examples

Next to each **objective**, list the examples that are best described by that objective.

A	
B	
C	
D	
E	
F	

Getting Ready for Class

After reading through the preceding section, respond in your own words and in complete sentences.

1. For each of the following expressions, give an example of an everyday situation that is modeled by the expression.
   a. $\$35 - \$12 = \$23$
   b. $-\$35 - \$12 = -\$47$
2. For each of the following expressions, give an example of an everyday situation that is modeled by the expression.
   a. $3(-\$25) = -\$75$
   b. $(-\$100) \div 5 = -\$20$
3. Why is division by 0 not allowed?
4. Why isn't the statement "two negatives make a positive" true?

Answer

27. a. -70
b. -70
c. -34

Problem Set 1.4 INTERMEDIATE Algebra Now™

Find each of the following sums.

1. $6 + (-2)$

2. $11 - 5$

3. $-6 + 2$

4. $-11 + 5$

Find each of the following differences.

5. $-7 - 3$

6. $-6 - 9$

7. $-7 - (-3)$

8. $-6 - (-9)$

9. $\frac{3}{4} - \left(-\frac{5}{6}\right)$

10. $\frac{2}{3} - \left(-\frac{7}{5}\right)$

11. $\frac{11}{42} - \frac{17}{30}$

12. $\frac{13}{70} - \frac{19}{42}$

13. Subtract 5 from -3.

14. Subtract -3 from 5.

15. Find the difference of -4 and 8.

16. Find the difference of 8 and -4.

17. Subtract $4x$ from $-3x$.

18. Subtract $-5x$ from $7x$.

19. What number do you subtract from 5 to get -8?

20. What number do you subtract from -3 to get 9?

21. Add -7 to the difference of 2 and 9.

22. Add -3 to the difference of 9 and 2.

23. Subtract $3a$ from the sum of $8a$ and a.

24. Subtract $-3a$ from the sum of $3a$ and $5a$.

Find the following products.

25. $3(-5)$

26. $-3(5)$

27. $-3(-5)$

28. $4(-6)$

29. $2(-3)(4)$

30. $-2(3)(-4)$

31. $-2(5x)$

32. $-5(4x)$

33. $-\frac{1}{3}(-3x)$

34. $-\frac{1}{6}(-6x)$

35. $-\frac{2}{3}\left(-\frac{3}{2}y\right)$

36. $-\frac{2}{5}\left(-\frac{5}{2}y\right)$

37. $-2(4x - 3)$

38. $-2(-5t + 6)$

39. $-\frac{1}{2}(6a - 8)$

40. $-\frac{1}{3}(6a - 9)$

Simplify each expression as much as possible.

41. $3(-4) - 2$

42. $-3(-4) - 2$

43. $4(-3) - 6(-5)$

44. $-6(-3) - 5(-7)$

45. $2 - 5(-4) - 6$

46. $3 - 8(-1) - 7$

47. $4 - 3(7 - 1) - 5$

48. $8 - 5(6 - 3) - 7$

49. $2(-3)^2 - 4(-2)^3$

50. $5(-2)^2 - 2(-3)^3$

51. $(2 - 8)^2 - (3 - 7)^2$

52. $(5 - 8)^2 - (4 - 8)^2$

53. $7(3 - 5)^3 - 2(4 - 7)^3$

54. $3(-7 + 9)^3 - 5(-2 + 4)^3$

55. $-3(2 - 9) - 4(6 - 1)$

56. $-5(5 - 6) - 7(2 - 8)$

57. $2 - 4[3 - 5(-1)]$

58. $6 - 5[2 - 4(-8)]$

59. $(8 - 7)[4 - 7(-2)]$

60. $(6 - 9)[15 - 3(-4)]$

61. $-3 + 4[6 - 8(-3 - 5)]$

62. $-2 + 7[2 - 6(-3 - 4)]$

63. $5 - 6[-3(2 - 9) - 4(8 - 6)]$

64. $9 - 4[-2(4 - 8) - 5(3 - 1)]$

The problems below are problems you will see later in the book. Simplify.

65. $1(-2) - 2(-16) + 1(9)$

66. $6(1) - 1(-5) + 1(2)$

67. $1(1) - 3(-2) + (-2)(-2)$

68. $-2(-14) + 3(-4) - 1(-10)$

69. $-4(0)(-2) - (-1)(1)(1) - 1(2)(3)$

70. $1(0)(1) + 3(1)(4) + (-2)(2)(-1)$

71. $1[0 - (-1)] - 3(2 - 4) + (-2)(-2 - 0)$

72. $-3(-1 - 1) + 4(-2 + 2) - 5[2 - (-2)]$

73. $3(-2)^2 + 2(-2) - 1$

74. $4(-1)^2 + 3(-1) - 2$

75. $2(-2)^3 - 3(-2)^2 + 4(-2) - 8$

76. $5 \cdot 2^3 - 3 \cdot 2^2 + 4 \cdot 2 - 5$

77. $\dfrac{0 - 4}{0 - 2}$

78. $\dfrac{0 + 6}{0 - 3}$

79. $\dfrac{-4 - 4}{-4 - 2}$

80. $\dfrac{6 + 6}{6 - 3}$

81. $\dfrac{-6 + 6}{-6 - 3}$

82. $\dfrac{4 - 4}{4 - 2}$

83. $\dfrac{2 - 4}{2 - 2}$

84. $\dfrac{3 + 6}{3 - 3}$

85. $\dfrac{3 - (-1)}{-3 - 3}$

86. $\dfrac{-1 - 3}{3 - (-3)}$

The problems below are problems you will see later in the book. Apply the distributive property, then simplify if possible.

87. $-1(5 - x)$

88. $-1(a - b)$

89. $-1(7 - x)$

90. $-1(6 - y)$

91. $-3(2x - 3y)$

92. $-1(x - 2z)$

93. $6\left(\dfrac{x}{2} - 3\right)$

94. $6\left(\dfrac{x}{3} + 1\right)$

95. $12\left(\dfrac{a}{4} + \dfrac{1}{2}\right)$

96. $15\left(\dfrac{a}{3} + 2\right)$

97. $15\left(\dfrac{x}{5} + 4\right)$

98. $10\left(\dfrac{x}{2} - 9\right)$

99. $12\left(\dfrac{y}{2} + \dfrac{y}{4} + \dfrac{y}{6}\right)$

100. $12\left(\dfrac{y}{3} - \dfrac{y}{6} + \dfrac{y}{2}\right)$

Simplify each expression.

101. $3(5x + 4) - x$

102. $4(7x + 3) - x$

103. $6 - 7(3 - m)$

104. $3 - 5(5 - m)$

105. $7 - 2(3x - 1) + 4x$

106. $8 - 5(2x - 3) + 4x$

107. $5(3y + 1) - (8y - 5)$

108. $4(6y + 3) - (6y - 6)$

109. $4(2 - 6x) - (3 - 4x)$

110. $7(1 - 2x) - (4 - 10x)$

111. $10 - 4(2x + 1) - (3x - 4)$

112. $7 - 2(3x + 5) - (2x - 3)$

Use the definition of division to write each division problem as a multiplication problem, then simplify.

113. $\frac{8}{-4}$

114. $\frac{-8}{4}$

115. $\frac{4}{0}$

116. $\frac{-7}{0}$

117. $\frac{0}{-3}$

118. $\frac{0}{5}$

119. $-\frac{3}{4} \div \frac{9}{8}$

120. $-\frac{2}{3} \div \frac{4}{9}$

121. $-8 \div \left(-\frac{1}{4}\right)$

122. $-12 \div \left(-\frac{2}{3}\right)$

123. $-40 \div \left(-\frac{5}{8}\right)$

124. $-30 \div \left(-\frac{5}{6}\right)$

125. $\frac{4}{9} \div (-8)$

126. $\frac{3}{7} \div (-6)$

Simplify as much as possible.

127. $\frac{3(-1) - 4(-2)}{8 - 5}$

128. $\frac{6(-4) - 5(-2)}{7 - 6}$

129. $8 - (-6)\left[\frac{2(-3) - 5(4)}{-8(6) - 4}\right]$

130. $-9 - 5\left[\frac{11(-1) - 9}{4(-3) + 2(5)}\right]$

131. $6 - (-3)\left[\frac{2 - 4(3 - 8)}{1 - 5(1 - 3)}\right]$

132. $8 - (-7)\left[\frac{6 - 1(6 - 10)}{4 - 3(5 - 7)}\right]$

Complete each of the following tables.

133.

a	b	SUM $a+b$	DIFFERENCE $a-b$	PRODUCT ab	QUOTIENT $\frac{a}{b}$
3	12				
−3	12				
3	−12				
−3	−12				

134.

a	b	SUM $a+b$	DIFFERENCE $a-b$	PRODUCT ab	QUOTIENT $\frac{a}{b}$
8	2				
−8	2				
8	−2				
−8	−2				

135.

x	$3(5x-2)$	$15x-6$	$15x-2$
−2			
−1			
0			
1			
2			

136.

x	$(x+1)^2$	x^2+1	x^2+2x+1
−2			
−1			
0			
1			
2			

137. Find the value of $-\frac{b}{2a}$ when

a. $a = 3, b = -6$
b. $a = -2, b = 6$
c. $a = -1, b = -2$
d. $a = -0.1, b = 27$

138. Find the value of $b^2 - 4ac$ when

a. $a = 3$, $b = -2$, and $c = 4$
b. $a = 1$, $b = -3$, and $c = -28$
c. $a = 1$, $b = -6$, and $c = 9$
d. $a = 0.1$, $b = -27$, and $c = 1{,}700$

Use a calculator to simplify each expression. If rounding is necessary, round your answers to the nearest ten thousandth (4 places past the decimal point). You will see these problems later in the book.

139. $\frac{1.3802}{0.9031}$

140. $\frac{1.0792}{0.6990}$

141. $\frac{1}{2}(-0.1587)$

142. $\frac{1}{2}(-0.7948)$

143. $\frac{1}{2}\left(\frac{1.2}{1.4} - 1\right)$

144. $\frac{1}{2}\left(\frac{1.3}{1.1} - 1\right)$

145. $\frac{(6.8)(3.9)}{7.8}$

146. $\frac{(2.4)(1.8)}{1.2}$

147. $\frac{0.0005(200)}{(0.25)^2}$

148. $\frac{0.0006(400)}{(0.25)^2}$

149. $-500 + 27(100) - 0.1(100)^2$

150. $-500 + 27(170) - 0.1(170)^2$

151. $-0.05(130)^2 + 9.5(130) - 200$

152. $-0.04(130)^2 + 8.5(130) - 210$

Applying the Concepts

153. Time Zones The continental United States is divided into four time zones. When it is 4:00 in the Pacific zone, it is 5:00 in the Mountain zone, 6:00 in the Central zone, and 7:00 in the Eastern zone.

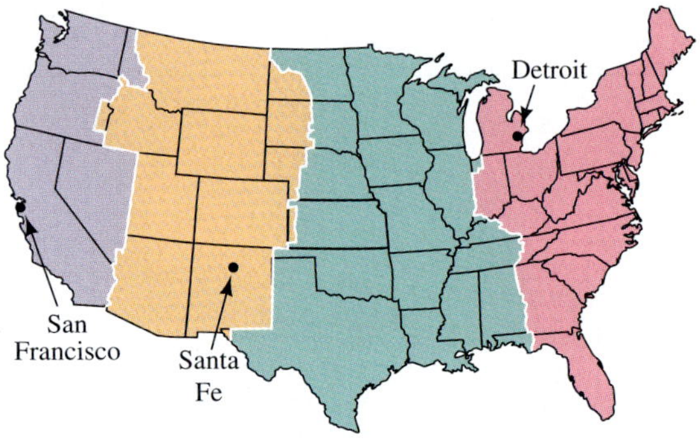

You board a plane at 6:55 P.M. in California (Pacific zone) and take 2 hours and 15 minutes to arrive in Santa Fe, New Mexico (Mountain zone) and another 3 hours and 20 minutes to arrive at your final destination, Detroit, Michigan (Eastern zone). At what local times did you arrive at Santa Fe and Detroit?

154. Oceans and Mountains The deepest ocean depth is 35,840 feet, found in the Pacific Ocean's Mariana Trench. The tallest mountain is Mount Everest, with a height of 29,028 feet. What is the difference between the highest point on earth and the lowest point on earth?

Royalty-Free/Corbis

Improving Your Quantitative Literacy

155. Ironman Triathlon In our number system, everything is in terms of powers of 10. With minutes and seconds, we think in terms of 60s. The chart appeared in *USA Today* in July 2005. The format for the times in the chart is:

hours:minutes:seconds

Find the difference between each of the following Triathlon times:

a. Kirill Litovtsenko and Cam Widoff
b. Ryan Bolton and Thomas Hellriegel
c. Cam Widoff and Simon Lessing

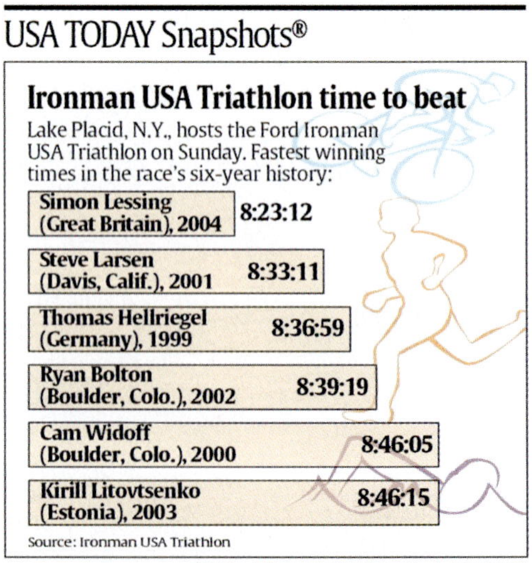

By Ellen J. Horrow and Adrienne Lewis, USA TODAY

Recognizing Patterns

1.5

Much of what we do in mathematics is concerned with recognizing patterns and classifying together groups of numbers that share a common characteristic. For instance, suppose you were asked to give the next number in this sequence:

3, 5, 7, . . .

Looking for a pattern, you may observe that each number is 2 more than the number preceding it. That being the case, the next number in the sequence will be 9 because 9 is 2 more than 7. Reasoning in this manner is called *inductive reasoning.* In mathematics, we use inductive reasoning when we notice a pattern to a sequence of numbers and then extend the sequence using the pattern.

OBJECTIVES	
A	Recognize a pattern in a sequence of numbers.
B	Extend an arithmetic sequence.
C	Extend a geometric sequence.
D	Recognize and extend the Fibonacci sequence.

EXAMPLE 1 Use inductive reasoning to find the next term in each sequence.

a. 5, 8, 11, 14, . . .
b. △, ▷, ▽, ◁, . . .
c. 1, 4, 9, 16, . . .

Solution In each case we use the pattern we observe in the first few terms to write the next term.

a. Each term comes from the previous term by adding 3. Therefore, the next term would be 17.
b. The triangles rotate a quarter turn to the right each time. The next term would be a triangle that points up, △.
c. This looks like the sequence of squares, $1^2, 2^2, 3^2, 4^2, \ldots$. The next term is $5^2 = 25$.

PRACTICE PROBLEMS

1. Use inductive reasoning to find the next term in each sequence below.
   a. 5, 15, 45, 135, . . .
   b. ▷, ▽, ◁, △, . . .
   c. 1, 8, 27, 64, . . .

Now that we have an intuitive idea of inductive reasoning, here is a formal definition.

DEFINITION

Inductive reasoning is reasoning in which a conclusion is drawn based on evidence and observations that support that conclusion. In mathematics this usually involves noticing that a few items in a group have a trait or characteristic in common and then concluding that all items in the group have that same trait.

Arithmetic Sequences

We can extend our work with sequences by classifying together sequences that share a common characteristic. Our first classification is for sequences that are constructed by adding the same number each time.

DEFINITION

An **arithmetic sequence** is a sequence of numbers in which each number (after the first number) comes from adding the same amount to the number before it.

Answer
1. **a.** 405 **b.** ▷ **c.** 125

The sequence

$$4, 7, 10, 13, \ldots$$

is an example of an arithmetic sequence, because each term is obtained from the preceding term by adding 3 each time. The number we add each time—in this case, 3—is the *common difference* because it can be obtained by subtraction.

2. Each sequence below is an arithmetic sequence. Find the next two numbers in each sequence.

a. $6, 16, 26, \ldots$

b. $-\frac{1}{2}, 0, \frac{1}{2}, \ldots$

c. $-5, 0, 5, \ldots$

EXAMPLE 2 Each sequence shown here is an arithmetic sequence. Find the next two numbers in each sequence.

a. $10, 16, 22, \ldots$

b. $\frac{1}{2}, 1, \frac{3}{2}, \ldots$

c. $5, 0, -5, \ldots$

Solution Because we know that each sequence is arithmetic, we know how to look for the number that is added to each term to produce the next consecutive term.

a. $10, 16, 22, \ldots$: Each term is found by adding 6 to the term before it. Therefore, the next two terms will be 28 and 34.

b. $\frac{1}{2}, 1, \frac{3}{2}, \ldots$: Each term comes from the term before it by adding $\frac{1}{2}$. The fourth term will be $\frac{3}{2} + \frac{1}{2} = 2$, while the fifth term will be $2 + \frac{1}{2} = \frac{5}{2}$.

c. $5, 0, -5, \ldots$: Each term comes from adding -5 to the term before it. Therefore, the next two terms will be $-5 + (-5) = -10$, and $-10 + (-5) = -15$.

Geometric Sequences

Our second classification of sequences with a common characteristic involves sequences that are constructed using multiplication.

DEFINITION

A **geometric sequence** is a sequence of numbers in which each number (after the first number) comes from the number before it by multiplying by the same amount each time.

The sequence

$$4, 12, 36, 108, \ldots$$

is a geometric sequence. Each term is obtained from the previous term by multiplying by 3. The amount by which we multiply each term to obtain the next term—in this case, 3—is called the *common ratio.*

3. Each sequence below is a geometric sequence. Find the next number in each sequence.

a. $10, 20, 40, \ldots$

b. $5, -15, 45, \ldots$

c. $2, 1, \frac{1}{2}, \ldots$

EXAMPLE 3 Each sequence shown here is a geometric sequence. Find the next number in each sequence.

a. $2, 10, 50, \ldots$

b. $3, -15, 75, \ldots$

c. $\frac{1}{8}, \frac{1}{4}, \frac{1}{2}, \ldots$

Solution Because each sequence is a geometric sequence, we know that each term is obtained from the previous term by multiplying by the same number each time.

Answer

2. a. 36, 46 **b.** $1, \frac{3}{2}$ **c.** 10, 15

a. 2, 10, 50, . . . : Starting with 2, each number is obtained from the previous number by multiplying by 5 each time. The next number will be $50 \cdot 5 = 250$.

b. 3, −15, 75, . . . : The sequence starts with 3. After that, each number is obtained by multiplying by −5 each time. The next number will be $75(-5) = -375$.

c. $\frac{1}{8}, \frac{1}{4}, \frac{1}{2}, \ldots$: This sequence starts with $\frac{1}{8}$. Multiplying each number in the sequence by 2 produces the next number in the sequence. To extend the sequence we multiply $\frac{1}{2}$ by 2:

$$\frac{1}{2} \cdot 2 = 1$$

The next number in the sequence is 1. ■

The Fibonacci Sequence

In the Study Skills introduction to this chapter we quoted the mathematician Fibonacci. There is a special sequence in mathematics named for Fibonacci.

Fibonacci sequence: 1, 1, 2, 3, 5, 8, . . .

To construct the Fibonacci sequence we start with two 1s. The rest of the numbers in the sequence are found by adding the two previous terms. Adding the first two terms, 1 and 1, we have 2. Then, adding 1 and 2 we have 3. In general, adding any two consecutive terms of the Fibonacci sequence gives us the next term.

A Mathematical Model

One of the reasons we study number sequences is because they can be used to model some of the patterns and events we see in the world around us. The discussion that follows shows how the Fibonacci sequence can be used to predict the number of bees in each generation of the family tree of a male honeybee. It is based on an example from Chapter 2 of the book *Mathematics: A Human Endeavor* by Harold Jacobs. If you find that you enjoy discovering patterns in mathematics, Mr. Jacobs' book has many interesting examples and problems involving patterns in mathematics.

A male honeybee has one parent, its mother, whereas a female honeybee has two parents, a mother and a father. (A male honeybee comes from an unfertilized egg; a female honeybee comes from a fertilized egg.) Using these facts, we construct the family tree of a male honeybee using ♂ to represent a male honeybee and ♀ to represent a female honeybee.

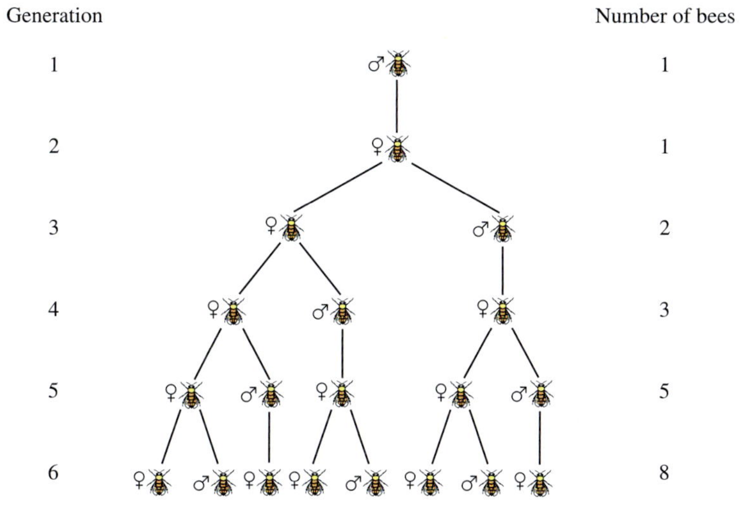

Answer
3. **a.** 80 **b.** −135 **c.** $\frac{1}{4}$

Looking at the numbers in the right column in our diagram, the sequence that gives us the number of bees in each generation of the family tree of a male honeybee is

1 1 2 3 5 8

As you can see, this is the Fibonacci sequence. We have taken our original diagram (the family tree of the male honeybee) and reduced it to a mathematical model (the Fibonacci sequence). The model can be used in place of the diagram to find the number of bees in any generation back from our first bee.

4. Find the number of bees in the 11th generation of the family tree of a male honeybee.

EXAMPLE 4 Find the number of bees in the tenth generation of the family tree of a male honeybee.

Solution We can continue the previous diagram and simply count the number of bees in the tenth generation, or we can use inductive reasoning to conclude that the number of bees in the tenth generation will be the tenth term of the Fibonacci sequence. Let's make it easy on ourselves and find the first ten terms of the Fibonacci sequence.

Generation:	1	2	3	4	5	6	7	8	9	10
Number of bees:	1	1	2	3	5	8	13	21	34	55

As you can see, the number of bees in the tenth generation is 55.

Connecting Two Sequences: Paired Data

David Young-Wolff/PhotoEdit

We can use the discussion that opened this chapter to illustrate how we connect sequences. Suppose an antidepressant has a half-life of 5 days, and its concentration in a patient's system is 80 ng/mL (nanograms/milliliter) when the patient stops taking the antidepressant. To find the concentration after a half-life passes, we multiply the previous concentration by $\frac{1}{2}$. Here are two sequences that together summarize this information.

Days: 0, 5, 10, 15, . . .

Concentration: 80, 40, 20, 10, . . .

The sequence of days is an arithmetic sequence in which we start with 0 and add 5 each time. The sequence of concentrations is a geometric sequence in which we start with 80 and multiply by $\frac{1}{2}$ each time. Here is the same information displayed in a table.

TABLE 1
CONCENTRATION OF AN ANTIDEPRESSANT

DAYS SINCE DISCONTINUING	CONCENTRATION (NG/ML)
0	80
5	40
10	20
15	10

The information in Table 1 is shown visually in Figures 1 and 2. The diagram in Figure 1 is called a *scatter diagram.* If we connect the points in the scatter diagram with straight lines, we have the diagram in Figure 2, which is called a *line graph.* In both figures, the horizontal line that shows days is called the *horizontal axis,* whereas the vertical line that shows concentration is called the *vertical axis.*

Answer
4. 89

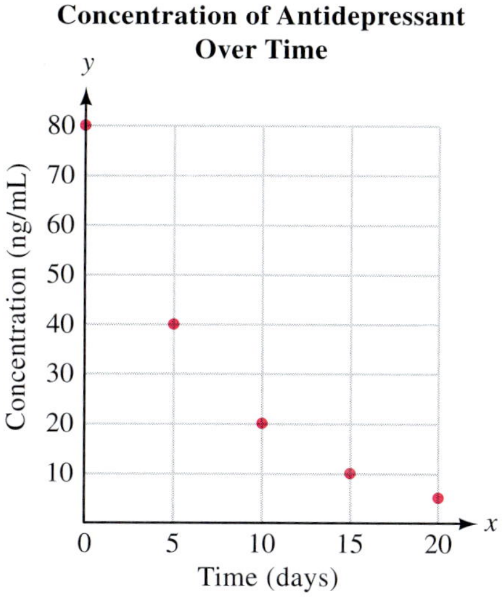

FIGURE 1 **Scatter diagram.**

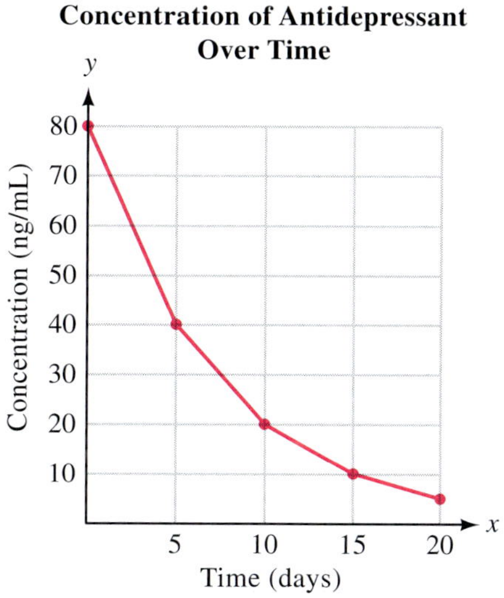

FIGURE 2 **Line graph.**

The data in Table 1 is called *paired data* because each number in the days column is paired with a specific number in the concentration column. For instance, after 15 days, the concentration is 10 ng/mL. Each pair of numbers from Table 1 is associated with one of the dots in the scatter diagram and line graph.

Table 1, Figure 1, and Figure 2 each describe the same relationship. The figures are visual descriptions, whereas the table is a numerical description. The discussion that precedes Table 1 is a written description of that relationship. This gives us three ways to describe the relationship: written, numerical, and visual. As we proceed further into our study of algebra, you will see this same relationship described a fourth way when we give the equation that relates time and concentration: $C = 80 \cdot 2^{-t/5}$.

Getting Ready for Class

After reading through the preceding section, respond in your own words and in complete sentences.

1. What is inductive reasoning?
2. If you were to describe the Fibonacci sequence in words, you would start this way: "The first two numbers are ones. After that, each number is found by . . . ." Finish the sentence so that someone reading it will know how to find the numbers in the Fibonacci sequence.
3. Create an arithmetic sequence and explain how it was formed.
4. Create a geometric sequence and explain how it was formed.

Linking Objectives and Examples

Next to each **objective**, list the examples that are best described by that objective.

A	
B	
C	
D	

Problem Set 1.5

INTERMEDIATE Algebra $f(x)$ Now™

Here are some sequences that we will be referring to throughout the book. Find the next number in each sequence.

1. 1, 2, 3, 4, . . . (The sequence of counting numbers.)

2. 0, 1, 2, 3, . . . (The sequence of whole numbers.)

3. 2, 4, 6, 8, . . . (The sequence of even numbers.)

4. 1, 3, 5, 7, . . . (The sequence of odd numbers.)

5. 1, 4, 9, 16, . . . (The sequence of squares.)

6. 1, 8, 27, 64, . . . (The sequence of cubes.)

Find the next number in each sequence.

7. 1, 8, 15, 22, . .

8. 1, 8, 64, 512, . . .

9. 1, 8, 14, 19, . . .

10. 1, 8, 16, 25, . . .

Give one possibility for the next term in each sequence.

11. △, ◁, ▽, ▷, . . .

12. →, ↓, ←, ↑, . . .

13. △, □, ○, ◬, ⊡, . . .

14. 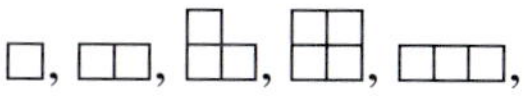, . . .

Each sequence shown here is an arithmetic sequence. In each case, find the next two numbers in the sequence.

15. 1, 5, 9, 13, . . .

16. 10, 16, 22, 28, . . .

17. 1, 0, -1, . . .

18. 6, 0, -6, . . .

19. 5, 2, -1, . . .

20. 8, 4, 0, . . .

21. $\frac{1}{4}, 0, -\frac{1}{4}, \ldots$

22. $\frac{2}{5}, 0, -\frac{2}{5}, \ldots$

23. $1, \frac{3}{2}, 2, \ldots$

24. $\frac{1}{3}, 1, \frac{5}{3}, \ldots$

Each sequence shown here is a geometric sequence. In each case, find the next number in the sequence.

25. 1, 3, 9, . . .

26. 1, 7, 49, . . .

27. 10, −30, 90, . . .

28. 10, −20, 40, . . .

29. $1, \frac{1}{2}, \frac{1}{4}, \ldots$

30. $1, \frac{1}{3}, \frac{1}{9}, \ldots$

31. 20, 10, 5, . . .

32. 8, 4, 2, . . .

33. 5, −25, 125, . . .

34. −4, 16, −64, . . .

35. $1, -\frac{1}{5}, \frac{1}{25}, \ldots$

36. $1, -\frac{1}{2}, \frac{1}{4}, \ldots$

37. Find the next number in the sequence 4, 8, . . . if the sequence is

a. An arithmetic sequence

b. A geometric sequence

38. Find the next number in the sequence 1, −4, . . . if the sequence is

a. An arithmetic sequence

b. A geometric sequence

39. Find the 12th term of the Fibonacci sequence.

40. Find the 13th term of the Fibonacci sequence.

Any number in the Fibonacci sequence is a *Fibonacci number.*

41. Name three Fibonacci numbers that are prime numbers.

42. Name three Fibonacci numbers that are composite numbers.

43. In the first ten terms of the Fibonacci sequence, which ones are even numbers?

44. In the first ten terms of the Fibonacci sequence, which ones are odd numbers?

The patterns in the tables below will become important when we do factoring of trinomials later in the book. Complete each table.

45.

TWO NUMBERS a AND b	THEIR PRODUCT ab	THEIR SUM $a+b$
1, −24		
−1, 24		
2, −12		
−2, 12		
3, −8		
−3, 8		
4, −6		
−4, 6		

46.

TWO NUMBERS a AND b	THEIR PRODUCT ab	THEIR SUM $a+b$
1, −54		
−1, 54		
2, −27		
−2, 27		
3, −18		
−3, 18		
6, −9		
−6, 9		

Applying the Concepts

47. Temperature and Altitude A pilot checks the weather conditions before flying and finds that the air temperature drops 3.5°F every 1,000 feet above the surface of the Earth. (The higher he flies, the colder the air.) If the air temperature is 41°F when the plane reaches 10,000 feet, write a sequence of numbers that gives the air temperature every 1,000 feet as the plane climbs from 10,000 feet to 15,000 feet. Is this sequence an arithmetic sequence?

48. Temperature and Altitude For the plane mentioned in Problem 47, at what altitude will the air temperature be 20°F?

Paul S. Bowen

49. Temperature and Altitude The weather conditions on a certain day are such that the air temperature drops 4.5°F every 1,000 feet above the surface of the Earth. If the air temperature is 41°F at 10,000 feet, write a sequence of numbers that gives the air temperature every 1,000 feet starting at 10,000 feet and ending at 5,000 feet. Is this sequence an arithmetic sequence?

50. Value of a Painting Suppose you own a painting that doubles in value every 5 years. If you bought the painting for $125 in 1990, write a sequence of numbers that gives the value of the painting every 5 years from the time you purchased it until the year 2010. Is this sequence a geometric sequence?

David Young-Wolff/PhotoEdit

Reread the introduction to this chapter and the discussion of paired data at the end of this section, then work the following problems.

51. Half-Life The half-lives of two antidepressants are given below. A patient taking Antidepressant 1 tells his doctor that he begins to feel sick if he misses his morning dose, and a patient taking Antidepressant 2 tells his doctor that he doesn't notice a difference if he misses a day of taking the medication. Explain these situations in terms of the half-life of the medication.

HALF-LIFE	
ANTIDEPRESSANT 1	ANTIDEPRESSANT 2
11 hours	5 days

52. Half-Life Two patients are taking the antidepressants mentioned in Problem 51. Both decide to stop their current medications and take another medication. The physician tells the patient on Antidepressant 2 to simply stop taking it but instructs the patient on Antidepressant 1 to take half the normal dose for 3 days, then one-fourth the normal dose for another 3 days before stopping the medication altogether. Use half-life to explain why the doctor uses two different methods of taking the patients off their medications.

53. Half-Life The half-life of a drug is 4 hours. A patient has been taking the drug on a regular basis for a few months and then discontinues taking it. The concentration of the drug in a patient's system is 60 ng/mL when the patient stops taking the medication. Complete the table, and then use the results to construct a line graph of that data (a template is provided in Figure 3).

HOURS SINCE DISCONTINUING	CONCENTRATION (ng/mL)
0	60
4	
8	
12	
16	

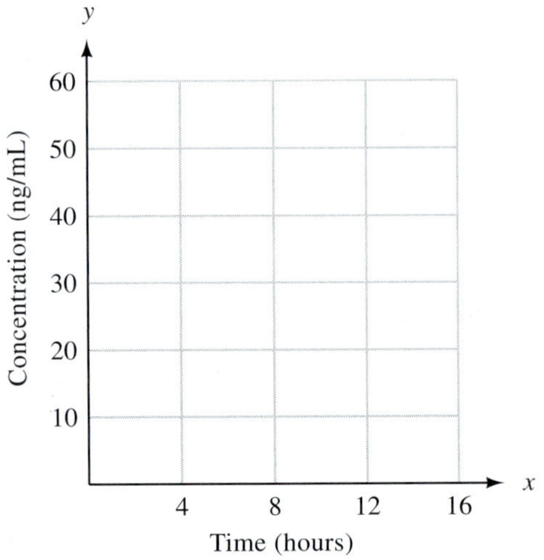

FIGURE 3 **Template for line graph.**

54. Half-Life The half-life of a drug is 8 hours. A patient has been taking the drug on a regular basis for a few months and then discontinues taking it. The concentration of the drug in a patient's system is 120 ng/mL when the patient stops taking the medication. Complete the table, and then use the results to construct a line graph of that data (a template is provided in Figure 4).

HOURS SINCE DISCONTINUING	CONCENTRATION (ng/mL)
0	120
8	
16	
24	
32	

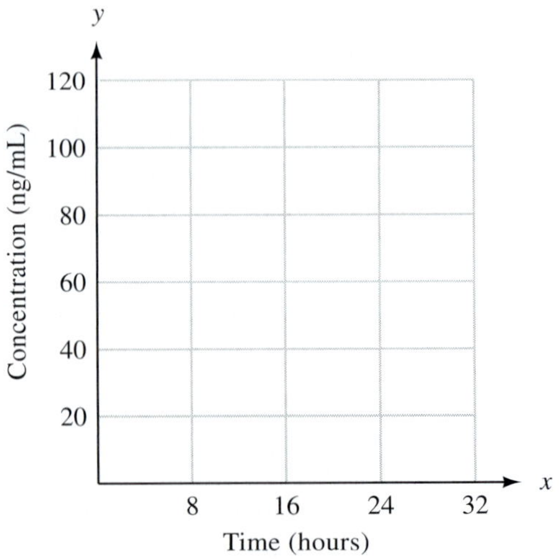

FIGURE 4 **Template for line graph.**

55. Boiling Point The boiling point of water at sea level is 212°F. The boiling point of water drops 1.8°F every 1,000 feet above sea level, and it rises 1.8°F every 1,000 feet below sea level. Complete the following table to write the sequence that gives the boiling points of water from 2,000 feet below sea level to 3,000 feet above sea level.

ELEVATION (ft)	BOILING POINT (°F)
−2,000	
−1,000	
0	
1,000	
2,000	
3,000	

56. Chlorine Level A swimming pool needs 1 ppm (part per million) of chlorine to keep bacteria growth to a minimum. A pool contains 3 ppm chlorine and loses 20% of that amount each day if no more chlorine is added. Fill in the following table, assuming that no additional chlorine is added to a pool containing 3 ppm chlorine. (Round to the nearest hundredth.)

DAY	1	2	3	4	5	6
CHLORINE LEVEL	3					

Extending the Concepts

57. Reading Graphs A patient is taking a prescribed dose of a medication every 4 hours during the day to relieve the symptoms of a cold. Figure 5 shows how the concentration of that medication in the patient's system changes over time. The 0 on the horizontal axis corresponds to the time the patient takes the first dose of the medication.

a. Explain what the steep straight line segments show with regard to the patient and his medication.

b. What has happened to make the graph fall off on the right?

c. What is the maximum concentration of the medication in the patient's system during the time period shown in Figure 5?

d. Find the values of A, B, and C.

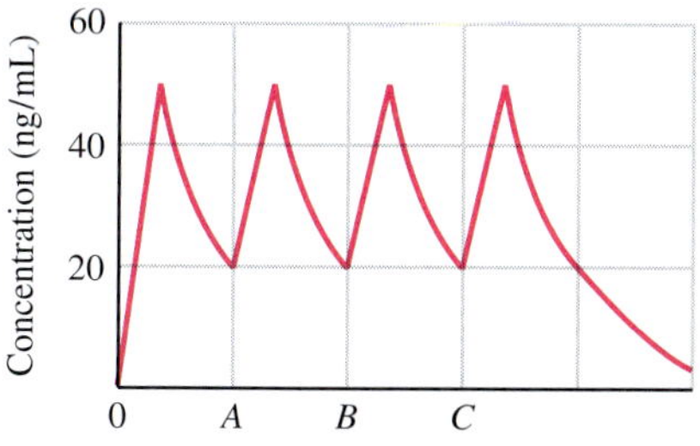

FIGURE 5

58. Reading Graphs Figure 6 shows the number of people in line at a theater box office to buy tickets for a movie that starts at 7:30. The box office opens at 6:45.

a. How many people are in line at 6:30?

b. How many people are in line when the box office opens?

c. How many people are in line when the show starts?

d. At what times are there 60 people in line?

e. How long after the show starts is there no one left in line?

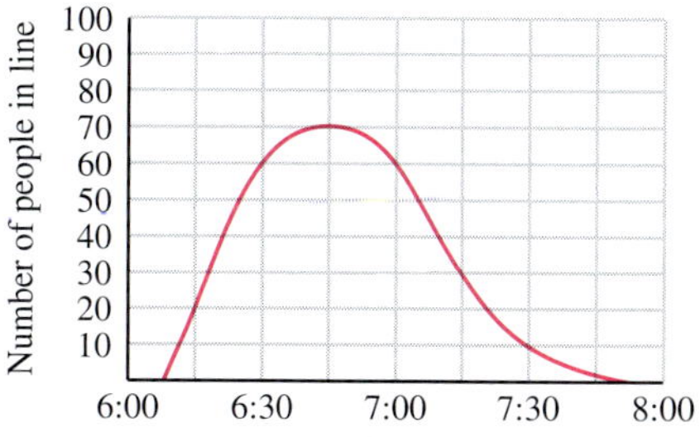

FIGURE 6

59. Shelf Life of Milk The diagram below shows the relationship between the temperature at which the milk is stored and the length of time before it spoils and cannot be consumed. Use the information in the diagram to complete the table. Then use the information in the table to construct a line graph.

SHELF LIFE OF MILK

TEMPERATURE (FAHRENHEIT)	SHELF-LIFE (DAYS)
32°	24
40°	
50°	
60°	
70°	

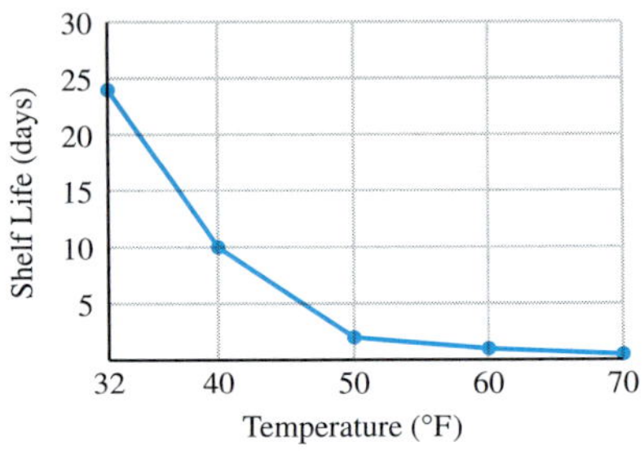

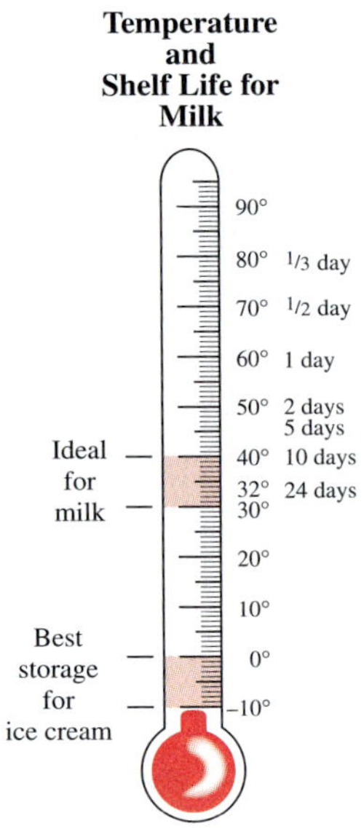

Improving Your Quantitative Literacy

60. Reading a Graph Information that comes to us from the media can answer questions and also raise questions. Sometimes the answers are contained in the data itself, and sometimes the answers come from outside the data.

a. How many launches were there in 1982?

b. How many launches were there in 1985?

c. During which years were there 7 launches?

d. Why were there no shuttle launches in 1987 and 2004?

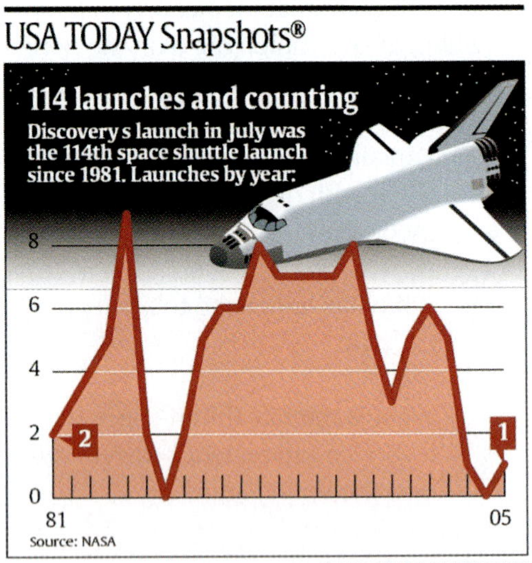

By Marcy E. Mullins, USA TODAY

From *USA Today*. Copyright 2005. Reprinted with permission.

Chapter 1 Summary

The numbers in brackets refer to the section(s) in which the topic can be found.

Exponents [1.1]

Exponents represent notation used to indicate repeated multiplication. In the expression 3^4, 3 is the *base* and 4 is the *exponent.*

$$3^4 = 3 \cdot 3 \cdot 3 \cdot 3 = 81$$

The expression 3^4 is said to be in *exponential form,* whereas the expression $3 \cdot 3 \cdot 3 \cdot 3$ is in *expanded form.*

EXAMPLES

The margins of the chapter summaries will be used for brief examples of the topics being reviewed, whenever it is convenient.

1. $2^5 = 2 \cdot 2 \cdot 2 \cdot 2 \cdot 2 = 32$
   $5^2 = 5 \cdot 5 = 25$
   $10^3 = 10 \cdot 10 \cdot 10 = 1{,}000$
   $1^4 = 1 \cdot 1 \cdot 1 \cdot 1 = 1$

Order of Operations [1.1]

When evaluating a mathematical expression, we will perform the operations in the following order, beginning with the expression in the innermost parentheses or brackets and working our way out.

1. Simplify all numbers with exponents, working from left to right if more than one of these numbers is present.
2. Then, do all multiplications and divisions left to right.
3. Finally, perform all additions and subtractions left to right.

2. $10 + (2 \cdot 3^2 - 4 \cdot 2)$
   $= 10 + (2 \cdot 9 - 4 \cdot 2)$
   $= 10 + (18 - 8)$
   $= 10 + 10$
   $= 20$

Sets [1.1]

A *set* is a collection of objects or things.

The *union* of two sets A and B, written $A \cup B$, is all the elements that are in A *or* are in B.

The *intersection* of two sets A and B, written $A \cap B$, is the set consisting of all elements common to both A *and* B.

Set A is a *subset* of set B, written $A \subset B$, if all elements in set A are also in set B.

3. If $A = \{0, 1, 2\}$ and $B = \{2, 3\}$, then $A \cup B = \{0, 1, 2, 3\}$ and $A \cap B = \{2\}$.

Special Sets [1.2]

Counting numbers = $\{1, 2, 3, \ldots\}$

Whole numbers = $\{0, 1, 2, 3, \ldots\}$

Integers = $\{\ldots, -3, -2, -1, 0, 1, 2, 3, \ldots\}$

Rational numbers = $\left\{\frac{a}{b} \mid a \text{ and } b \text{ are integers}, b \neq 0\right\}$

Irrational numbers = $\{x \mid x \text{ is real, but not rational}\}$

Real numbers = $\{x \mid x \text{ is rational or } x \text{ is irrational}\}$

Prime numbers = $\{2, 3, 5, 7, 11, \ldots\} = \{x \mid x \text{ is a positive integer greater than 1 whose only positive divisors are itself and 1}\}$

4. 5 is a counting number, a whole number, an integer, a rational number, and a real number.
   $\frac{3}{4}$ is a rational number and a real number.
   $\sqrt{2}$ is an irrational number and a real number.

Opposites [1.3, 1.4]

5. The numbers 5 and -5 are opposites; their sum is 0.

$$5 + (-5) = 0$$

Any two real numbers the same distance from 0 on the number line, but in opposite directions from 0, are called *opposites,* or *additive inverses.* Opposites always add to 0.

Reciprocals [1.3, 1.4]

6. The numbers 3 and $\frac{1}{3}$ are reciprocals; their product is 1.

$$3\left(\frac{1}{3}\right) = 1$$

Any two real numbers whose product is 1 are called *reciprocals.* Every real number has a reciprocal except 0.

Absolute Value [1.3]

7. $|5| = 5$
$|-5| = 5$

The *absolute value* of a real number is its distance from 0 on the number line. If $|x|$ represents the absolute value of x, then

$$|x| = \begin{cases} x & \text{if} \quad x \geq 0 \\ -x & \text{if} \quad x < 0 \end{cases}$$

The absolute value of a real number is never negative.

Inequalities [1.2]

8. Graph each inequality mentioned at the right.

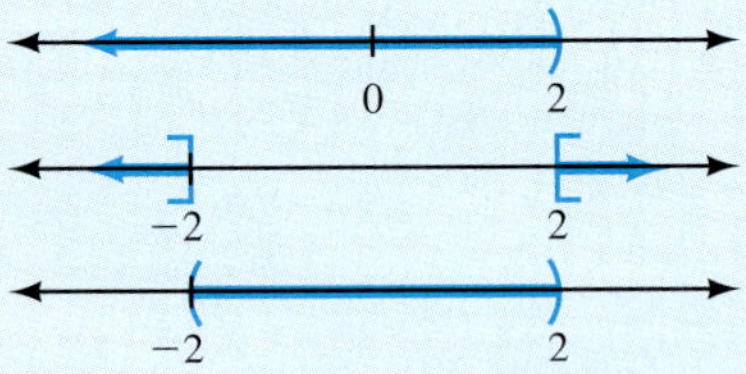

The set $\{x \mid x < 2\}$ is the set of all real numbers that are less than 2. To graph this set we place a right parenthesis at 2 on the real number line and then draw an arrow that starts at 2 and points to the left.

The set $\{x \mid x \leq -2 \text{ or } x \geq 2\}$ is the set of all real numbers that are either less than or equal to -2 or greater than or equal to 2.

The set $\{x \mid -2 < x < 2\}$ is the set of all real numbers that are between -2 and 2; that is, the real numbers that are greater than -2 and less than 2.

Properties of Real Numbers [1.3]

	For Addition	*For Multiplication*
Commutative	$a + b = b + a$	$ab = ba$
Associative	$a + (b + c) = (a + b) + c$	$a(bc) = (ab)c$
Identity	$a + 0 = a$	$a \cdot 1 = a$
Inverse	$a + (-a) = 0$	$a\left(\frac{1}{a}\right) = 1$
Distributive	$a(b + c) = ab + ac$	

Addition [1.4]

To add two real numbers with

1. *The same sign:* Simply add absolute values and use the common sign.
2. *Different signs:* Subtract the smaller absolute value from the larger absolute value. The answer has the same sign as the number with the larger absolute value.

9.
$$5 + 3 = 8$$
$$5 + (-3) = 2$$
$$-5 + 3 = -2$$
$$-5 + (-3) = -8$$

Subtraction [1.4]

If a and b are real numbers,

$$a - b = a + (-b)$$

To subtract b, add the opposite of b.

10.
$$6 - 2 = 6 + (-2) = 4$$
$$6 - (-2) = 6 + 2 = 8$$

Multiplication [1.4]

To multiply two real numbers, simply multiply their absolute values. Like signs give a positive answer. Unlike signs give a negative answer.

11.
$$5(4) = 20$$
$$5(-4) = -20$$
$$-5(4) = -20$$
$$-5(-4) = 20$$

Division [1.4]

If a and b are real numbers and $b \neq 0$, then

$$\frac{a}{b} = a \cdot \left(\frac{1}{b}\right)$$

To divide by b, multiply by the reciprocal of b.

12.
$$\frac{12}{-3} = -4$$
$$\frac{-12}{-3} = 4$$

Inductive Reasoning [1.5]

Inductive reasoning is reasoning in which a conclusion is drawn based on evidence and observations that support that conclusion. In mathematics this usually involves noticing that a few items in a group have a trait or characteristic in common and then concluding that all items in the group have that same trait.

13. We use inductive reasoning when we conclude that the next number in the sequence below is 25.

$$1, 4, 9, 16, \ldots$$

Arithmetic Sequence [1.5]

An *arithmetic sequence* is a sequence of numbers in which each number (after the first number) comes from adding the same amount to the number before it. The number we add to each term to obtain the next term is called the *common difference.*

14. The following sequence is an arithmetic sequence because each term is obtained from the preceding term by adding 3 each time.

$$4, 7, 10, 13, \ldots$$

Geometric Sequence [1.5]

15. The following sequence is a geometric sequence because each term is obtained from the previous term by multiplying by 3 each time.

$$4, 12, 36, 108, \ldots$$

A *geometric sequence* is a sequence of numbers in which each number (after the first number) comes from the number before it by multiplying by the same amount each time. The amount by which we multiply each term to obtain the next term is called the *common ratio.*

COMMON MISTAKES

1. Interpreting absolute value as changing the sign of the number inside the absolute value symbols; that is, $|-5| = +5$, $|+5| = -5$. To avoid this mistake, remember absolute value is defined as a distance and distance is always measured in positive units.
2. Confusing $-(-5)$ with $-|-5|$. The first answer is $+5$, but the second answer is -5.

Chapter 1 Review/Test

The problems below form a comprehensive review of the material in this chapter. They can be used to study for exams. If you would like to take a practice test on this chapter, you can use the odd-numbered problems. Give yourself an hour and work as many of the odd-numbered problems as possible. When you are finished, or when an hour has passed, check your answers with the answers in the back of the book. You can use the even-numbered problems for a second practice test.

The numbers in brackets refer to the section(s) in the text where similar problems can be found.

Translate each expression into symbols. [1.1]

1. The sum of x and 2.

2. The difference of x and 2.

3. The quotient of x and 2.

4. The product of 2 and x.

5. Twice the sum of x and y.

6. The sum of twice x and y.

Expand and multiply. [1.1]

7. 3^3 **8.** 5^3 **9.** 8^2 **10.** 1^8 **11.** 2^5 **12.** 3^4

Simplify each expression. [1.1]

13. $2 + 3 \cdot 5$ **14.** $10 - 2 \cdot 3$ **15.** $20 \div 2 + 3$ **16.** $30 \div 6 + 4 \div 2$

17. $3 + 2(5 - 2)$ **18.** $(10 - 2)(7 - 3)$ **19.** $3 \cdot 4^2 - 2 \cdot 3^2$ **20.** $3 + 5(2 \cdot 3^2 - 10)$

Let $A = \{1, 3, 5\}$, $B = \{2, 4, 6\}$, and $C = \{0, 1, 2, 3, 4\}$, and find each of the following. [1.1]

21. $A \cup B$

22. $A \cap C$

23. $\{x \mid x \in A \quad \text{and} \quad x \notin C\}$

24. $\{x \mid x \in B \quad \text{and} \quad x > 4\}$

25. Locate the numbers $-4, -2.5, -1, 0, 1.5, 3.1, 4.75$ on the number line. [1.2]

26. Locate the numbers $-4.75, -3, -1, 0, 1.5, 2.3, 3$ on the number line. [1.2]

Give the opposite and reciprocal of each number. [1.3]

27. 2

28. $-\frac{2}{5}$

Simplify. [1.3]

29. $|-3|$ **30.** $-|-5|$ **31.** $|-4|$ **32.** $|10 - 16|$

For the set $\{-7, -4.2, -\sqrt{3}, 0, \frac{3}{4}, \pi, 5\}$ list all the elements that are in the following sets: [1.2]

33. Integers **34.** Rational numbers **35.** Irrational numbers

36. Factor 4,356 into the product of prime factors. [1.2]

37. Reduce $\frac{4{,}356}{5{,}148}$ to lowest terms. [1.2]

Multiply. [1.3]

38. $\frac{3}{4} \cdot \frac{8}{5} \cdot \frac{5}{6}$ **39.** $\left(\frac{3}{4}\right)^3$ **40.** $\frac{1}{4} \cdot 8$

Graph each inequality. [1.2]

41. $\{x \mid x < -2 \quad \text{or} \quad x > 3\}$

42. $\{x \mid x > 2 \quad \text{and} \quad x < 5\}$

43. $\{x \mid -3 \le x \le 4\}$

44. $\{x \mid 0 \le x \le 5 \quad \text{or} \quad x > 10\}$

Translate each statement into an equivalent inequality. [1.2]

45. x is at least 4

46. x is no more than 5

47. x is between 0 and 8

48. x is between 0 and 8, inclusive

Combine similar terms. [1.4]

49. $-2y + 4y$

50. $-3x - x + 7x$

51. $3x - 2 + 5x + 7$

52. $2y + 4 - y - 2$

Match each expression on the left with the letter of the appropriate property (or properties) on the right. [1.3]

53. $x + 3 = 3 + x$

54. $(x + 2) + 3 = x + (2 + 3)$

55. $3(x + 4) = 3(4 + x)$

56. $(5x)y = x(5y)$

57. $(x + 2) + y = (x + y) + 2$

58. $3(1) = 3$

59. $5 + 0 = 5$

60. $5 + (-5) = 0$

a. Commutative property of addition
b. Commutative property of multiplication
c. Associative property of addition
d. Associative property of multiplication
e. Additive identity
f. Multiplicative identity
g. Additive inverse
h. Multiplicative inverse

Find the following sums and differences. [1.4]

61. $5 - 3$

62. $-5 - (-3)$

63. $7 + (-2) - 4$

64. $6 - (-3) + 8$

65. $|-4| - |-3| + |-2|$

66. $|7 - 9| - |-3 - 5|$

67. $6 - (-3) - 2 - 5$

68. $2 \cdot 3^2 - 4 \cdot 2^3 + 5 \cdot 4^2$

69. $-\frac{1}{12} - \frac{1}{6} - \frac{1}{4} - \frac{1}{3}$

70. $-\frac{1}{3} - \frac{1}{4} - \frac{1}{6} - \frac{1}{12}$

Find the following products. [1.4]

71. $6(-7)$

72. $-3(5)(-2)$

73. $7(3x)$

74. $-3(2x)$

Apply the distributive property. [1.4]

75. $-2(3x - 5)$

76. $-3(2x - 7)$

77. $-\frac{1}{2}(2x - 6)$

78. $-3(5x - 1)$

Divide. [1.4]

79. $-\frac{5}{8} \div \frac{3}{4}$

80. $-12 \div \frac{1}{3}$

81. $\frac{3}{5} \div 6$

82. $\frac{4}{7} \div (-2)$

Simplify each expression as much as possible. [1.4]

83. $2(-5) - 3$

84. $3(-4) - 5$

85. $6 + 3(-2)$

86. $7 + 2(-4)$

87. $-3(2) - 5(6)$

88. $-4(3)^2 - 2(-1)^3$

89. $8 - 2(6 - 10)$

90. $(8 - 2)(6 - 10)$

91. $\dfrac{3(-4) - 8}{-5 - 5}$

92. $\dfrac{9(-1)^3 - 3(-6)^2}{6 - 9}$

93. $4 - (-2)\left[\dfrac{6 - 3(-4)}{1 + 5(-2)}\right]$

Simplify. [1.3, 1.4]

94. $7 - 2(3y - 1) + 4y$

95. $4(3x - 1) - 5(6x + 2)$

96. $4(2a - 5) - (3a + 2)$

Find the next number in each sequence. Identify any sequences that are arithmetic and any that are geometric. [1.5]

97. 11, 8, 5, 2, . . .

98. 1, 1, 2, 3, 5, . . .

99. $1, \frac{1}{2}, 0, -\frac{1}{2}, \ldots$

100. $1, -\frac{1}{2}, \frac{1}{4}, -\frac{1}{8}, \ldots$

Chapter 1 Projects

Basic Properties and Definitions

Students and Instructors: The end of each chapter in this book will contain a section like this one containing two projects. The group projects are intended to be done in class. The research projects are to be completed outside of class; they can be done in groups or individually. In my classes, I use the research projects for extra credit. I require all research projects to be done on a word processor and to be free of spelling errors.

GAUSS'S METHOD FOR ADDING CONSECUTIVE INTEGERS

Group Project

Number of People 3

Time Needed 15–20 minutes

Equipment Paper and pencil

Background There is a popular story about famous mathematician Karl Friedrich Gauss (1777–1855). As the story goes, a 9-year-old Gauss and the rest of his class were given what the teacher had hoped was busy work. They were told to find the sum of all the whole numbers from 1 to 100. While the rest of the class labored under the assignment, Gauss found the answer within a few moments. He may have set up the problem like this:

Karl Gauss, 1777–1855

1	2	3	4	. . .	98	99	100
100	99	98	97	. . .	3	2	1

Procedure Copy the preceding illustration to get started.

1. Add the numbers in the columns and write your answers below the line.
2. How many numbers do you have below the line?
3. Suppose you add all the numbers below the line. How will your result be related to the sum of all the whole numbers from 1 to 100?
4. Use multiplication to add all the numbers below the line.
5. What is the sum of all the whole numbers from 1 to 100?
6. Use Gauss's method to add all the whole numbers from 1 to 10. (The answer is 55.)
7. Explain, in your own words, Gauss's method for finding this sum so quickly.

SOFIA KOVALEVSKAYA

Research Project

Sofia Kovalevskaya (1850–1891) was born in Moscow, Russia, around the time that Gauss died. Both Gauss and Kovalevskaya showed talent in mathematics early in their lives, but unlike Gauss, Kovalevskaya had to struggle to receive an education and have her work recognized. The Russian people honored her with the commemorative postage stamp shown here. Research the life of Sofia Kovalevskaya, noting the obstacles she faced in pursuing her dream of studying and producing research in mathematics. Then summarize your results in an essay.

2

Yellow Dog Productions/Getty Images

CHAPTER OUTLINE

INTERMEDIATE Algebra $f(x)$ Now™

Equations and Inequalities in One Variable

A recent newspaper article gave the following guideline for college students taking out loans to finance their education: The maximum monthly payment on the amount borrowed should not exceed 8% of their monthly starting salary. In this situation, the maximum monthly payment can be described mathematically with the formula

$$y = 0.08x$$

Using this formula, we can construct a table and a line graph that show the maximum student loan payment that can be made for a variety of starting salaries.

MAXIMUM STUDENT LOAN PAYMENTS

MONTHLY STARTING SALARY	MAXIMUM LOAN PAYMENT
\$2,000	\$160
\$2,500	\$200
\$3,000	\$240
\$3,500	\$280
\$4,000	\$320
\$4,500	\$360
\$5,000	\$400

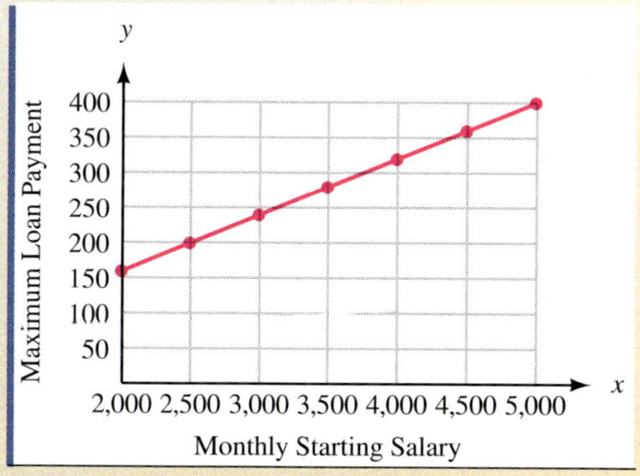

FIGURE 1

In this chapter, we begin our work with connecting tables and line graphs with algebraic formulas.

Visit our web site at http://www.brookscole.com

Study Skills

If you have successfully completed Chapter 1, then you have made a good start at developing the study skills necessary to succeed in all math classes. Here is the list of study skills for this chapter.

1 Imitate Success

Your work should look like the work you see in this book and the work your instructor shows. The steps shown in solving problems in this book were written by someone who has been successful in mathematics. The same is true of your instructor. Your work should imitate the work of people who have been successful in mathematics.

2 List Difficult Problems

Begin to make lists of problems that give you the most difficulty. These are the problems in which you are repeatedly making mistakes.

3 Begin to Develop Confidence with Word Problems

It seems that the main difference between people who are good at working word problems and those who are not is confidence. People with confidence know that no matter how long it takes them, they will eventually be able to solve the problem. Those without confidence begin by saying to themselves, "I'll never be able to work this problem." Are you like that? If you are, what you need to do is put your old ideas about you and word problems aside for a while and make a decision to be successful. Sometimes that's all it takes. Instead of telling yourself that you can't do word problems, that you don't like them, or that they're not good for anything anyway, decide to do whatever it takes to master them.

Getting Ready for Chapter 2

Simplify each expression.

1. $-6\left(\frac{2}{3}\right)$

2. $-24\left(\frac{3}{8}\right)$

3. $\frac{4}{3}(-9)$

4. $\frac{2}{3}\left(-\frac{21}{16}\right)$

5. $-\frac{3}{8}+\left(-\frac{1}{2}\right)$

6. $-\frac{7}{8}+\frac{1}{2}$

7. $\frac{1}{2}(2x)$

8. $\frac{4}{3}\left(\frac{3}{4}x\right)$

9. $24\left(\frac{2}{3}x\right)$

10. $24\left(\frac{2}{3}x+\frac{1}{2}\right)$

11. $100(0.05x)$

12. $0.06x + 0.05(10{,}000 - x)$

13. $3(2y-1)+y$

14. $8-3(4x-2)+5x$

15. Find the value of $2x-1$ when x is

a. 2 **b.** 3 **c.** 5

16. Find the value of $\frac{1}{x+1}$ when x is

a. 1 **b.** 2 **c.** 3

Translate into symbols.

17. Three less than twice a number

18. Ten less than four times a number

19. The sum of x and y is 180.

20. The sum of a and b is 90.

Linear Equations in One Variable

2.1

A ***linear equation in one variable*** is any equation that can be put in the form

$$ax + b = c$$

where a, b, and c are constants and $a \neq 0$. For example, each of the equations

$$5x + 3 = 2 \qquad 2x = 7 \qquad 2x + 5 = 0$$

are linear because they can be put in the form $ax + b = c$. In the first equation, $5x$, 3, and 2 are called *terms* of the equation: $5x$ is a variable term; 3 and 2 are constant terms.

OBJECTIVES

A Solve a linear equation in one variable.

DEFINITION

The **solution set** for an equation is the set of all numbers that, when used in place of the variable, make the equation a true statement.

EXAMPLE 1 The solution set for $2x - 3 = 9$ is $\{6\}$ since replacing x with 6 makes the equation a true statement.

If	$x = 6$	
then	$2x - 3 = 9$	
becomes	$2(6) - 3 = 9$	
	$12 - 3 = 9$	
	$9 = 9$	**A true statement**

DEFINITION

Two or more equations with the same solution set are called **equivalent equations.**

EXAMPLE 2 The equations $2x - 5 = 9$, $x - 1 = 6$, and $x = 7$ are all equivalent equations because the solution set for each is $\{7\}$.

PRACTICE PROBLEMS

1. Show that $x = 3$ is a solution to the equation $4x - 2 = 10$.
2. Show that the equations $3x + 1 = 16$ and $4x - 6 = 14$ both have solution set $\{5\}$.

Properties of Equality

The first property states that adding the same quantity to both sides of an equation preserves equality. Or, more importantly, adding the same amount to both sides of an equation *never changes* the solution set. This property is called the *addition property of equality* and is stated in symbols as follows.

Addition Property of Equality

For any three algebraic expressions, A, B, and C,

if $\quad A = B$

then $\quad A + C = B + C$

In words: Adding the same quantity to both sides of an equation will not change the solution set.

Our second new property is called the *multiplication property of equality* and is stated as follows.

Answers

1–2. See Solutions to Selected Practice Problems.

NOTE

Because subtraction is defined in terms of addition and division is defined in terms of multiplication, we do not need to introduce separate properties for subtraction and division. The solution set for an equation will never be changed by subtracting the same amount from both sides or by dividing both sides by the same nonzero quantity.

3. Solve $\frac{2}{3}x + 4 = -8$

Multiplication Property of Equality

For any three algebraic expressions A, B, and C, where $C \neq 0$,

if $A = B$

then $AC = BC$

In words: Multiplying both sides of an equation by the same nonzero quantity will not change the solution set.

EXAMPLE 3 Solve $\frac{3}{4}x + 5 = -4$.

Solution We begin by adding -5 to both sides of the equation. Once this has been done, we multiply both sides by the reciprocal of $\frac{3}{4}$, which is $\frac{4}{3}$.

$$\frac{3}{4}x + 5 = -4$$

$$\frac{3}{4}x + 5 + (\mathbf{-5}) = -4 + (\mathbf{-5}) \qquad \textbf{Add } -5 \textbf{ to both sides}$$

$$\frac{3}{4}x = -9$$

$$\mathbf{\frac{4}{3}}\left(\frac{3}{4}x\right) = \mathbf{\frac{4}{3}}(-9) \qquad \textbf{Multiply both sides by } \tfrac{4}{3}$$

$$x = -12 \qquad \tfrac{4}{3}(-9) = \tfrac{4}{3}(-\tfrac{9}{1}) = -\tfrac{36}{3} = -12$$

4. Find the solution set for

$3a - 3 = -5a + 9$

EXAMPLE 4 Find the solution set for $3a - 5 = -6a + 1$.

Solution To solve for a we must isolate it on one side of the equation. Let's decide to isolate a on the left side by adding $6a$ to both sides of the equation.

$$3a - 5 = -6a + 1$$

$$3a + \mathbf{6a} - 5 = -6a + \mathbf{6a} + 1 \qquad \textbf{Add } 6a \textbf{ to both sides}$$

$$9a - 5 = 1$$

$$9a - 5 + \mathbf{5} = 1 + \mathbf{5} \qquad \textbf{Add 5 to both sides}$$

$$9a = 6$$

$$\mathbf{\frac{1}{9}}(9a) = \mathbf{\frac{1}{9}}(6) \qquad \textbf{Multiply both sides by } \tfrac{1}{9}$$

$$a = \frac{2}{3} \qquad \tfrac{1}{9}(6) = \tfrac{6}{9} = \tfrac{2}{3}$$

The solution set is $\left\{\frac{2}{3}\right\}$.

NOTE

From the previous chapter we know that multiplication by a number and division by its reciprocal always produce the same result. Because of this fact, instead of multiplying each side of our equation by $\frac{1}{9}$, we could just as easily divide each side by 9. If we did so, the last two lines in our solution would look like this:

$$\frac{9a}{9} = \frac{6}{9}$$

$$a = \frac{2}{3}$$

We can check our solution in Example 4 by replacing a in the original equation with $\frac{2}{3}$.

When $a = \frac{2}{3}$

the equation $3a - 5 = -6a + 1$

becomes $3\left(\frac{2}{3}\right) - 5 = -6\left(\frac{2}{3}\right) + 1$

$$2 - 5 = -4 + 1$$

$$-3 = -3 \qquad \textbf{A true statement}$$

Answers

3. -18 **4.** $\left\{\frac{3}{2}\right\}$

There will be times when we solve equations and end up with a negative sign in front of the variable. The next example shows how to handle this situation.

EXAMPLE 5 Solve each equation.

a. $-x = 4$ **b.** $-y = -8$

Solution Neither equation can be considered solved because of the negative sign in front of the variable. To eliminate the negative signs we simply multiply both sides of each equation by -1.

a. $-x = 4$ **b.** $-y = -8$

$\mathbf{-1}(-x) = \mathbf{-1}(4)$ $\mathbf{-1}(-y) = \mathbf{-1}(-8)$ **Multiply each side by −1**

$x = -4$ $y = 8$

EXAMPLE 6 Solve $\frac{2}{3}x + \frac{1}{2} = -\frac{3}{8}$.

Solution We can solve this equation by applying our properties and working with fractions, or we can begin by eliminating the fractions. Let's use both methods.

Method 1 Working with the fractions.

$$\frac{2}{3}x + \frac{1}{2} + \left(-\frac{\mathbf{1}}{\mathbf{2}}\right) = -\frac{3}{8} + \left(-\frac{\mathbf{1}}{\mathbf{2}}\right) \quad \textbf{Add } -\tfrac{1}{2} \textbf{ to each side}$$

$$\frac{2}{3}x = -\frac{7}{8} \quad -\tfrac{3}{8} + \left(-\tfrac{1}{2}\right) = -\tfrac{3}{8} + \left(-\tfrac{4}{8}\right)$$

$$\frac{\mathbf{3}}{\mathbf{2}}\left(\frac{2}{3}x\right) = \frac{\mathbf{3}}{\mathbf{2}}\left(-\frac{7}{8}\right) \quad \textbf{Multiply each side by } \tfrac{3}{2}$$

$$x = -\frac{21}{16}$$

Method 2 Eliminating the fractions in the beginning.

Our original equation has denominators of 3, 2, and 8. The least common denominator, abbreviated LCD, for these three denominators is 24, and it has the property that all three denominators will divide it evenly. If we multiply both sides of our equation by 24, each denominator will divide into 24, and we will be left with an equation that does not contain any denominators other than 1.

$$\mathbf{24}\left(\frac{2}{3}x + \frac{1}{2}\right) = \mathbf{24}\left(-\frac{3}{8}\right) \quad \textbf{Multiply each side by the LCD 24}$$

$$24\left(\frac{2}{3}x\right) + 24\left(\frac{1}{2}\right) = 24\left(-\frac{3}{8}\right) \quad \textbf{Distributive property on the left side}$$

$$16x + 12 = -9 \quad \textbf{Multiply}$$

$$16x = -21 \quad \textbf{Add } -12 \textbf{ to each side}$$

$$x = -\frac{21}{16} \quad \textbf{Multiply each side by } \tfrac{1}{16}$$

Check To check our solution, we substitute $x = -\frac{21}{16}$ back into our original equation to obtain

$$\frac{2}{3}\left(-\frac{21}{16}\right) + \frac{1}{2} \stackrel{?}{=} -\frac{3}{8}$$

$$-\frac{7}{8} + \frac{1}{2} \stackrel{?}{=} -\frac{3}{8}$$

$$-\frac{7}{8} + \frac{4}{8} \stackrel{?}{=} -\frac{3}{8}$$

$$-\frac{3}{8} = -\frac{3}{8} \quad \textbf{A true statement}$$

5. Solve each equation.

a. $-x = \frac{2}{3}$

b. $-y = -4$

6. Solve $\frac{3}{5}x + \frac{1}{3} = -\frac{5}{6}$.

NOTE

We are placing a question mark over the equal sign because we don't know yet if the expression on the left will be equal to the expression on the right.

Answers

5. a. $x = -\frac{2}{3}$ **b.** $y = 4$

6. $-\frac{35}{18}$

7. Solve.
$0.08x + 0.10(8{,}000 - x) = 680$

EXAMPLE 7

Solve the equation $0.06x + 0.05(10{,}000 - x) = 560$.

Solution We can solve the equation in its original form by working with the decimals, or we can eliminate the decimals first by using the multiplication property of equality and solve the resulting equation. Here are both methods.

Method 1 Working with the decimals.

$0.06x + 0.05(10{,}000 - x) = 560$	**Original equation**
$0.06x + 0.05(10{,}000) - 0.05x = 560$	**Distributive property**
$0.01x + 500 = 560$	**Simplify the left side**
$0.01x + 500 + (\mathbf{-500}) = 560 + (\mathbf{-500})$	**Add −500 to each side**
$0.01x = 60$	
$\dfrac{0.01x}{\mathbf{0.01}} = \dfrac{60}{\mathbf{0.01}}$	**Divide each side by 0.01**
$x = 6{,}000$	

Method 2 Eliminating the decimals in the beginning: To move the decimal point two places to the right in $0.06x$ and 0.05, we multiply each side of the equation by 100.

$0.06x + 0.05(10{,}000 - x) = 560$	**Original equation**
$0.06x + 500 - 0.05x = 560$	**Distributive property**
$\mathbf{100}(0.06x) + \mathbf{100}(500) - \mathbf{100}(0.05x) = \mathbf{100}(560)$	**Multiply each side by 100**
$6x + 50{,}000 - 5x = 56{,}000$	
$x + 50{,}000 = 56{,}000$	**Simplify the left side**
$x = 6{,}000$	**Add −50,000 to each side**

Using either method, the solution to our equation is 6,000.

Check We check our work (to be sure we have not made a mistake in applying the properties or in arithmetic) by substituting 6,000 into our original equation and simplifying each side of the result separately, as the following shows.

$0.06(\mathbf{6{,}000}) + 0.05(10{,}000 - \mathbf{6{,}000}) \stackrel{?}{=} 560$	
$0.06(6{,}000) + 0.05(4{,}000) \stackrel{?}{=} 560$	
$360 + 200 \stackrel{?}{=} 560$	
$560 = 560$	**A true statement**

Answer
7. 6,000

Here is a list of steps to use as a guideline for solving linear equations in one variable.

Strategy for Solving Linear Equations in One

Step 1: **a.** Use the distributive property to separate terms, if necessary.

b. If fractions are present, consider multiplying both sides by the LCD to eliminate the fractions. If decimals are present, consider multiplying both sides by a power of 10 to clear the equation of decimals.

c. Combine similar terms on each side of the equation.

Step 2: Use the addition property of equality to get all variable terms on one side of the equation and all constant terms on the other side. A **variable term** is a term that contains the variable. A **constant term** is a term that does not contain the variable (the number 3, for example).

Step 3: Use the multiplication property of equality to get the variable by itself on one side of the equation.

Step 4: Check your solution in the original equation to be sure that you have not made a mistake in the solution process.

As you will see as you work through the problems in the problem set, it is not always necessary to use all four steps when solving equations. The number of steps used depends on the equation. In Example 8 there are no fractions or decimals in the original equation, so Step 1b will not be used.

EXAMPLE 8

Solve $3(2y - 1) + y = 5y + 3$.

Solution Applying the steps outlined in the preceding strategy, we have

Step 1: **a.** $3(2y - 1) + y = 5y + 3$ **Original equation**

$6y - 3 + y = 5y + 3$ **Distributive property**

c. $7y - 3 = 5y + 3$ **Simplify**

Step 2: $7y + (\mathbf{-5y}) - 3 = 5y + (\mathbf{-5y}) + 3$ **Add $-5y$ to both sides**

$2y - 3 = 3$

$2y - 3 + \mathbf{3} = 3 + \mathbf{3}$ **Add +3 to both sides**

$2y = 6$

Step 3: $\frac{1}{2}(2y) = \frac{1}{2}(6)$ **Multiply by $\frac{1}{2}$**

$y = 3$

Step 4: *Check* When $y = 3$

the equation $3(2y - 1) + y = 5y + 3$

becomes $3(2 \cdot \mathbf{3} - 1) + \mathbf{3} \stackrel{?}{=} 5 \cdot \mathbf{3} + 3$

$3(5) + 3 \stackrel{?}{=} 15 + 3$

$18 = 18$ **A true statement**

8. Solve
$2(5y - 3) + 4 = 9y - 6$

Answer
8. -4

9. Solve for x.
$6 - 2(5x - 1) + 4x = 20$

NOTE

It would be a mistake to subtract 3 from 8 first because the rule for order of operations indicates we are to do multiplication before subtraction.

EXAMPLE 9 Solve the equation $8 - 3(4x - 2) + 5x = 35$.

Solution We must begin by distributing the -3 across the quantity $4x - 2$.

Step 1:	**a.**	$8 - 3(4x - 2) + 5x = 35$	**Original equation**
	b.	$8 - 12x + 6 + 5x = 35$	**Distributive property**
	c.	$-7x + 14 = 35$	**Simplify**
Step 2:		$-7x = 21$	**Add −14 to each side**
Step 3:		$x = -3$	**Multiply by** $-\frac{1}{7}$

Step 4: When x is replaced by -3 in the original equation, a true statement results. Therefore, -3 is the solution to our equation.

Identities and Equations with No Solution

Two special cases are associated with solving linear equations in one variable, each of which is illustrated in the following examples.

10. Solve.
$3(5x + 1) = 10 + 15x$

EXAMPLE 10 Solve for x: $2(3x - 4) = 3 + 6x$.

Solution Applying the distributive property to the left side gives us

$6x - 8 = 3 + 6x$ **Distributive property**

Now, if we add $-6x$ to each side, we are left with the following

$$-8 = 3$$

which is a false statement. This means that there is no solution to our equation. Any number we substitute for x in the original equation will lead to a similar false statement.

11. Solve.
$-4 + 8x = 2(4x - 2)$

EXAMPLE 11 Solve for x: $-15 + 3x = 3(x - 5)$.

Solution We start by applying the distributive property to the right side.

$-15 + 3x = 3x - 15$ **Distributive property**

If we add $-3x$ to each side, we are left with the true statement

$$-15 = -15$$

In this case, our result tells us that any number we use in place of x in the original equation will lead to a true statement. Therefore, all real numbers are solutions to our equation. We say the original equation is an *identity* because the left side is always identically equal to the right side.

Linking Objectives and Examples

Next to each **objective**, list the examples that are best described by that objective.

A	

Getting Ready for Class

After reading through the preceding section, respond in your own words and in complete sentences.

1. What is a solution to an equation?
2. What are equivalent equations?
3. Describe how to eliminate fractions in an equation.
4. Suppose when solving an equation your result is the statement "$3 = -3$." What would you conclude about the solution to the equation?

Answers
9. -2 10. No solution
11. Identity

Problem Set 2.1

INTERMEDIATE Algebra $f(x)$ Now™

Solve each of the following equations.

1. $x - 5 = 3$
2. $x + 2 = 7$
3. $2x - 4 = 6$
4. $3x - 5 = 4$
5. $7 = 4a - 1$
6. $10 = 3a - 5$
7. $3 - y = 10$
8. $5 - 2y = 11$
9. $-3 - 4x = 15$
10. $-8 - 5x = -6$
11. $-3 = 5 + 2x$
12. $-12 = 6 + 9x$
13. $-300y + 100 = 500$
14. $-20y + 80 = 30$
15. $160 = -50x - 40$
16. $110 = -60x - 50$
17. $-x = 2$
18. $-x = \frac{1}{2}$
19. $-a = -\frac{3}{4}$
20. $-a = -5$
21. $\frac{2}{3}x = 8$
22. $\frac{3}{2}x = 9$
23. $-\frac{3}{5}a + 2 = 8$
24. $-\frac{5}{3}a + 3 = 23$
25. $8 = 6 + \frac{2}{7}y$
26. $1 = 4 + \frac{3}{7}y$
27. $2x - 5 = 3x + 2$
28. $5x - 1 = 4x + 3$
29. $-3a + 2 = -2a - 1$
30. $-4a - 8 = -3a + 7$

31. $5 - 2x = 3x + 1$

32. $7 - 3x = 8x - 4$

33. $2x - 3 - x = 3x + 5$

34. $3x - 5 - 2x = 2x - 3$

35. $11x - 5 + 4x - 2 = 8x$

36. $2x + 7 - 3x + 4 = -2x$

37. $5(y + 2) - 4(y + 1) = 3$

38. $6(y - 3) - 5(y + 2) = 8$

39. $6 - 7(m - 3) = -1$

40. $3 - 5(2m - 5) = -2$

Solve each equation.

41. $4(a - 3) + 5 = 7(3a - 1)$

42. $6(a - 4) + 6 = 2(5a + 2)$

43. $7 + 3(x + 2) = 4(x - 1)$

44. $5 + 2(4x - 4) = 3(2x - 1)$

45. $5 = 7 - 2(3x - 1) + 4x$

46. $20 = 8 - 5(2x - 3) + 4x$

47. $\frac{1}{2}x + \frac{1}{4} = \frac{1}{3}x + \frac{5}{4}$

48. $\frac{2}{3}x - \frac{3}{4} = \frac{1}{6}x + \frac{21}{4}$

49. $-\frac{2}{5}x + \frac{2}{15} = \frac{2}{3}$

50. $-\frac{1}{6}x + \frac{2}{3} = \frac{1}{4}$

51. $\frac{3}{4}(8x - 4) = \frac{2}{3}(6x - 9)$

52. $\frac{3}{5}(5x + 10) = \frac{5}{6}(12x - 18)$

53. $\frac{1}{4}(12a + 1) - \frac{1}{4} = 5$

54. $\frac{2}{3}(6x - 1) + \frac{2}{3} = 4$

55. $\frac{1}{2}x + \frac{1}{3}x + \frac{1}{4}x = 13$

56. $\frac{1}{3}x + \frac{1}{4}x + \frac{1}{5}x = 47$

57. $0.08x + 0.09(9{,}000 - x) = 750$

58. $0.08x + 0.09(9{,}000 - x) = 500$

59. $0.12x + 0.10(15{,}000 - x) = 1{,}600$

60. $0.09x + 0.11(11{,}000 - x) = 1{,}150$

61. $0.35x - 0.2 = 0.15x + 0.1$

62. $0.25x - 0.05 = 0.2x + 0.15$

63. $0.42 - 0.18x = 0.48x - 0.24$

64. $0.3 - 0.12x = 0.18x + 0.06$

Solve each equation, if possible.

65. $3x - 6 = 3(x + 4)$

66. $7x - 14 = 7(x - 2)$

67. $4y + 2 - 3y + 5 = 3 + y + 4$

68. $7y + 5 - 2y - 3 = 6 + 5y - 4$

69. $2(4t - 1) + 3 = 5t + 4 + 3t$

70. $5(2t - 1) + 1 = 2t - 4 + 8t$

Now that you have practiced solving a variety of equations, we can turn our attention to the type of equation you will see as you progress through the book. Each equation appears later in the book exactly as you see it below.

Solve each equation.

71. $x - 6 = 0$

72. $x + 4 = 0$

73. $2x + 1 = 0$

74. $3x + 1 = 0$

75. $3x + 2 = 0$

76. $5x - 4 = 0$

77. $0 = 6{,}400a + 70$

78. $0 = 6{,}400a + 60$

Solve each equation.

79. $x + 2 = 2x$

80. $x + 2 = 7x$

81. $0.07x = 1.4$

82. $0.02x = 0.3$

83. $5(2x + 1) = 12$

84. $4(3x - 2) = 21$

85. $50 = \frac{K}{48}$

86. $50 = \frac{K}{24}$

87. $100P = 2{,}400$

88. $3.5d = 16(3.5)^2$

89. $x + (3x + 2) = 26$

90. $2(1) + y = 4$

91. $2x - 3(3x - 5) = -6$

92. $2(2y + 6) + 3y = 5$

93. $5\left(\frac{19}{15}\right) + 5y = 9$

94. $4\left(\frac{19}{15}\right) - 2y = 4$

95. $2\left(\frac{29}{22}\right) - 3y = 4$

96. $2x - 3\left(-\frac{5}{11}\right) = 4$

Applying the Concepts

97. Cost of a Taxi Ride The taximeter was invented in 1891 by Wilhelm Bruhn. The city of Chicago charges \$1.80 plus \$0.40 per mile for a taxi ride.

Jan Halaska/Index Stock Imagery/PictureQuest

a. A woman paid a fare of \$6.60. Write an equation that connects the fare the woman paid, the miles she traveled, n, and the charges the taximeter computes.

b. Solve the equation from part (a) to determine how many miles the woman traveled.

98. Coughs and Earaches In 1992, twice as many people visited their doctor because of a cough than an earache. The total number of doctor's visits for these two ailments was reported to be 45 million.

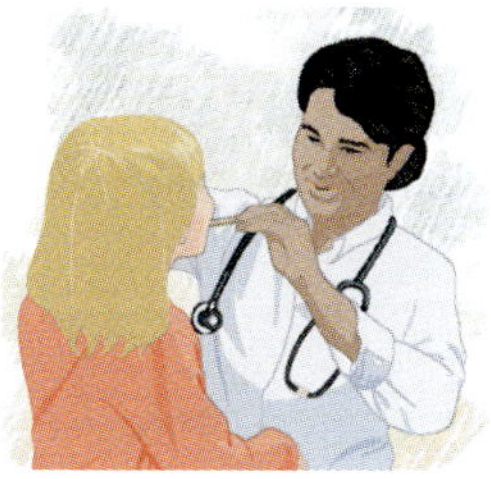

a. Let x represent the number of earaches reported in 1992, then write an expression using x for the number of coughs reported in 1992.

b. Write an equation that relates 45 million to the variable x.

c. Solve the equation from part (b) to determine the number of people who visited their doctor in 1992 to report an earache.

99. Population Density In July 2001 the population of Puerto Rico was estimated to be 3,937,000 people, with a population density of 1,125 people per square mile.

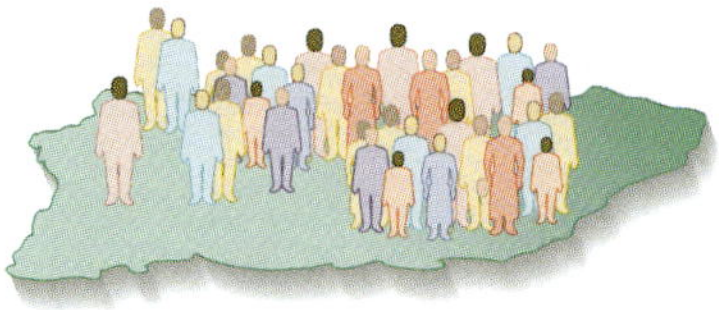

a. Let A represent the area of Puerto Rico in square miles, and write an equation that shows that the population is equal to the product of the area and the population density.

b. Solve the equation from Part (a), rounding your solution to the nearest square mile.

100. Solving Equations by Trial and Error Sometimes equations can be solved most easily by trial and error. Solve the following equations by trial and error.

a. Find x and y if $x \cdot y + 1 = 36$, and both x and y are prime.

b. Find w, t, and z if $w + t + z + 10 = 52$, and w, t, and z are consecutive terms of a Fibonacci sequence.

c. Find x and y if $x \neq y$ and $x^y = y^x$.

Getting Ready for the Next Section

Problems under this heading *Getting Ready for the Next Section* are problems that you must be able to work in order to understand the material in the next section. In this case, the problems below are variations on the types of problems you have already worked in this problem set. They are exactly the type of problems you will see in explanations and examples in the next section.

Solve each equation.

101. $x \cdot 42 = 21$

102. $x \cdot 84 = 21$

103. $25 = 0.4x$

104. $35 = 0.4x$

105. $12 - 4y = 12$

106. $-6 - 3y = 6$

107. $525 = 900 - 300p$

108. $375 = 900 - 300p$

109. $486.7 = 78.5 + 31.4h$

110. $486.7 = 113.0 + 37.7h$

111. Find the value of $2x - 1$ when x is

a. 2 **b.** 3 **c.** 5

112. Find the value of $\frac{1}{x+1}$ when x is

a. 1 **b.** 2 **c.** 3

Maintaining Your Skills

From this point on, each problem set will contain a number of problems under the heading *Maintaining Your Skills.* These problems cover the most important skills you have learned in previous sections and chapters. Hopefully, by working these problems on a regular basis, you will keep yourself current on all the topics we have covered, and possibly need less time to study for tests and quizzes.

Identify the property (or properties) that justifies each of the following statements.

113. $ax = xa$

114. $5\left(\frac{1}{5}\right) = 1$

115. $3 + (x + y) = (3 + x) + y$

116. $3 + (x + y) = (x + y) + 3$

117. $3 + (x + y) = (3 + y) + x$

118. $7(3x - 5) = 21x - 35$

119. $5(1) = 5$

120. $5 + 0 = 5$

121. $4(xy) = 4(yx)$

122. $4(xy) = (4y)x$

123. $2 + 0 = 2$

124. $2 + (-2) = 0$

Formulas 2.2

A *formula* in mathematics is an equation that contains more than one variable. Some formulas are probably already familiar to you—for example, the formula for the area (A) of a rectangle with length l and width w is $A = lw$.

To begin our work with formulas, we will consider some examples in which we are given numerical replacements for all but one of the variables.

OBJECTIVES	
A	Solve a formula with numerical replacements for all but one of its variables.
B	Solve formulas for the indicated variable.
C	Solve basic percent problems by translating them into equations.
D	Find terms of a sequence using the general term.

EXAMPLE 1 Find y when x is 4 in the formula $3x - 4y = 2$.

Solution We substitute 4 for x in the formula and then solve for y:

When $x = 4$

the formula $3x - 4y = 2$

becomes $3(4) - 4y = 2$

$12 - 4y = 2$ **Multiply 3 and 4**

$-4y = -10$ **Add −12 to each side**

$y = \frac{5}{2}$ **Divide each side by −4**

Note that in the last line of Example 1 we divided each side of the equation by −4. Remember that this is equivalent to multiplying each side of the equation by $-\frac{1}{4}$.

PRACTICE PROBLEMS

1. Find y when x is −3 in $2x - 3y = 6$.

EXAMPLE 2 A store selling art supplies finds that they can sell x sketch pads each week at a price of p dollars each, according to the formula $x = 900 - 300p$. What price should they charge for each sketch pad if they want to sell 525 pads each week?

Solution Here we are given a formula, $x = 900 - 300p$, and asked to find the value of p if x is 525. To do so, we simply substitute 525 for x and solve for p:

When $x = 525$

the formula $x = 900 - 300p$

becomes $525 = 900 - 300p$

$-375 = -300p$ **Add −900 to each side**

$1.25 = p$ **Divide each side by −300**

To sell 525 sketch pads, the store should charge \$1.25 for each pad.

2. Repeat Example 2 if they want to sell 375 pads each week.

Our next example involves a formula that contains the number π. In this book, you can use either 3.14 or $\frac{22}{7}$ as an approximation for π. Generally speaking, if the problem contains decimals, use 3.14 to approximate π.

Answers
1. −4 2. \$1.75

3. Repeat Example 3 using 6 centimeters for the radius of each can. Round your answer to the nearest tenth.

EXAMPLE 3 A company manufacturing coffee cans uses 486.7 square centimeters of material to make each can. Because of the way the cans are to be placed in boxes for shipping, the radius of each can must be 5 centimeters. If the formula for the surface area of a can is $S = \pi r^2 + 2\pi rh$ (the cans are open at the top), find the height of each can. (Use 3.14 as an approximation for π.)

Solution Substituting 486.7 for S, 5 for r, and 3.14 for π into the formula $S = \pi r^2 + 2\pi rh$, we have

Surface area = 486.7 cm²

$$486.7 = (3.14)(5^2) + 2(3.14)(5)h$$

$$486.7 = 78.5 + 31.4h$$

$$408.2 = 31.4h \qquad \textbf{Add } -78.5 \textbf{ to each side}$$

$$13 = h \qquad \textbf{Divide each side by 31.4}$$

The height of each can is 13 centimeters.

Facts From Geometry: Formulas for Area and Perimeter

To review, here are the formulas for the area and perimeter of some common geometric objects.

NOTE
The pink line labeled h in the triangle is its height, or altitude. It extends from the top of the triangle down to the base, meeting the base at an angle of 90°. The altitude of a triangle is always perpendicular to the base. The small square shown where the altitude meets the base is used to indicate that the angle formed is 90°.

A Square

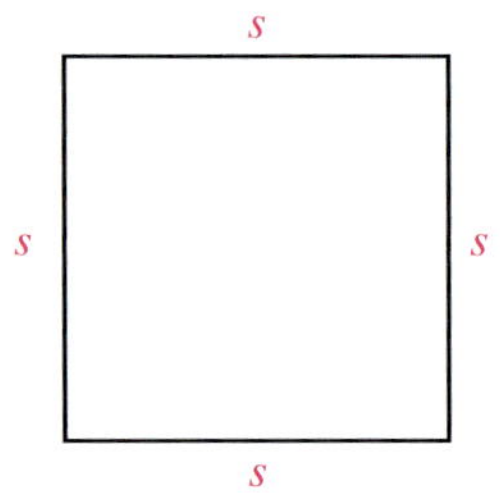

Perimeter $= 4s$
Area $= s^2$

A Rectangle

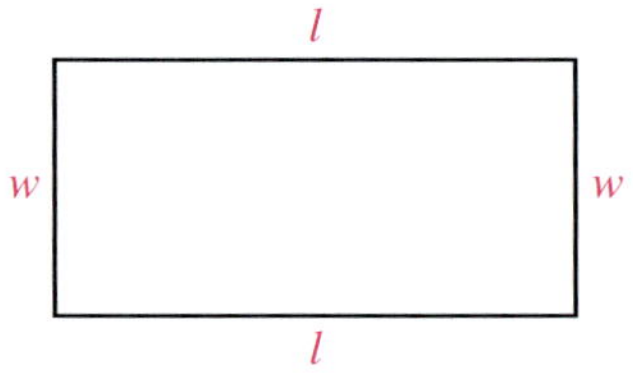

Perimeter $= 2l + 2w$
Area $= lw$

A Triangle

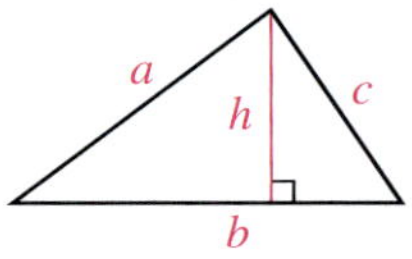

Perimeter $= a + b + c$
Area $= \frac{1}{2}bh$

The formula for perimeter gives us the distance around the outside of the object along its sides, whereas the formula for area gives us a measure of the amount of surface the object covers.

4. Solve $P = 2w + 2l$ for l.

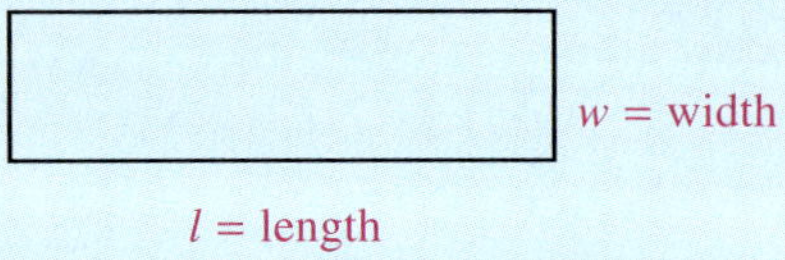

EXAMPLE 4 Given the formula $P = 2w + 2l$, solve for w.

Solution To solve for w, we must isolate it on one side of the equation. We can accomplish this if we delete the $2l$ term and the coefficient 2 from the right side of the equation.

To begin, we add $-2l$ to both sides:

$$P + (\mathbf{-2l}) = 2w + 2l + (\mathbf{-2l})$$

$$P - 2l = 2w$$

To delete the 2 from the right side, we can multiply both sides by $\frac{1}{2}$:

$$\frac{\mathbf{1}}{\mathbf{2}}(P - 2l) = \frac{\mathbf{1}}{\mathbf{2}}(2w)$$

$$\frac{P - 2l}{2} = w$$

Answer
3. 9.9 cm

The two formulas

$$P = 2w + 2l \qquad \text{and} \qquad w = \frac{P - 2l}{2}$$

give the relationship between P, l, and w. They look different, but they both say the same thing about P, l, and w. The first formula gives P in terms of l and w, and the second formula gives w in terms of P and l.

NOTE

We know we are finished solving a formula for a specified variable when that variable appears alone on one side of the equal sign and not on the other.

EXAMPLE 5

Solve for x: $ax - 3 = bx + 5$.

Solution In this example, we must begin by collecting all the variable terms on the left side of the equation and all the constant terms on the other side (just like we did when we were solving linear equations):

$ax - 3 = bx + 5$	
$ax - bx - 3 = 5$	**Add $-bx$ to each side**
$ax - bx = 8$	**Add 3 to each side**

At this point we need to apply the distributive property to write the left side as $(a - b)x$. After that, we divide each side by $a - b$:

$(a - b)x = 8$	**Distributive property**
$x = \frac{8}{a - b}$	**Divide each side by $a - b$**

5. Solve for x: $ax + 5 = cx + 3$.

NOTE

We are applying the distributive property in the same way we applied it when we first learned how to simplify $7x - 4x$. Recall that $7x - 4x = 3x$ because

$$7x - 4x = (7 - 4)x = 3x$$

We are using the same type of reasoning when we write

$$ax - bx = (a - b)x$$

EXAMPLE 6

Solve for y: $\frac{y - b}{x - 0} = m$.

Solution Although we will do more extensive work with formulas of this form later in the book, we need to know how to solve this particular formula for y in order to understand some things in the next chapter. We begin by simplifying the denominator on the left side and then multiplying each side of the formula by x. Doing so makes the rest of the solution process simple.

$\frac{y - b}{x - 0} = m$	**Original formula**
$\frac{y - b}{x} = m$	**$x - 0 = x$**
$x \cdot \frac{y - b}{x} = m \cdot x$	**Multiply each side by x**
$y - b = mx$	**Simplify each side**
$y = mx + b$	**Add b to each side**

This is our solution. If we look back to the first step, we can justify our result on the left side of the equation this way: Dividing by x is equivalent to multiplying by its reciprocal $\frac{1}{x}$. Here is what it looks like when written out completely:

$$x \cdot \frac{y - b}{x} = x \cdot \frac{1}{x} \cdot (y - b) = 1(y - b) = y - b$$

6. Solve for y: $\frac{y - 2}{x - 0} = \frac{4}{3}$.

Answers

4. $l = \frac{P - 2w}{2}$ 5. $x = \frac{-2}{a - c}$

6. $y = \frac{4}{3}x + 2$

7. Solve for y: $\frac{y+8}{x-7} = 6$.

EXAMPLE 7

Solve for y: $\frac{y-4}{x-5} = 3$.

Solution We proceed as we did in the previous example, but this time we clear the formula of fractions by multiplying each side of the formula by $x - 5$.

$$\frac{y-4}{x-5} = 3 \quad \text{Original formula}$$

$$(x-5)\cdot\frac{y-4}{x-5} = 3\cdot(x-5) \quad \text{Multiply each side by } (x-5)$$

$$y - 4 = 3x - 15 \quad \text{Simplify each side}$$

$$y = 3x - 11 \quad \text{Add 4 to each side}$$

We have solved for y. We can justify our result on the left side of the equation this way: Dividing by $x - 5$ is equivalent to multiplying by its reciprocal $\frac{1}{x-5}$. Here are the details:

$$(x-5)\cdot\frac{y-4}{x-5} = (x-5)\cdot\frac{1}{x-5}\cdot(y-4) = 1(y-4) = y-4$$

Basic Percent Problems

The next examples in this section show how basic percent problems can be translated directly into equations. To understand these examples, we must recall that percent means "per hundred." That is, 75% is the same as 75/100, 0.75, and, in reduced fraction form, $\frac{3}{4}$. Likewise, the decimal 0.25 is equivalent to 25%. To change a decimal to a percent, we move the decimal point two places to the right and write the % symbol. To change from a percent to a decimal, we drop the % symbol and move the decimal point two places to the left. The table that follows gives some of the most commonly used fractions and decimals and their equivalent percents.

FRACTION	DECIMAL	PERCENT
$\frac{1}{2}$	0.5	50%
$\frac{1}{4}$	0.25	25%
$\frac{3}{4}$	0.75	75%
$\frac{1}{3}$	$0.\overline{3}$	$33\frac{1}{3}\%$
$\frac{2}{3}$	$0.\overline{6}$	$66\frac{2}{3}\%$
$\frac{1}{5}$	0.2	20%
$\frac{2}{5}$	0.4	40%
$\frac{3}{5}$	0.6	60%
$\frac{4}{5}$	0.8	80%

8. What number is 25% of 74?

NOTE

We write 0.15 instead of 15% when we translate the sentence into an equation because we cannot do calculations with the % symbol. Because percent (%) means per hundred, we think of 15% as 15 hundredths, or 0.15.

Answers

7. $y = 6x - 50$ 8. 18.5

EXAMPLE 8

What number is 15% of 63?

Solution To solve a problem like this, we let x = the number in question and then translate the sentence directly into an equation. Here is how it is done:

What number is 15% of 63?

$$\underbrace{\text{What number}}_{x}\ \underset{=}{\text{is}}\ \underset{0.15}{15\%}\ \underset{\cdot}{\text{of}}\ \underset{63}{63?}$$

$$x = 0.15 \cdot 63 = 9.45$$

The number 9.45 is 15% of 63.

EXAMPLE 9 What percent of 42 is 21?

Solution We translate the sentence as follows:

$$\underbrace{\text{What percent}}_{x} \text{ of } 42 \text{ is } 21?$$
$$x \cdot 42 = 21$$

Next, we divide each side by 42:

$$x = \frac{21}{42}$$
$$= 0.50 \text{ or } 50\%$$

EXAMPLE 10 25 is 40% of what number?

Solution Again, we translate the sentence directly:

$$25 \text{ is } 40\% \text{ of } \underbrace{\text{what number}}_{x}?$$
$$25 = 0.40 \cdot x$$

We solve the equation by dividing both sides by 0.40:

$$\frac{25}{0.40} = \frac{0.40 \cdot x}{0.40}$$
$$62.5 = x$$

25 is 40% of 62.5.

Formulas for Number Sequences

In Chapter 1 we used inductive reasoning to find the next number in a sequence of numbers. What we are going to do now is extend our work with number sequences so that we can find any number in the sequence. To begin, we introduce notation that allows us to associate each number in a number sequence with its position in the sequence. To illustrate, consider the sequence of even numbers:

Even numbers: 2, 4, 6, 8, 10, . . .

The number 8 is the fourth number in sequence. Its *position* in the sequence is 4, and its *value* is 8. Here is the sequence of even numbers again, this time written so that the position of each term is noted.

Position: 1, 2, 3, 4, 5, . . .

Value: 2, 4, 6, 8, 10, . . .

The notation we use to indicate the position of a number in a number sequence involves *subscripts*. Subscripts are numbers written to the right and just below another number or variable. In the expression a_4, the 4 is a subscript. We use subscripts to designate the terms of a sequence this way:

The *first term* of a sequence is designated by a_1.

The *second term* of a sequence is designated by a_2.

The *third term* of a sequence is designated by a_3.

The *fourth term* of a sequence is designated by a_4.

The expression a_n is called the nth term, or the *general term* of the sequence. The general term is used to define the other terms of the sequence; that is, if we are given the formula for the general term a_n, we can find any other term in the sequence. The following examples illustrate the concept.

9. What percent of 84 is 21?

10. 35 is 40% of what number?

Answers

9. 25% **10.** 87.5

11. Find the first four terms of the sequence whose general term is given by $a_n = 3n - 2$.

EXAMPLE 11 Find the first four terms of the sequence whose general term is given by $a_n = 2n - 1$.

Solution To find the first, second, third, and fourth terms of this sequence, we simply substitute 1, 2, 3, and 4 for n in the formula $2n - 1$:

If The general term is $a_n = 2n - 1$

then The first term is $a_1 = 2(1) - 1 = 1$

The second term is $a_2 = 2(2) - 1 = 3$

The third term is $a_3 = 2(3) - 1 = 5$

The fourth term is $a_4 = 2(4) - 1 = 7$

The first four terms of this sequence are the odd numbers 1, 3, 5, and 7. The whole sequence can be written as

$$1, 3, 5, \ldots, 2n - 1, \ldots$$

Because each term in this sequence is larger than the preceding term, we say the sequence is an *increasing sequence.*

12. Write the first four terms of the sequence defined by

$$a_n = \frac{n}{n+1}$$

EXAMPLE 12 Write the first four terms of the sequence defined by

$$a_n = \frac{1}{n+1}$$

Solution Replacing n with 1, 2, 3, and 4, we have, respectively, the first four terms:

$$\text{First term} = a_1 = \frac{1}{1+1} = \frac{1}{2}$$

$$\text{Second term} = a_2 = \frac{1}{2+1} = \frac{1}{3}$$

$$\text{Third term} = a_3 = \frac{1}{3+1} = \frac{1}{4}$$

$$\text{Fourth term} = a_4 = \frac{1}{4+1} = \frac{1}{5}$$

The sequence defined by

$$a_n = \frac{1}{n+1}$$

can be written as

$$\frac{1}{2}, \frac{1}{3}, \frac{1}{4}, \ldots, \frac{1}{n+1}, \ldots$$

Because each term in the sequence is smaller than the term preceding it, the sequence is said to be a *decreasing sequence.*

Linking Objectives and Examples

Next to each **objective**, list the examples that are best described by that objective.

A	
B	
C	
D	

Getting Ready for Class

After reading through the preceding section, respond in your own words and in complete sentences.

1. What is a formula in mathematics?
2. Create a formula for a sequence and explain how to obtain the first four terms of your sequence.
3. Write a percent problem that can be solved by the equation $30 = 0.25x$.
4. Explain in words the formula for the perimeter of a rectangle.

Answers

11. 1, 4, 7, 10

12. $\frac{1}{2}, \frac{2}{3}, \frac{3}{4}, \frac{4}{5}$

Problem Set 2.2

INTERMEDIATE Algebra $f(x)$ Now™

Use the formula $3x - 4y = 12$ to find y if

1. x is 0

2. x is -2

3. x is 4

4. x is -4

Use the formula $y = 2x - 3$ to find x when

5. y is 0

6. y is -3

7. y is 5

8. y is -5

Problems 9 through 26 are all problems that you will see later in the text.

9. If $x - 2y = 4$ and $y = -\frac{6}{5}$, find x.

10. If $x - 2y = 4$ and $x = \frac{8}{5}$, find y.

11. If $2x + 3y = 6$, find y when x is 0.

12. If $2x + 3y = 6$, find x when y is 0.

13. Let $x = 160$ and $y = 0$ in $y = a(x - 80)^2 + 70$ and solve for a.

14. Let $x = 0$ and $y = 0$ in $y = a(x - 80)^2 + 70$ and solve for a.

15. Find R if $p = 1.5$ and $R = (900 - 300p)p$.

16. Find R if $p = 2.5$ and $R = (900 - 300p)p$.

17. Find P if $P = -0.1x^2 + 27x - 1{,}700$ and

a. $x = 100$ **b.** $x = 170$

18. Find P if $P = -0.1x^2 + 27x - 1{,}820$ and

a. $x = 130$ **b.** $x = 140$

19. Find h if $h = 16 + 32t - 16t^2$

a. $t = \frac{1}{4}$ **b.** $t = \frac{7}{4}$

20. Find h if $h = 64t - 16t^2$

a. $t = 1$ **b.** $t = 3$

21. Find y if $x = \frac{3}{2}$ and $y = -2x^2 + 6x - 5$.

22. Find y if $x = \frac{1}{2}$ and $y = -2x^2 + 6x - 5$.

23. If $y = Kx$, find K if $x = 5$ and $y = 15$.

24. If $d = Kt^2$, find K if $t = 2$ and $d = 64$.

25. If $V = \frac{K}{P}$, find K if $P = 48$ and $V = 50$.

26. If $y = Kxz^2$, find K if $x = 5$, $z = 3$, and $y = 180$.

Solve each of the following formulas for the indicated variable.

27. $A = lw$ for l

28. $A = \frac{1}{2}bh$ for b

29. $I = prt$ for t

30. $I = prt$ for r

31. $PV = nRT$ for T

32. $PV = nRT$ for R

33. $y = mx + b$ for x

34. $A = P + Prt$ for t

35. $\mathrm{C} = \frac{5}{9}(\mathrm{F} - 32)$ for F

36. $\mathrm{F} = \frac{9}{5}\mathrm{C} + 32$ for C

37. $h = vt + 16t^2$ for v

38. $h = vt - 16t^2$ for v

39. $A = a + (n - 1)d$ for d

40. $A = a + (n - 1)d$ for n

41. $2x + 3y = 6$ for y

42. $2x - 3y = 6$ for y

43. $-3x + 5y = 15$ for y

44. $-2x - 7y = 14$ for y

45. $2x - 6y + 12 = 0$ for y

46. $7x - 2y - 6 = 0$ for y

47. $ax + 4 = bx + 9$ for x

48. $ax - 5 = cx - 2$ for x

49. $A = P + Prt$ for P

50. $ax + b = cx + d$ for x

Solve for y.

51. $\frac{x}{8} + \frac{y}{2} = 1$

52. $\frac{x}{7} + \frac{y}{9} = 1$

53. $\frac{x}{5} + \frac{y}{-3} = 1$

54. $\frac{x}{16} + \frac{y}{-2} = 1$

Problems 55 through 60 are all problems that you will see later in the text. Solve each formula for y.

55. $x = 2y - 3$

56. $x = 4y + 1$

57. $y - 3 = -2(x + 4)$

58. $y + 1 = -\frac{2}{3}(x - 3)$

59. Solve for y.

a. $\frac{y+1}{x-0} = 4$

b. $\frac{y+2}{x-4} = -\frac{1}{2}$

c. $\frac{y+3}{x-7} = 0$

60. Solve for y.

a. $\frac{y-1}{x-0} = -3$

b. $\frac{y-2}{x-6} = \frac{2}{3}$

c. $\frac{y-3}{x-1} = 0$

Translate each of the following into a linear equation and then solve the equation.

61. What number is 54% of 38?

62. What number is 11% of 67?

63. What percent of 36 is 9?

64. What percent of 50 is 5?

65. 37 is 4% of what number?

66. 8 is 2% of what number?

Write the first five terms of the sequences with the following general terms.

67. $a_n = 3n + 1$

68. $a_n = 4n - 1$

69. $a_n = n^2 + 3$

70. $a_n = n^3 + 1$

71. $a_n = \frac{n}{n+3}$

72. $a_n = \frac{n}{n+2}$

73. $a_n = \frac{1}{n^2}$

74. $a_n = \frac{1}{n^3}$

75. $a_n = 2^n$

76. $a_n = 3^n$

77. $a_n = 1 + \frac{1}{n}$

78. $a_n = 1 - \frac{1}{n}$

Applying the Concepts

79. Air Pollution Air pollution in the United States has become a concern in recent years. The formula below models the sulfur dioxide emissions during a given year between 1970 and 1997.

$$E = 30{,}366 - 395x$$

In the formula, E is the emission of sulfur dioxide in thousands of tons, and x is the number of years after 1970; that is, $x = 0$ on January 1, 1970. In what year were sulfur dioxide emissions 22,071 thousand tons? (U.S. Environmental Protection Agency, *National Air Quality and Emissions Trend Report.)*

80. Fermat's Last Theorem As we mentioned in the previous chapter, the postage stamp shows Fermat's last theorem, which states that if n is an integer greater than 2, then there are no positive integers x, y, and z that will make the formula $x^n + y^n = z^n$ true.

Use the formula $x^n + y^n = z^n$ to

a. Find x if $n = 1$, $y = 7$, and $z = 15$.

b. Find y if $n = 1$, $x = 23$, and $z = 37$.

81. Estimating Vehicle Weight If you can measure the area that the tires on your car contact the ground and know the air pressure in the tires, then you can estimate the weight of your car with the following formula:

$$W = \frac{APN}{2{,}000}$$

where W is the vehicle's weight in tons, A is the average tire contact area with a hard surface in square inches, P is the air pressure in the tires in pounds per square inch (psi, or lb/in^2), and N is the number of tires.

a. What is the approximate weight of a car if the average tire contact area is a rectangle 6 inches by 5 inches and if the air pressure in the tires is 30 psi?

b. What is the approximate weight of a car if the average tire contact area is a rectangle 5 inches by 4 inches and the tire pressure is 30 psi?

82. Digital Video The biggest video download of all time was a *Star Wars* movie trailer. The video was compressed so it would be small enough for people to download over the Internet. A formula for estimating the size, in kilobytes, of a compressed video is

$$S = \frac{height \cdot width \cdot fps \cdot time}{35{,}000}$$

where *height* and *width* are in pixels, *fps* is the number of frames per second the video is to play (television plays at 30 fps), and *time* is given in seconds.

a. Estimate the size in kilobytes of the *Star Wars* trailer that has a height of 480 pixels, has a width of 216 pixels, plays at 30 fps, and runs for 150 seconds.

b. Estimate the size in kilobytes of the *Star Wars* trailer that has a height of 320 pixels, has a width of 144 pixels, plays at 15 fps, and runs for 150 seconds.

Pricing A company that manufactures ink cartridges finds that they can sell x cartridges each week at a price of p dollars each, according to the formula $x = 1{,}300 - 100p$. What price should they charge for each cartridge if they want to sell

83. 800 cartridges each week?

84. 400 cartridges each week?

85. 300 cartridges each week?

86. 900 cartridges each week?

Volume The volume of a cylinder is given by the formula $V = \pi r^2 h$, where r is the radius and h is the height. Find the height if

87. the volume is 308 cubic centimeters and the radius is 7 centimeters. (Use $\frac{22}{7}$ for π.)

88. the volume is 308 cubic centimeters and the radius is $\frac{7}{2}$ centimeters.

89. the volume is 628 cubic inches and the radius is 10 inches. (Use 3.14 for π.)

90. the volume is 12.56 cubic inches and the radius is 5 inches.

Getting Ready for the Next Section

To understand all of the explanations and examples in the next section, you must be able to work the problems below.

Translate into symbols.

91. Three less than twice a number

92. Ten less than four times a number

93. The sum of x and y is 180.

94. The sum of a and b is 90.

Solve each equation.

95. $x + 2x = 90$

96. $x + 5x = 180$

97. $2(2x - 3) + 2x = 45$

98. $2(4x - 10) + 2x = 12.5$

99. $6x + 5(10{,}000 - x) = 56{,}000$

100. $x + 0.0725x = 17{,}481.75$

Maintaining Your Skills

Simplify using the rule for order of operations.

101. $38 - 19 + 1$

102. $200 - 150 + 20$

103. $57 - 18 - 8$

104. $71 - 11 - 1$

105. $28 \div 7 \cdot 2$

106. $48 \div 3 \cdot 2$

107. $125 \div 25 \div 5$

108. $36 \div 12 \div 4$

Improving Your Quantitative Literacy

Exercise Physiology In exercise physiology, a person's maximum heart rate, in beats per minute, is found by subtracting his age, in years, from 220. So, if A represents your age in years, then your maximum heart rate is

$$M = 220 - A$$

A person's training heart rate, in beats per minute, is her resting heart rate plus 60% of the difference between her maximum heart rate and her resting heart rate. If resting heart rate is R and maximum heart rate is M, then the formula that gives training heart rate is

$$T = R + 0.6(M - R)$$

109. **Training Heart Rate** Shar is 46 years old. Her daughter, Sara, is 26 years old. If they both have a resting heart rate of 60 beats per minute, find the training heart rate for each.

110. **Training Heart Rate** Shane is 30 years old and has a resting heart rate of 68 beats per minute. Her mother, Carol, is 52 years old and has the same resting heart rate. Find the training heart rate for Shane and for Carol.

Tom & Dee Ann McCarthy/Corbis

Applications

2.3

OBJECTIVES

A	Apply the Blueprint for Problem Solving to a variety of application problems.
B	Use a formula to construct a table of paired data.

In this section we use the skills we have developed for solving equations to solve problems written in words. You may find that some of the examples and problems are more realistic than others. Since we are just beginning our work with application problems, even the ones that seem unrealistic are good practice. What is important in this section is the *method* we use to solve application problems, not the applications themselves. The method, or strategy, that we use to solve application problems is called the *Blueprint for Problem Solving*. It is an outline that will overlay the solution process we use on all application problems.

Blueprint for Problem Solving

Step 1: ***Read*** the problem, and then mentally ***list*** the items that are known and the items that are unknown.

Step 2: ***Assign a variable*** to one of the unknown items. (In most cases this will amount to letting $x =$ the item that is asked for in the problem.) Then ***translate*** the other ***information*** in the problem to expressions involving the variable.

Step 3: ***Reread*** the problem, and then ***write an equation,*** using the items and variable listed in steps 1 and 2, that describes the situation.

Step 4: ***Solve the equation*** found in step 3.

Step 5: ***Write your answer*** using a complete sentence.

Step 6: ***Reread*** the problem, and ***check*** your solution with the original words in the problem.

A number of substeps occur within each of the steps in our blueprint. For instance, with steps 1 and 2 it is always a good idea to draw a diagram or picture if it helps you visualize the relationship between the items in the problem.

EXAMPLE 1

The length of a rectangle is 3 inches less than twice the width. The perimeter is 45 inches. Find the length and width.

Solution When working problems that involve geometric figures, a sketch of the figure helps organize and visualize the problem.

Step 1: ***Read and list.***
Known items: The figure is a rectangle. The length is 3 inches less than twice the width. The perimeter is 45 inches.
Unknown items: The length and the width

Step 2: ***Assign a variable and translate information.***
Since the length is given in terms of the width (the length is 3 less than twice the width), we let $x =$ the width of the rectangle. The length is 3 less than twice the width, so it must be $2x - 3$. The diagram in Figure 1 is a visual description of the relationships we have listed so far.

PRACTICE PROBLEMS

1. The width of a rectangle is 10 feet less than 4 times the length. If the perimeter is 12.5 feet, find the length and width.

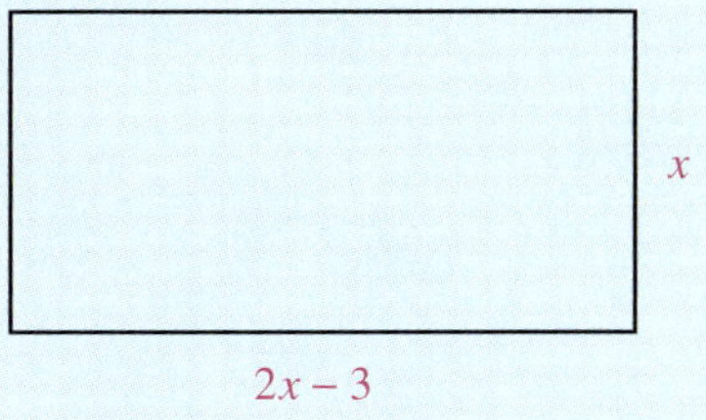

FIGURE 1

Step 3: ***Reread and write an equation.***

The equation that describes the situation is

$$\text{Twice the length} + \text{twice the width} = \text{perimeter}$$

$$2(2x - 3) \quad + \quad 2x \quad = \quad 45$$

Step 4: ***Solve the equation.***

$$\begin{aligned} 2(2x - 3) + 2x &= 45 \\ 4x - 6 + 2x &= 45 \\ 6x - 6 &= 45 \\ 6x &= 51 \\ x &= 8.5 \end{aligned}$$

Step 5: ***Write the answer.***

The width is 8.5 inches. The length is $2x - 3 = 2(8.5) - 3 = 14$ inches.

Step 6: ***Reread and check.***

If the length is 14 inches and the width is 8.5 inches, then the perimeter must be $2(14) + 2(8.5) = 28 + 17 = 45$ inches. Also, the length, 14, is 3 less than twice the width. ■

Remember as you read through the steps in the solutions to the examples in this section that step 1 is done mentally. Read the problem and then *mentally* list the items that you know and the items that you don't know. The purpose of step 1 is to give you direction as you begin to work application problems. Finding the solution to an application problem is a process; it doesn't happen all at once. The first step is to read the problem with a purpose in mind. That purpose is to mentally note the items that are known and the items that are unknown.

EXAMPLE 2 In April 1998, Pat bought a new Ford Mustang with a 5.0-liter engine. The total price, which includes the price of the car plus sales tax, was $17,481.75. If the sales tax rate is 7.25%, what was the price of the car?

2. If Pat had purchased the convertible version of the Ford Mustang discussed in Example 2, the total price (the price of the car plus sales tax) would have been $23,466.30. Find the price of the convertible if the sales tax rate is 7.25%.

Solution

Step 1: ***Read and list.***

Known items: The total price is $17,481.75. The sales tax rate is 7.25%, which is 0.0725 in decimal form.

Unknown item: The price of the car.

Step 2: ***Assign a variable and translate information.***

If we let x = the price of the car, then to calculate the sales tax, we multiply the price of the car x by the sales tax rate:

$$\begin{aligned} \text{Sales tax} &= (\text{sales tax rate})(\text{price of the car}) \\ &= 0.0725x \end{aligned}$$

Step 3: ***Reread and write an equation.***

$$\text{Car price} + \text{sales tax} = \text{total price}$$

$$x \quad + \quad 0.0725x \quad = \quad 17{,}481.75$$

Step 4: ***Solve the equation.***

$$\begin{aligned} x + 0.0725x &= 17{,}481.75 \\ 1.0725x &= 17{,}481.75 \\ x &= \frac{17{,}481.75}{1.0725} \\ &= 16{,}300.00 \end{aligned}$$

Answer

1. length = 3.25 feet, width = 3 feet

Step 5: ***Write the answer.***
The price of the car is $16,300.00

Step 6: ***Reread and check.***
The price of the car is $16,300.00. The tax is 0.0725(16,300) = $1,181.75. Adding the retail price and the sales tax we have a total bill of $17,481.75.

Facts From Geometry: Angles in General

An angle is formed by two rays with the same endpoint. The common endpoint is called the *vertex* of the angle, and the rays are called the *sides* of the angle.

In Figure 2, angle θ (theta) is formed by the two rays *OA* and *OB*. The vertex of θ is *O*. Angle θ is also denoted as angle *AOB*, where the letter associated with the vertex is always the middle letter in the three letters used to denote the angle.

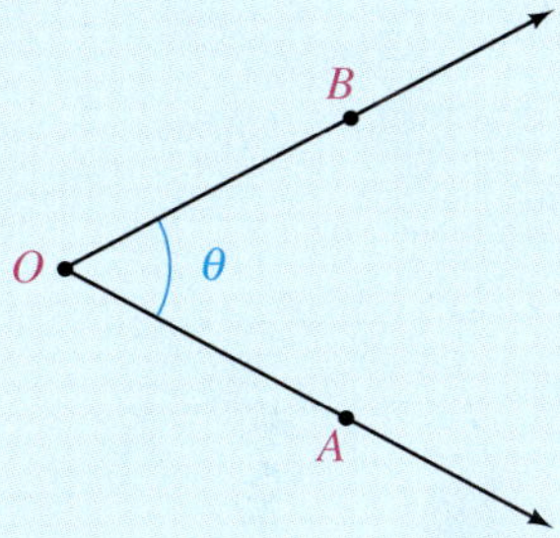

FIGURE 2

Degree Measure

The angle formed by rotating a ray through one complete revolution about its endpoint (Figure 3) has a measure of 360 degrees, which we write as 360°.

One degree of angle measure, written 1°, is $\frac{1}{360}$ of a complete rotation of a ray about its endpoint; there are 360° in one full rotation. (The number 360 was decided on by early civilizations because it was believed that the earth was at the center of the universe and the sun would rotate once around the earth every 360 days.) Similarly, 180° is half of a complete rotation, and 90° is a quarter of a full rotation. Angles that measure 90° are called *right angles,* and angles that measure 180° are called *straight angles.* If an angle measures between 0° and 90° it is called an *acute angle,* and an angle that measures between 90° and 180° is an *obtuse angle.* Figure 4 illustrates further.

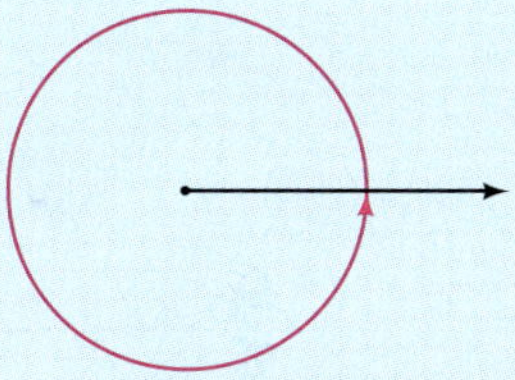
One complete revolution = 360°

FIGURE 3

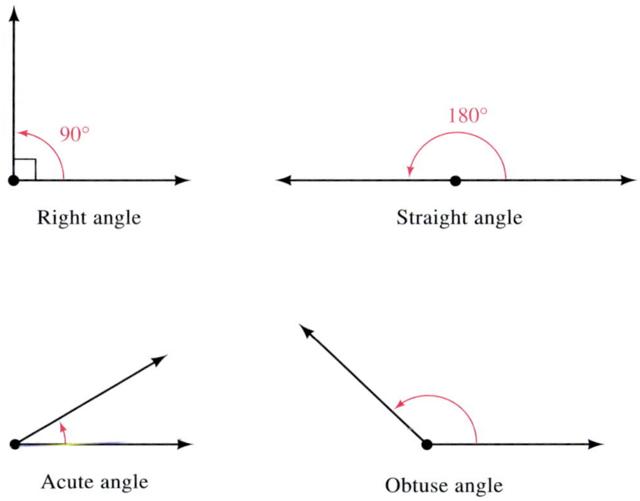

FIGURE 4

Complementary Angles and Supplementary Angles

If two angles add up to 90°, we call them *complementary angles,* and each is called the *complement* of the other. If two angles have a sum of 180°, we call them *supplementary angles,* and each is called the *supplement* of the other. Figure 5 illustrates the relationship between angles that are complementary and angles that are supplementary.

Answer
2. $21,880.00

Complementary angles: $\alpha + \beta = 90°$ Supplementary angles: $\alpha + \beta = 180°$

FIGURE 5

Special Triangles

It is not unusual to have the terms we use in mathematics show up in the descriptions of things we find in the world around us. The flag of Puerto Rico shown here is described on the government website as "Five equal horizontal bands of red (top and bottom) alternating with white; a blue isosceles triangle based on the hoist side bears a large white five-pointed star in the center." An *isosceles triangle* as shown here and in Figure 6, is a triangle with two sides of equal length.

Angles A and B in the isosceles triangle in Figure 6 are called the *base angles:* they are the angles opposite the two equal sides. In every isosceles triangle, the base angles are equal.

NOTE

As you can see from Figures 6 and 7, one way to label the important parts of a triangle is to label the vertices with capital letters and the sides with small letters: side a is opposite vertex A, side b is opposite vertex B, and side c is opposite vertex C.

Also, because each vertex is the vertex of one of the angles of the triangle, we refer to the three interior angles as A, B, and C.

Finally, in any triangle, the sum of the interior angles is 180°. For the triangles shown in Figures 6 and 7, the relationship is written

$$A + B + C = 180°$$

Isosceles Triangle

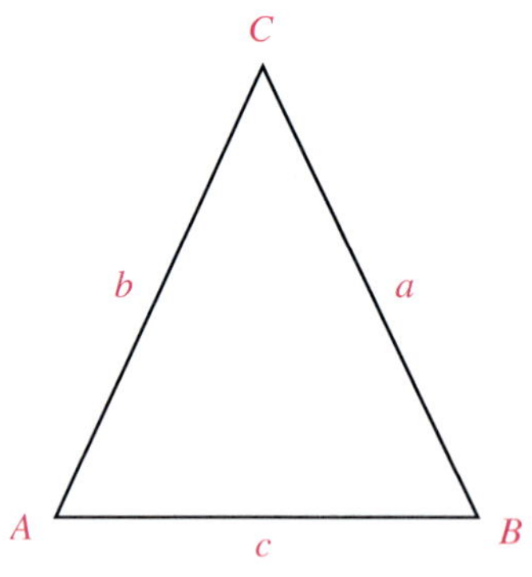

$a = b$

FIGURE 6

Equilateral Triangle

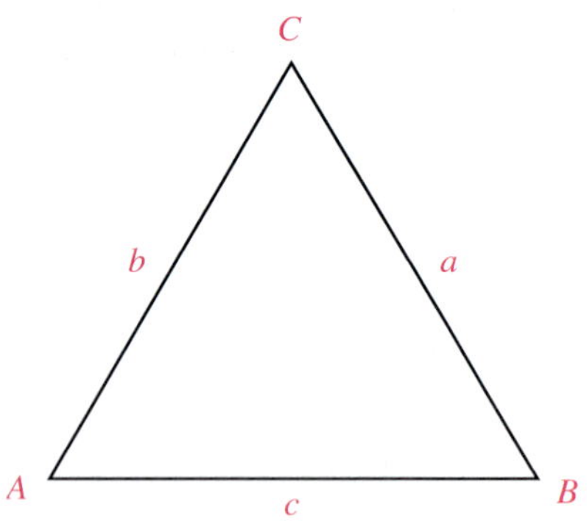

$a = b = c$

FIGURE 7

An *equilateral triangle* (Figure 7) is a triangle with three sides of equal length. If all three sides in a triangle have the same length, then the three interior angles in the triangle also must be equal. Because the sum of the interior angles in a triangle is always 180°, the three interior angles in any equilateral triangle must be 60°.

EXAMPLE 3 Two complementary angles are such that one is twice as large as the other. Find the two angles.

Solution Applying the Blueprint for Problem Solving, we have:

Step 1: ***Read and list.***
Known items: Two complementary angles. One is twice as large as the other.
Unknown items: The size of the angles

Step 2: ***Assign a variable and translate information.***
Let x = the smaller angle. The larger angle is twice the smaller so we represent the larger angle with $2x$.

Step 3: ***Reread and write an equation.***
Because the two angles are complementary, their sum is 90. Therefore,

$$x + 2x = 90$$

Step 4: ***Solve the equation.***

$$x + 2x = 90$$
$$3x = 90$$
$$x = 30$$

Step 5: ***Write the answer.***
The smaller angle is 30°, and the larger angle is $2 \cdot 30 = 60°$.

Step 6: ***Reread and check.***
The larger angle is twice the smaller angle, and their sum is 90°.

Suppose we know that the sum of two numbers is 50. If we let x represent one of the two numbers, how can we represent the other? Let's suppose for a moment that x turns out to be 30. Then the other number will be 20 because their sum is 50; that is, if two numbers add up to 50, and one of them is 30, then the other must be $50 - 30 = 20$. Generalizing this to any number x, we see that if two numbers have a sum of 50 and one of the numbers is x, then the other must be $50 - x$. The following table shows some additional examples:

IF TWO NUMBERS HAVE A SUM OF	AND ONE OF THEM IS	THEN THE OTHER MUST BE
50	x	$50 - x$
10	y	$10 - y$
12	n	$12 - n$

Interest Problem

EXAMPLE 4 Suppose a person invests a total of $10,000 in two accounts. One account earns 5% annually, and the other earns 6% annually. If the total interest earned from both accounts in a year is $560, how much is invested in each account?

Solution

Step 1: ***Read and list.***
Known items: Two accounts. One pays interest of 5%, and the other pays 6%. The total dollars invested is $10,000.
Unknown items: The number of dollars invested in each individual account.

3. Two supplementary angles are such that one is five times as large as the other. Find the two angles.

4. Howard invests a total of $8,000 in two accounts. One account earns 8% annually, and the other earns 10% annually. If the total interest earned from both accounts in a year is $680, how much is invested in each account?

Answer
3. 30° and 150°

Step 2: ***Assign a variable and translate information.***
If we let x equal the amount invested at 6%, then $10{,}000 - x$ is the amount invested at 5%. The total interest earned from both accounts is \$560. The amount of interest earned on x dollars at 6% is $0.06x$, whereas the amount of interest earned on $10{,}000 - x$ dollars at 5% is $0.05(10{,}000 - x)$.

	DOLLARS AT 6%	DOLLARS AT 5%	TOTAL
Number of	x	$10{,}000 - x$	10,000
Interest on	$0.06x$	$0.05(10{,}000 - x)$	560

Step 3: ***Reread and write an equation.***
The last line gives us the equation we are after:

$$0.06x + 0.05(10{,}000 - x) = 560$$

Step 4: ***Solve the equation.***
To make this equation a little easier to solve, we begin by multiplying both sides by 100 to move the decimal point two places to the right:

$$\begin{aligned} 6x + 5(10{,}000 - x) &= 56{,}000 \\ 6x + 50{,}000 - 5x &= 56{,}000 \\ x + 50{,}000 &= 56{,}000 \\ x &= 6{,}000 \end{aligned}$$

Step 5: ***Write the answer.***
The amount of money invested at 6% is \$6,000. The amount of money invested at 5% is \$10,000 − \$6,000 = \$4,000.

Step 6: ***Reread and check.***
To check our results, we find the total interest from the two accounts:

The interest on \$6,000 at 6% is 0.06(6,000) = 360
The interest on \$4,000 at 5% is 0.05(4,000) = 200
The total interest = \$560

Table Building

We can use our knowledge of formulas to build tables of paired data. As you will see, equations or formulas that contain exactly two variables produce pairs of numbers that can be used to construct tables.

5. Rework Example 5 if the string is 14 inches long and the rectangle has a width of 1, 2, 3, 4, 5, or 6 inches.

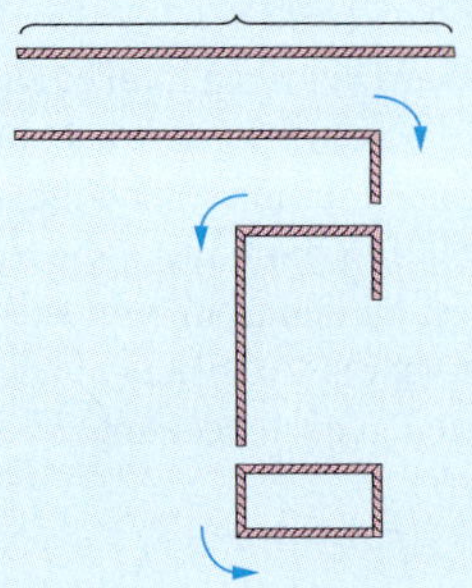

EXAMPLE 5 A piece of string 12 inches long is to be formed into a rectangle. Build a table that gives the length of the rectangle if the width is 1, 2, 3, 4, or 5 inches. Then find the area of each of the rectangles formed.

Solution Because the formula for the perimeter of a rectangle is $P = 2l + 2w$ and our piece of string is 12 inches long, then the formula we will use to find the lengths for the given widths is $12 = 2l + 2w$. To solve this formula for l, we divide each side by 2 and then subtract w. The result is $l = 6 - w$. Table 1 organizes our work so that the formula we use to find l for a given value of w is shown, and we have added a last column to give us the areas of the rectangles formed. The units for the first three columns are inches, and the units for the numbers in the last column are square inches.

Answer
4. \$6,000 at 8%, \$2,000 at 10%

TABLE 1			
LENGTH, WIDTH, AND AREA			
WIDTH (IN.)	LENGTH (IN.)	AREA (IN.^{2})	
w	$l = 6 - w$	l	$A = lw$
1	$l = 6 - 1$	5	5
2	$l = 6 - 2$	4	8
3	$l = 6 - 3$	3	9
4	$l = 6 - 4$	2	8
5	$l = 6 - 5$	1	5

Figures 8 and 9 show two bar charts constructed from the information in Table 1.

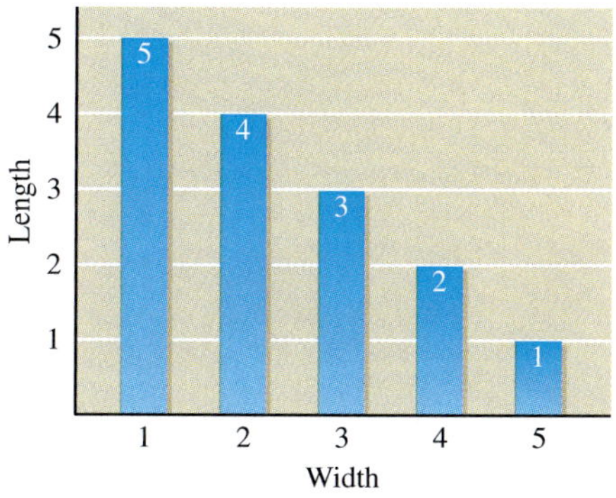

FIGURE 8 **Length and width of rectangles with perimeters fixed at 12 inches.**

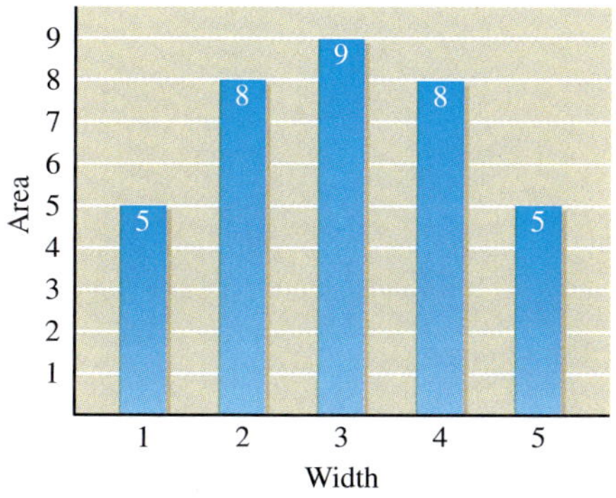

FIGURE 9 **Area and width of rectangles with perimeters fixed at 12 inches.**

USING TECHNOLOGY

A number of graphing calculators have table-building capabilities. We can let the calculator variable X represent the widths of the rectangles in Example 5. To find the lengths, we set variable Y_1 equal to $6 - X$. The area of each rectangle can be found by setting variable Y_2 equal to $X * Y_1$. To have the calculator produce the table automatically, we use a table minimum of 0 and a table increment of 1. Here is a summary of how the graphing calculator is set up:

Table Setup
Table minimum = 0
Table increment = 1
Independent variable: Auto
Dependent variable: Auto

Y Variables Setup
$Y_1 = 6 - X$
$Y_2 = X * Y_1$

The table will look like this:

X	Y_1	Y_2
0	6	0
1	5	5
2	4	8
3	3	9
4	2	8
5	1	5
6	0	0

Answer

5. See Solutions to Practice Problems Section

Linking Objectives and Examples

Next to each **objective**, list the examples that are best described by that objective.

A	
B	

Getting Ready for Class

After reading through the preceding section, respond in your own words and in complete sentences.

1. What is the first step in solving an application problem?
2. What is the biggest obstacle between you and success in solving application problems?
3. Write an application problem for which the solution depends on solving the equation $2x + 2 \cdot 3 = 18$.
4. What is the last step in solving an application problem? Why is this step important?

Problem Set 2.3

INTERMEDIATE Algebra $f(x)$ Now™

Solve each application problem. Be sure to follow the steps in the Blueprint for Problem Solving.

Geometry Problems

1. **Rectangle** A rectangle is twice as long as it is wide. The perimeter is 60 feet. Find the dimensions.

2. **Rectangle** The length of a rectangle is 5 times the width. The perimeter is 48 inches. Find the dimensions.

3. **Square** A square has a perimeter of 28 feet. Find the length of each side.

4. **Square** A square has a perimeter of 36 centimeters. Find the length of each side.

5. **Triangle** A triangle has a perimeter of 23 inches. The medium side is 3 inches more than the shortest side, and the longest side is twice the shortest side. Find the shortest side.

6. **Triangle** The longest side of a triangle is two times the shortest side, whereas the medium side is 3 meters more than the shortest side. The perimeter is 27 meters. Find the dimensions.

7. **Rectangle** The length of a rectangle is 3 meters less than twice the width. The perimeter is 18 meters. Find the width.

8. **Rectangle** The length of a rectangle is 1 foot more than twice the width. The perimeter is 20 feet. Find the dimensions.

9. **Livestock Pen** A livestock pen is built in the shape of a rectangle that is twice as long as it is wide. The perimeter is 48 feet. If the material used to build the pen is \$1.75 per foot for the longer sides and \$2.25 per foot for the shorter sides (the shorter sides have gates, which increase the cost per foot), find the cost to build the pen.

$p = 48$ ft.

10. **Garden** A garden is in the shape of a square with a perimeter of 42 feet. The garden is surrounded by two fences. One fence is around the perimeter of the garden, whereas the second fence is 3 feet from the first fence, as Figure 10 indicates. If the material used to build the two fences is \$1.28 per foot, what was the total cost of the fences?

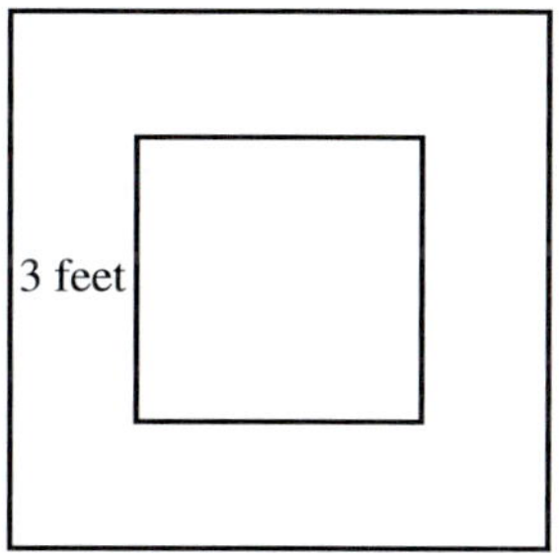

FIGURE 10

Percent Problems

11. **Money** Shane returned from a trip to Las Vegas with \$300.00, which was 50% more money than he had at the beginning of the trip. How much money did Shane have at the beginning of his trip?

12. **Items Sold** Every item in the Just a Dollar store is priced at \$1.00. When Mary Jo opens the store, there is \$125.50 in the cash register. When she counts the money in the cash register at the end of the day, the total is \$1,058.60. If the sales tax rate is 8.5%, how many items were sold that day?

13. Monthly Salary An accountant earns $3,440 per month after receiving a 5.5% raise. What was the accountant's monthly income before the raise? Round your answer to the nearest cent.

14. Textbook Price Suppose a college bookstore buys a textbook from a publishing company and then marks up the price they paid for the book 33% and sells it to a student at the marked-up price. If the student pays $75.00 for the textbook, what did the bookstore pay for it? Round your answer to the nearest cent.

15. Movies *Batman Forever* grossed $52.8 million on its opening weekend and had one of the most successful movie launches in history. If *Batman Forever* accounted for approximately 53% of all box office receipts that weekend, what were the total box office receipts?

16. Hourly Wage A sheet metal worker earns $26.80 per hour after receiving a 4.5% raise. What was the sheet metal worker's hourly pay before the raise? Round your answer to the nearest cent.

17. Calories in Milk The information on the milk carton in Figure 11 states that it contains 20% fewer calories than whole milk. If one serving of this milk contains 120 calories, how many calories are in an equivalent serving of whole milk?

FIGURE 11

18. Movies The movie with the biggest opening weekend in 1999 was *Star Wars Episode I: The Phantom Menace* with opening weekend receipts of $64.8 million. In 2001 the movie with the biggest opening weekend was *Harry Potter and the Sorcerer's Stone* with opening weekend receipts of $113.4 million. What was the percent increase in opening weekend receipts from the top weekend movie in 1999 to the top weekend movie of 2001?

More Geometry Problems

19. Angles Two supplementary angles are such that one is eight times larger than the other. Find the two angles.

20. Angles Two complementary angles are such that one is five times larger than the other. Find the two angles.

21. Angles One angle is 12° less than four times another. Find the measure of each angle if

a. they are complements of each other.

b. they are supplements of each other.

22. Angles One angle is 4° more than three times another. Find the measure of each angle if

a. they are complements of each other.

b. they are supplements of each other.

23. Triangle A triangle is such that the largest angle is three times the smallest angle. The third angle is 9° less than the largest angle. Find the measure of each angle.

24. Triangle The smallest angle in a triangle is half of the largest angle. The third angle is 15° less than the largest angle. Find the measure of all three angles.

25. Triangle The smallest angle in a triangle is one-third of the largest angle. The third angle is 10° more than the smallest angle. Find the measure of all three angles.

26. Triangle The third angle in an isosceles triangle is half as large as each of the two base angles. Find the measure of each angle.

27. Isosceles Triangle The third angle in an isosceles triangle is 8° more than twice as large as each of the two base angles. Find the measure of each angle.

28. Isosceles Triangle The third angle in an isosceles triangle is 4° more than one fifth of each of the two base angles. Find the measure of each angle.

Interest Problems

29. Investing A woman has a total of $9,000 to invest. She invests part of the money in an account that pays 8% per year and the rest in an account that pays 9% per year. If the interest earned in the first year is $750, how much did she invest in each account?

	DOLLARS AT 8%	DOLLARS AT 9%	TOTAL
Number of			
Interest on			

30. Investing A man invests $12,000 in two accounts. If one account pays 10% per year and the other pays 7% per year, how much was invested in each account if the total interest earned in the first year was $960?

	DOLLARS AT 10%	DOLLARS AT 7%	TOTAL
Number of			
Interest on			

31. Investing A total of $15,000 is invested in two accounts. One of the accounts earns 12% per year, and the other earns 10% per year. If the total interest earned in the first year is $1,600, how much was invested in each account?

32. Investing A total of $11,000 is invested in two accounts. One of the two accounts pays 9% per year, and the other account pays 11% per year. If the total interest paid in the first year is $1,150, how much was invested in each account?

33. Investing Stacy has a total of $6,000 in two accounts. The total amount of interest she earns from both accounts in the first year is $500. If one of the accounts earns 8% interest per year and the other earns 9% interest per year, how much did she invest in each account?

34. Investing Travis has a total of $6,000 invested in two accounts. The total amount of interest he earns from the accounts in the first year is $410. If one account pays 6% per year and the other pays 8% per year, how much did he invest in each account?

35. Credit Cards The Shell MasterCard lets customers earn free gasoline based on purchases made with the card. On purchases of Shell products, including gasoline, 3% of the purchase price goes toward free gasoline. On all other purchases, 2% of the purchase price goes toward the free gasoline. Last year Melissa spent $2,850 with her Shell MasterCard and accumulated $64.00 worth of free gasoline. How much were her Shell purchases? How much were her other purchases?

36. Credit Cards Angela owns the same type of Shell MasterCard as Melissa in Problem 35. Last year Angela spent $5,625 with her Shell Master Card and accumulated $134.15 worth of free gasoline. How much were her Shell purchases? How much were her other purchases?

Miscellaneous Problems

37. **Tickets** Tickets for the father-and-son breakfast were \$2.00 for fathers and \$1.50 for sons. If a total of 75 tickets were sold for \$127.50, how many fathers and how many sons attended the breakfast?

38. **Tickets** A Girl Scout troop sells 62 tickets to their mother-and-daughter dinner, for a total of \$216. If the tickets cost \$4.00 for mothers and \$3.00 for daughters, how many of each ticket did they sell?

39. **Sales Tax** A woman owns a small, cash-only business in a state that requires her to charge 6% sales tax on each item she sells. At the beginning of the day, she has \$250 in the cash register. At the end of the day, she has \$1,204 in the register. How much money should she send to the state government for the sales tax she collected?

40. **Sales Tax** A store is located in a state that requires 6% tax on all items sold. If the store brings in a total of \$3,392 in one day, how much of that total was sales tax?

41. **Long-Distance Phone Calls** Patrick goes away to college. The first week he is away from home he calls his girlfriend, using his parents' telephone credit card, and talks for a long time. The telephone company charges 40 cents for the first minute and 30 cents for each additional minute, and then adds on a 50-cent service charge for using the credit card. If his parents receive a bill for \$13.80 for Patrick's call, how long did he talk?

42. **Long-Distance Phone Calls** A person makes a long-distance person-to-person call to Santa Barbara, California. The telephone company charges 41 cents for the first minute and 32 cents for each additional minute. Because the call is person-to-person, there is also a service charge of \$3.00. If the cost of the call is \$6.29, how many minutes did the person talk?

43. Use $h = 32t - 16t^2$ to complete the table.

t	0	$\frac{1}{4}$	1	$\frac{7}{4}$	2
h					

44. Use $s = \frac{60}{t}$ to complete the table.

t	4	6	8	10
s				

Coffee Sales The chart appeared in *USA Today* in October 2005. Use the information to complete the tables below. Round to the nearest tenth of a billion dollars.

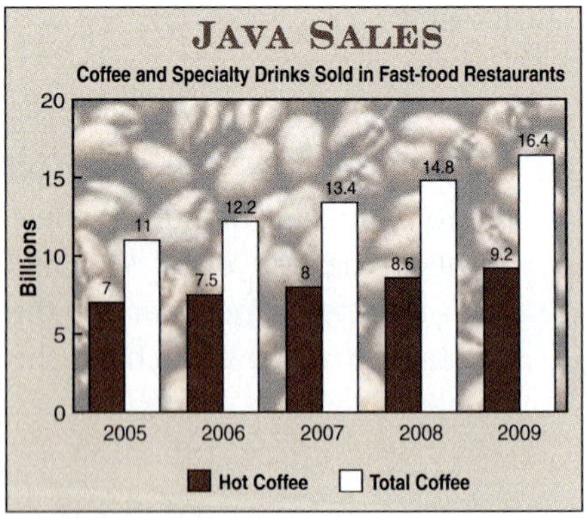

Data from *USA Today*. Copyright 2005.

45.

HOT COFFEE SALES	
YEAR	SALES (BILLIONS OF DOLLARS)
2005	
2006	
2007	
2008	
2009	

46.

TOTAL COFFEE SALES	
YEAR	SALES (BILLIONS OF DOLLARS)
2005	
2006	
2007	
2008	
2009	

Table Building

Maximum Heart Rate In exercise physiology, a person's maximum heart rate, in beats per minute, is found by subtracting his age in years from 220. So, if A represents your age in years, then your maximum heart rate is

$$M = 220 - A$$

Use this formula to complete the following tables.

47.

AGE (YEARS)	MAXIMUM HEART RATE (BEATS PER MINUTE)
18	
19	
20	
21	
22	
23	

48.

AGE (YEARS)	MAXIMUM HEART RATE (BEATS PER MINUTE)
15	
20	
25	
30	
35	
40	

Training Heart Rate A person's training heart rate, in beats per minute, is the person's resting heart rate plus 60% of the difference between maximum heart rate and their resting heart rate. If resting heart rate is R and maximum heart rate is M, then the formula that gives training heart rate is

$$T = R + 0.6(M - R)$$

Use this formula along with the results of Problems 49 and 50 to fill in the following two tables.

49. For a 20-year-old person

RESTING HEART RATE (BEATS PER MINUTE)	TRAINING HEART RATE (BEATS PER MINUTE)
60	
62	
64	
68	
70	
72	

50. For a 40-year-old person

RESTING HEART RATE (BEATS PER MINUTE)	TRAINING HEART RATE (BEATS PER MINUTE)
60	
62	
64	
68	
70	
72	

Problems 51–52 may be solved using a graphing calculator.

51. Livestock Pen A farmer buys 48 feet of fencing material to build a rectangular livestock pen. Fill in the second column of the table to find the length of the pen if the width is 2, 4, 6, 8, 10, or 12 feet. Then find the area of each of the pens formed.

w	l	A
2		
4		
6		
8		
10		
12		

52. Model Rocket A small rocket is projected straight up into the air with a velocity of 128 feet per second. The formula that gives the height h of the rocket t seconds after it is launched is

$$h = -16t^2 + 128t$$

Use this formula to find the height of the rocket after 1, 2, 3, 4, 5, and 6 seconds.

TIME (SECONDS)	HEIGHT (FEET)
1	
2	
3	
4	
5	
6	

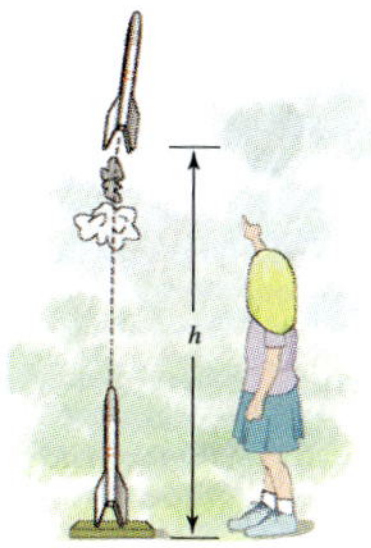

Getting Ready for the Next Section

To understand all of the explanations and examples in the next section, you must be able to work the problems below.

Graph each inequality.

53. $x < 2$

54. $x \le 2$

55. $x \ge -3$

56. $x > -3$

Solve each equation.

57. $-2x - 3 = 7$

58. $3x + 3 = 2x - 1$

59. $3(2x - 4) - 7x = -3x$

60. $3(2x + 5) = -3x$

Maintaining Your Skills

Graph each of the following on the number line provided.

61. $\{x \mid x > -5\}$

62. $\{x \mid x \le 4\}$

63. $\{x \mid x \le -2 \text{ or } x > 5\}$

64. $\{x \mid x < 3 \text{ or } x \ge 5\}$

65. $\{x \mid x > -4 \text{ and } x < 0\}$

66. $\{x \mid x \ge 0 \text{ and } x \le 2\}$

67. $\{x \mid 1 \le x \le 4\}$

68. $\{x \mid -4 < x < -2\}$

Linear Inequalities in One Variable

2.4

A *linear inequality in one variable* is any inequality that can be put in the form

$$ax + b < c \qquad (a, b, \text{and } c \text{ constants}, a \neq 0)$$

where the inequality symbol ($<$) can be replaced with any of the other three inequality symbols ($\leq$, $>$, or $\geq$).

Some examples of linear inequalities are

$$3x - 2 \geq 7 \qquad -5y < 25 \qquad 3(x - 4) > 2x$$

Our first property for inequalities is similar to the addition property we used when solving equations.

Addition Property for Inequalities

For any algebraic expressions, A, B, and C,

if $\quad A < B$

then $\quad A + C < B + C$

In words: Adding the same quantity to both sides of an inequality will not change the solution set.

EXAMPLE 1 Solve $3x + 3 < 2x - 1$, and graph the solution.

Solution We use the addition property for inequalities to write all the variable terms on one side and all constant terms on the other side:

$3x + 3 < 2x - 1$	
$3x + (\mathbf{-2x}) + 3 < 2x + (\mathbf{-2x}) - 1$	**Add $-2x$ to each side**
$x + 3 < -1$	
$x + 3 + (\mathbf{-3}) < -1 + (\mathbf{-3})$	**Add -3 to each side**
$x < -4$	

The solution set is all real numbers that are less than -4. To show this we can use set notation and write

$$\{x \mid x < -4\}$$

Or we can graph the solution set on the number line using a left-opening parenthesis at -4 to show that -4 is not part of the solution set:

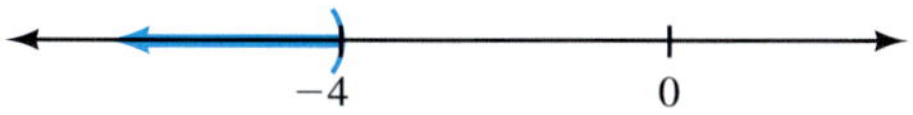

This graph gives rise to the following notation, called *interval notation,* that is an alternative way to write the solution set.

$$(-\infty, -4)$$

The preceding expression indicates that the solution set is all real numbers from negative infinity up to, but not including, -4.

We have three equivalent representations for the solution set to our original inequality. Here are all three together.

Set Notation	Line Graph	Interval Notation
$\{x \mid x < -4\}$		$(-\infty, -4)$

OBJECTIVES

A	Solve a linear inequality in one variable and graph the solution set.
B	Solve a compound inequality and graph the solution set.
C	Write solutions to inequalities using interval notation.
D	Solve application problems using inequalities.

NOTE

Since subtraction is defined as addition of the opposite, our new property holds for subtraction as well as addition; that is, we can subtract the same quantity from each side of an inequality and always be sure that we have not changed the solution.

PRACTICE PROBLEMS

1. Solve $4x - 2 > 3x + 4$ and graph the solution.

NOTE

The English mathematician John Wallis (1616–1703) was the first person to use the ∞ symbol to represent infinity. When we encounter the interval $(3, \infty)$, we read it as "the interval from 3 to infinity," and we mean the set of real numbers that are greater than three. Likewise, the interval $(-\infty, -4)$ is read "the interval from negative infinity to -4," which is all real numbers less than -4.

Answers

1–5. See Solutions to Selected Practice Problems for this section.

Interval Notation and Graphing

The table below shows the connection between inequalities, interval notation, and number line graphs. We have included the graphs with open and closed circles for those of you who have used this type of graph previously. In this book, we will continue to show our graphs using the parentheses/brackets method.

Inequality notation	Interval notation	Graph using parentheses/brackets	Graph using open and closed circles
$x < 2$	$(-\infty, 2)$		
$x \le 2$	$(-\infty, 2]$		
$x \ge -3$	$[-3, \infty)$		
$x > -3$	$(-3, \infty)$		

> **NOTE**
> The purpose of this discussion is to justify the multiplication property for inequalities. If you are having trouble understanding it, you may want to read the property itself first and then read through the discussion here. What is important is that you understand and can use the multiplication property for inequalities.

Before we state the multiplication property for inequalities, we will take a look at what happens to an inequality statement when we multiply both sides by a positive number and what happens when we multiply by a negative number.

We begin by writing three true inequality statements:

$$3 < 5 \qquad -3 < 5 \qquad -5 < -3$$

We multiply both sides of each inequality by a positive number—say, 4:

$$\begin{array}{ccc} 4(3) < 4(5) & 4(-3) < 4(5) & 4(-5) < 4(-3) \\ 12 < 20 & -12 < 20 & -20 < -12 \end{array}$$

Notice in each case that the resulting inequality symbol points in the same direction as the original inequality symbol. Multiplying both sides of an inequality by a positive number preserves the *sense* of the inequality.

Let's take the same three original inequalities and multiply both sides by -4:

$$\begin{array}{ccc} 3 < 5 & -3 < 5 & -5 < -3 \\ -4(3) > -4(5) & -4(-3) > -4(5) & -4(-5) > -4(-3) \\ -12 > -20 & 12 > -20 & 20 > 12 \end{array}$$

Notice in this case that the resulting inequality symbol always points in the opposite direction from the original one. Multiplying both sides of an inequality by a negative number *reverses* the sense of the inequality. Keeping this in mind, we will now state the multiplication property for inequalities.

> **NOTE**
> Because division is defined as multiplication by the reciprocal, we can apply our new property to division as well as to multiplication. We can divide both sides of an inequality by any nonzero number as long as we reverse the direction of the inequality when the number we are dividing by is negative.

Multiplication Property for Inequalities

Let A, B, and C represent algebraic expressions.

If	$A < B$		
then	$AC < BC$	if	C is positive ($C > 0$)
or	$AC > BC$	if	C is negative ($C < 0$)

In words: Multiplying both sides of an inequality by a positive number always produces an equivalent inequality. Multiplying both sides of an inequality by a negative number reverses the sense of the inequality.

The multiplication property for inequalities does not limit what we can do with inequalities. We are still free to multiply both sides of an inequality by any nonzero number we choose. If the number we multiply by happens to be *negative,* then we *must also reverse* the direction of the inequality.

EXAMPLE 2 Find the solution set for $-2y - 3 \le 7$.

Solution We begin by adding 3 to each side of the inequality:

$$-2y - 3 \le 7$$

$$-2y \le 10 \qquad \textbf{Add 3 to both sides}$$

$$-\frac{1}{2}(-2y) \ge -\frac{1}{2}(10) \qquad \textbf{Multiply by } -\tfrac{1}{2} \textbf{ and reverse the direction of the inequality symbol}$$

$$y \ge -5$$

The solution set is all real numbers that are greater than or equal to -5. Below are three equivalent ways to represent this solution set.

Set Notation	Line Graph	Interval Notation
$\{y \mid y \ge -5\}$		$[-5, \infty)$

Notice how a bracket is used with interval notation to show that -5 is part of the solution set.

When our inequalities become more complicated, we use the same basic steps we used in Section 2.1 when we were solving equations; that is, we simplify each side of the inequality before we apply the addition property or multiplication property. When we have solved the inequality, we graph the solution on a number line.

EXAMPLE 3 Solve $3(2x - 4) - 7x \le -3x$.

Solution We begin by using the distributive property to separate terms. Next, simplify both sides.

$$3(2x - 4) - 7x \le -3x \qquad \textbf{Original inequality}$$

$$6x - 12 - 7x \le -3x \qquad \textbf{Distributive property}$$

$$-x - 12 \le -3x \qquad \mathbf{6x - 7x = (6 - 7)x = -x}$$

$$-12 \le -2x \qquad \textbf{Add } x \textbf{ to both sides}$$

$$-\frac{1}{2}(-12) \ge -\frac{1}{2}(-2x) \qquad \textbf{Multiply both sides by } -\tfrac{1}{2} \textbf{ and reverse the direction of the inequality symbol}$$

$$6 \ge x$$

This last line is equivalent to $x \le 6$. The solution set can be represented with any of the three following items.

Set Notation	Line Graph	Interval Notation
$\{x \mid x \le 6\}$		$(-\infty, 6]$

2. Find the solution set for $-3y - 2 < 7$

3. Solve $3(2x + 5) \le -3x$.

NOTE

In Examples 2 and 3, notice that each time we multiplied both sides of the inequality by a negative number we also reversed the direction of the inequality symbol. Failure to do so would cause our graph to lie on the wrong side of the endpoint.

4. Solve and graph the solution.
$-7 \le 2x + 1 \le 7$

EXAMPLE 4 Solve and graph $-3 \le 2x - 5 \le 3$.

Solution We can extend our properties for addition and multiplication to cover this situation. If we add a number to the middle expression, we must add the same number to the outside expressions. If we multiply the center expression by a number, we must do the same to the outside expressions, remembering to reverse the direction of the inequality symbols if we multiply by a negative number. We begin by adding 5 to all three parts of the inequality:

$$-3 \le 2x - 5 \le 3$$
$$2 \le 2x \le 8 \quad \textbf{Add 5 to all three members}$$
$$1 \le x \le 4 \quad \textbf{Multiply through by } \tfrac{1}{2}$$

Here are three ways to write this solution set:

Set Notation	Line Graph	Interval Notation
$\{x \mid 1 \le x \le 4\}$		$[1, 4]$

Interval Notation and Graphing

The table below shows the connection between interval notation and number line graphs for a variety of continued inequalities. Again, we have included the graphs with open and closed circles for those of you who have used this type of graph previously. Remember, however, that in this book we will be using the parentheses/brackets method of graphing.

Inequality notation	Interval notation	Graph using parentheses/brackets	Graph using open and closed circles
$-3 < x < 2$	$(-3, 2)$		
$-3 \le x \le 2$	$[-3, 2]$		
$-3 \le x < 2$	$[-3, 2)$		
$-3 < x \le 2$	$(-3, 2]$		

5. Graph the solution for the compound inequality.
$3t - 6 \le -3$ or $3t - 6 \ge 3$

EXAMPLE 5 Solve the compound inequality.

$$3t + 7 \le -4 \quad \text{or} \quad 3t + 7 \ge 4$$

Solution We solve each half of the compound inequality separately, then we graph the solution set:

$$3t + 7 \le -4 \quad \text{or} \quad 3t + 7 \ge 4$$
$$3t \le -11 \quad \text{or} \quad 3t \ge -3 \quad \textbf{Add } -7$$
$$t \le -\frac{11}{3} \quad \text{or} \quad t \ge -1 \quad \textbf{Multiply by } \tfrac{1}{3}$$

The solution set can be written in any of the following ways:

Set Notation	Line Graph	Interval Notation
$\{t \mid t \le -\frac{11}{3} \text{ or } t \ge -1\}$		$(-\infty, -\frac{11}{3}] \cup [-1, \infty)$

EXAMPLE 6

A company that manufactures ink cartridges for printers finds that they can sell x cartridges each week at a price of p dollars each, according to the formula $x = 1{,}300 - 100p$. What price should they charge for each cartridge if they want to sell at least 300 cartridges a week?

Solution Because x is the number of cartridges they sell each week, an inequality that corresponds to selling at least 300 cartridges a week is

$$x \ge 300$$

Substituting $1{,}300 - 100p$ for x gives us an inequality in the variable p.

$$1{,}300 - 100p \ge 300$$

$$-100p \ge -1{,}000 \qquad \textbf{Add } -1{,}300 \textbf{ to each side}$$

$$p \le 10 \qquad \textbf{Divide each side by } -100\textbf{, and reverse the direction of the inequality symbol}$$

To sell at least 300 cartridges each week, the price per cartridge should be no more than \$10; that is, selling the cartridges for \$10 or less will produce weekly sales of 300 or more cartridges.

6. Use the information in Example 6 to find the price they must charge to sell at least 800 cartridges a week.

EXAMPLE 7

The formula $F = \frac{9}{5}C + 32$ gives the relationship between the Celsius and Fahrenheit temperature scales. If the temperature range on a certain day is 86° to 104° Fahrenheit, what is the temperature range in degrees Celsius?

Solution From the given information we can write $86 \le F \le 104$. But, because F is equal to $\frac{9}{5}C + 32$, we can also write

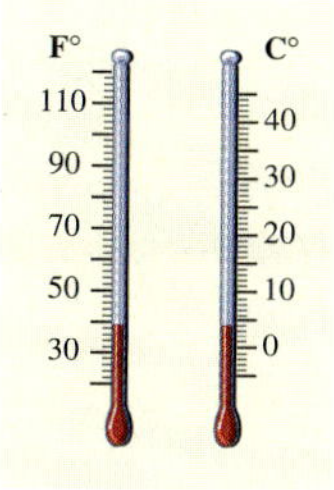

$$86 \le \frac{9}{5}C + 32 \le 104$$

$$54 \le \frac{9}{5}C \le 72 \qquad \textbf{Add } -32 \textbf{ to each member}$$

$$\frac{5}{9}(54) \le \frac{5}{9}\left(\frac{9}{5}C\right) \le \frac{5}{9}(72)$$

$$30 \le C \le 40$$

A temperature range of 86° to 104° Fahrenheit corresponds to a temperature range of 30° to 40° Celsius.

7. During one 24-hour period, the temperature in Provo, Utah, ranged from a low of 18° below zero to a high of 22° above zero Fahrenheit. Find the corresponding temperatures in degrees Celsius. Round your answer to the nearest tenth of a degree.

Getting Ready for Class

After reading through the preceding section, respond in your own words and in complete sentences.

1. What is the addition property for inequalities?
2. When we use interval notation to denote a section of the real number line, when do we use parentheses ( ) and when do we use brackets [ ]?
3. Explain the difference between the multiplication property of equality and the multiplication property for inequalities.
4. When solving an inequality, when do we change the direction of the inequality symbol?

Linking Objectives and Examples

Next to each **objective**, list the examples that are best described by that objective.

A	
B	
C	
D	

Answers

6. \$5 or less 7. $-27.8° \le C \le -5.6°$

Problem Set 2.4

INTERMEDIATE Algebra f(x) Now™

Solve each of the following inequalities and graph each solution.

1. $2x \le 3$

2. $5x \ge -115$

3. $\frac{1}{2}x > 2$

4. $\frac{1}{3}x > 4$

5. $-5x \le 25$

6. $-7x \ge 35$

7. $-\frac{3}{2}x > -6$

8. $-\frac{2}{3}x < -8$

9. $-12 \le 2x$

10. $-20 \ge 4x$

11. $-1 \ge -\frac{1}{4}x$

12. $-1 \le -\frac{1}{5}x$

13. $-3x + 1 > 10$

14. $-2x - 5 \le 15$

15. $\frac{1}{2} - \frac{m}{12} \le \frac{7}{12}$

16. $\frac{1}{2} - \frac{m}{10} > -\frac{1}{5}$

17. $\frac{1}{2} \geq -\frac{1}{6} - \frac{2}{9}x$

18. $\frac{9}{5} > -\frac{1}{5} - \frac{1}{2}x$

19. $-40 \leq 30 - 20y$

20. $-20 > 50 - 30y$

21. $\frac{2}{3}x - 3 < 1$

22. $\frac{3}{4}x - 2 > 7$

23. $10 - \frac{1}{2}y \leq 36$

24. $8 - \frac{1}{3}y \geq 20$

Simplify each side first, then solve the following inequalities. Write your answers with interval notation.

25. $2(3y + 1) \leq -10$

26. $3(2y - 4) > 0$

27. $-(a + 1) - 4a \leq 2a - 8$

28. $-(a - 2) - 5a \leq 3a + 7$

29. $\frac{1}{3}t - \frac{1}{2}(5 - t) < 0$

30. $\frac{1}{4}t - \frac{1}{3}(2t - 5) < 0$

31. $-2 \leq 5 - 7(2a + 3)$

32. $1 < 3 - 4(3a - 1)$

33. $-\frac{1}{3}(x + 5) \leq -\frac{2}{9}(x - 1)$

34. $-\frac{1}{2}(2x + 1) \leq -\frac{3}{8}(x + 2)$

Solve each inequality. Write your answer using inequality notation.

35. $20x + 9{,}300 > 18{,}000$

36. $20x + 4{,}800 > 18{,}000$

Solve the following continued inequalities. Use both a line graph and interval notation to write each solution set.

37. $-2 \leq m - 5 \leq 2$

38. $-3 \leq m + 1 \leq 3$

39. $-60 < 20a + 20 < 60$

40. $-60 < 50a - 40 < 60$

41. $0.5 \le 0.3a - 0.7 \le 1.1$

42. $0.1 \le 0.4a + 0.1 \le 0.3$

43. $3 < \frac{1}{2}x + 5 < 6$

44. $5 < \frac{1}{4}x + 1 < 9$

45. $4 < 6 + \frac{2}{3}x < 8$

46. $3 < 7 + \frac{4}{5}x < 15$

Graph the solution sets for the following compound inequalities. Then write each solution set using interval notation.

47. $x + 5 \le -2$ or $x + 5 \ge 2$

48. $3x + 2 < -3$ or $3x + 2 > 3$

49. $5y + 1 \le -4$ or $5y + 1 \ge 4$

50. $7y - 5 \le -2$ or $7y - 5 \ge 2$

51. $2x + 5 < 3x - 1$ or $x - 4 > 2x + 6$

52. $3x - 1 > 2x + 4$ or $5x - 2 < 3x + 4$

Applying the Concepts

A store selling art supplies finds that they can sell x sketch pads each week at a price of p dollars each, according to the formula $x = 900 - 300p$. What price should they charge if they want to sell

53. at least 300 pads each week?

54. more than 600 pads each week?

55. less than 525 pads each week?

56. at most 375 pads each week?

57. Amtrak The average number of passengers carried by Amtrak declined each year for the years 1990–1996 (American Association of Railroads, Washington, DC, *Railroad Facts, Statistics of Railroads of Class 1*.) The linear model for the number of passengers carried each year by Amtrak is given by $P = 22{,}419 - 399x$, where P is the number of passengers, in millions, and x is the number of years after January 1, 1990. In what years did Amtrak have more than 20,500 million passengers?

Courtesy of Amtrak

58. Health Insurance Income The revenues from health insurance premiums collected by life insurance companies in the United States has been increasing since 1980 (American Council of Life Insurance, *Life Insurance Fact Book,* annual.) A model for this growth during 1980–1997 is

$$P = 4x + 23$$

where P is the annual health insurance premiums collected in billions of dollars, and x is the number of years after January 1, 1980. In what years did the insurance companies collect between \$60 and \$80 billion in health insurance premiums each year?

59. Cell Phone Cost Suppose you have a cellular phone in which you have committed to spend less than \$30 on each monthly bill. Your rate plan costs \$20 a month and includes 30 free minutes. Each additional minute costs \$0.40. Set up and solve an inequality to determine how many minutes you may use each month to keep your bill under \$30.

60. Student Loan When considering how much debt to incur in student loans, you learn that it is wise to keep your student loan payment to 8% or less of your starting monthly income. Suppose you anticipate a starting annual salary of \$24,000. Set up and solve an inequality that represents the amount of monthly debt for student loans that would be considered manageable.

Temperature Each of the following temperature ranges is in degrees Fahrenheit. Use the formula $F = \frac{9}{5}C + 32$ to find the corresponding temperature range in degrees Celsius.

61. 95° to 113°

62. 68° to 86°

63. −13° to 14°

64. −4° to 23°

Getting Ready for the Next Section

To understand all of the explanations and examples in the next section, you must be able to work the problems below.

Solve each equation.

65. $2a - 1 = -7$

66. $3x - 6 = 9$

67. $\frac{2}{3}x - 3 = 7$

68. $\frac{2}{3}x - 3 = -7$

69. $x - 5 = x - 7$

70. $x + 3 = x + 8$

71. $x - 5 = -x - 7$

72. $x + 3 = -x + 8$

Maintaining Your Skills

The problems that follow review some of the more important skills you have learned in previous sections and chapters. You can consider the time you spend working these problems as time spent studying for exams.

Simplify.

73. $|-3|$

74. $|3|$

75. $-|-3|$

76. $-(-3)$

77. Give a definition for the absolute value of x that involves the number line. (This is the geometric definition.)

78. Give a definition of the absolute value of x that does not involve the number line. (This is the algebraic definition.)

Extending the Concepts

Assume a, b, and c are positive, and solve each formula for x.

79. $ax + b < c$

80. $\frac{x}{a} + \frac{y}{b} < 1$

81. $-c < ax + b < c$

82. $-1 < \frac{ax + b}{c} < 1$

Improving Your Quantitative Literacy

83. What Is Luxury? As you can see, the percents shown in the graph are from a survey of 770 families.

a. How many families responded that luxury is happiness and satisfaction f rom being with friends and family and enjoying good times?

b. If x represents the annual income of any of the families surveyed, write an inequality using x that indicates their level of income.

c. The margin of error for the survey is $\pm 4\%$. What does this mean in relation to the 19% category?

Data from *USA Today.* Copyright 2005. Reprinted with permission.

Equations with Absolute Value

2.5

Previously we defined the absolute value of x, $|x|$, to be the distance between x and 0 on the number line. The absolute value of a number measures its distance from 0.

OBJECTIVES

A Solve equations with absolute value symbols.

EXAMPLE 1 Solve for x: $|x| = 5$.

Solution Using the definition of absolute value, we can read the equation as, "The distance between x and 0 on the number line is 5." If x is 5 units from 0, then x can be 5 or -5:

$$\text{If } |x| = 5 \quad \text{then } x = 5 \quad \text{or} \quad x = -5$$

In general, then, we can see that any equation of the form $|a| = b$ is equivalent to the equations $a = b$ or $a = -b$, as long as $b > 0$.

PRACTICE PROBLEMS

1. Solve for x: $|x| = 3$.

EXAMPLE 2 Solve $|2a - 1| = 7$.

Solution We can read this question as "$2a - 1$ is 7 units from 0 on the number line." The quantity $2a - 1$ must be equal to 7 or -7:

$$|2a - 1| = 7$$
$$2a - 1 = 7 \quad \text{or} \quad 2a - 1 = -7$$

We have transformed our absolute value equation into two equations that do not involve absolute value. We can solve each equation separately.

$$\begin{aligned} 2a - 1 &= 7 & \text{or} \quad 2a - 1 &= -7 & \\ 2a &= 8 & \text{or} \quad 2a &= -6 & \textbf{Add 1 to both sides} \\ a &= 4 & \text{or} \quad a &= -3 & \textbf{Multiply by } \tfrac{1}{2} \end{aligned}$$

Our solution set is $\{4, -3\}$.

To check our solutions, we put them into the original absolute value equation:

When	$a = 4$	When	$a = -3$
the equation	$\lvert 2a - 1\rvert = 7$	the equation	$\lvert 2a - 1\rvert = 7$
becomes	$\lvert 2(4) - 1\rvert = 7$	becomes	$\lvert 2(-3) - 1\rvert = 7$
	$\lvert 7\rvert = 7$		$\lvert -7\rvert = 7$
	$7 = 7$		$7 = 7$

2. Solve $|3x - 6| = 9$.

EXAMPLE 3 Solve $\left|\frac{2}{3}x - 3\right| + 5 = 12$.

Solution To use the definition of absolute value to solve this equation, we must isolate the absolute value on the left side of the equal sign. To do so, we add -5 to both sides of the equation to obtain

$$\left|\frac{2}{3}x - 3\right| = 7$$

Now that the equation is in the correct form, we can write

$$\begin{aligned} \frac{2}{3}x - 3 &= 7 & \text{or} \quad \frac{2}{3}x - 3 &= -7 & \\ \frac{2}{3}x &= 10 & \text{or} \quad \frac{2}{3}x &= -4 & \textbf{Add 3 to both sides} \\ x &= 15 & \text{or} \quad x &= -6 & \textbf{Multiply by } \tfrac{3}{2} \end{aligned}$$

The solution set is $\{15, -6\}$.

3. Solve $|4x - 3| + 2 = 3$.

Answers

1. $-3, 3$ 2. $-1, 5$ 3. $1, \frac{1}{2}$

4. Solve $|7a - 1| = -2$.

NOTE

Recall that ∅ is the symbol we use to denote the empty set. When we use it to indicate the solutions to an equation, then we are saying the equation has no solution.

EXAMPLE 4 Solve $|3a - 6| = -4$.

Solution The solution set is ∅ because the left side cannot be negative and the right side is negative. No matter what we try to substitute for the variable a, the quantity $|3a - 6|$ will always be positive or zero. It can never be -4.

Consider the statement $|a| = |b|$. What can we say about a and b? We know they are equal in absolute value. By the definition of absolute value, they are the same distance from 0 on the number line. They must be equal to each other or opposites of each other. In symbols, we write

$$\underbrace{|a| = |b|}_{\text{Equal in absolute value}} \Leftrightarrow \underbrace{a = b}_{\text{Equals}} \quad \text{or} \quad \underbrace{a = -b}_{\text{Opposites}}$$

5. Solve $|x + 3| = |x + 8|$.

EXAMPLE 5 Solve $|x - 5| = |x - 7|$.

Solution The quantities $x - 5$ and $x - 7$ must be equal or they must be opposites because their absolute values are equal:

Equals		*Opposites*
$x - 5 = x - 7$	or	$x - 5 = -(x - 7)$
$-5 = -7$		$x - 5 = -x + 7$
No solution here		$2x - 5 = 7$
		$2x = 12$
		$x = 6$

Because the first equation leads to a false statement, it will not give us a solution. (If either of the two equations were to reduce to a true statement, it would mean all real numbers would satisfy the original equation.) In this case, our only solution is $x = 6$.

Linking Objectives and Examples

Next to each **objective**, list the examples that are best described by that objective.

A	

Getting Ready for Class

After reading through the preceding section, respond in your own words and in complete sentences.

1. Why do some of the equations in this section have two solutions instead of one?
2. Translate $|x| = 6$ into words using the definition of absolute value.
3. Explain in words what the equation $|x - 3| = 4$ means with respect to distance on the number line.
4. When is the statement $|x| = x$ true?

Answers

4. No solution 5. $-\frac{11}{2}$

Problem Set 2.5

INTERMEDIATE Algebra Now™

Use the definition of absolute value to solve each of the following equations.

1. $|x| = 4$
2. $|x| = 7$
3. $2 = |a|$
4. $5 = |a|$
5. $|x| = -3$
6. $|x| = -4$
7. $|a| + 2 = 3$
8. $|a| - 5 = 2$
9. $|y| + 4 = 3$
10. $|y| + 3 = 1$
11. $4 = |x| - 2$
12. $3 = |x| - 5$
13. $|x - 2| = 5$
14. $|x + 1| = 2$
15. $|a - 4| = \frac{5}{3}$
16. $|a + 2| = \frac{7}{5}$
17. $1 = |3 - x|$
18. $2 = |4 - x|$
19. $\left|\frac{3}{5}a + \frac{1}{2}\right| = 1$
20. $\left|\frac{2}{7}a + \frac{3}{4}\right| = 1$
21. $60 = |20x - 40|$
22. $800 = |400x - 200|$
23. $|2x + 1| = -3$
24. $|2x - 5| = -7$
25. $\left|\frac{3}{4}x - 6\right| = 9$
26. $\left|\frac{4}{5}x - 5\right| = 15$
27. $\left|1 - \frac{1}{2}a\right| = 3$
28. $\left|2 - \frac{1}{3}a\right| = 10$

Solve each equation.

29. $|3x + 4| + 1 = 7$

30. $|5x - 3| - 4 = 3$

31. $|3 - 2y| + 4 = 3$

32. $|8 - 7y| + 9 = 1$

33. $3 + |4t - 1| = 8$

34. $2 + |2t - 6| = 10$

35. $\left|9 - \frac{3}{5}x\right| + 6 = 12$

36. $\left|4 - \frac{2}{7}x\right| + 2 = 14$

37. $5 = \left|\frac{2x}{7} + \frac{4}{7}\right| - 3$

38. $7 = \left|\frac{3x}{5} + \frac{1}{5}\right| + 2$

39. $2 = -8 + \left|4 - \frac{1}{2}y\right|$

40. $1 = -3 + \left|2 - \frac{1}{4}y\right|$

41. $|3a + 1| = |2a - 4|$

42. $|5a + 2| = |4a + 7|$

43. $\left|x - \frac{1}{3}\right| = \left|\frac{1}{2}x + \frac{1}{6}\right|$

44. $\left|\frac{1}{10}x - \frac{1}{2}\right| = \left|\frac{1}{5}x + \frac{1}{10}\right|$

45. $|y - 2| = |y + 3|$

46. $|y - 5| = |y - 4|$

47. $|3x - 1| = |3x + 1|$

48. $|5x - 8| = |5x + 8|$

49. $|3 - m| = |m + 4|$

50. $|5 - m| = |m + 8|$

51. $|0.03 - 0.01x| = |0.04 + 0.05x|$

52. $|0.07 - 0.01x| = |0.08 - 0.02x|$

53. $|x - 2| = |2 - x|$

54. $|x - 4| = |4 - x|$

55. $\left|\frac{x}{5} - 1\right| = \left|1 - \frac{x}{5}\right|$

56. $\left|\frac{x}{3} - 1\right| = \left|1 - \frac{x}{3}\right|$

Applying the Concepts

57. Amtrak Amtrak's annual passenger revenue for the years 1985–1995 is modeled approximately by the formula

$$R = -60|x - 11| + 962$$

where R is the annual revenue in millions of dollars and x is the number of years since January 1, 1980 (Association of American Railroads, Washington, DC, *Railroad Facts, Statistics of Railroads of Class 1,* annual). In what years was the passenger revenue \$722 million?

58. Corporate Profits The corporate profits for various U.S. industries vary from year to year. An approximate model for profits of U.S. "communications companies" during a given year between 1990 and 1997 is given by

$$P = -3{,}400|x - 5.5| + 36{,}000$$

where P is the annual profits (in millions of dollars) and x is the number of years since January 1, 1990 (U.S. Bureau of Economic Analysis, Income and Product Accounts of the U.S. (1929–1994), *Survey of Current Business,* September 1998). Use the model to determine the years in which profits of "communication companies" were \$31.5 billion (\$31,500 million).

Getting Ready for the Next Section

To understand all of the explanations and examples in the next section, you must be able to work the problems below.

Solve each inequality. Do not graph the solution set.

59. $2x - 5 < 3$

60. $-3 < 2x - 5$

61. $-4 \le 3a + 7$

62. $3a + 7 \le 4$

63. $4t - 3 \le -9$

64. $4t - 3 \ge 9$

Maintaining Your Skills

The problems that follow review some of the more important skills you have learned in previous sections and chapters. You can consider the time you spend working these problems as time spent studying for exams.

Graph each inequality.

65. $x < -2$ or $x > 8$

66. $1 < x < 4$

67. $-2 \le x \le 1$

68. $x \le -\frac{3}{2}$ or $x \ge 3$

Simplify each expression.

69. $\frac{38}{30}$

70. $\frac{10}{25}$

71. $\frac{240}{6}$

72. $\frac{39}{13}$

73. $\frac{0 + 6}{0 - 3}$

74. $\frac{6 + 6}{6 - 3}$

75. $\frac{4 - 4}{4 - 2}$

76. $\frac{3 + 6}{3 - 3}$

77. $\frac{-1 - 3}{3 - (-3)}$

2.6 Inequalities Involving Absolute Value

In this section we will again apply the definition of absolute value to solve inequalities involving absolute value. Again, the absolute value of x, which is $|x|$, represents the distance that x is from 0 on the number line. We will begin by considering three absolute value equations/inequalities and their English translations:

Expression	*In Words*
$\lvert x \rvert = 7$	x is exactly 7 units from 0 on the number line
$\lvert a \rvert < 5$	a is less than 5 units from 0 on the number line
$\lvert y \rvert \geq 4$	y is greater than or equal to 4 units from 0 on the number line

Once we have translated the expression into words, we can use the translation to graph the original equation or inequality. The graph then is used to write a final equation or inequality that does not involve absolute value.

Original Expression	*Graph*	*Final Expression*
$\lvert x \rvert = 7$	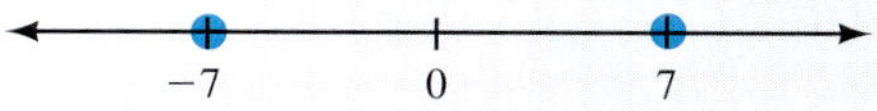	$x = -7$ or $x = 7$
$\lvert a \rvert < 5$	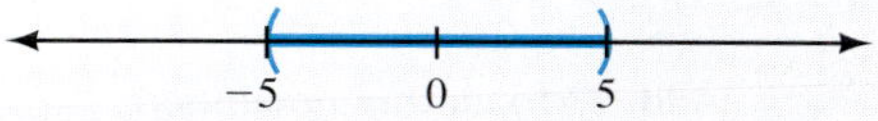	$-5 < a < 5$
$\lvert y \rvert \geq 4$	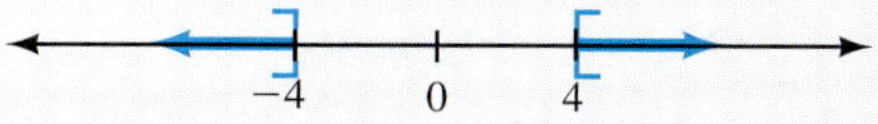	$y \leq -4$ or $y \geq 4$

Although we will not always write out the English translation of an absolute value inequality, it is important that we understand the translation. Our second expression, $|a| < 5$, means a is within 5 units of 0 on the number line. The graph of this relationship is

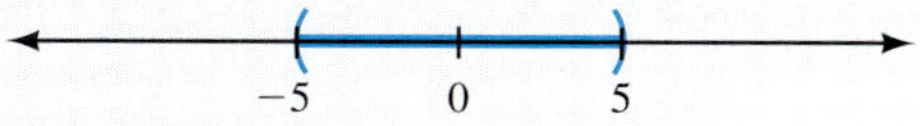

which can be written with the following continued inequality:

$$-5 < a < 5$$

We can follow this same kind of reasoning to solve more complicated absolute value inequalities.

OBJECTIVES

A Solve inequalities with absolute value and graph the solution set.

EXAMPLE 1

Graph the solution set: $|2x - 5| < 3$.

Solution The absolute value of $2x - 5$ is the distance that $2x - 5$ is from 0 on the number line. We can translate the inequality as "$2x - 5$ is less than 3 units from 0 on the number line"; that is, $2x - 5$ must appear between -3 and 3 on the number line.

A picture of this relationship is

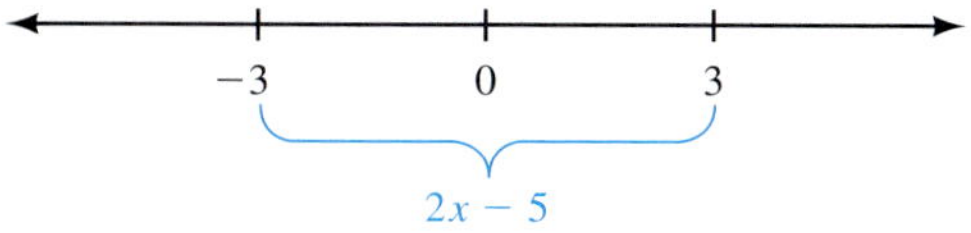

Using the picture, we can write an inequality without absolute value that describes the situation:

$$-3 < 2x - 5 < 3$$

PRACTICE PROBLEMS

1. Graph the solution set.
$$|2x + 1| \leq 7$$

Answers

1–8. See Solutions to Selected Practice Problems for this section.

Next, we solve the continued inequality by first adding 5 to all three members and then multiplying all three by $\frac{1}{2}$.

$$-3 < 2x - 5 < 3$$

$$2 < 2x < 8 \qquad \textbf{Add 5 to all three members}$$

$$1 < x < 4 \qquad \textbf{Multiply each member by } \tfrac{1}{2}$$

The graph of the solution set is

We can see from the solution that for the absolute value of $2x - 5$ to be within 3 units of 0 on the number line, x must be between 1 and 4.

2. Solve and graph $|4a - 3| < 5$.

EXAMPLE 2

Solve and graph $|3a + 7| \le 4$.

Solution We can read the inequality as, "The distance between $3a + 7$ and 0 is less than or equal to 4." Or, "$3a + 7$ is within 4 units of 0 on the number line." This relationship can be written without absolute value as

$$-4 \le 3a + 7 \le 4$$

Solving as usual, we have

$$-4 \le 3a + 7 \le 4$$

$$-11 \le 3a < -3 \qquad \textbf{Add } -7 \textbf{ to all three members}$$

$$-\frac{11}{3} \le a \le -1 \qquad \textbf{Multiply each member by } \tfrac{1}{3}$$

We can see from Examples 1 and 2 that to solve an inequality involving absolute value, we must be able to write an equivalent expression that does not involve absolute value.

3. Solve $|x + 2| > 7$ and graph the solution.

EXAMPLE 3

Solve $|x - 3| > 5$ and graph the solution.

Solution We interpret the absolute value inequality to mean that $x - 3$ is more than 5 units from 0 on the number line. The quantity $x - 3$ must be either above +5 or below −5. Here is a picture of the relationship:

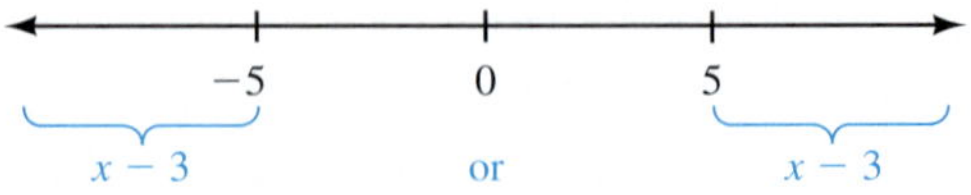

An inequality without absolute value that also describes this situation is

$$x - 3 < -5 \quad \text{or} \quad x - 3 > 5$$

Adding 3 to both sides of each inequality we have

$$x < -2 \quad \text{or} \quad x > 8$$

The graph of which is

EXAMPLE 4

Graph the solution set: $|4t - 3| \geq 9$.

Solution The quantity $4t - 3$ is greater than or equal to 9 units from 0. It must be either above $+9$ or below -9.

$$4t - 3 \leq -9 \quad \text{or} \quad 4t - 3 \geq 9$$

$$4t \leq -6 \quad \text{or} \quad 4t \geq 12 \qquad \textbf{Add 3}$$

$$t \leq -\frac{6}{4} \quad \text{or} \quad t \geq \frac{12}{4} \qquad \textbf{Multiply by } \tfrac{1}{4}$$

$$t \leq -\frac{3}{2} \quad \text{or} \quad t \geq 3$$

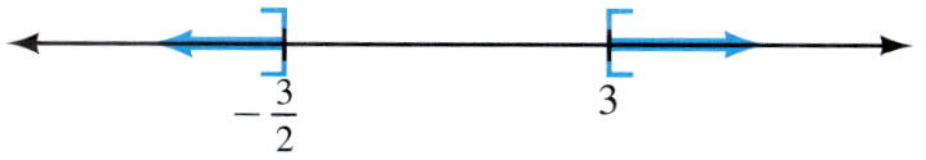

4. Graph the solution set.
$|3t - 6| \geq 3$

We can use the results of our first few examples and the material in the previous section to summarize the information we have related to absolute value equations and inequalities.

Rewriting Absolute Value Equations and Inequalities

If c is a positive real number, then each of the following statements on the left is equivalent to the corresponding statement on the right.

With Absolute Value	*Without Absolute Value*
$\lvert x \rvert = c$	$x = -c$ or $x = c$
$\lvert x \rvert < c$	$-c < x < c$
$\lvert x \rvert > c$	$x < -c$ or $x > c$
$\lvert ax + b \rvert = c$	$ax + b = -c$ or $ax + b = c$
$\lvert ax + b \rvert < c$	$-c < ax + b < c$
$\lvert ax + b \rvert > c$	$ax + b < -c$ or $ax + b > c$

EXAMPLE 5

Solve and graph $|2x + 3| + 4 < 9$.

Solution Before we can apply the method of solution we used in the previous examples, we must isolate the absolute value on one side of the inequality. To do so, we add $\mathbf{-4}$ to each side.

$$|2x + 3| + 4 < 9$$

$$|2x + 3| + 4 + (\mathbf{-4}) < 9 + (\mathbf{-4})$$

$$|2x + 3| < 5$$

From this last line we know that $2x + 3$ must be between -5 and $+5$.

$$-5 < 2x + 3 < 5$$

$$-8 < 2x < 2 \qquad \textbf{Add } \mathbf{-3} \textbf{ to each member}$$

$$-4 < x < 1 \qquad \textbf{Multiply each member by } \tfrac{1}{2}$$

The graph is

5. Solve and graph.
$|2x + 5| - 2 < 9$

6. Solve and graph.
$|5 - 2t| > 3$

EXAMPLE 6 Solve and graph $|4 - 2t| > 2$.

Solution The inequality indicates that $4 - 2t$ is less than -2 or greater than $+2$. Writing this without absolute value symbols, we have

$$4 - 2t < -2 \quad \text{or} \quad 4 - 2t > 2$$

To solve these inequalities we begin by adding $\mathbf{-4}$ to each side.

$$4 + (\mathbf{-4}) - 2t < -2 + (\mathbf{-4}) \quad \text{or} \quad 4 + (\mathbf{-4}) - 2t > 2 + (\mathbf{-4})$$

$$-2t < -6 \quad \text{or} \quad -2t > -2$$

Next we must multiply both sides of each inequality by $\mathbf{-\frac{1}{2}}$. When we do so, we must also reverse the direction of each inequality symbol.

$$-2t < -6 \quad \text{or} \quad -2t > -2$$

$$-\tfrac{1}{2}(-2t) > -\tfrac{1}{2}(-6) \quad \text{or} \quad -\tfrac{1}{2}(-2t) < -\tfrac{1}{2}(-2)$$

$$t > 3 \quad \text{or} \quad t < 1$$

Although in situations like this we are used to seeing the "less than" symbol written first, the meaning of the solution is clear. We want to graph all real numbers that are either greater than 3 or less than 1. Here is the graph.

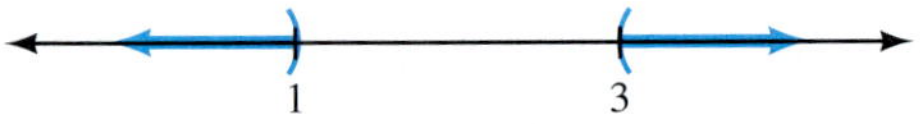

NOTE

Remember, the multiplication property for inequalities requires that we reverse the direction of the inequality symbol *every* time we multiply both sides of an inequality by a negative number.

Because absolute value always results in a nonnegative quantity, we sometimes come across special solution sets when a negative number appears on the right side of an absolute value inequality.

7. Solve $|8y + 3| \le -5$.

EXAMPLE 7 Solve $|7y - 1| < -2$.

Solution The *left* side is never negative because it is an absolute value. The *right* side is negative. We have a positive quantity (or zero) less than a negative quantity, which is impossible. The solution set is the empty set, $\varnothing$. There is no real number to substitute for y to make this inequality a true statement.

8. Solve $|2x - 9| \ge -7$.

EXAMPLE 8 Solve $|6x + 2| > -5$.

Solution This is the opposite case from that in Example 7. No matter what real number we use for x on the *left* side, the result will always be positive, or zero. The *right* side is negative. We have a positive quantity (or zero) greater than a negative quantity. Every real number we choose for x gives us a true statement. The solution set is the set of all real numbers.

Linking Objectives and Examples

Next to each **objective**, list the examples that are best described by that objective.

A	

Getting Ready for Class

After reading through the preceding section, respond in your own words and in complete sentences.

1. Write an inequality containing absolute value, the solution to which is all the numbers between -5 and 5 on the number line.
2. Translate $|x| \ge 3$ into words using the definition of absolute value.
3. Explain in words what the inequality $|x - 5| < 2$ means with respect to distance on the number line.
4. Why is there no solution to the inequality $|2x - 3| < 0$?

Problem Set 2.6

INTERMEDIATE Algebra $f(x)$ Now™

Solve each of the following inequalities using the definition of absolute value. Graph the solution set in each case.

1. $|x| < 3$

2. $|x| \le 7$

3. $|x| \ge 2$

4. $|x| > 4$

5. $|x| + 2 < 5$

6. $|x| - 3 < -1$

7. $|t| - 3 > 4$

8. $|t| + 5 > 8$

9. $|y| < -5$

10. $|y| > -3$

11. $|x| \ge -2$

12. $|x| \le -4$

13. $|x - 3| < 7$

14. $|x + 4| < 2$

15. $|a + 5| \ge 4$

16. $|a - 6| \ge 3$

Solve each inequality and graph the solution set.

17. $|a - 1| < -3$

18. $|a + 2| \geq -5$

19. $|2x - 4| < 6$

20. $|2x + 6| < 2$

21. $|3y + 9| \geq 6$

22. $|5y - 1| \geq 4$

23. $|2k + 3| \geq 7$

24. $|2k - 5| \geq 3$

25. $|x - 3| + 2 < 6$

26. $|x + 4| - 3 < -1$

27. $|2a + 1| + 4 \geq 7$

28. $|2a - 6| - 1 \geq 2$

29. $|3x + 5| - 8 < 5$

30. $|6x - 1| - 4 \leq 2$

Solve each inequality, and graph the solution set. Keep in mind that if you multiply or divide both sides of an inequality by a negative number, you must reverse the inequality sign.

31. $|5 - x| > 3$

32. $|7 - x| > 2$

33. $\left|3 - \frac{2}{3}x\right| \geq 5$

34. $\left|3 - \frac{3}{4}x\right| \geq 9$

35. $\left|2 - \frac{1}{2}x\right| > 1$

36. $\left|3 - \frac{1}{3}x\right| > 1$

Solve each inequality.

37. $|x - 1| < 0.01$

38. $|x + 1| < 0.01$

39. $|2x + 1| \geq \frac{1}{5}$

40. $|2x - 1| \geq \frac{1}{8}$

41. $\left|\frac{3x - 2}{5}\right| \leq \frac{1}{2}$

42. $\left|\frac{4x - 3}{2}\right| \leq \frac{1}{3}$

43. $\left|2x - \frac{1}{5}\right| < 0.3$

44. $\left|3x - \frac{3}{5}\right| < 0.2$

45. Write the continued inequality $-4 \leq x \leq 4$ as a single inequality involving absolute value.

46. Write the continued inequality $-8 \leq x \leq 8$ as a single inequality involving absolute value.

47. Write $-1 \leq x - 5 \leq 1$ as a single inequality involving absolute value.

48. Write $-3 \leq x + 2 \leq 3$ as a single inequality involving absolute value.

49. Speed Limits The interstate speed limit for cars is 75 miles per hour in Nebraska, Nevada, New Mexico, Oklahoma, South Dakota, Utah, and Wyoming and is the highest in the United States. To discourage passing, minimum speeds are also posted, so that the difference between the fastest and slowest moving traffic is no more than 20 miles per hour. Therefore, the speed x of a car must satisfy the inequality $55 \le x \le 75$. Write this inequality as an absolute value inequality.

50. Wavelengths of Light When white light from the sun passes through a prism, it is broken down into bands of light that form colors. The wavelength, v, (in nanometers) of some common colors are:

Blue:	$424 < v < 491$
Green:	$491 < v < 575$
Yellow:	$575 < v < 585$
Orange:	$585 < v < 647$
Red:	$647 < v < 700$

When a fireworks display made of copper is burned, it lets out light with wavelengths, v, that satisfy the relationship $|v - 455| < 23$. Write this inequality without absolute values; find the range of possible values for v; and then, using the preceding list of wavelengths, determine the color of that copper fireworks display.

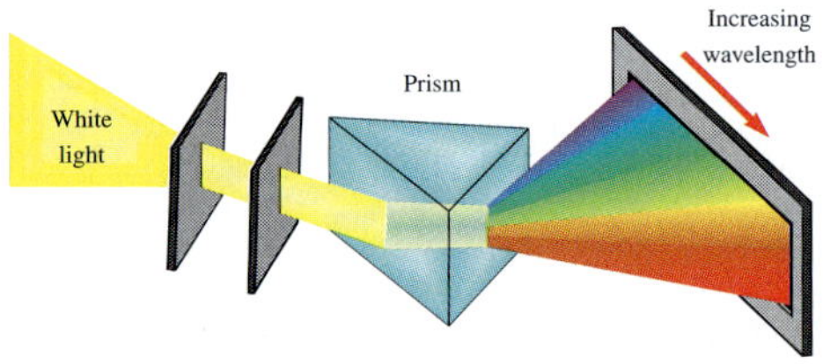

Maintaining Your Skills

The problems that follow review some of the more important skills you have learned in previous sections and chapters. You can consider the time you spend working these problems as time spent studying for exams.

Simplify each expression as much as possible.

51. $-9 \div \frac{3}{2}$

52. $-\frac{4}{5} \div (-4)$

53. $3 - 7(-6 - 3)$

54. $(3 - 7)(-6 - 3)$

55. $-4(-2)^3 - 5(-3)^2$

56. $4(2 - 5)^3 - 3(4 - 5)^5$

57. $-2(-3 + 8) - 7(-9 + 6)$

58. $-3 - 6[5 - 2(-3 - 1)]$

59. $\frac{2(-3) - 5(-6)}{-1 - 2 - 3}$

60. $\frac{4 - 8(3 - 5)}{2 - 4(3 - 5)}$

61. $6(1) - 1(-5) + 1(2)$

62. $-2(-14) + 3(-4) - 1(-10)$

63. $1(0)(1) + 3(1)(4) + (-2)(2)(-1)$

64. $-3(-1 - 1) + 4(-2 + 2) - 5[2 - (-2)]$

65. $4(-1)^2 + 3(-1) - 2$

66. $5 \cdot 2^3 - 3 \cdot 2^2 + 4 \cdot 2 - 5$

Chapter 2 Summary

Addition Property of Equality [2.1]

For algebraic expressions A, B, and C,

$$\text{if} \quad A = B$$
$$\text{then} \quad A + C = B + C$$

This property states that we can add the same quantity to both sides of an equation without changing the solution set.

EXAMPLES

1 We can solve

$$x + 3 = 5$$

by adding $\mathbf{-3}$ to both sides:

$$x + 3 + (\mathbf{-3}) = 5 + (\mathbf{-3})$$
$$x = 2$$

Multiplication Property of Equality [2.1]

For algebraic expressions A, B, and C,

$$\text{if} \quad A = B$$
$$\text{then} \quad AC = BC \quad C \neq 0$$

Multiplying both sides of an equation by the same nonzero quantity never changes the solution set.

2. We can solve $3x = 12$ by multiplying both sides by $\frac{1}{3}$.

$$3x = 12$$
$$\frac{1}{3}(3x) = \frac{1}{3}(12)$$
$$x = 4$$

Strategy for Solving Linear Equations in One Variable [2.1]

Step 1: **a.** Use the distributive property to separate terms, if necessary.

b. If fractions are present, consider multiplying both sides by the LCD to eliminate the fractions. If decimals are present, consider multiplying both sides by a power of 10 to clear the equation of decimals.

c. Combine similar terms on each side of the equation.

Step 2: Use the addition property of equality to get all variable terms on one side of the equation and all constant terms on the other side. A variable term is a term that contains the variable (for example, $5x$). A constant term is a term that does not contain the variable (the number 3, for example).

Step 3: Use the multiplication property of equality to get x (that is, $1x$) by itself on one side of the equation.

Step 4: Check your solution in the original equation to be sure that you have not made a mistake in the solution process.

3. Solve: $3(2x - 1) = 9$.

$$3(2x - 1) = 9$$
$$6x - 3 = 9$$
$$6x - 3 + \mathbf{3} = 9 + \mathbf{3}$$
$$6x = 12$$
$$\frac{1}{6}(6x) = \frac{1}{6}(12)$$
$$x = 2$$

Formulas [2.2]

A ***formula*** in algebra is an equation involving more than one variable. To solve a formula for one of its variables, simply isolate that variable on one side of the equation.

4. Solve for w:

$$P = 2l + 2w$$
$$P - 2l = 2w$$
$$\frac{P - 2l}{2} = w$$

Blueprint for Problem Solving [2.3]

Step 1: ***Read*** the problem, and then mentally ***list*** the items that are known and the items that are unknown.

Step 2: ***Assign a variable*** to one of the unknown items. (In most cases this will amount to letting x = the item that is asked for in the problem.) Then ***translate*** the other ***information*** in the problem to expressions involving the variable.

Step 3: ***Reread*** the problem, and then ***write an equation,*** using the items and variables listed in steps 1 and 2, that describes the situation.

Step 4: ***Solve the equation*** found in step 3.

Step 5: ***Write*** your ***answer*** using a complete sentence.

Step 6: ***Reread*** the problem, and ***check*** your solution with the original words in the problem.

5. The perimeter of a rectangle is 32 inches. If the length is 3 times the width, find the dimensions.

Step 1: This step is done mentally.

Step 2: Let x = the width. Then the length is $3x$.

Step 3: The perimeter is 32; therefore

$$2x + 2(3x) = 32$$

Step 4:

$$8x = 32$$
$$x = 4$$

Step 5: The width is 4 inches. The length is 3(4) = 12 inches.

Step 6: The perimeter is 2(4) + 2(12), which is 32. The length is 3 times the width.

Addition Property for Inequalities [2.4]

For expressions A, B, and C,

if $\quad A < B$

then $\quad A + C < B + C$

Adding the same quantity to both sides of an inequality never changes the solution set.

6. Adding 5 to both sides of the inequality $x - 5 < -2$ gives

$$x - 5 + \mathbf{5} < -2 + \mathbf{5}$$
$$x < 3$$

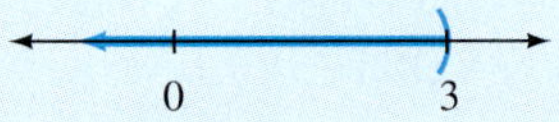

Multiplication Property for Inequalities [2.4]

For expressions A, B, and C,

if $\quad A < B$

then $\quad AC < BC \quad$ if $\quad C > 0$ (C is positive)

or $\quad AC > BC \quad$ if $\quad C < 0$ (C is negative)

We can multiply both sides of an inequality by the same nonzero number without changing the solution set as long as each time we multiply by a negative number we also reverse the direction of the inequality symbol.

7. Multiplying both sides of $-2x \geq 6$ by $-\frac{1}{2}$ gives

$$-2x \geq 6$$
$$-\tfrac{1}{2}(-2x) \leq -\tfrac{1}{2}(6)$$
$$x \leq -3$$

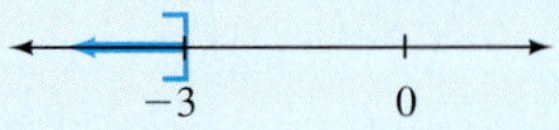

Absolute Value Equations [2.5]

To solve an equation that involves absolute value, we isolate the absolute value on one side of the equation and then rewrite the absolute value equation as two separate equations that do not involve absolute value. In general, if b is a positive real number, then

$$|a| = b \quad \text{is equivalent to} \quad a = b \quad \text{or} \quad a = -b$$

8. To solve

$$|2x - 1| + 2 = 7$$

we first isolate the absolute value on the left side by adding -2 to each side to obtain

$$|2x - 1| = 5$$

$$2x - 1 = 5 \quad \text{or} \quad 2x - 1 = -5$$

$$2x = 6 \quad \text{or} \quad 2x = -4$$

$$x = 3 \quad \text{or} \quad x = -2$$

Absolute Value Inequalities [2.6]

To solve an inequality that involves absolute value, we first isolate the absolute value on the left side of the inequality symbol. Then we rewrite the absolute value inequality as an equivalent continued or compound inequality that does not contain absolute value symbols. In general, if b is a positive real number, then

$$|a| < b \quad \text{is equivalent to} \quad -b < a < b$$

and

$$|a| > b \quad \text{is equivalent to} \quad a < -b \quad \text{or} \quad a > b$$

9. To solve

$$|x - 3| + 2 < 6$$

we first add -2 to both sides to obtain

$$|x - 3| < 4$$

which is equivalent to

$$-4 < x - 3 < 4$$

$$-1 < x < 7$$

COMMON MISTAKES

A very common mistake in solving inequalities is to forget to reverse the direction of the inequality symbol when multiplying both sides by a negative number. When this mistake occurs, the graph of the solution set is always drawn on the wrong side of the endpoint.

Chapter 2 Review/Test

The problems below form a comprehensive review of the material in this chapter. They can be used to study for exams. If you would like to take a practice test on this chapter, you can use the odd-numbered problems. Give yourself an hour and work as many of the odd-numbered problems as possible. When you are finished, or when an hour has passed, check your answers with the answers in the back of the book. You can use the even-numbered problems for a second practice test.

Solve each equation [2.1]

1. $x - 3 = 7$

2. $5x - 2 = 8$

3. $400 - 100a = 200$

4. $5 - \frac{2}{3}a = 7$

5. $4x - 2 = 7x + 7$

6. $\frac{3}{2}x - \frac{1}{6} = -\frac{7}{6}x - \frac{1}{6}$

7. $7y - 5 - 2y = 2y - 3$

8. $\frac{3y}{4} - \frac{1}{2} + \frac{3y}{2} = 2 - y$

9. $3(2x + 1) = 18$

10. $-\frac{1}{2}(4x - 2) = -x$

11. $8 - 3(2t + 1) = 5(t + 2)$

12. $8 + 4(1 - 3t) = -3(t - 4) + 2$

Substitute the given values in each formula and then solve for the variable that does not have a numerical replacement. [2.2]

13. $P = 2b + 2h$: $P = 40$, $b = 3$

14. $A = P + Prt$: $A = 2{,}000$, $P = 1{,}000$, $r = 0.05$

Solve each formula for the indicated variable. [2.2]

15. $I = prt$ for p

16. $y = mx + b$ for x

17. $4x - 3y = 12$ for y

18. $d = vt + 16t^2$ for v

For Problems 19–22, the general term of a sequence is given. Use the general term to find the first four terms of each sequence. Identify sequences that are increasing or decreasing. [2.2]

19. $a_n = 3n$

20. $a_n = n + 3$

21. $a_n = (-2)^n$

22. $a_n = \frac{1}{n}$

Solve each application. In each case, be sure to show the equation that describes the situation. [2.3]

23. Geometry The length of a rectangle is 3 times the width. The perimeter is 32 feet. Find the length and width.

24. Geometry The three sides of a triangle are given by three consecutive integers. If the perimeter is 12 meters, find the length of each side.

25. Brick Laying The formula $N = 7 \cdot L \cdot H$ gives the number of standard bricks needed in a wall of L feet long and H feet high and is called the *brick-layer's formula.*

a. How many bricks would be required in a wall 45 feet long by 12 feet high?

b. An 8-foot-high wall is to be built from a load of 35,000 bricks. What length wall can be built?

26. Salary A teacher has a salary of $25,920 for her second year on the job. If this is 4.2% more than her first year salary, how much did she earn her first-year?

Solve each inequality. Write your answer using interval notation. [2.4]

27. $-8a > -4$

28. $6 - a \geq -2$

29. $\frac{3}{4}x + 1 \leq 10$

30. $800 - 200x < 1{,}000$

31. $\frac{1}{3} \leq \frac{1}{6}x \leq 1$

32. $-0.01 \leq 0.02x - 0.01 \leq 0.01$

33. $5t + 1 \leq 3t - 2$ or $-7t \leq -21$

34. $3(x + 1) < 2(x + 2)$ or $2(x - 1) \geq x + 2$

Solve each equation. [2.5]

35. $|x| = 2$

36. $|a| - 3 = 1$

37. $|x - 3| = 1$

38. $|2y - 3| = 5$

39. $|4x - 3| + 2 = 11$

40. $\left|\frac{7}{3} - \frac{x}{3}\right| + \frac{4}{3} = 2$

41. $|5t - 3| = |3t - 5|$

42. $\left|\frac{1}{2} - x\right| = \left|x + \frac{1}{2}\right|$

Solve each inequality and graph the solution set. [2.6]

43. $|x| < 5$

44. $|0.01a| \geq 5$

45. $|x| < 0$

46. $|2t + 1| - 3 < 2$

Solve each equation or inequality, if possible. [2.1, 2.4, 2.5, 2.6]

47. $2x - 3 = 2(x - 3)$

48. $3(5x - \frac{1}{2}) = 15x + 2$

49. $|4y + 8| = -1$

50. $|x| > 0$

51. $|5 - 8t| + 4 \leq 1$

52. $|2x + 1| \geq -4$

Chapter 2 Projects

Linear Equations and Inequalities in One Variable

FINDING THE MAXIMUM HEIGHT OF A MODEL ROCKET

Group Project

Number of People 3

Time Needed 20 minutes

Equipment Paper and pencil

Background In this chapter, we used formulas to do some table building. Once we have a table, it is sometimes possible to use just the table information to extend what we know about the situation described by the table. In this project, we take some basic information from a table and then look for patterns among the table entries. Once we have established the patterns, we continue them and, in so doing, solve a realistic application problem.

Procedure A model rocket is launched into the air. Table 1 gives the height of the rocket every second after takeoff for the first 5 seconds. Figure 1 is a graphical representation of the information in Table 1.

TABLE 1
HEIGHT OF A MODEL ROCKET

TIME (SECONDS)	HEIGHT (FEET)
0	0
1	176
2	320
3	432
4	512
5	560

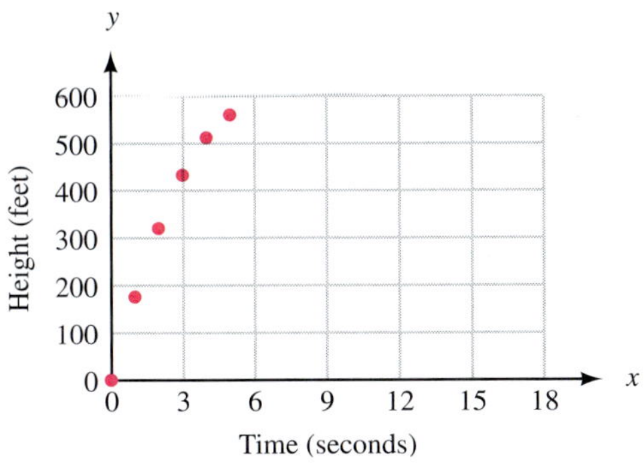

FIGURE 1

TABLE 1
HEIGHT OF A MODEL ROCKET

TIME (SECONDS)	HEIGHT (FEET)
0	0
1	176
2	320
3	432
4	512
5	560

FIRST DIFFERENCES

SECOND DIFFERENCES

1. Table 1 is shown again at the top of the second column with two new columns. Fill in the first five entries in the First Differences column by finding the difference of consecutive heights. For example, the second entry in the First Differences column will be the difference of 320 and 176, which is 144.

2. Start filling in the Second Differences column by finding the differences of the First Differences.

3. Once you see the pattern in the Second Differences table, fill in the rest of the entries.

4. Now, using the results in the Second Differences table, go back and complete the First Differences table.

5. Now, using the results in the First Differences table, go back and complete the Heights column in the original table.

6. Plot the rest of the points from Table 1 on the graph in Figure 1.

7. What is the maximum height of the rocket?

8. How long was the rocket in the air?

THE EQUAL SIGN

Research Project

We have been using the equal sign, =, for some time now. The first published use of the symbol was in 1557, with the publication of *The Whetstone of Witte* by the English mathematician and physician Robert Recorde. Research the first use of the symbols we use for addition, subtraction, multiplication, and division and then write an essay on the subject from your results.

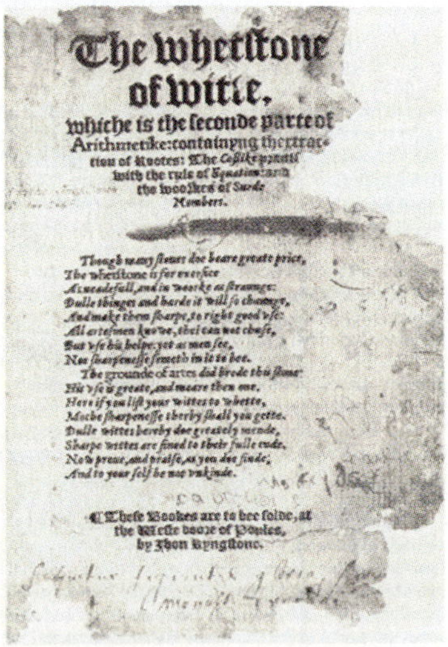
The whetstone of witte,
whiche is the seconde parte of
Arithmetike: containyng the extraction of Rootes: The Cossike practise, with the rule of Equation: and the woorkes of Surde Nombers.

Though many stones doe beare greate price,
The whetstone is for exercise
As neadefull, and in woorke as straunge:
Dulle thinges and harde it will so chaunge,
And make them sharpe, to right good vse:
All artesmen knowe, thei can not chuse,
But vse his helpe: yet as men see,
Noe sharpenesse semeth in it to bee.
The grounde of artes did brede this stone:
His vse is greate, and moare then one.
Here if you lift your wittes to whette,
Moche sharpenesse therby shall you gette.
Dulle wittes hereby doe greately mende,
Sharpe wittes are fined to their fulle ende.
Now proue, and praise, as you doe finde,
And to your self be not vnkinde.

These Bookes are to bee solde, at the Weste doore of Poules, by Ihon Kyngstone.

Smithsonian Institution Libraries

3

David Young-Wolff/PhotoEdit

CHAPTER OUTLINE

INTERMEDIATE Algebra $f(x)$ Now™

Equations and Inequalities in Two Variables

A student is heating water in a chemistry lab. As the water heats, she records the temperature readings from two thermometers, one giving temperature in degrees Fahrenheit and the other in degrees Celsius. The table below shows some of the data she collects. Figure 1 is a scatter diagram that gives a visual representation of the data in the table.

CORRESPONDING TEMPERATURES

IN DEGREES FAHRENHEIT	IN DEGREES CELSIUS
77	25
95	35
167	75
212	100

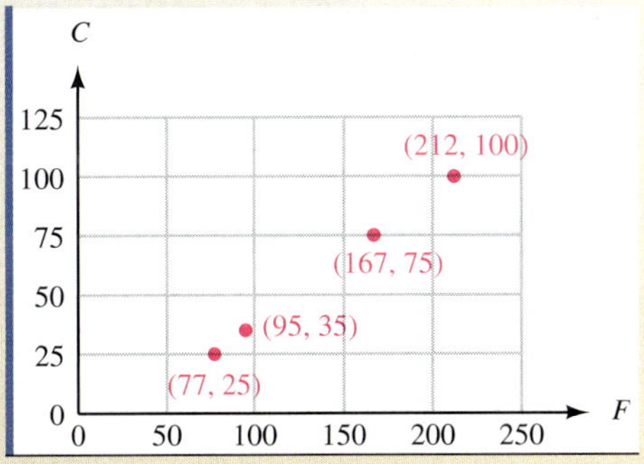

FIGURE 1

The exact relationship between the Fahrenheit and Celsius temperature scales is given by the formula

$$C = \frac{5}{9}(F - 32)$$

We have three ways to describe the relationship between the two temperature scales: a table, a graph, and an equation. But, most important to us, we don't need to accept this formula on faith, because the material we will present in this chapter gives us the ability to derive the formula from the data in the table above.

Visit our web site at http://www.brookscole.com

Study Skills

Try to arrange your daily study habits so that you have very little studying to do the night before your next exam. The next two goals will help you achieve this.

1 Review with the Exam in Mind

Each day you should review material that will be covered on the next exam. Your review should consist of working problems. Preferably, the problems you work should be problems from your list of difficult problems.

2 Pay Attention to Instructions

Each of the following is a valid instruction with respect to the equation $y = 3x - 2$, and the result of applying the instructions will be different in each case:

Find x when y is 10.
Solve for x.
Graph the equation.
Find the intercepts.

There are many things to do with the equation $y = 3x - 2$. If you train yourself to pay attention to the instructions that accompany a problem as you work through the assigned problems, you will not find yourself confused about what to do with a problem when you see it on a test.

Getting Ready for Chapter 3

1. $16(4.5)^2$

2. $\dfrac{0.2(0.5)^2}{100}$

3. $-\dfrac{1}{3}(-3) + 2$

4. $4(-2) - 1$

5. $\dfrac{4 - (-2)}{3 - 1}$

6. $\dfrac{-3 - (-3)}{2 - (-1)}$

7. $\dfrac{4}{3}(3.14) \cdot 3^3$ (round to the nearest whole number)

8. $3(-2)^2 + 2(-2) - 1$

9. If $2x + 3y = 6$, find x when $y = 0$.

10. If $s = \dfrac{60}{t}$, find s when $t = 15$.

11. Write -0.06 as a fraction in lowest terms.

12. Write -0.6 as a fraction in lowest terms.

13. Solve $2x - 3y = 5$ for y.

14. Solve $y - 4 = -\dfrac{1}{2}(x + 1)$ for y.

15. Which of the following are solutions to $x + y \geq 5$?
(0, 0) (5, 0) (4, 3)

16. Which of the following are solutions to $y < \dfrac{1}{2}x + 3$?
(0, 0) (2, 0) (−2, 4)

17. Use $h = 64t - 16t^2$ to complete the table.

t	0	1	2	3	4
h					

18. Use $s = \dfrac{60}{t}$ to complete the table.

t	4	6	8	10
s				

19. If $y = \dfrac{K}{x}$, find K when $x = 3$ and $y = 45$.

20. If $y = Kxz^2$, find K when $x = 2$, $y = 81$, and $z = 9$.

Paired Data and the Rectangular Coordinate System

3.1

OBJECTIVES	
A	Graph ordered pairs on a rectangular coordinate system.
B	Graph linear equations by finding intercepts or by making a table.
C	Graph horizontal and vertical lines.

In this section, we begin our work with the visual component of algebra. We are setting the stage for two important topics: *functions* and *graphs*. Both topics have a wide variety of applications and are found in all the math classes that follow this one.

Table 1 gives the net price of a popular intermediate algebra text at the beginning of each year in which a new edition was published. (The net price is the price the bookstore pays for the book, not the price you pay for it.)

TABLE 1

PRICE OF A TEXTBOOK

EDITION	YEAR PUBLISHED	NET PRICE ($)
First	1982	16.95
Second	1985	23.50
Third	1989	30.50
Fourth	1993	39.25
Fifth	1997	47.50
Sixth	2001	55.00

The information in Table 1 is shown visually in Figures 1 and 2. The diagram in Figure 1 is called a *bar chart.* The diagram in Figure 2 is called a *line graph.* In both figures, the horizontal line that shows years is called the *horizontal axis,* and the vertical line that shows prices is called the *vertical axis.*

Recall from Chapter 1 that the data in Table 1 are called *paired data* because each number in the years column is paired with a specific number in the price column. Each of these pairs of numbers from Table 1 is associated with one of the solid bars in the bar chart (Figure 1) and one of the dots in the line graph (Figure 2).

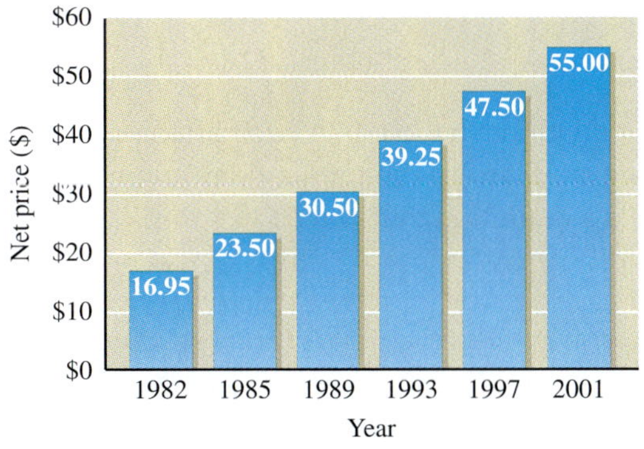

FIGURE 1

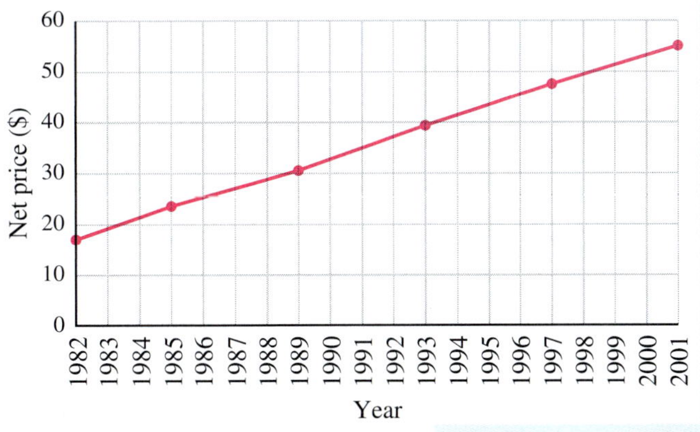

FIGURE 2

USING TECHNOLOGY

Spreadsheet Programs

When I put together the manuscript for this book, I used a spreadsheet program to draw the bar chart and line graph shown in Figures 1 and 2.

Figure 3 shows what the screen of my computer looked like when I was preparing Figure 1. The bar chart was drawn by the computer from the data in the table. This was just one of many ways I could have chosen to display the data.

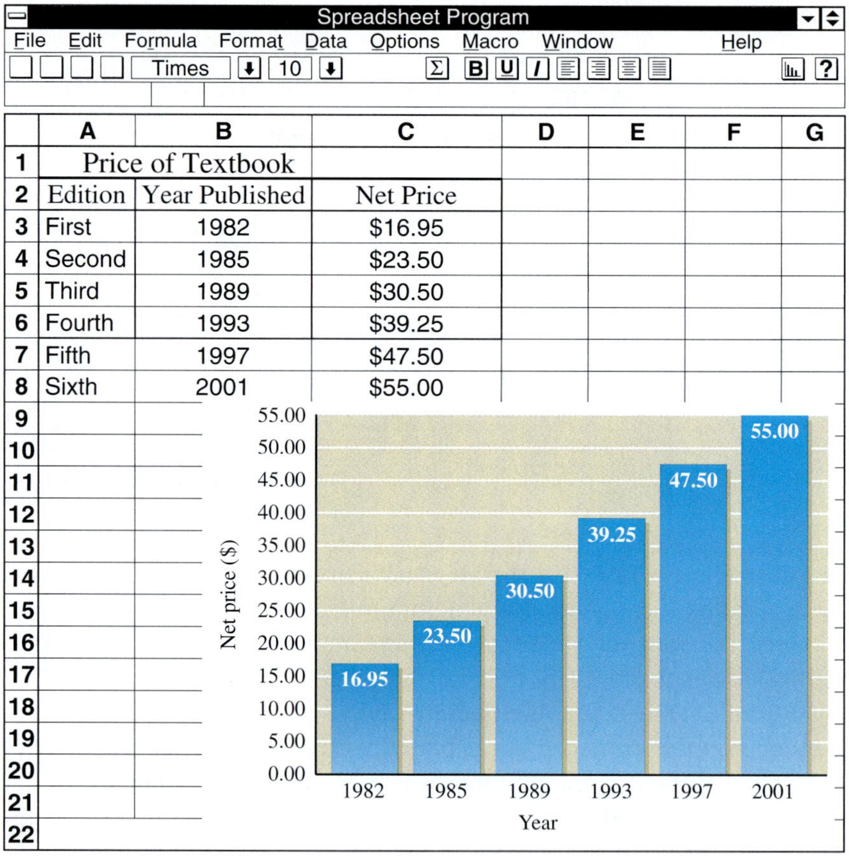

FIGURE 3

Ordered Pairs

Paired data play an important role in equations that contain two variables. Working with these equations is easier if we standardize the terminology and notation associated with paired data. So here is a definition that will do just that.

DEFINITION

A pair of numbers enclosed in parentheses and separated by a comma, such as $(-2, 1)$, is called an **ordered pair** of numbers. The first number in the pair is called the ***x*-coordinate** of the ordered pair; the second number is called the ***y*-coordinate.** For the ordered pair $(-2, 1)$, the x-coordinate is -2 and the y-coordinate is 1.

To standardize the way in which we display paired data visually, we use a rectangular coordinate system. A *rectangular coordinate system* is made by drawing two real number lines at right angles to each other. The two number lines, called *axes,* cross each other at 0. This point is called the *origin.* Positive directions are to the right and up. Negative directions are down and to the left. The rectangular coordinate system is shown in Figure 4.

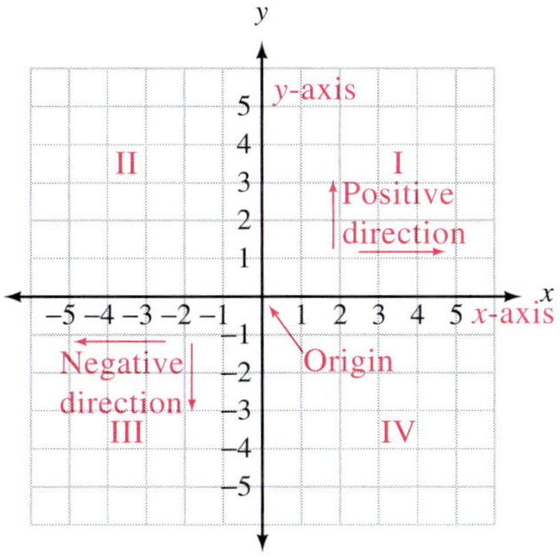

FIGURE 4

The horizontal number line is called the *x-axis* and the vertical number line is called the *y-axis.* The two number lines divide the coordinate system into four quadrants, which we number I through IV in a counterclockwise direction. Points on the axes are not considered as being in any quadrant.

To graph the ordered pair (a, b) on a rectangular system, we start at the origin and move a units right or left (right if a is positive, left if a is negative). Then we move b units up or down (up if b is positive and down if b is negative). The point where we end up is the graph of the ordered pair (a, b).

EXAMPLE 1 Plot (graph) the ordered pairs (2, 5), (−2, 5), (−2, −5), and (2, −5).

Solution To graph the ordered pair (2, 5), we start at the origin and move 2 units to the right, then 5 units up. We are now at the point whose coordinates are (2, 5). We graph the other three ordered pairs in a similar manner (see Figure 5).

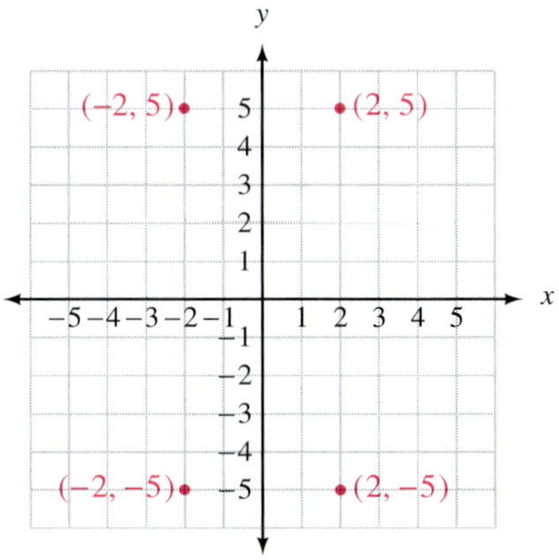

FIGURE 5

NOTE

A rectangular coordinate system allows us to connect algebra and geometry by associating geometric shapes (the curves shown in the diagrams) with algebraic equations. The French philosopher and mathematician René Descartes (1596–1650) usually is credited with the invention of the rectangular coordinate system, which often is referred to as the *Cartesian coordinate system* in his honor. As a philosopher, Descartes is responsible for the statement, "I think, therefore, I am." Until Descartes invented his coordinate system in 1637, algebra and geometry were treated as separate subjects.

PRACTICE PROBLEMS

1. Plot the ordered pairs (1,3), (−1, 3), (−1, −3), and (1, −3).

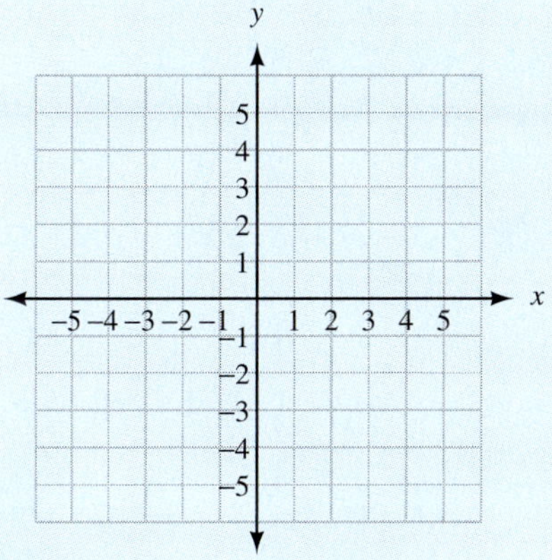

NOTE

From Example 1 we see that any point in quadrant I has both its *x*- and *y*-coordinates positive (+, +). Points in quadrant II have negative *x*-coordinates and positive *y*-coordinates (−, +). In quadrant III both coordinates are negative (−, −). In quadrant IV the form is (+, −).

Answers

See Solutions to Selected Practice Problems for all answers in this section.

2. Plot the ordered pairs $(4, -2)$, $(-\frac{1}{2}, 3)$, $(1, 0)$, $(0, -5)$, $(-6, 0)$, and $(0, 4)$.

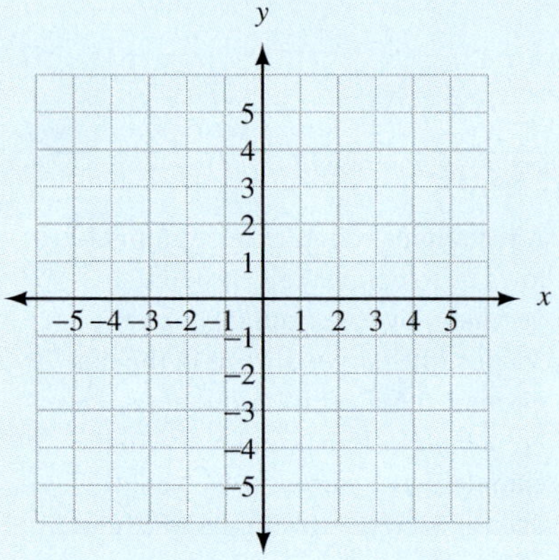

EXAMPLE 2

Graph the ordered pairs $(1, -3)$, $(\frac{1}{2}, 2)$, $(3, 0)$, $(0, -2)$, $(-1, 0)$, and $(0, 5)$.

Solution

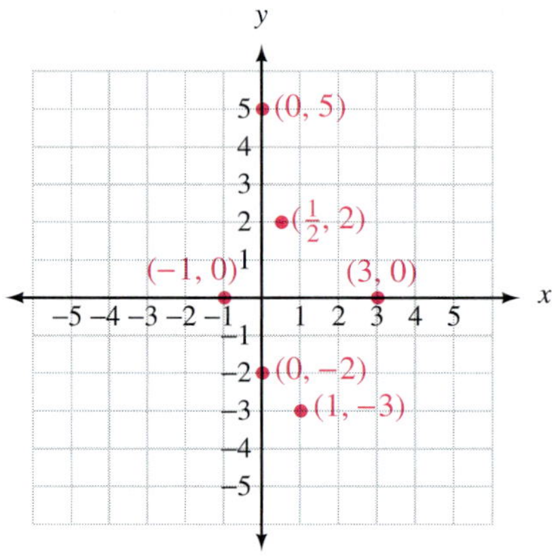

FIGURE 6

From Figure 6 we see that any point on the x-axis has a y-coordinate of 0 (it has no vertical displacement), and any point on the y-axis has an x-coordinate of 0 (no horizontal displacement).

Linear Equations

We can plot a single point from an ordered pair, but to draw a line, we need two points or an equation in two variables.

DEFINITION

Any equation that can be put in the form $ax + by = c$, where a, b, and c are real numbers and a and b are not both 0, is called a **linear equation in two variables.** The graph of any equation of this form is a straight line (that is why these equations are called "linear"). The form $ax + by = c$ is called **standard form.**

To graph a linear equation in two variables, we simply graph its solution set; that is, we draw a line through all the points whose coordinates satisfy the equation.

3. Graph $y = \frac{1}{2}x + 3$.

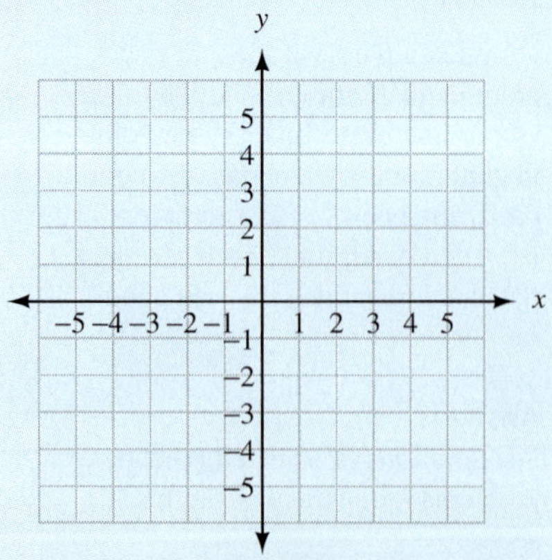

EXAMPLE 3

Graph the equation $y = -\frac{1}{3}x + 2$.

Solution We need to find three ordered pairs that satisfy the equation. To do so, we can let x equal any numbers we choose and find corresponding values of y. But since every value of x we substitute into the equation is going to be multiplied by $-\frac{1}{3}$, let's use numbers for x that are divisible by 3, like -3, 0, and 3. That way, when we multiply them by $-\frac{1}{3}$, the result will be an integer.

$$\text{Let } x = -3; \quad y = -\frac{1}{3}(-3) + 2$$
$$y = 1 + 2$$
$$y = 3$$

The ordered pair $(-3, 3)$ is one solution.

Let $x = 0$; $y = -\frac{1}{3}(0) + 2$

$y = 0 + 2$

$y = 2$

The ordered pair (0, 2) is a second solution.

Let $x = 3$; $y = -\frac{1}{3}(3) + 2$

$y = -1 + 2$

$y = 1$

In table form

x	y
−3	3
0	2
3	1

The ordered pair (3, 1) is a third solution.

Plotting the ordered pairs (−3, 3), (0, 2), and (3, 1) and drawing a straight line through their graphs, we have the graph of the equation $y = -\frac{1}{3}x + 2$, as shown in Figure 7.

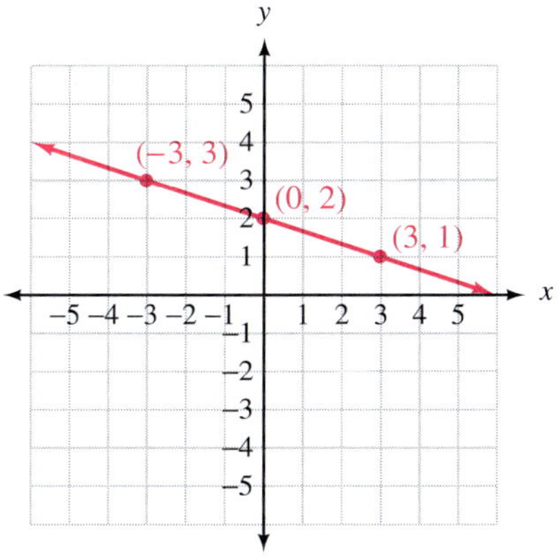

FIGURE 7

NOTE

It takes only two points to determine a straight line. We have included a third point for "insurance." If all three points do not line up in a straight line, we have made a mistake.

Example 3 illustrates again the connection between algebra and geometry that we mentioned earlier in this section. Descartes' rectangular coordinate system allows us to associate the equation $y = -\frac{1}{3}x + 2$ (an algebraic concept) with a specific straight line (a geometric concept). The study of the relationship between equations in algebra and their associated geometric figures is called *analytic geometry*.

Intercepts

Two important points on the graph of a straight line, if they exist, are the points where the graph crosses the axes.

DEFINITION

The ***x*-intercept** of the graph of an equation is the x-coordinate of the point where the graph crosses the x-axis. The ***y*-intercept** is defined similarly.

Because any point on the x-axis has a y-coordinate of 0, we can find the x-intercept by letting $y = 0$ and solving the equation for x. We find the y-intercept by letting $x = 0$ and solving for y.

4. Find the x- and y-intercepts for $2x - 3y = 6$, then graph the solution set.

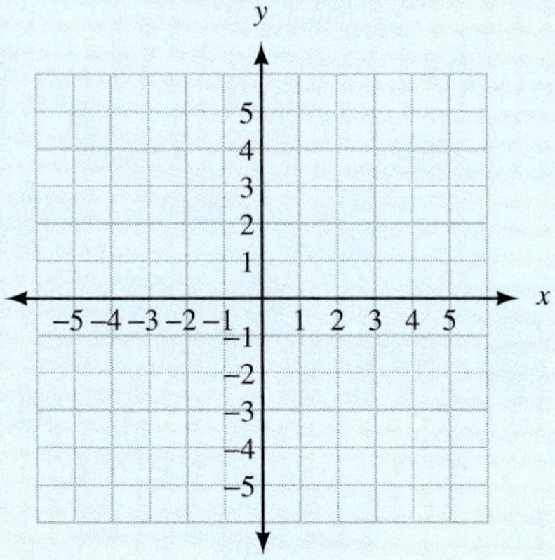

EXAMPLE 4

Find the x- and y-intercepts for $2x + 3y = 6$; then graph the solution set.

Solution To find the y-intercept we let $x = 0$.

When $x = 0$

we have $2(0) + 3y = 6$

$3y = 6$

$y = 2$

The y-intercept is 2, and the graph crosses the y-axis at the point (0, 2).

When $y = 0$

we have $2x + 3(0) = 6$

$2x = 6$

$x = 3$

The x-intercept is 3, so the graph crosses the x-axis at the point (3, 0). We use these results to graph the solution set for $2x + 3y = 6$. The graph is shown in Figure 8.

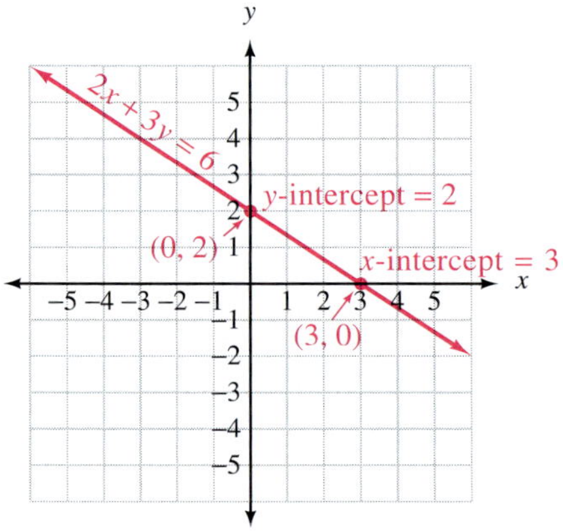

FIGURE 8

NOTE

Graphing straight lines by finding the intercepts works best when the coefficients of x and y are factors of the constant term.

5. Graph the line $x = -1$ and the line $y = 4$.

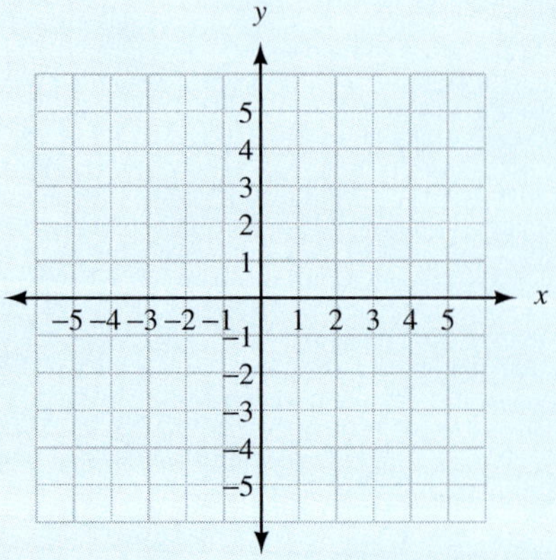

EXAMPLE 5

Graph each of the following lines.

a. $y = \frac{1}{2}x$ **b.** $x = 3$ **c.** $y = -2$

Solution

a. The line $y = \frac{1}{2}x$ passes through the origin because (0, 0) satisfies the equation. To sketch the graph we need at least one more point on the line. When x is 2, we obtain the point (2, 1), and when x is -4, we obtain the point $(-4, -2)$. The graph of $y = \frac{1}{2}x$ is shown in Figure 9a.

b. The line $x = 3$ is the set of all points whose x-coordinate is 3. The variable y does not appear in the equation, so the y-coordinate can be any number. Note that we can write our equation as a linear equation in two variables by writing it as $x + 0y = 3$. Because the product of 0 and y will always be 0, y can be any number. The graph of $x = 3$ is the vertical line shown in Figure 9b.

c. The line $y = -2$ is the set of all points whose y-coordinate is -2. The variable x does not appear in the equation, so the x-coordinate can be any number. Again, we can write our equation as a linear equation in two variables by writing it as $0x + y = -2$. Because the product of 0 and x will always be 0, x can be any number. The graph of $y = -2$ is the horizontal line shown in Figure 9c.

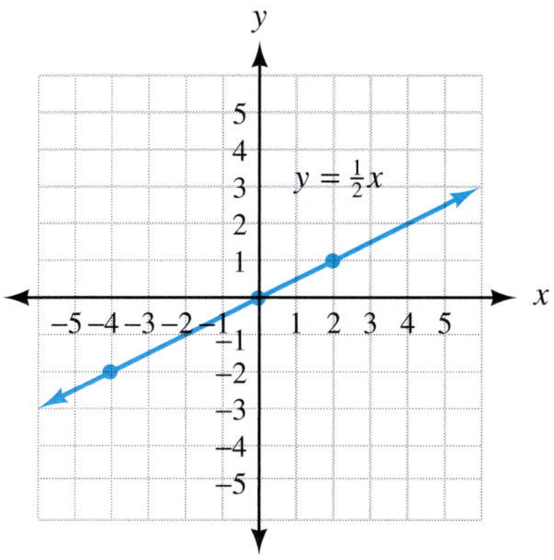

FIGURE 9A

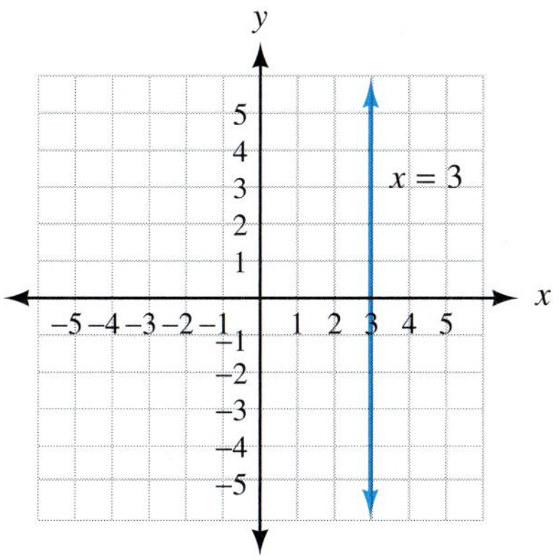

FIGURE 9B

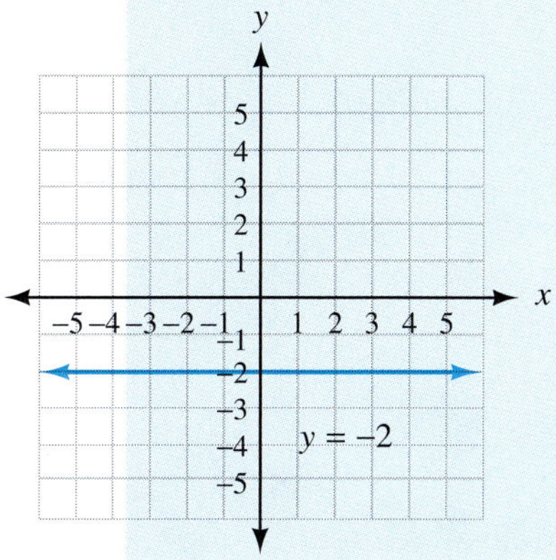

FIGURE 9C

Facts From Geometry: Special Equations and Their Graphs

For the equations below, m, a, and b are real numbers.

Through the Origin

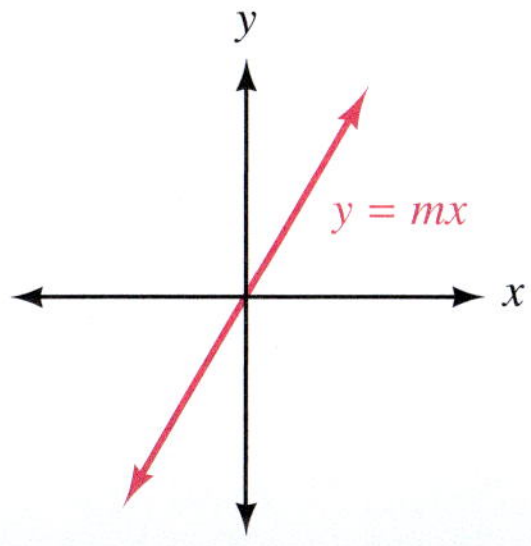

FIGURE 10A **Any equation of the form $y = mx$ has a graph that passes through the origin.**

Vertical Line

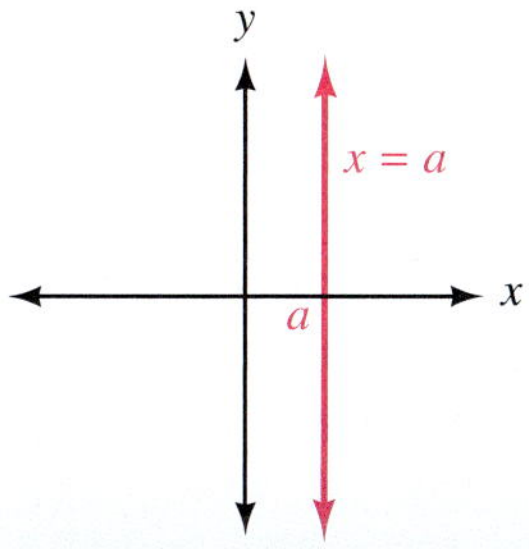

FIGURE 10B **Any equation of the form $x = a$ has a vertical line for its graph.**

Horizontal Line

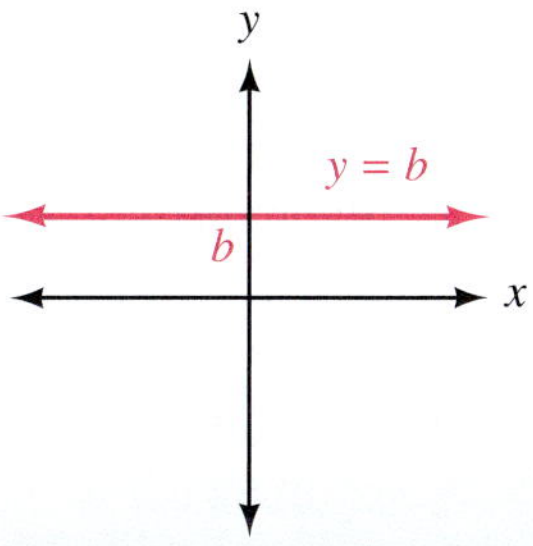

FIGURE 10C **Any equation of the form $y = b$ has a horizontal line for its graph.**

USING TECHNOLOGY

Graphing Calculators and Computer Graphing Programs

A variety of computer programs and graphing calculators are currently available to help us graph equations and then obtain information from those graphs much faster than we could with paper and pencil. We will not give instructions for all the available calculators. Most of the instructions we give are generic in form. You will have to use the manual that came with your calculator to find the specific instructions for your calculator.

Graphing with Trace and Zoom

All graphing calculators have the ability to graph a function and then trace over the points on the graph, giving their coordinates. Furthermore, all graphing calculators can zoom in and out on a graph that has been drawn. To graph a linear equation on a graphing calculator, we first set the graph window. Most calculators call the smallest value of x Xmin and the largest value of x Xmax. The counterpart values of y are Ymin and Ymax. We will use the notation

Window: X from -5 to 4, Y from -3 to 2

to stand for a window in which

Xmin $= -5$ Ymin $= -3$
Xmax $= 4$ Ymax $= 2$

Set your calculator with the following window:

Window: X from -10 to 10, Y from -10 to 10

Graph the equation $Y = -X + 8$. On the TI-82/83, you use the [Y=] key to enter the equation; you enter a negative sign with the [(−)] key, and a subtraction sign with the [−] key. The graph will be similar to the one shown in Figure 11.

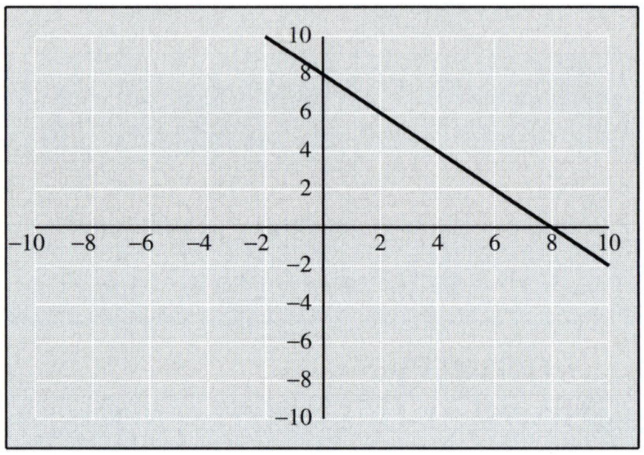

FIGURE 11

Use the Trace feature of your calculator to name three points on the graph. Next, use the Zoom feature of your calculator to zoom out so your window is twice as large.

Solving for *y* First

To graph the equation from Example 4, $2x + 3y = 6$, on a graphing calculator, you must first solve it for y. When you do so, you will get $y = -\frac{2}{3}x + 2$, which you enter into your calculator as $Y = -(2/3)X + 2$. Graph this equation in the window described here, and compare your results with the graph in Figure 8.

Window: X from -6 to 6, Y from -6 to 6

Hint on Tracing

If you are going to use the Trace feature and you want the x-coordinates to be exact numbers, set your window so that the range of X inputs is a multiple of the number of horizontal pixels on your calculator screen. On the TI-82/83, the screen has 94 pixels across. Here are a few convenient trace windows:

X from -4.7 to 4.7	To trace to the nearest tenth
X from -47 to 47	To trace to the nearest integer
X from 0 to 9.4	To trace to the nearest tenth
X from 0 to 94	To trace to the nearest integer
X from -94 to 94	To trace to the nearest even integer

Linking Objectives and Examples

Next to each **objective**, list the examples that are best described by that objective.

A	
B	
C	

Getting Ready for Class

After reading through the preceding section, respond in your own words and in complete sentences.

1. Explain how you would construct a rectangular coordinate system from two real number lines.
2. Explain in words how you would graph the ordered pair $(2, -3)$.
3. How can you tell if an ordered pair is a solution to the equation $y = 2x - 5$?
4. If you were looking for solutions to the equation $y = \frac{1}{3}x + 5$, why would it be easier to substitute 6 for x than to substitute 5 for x?

Problem Set 3.1

INTERMEDIATE Algebra $f(x)$ Now™

Graph each of the following ordered pairs on a rectangular coordinate system.

1. **a.** (1, 2)
b. (−1, −2)
c. (5, 0)
d. (0, 2)
e. (−5, −5)
f. $\left(\frac{1}{2}, 2\right)$

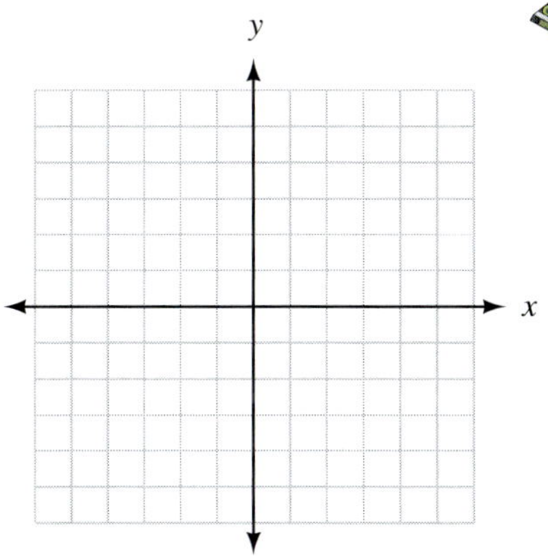

2. **a.** (−1, 2)
b. (1, −2)
c. (0, −3)
d. (4, 0)
e. (−4, −1)
f. $\left(3, \frac{1}{4}\right)$

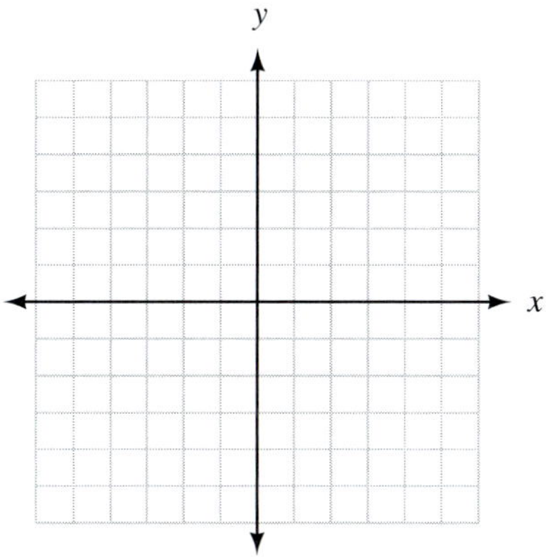

Give the coordinates of each point.

3.

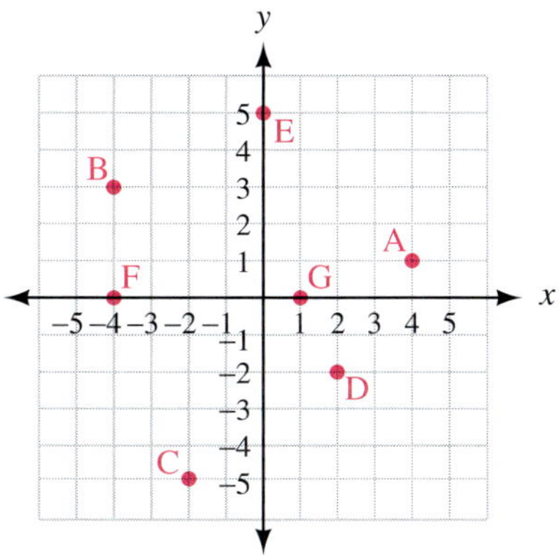

4.

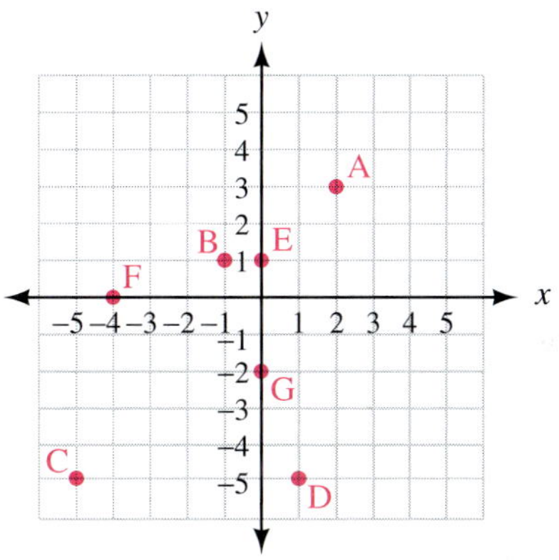

Graph each of the following linear equations by first finding the intercepts.

5. $2x - 3y = 6$

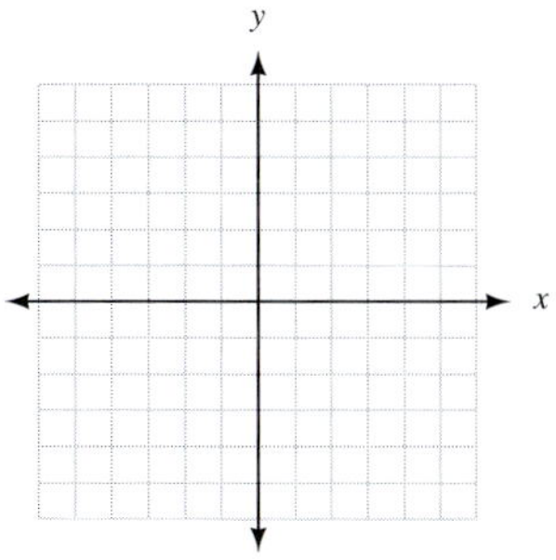

6. $y - 2x = 4$

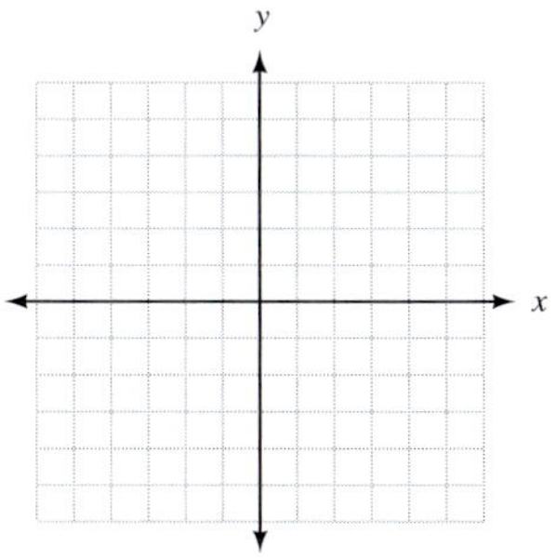

7. $4x - 5y = 20$

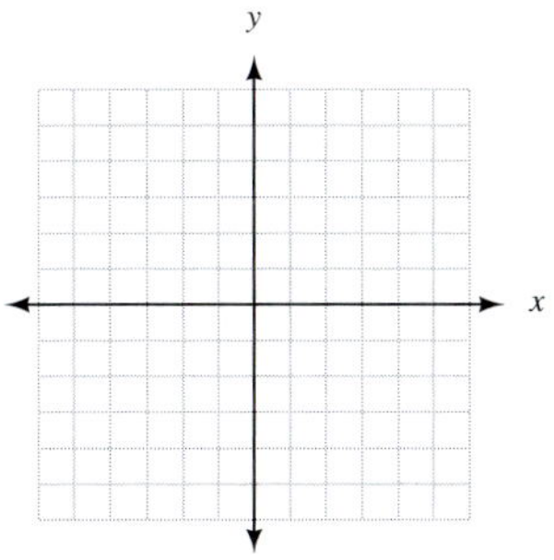

8. $5x - 3y - 15 = 0$

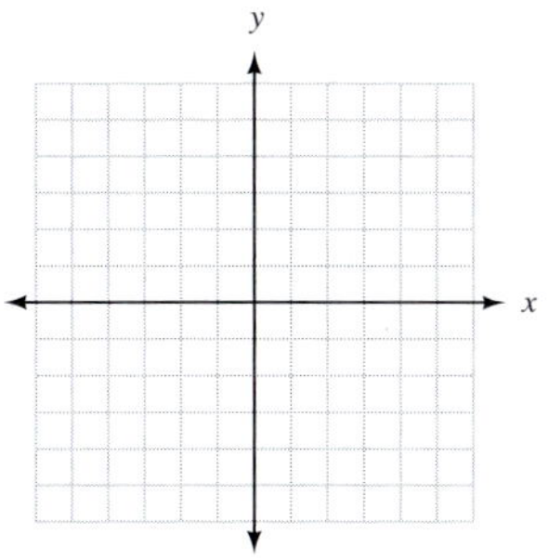

9. $y = 2x + 3$

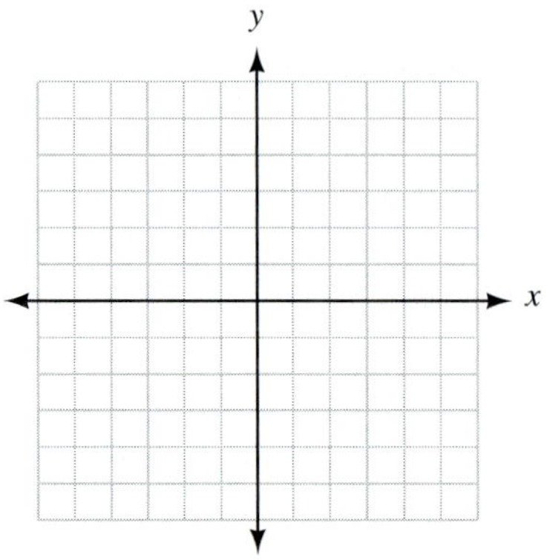

10. $y = 3x - 2$

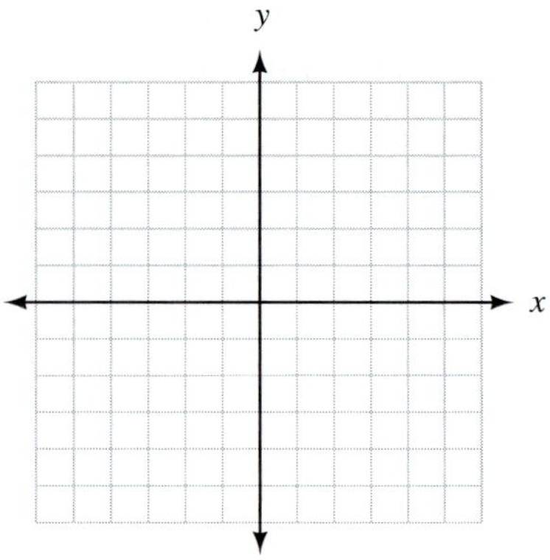

11. Which of the following tables could be produced from the equation $y = 2x - 6$?

a.

x	y
0	6
1	4
2	2
3	0

b.

x	y
0	−6
1	−4
2	−2
3	0

c.

x	y
0	−6
1	−5
2	−4
3	−3

12. Which of the following tables could be produced from the equation $3x - 5y = 15$?

a.

x	y
0	5
−3	0
10	3

b.

x	y
0	−3
5	0
10	3

c.

x	y
0	−3
−5	0
10	−3

Graph each of the following lines.

13. $y = \frac{1}{3}x$

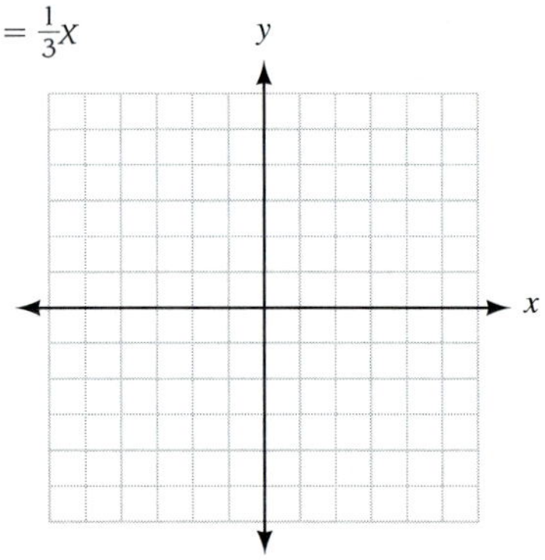

14. $y = \frac{1}{2}x$

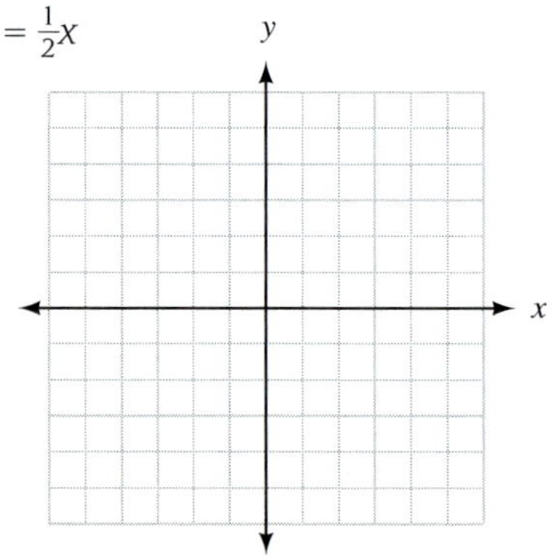

15. $-2x + y = -3$

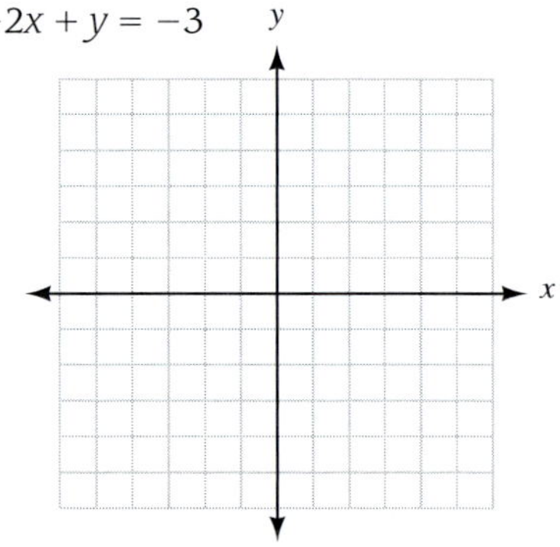

16. $-3x + y = -2$

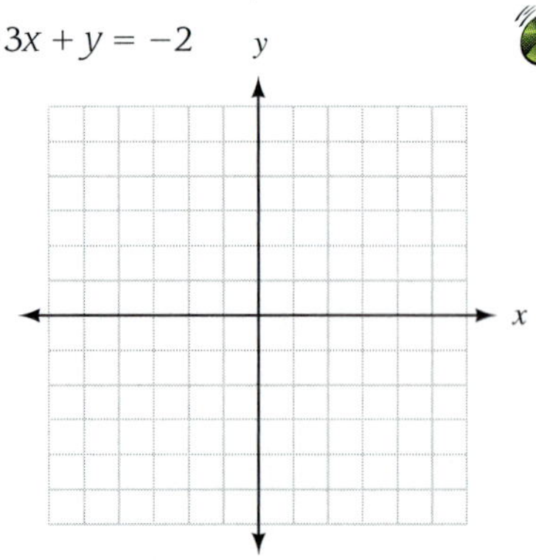

17. $y = -\frac{2}{3}x + 1$

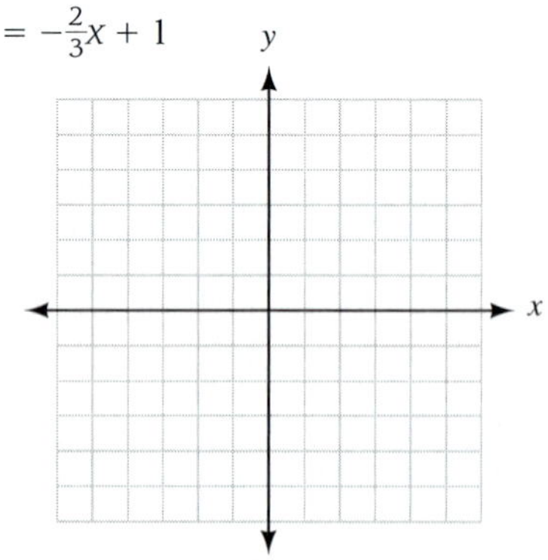

18. $y = -\frac{2}{3}x - 1$

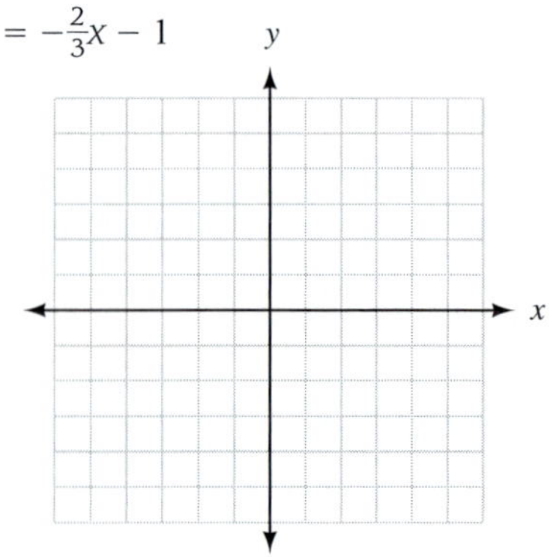

19. $\dfrac{x}{3} + \dfrac{y}{4} = 1$

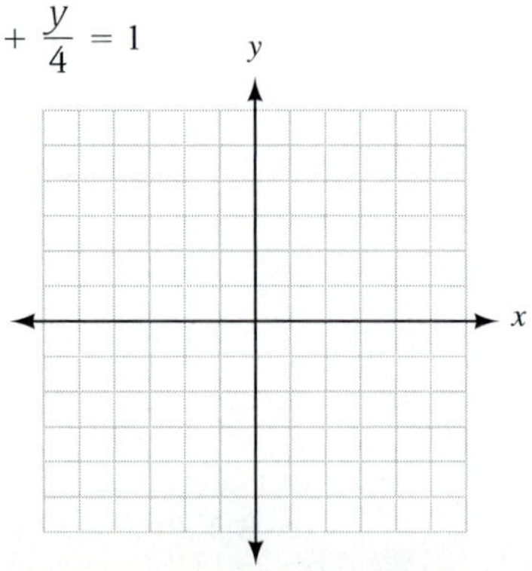

20. $\dfrac{x}{-2} + \dfrac{y}{3} = 1$

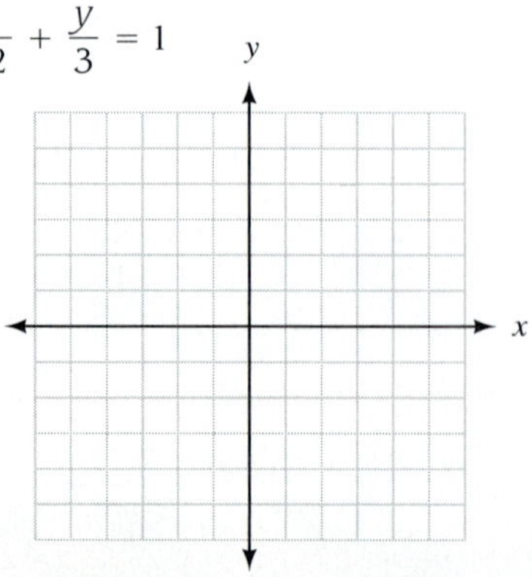

21. The graph shown here is the graph of which of the following equations?

a. $3x - 2y = 6$

b. $2x - 3y = 6$

c. $2x + 3y = 6$

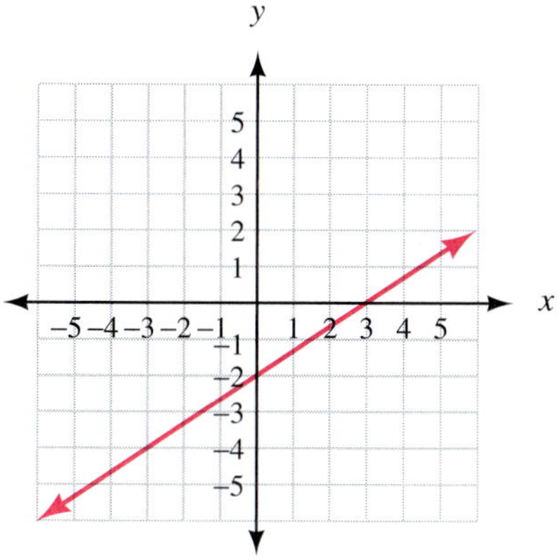

22. The graph shown here is the graph of which of the following equations?

a. $3x - 2y = 8$

b. $2x - 3y = 8$

c. $2x + 3y = 8$

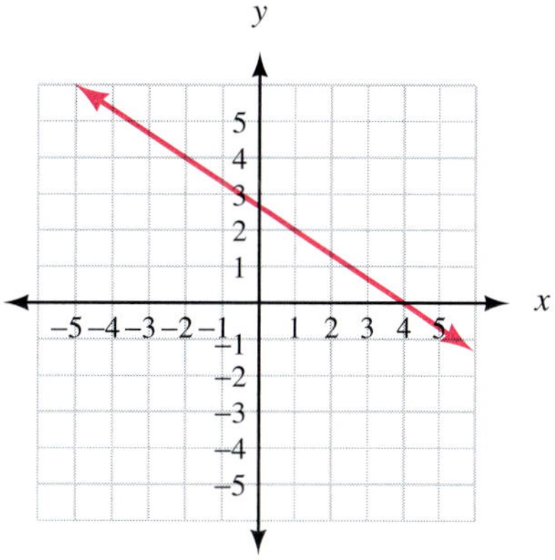

23. Graph each of the following lines.

a. $y = 2x$ **b.** $x = -3$ **c.** $y = 2$

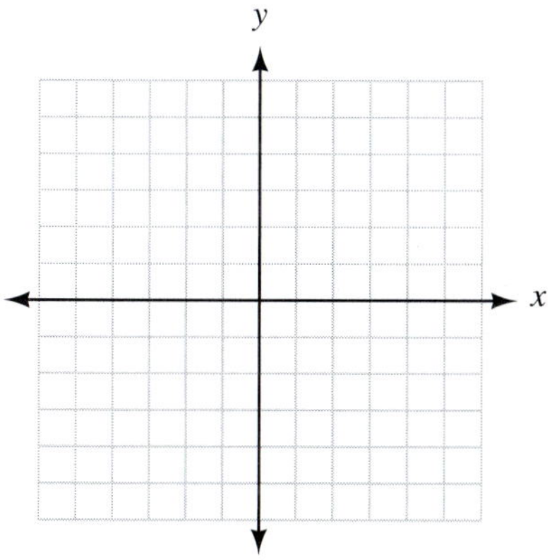

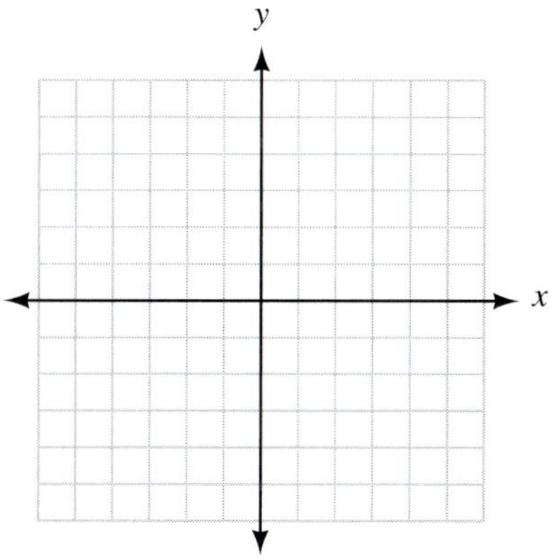

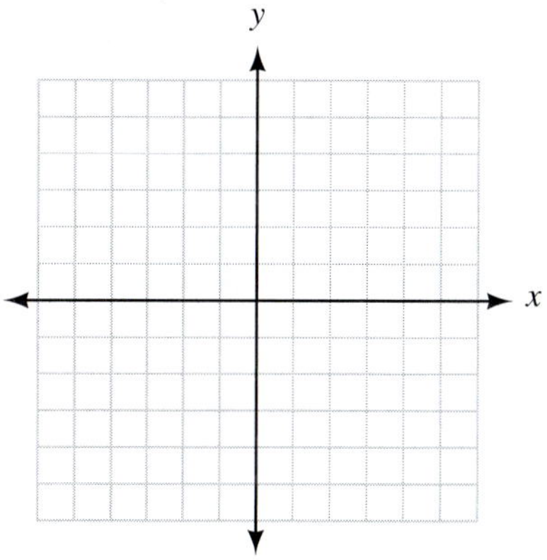

24. Graph each of the following lines.

a. $y = 3x$ **b.** $x = -2$ **c.** $y = 4$

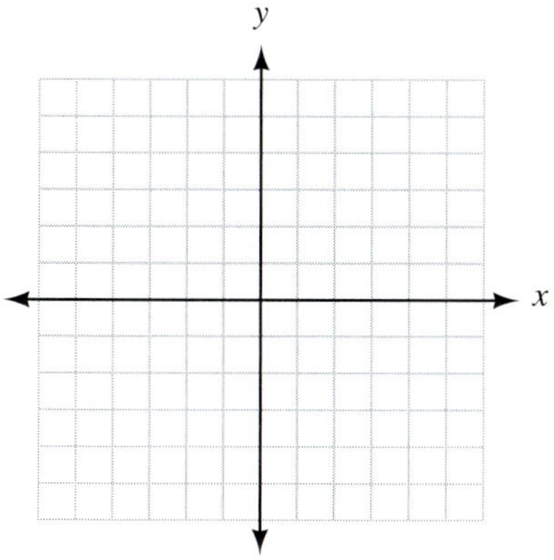

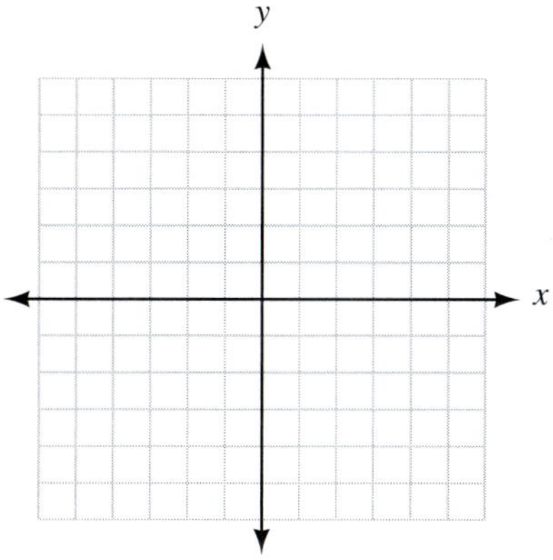

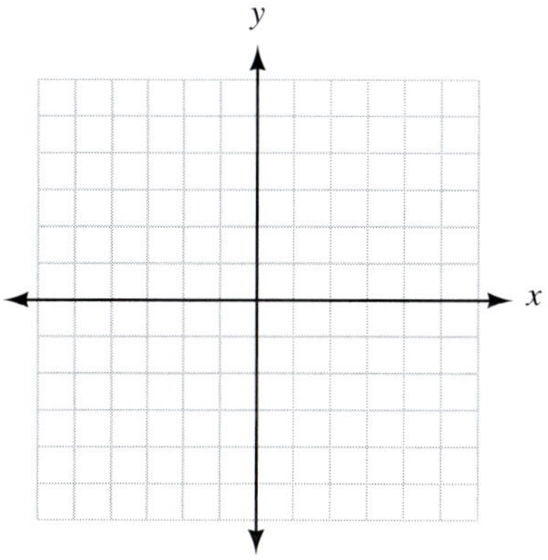

25. Graph each of the following lines.

a. $y = -\frac{1}{2}x$ **b.** $x = 4$ **c.** $y = -3$

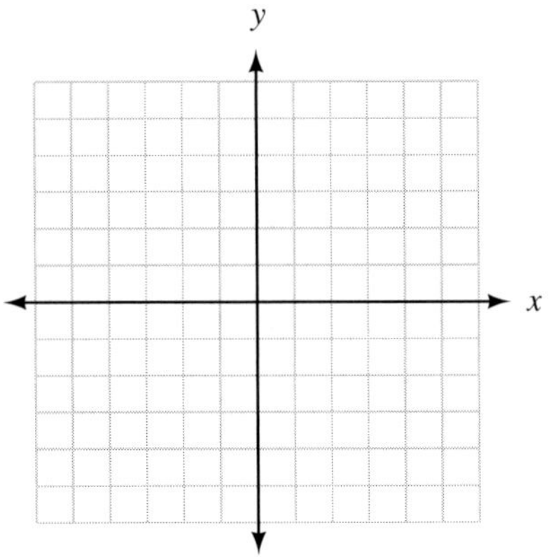

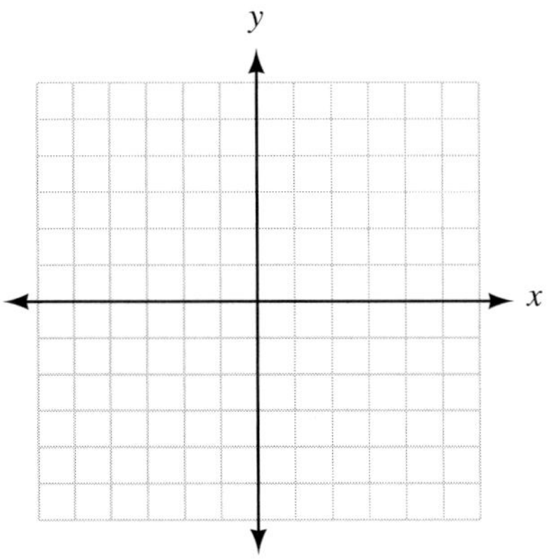

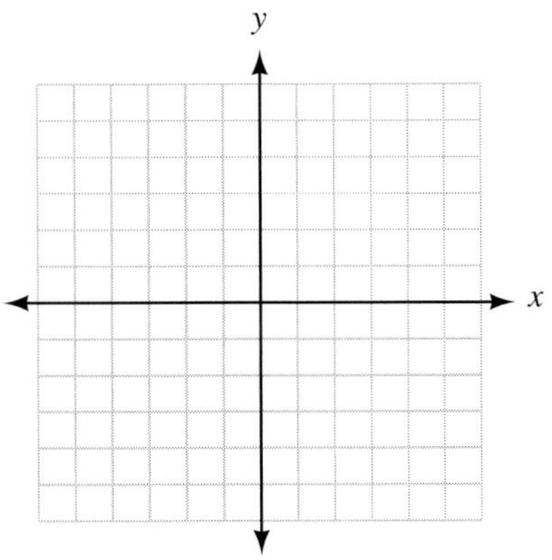

26. Graph each of the following lines.

a. $y = -\frac{1}{3}x$ **b.** $x = 1$ **c.** $y = -5$

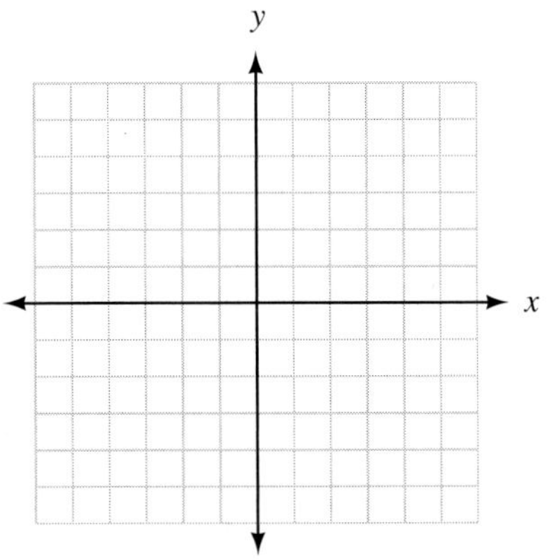

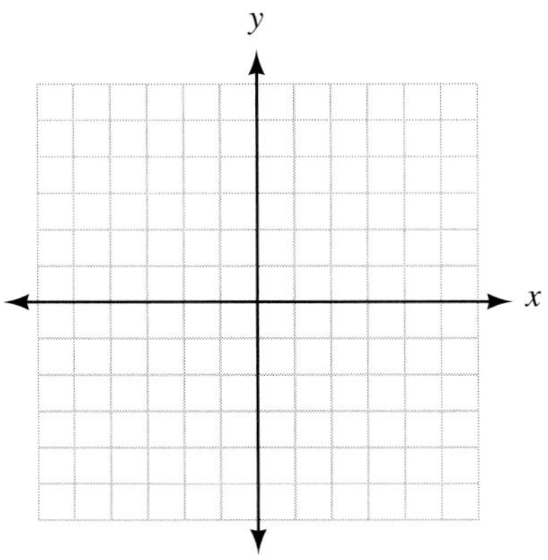

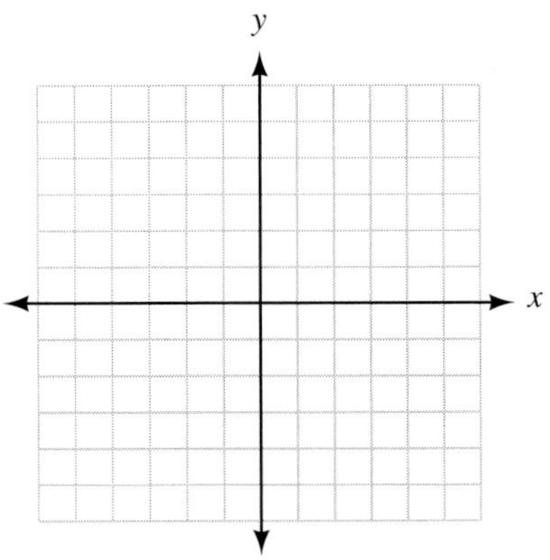

27. Graph the line $0.02x + 0.03y = 0.06$.

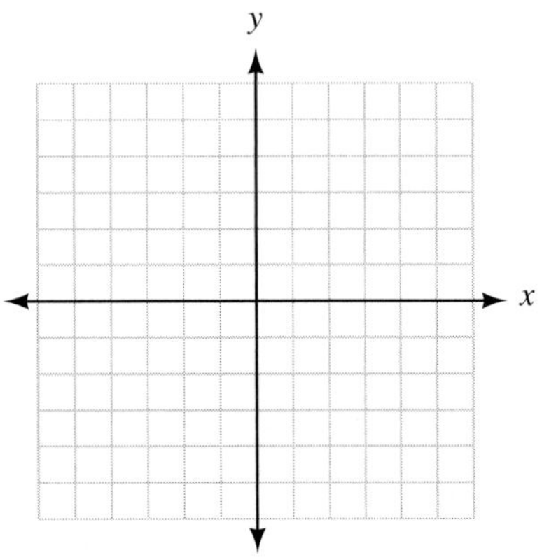

28. Graph the line $0.05x - 0.03y = 0.15$.

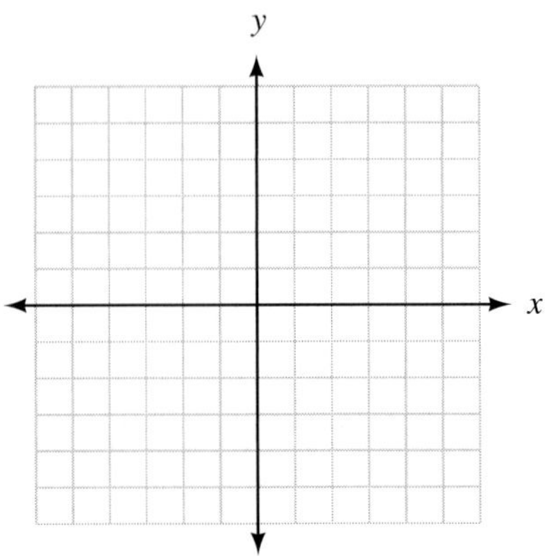

29. The ordered pairs that satisfy the equation $y = 3x$ all have the form $(x, 3x)$ because y is always 3 times x. Graph all ordered pairs of the form $(x, 3x)$.

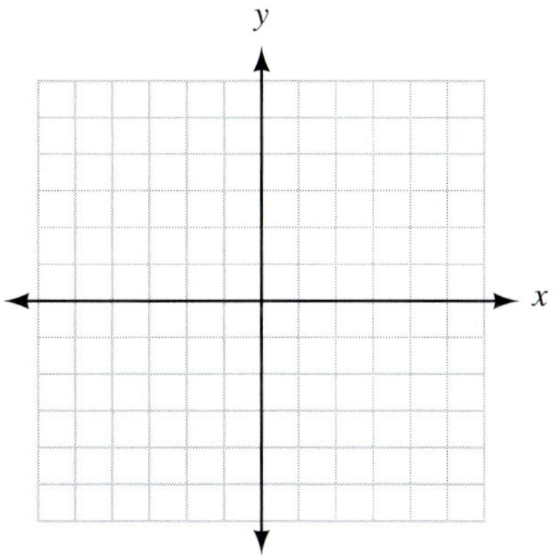

30. Graph all ordered pairs of the form $(x, -3x)$.

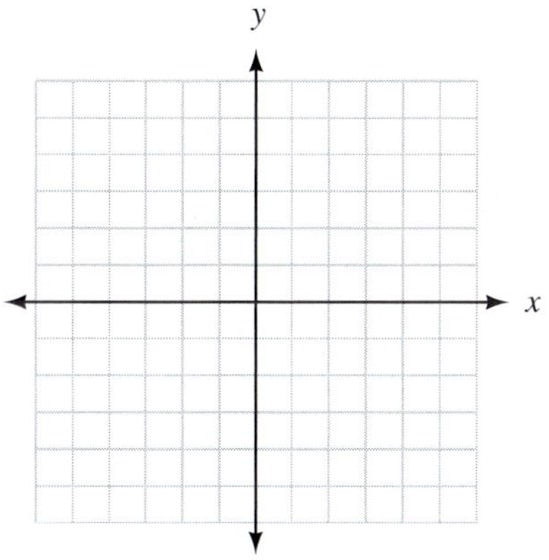

Applying the Concepts

31. A Snail's Pace A snail is climbing straight up a brick wall at a constant rate. It takes the snail 4 hours to climb up 6 feet. After climbing for 4 hours, the snail rests for 4 hours, during which time it slides 2 feet down the wall. If the snail starts at the bottom of the wall and repeats this climbing up/sliding down process continuously, construct a table that gives the snail's height above the ground every 4 hours, starting at 0 and ending at 24 hours. Then construct a line graph from the information in the table.

TIME (HR)	DISTANCE (FT)

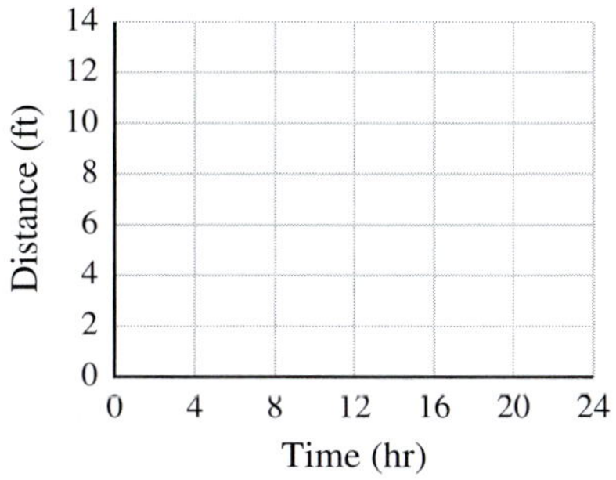

32. Air Temperature A pilot checks the weather conditions before flying and finds that the air temperature drops 3.5°F every 1,000 feet. If the air temperature is 41°F when the plane reaches 10,000 feet, construct a table that gives the air temperature every 2,000 feet, starting at 10,000 feet and ending at 20,000 feet. Then construct a line graph from the information in the table.

ALTITUDE (FT)	TEMPERATURE (°F)

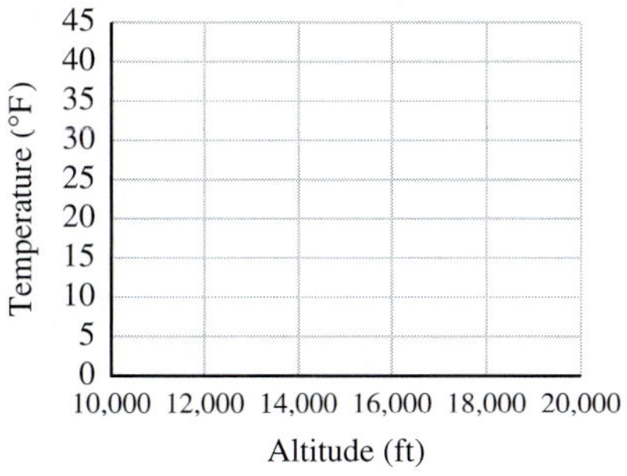

33. Hourly Wages Suppose you have a job that pays \$7.50 per hour, and you work anywhere from 0 to 40 hours per week. Table 2 gives the amount of money you will earn in 1 week for working various hours. Construct a line graph from the information in Table 2. You can copy the grid in Figure 12 to use as a guide.

TABLE 2

WEEKLY WAGES

HOURS WORKED	PAY (\$)
0	0
10	75
20	150
30	225
40	300

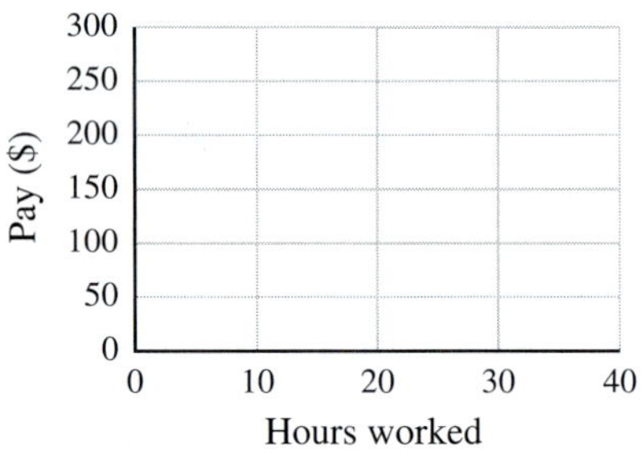

FIGURE 12 **Template for constructing a line graph.**

34. Softball Toss Chaudra is tossing a softball into the air with an underhand motion. It takes exactly 2 seconds for the ball to come back to her. Table 3 shows the distance the ball is above her hand at quarter-second intervals. Construct a line graph from the information in the table. Use the grid in Figure 13 as a guide.

TABLE 3

TOSSING A SOFTBALL INTO THE AIR

TIME (SEC)	DISTANCE (FT)
0	0
0.25	7
0.5	12
0.75	15
1	16
1.25	15
1.5	12
1.75	7
2	0

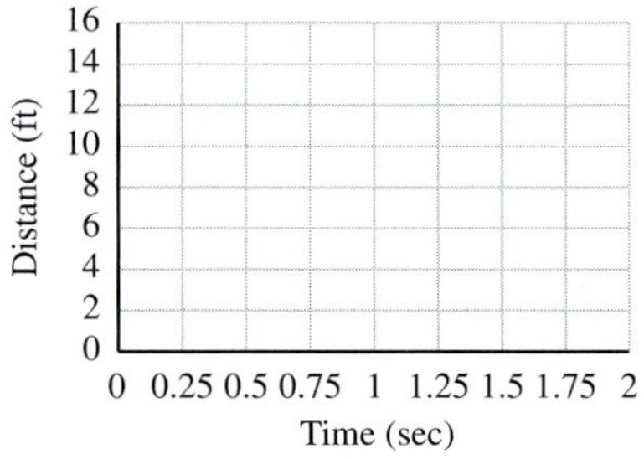

FIGURE 13 **Template for constructing a line graph.**

35. Intensity of Light Table 4 gives the intensity of light that falls on a surface at various distances from a 100-watt light bulb. Use the template shown in Figure 14 to construct a bar chart from the information in Table 4.

TABLE 4

LIGHT INTENSITY FROM A 100-WATT LIGHT BULB

DISTANCE ABOVE SURFACE (FT)	INTENSITY (LUMENS/SQ FT)
1	120.0
2	30.0
3	13.3
4	7.5
5	4.8
6	3.3

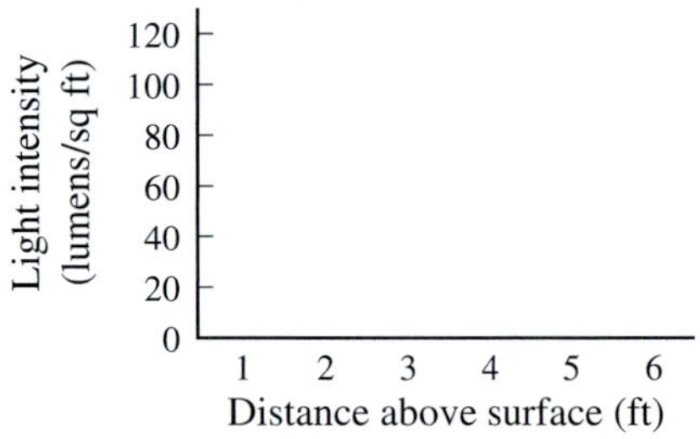

FIGURE 14 **Template for constructing a bar chart.**

36. Value of a Painting A piece of abstract art was purchased in 1990 for \$125. Table 5 shows the value of the painting at various times, assuming that it doubles in value every 5 years. Construct a bar chart from the information in the table, using the template in Figure 15.

TABLE 5

VALUE OF A PAINTING

YEAR	VALUE (\$)
1990	125
1995	250
2000	500
2005	1,000
2010	2,000

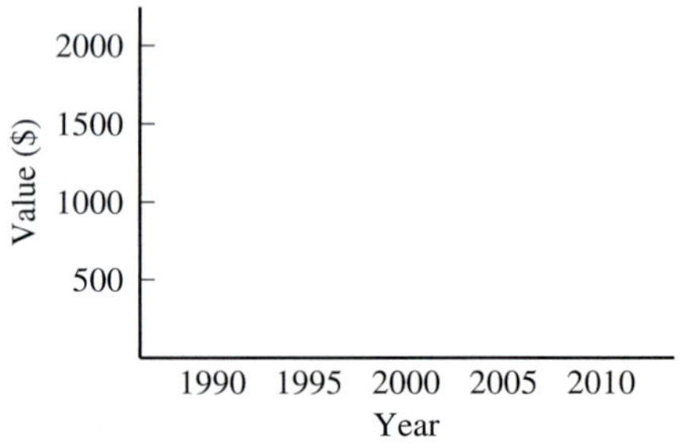

FIGURE 15 **Template for constructing a bar chart.**

37. **Reading Graphs** When we see graphs in the media, the axes are not always labeled completely. Use the chart to answer the following questions.

 a. Approximately what percentage of people were online in 1998?

 b. Approximately what percentage of people were online in 2003?

 c. Name the year or years during which 65% of adults were online.

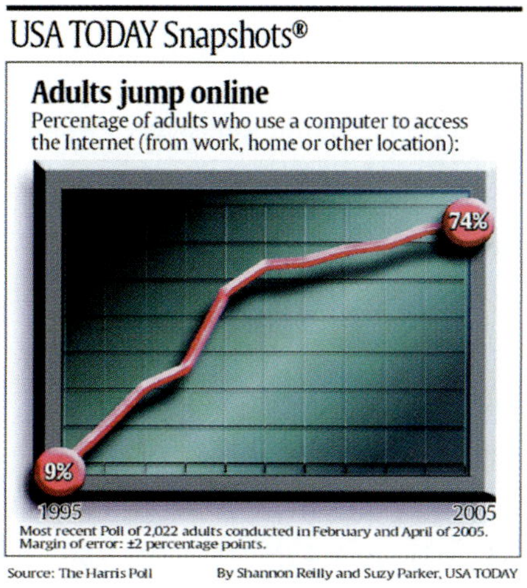

38. **Bridal Books** Use the chart to make a table of values of x and y, where x is years and y is the number of bridal books published.

x	y

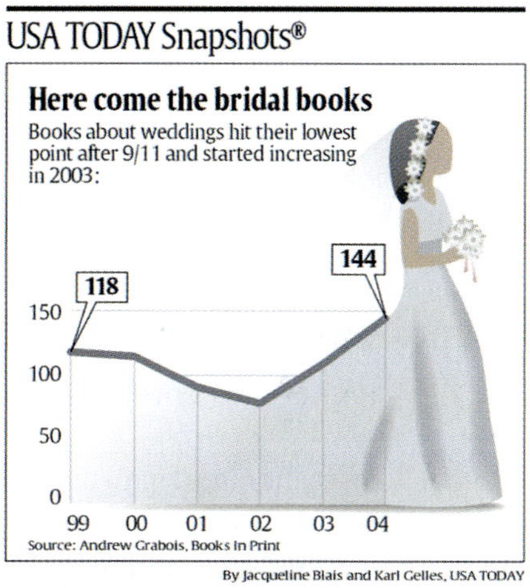

Getting Ready for the Next Section

39. Write -0.06 as a fracton with denominator 100.

40. Write -0.07 as a fraction with denominator 100.

41. If $y = 2x - 3$, find y when $x = 2$.

42. If $y = 2x - 3$, find x when $y = 5$.

Simplify.

43. $\dfrac{1-(-3)}{-5-(-2)}$

44. $\dfrac{-3-1}{-2-(-5)}$

45. $\dfrac{-1-4}{3-3}$

46. $\dfrac{-3-(-3)}{2-(-1)}$

47. The product of $\frac{2}{3}$ and what number will result in

 a. 1? **b.** -1?

48. The product of 3 and what number will result in

 a. 1? **b.** -1?

Maintaining Your Skills

The problems that follow review some of the more important skills you have learned in previous sections and chapters.

Solve each equation.

49. $5x - 4 = -3x + 12$

50. $\frac{1}{2} - \frac{y}{5} = -\frac{9}{10} + \frac{y}{2}$

51. $\frac{1}{2} - \frac{1}{8}(3t - 4) = -\frac{7}{8}t$

52. $3(5t - 1) - (3 - 2t) = 5t - 8$

53. $50 = \frac{K}{24}$

54. $3.5d = 16(3.5)^2$

55. $2(1) + y = 4$

56. $2(2y + 6) + 3y = 5$

57. $4\left(\frac{19}{15}\right) - 2y = 4$

58. $2x - 3\left(-\frac{5}{11}\right) = 4$

Improving Your Quantitative Literacy

Now that you have had a chance to look at some of the quantitative literacy problems in Chapters 1 and 2, we can give a more detailed definition. Here is what the *Mathematical Association of America* has to say about quantitative literacy:

> . . . every college graduate should be able to apply simple mathematical methods to the solution of real-world problems. A quantitatively literate college graduate should be able to:
>
> **1.** Interpret mathematical models such as formulas, graphs, tables, and schematics, and draw inferences from them.
> **2.** Represent mathematical information symbolically, visually, numerically, and verbally.
> **3.** Estimate and check answers to mathematical problems in order to determine reasonableness, identify alternatives, and select optimal results.
> **4.** Recognize that mathematical and statistical methods have limits.

List two specific skills that you would like to acquire from this course that will improve your quantitative literacy. For example, you may want to better understand the financing options available for purchasing a new car.

59. Misleading Chart One of our goals is for you to read graphs and charts that you see outside of class, and to recognize when the information in the chart is displayed incorrectly or in a misleading way. Here is a chart that appeared on the CNN.com website. It shows the results of a survey that asked people if the Florida state court was right to order a feeding tube removed from Terri Schiavo.

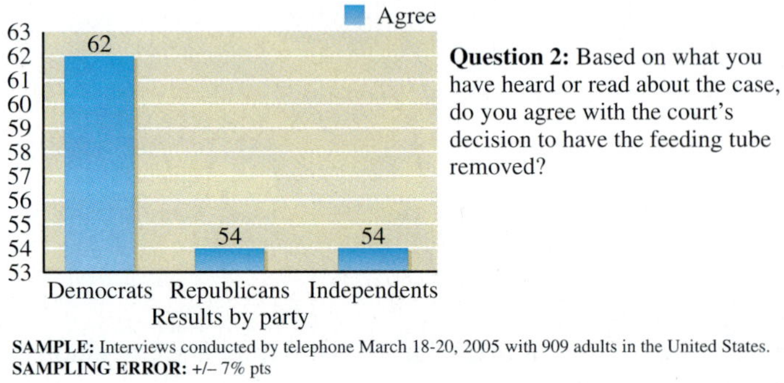

a. What is the impression given by the chart?
b. How does the margin of error affect the results of the survey?
c. Draw the chart again so it more accurately reflects the results of the survey.

The Slope of a Line

3.2

A highway sign tells us we are approaching a 6% downgrade. As we drive down this hill, each 100 feet we travel horizontally is accompanied by a 6-foot drop in elevation.

Highway sign

Mathematical model

OBJECTIVES	
A	Find the slope of a line from its graph.
B	Find the slope of a line given two points on the line.

In mathematics we say the slope of the highway is $-0.06 = -\frac{6}{100} = -\frac{3}{50}$. The *slope* is the ratio of the vertical change to the accompanying horizontal change.

In defining the slope of a straight line, we are looking for a number to associate with a straight line that does two things. First, we want the slope of a line to measure the "steepness" of the line; that is, in comparing two lines, the slope of the steeper line should have the larger numerical value. Second, we want a line that *rises* going from left to right to have a *positive* slope. We want a line that *falls* going from left to right to have a *negative* slope. (A line that neither rises nor falls going from left to right must, therefore, have 0 slope.)

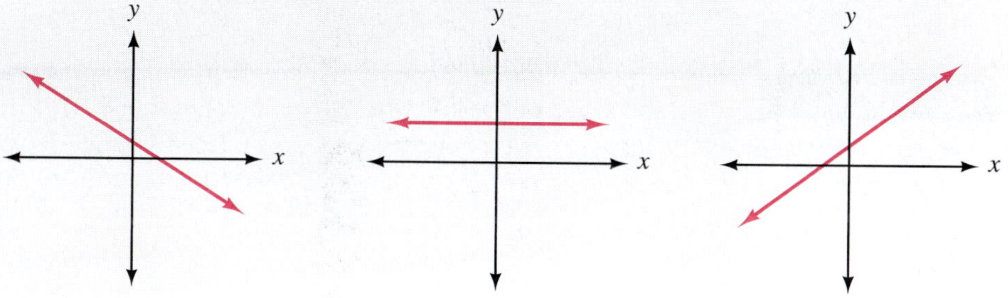

Negative slope　　Zero slope　　Positive slope

Geometrically, we can define the *slope* of a line as the ratio of the vertical change to the horizontal change encountered when moving from one point to another on the line. The vertical change is sometimes called the *rise*. The horizontal change is called the *run*.

EXAMPLE 1

Find the slope of the line $y = 2x - 3$.

Solution To use our geometric definition, we first graph $y = 2x - 3$ (Figure 1). We then pick any two convenient points and find the ratio of rise to run. By convenient points we mean points with integer coordinates. If we let $x = 2$ in the equation, then $y = 1$. Likewise if we let $x = 4$, then y is 5.

PRACTICE PROBLEMS

1. Graph the line $y = 3x - 5$ and then find the slope.

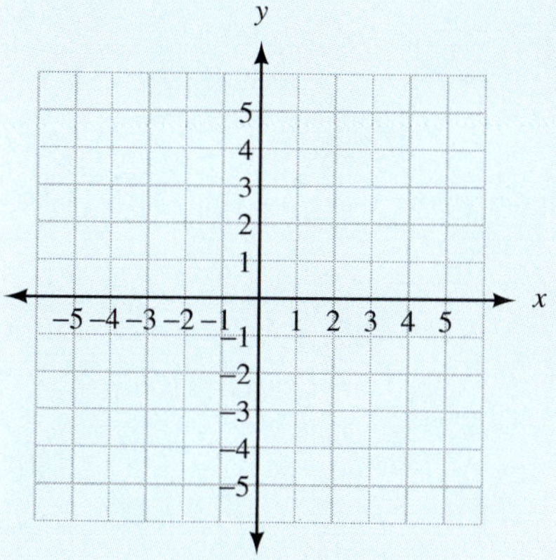

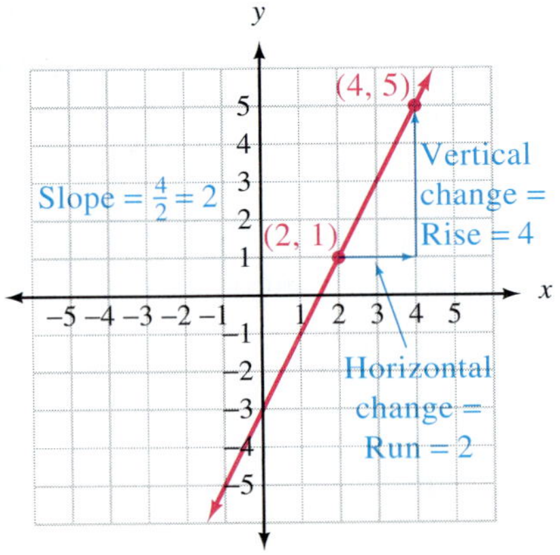

FIGURE 1

Our line has a slope of 2.

Notice that we can measure the vertical change (rise) by subtracting the y-coordinates of the two points shown in Figure 1: $5 - 1 = 4$. The horizontal change (run) is the difference of the x-coordinates: $4 - 2 = 2$. This gives us a second way of defining the slope of a line.

DEFINITION

The **slope** of the line between two points (x_1, y_1) and (x_2, y_2) is given by

$$\text{Slope} = m = \frac{\text{Rise}}{\text{Run}} = \frac{y_2 - y_1}{x_2 - x_1}$$

↑ Geometric form ↑ Algebraic form

EXAMPLE 2

Find the slope of the line through $(-2, -3)$ and $(-5, 1)$.

2. Find the slope of the line through $(3, 4)$ and $(1, -2)$.

Solution

$$m = \frac{y_2 - y_1}{x_2 - x_1} = \frac{1 - (-3)}{-5 - (-2)} = \frac{4}{-3} = -\frac{4}{3}$$

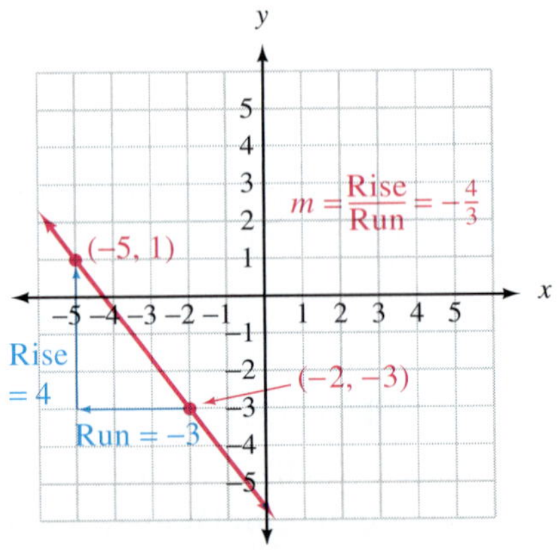

FIGURE 2

Looking at the graph of the line between the two points (Figure 2), we can see our geometric approach does not conflict with our algebraic approach.

Answer
1. 3

We should note here that it does not matter which ordered pair we call (x_1, y_1) and which we call (x_2, y_2). If we were to reverse the order of subtraction of both the x- and y-coordinates in the preceding example, we would have

$$m = \frac{-3 - 1}{-2 - (-5)} = \frac{-4}{3} = -\frac{4}{3}$$

which is the same as our previous result.

Note: The two most common mistakes students make when first working with the formula for the slope of a line are

1. Putting the difference of the x-coordinates over the difference of the y-coordinates.
2. Subtracting in one order in the numerator and then subtracting in the opposite order in the denominator. You would make this mistake in Example 2 if you wrote $1 - (-3)$ in the numerator and then $-2 - (-5)$ in the denominator.

EXAMPLE 3 Find the slope of the line containing $(3, -1)$ and $(3, 4)$.

Solution Using the definition for slope, we have

$$m = \frac{-1 - 4}{3 - 3} = \frac{-5}{0}$$

The expression $\frac{-5}{0}$ is undefined; that is, there is no real number to associate with it. In this case, we say the line *has no slope.*

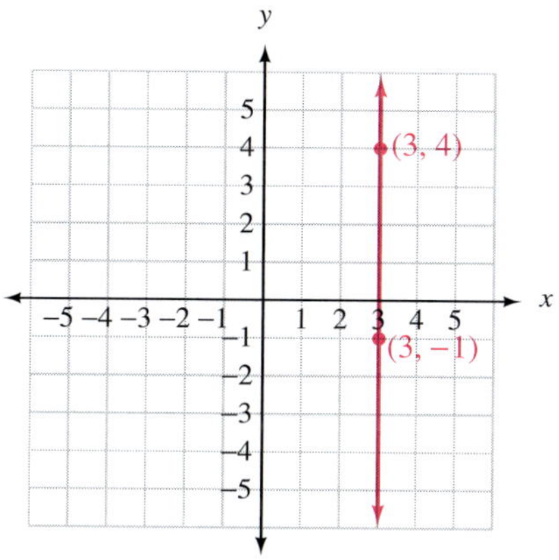

FIGURE 3

The graph of our line is shown in Figure 3. Our line with no slope is a vertical line. All vertical lines have no slope. (And all horizontal lines, as we mentioned earlier, have 0 slope.)

3. Find the slope of the line through $(2, -3)$ and $(-1, -3)$.

Slopes of Parallel and Perpendicular Lines

In geometry we call lines in the same plane that never intersect parallel. For two lines to be nonintersecting, they must rise or fall at the same rate. In other words, two lines are *parallel* if and only if they have the *same slope.*

Although it is not as obvious, it is also true that two nonvertical lines are *perpendicular* if and only if the *product of their slopes is* -1. This is the same as saying their slopes are negative reciprocals.

Answers
2. 3 **3.** 0

We can state these facts with symbols as follows: If line l_1 has slope m_1 and line l_2 has slope m_2, then

$$l_1 \text{ is parallel to } l_2 \Leftrightarrow m_1 = m_2$$

and

$$l_1 \text{ is perpendicular to } l_2 \Leftrightarrow m_1 \cdot m_2 = -1$$

$$\left(\text{or } m_1 = \frac{-1}{m_2}\right)$$

For example, if a line has a slope of $\frac{2}{3}$, then any line parallel to it has a slope of $\frac{2}{3}$. Any line perpendicular to it has a slope of $-\frac{3}{2}$ (the negative reciprocal of $\frac{2}{3}$).

Although we cannot give a formal proof of the relationship between the slopes of perpendicular lines at this level of mathematics, we can offer some justification for the relationship. Figure 4 shows the graphs of two lines. One of the lines has a slope of $\frac{2}{3}$; the other has a slope of $-\frac{3}{2}$. As you can see, the lines are perpendicular.

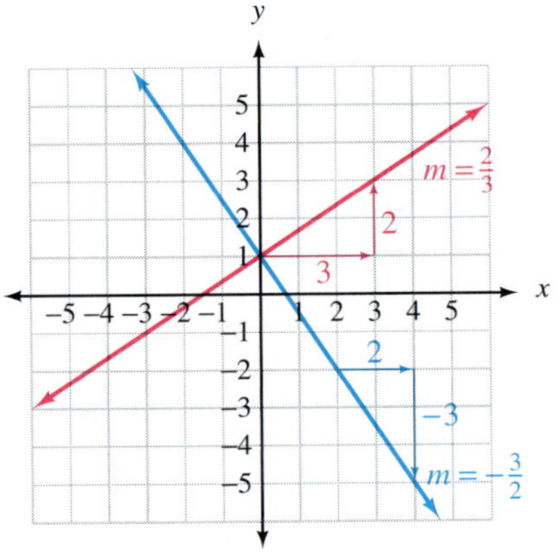

FIGURE 4

Slope and Rate of Change

So far, the slopes we have worked with represent the ratio of the change in y to a corresponding change in x, or, on the graph of the line, the slope is the ratio of vertical change to horizontal change in moving from one point on the line to another. However, when our variables represent quantities from the world around us, slope can have additional interpretations.

EXAMPLE 4 On the chart below, find the slope of the line connecting the first point (1955, 0.29) with the last point (2005, 2.93). Explain the significance of the result.

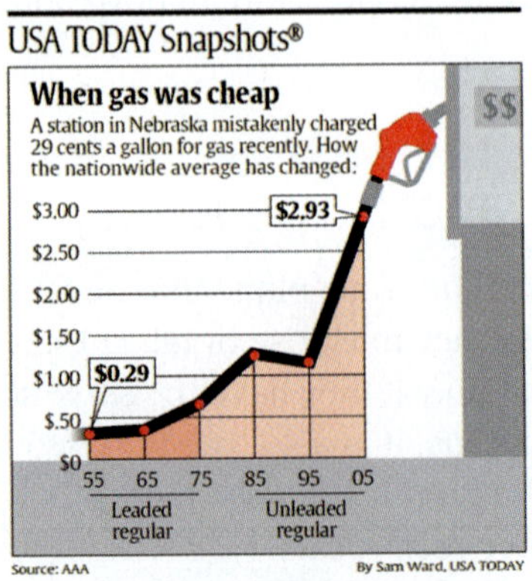

From *USA Today*. Copyright 2005. Reprinted with permission.

4. On the chart, find the slope of the line connecting the point (1985, 1.25) with the last point (2005, 2.93). Explain in words what your result represents.

Solution The slope of the line connecting the first point (1955, 0.29) with the last point (2005, 2.93), is

$$m = \frac{2.93 - 0.29}{2005 - 1955} = \frac{2.64}{50} = 0.0528$$

The units are dollars/year. If we write this in terms of cents we have

$$m = 5.28 \text{ cents/year}$$

which is the average change in the price of a gallon of gasoline over a 50-year period of time.

Likewise, if we connect the points (1995, 1.10) and (2005, 2.93), the line that results has a slope of

$$m = \frac{2.93 - 1.10}{2005 - 1995} = \frac{1.83}{10} = 0.183 \text{ dollars/year} = 18.3 \text{ cents/year}$$

which is the average change in the price of a gallon of gasoline over a 10-year period. As you can imagine by looking at the chart, the line connecting the first and last point is not as steep as the line connecting the points from 1995 and 2005, and this is what we are seeing numerically with our slope calculations. If we were summarizing this information for an article in the newspaper, we could say, although the price of a gallon of gasoline has increased only 5.28 cents per year over the last 50 years, in the last 10 years the average annual rate of increase has more than tripled, to 18.3 cents per year. ■

USING TECHNOLOGY

Families of Curves

We can use a graphing calculator to investigate the effects of the numbers a and b on the graph of $y = ax + b$. To see how the number b affects the graph, we can hold a constant and let b vary. Doing so will give us a *family* of curves. Suppose we set $a = 1$ and then let b take on integer values from -3 to 3. The equations we obtain are

$$\begin{aligned} y &= x - 3 \\ y &= x - 2 \\ y &= x - 1 \\ y &= x \\ y &= x + 1 \\ y &= x + 2 \\ y &= x + 3 \end{aligned}$$

We will give three methods of graphing this set of equations on a graphing calculator.

Method 1: Y-Variables List

To use the Y-variables list, enter each equation at one of the Y variables, set the graph window, then graph. The calculator will graph the equations in order, starting with Y_1 and ending with Y_7. Following is the Y-variables list, an appropriate window, and a sample of the type of graph obtained (Figure 5).

Answer

4. 8.4 cents/year, which is the average rate of change of the price of a gallon of gasoline over a 20-year period of time.

$Y_1 = X - 3$
$Y_2 = X - 2$
$Y_3 = X - 1$
$Y_4 = X$
$Y_5 = X + 1$
$Y_6 = X + 2$
$Y_7 = X + 3$

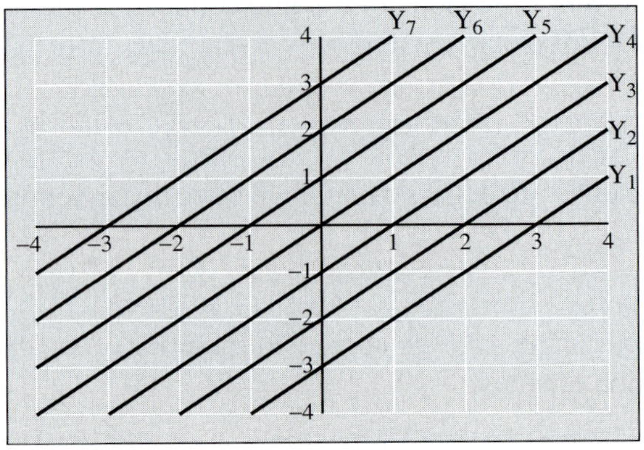

FIGURE 5

Window: X from −4 to 4, Y from −4 to 4

Method 2: Programming

The same result can be obtained by programming your calculator to graph $y = x + b$ for $b = -3, -2, -1, 0, 1, 2$, and 3. Here is an outline of a program that will do this. Check the manual that came with your calculator to find the commands for your calculator.

Step 1: Clear screen
Step 2: Set window for X from −4 to 4 and Y from −4 to 4
Step 3: $-3 \to B$
Step 4: Label 1
Step 5: Graph $Y = X + B$
Step 6: $B + 1 \to B$
Step 7: If $B < 4$, go to 1
Step 8: End

Method 3: Using Lists

On the TI-82/83 you can set Y_1 as follows

$$Y_1 = X + \{-3, -2, -1, 0, 1, 2, 3\}$$

When you press [GRAPH], the calculator will graph each line from $y = x + (-3)$ to $y = x + 3$.

Each of the three methods will produce graphs similar to those in Figure 5.

Linking Objectives and Examples

Next to each **objective**, list the examples that are best described by that objective.

A	
B	

Getting Ready for Class

After reading through the preceding section, respond in your own words and in complete sentences.

1. If you were looking at a graph that described the performance of a stock you had purchased, why would it be better if the slope of the line were positive, rather than negative?
2. Describe the behavior of a line with a negative slope.
3. Would you rather climb a hill with a slope of $\frac{1}{2}$ or a slope of 3? Explain why.
4. Describe how to obtain the slope of a line if you know the coordinates of two points on the line.

Problem Set 3.2

INTERMEDIATE Algebra $f(x)$ Now™

Find the slope of each of the following lines from the given graph.

1.

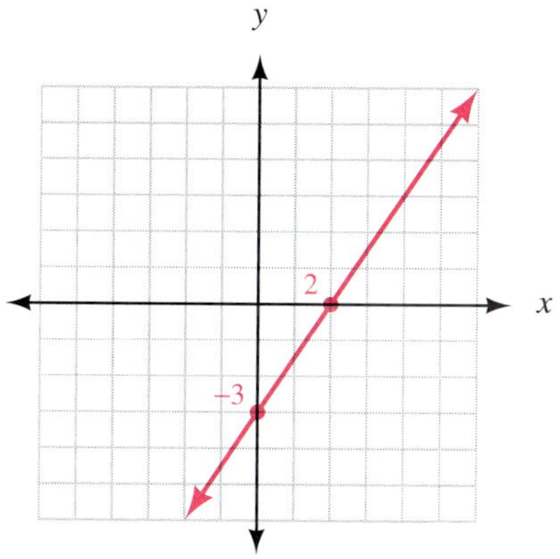

2.

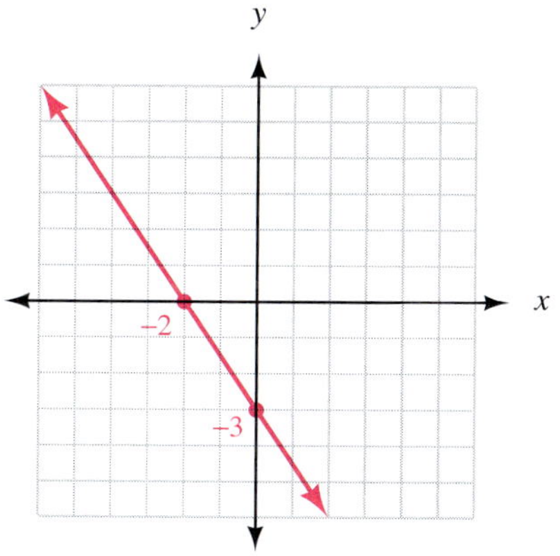

3.

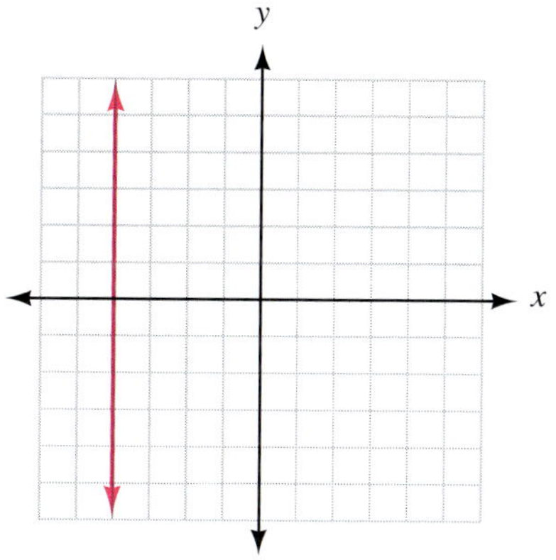

4.

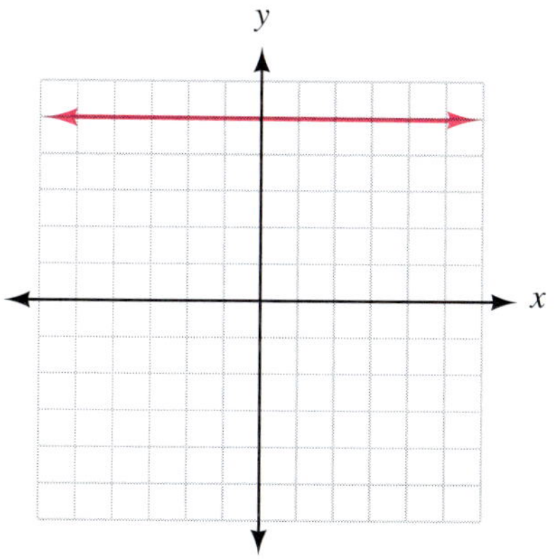

5.

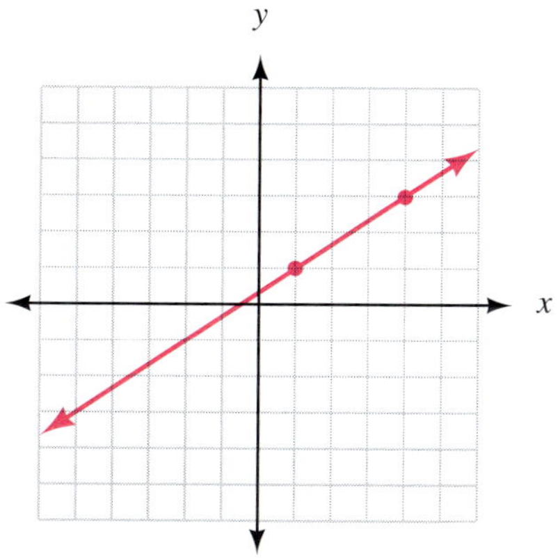

6.

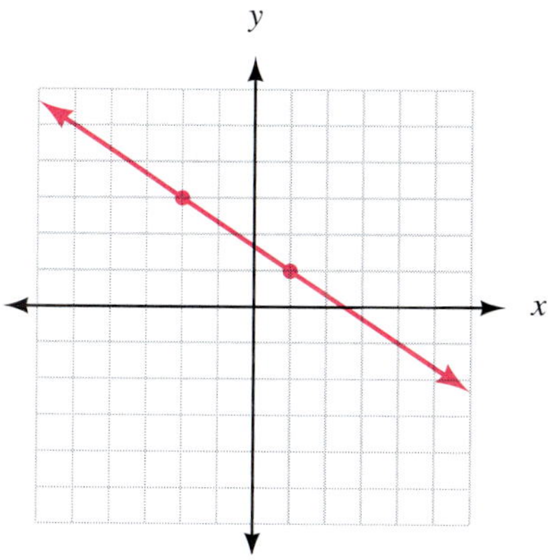

Find the slope of the line through the following pairs of points. Then, plot each pair of points, draw a line through them, and indicate the rise and run in the graph in the manner shown in Example 2.

7. (2, 1), (4, 4)

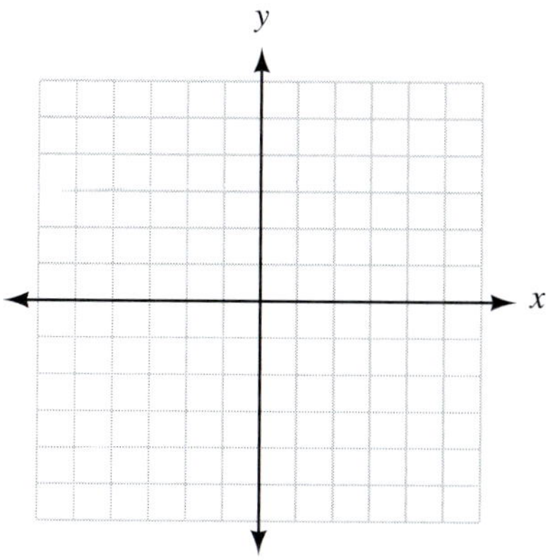

8. (3, 1), (5, 4)

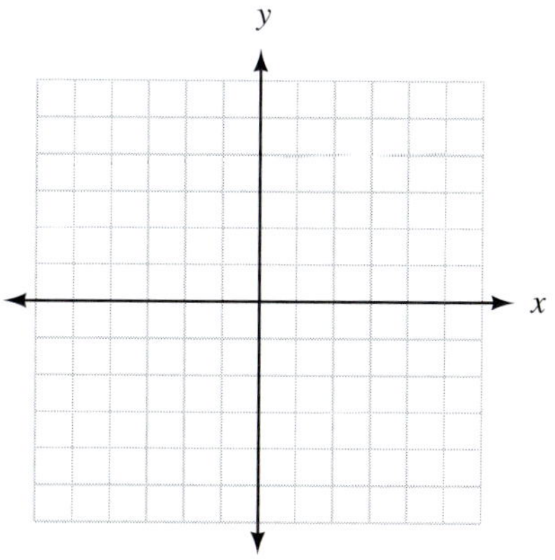

9. (1, 4), (5, 2)

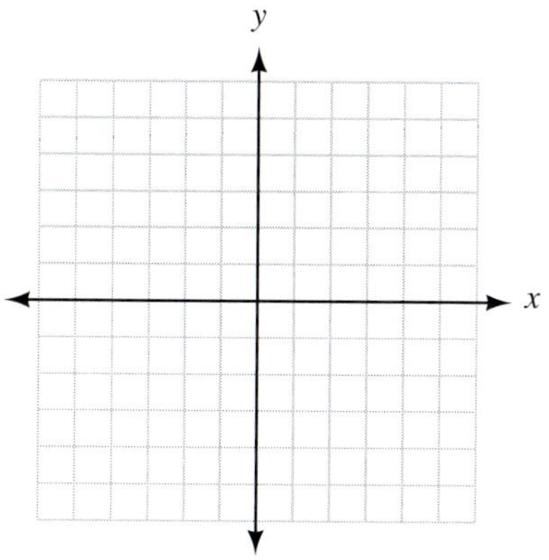

10. (1, 3), (5, 2)

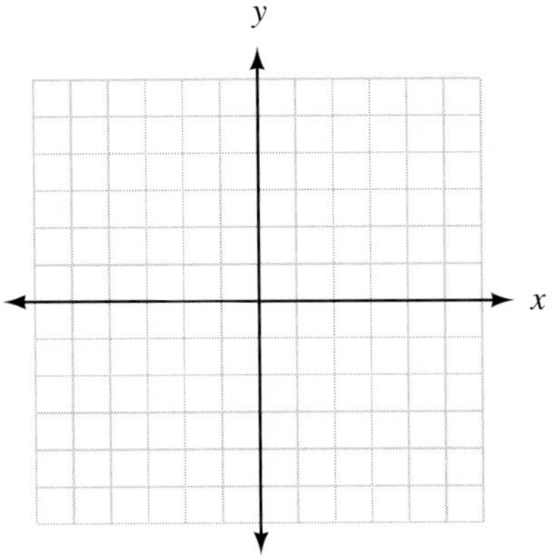

11. (1, −3), (4, 2)

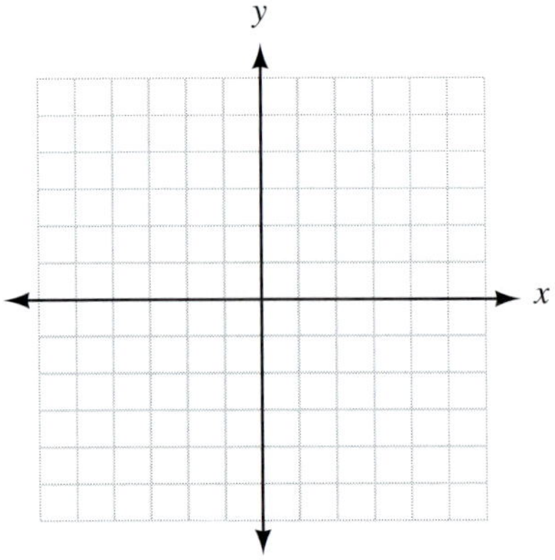

12. (2, −3), (5, 2)

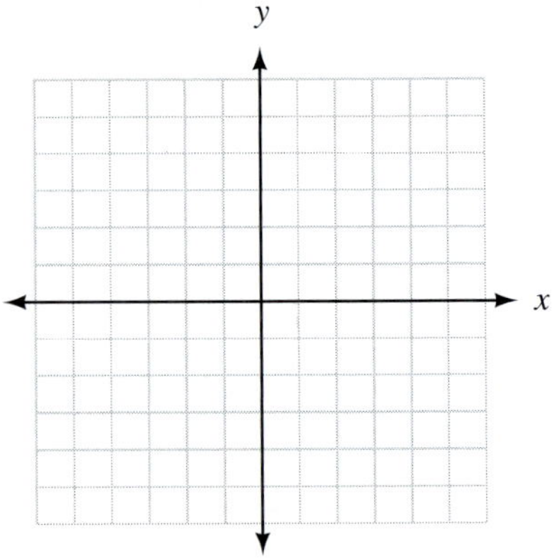

13. (−3, −2), (1, 3)

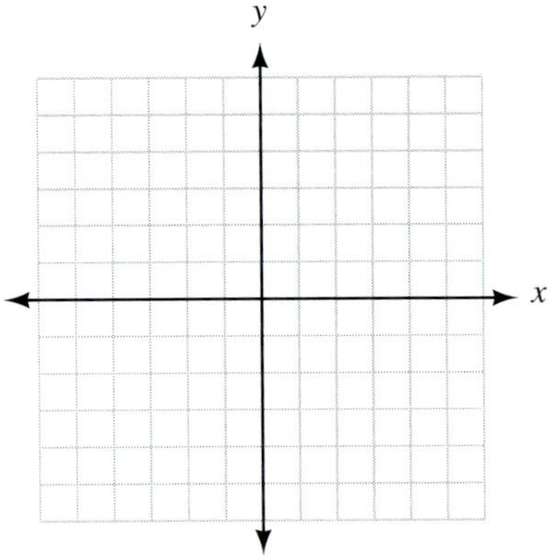

14. (−3, −1), (1, 4)

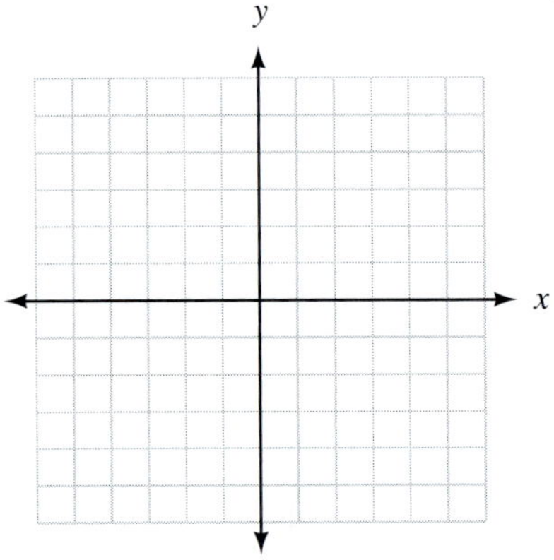

15. (−3, 2), (3, −2)

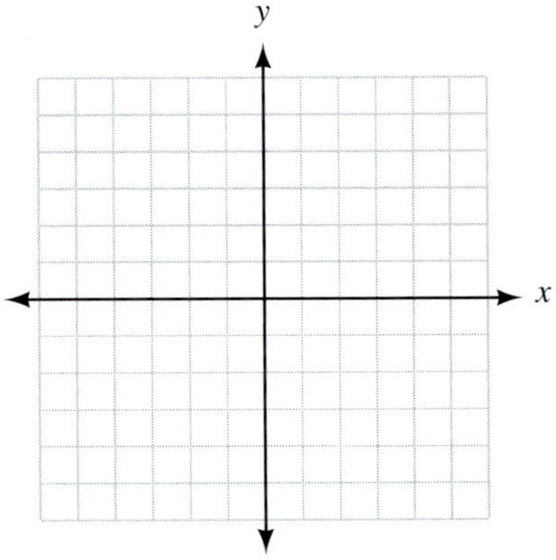

16. (−3, 3), (3, −1)

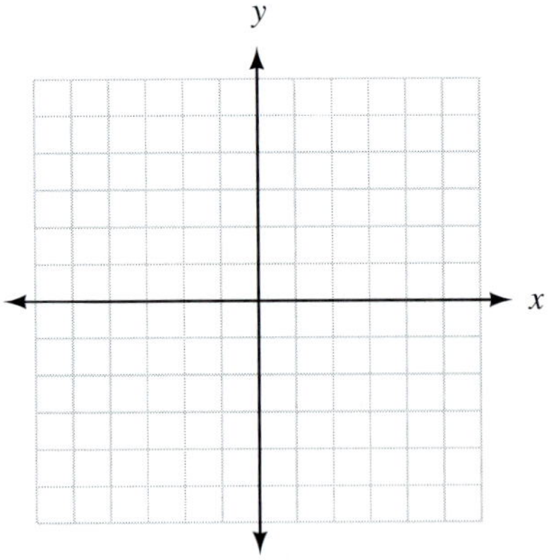

17. (2, −5), (3, −2)

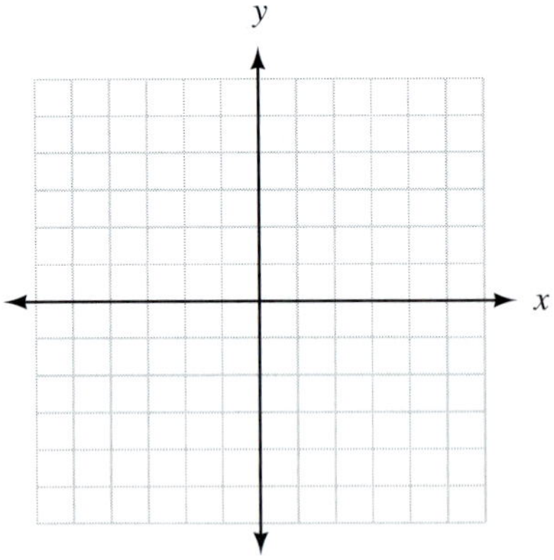

18. (2, −4), (3, −1)

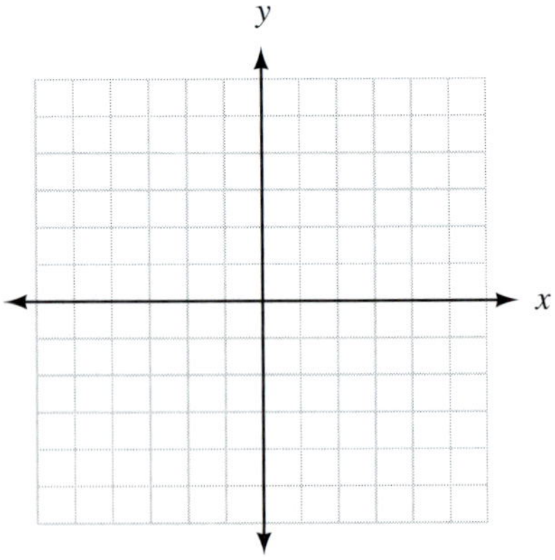

For each equation below, complete the table, and then use the results to find the slope of the graph of the equation.

19. $2x + 3y = 6$

x	y
0	
	0

20. $3x - 2y = 6$

x	y
0	
	0

21. $y = \frac{2}{3}x - 5$

x	y
0	
3	

22. $y = -\frac{3}{4}x + 2$

x	y
0	
4	

23. Graph the line that has an x-intercept of 3 and a y-intercept of -2. What is the slope of this line?

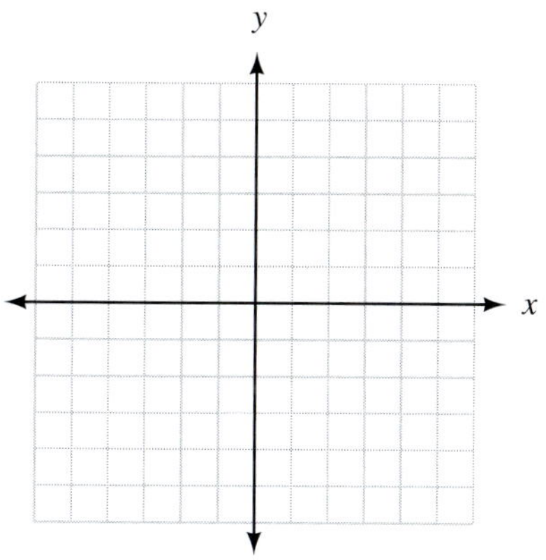

24. Graph the line that has an x-intercept of 2 and a y-intercept of -3. What is the slope of this line?

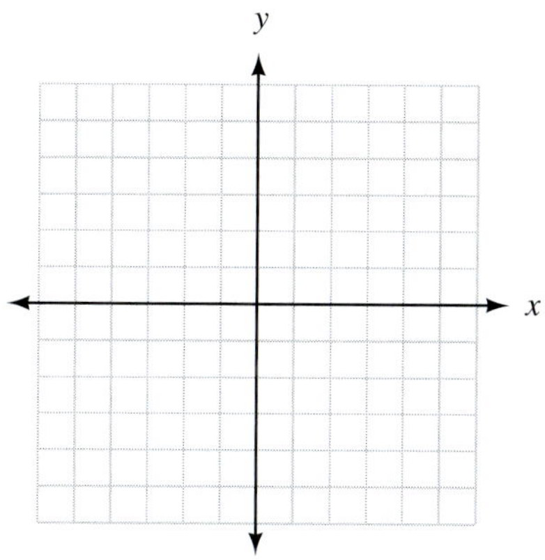

25. Graph the line with x-intercept 4 and y-intercept 2. What is the slope of this line?

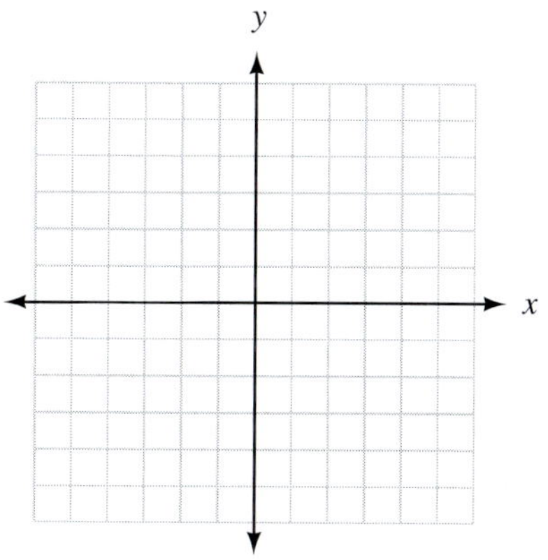

26. Graph the line with x-intercept -4 and y-intercept -2. What is the slope of this line?

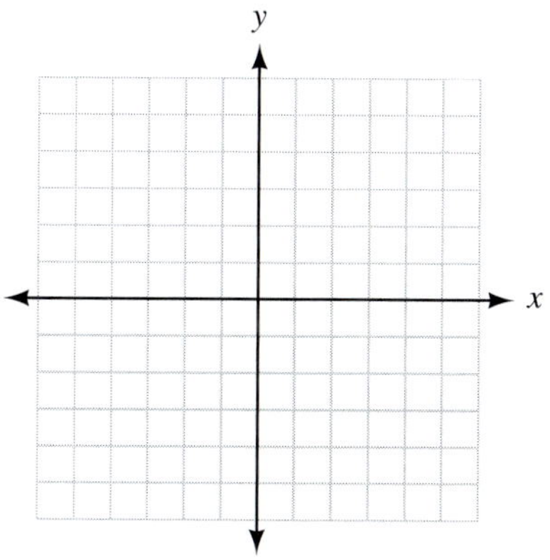

27. Determine if each of the following tables could represent ordered pairs from an equation of a line.

a.

x	y
0	5
1	7
2	9
3	11

b.

x	y
-2	-5
0	-2
2	0
4	1

28. The following lines have slope 2, $\frac{1}{2}$, 0, and -1. Match each line to its slope value.

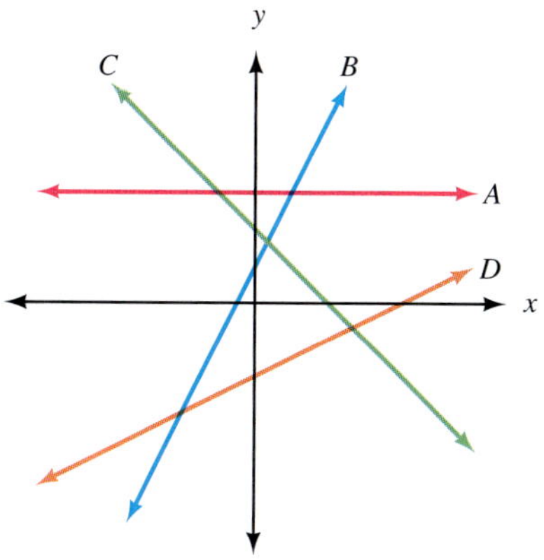

29. Find the slope of any line parallel to the line through (2, 3) and (−8, 1).

30. Find the slope of any line parallel to the line through (2, 5) and (5, −3).

31. Line l contains the points (5, −6) and (5, 2). Give the slope of any line perpendicular to l.

32. Line l contains the points (3, 4) and (−3, 1). Give the slope of any line perpendicular to l.

33. Line l has a slope of $\frac{2}{3}$. A horizontal change of 12 will always be accompanied by how much of a vertical change?

34. For any line with slope $\frac{4}{5}$, a vertical change of 8 is always accompanied by how much of a horizontal change?

Applying the Concepts

35. Slope of a Highway A sign at the top of the Cuesta Grade, outside of San Luis Obispo, reads "7% downgrade next 3 miles." The diagram shown here is a model of the Cuesta Grade that takes into account the information on that sign.

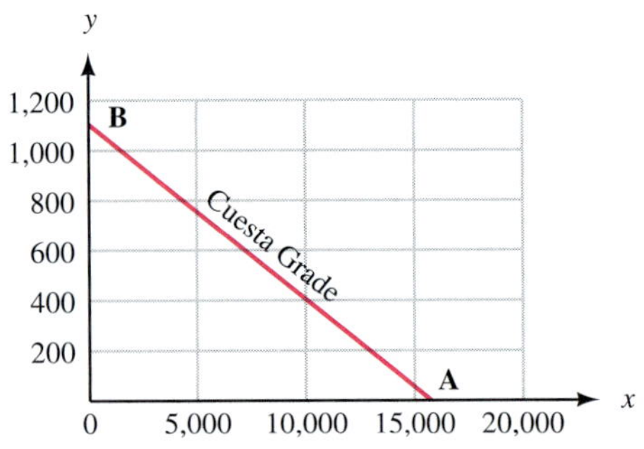

a. At point B, the graph crosses the y-axis at 1,106 feet. How far is it from the origin to point A?

b. What is the slope of the Cuesta Grade?

36. Heating a Block of Ice A block of ice with an initial temperature of −20°C is heated at a steady rate. The graph shows how the temperature changes as the ice melts to become water and the water boils to become steam and water.

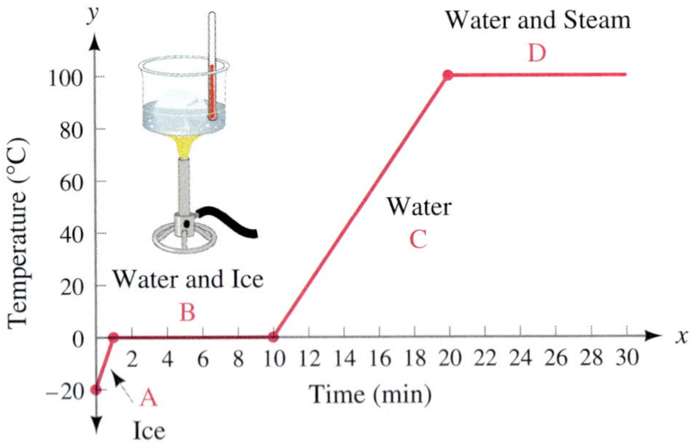

a. How long does it take all the ice to melt?

b. From the time the heat is applied to the block of ice, how long is it before the water boils?

c. Find the slope of the line segment labeled A. What units would you attach to this number?

d. Find the slope of the line segment labeled C. Be sure to attach units to your answer.

e. Is the temperature changing faster during the 1st minute or the 16th minute?

37. Slope and Rate of Change Find the slope of the line connecting the first and last points on the chart. Explain in words what the slope represents.

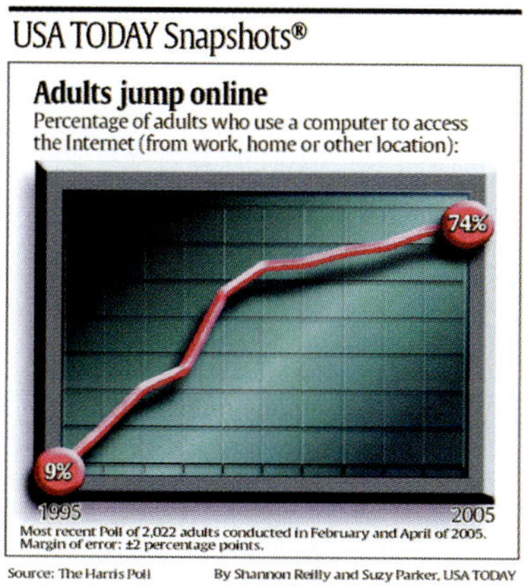

From *USA Today.* Copyright 2005. Reprinted with permission.

38. Bridal Books Find the slope of the line that connects the points (1999, 118) and (2002, 75). Round to the nearest tenth. Explain in words what the slope represents.

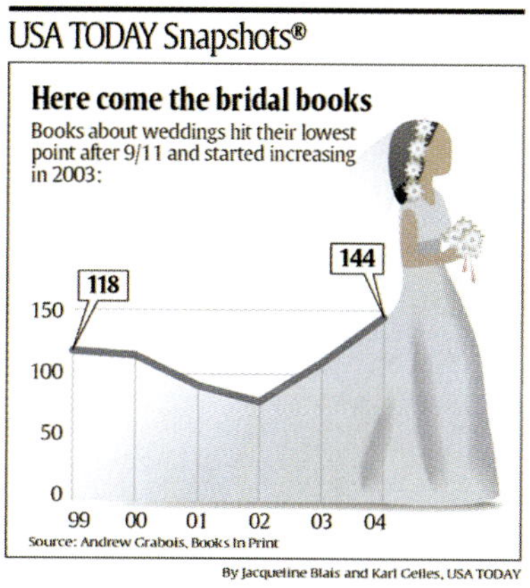

From *USA Today.* Copyright 2005. Reprinted with permission.

39. Health Care Use the chart to work the following problems involving slope.

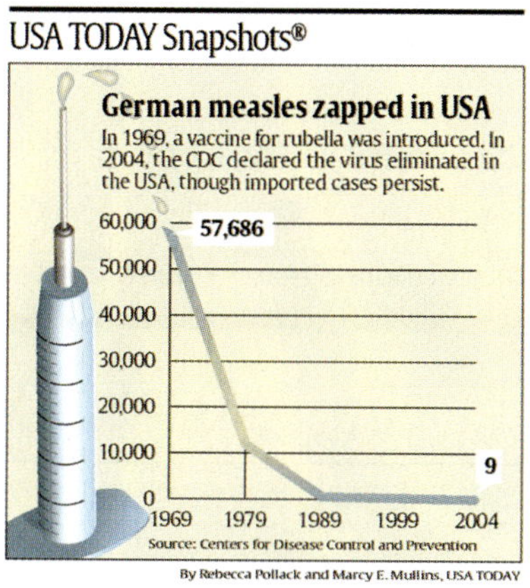

From *USA Today.* Copyright 2005. Reprinted with permission.

a. Find the slope of the line from 1969 to 1979, and then give a written explanation of the significance of that number.

b. Find the slope of the line from 1979 to 1989, and then give a written explanation of the significance of that number.

c. What is the slope of the line connecting (1989, 9) and (2004, 9)? What does this tell us about German measles?

40. Triathlon Entries Use the chart to work the following problems.

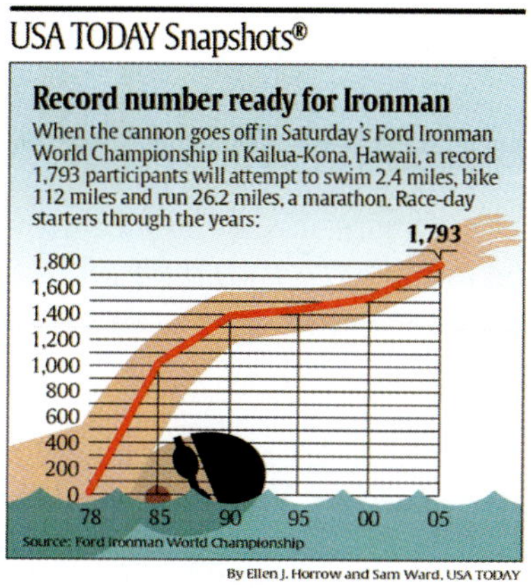

From *USA Today.* Copyright 2005. Reprinted with permission.

a. Find the slope of the line that connects the points (1978, 0) and (2005, 1,793). Round your answer to the nearest tenth. Explain in words what the slope represents.

b. Of the line segments shown in the chart, one has a larger slope than the rest. Find that slope and give an interpretation of its meaning.

Getting Ready for the Next Section

Simplify.

41. $2\left(-\frac{1}{2}\right)$

42. $\frac{3-(-1)}{-3-3}$

43. $\frac{5-(-3)}{2-6}$

44. $3\left(-\frac{2}{3}x+1\right)$

Solve for y.

45. $\frac{y-b}{x-0}=m$

46. $2x+3y=6$

47. $y-3=-2(x+4)$

48. $y+1=-\frac{2}{3}(x-3)$

49. If $y=-\frac{4}{3}x+5$, find y when x is 0.

50. If $y=-\frac{4}{3}x+5$, find y when x is 3.

Maintaining Your Skills

The problems that follow review some of the more important skills you have learned in previous sections and chapters.

51. If $3x+2y=12$, find y when x is 4.

52. If $y=3x-1$, find x when y is 0.

53. Solve the formula $3x+2y=12$ for y.

54. Solve the formula $y=3x-1$ for x.

55. Solve the formula $A=P+Prt$ for t.

56. Solve the formula $S=\pi r^2+2\pi rh$ for h.

Improving Your Quantitative Literacy

57. Cycling A cyclist is traveling at a constant speed. The graph shows the distance the cyclist travels over time when there is no wind present, when she travels against the wind, and when she travels with the wind.

a. Use the concept of slope to find the speed of the cyclist under each of the three conditions.

b. Compare the speed of the cyclist when she is traveling without any wind to when she is riding against the wind. How do the two speeds differ?

c. Compare the speed of the cyclist when she is traveling without any wind to when she is riding with the wind. How do the two speeds differ?

d. What is the speed of the wind?

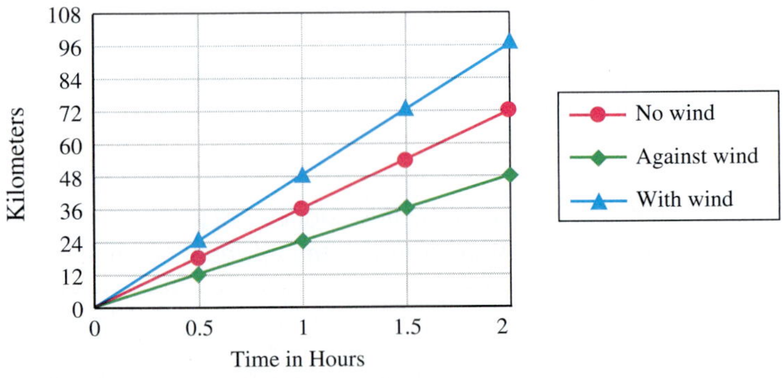

3.3 The Equation of a Line

The table and illustrations show some corresponding temperatures on the Fahrenheit and Celsius temperature scales. For example, water freezes at 32°F and 0°C, and boils at 212°F and 100°C.

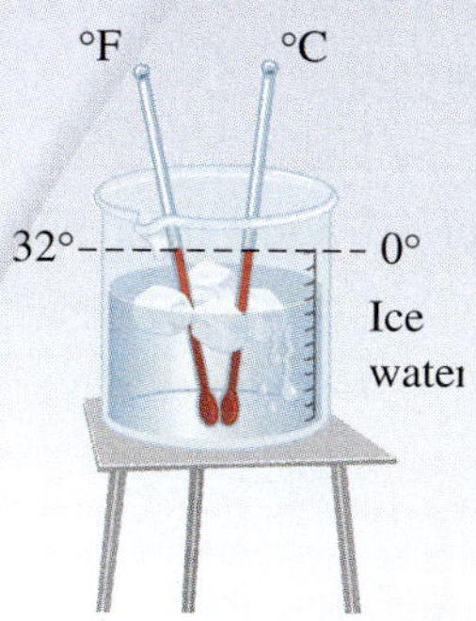

DEGREES CELSIUS	DEGREES FAHRENHEIT
0	32
25	77
50	122
75	167
100	212

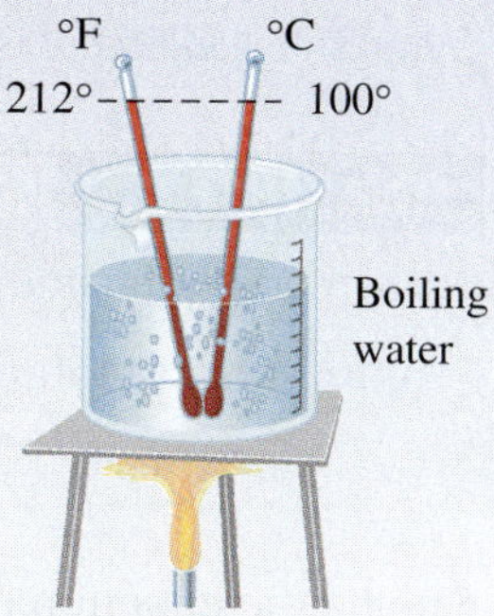

OBJECTIVES

A	Find the equation of a line given its slope and *y*-intercept.
B	Find the slope and *y*-intercept from the equation of a line.
C	Find the equation of a line given the slope and a point on the line.
D	Find the equation of a line given two points on the line.

If we plot all the points in the table using the *x*-axis for temperatures on the Celsius scale and the *y*-axis for temperatures on the Fahrenheit scale, we see that they line up in a straight line (Figure 1). This means that a linear equation in two variables will give a perfect description of the relationship between the two scales. That equation is

$$F = \frac{9}{5}C + 32$$

The techniques we use to find the equation of a line from a set of points is what this section is all about.

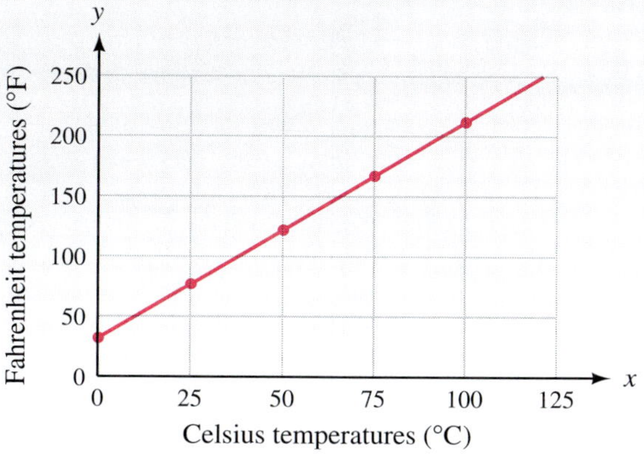

FIGURE 1

Suppose line l has slope m and y-intercept b. What is the equation of l? Because the y-intercept is b, we know the point $(0, b)$ is on the line. If (x, y) is any other point on l, then using the definition for slope, we have

$\frac{y - b}{x - 0} = m$	**Definition of slope**
$y - b = mx$	**Multiply both sides by x**
$y = mx + b$	**Add b to both sides**

This last equation is known as the *slope-intercept form* of the equation of a straight line.

Slope-Intercept Form of the Equation of a Line

The equation of any line with slope m and y-intercept b is given by

$$y = mx + b$$

Slope (m) y-intercept (b)

When the equation is in this form, the *slope* of the line is always the *coefficent of* x, and the *y-intercept* is always the *constant term.*

PRACTICE PROBLEMS

1. Find the equation of the line with slope $\frac{2}{3}$ and y-intercept 1. Then graph the line.

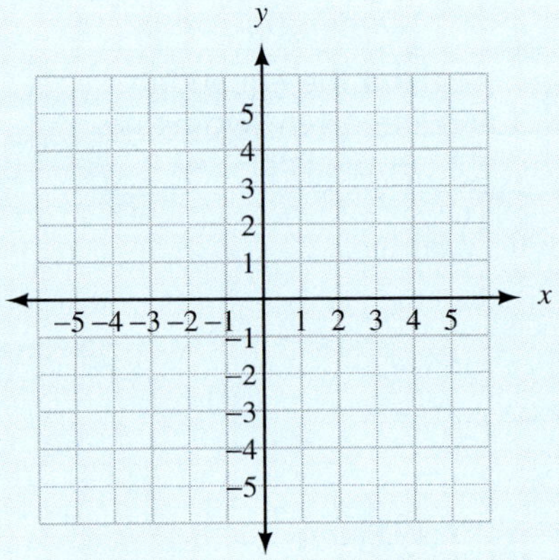

2. Give the slope and y-intercept for the line $4x - 5y = 7$.

EXAMPLE 1 Find the equation of the line with slope $-\frac{4}{3}$ and y-intercept 5. Then graph the line.

Solution Substituting $m = -\frac{4}{3}$ and $b = 5$ into the equation $y = mx + b$, we have

$$y = -\frac{4}{3}x + 5$$

Finding the equation from the slope and y-intercept is just that easy. If the slope is m and the y-intercept is b, then the equation is always $y = mx + b$. Now, let's graph the line.

Because the y-intercept is 5, the graph goes through the point (0, 5). To find a second point on the graph, we start at (0, 5) and move 4 units down (that's a rise of −4) and 3 units to the right (a run of 3). The point we end up at is (3, 1). Drawing a line that passes through (0, 5) and (3, 1), we have the graph of our equation. (Note that we could also let the rise = 4 and the run = −3 and obtain the same graph.) The graph is shown in Figure 2.

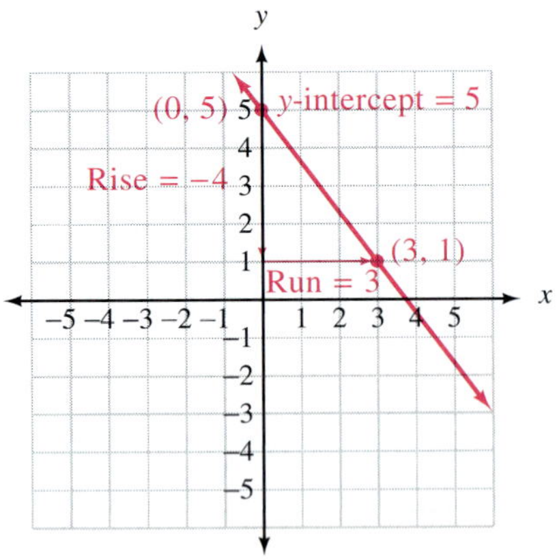

FIGURE 2

EXAMPLE 2 Give the slope and y-intercept for the line $2x - 3y = 5$.

Solution To use the slope-intercept form, we must solve the equation for y in terms of x.

$$2x - 3y = 5$$

$$-3y = -2x + 5 \quad \textbf{Add } -2x \textbf{ to both sides}$$

$$y = \frac{2}{3}x - \frac{5}{3} \quad \textbf{Divide by } -3$$

The last equation has the form $y = mx + b$. The slope must be $m = \frac{2}{3}$, and the y-intercept is $b = -\frac{5}{3}$.

Answers

1. See Solutions to Selected Practice Problems.
2. $m = \frac{4}{5}$, $b = -\frac{7}{5}$

EXAMPLE 3 Graph the equation $2x + 3y = 6$ using the slope and y-intercept.

Solution Although we could graph this equation by finding ordered pairs that are solutions to the equation and drawing a line through their graphs, it is sometimes easier to graph a line using the slope-intercept form of the equation.

Solving the equation for y, we have,

$$2x + 3y = 6$$

$$3y = -2x + 6 \qquad \textbf{Add } -2x \textbf{ to both sides}$$

$$y = -\frac{2}{3}x + 2 \qquad \textbf{Divide by 3}$$

The slope is $m = -\frac{2}{3}$ and the y-intercept is $b = 2$. Therefore, the point (0, 2) is on the graph and the ratio of rise to run going from (0, 2) to any other point on the line is $-\frac{2}{3}$. If we start at (0, 2) and move 2 units up (that's a rise of 2) and 3 units to the left (a run of -3), we will be at another point on the graph. (We could also go down 2 units and right 3 units and still be assured of ending up at another point on the line because $\frac{2}{-3}$ is the same as $\frac{-2}{3}$.)

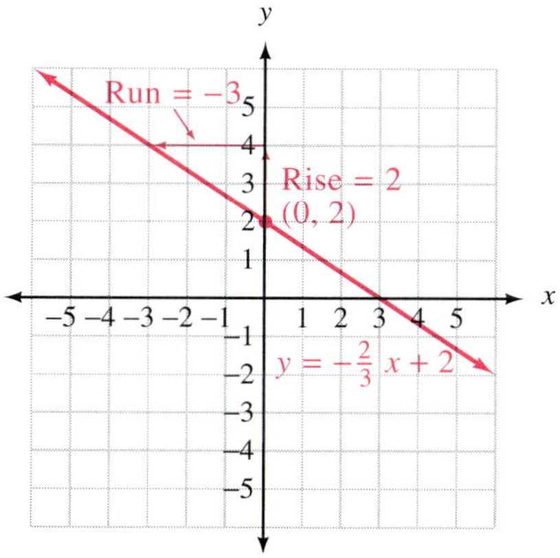

FIGURE 3

A second useful form of the equation of a straight line is the point-slope form.

Let line l contain the point (x_1, y_1) and have slope m. If (x, y) is any other point on l, then by the definition of slope we have

$$\frac{y - y_1}{x - x_1} = m$$

Multiplying both sides by $(x - x_1)$ gives us

$$\mathbf{(x - x_1)} \cdot \frac{y - y_1}{x - x_1} = m\mathbf{(x - x_1)}$$

$$y - y_1 = m(x - x_1)$$

This last equation is known as the *point-slope form* of the equation of a straight line.

Point-Slope Form of the Equation of a Line

The equation of the line through (x_1, y_1) with slope m is given by

$$y - y_1 = m(x - x_1)$$

This form of the equation of a straight line is used to find the equation of a line, either given one point on the line and the slope, or given two points on the line.

3. Graph the line $-3x + 2y = -6$ on the following coordinate system. Use the method shown in Example 3.

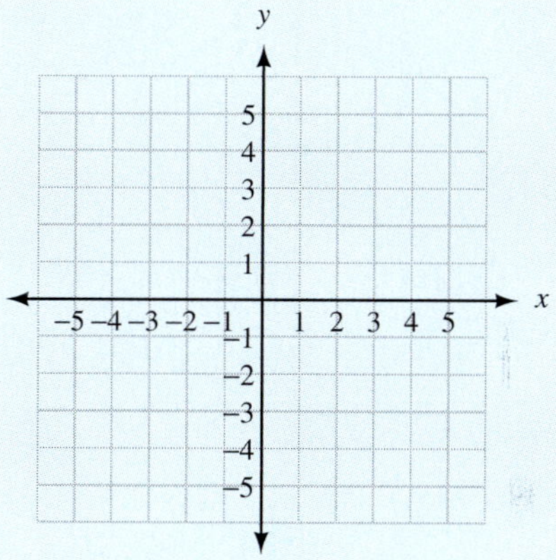

NOTE

As we mentioned earlier, the rectangular coordinate system is the tool we use to connect algebra and geometry. Example 3 illustrates this connection, as do the many other examples in this chapter. In Example 3, Descartes' rectangular coordinate system allows us to associate the equation $2x + 3y = 6$ (an algebraic concept) with the straight line (a geometric concept) shown in Figure 3.

Answer

3. See Solutions to Selected Practice Problems.

4. Find the equation of the line with slope 3 that contains the point $(-1, 2)$.

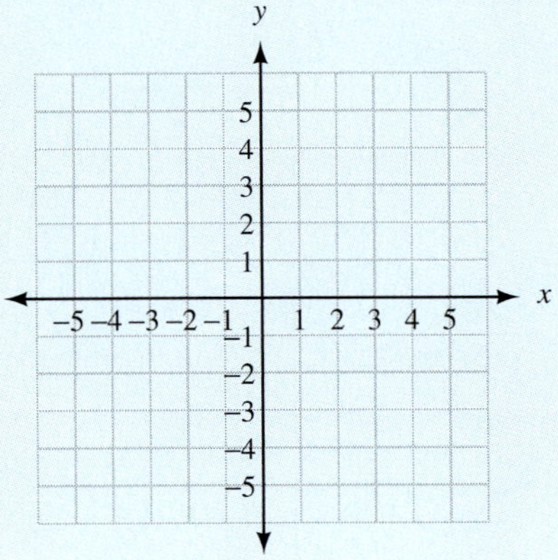

5. Find the equation of the line through $(2, 5)$, and $(6, -3)$.

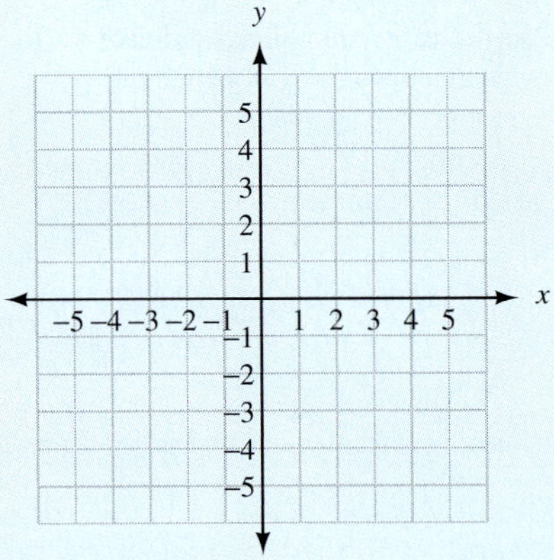

EXAMPLE 4 Find the equation of the line with slope -2 that contains the point $(-4, 3)$. Write the answer in slope-intercept form.

Solution

Using	$(x_1, y_1) = (-4, 3)$	and	$m = -2$
in	$y - y_1 = m(x - x_1)$		**Point-slope form**
gives us	$y - 3 = -2(x + 4)$		***Note:*** **$x - (-4) = x + 4$**
	$y - 3 = -2x - 8$		**Multiply out right side**
	$y = -2x - 5$		**Add 3 to each side**

Figure 4 is the graph of the line that contains $(-4, 3)$ and has a slope of -2. Notice that the y-intercept on the graph matches that of the equation we found.

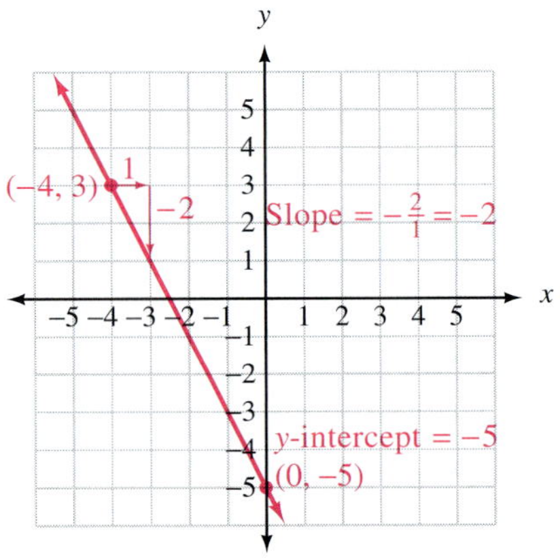

FIGURE 4

EXAMPLE 5 Find the equation of the line that passes through the points $(-3, 3)$ and $(3, -1)$.

Solution We begin by finding the slope of the line:

$$m = \frac{3 - (-1)}{-3 - 3} = \frac{4}{-6} = -\frac{2}{3}$$

Using $(x_1, y_1) = (3, -1)$ and $m = -\frac{2}{3}$ in $y - y_1 = m(x - x_1)$ yields

$$y + 1 = -\frac{2}{3}(x - 3)$$

$$y + 1 = -\frac{2}{3}x + 2 \quad \textbf{Multiply out right side}$$

$$y = -\frac{2}{3}x + 1 \quad \textbf{Add } -1 \textbf{ to each side}$$

Figure 5 shows the graph of the line that passes through the points $(-3, 3)$ and $(3, -1)$. As you can see, the slope and y-intercept are $-\frac{2}{3}$ and 1, respectively.

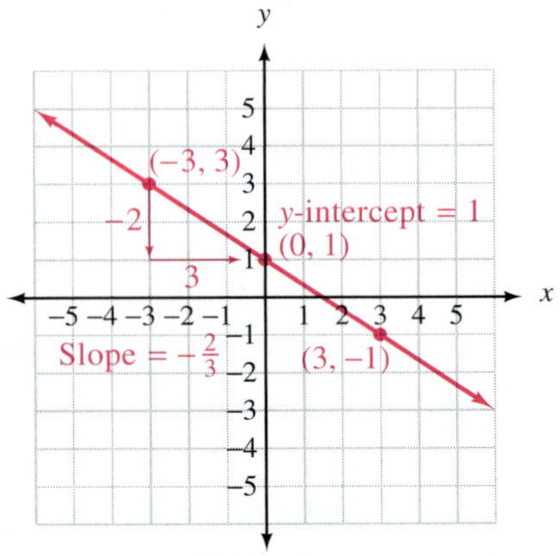

FIGURE 5

NOTE

In Example 5 we could have used the point $(-3, 3)$ instead of $(3, -1)$ and obtained the same equation; that is, using $(x_1, y_1) = (-3, 3)$ and $m = -\frac{2}{3}$ in $y - y_1 = m(x - x_1)$ gives us

$$y - 3 = -\frac{2}{3}(x + 3)$$

$$y - 3 = -\frac{2}{3}x - 2$$

$$y = -\frac{2}{3}x + 1$$

which is the same result we obtained using $(3, -1)$.

Answers

4. $y = 3x + 5$ 5. $y = -2x + 9$

The last form of the equation of a line that we will consider in this section is called the standard form. It is used mainly to write equations in a form that is free of fractions and is easy to compare with other equations.

Standard Form for the Equation of a Line

If a, b, and c are integers, then the equation of a line is in standard form when it has the form

$$ax + by = c$$

If we were to write the equation

$$y = -\frac{2}{3}x + 1$$

in standard form, we would first multiply both sides by 3 to obtain

$$3y = -2x + 3$$

Then we would add $2x$ to each side, yielding

$$2x + 3y = 3$$

which is a linear equation in standard form.

EXAMPLE 6 Give the equation of the line through $(-1, 4)$ whose graph is perpendicular to the graph of $2x - y = -3$. Write the answer in standard form.

Solution To find the slope of $2x - y = -3$, we solve for y:

$$2x - y = -3$$
$$y = 2x + 3$$

The slope of this line is 2. The line we are interested in is perpendicular to the line with slope 2 and must, therefore, have a slope of $-\frac{1}{2}$.

Using $(x_1, y_1) = (-1, 4)$ and $m = -\frac{1}{2}$, we have

$$y - y_1 = m(x - x_1)$$
$$y - 4 = -\frac{1}{2}(x + 1)$$

Because we want our answer in standard form, we multiply each side by 2.

$$2y - 8 = -1(x + 1)$$
$$2y - 8 = -x - 1$$
$$x + 2y - 8 = -1$$
$$x + 2y = 7$$

The last equation is in standard form. ■

6. Find the equation of the line through (3, 2) that is perpendicular to the graph of $3x - y = 2$. Write your answer in standard form.

As a final note, the summary reminds us that all horizontal lines have equations of the form $y = b$ and slopes of 0. Because they cross the y-axis at b, the y-intercept is b; there is no x-intercept. Vertical lines have no slope and equations of the form $x = a$. Each will have an x-intercept at a and no y-intercept. Finally, equations of the form $y = mx$ have graphs that pass through the origin. The slope is always m and both the x-intercept and the y-intercept are 0.

Answer

6. $x + 3y = 9$

Facts From Geometry: Special Equations and Their Graphs, Slopes, and Intercepts

For the equations below, m, a, and b are real numbers.

Through the Origin

Equation: $y = mx$
Slope $= m$
x-intercept $= 0$
y-intercept $= 0$

Vertical Line

Equation: $x = a$
No slope
x-intercept $= a$
No y-intercept

Horizontal Line

Equation: $y = b$
Slope $= 0$
No x-intercept
y-intercept $= b$

Through the Origin

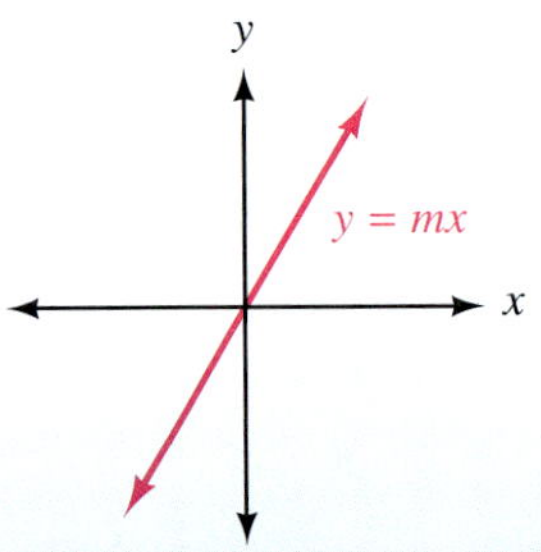

FIGURE 6A

Vertical Line

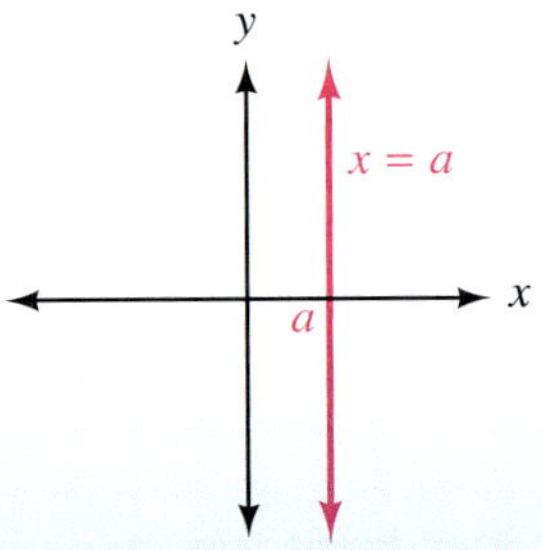

FIGURE 6B

Horizontal Line

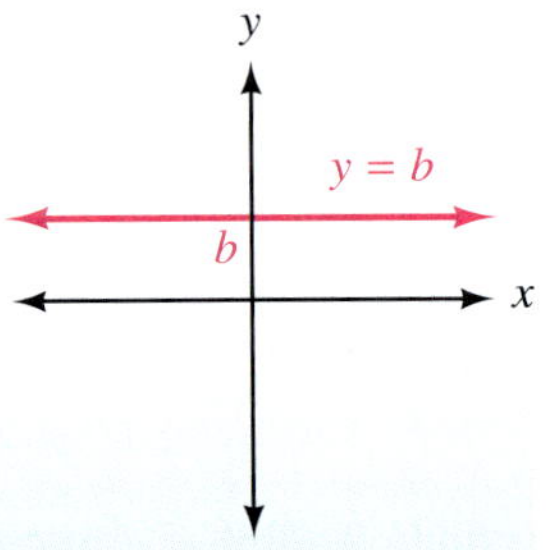

FIGURE 6C

USING TECHNOLOGY

Graphing Calculators

One advantage of using a graphing calculator to graph lines is that a calculator does not care whether the equation has been simplified or not. To illustrate, in Example 5 we found that the equation of the line with slope $-\frac{2}{3}$ that passes through the point $(3, -1)$ is

$$y + 1 = -\frac{2}{3}(x - 3)$$

Normally, to graph this equation we would simplify it first. With a graphing calculator, we add -1 to each side and enter the equation this way:

$$Y_1 = -(2/3)(X - 3) - 1$$

Linking Objectives and Examples

Next to each **objective**, list the examples that are best described by that objective.

A	
B	
C	
D	

Getting Ready for Class

After reading through the preceding section, respond in your own words and in complete sentences.

1. How would you graph the line $y = \frac{1}{2}x + 3$?
2. What is the slope-intercept form of the equation of a line?
3. Describe how you would find the equation of a line if you knew the slope and the y-intercept of the line.
4. If you had the graph of a line, how would you use it to find the equation of the line?

Problem Set 3.3

INTERMEDIATE Algebra $f(x)$ Now™

Write the equation of the line with the given slope and y-intercept.

1. $m = 2, b = 3$

2. $m = -4, b = 2$

3. $m = 1, b = -5$

4. $m = -5, b = -3$

5. $m = \frac{1}{2}, b = \frac{3}{2}$

6. $m = \frac{2}{3}, b = \frac{5}{6}$

7. $m = 0, b = 4$

8. $m = 0, b = -2$

Give the slope and y-intercept for each of the following equations. Sketch the graph using the slope and y-intercept. Give the slope of any line perpendicular to the given line.

9. $y = 3x - 2$

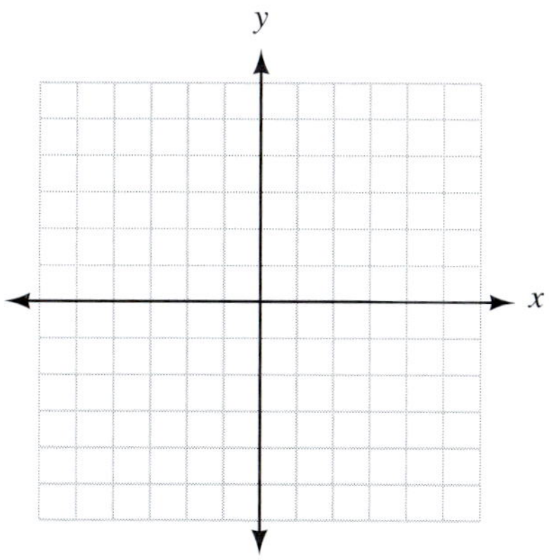

10. $y = 2x + 3$

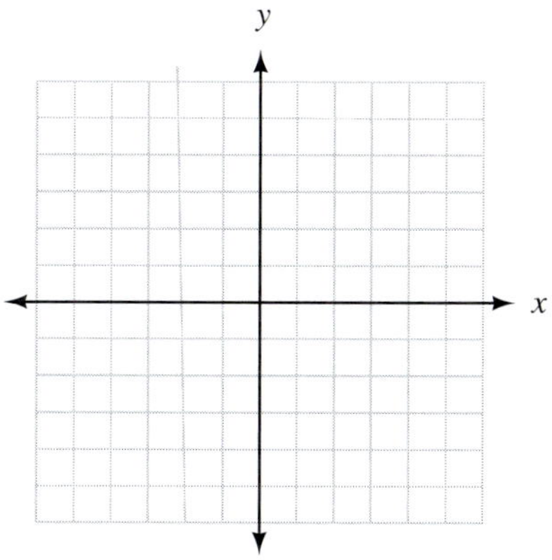

11. $2x - 3y = 12$

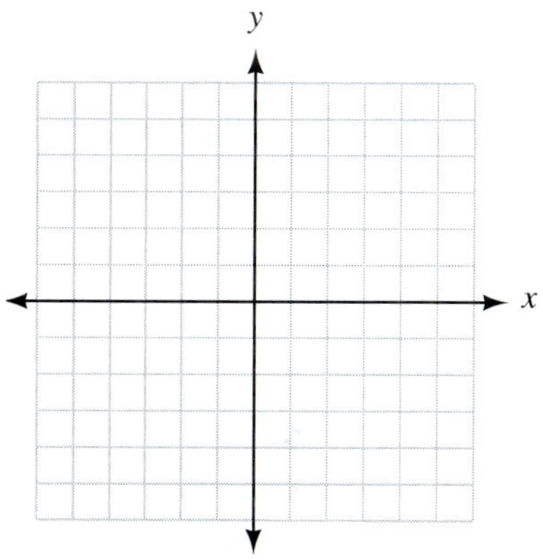

12. $3x - 2y = 12$

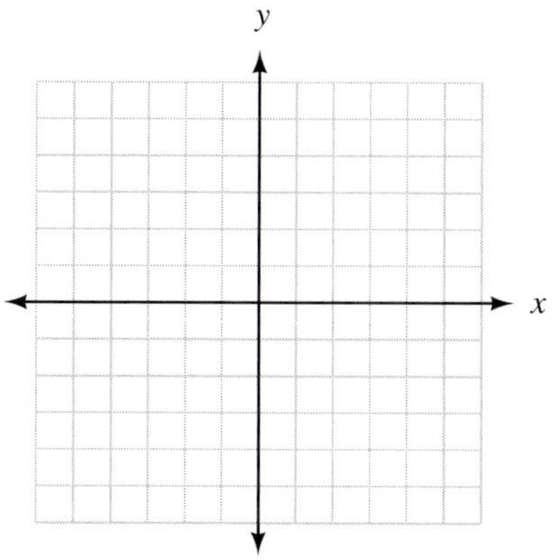

13. $4x + 5y = 20$

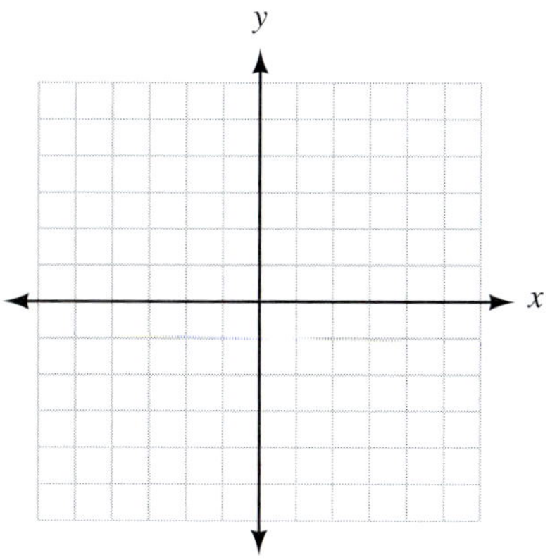

14. $5x - 4y = 20$

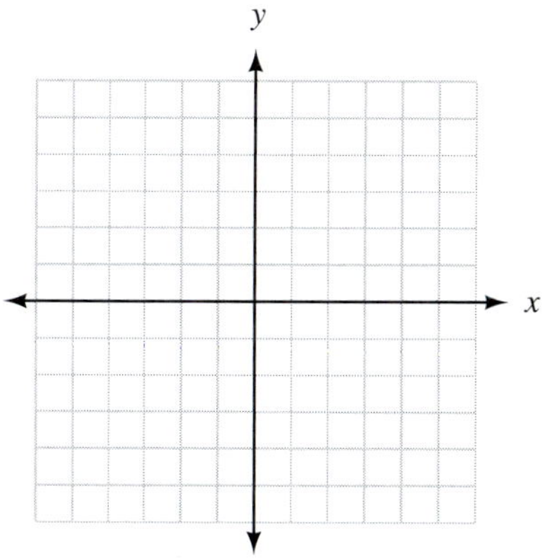

For each of the following lines, name the slope and y-intercept. Then write the equation of the line in slope-intercept form.

15.

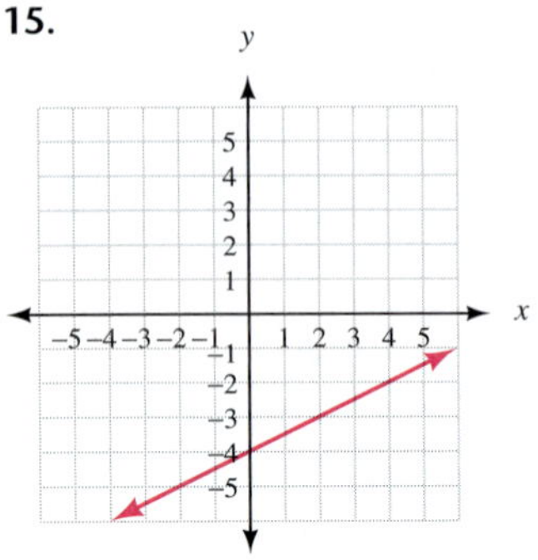

16.

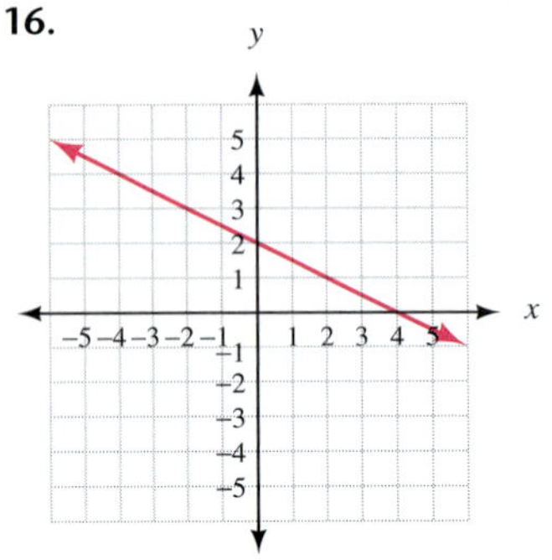

17.

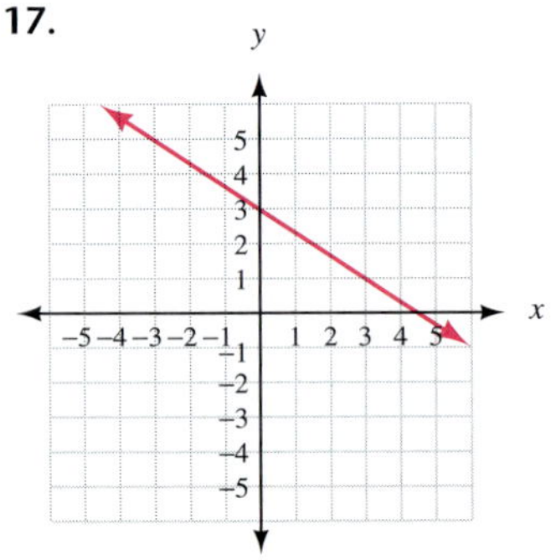

18.

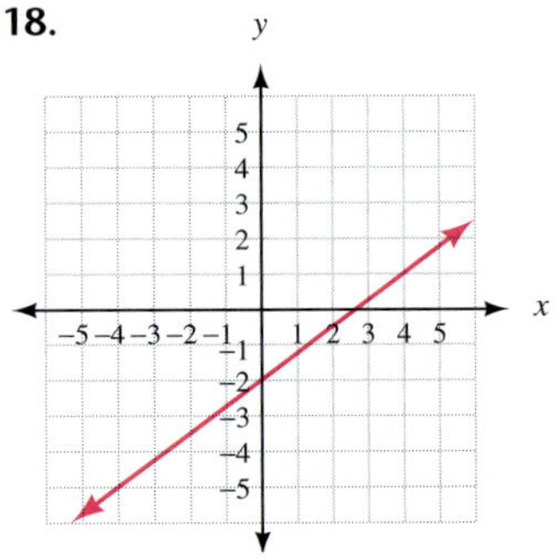

For each of the following problems, the slope and one point on a line are given. In each case, find the equation of that line. (Write the equation for each line in slope-intercept form.)

19. $(-2, -5)$; $m = 2$

20. $(-1, -5)$; $m = 2$

21. $(-4, 1)$; $m = -\frac{1}{2}$

22. $(-2, 1)$; $m = -\frac{1}{2}$

23. $\left(-\frac{1}{3}, 2\right)$; $m = -3$

24. $\left(-\frac{2}{3}, 5\right)$; $m = -3$

Find the equation of the line that passes through each pair of points. Write your answers in standard form.

25. $(-2, -4)$, $(1, -1)$

26. $(2, 4)$, $(-3, -1)$

27. $(-1, -5)$, $(2, 1)$

28. $(-1, 6)$, $(1, 2)$

29. $\left(\frac{1}{3}, -\frac{1}{5}\right)$, $\left(-\frac{1}{3}, -1\right)$

30. $\left(-\frac{1}{2}, -\frac{1}{2}\right)$, $\left(\frac{1}{2}, \frac{1}{10}\right)$

EXAMPLE 6 Graph $y = |x|$. Use the graph to determine whether we have the graph of a function. State the domain and range.

Solution We let x take on values of -4, -3, -2, -1, 0, 1, 2, 3, and 4. The corresponding values of y are shown in the table. The graph is shown in Figure 6.

x	y
−4	4
−3	3
−2	2
−1	1
0	0
1	1
2	2
3	3
4	4

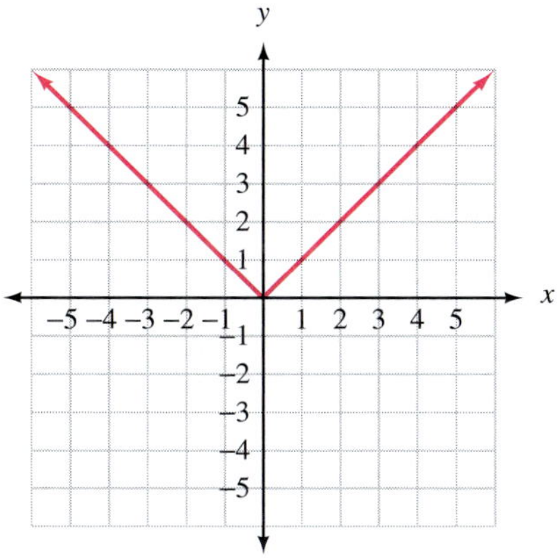

FIGURE 6

Because no vertical line can be found that crosses the graph in more than one place, $y = |x|$ is a function. The domain is all real numbers. The range is $\{y \mid y \geq 0\}$.

Getting Ready for Class

After reading through the preceding section, respond in your own words and in complete sentences.

1. What is a function?
2. What is the vertical line test?
3. Is every line the graph of a function? Explain.
4. Which variable is usually associated with the domain of a function?

6. Graph $y = |x| - 3$.

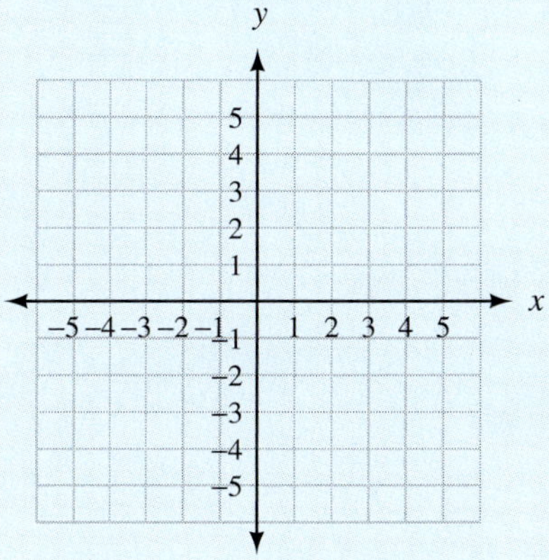

Linking Objectives and Examples

Next to each **objective**, list the examples that are best described by that objective.

A	
B	
C	

Answer

6. See Solutions to Selected Practice Problems.

Problem Set 3.5

INTERMEDIATE Algebra $f(x)$ Now™

For each of the following relations, give the domain and range, and indicate which are also functions.

1. {(1, 3), (2, 5), (4, 1)}

2. {(3, 1), (5, 7), (2, 3)}

3. {(−1, 3), (1, 3), (2, −5)}

4. {(3, −4), (−1, 5), (3, 2)}

5. {(7, −1), (3, −1), (7, 4)}

6. {(5, −2), (3, −2), (5, −1)}

State whether each of the following graphs represents a function.

7.

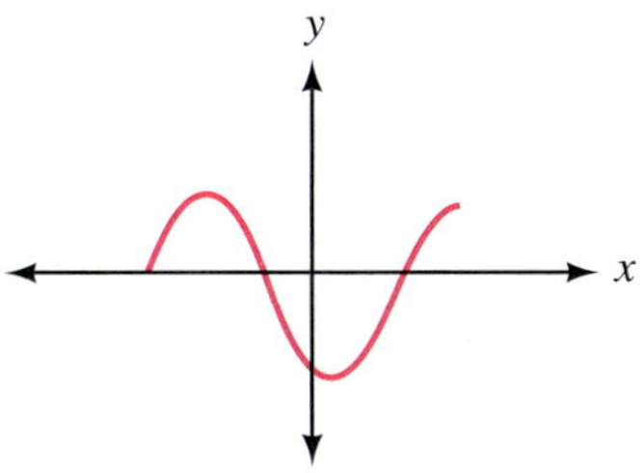

8.

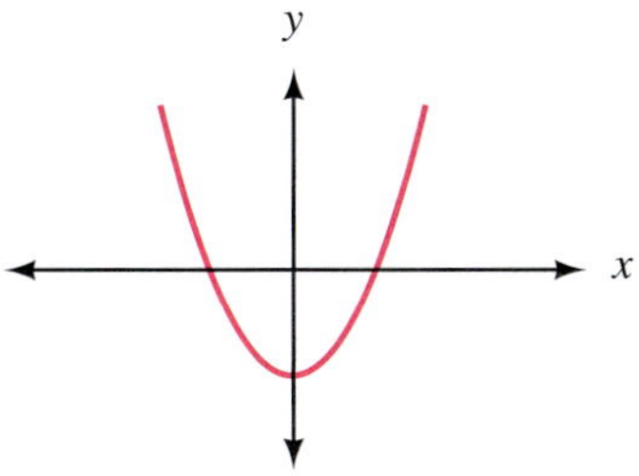

9.

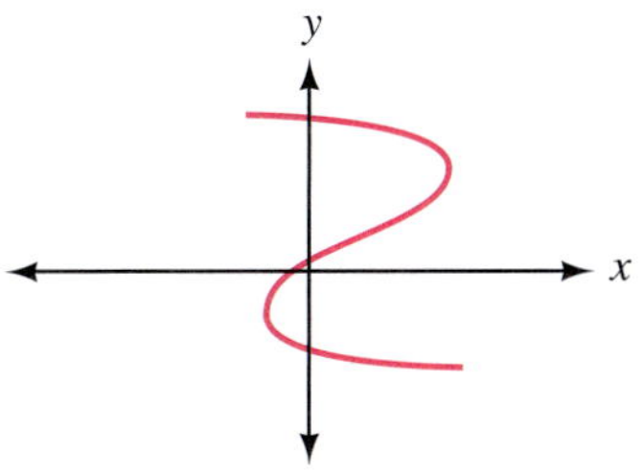

10.

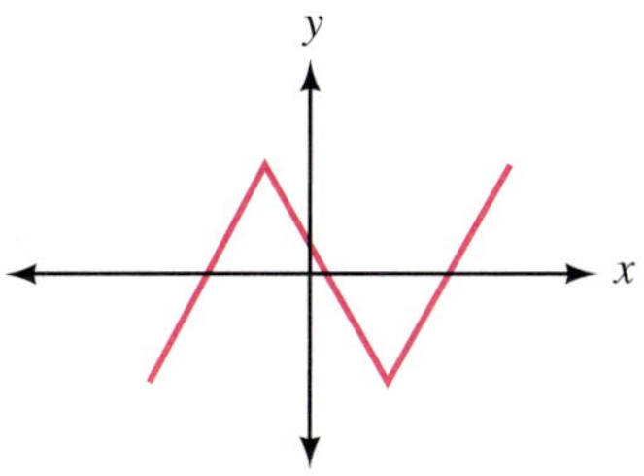

11.

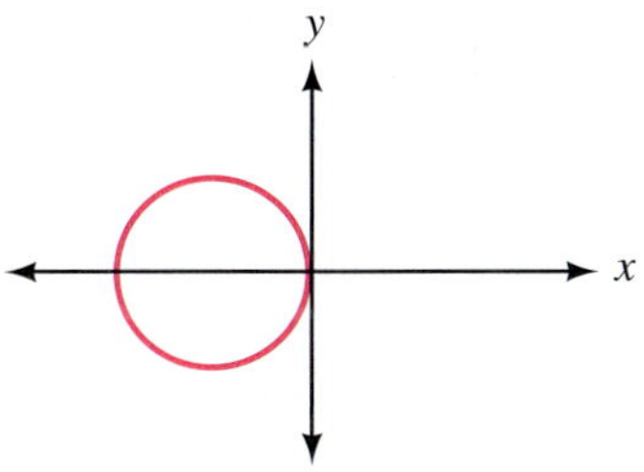

12.

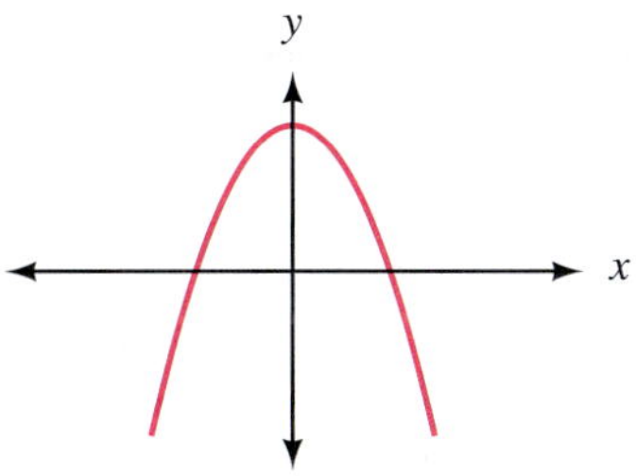

13.

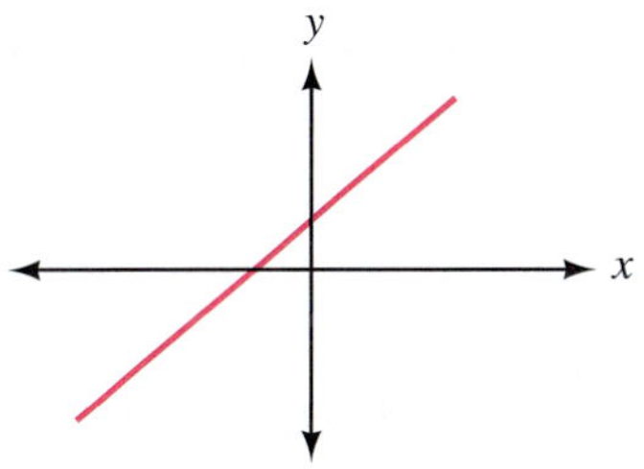

14.

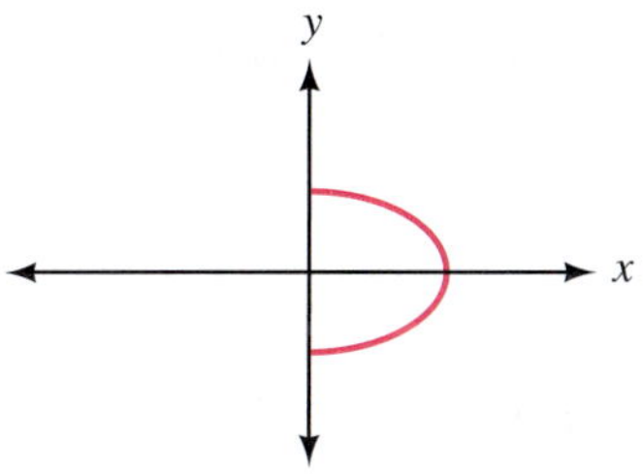

15.

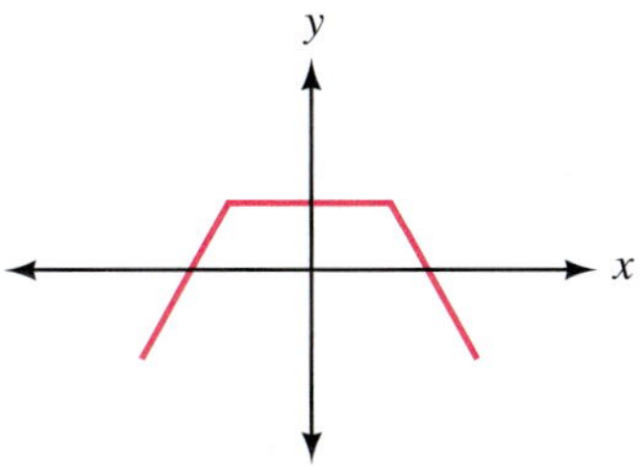

16.

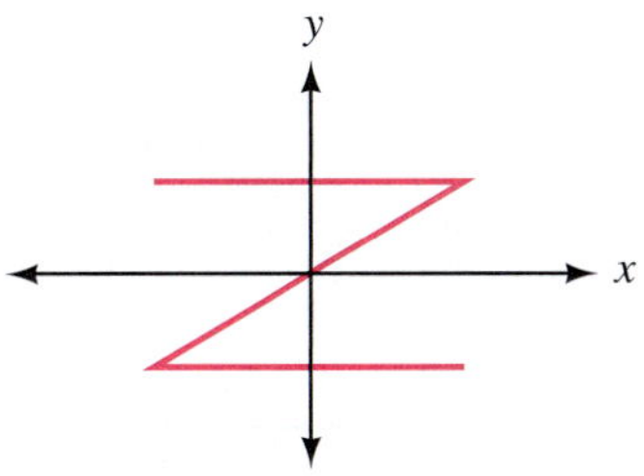

Graph each of the following relations. In each case, use the graph to find the domain and range, and indicate whether the graph is the graph of a function.

17. $y = x^2 - 1$

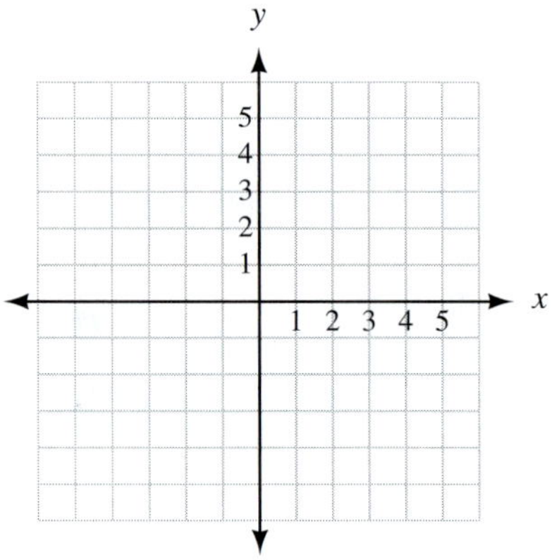

18. $y = x^2 + 1$

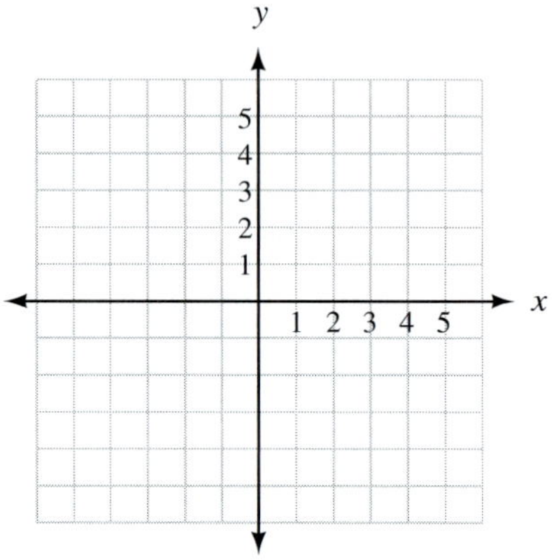

19. $y = x^2 + 4$

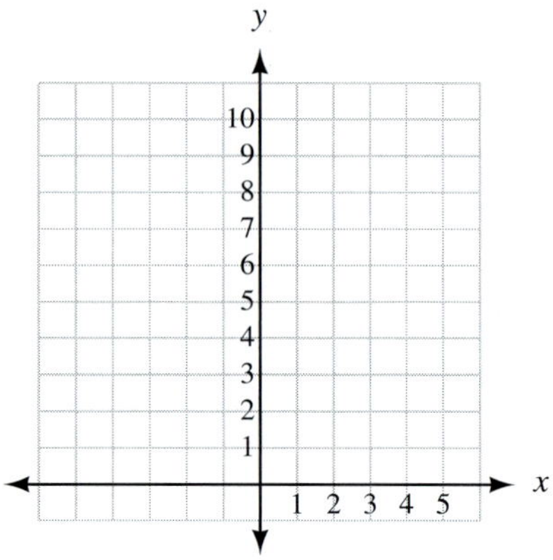

20. $y = x^2 - 9$

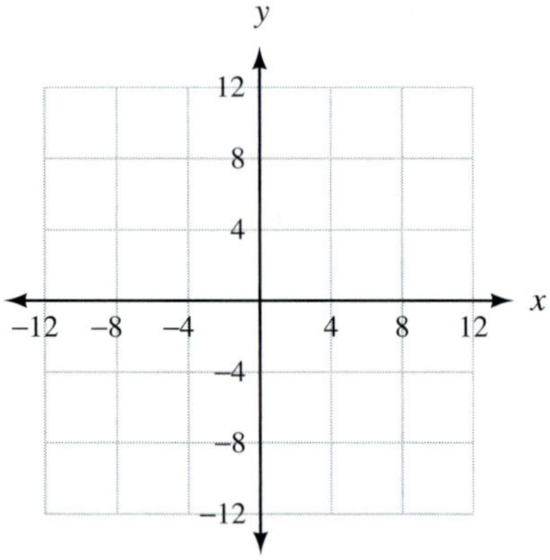

21. $x = y^2 - 1$

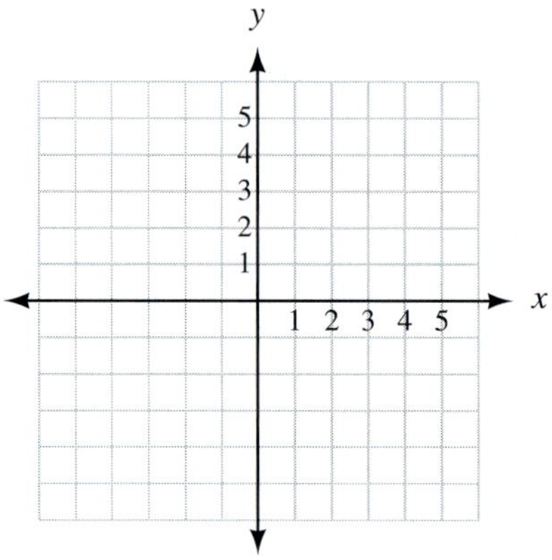

22. $x = y^2 + 1$

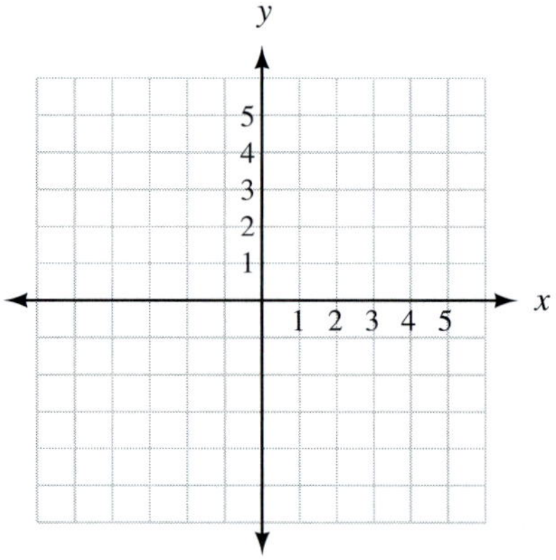

23. $x = y^2 + 4$

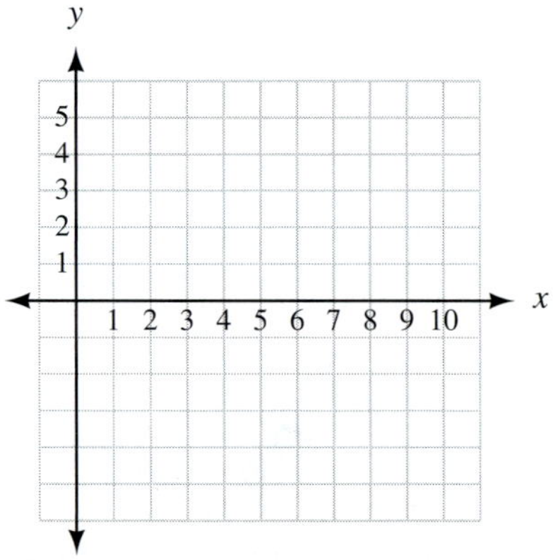

24. $x = y^2 - 9$

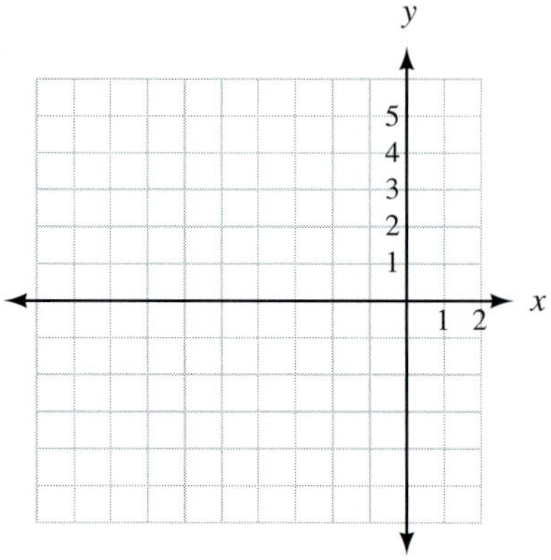

25. $y = |x - 2|$

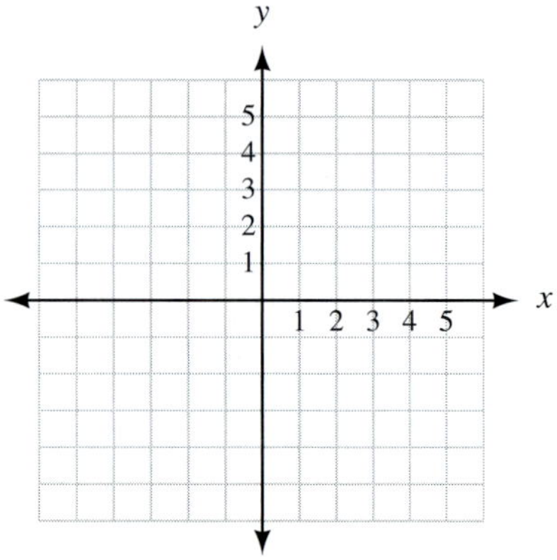

26. $y = |x + 2|$

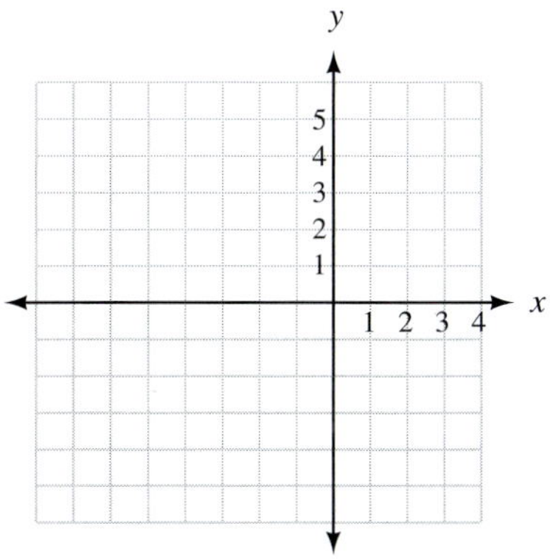

27. $y = |x| - 2$

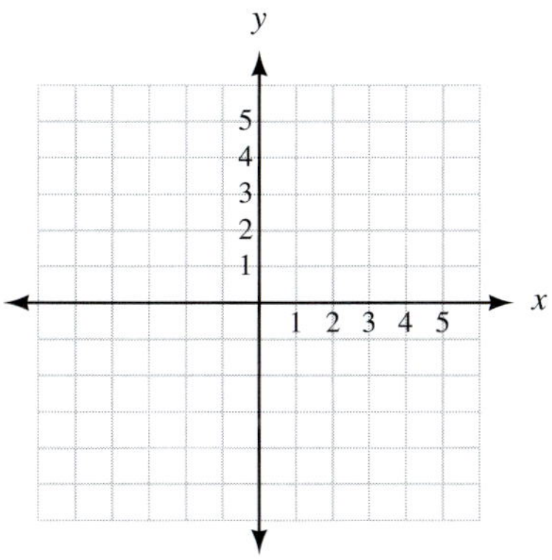

28. $y = |x| + 2$

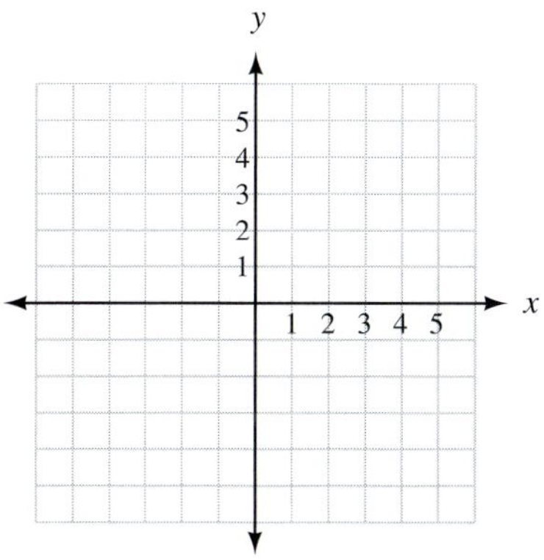

Applying the Concepts

29. Suppose you have a job that pays $8.50 per hour and you work anywhere from 10 to 40 hours per week.

a. Write an equation, with a restriction on the variable x, that gives the amount of money, y, you will earn for working x hours in one week.

b. Use the function rule to complete the table.

TABLE 4

	WEEKLY WAGES	
HOURS WORKED	FUNCTION RULE	GROSS PAY ($)
x		y
10		
20		
30		
40		

c. Use the data from the table to graph the function.

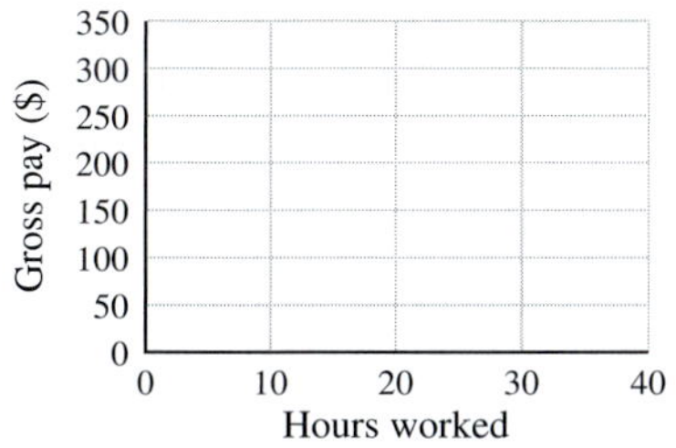

FIGURE 7 **Template for line graph.**

d. State the domain and range of this function.

e. What is the minimum amount you can earn in a week with this job? What is the maximum amount?

30. The ad shown here was in the local newspaper. Suppose you are hired for the job described in the ad.

312 Help Wanted

ESPRESSO BAR OPERATOR

Must be dependable, honest, service-oriented. Coffee exp desired. 15–30 hrs per wk. $5.25/hr. Start 5/31. Apply in person: Espresso Yourself, Central Coast Mall. Deadline 5/23.

a. If x is the number of hours you work per week and y is your weekly gross pay, write the equation for y. (Be sure to include any restrictions on the variable x that are given in the ad.)

b. Use the function rule to complete the table.

TABLE 5

	WEEKLY WAGES	
HOURS WORKED	FUNCTION RULE	GROSS PAY ($)
x		y
15		
20		
25		
30		

c. Use the data from the table to graph the function.

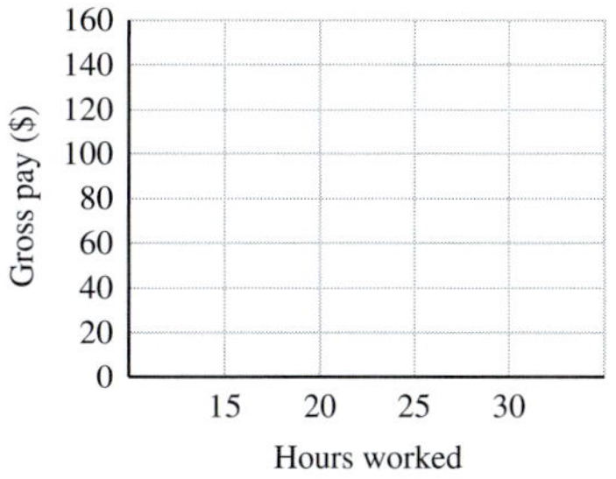

FIGURE 8 **Template for line graph.**

d. State the domain and range of this function.

e. What is the minimum amount you can earn in a week with this job? What is the maximum amount?

31. Tossing a Coin Hali is tossing a quarter into the air with an underhand motion. The distance the quarter is above her hand at any time is given by the function

$$h = 16t - 16t^2 \quad \text{for} \quad 0 \le t \le 1$$

where h is the height of the quarter in feet, and t is the time in seconds.

a. Fill in the table.

TIME (SEC) t	FUNCTION RULE $h = 16t - 16t^2$	DISTANCE (FT) h
0		
0.1		
0.2		
0.3		
0.4		
0.5		
0.6		
0.7		
0.8		
0.9		
1		

b. State the domain and range of this function.

c. Use the data from the table to graph the function.

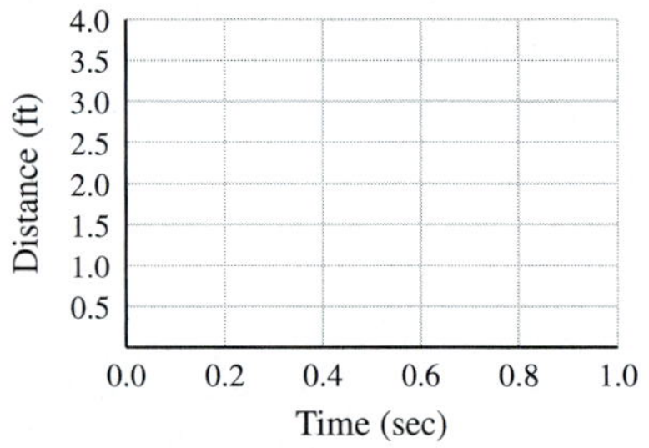

32. Intensity of Light The formula below gives the intensity of light that falls on a surface at various distances from a 100-watt light bulb:

$$I = \frac{120}{d^2} \quad \text{for} \quad d > 0$$

where I is the intensity of light (in lumens per square foot), and d is the distance (in feet) from the light bulb to the surface.

a. Fill in the table.

DISTANCE (FT) D	FUNCTION RULE $I = \frac{120}{d^2}$	INTENSITY I
1		
2		
3		
4		
5		
6		

b. Use the data from the table to graph the function.

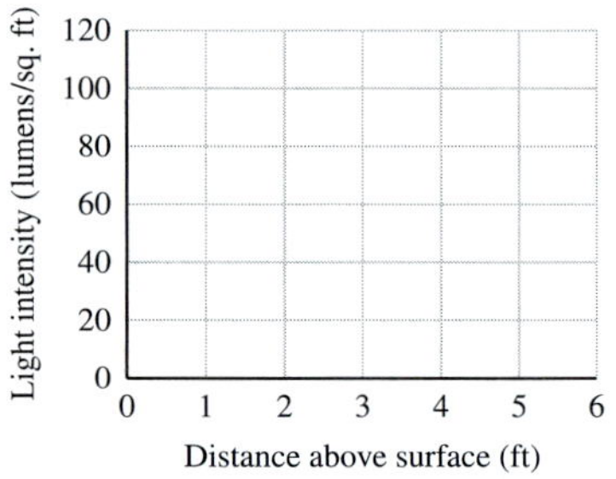

33. Area of a Circle The formula for the area A of a circle with radius r is given by $A = \pi r^2$. The formula shows that A is a function of r.

a. Graph the function $A = \pi r^2$ for $0 \le r \le 3$. (On the graph, let the horizontal axis be the r-axis, and let the vertical axis be the A-axis.)

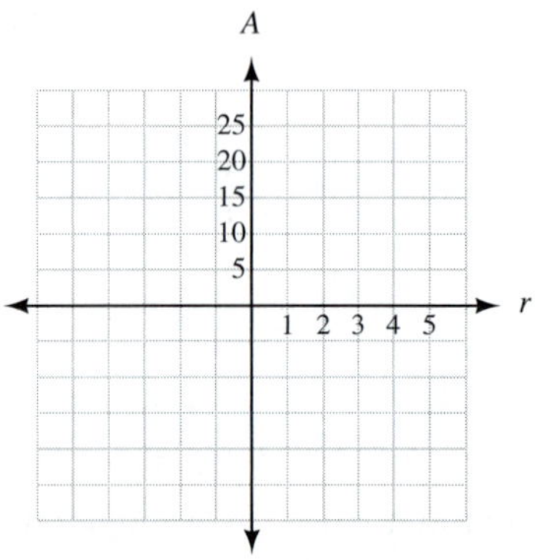

b. State the domain and range of the function $A = \pi r^2$, $0 \le r \le 3$.

34. Area and Perimeter of a Rectangle A rectangle is 2 inches longer than it is wide. Let x = the width, P = the perimeter, and A = the area of the rectangle.

a. Write an equation that will give the perimeter P in terms of the width x of the rectangle. Are there any restrictions on the values that x can assume?

b. Graph the relationship between P and x.

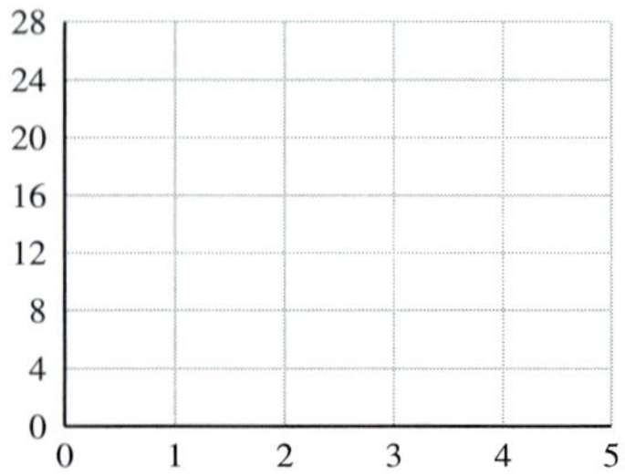

35. **Tossing a Ball** A ball is thrown straight up into the air from ground level. The relationship between the height h of the ball at any time t is illustrated by the following graph:

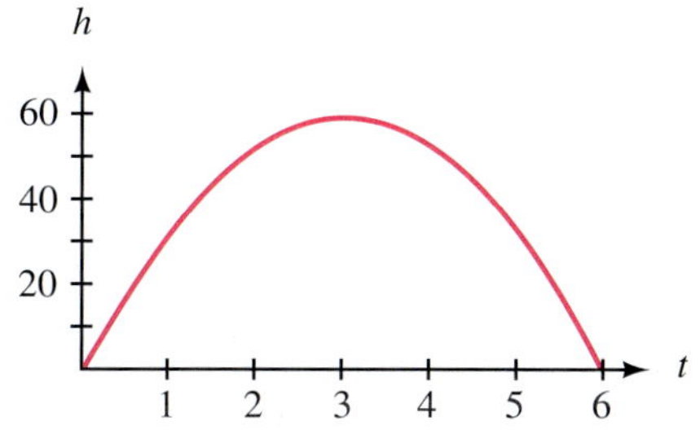

The horizontal axis represents time t, and the vertical axis represents height h.

a. Is this graph the graph of a function?

b. State the domain and range.

c. At what time does the ball reach its maximum height?

d. What is the maximum height of the ball?

e. At what time does the ball hit the ground?

36. **Company Profits** The amount of profit a company earns is based on the number of items it sells. The relationship between the profit P and number of items it sells x, is illustrated by the following graph:

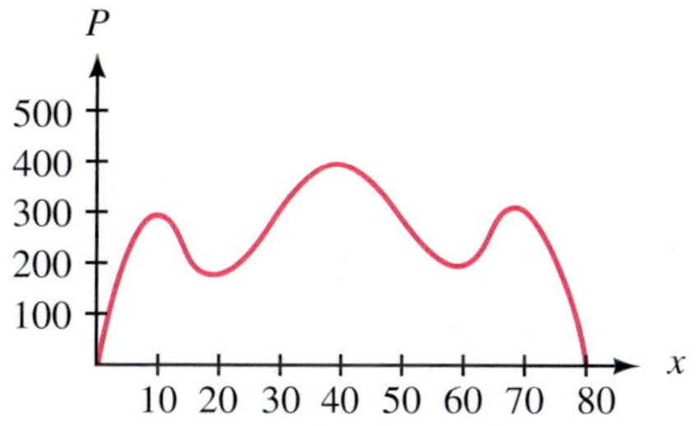

The horizontal axis represents items sold, x, and the vertical axis represents the profit, P.

a. Is this graph the graph of a function?

b. State the domain and range.

c. How many items must the company sell to make their maximum profit?

d. What is their maximum profit?

Getting Ready for the Next Section

Simplify. Round to the nearest whole number if necessary.

37. $7.5(20)$

38. $60 \div 7.5$

39. $4(3.14)(9)$

40. $\frac{4}{3}(3.14) \cdot 3^3$

41. $4(-2) - 1$

42. $3(3)^2 + 2(3) - 1$

43. If $s = \frac{60}{t}$, find s when

a. $t = 10$ **b.** $t = 8$

44. If $y = 3x^2 + 2x - 1$, find y when

a. $x = 0$ **b.** $x = -2$

45. Find the value of $x^2 + 2$ for

a. $x = 5$ **b.** $x = -2$

46. Find the value of $125 \cdot 2^t$ for

a. $t = 0$ **b.** $t = 1$

Maintaining Your Skills

The problems that follow review some of the more important skills you have learned in previous sections and chapters.

For the equation $y = 3x - 2$:

47. Find y if x is 4.

48. Find y if x is 0.

49. Find y if x is -4.

50. Find y if x is -2.

For the equation $y = x^2 - 3$:

51. Find y if x is 2.

52. Find y if x is -2.

53. Find y if x is 0.

54. Find y if x is -4.

55. If $x - 2y = 4$ and $x = \frac{8}{5}$, find y.

56. If $5x - 10y = 15$, find y when x is 3.

57. Let $x = 0$ and $y = 0$ in $y = a(x - 80)^2 + 70$ and solve for a.

58. Find R if $p = 2.5$ and $R = (900 - 300p)p$.

Extending the Concepts

59. Profits Match each of the following statements to the appropriate graph indicated by Figures 9–12.

a. Sarah works 25 hours in a week to earn \$250.

b. Justin works 35 hours in a week to earn \$560.

c. Rosemary works 30 hours in a week to earn \$360.

d. Marcus works 40 hours in a week to earn \$320.

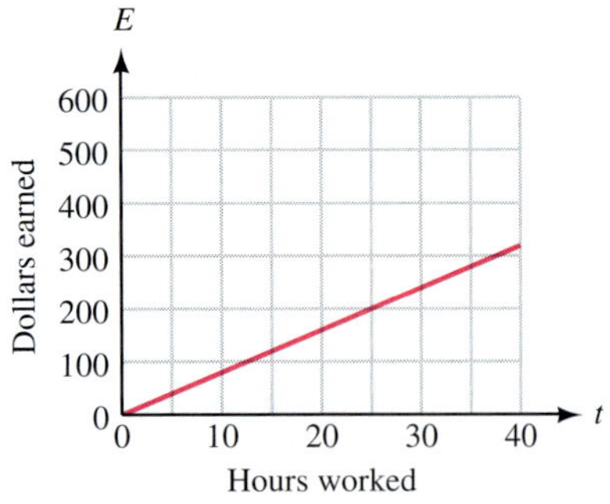

FIGURE 9

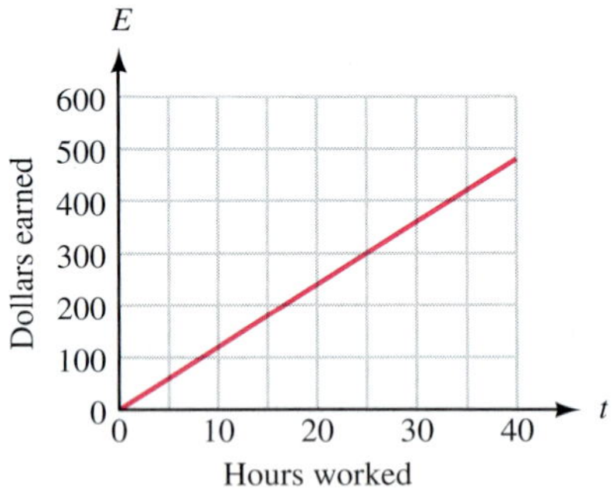

FIGURE 10

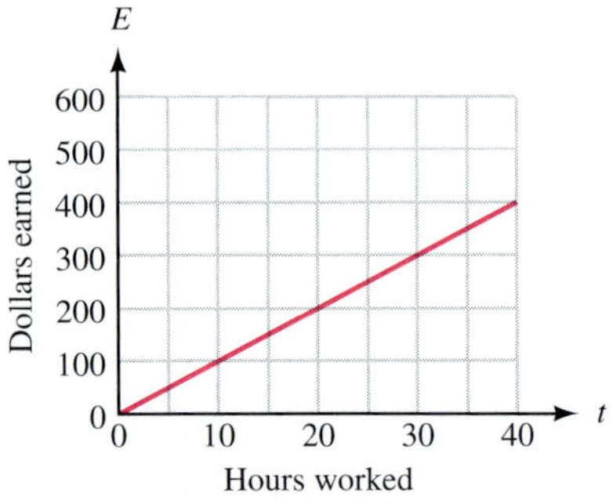

FIGURE 11

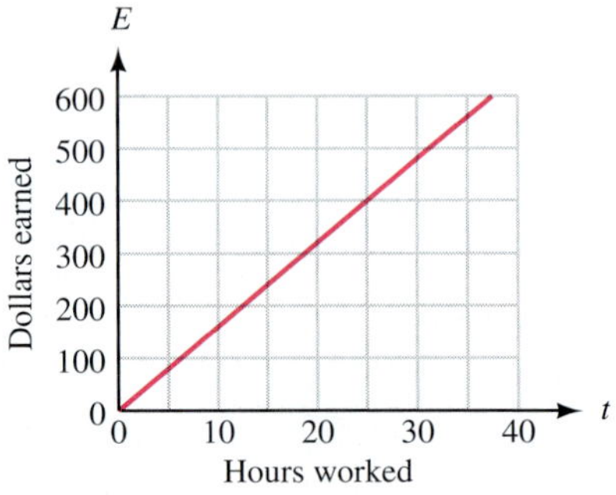

FIGURE 12

3.6 Function Notation

Let's return to the discussion that introduced us to functions. If a job pays \$7.50 per hour for working from 0 to 40 hours a week, then the amount of money y earned in 1 week is a function of the number of hours worked, x. The exact relationship between x and y is written

$$y = 7.5x \quad \text{for} \quad 0 \le x \le 40$$

Because the amount of money earned y depends on the number of hours worked x, we call y the *dependent variable* and x the *independent variable.* Furthermore, if we let f represent all the ordered pairs produced by the equation, then we can write

$$f = \{(x, y) \mid y = 7.5x \text{ and } 0 \le x \le 40\}$$

Once we have named a function with a letter, we can use an alternative notation to represent the dependent variable y. The alternative notation for y is $f(x)$. It is read "f of x" and can be used instead of the variable y when working with functions. The notation y and the notation $f(x)$ are equivalent—that is,

$$y = 7.5x \Leftrightarrow f(x) = 7.5x$$

When we use the notation $f(x)$ we are using *function notation.* The benefit of using function notation is that we can write more information with fewer symbols than we can by using just the variable y. For example, asking how much money a person will make for working 20 hours is simply a matter of asking for $f(20)$. Without function notation, we would have to say "find the value of y that corresponds to a value of $x = 20$." To illustrate further, using the variable y, we can say "y is 150 when x is 20." Using the notation $f(x)$, we simply say "$f(20) = 150$." Each expression indicates that you will earn \$150 for working 20 hours.

EXAMPLE 1 If $f(x) = 7.5x$, find $f(0)$, $f(10)$, and $f(20)$.

Solution To find $f(0)$ we substitute 0 for x in the expression $7.5x$ and simplify. We find $f(10)$ and $f(20)$ in a similar manner—by substitution.

$$\begin{aligned} \text{If} \quad f(x) &= 7.5x \\ \text{then} \quad f(\mathbf{0}) &= 7.5(\mathbf{0}) = 0 \\ f(\mathbf{10}) &= 7.5(\mathbf{10}) = 75 \\ f(\mathbf{20}) &= 7.5(\mathbf{20}) = 150 \end{aligned}$$

If we changed the example in the discussion that opened this section so that the hourly wage was \$6.50 per hour, we would have a new equation to work with:

$$y = 6.5x \quad \text{for} \quad 0 \le x \le 40$$

Suppose we name this new function with the letter g. Then

$$g = \{(x, y) \mid y = 6.5x \text{ and } 0 \le x \le 40\}$$

and

$$g(x) = 6.5x$$

If we want to talk about both functions in the same discussion, having two different letters, f and g, makes it easy to distinguish between them. For example,

OBJECTIVES

A	Use function notation to find the value of a function for a given value of the variable.

PRACTICE PROBLEMS

1. If $f(x) = 8x$, find
   a. $f(0)$
   b. $f(5)$
   c. $f(10.5)$

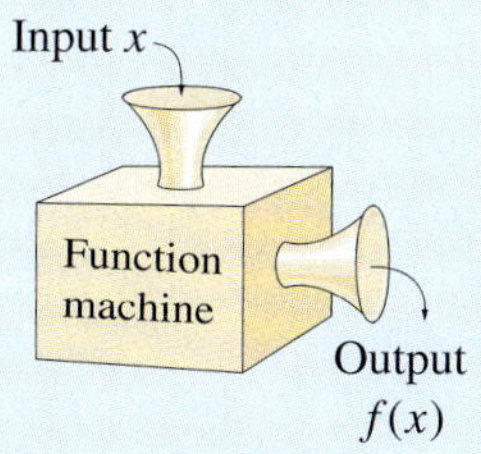

Some students like to think of functions as machines. Values of x are put into the machine, which transforms them into values of $f(x)$, which then are output by the machine.

Answer
1. a. 0 b. 40 c. 84

because $f(x) = 7.5x$ and $g(x) = 6.5x$, asking how much money a person makes for working 20 hours is simply a matter of asking for $f(20)$ or $g(20)$, avoiding any confusion over which hourly wage we are talking about.

The diagrams shown in Figure 1 further illustrate the similarities and differences between the two functions we have been discussing.

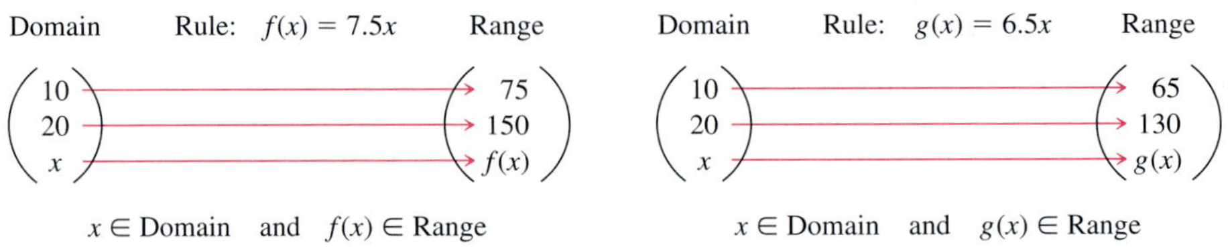

FIGURE 1 **Function maps**

Function Notation and Graphs

We can visualize the relationship between x and $f(x)$ or $g(x)$ on the graphs of the two functions. Figure 2 shows the graph of $f(x) = 7.5x$ along with two additional line segments. The horizontal line segment corresponds to $x = 20$, and the vertical line segment corresponds to $f(20)$. Figure 3 shows the graph of $g(x) = 6.5x$ along with the horizontal line segment that corresponds to $x = 20$, and the vertical line segment that corresponds to $g(20)$. (Note that the domain in each case is restricted to $0 \le x \le 40$.)

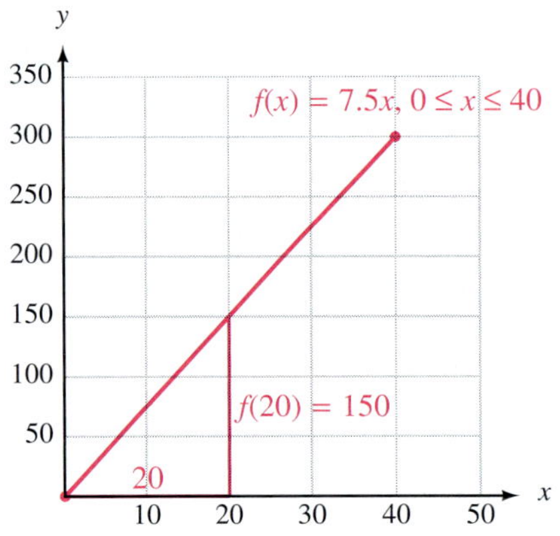

FIGURE 2

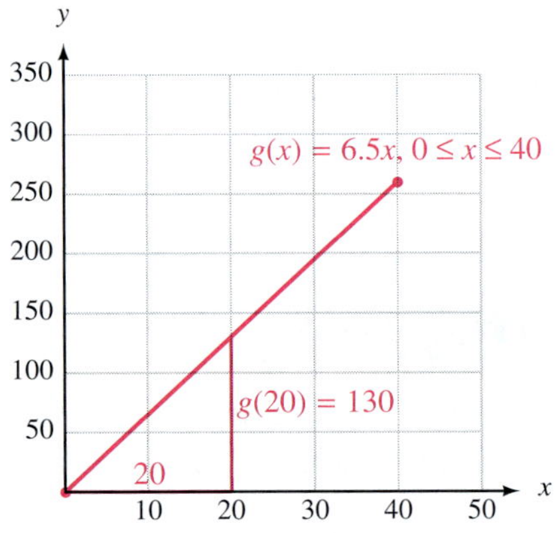

FIGURE 3

Using Function Notation

The remaining examples in this section show a variety of ways to use and interpret function notation.

EXAMPLE 2 If it takes Lorena t minutes to run a mile, then her average speed $s(t)$ in miles per hour is given by the formula

$$s(t) = \frac{60}{t} \quad \text{for} \quad t > 0$$

Find $s(10)$ and $s(8)$, and then explain what they mean.

Solution To find $s(10)$, we substitute 10 for t in the equation and simplify:

$$s(\mathbf{10}) = \frac{60}{\mathbf{10}} = 6$$

2. When Lorena runs a mile in t minutes, then her average speed in feet per seconds is given by $s(t) = \frac{88}{t}$, $t > 0$

a. Find $s(8)$ and explain what it means.

b. Find $s(11)$ and explain what it means.

In words: When Lorena runs a mile in 10 minutes, her average speed is 6 miles per hour.

We calculate $s(8)$ by substituting 8 for t in the equation. Doing so gives us

$$s(\mathbf{8}) = \frac{60}{\mathbf{8}} = 7.5$$

In words: Running a mile in 8 minutes is running at a rate of 7.5 miles per hour.

EXAMPLE 3

A painting is purchased as an investment for \$125. If its value increases continuously so that it doubles every 5 years, then its value is given by the function

$$V(t) = 125 \cdot 2^{t/5} \quad \text{for} \quad t \geq 0$$

where t is the number of years since the painting was purchased, and $V(t)$ is its value (in dollars) at time t. Find $V(5)$ and $V(10)$, and explain what they mean.

Solution The expression $V(5)$ is the value of the painting when $t = 5$ (5 years after it is purchased). We calculate $V(5)$ by substituting 5 for t in the equation $V(t) = 125 \cdot 2^{t/5}$. Here is our work:

$$V(\mathbf{5}) = 125 \cdot 2^{\mathbf{5}/5} = 125 \cdot 2^1 = 125 \cdot 2 = 250$$

In words: After 5 years, the painting is worth \$250.

The expression $V(10)$ is the value of the painting after 10 years. To find this number, we substitute 10 for t in the equation:

$$V(\mathbf{10}) = 125 \cdot 2^{\mathbf{10}/5} = 125 \cdot 2^2 = 125 \cdot 4 = 500$$

In words: The value of the painting 10 years after it is purchased is \$500.

EXAMPLE 4

A balloon has the shape of a sphere with a radius of 3 inches. Use the following formulas to find the volume and surface area of the balloon.

$$V(r) = \frac{4}{3}\pi r^3 \qquad S(r) = 4\pi r^2$$

Solution As you can see, we have used function notation to write the two formulas for volume and surface area because each quantity is a function of the radius. To find these quantities when the radius is 3 inches, we evaluate $V(3)$ and $S(3)$:

$$V(\mathbf{3}) = \frac{4}{3}\pi \mathbf{3}^3 = \frac{4}{3}\pi 27 = 36\pi \text{ cubic inches, or } 113 \text{ cubic inches}$$ **To the nearest whole number**

$$S(\mathbf{3}) = 4\pi \mathbf{3}^2 = 36\pi \text{ square inches, or } 113 \text{ square inches}$$ **To the nearest whole number**

The fact that $V(3) = 36\pi$ means that the ordered pair $(3, 36\pi)$ belongs to the function V. Likewise, the fact that $S(3) = 36\pi$ tells us that the ordered pair $(3, 36\pi)$ is a member of function S.

We can generalize the discussion at the end of Example 4 this way:

$$(a, b) \in f \quad \text{if and only if} \quad f(a) = b$$

3. A medication has a half-life of 5 days. If the concentration of the medication in a patient's system is 80 ng/mL, and the patient stops taking it, then t days later the concentration will be

$$C(t) = 80\left(\frac{1}{2}\right)^{t/5}$$

Find each of the following, and explain what they mean.

a. $C(5)$
b. $C(10)$

4. The following formulas give the circumference and area of a circle with a radius of r. Use the formulas to find the circumference and area of a circular plate if the radius is 5 inches.

$$C(r) = 2\pi r$$
$$A(r) = \pi r^2$$

Answers

2. **a.** $s(8) = 11$; runs a mile in 8 minutes, average speed is 11 feet per second. **b.** $s(11) = 8$; runs a mile in 11 minutes, average speed is 8 feet per second.
3. **a.** $C(5) = 40$ ng/mL; after 5 days the concentration is 40 ng per mL.
   b. $C(10) = 20$ ng/mL; after 10 days the concentration is 20 ng per mL.
4. $C(5) = 10\pi \approx 31.4$ inches
   $A(5) = 25\pi \approx 78.5$ in^2

USING TECHNOLOGY

More About Example 4

If we look back at Example 4, we see that when the radius of a sphere is 3, the numerical values of the volume and surface area are equal. How unusual is this? Are there other values of r for which $V(r)$ and $S(r)$ are equal? We can answer this question by looking at the graphs of both V and S.

To graph the function $V(r) = \frac{4}{3}\pi r^3$, set $Y_1 = 4\pi X^3/3$. To graph $S(r) = 4\pi r^2$, set $Y_2 = 4\pi X^2$. Graph the two functions in each of the following windows:

Window 1: X from −4 to 4, Y from −2 to 10

Window 2: X from 0 to 4, Y from 0 to 50

Window 3: X from 0 to 4, Y from 0 to 150

Then use the Trace and Zoom features of your calculator to locate the point in the first quadrant where the two graphs intersect. How do the coordinates of this point compare with the results in Example 4?

5. If $f(x) = 4x^2 - 3$, find
 a. $f(0)$
 b. $f(3)$
 c. $f(-2)$

EXAMPLE 5 If $f(x) = 3x^2 + 2x - 1$, find $f(0)$, $f(3)$, and $f(-2)$.

Solution Because $f(x) = 3x^2 + 2x - 1$, we have

$$f(\mathbf{0}) = 3(\mathbf{0})^2 + 2(\mathbf{0}) - 1 = 0 + 0 - 1 = -1$$

$$f(\mathbf{3}) = 3(\mathbf{3})^2 + 2(\mathbf{3}) - 1 = 27 + 6 - 1 = 32$$

$$f(\mathbf{-2}) = 3(\mathbf{-2})^2 + 2(\mathbf{-2}) - 1 = 12 - 4 - 1 = 7$$

In Example 5, the function f is defined by the equation $f(x) = 3x^2 + 2x - 1$. We could just as easily have said $y = 3x^2 + 2x - 1$; that is, $y = f(x)$. Saying $f(-2) = 7$ is exactly the same as saying y is 7 when x is −2.

6. If $f(x) = 2x + 1$ and $g(x) = x^2 - 3$, find
 a. $f(5)$
 b. $g(5)$
 c. $f(-2)$
 d. $g(-2)$
 e. $f(a)$
 f. $g(a)$

EXAMPLE 6 If $f(x) = 4x - 1$ and $g(x) = x^2 + 2$, then

$$f(\mathbf{5}) = 4(\mathbf{5}) - 1 = 19 \quad \text{and} \quad g(\mathbf{5}) = \mathbf{5}^2 + 2 = 27$$

$$f(\mathbf{-2}) = 4(\mathbf{-2}) - 1 = -9 \quad \text{and} \quad g(\mathbf{-2}) = (\mathbf{-2})^2 + 2 = 6$$

$$f(\mathbf{0}) = 4(\mathbf{0}) - 1 = -1 \quad \text{and} \quad g(\mathbf{0}) = \mathbf{0}^2 + 2 = 2$$

$$f(\boldsymbol{z}) = 4\boldsymbol{z} - 1 \quad \text{and} \quad g(\boldsymbol{z}) = \boldsymbol{z}^2 + 2$$

$$f(\boldsymbol{a}) = 4\boldsymbol{a} - 1 \quad \text{and} \quad g(\boldsymbol{a}) = \boldsymbol{a}^2 + 2$$

Answers

5. a. −3 b. 33 c. 13
6. a. 11 b. 22 c. −3 d. 1
 e. $2a + 1$ f. $a^2 - 3$

USING TECHNOLOGY

More About Example 6

Most graphing calculators can use tables to evaluate functions. To work Example 6 using a graphing calculator table, set Y_1 equal to $4X - 1$ and Y_2 equal to $X^2 + 2$. Then set the independent variable in the table to Ask instead of Auto. Go to your table and input 5, −2, and 0. Under Y_1 in the table, you will find $f(5)$, $f(-2)$, and $f(0)$. Under Y_2, you will find $g(5)$, $g(-2)$, and $g(0)$.

Table Setup
Table minimum = 0
Table increment = 1
Independent variable: Ask
Dependent variable: Ask

Y Variables Setup
$Y_1 = 4X - 1$
$Y_2 = X^2 + 2$

The table will look like this:

X	Y_1	Y_2
5	19	27
−2	−9	6
0	−1	2

Although the calculator asks us for a table increment, the increment doesn't matter since we are inputting the X values ourselves.

EXAMPLE 7

If the function f is given by

$$f = \{(-2, 0), (3, -1), (2, 4), (7, 5)\}$$

then $f(-2) = 0$, $f(3) = -1$, $f(2) = 4$, and $f(7) = 5$. ■

7. If $f = \{(-4, 1), (2, -3), (7, 9)\}$, find
   a. $f(-4)$
   b. $f(2)$
   c. $f(7)$

EXAMPLE 8

If $f(x) = 2x^2$ and $g(x) = 3x - 1$, find

a. $f[g(2)]$ **b.** $g[f(2)]$

Solution The expression $f[g(2)]$ is read "f of g of 2."

a. Since $g(2) = 3(2) - 1 = 5$,

$$f[g(2)] = f(5) = 2(5)^2 = 50$$

b. Since $f(2) = 2(2)^2 = 8$,

$$g[f(2)] = g(8) = 3(8) - 1 = 23$$ ■

8. If $f(x) = 3x^2$ and $g(x) = 4x + 1$, find
   a. $f[g(2)]$
   b. $g[f(2)]$

Getting Ready for Class

After reading through the preceding section, respond in your own words and in complete sentences.

1. Explain what you are calculating when you find $f(2)$ for a given function f.
2. If $s(t) = \frac{60}{t}$, how do you find $s(10)$?
3. If $f(2) = 3$ for a function f, what is the relationship between the numbers 2 and 3 and the graph of f?
4. If $f(6) = 0$ for a particular function f, then you can immediately graph one of the intercepts. Explain.

Linking Objectives and Examples

Next to each **objective**, list the examples that are best described by that objective.

A	

Answers
7. a. 1 b. −3 c. 9
8. a. 243 b. 49

Problem Set 3.6

INTERMEDIATE Algebra $f(x)$ Now™

Let $f(x) = 2x - 5$ and $g(x) = x^2 + 3x + 4$. Evaluate the following.

1. $f(2)$
2. $f(3)$
3. $f(-3)$
4. $g(-2)$
5. $g(-1)$
6. $f(-4)$
7. $g(-3)$
8. $g(2)$
9. $g(4) + f(4)$
10. $f(2) - g(3)$
11. $f(3) - g(2)$
12. $g(-1) + f(-1)$

Let $f(x) = 3x^2 - 4x + 1$ and $g(x) = 2x - 1$. Evaluate the following.

13. $f(0)$
14. $g(0)$
15. $g(-4)$
16. $f(1)$
17. $f(-1)$
18. $g(-1)$
19. $g(10)$
20. $f(10)$
21. $f(3)$
22. $g(3)$
23. $g\left(\frac{1}{2}\right)$
24. $g\left(\frac{1}{4}\right)$
25. $f(a)$
26. $g(b)$

If $f = \{(1, 4), (-2, 0), (3, \frac{1}{2}), (\pi, 0)\}$ and $g = \{(1, 1), (-2, 2), (\frac{1}{2}, 0)\}$, find each of the following values of f and g.

27. $f(1)$

28. $g(1)$

29. $g\left(\frac{1}{2}\right)$

30. $f(3)$

31. $g(-2)$

32. $f(\pi)$

Let $f(x) = 2x^2 - 8$ and $g(x) = \frac{1}{2}x + 1$. Evaluate each of the following.

33. $f(0)$

34. $g(0)$

35. $g(-4)$

36. $f(1)$

37. $f(a)$

38. $g(z)$

39. $f(b)$

40. $g(t)$

41. $f[g(2)]$

42. $g[f(2)]$

43. $g[f(-1)]$

44. $f[g(-2)]$

45. $g[f(0)]$

46. $f[g(0)]$

47. Graph the function $f(x) = \frac{1}{2}x + 2$. Then draw and label the line segments that represent $x = 4$ and $f(4)$.

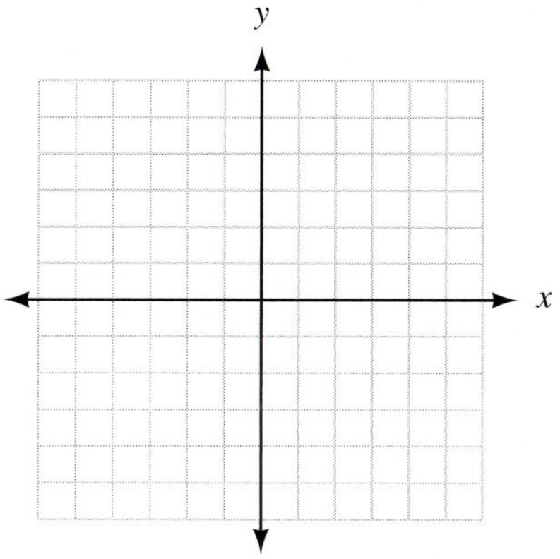

48. Graph the function $f(x) = -\frac{1}{2}x + 6$. Then draw and label the line segments that represent $x = 4$ and $f(4)$.

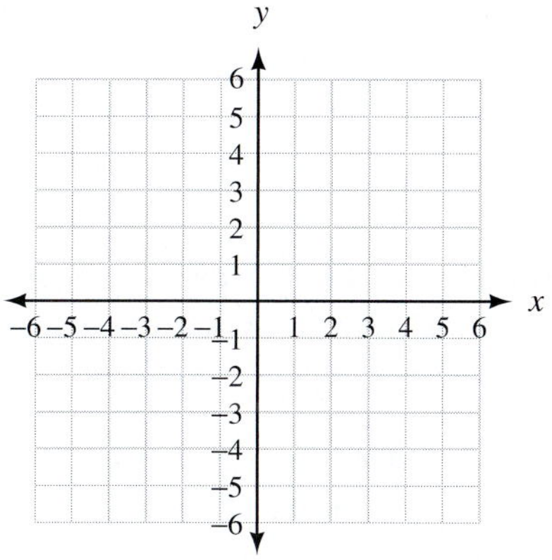

49. For the function $f(x) = \frac{1}{2}x + 2$, find the value of x for which $f(x) = x$.

50. For the function $f(x) = -\frac{1}{2}x + 6$, find the value of x for which $f(x) = x$.

51. Graph the function $f(x) = x^2$. Then draw and label the line segments that represent $x = 1$ and $f(1)$, $x = 2$ and $f(2)$, and, finally, $x = 3$ and $f(3)$.

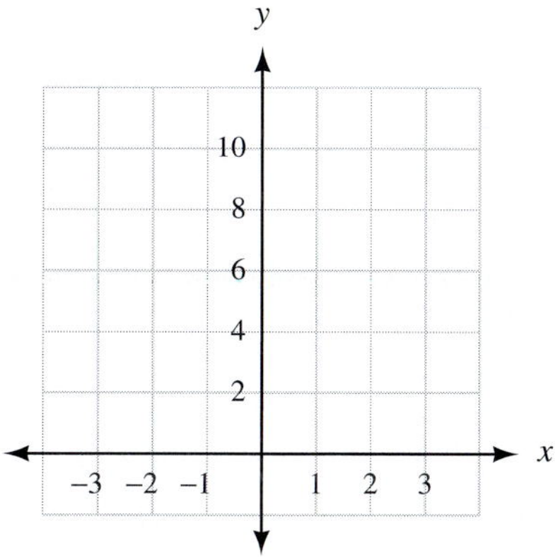

52. Graph the function $f(x) = x^2 - 2$. Then draw and label the line segments that represent $x = 2$ and $f(2)$, and the line segments corresponding to $x = 3$ and $f(3)$.

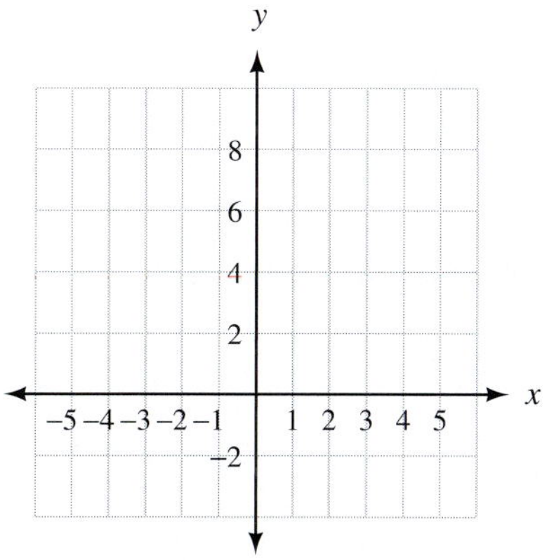

Applying the Concepts

53. Investing in Art A painting is purchased as an investment for \$150. If its value increases continuously so that it doubles every 3 years, then its value is given by the function

$$V(t) = 150 \cdot 2^{t/3} \quad \text{for} \quad t \geq 0$$

where t is the number of years since the painting was purchased, and $V(t)$ is its value (in dollars) at time t. Find $V(3)$ and $V(6)$, and then explain what they mean.

54. Average Speed If it takes Minke t minutes to run a mile, then her average speed $s(t)$, in miles per hour, is given by the formula

$$s(t) = \frac{60}{t} \quad \text{for} \quad t > 0$$

Find $s(4)$ and $s(5)$, and then explain what they mean.

55. Dimensions of a Rectangle The length of a rectangle is 3 inches more than twice the width. Let x represent the width of the rectangle and $P(x)$ represent the perimeter of the rectangle. Use function notation to write the relationship between x and $P(x)$, noting any restrictions on the variable x.

56. Dimensions of a Rectangle The length of a rectangle is 3 inches more than twice the width. Let x represent the width of the rectangle and $A(x)$ represent the area of the rectangle. Use function notation to write the relationship between x and $A(x)$, noting any restrictions on the variable x.

Straight-Line Depreciation Straight-line depreciation is an accounting method used to help spread the cost of new equipment over a number of years. It takes into account both the cost when new and the salvage value, which is the value of the equipment at the time it gets replaced.

57. Value of a Copy Machine The function $V(t) = -3{,}300t + 18{,}000$, where V is value and t is time in years, can be used to find the value of a large copy machine during the first 5 years of use.

a. What is the value of the copier after 3 years and 9 months?

b. What is the salvage value of this copier if it is replaced after 5 years?

c. State the domain of this function.

d. Sketch the graph of this function.

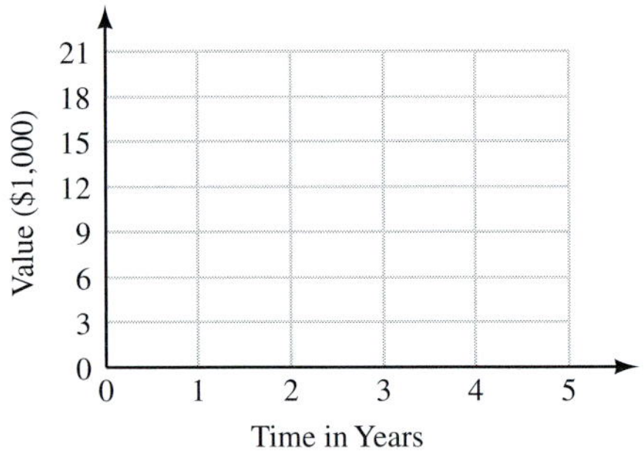

e. What is the range of this function?

f. After how many years will the copier be worth only $10,000?

58. Value of a Forklift The function $V = -16{,}500t + 125{,}000$, where V is value and t is time in years, can be used to find the value of an electric forklift during the first 6 years of use.

a. What is the value of the forklift after 2 years and 3 months?

b. What is the salvage value of this forklift if it is replaced after 6 years?

c. State the domain of this function.

d. Sketch the graph of this function.

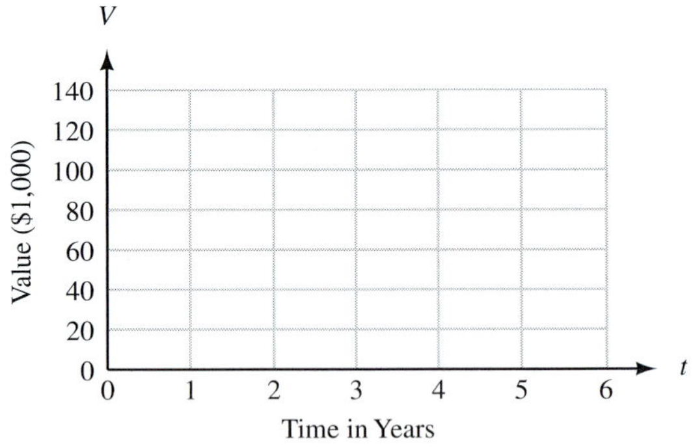

e. What is the range of this function?

f. After how many years will the forklift be worth only $45,000?

Photo Disc/Getty Images

Michael Grecco/The Image Bank/Getty Images

Getting Ready for the Next Section

Simplify.

59. $16(3.5)^2$

60. $\dfrac{2{,}400}{100}$

61. $\dfrac{180}{45}$

62. $4(2)(4)^2$

63. $\dfrac{0.0005(200)}{(0.25)^2}$

64. $\dfrac{0.2(0.5)^2}{100}$

65. If $y = Kx$, find K if $x = 5$ and $y = 15$.

66. If $d = Kt^2$, find K if $t = 2$ and $d = 64$.

67. If $V = \dfrac{K}{P}$, find K if $P = 48$ and $V = 50$.

68. If $y = Kxz^2$, find K if $x = 5$, $z = 3$, and $y = 180$.

Maintaining Your Skills

The problems that follow review some of the more important skills you have learned in previous sections and chapters.

Solve each equation.

69. $|3x - 5| = 7$

70. $|0.04 - 0.03x| = 0.02$

71. $|4y + 2| - 8 = -2$

72. $4 = |3 - 2y| - 5$

73. $5 + |6t + 2| = 3$

74. $7 + |3 - \frac{3}{4}t| = 10$

Extending the Concepts

The graphs of two functions are shown in Figures 4 and 5. Use the graphs to find the following.

75. **a.** $f(2)$
b. $f(-4)$
c. $g(0)$
d. $g(3)$

76. **a.** $g(2) - f(2)$
b. $f(1) + g(1)$
c. $f[g(3)]$
d. $g[f(3)]$

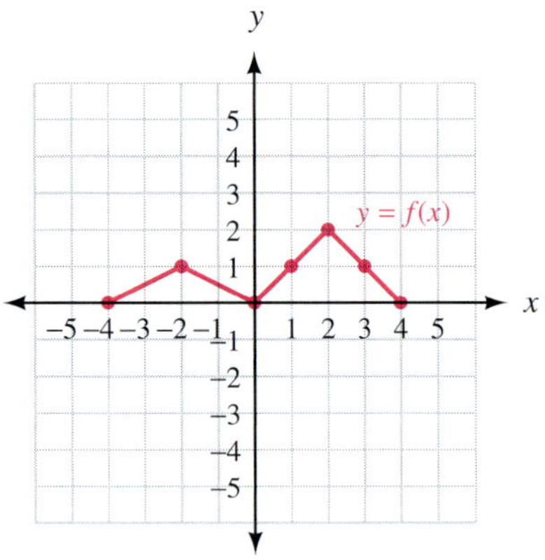

FIGURE 4

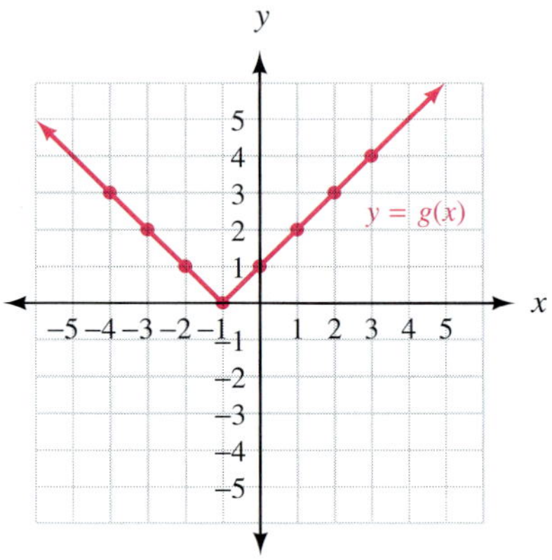

FIGURE 5

77. **Step Function** Figure 6 shows the graph of the step function C that was used to calculate the first-class postage on a letter weighing x ounces in 2006. Use this graph to answer questions (a) through (d).

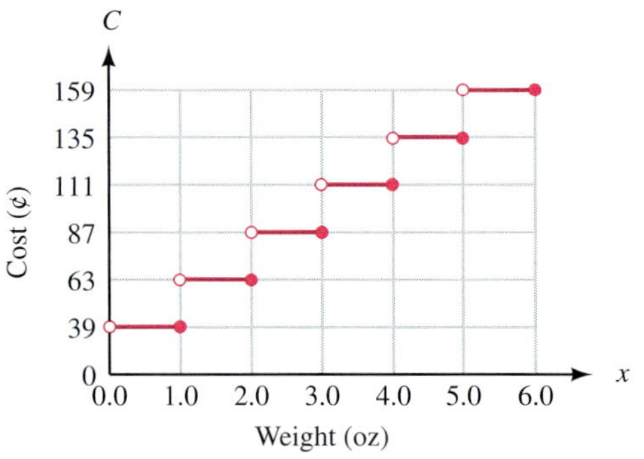

FIGURE 6 **The graph of $C(x)$**

a. Fill in the following table:

Weight (ounces)	0.6	1.0	1.1	2.5	3.0	4.8	5.0	5.3
Cost (cents)								

b. If a letter costs 87 cents to mail, how much does it weigh? State your answer in words. State your answer as an inequality.

c. If the entire function is shown in Figure 6, state the domain.

d. State the range of the function shown in Figure 6.

78. **Step Function** A taxi ride in Boston at the time I am writing this problem is \$1.50 for the first $\frac{1}{4}$ mile, and then \$0.25 for each additional $\frac{1}{8}$ of a mile. The following graph shows how much you will pay for a taxi ride of 1 mile or less.

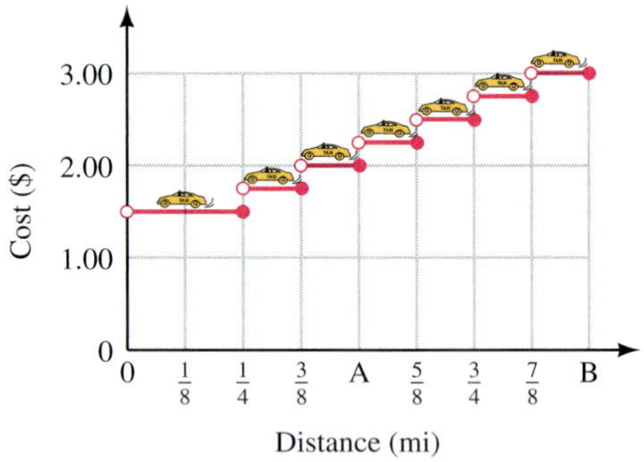

FIGURE 7

a. What is the most you will pay for this taxi ride?

b. How much does it cost to ride the taxi for $\frac{8}{10}$ of a mile?

c. Find the values of A and B on the horizontal axis.

d. If a taxi ride costs \$2.50, what distance was the ride?

e. If the complete function is shown in Figure 7, find the domain and range of the function.

Variation

3.7

If you are a runner and you average t minutes for every mile you run during one of your workouts, then your speed s in miles per hour is given by the equation and graph shown here. Figure 1 is shown in the first quadrant only because both t and s are positive.

$$s = \frac{60}{t}$$

INPUT t	OUTPUT s
4	15.0
6	10.0
8	7.5
10	6.0
12	5.0
14	4.3

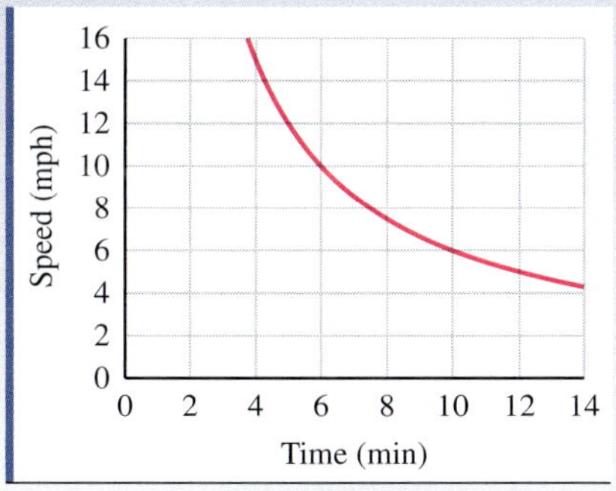

FIGURE 1

OBJECTIVES

A	Set up and solve problems with direct, inverse, or joint variation.

You know intuitively that as your average time per mile (t) increases, your speed (s) decreases. Likewise, lowering your time per mile will increase your speed. The equation and Figure 1 also show this to be true: increasing t decreases s, and decreasing t increases s. Quantities that are connected in this way are said to *vary inversely* with each other. Inverse variation is one of the topics we will study in this section.

There are two main types of variation: *direct variation* and *inverse variation*. Variation problems are most common in the sciences, particularly in chemistry and physics.

Direct Variation

When we say the variable *y varies directly* with the variable x, we mean that the relationship can be written in symbols as $y = Kx$, where K is a nonzero constant called the *constant of variation* (or *proportionality constant*). Another way of saying y varies directly with x is to say y is *directly proportional* to x.

Study the following list. It gives the mathematical equivalent of some direct variation statements.

ENGLISH PHRASE	ALGEBRAIC EQUATION
y varies directly with x	$y = Kx$
s varies directly with the square of t	$s = Kt^2$
y is directly proportional to the cube of z	$y = Kz^3$
u is directly proportional to the square root of v	$u = K\sqrt{v}$

EXAMPLE 1

y varies directly with x. If y is 15 when x is 5, find y when x is 7.

Solution The first sentence gives us the general relationship between x and y. The equation equivalent to the statement "y varies directly with x" is

$$y = Kx$$

PRACTICE PROBLEMS

1. y varies directly with x. If y is 24 when x is 8, find y when x is 2.

The graph of the direct variation statement $y = 3x$ is a line with slope 3 that passes through the origin.

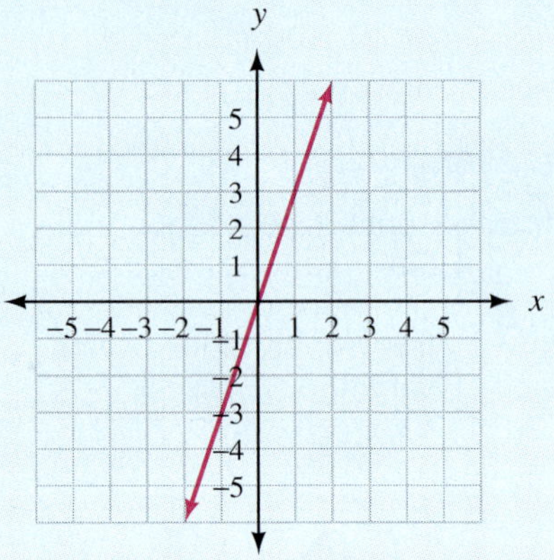

The first part of the second sentence in our example gives us the information necessary to evaluate the constant K:

When	$y = 15$
and	$x = 5$
the equation	$y = Kx$
becomes	$15 = K \cdot 5$
or	$K = 3$

The equation now can be written specifically as

$$y = 3x$$

Letting $x = 7$, we have

$$y = 3 \cdot 7$$

$$y = 21$$

2. Use the information in Example 2 to find how far the skydiver will fall in 2.5 seconds.

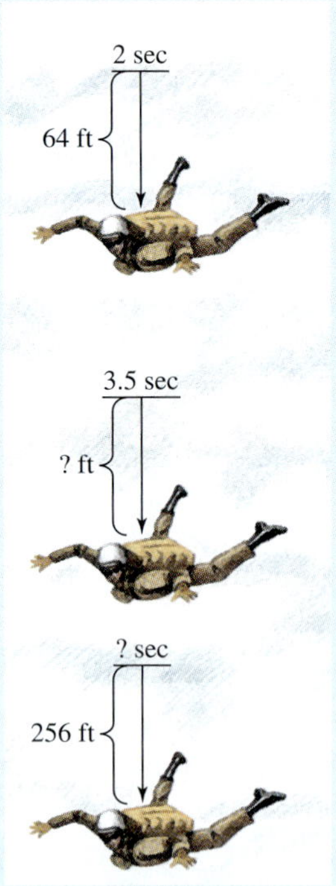

EXAMPLE 2 A skydiver jumps from a plane. Like any object that falls toward Earth, the distance the skydiver falls is directly proportional to the square of the time he has been falling until he reaches his terminal velocity. If the skydiver falls 64 feet in the first 2 seconds of the jump, then

a. How far will he have fallen after 3.5 seconds?
b. Graph the relationship between distance and time.
c. How long will it take him to fall 256 feet?

Solution We let t represent the time the skydiver has been falling, then we can let $d(t)$ represent the distance he has fallen.

a. Because $d(t)$ is directly proportional to the square of t, we have the general function that describes this situation:

$$d(t) = Kt^2$$

Next, we use the fact that $d(2) = 64$ to find K.

$$64 = K(2^2)$$

$$K = 16$$

The specific equation that describes this situation is

$$d(t) = 16t^2$$

To find how far a skydiver will fall after 3.5 seconds, we find $d(3.5)$:

$$d(3.5) = 16(3.5^2)$$

$$d(3.5) = 196$$

A skydiver will fall 196 feet after 3.5 seconds.

b. To graph this equation, we use a table:

INPUT t	OUTPUT $d(t)$
0	0
1	16
2	64
3	144
4	256
5	400

Answer
1. 6

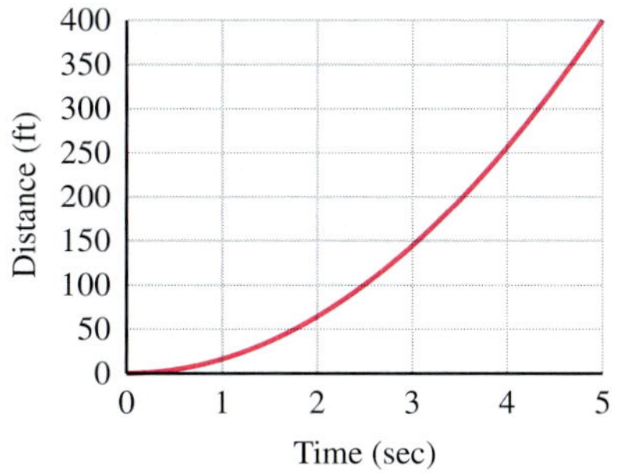

FIGURE 2

c. From the table or the graph (Figure 2), we see that it will take 4 seconds for the skydiver to fall 256 feet.

Inverse Variation

Running From the introduction to this section, we know that the relationship between the number of minutes (t) it takes a person to run a mile and his or her average speed in miles per hour (s) can be described with the following equation, table, and Figure 3.

$$s = \frac{60}{t}$$

INPUT t	OUTPUT s
4	15.0
6	10.0
8	7.5
10	6.0
12	5.0
14	4.3

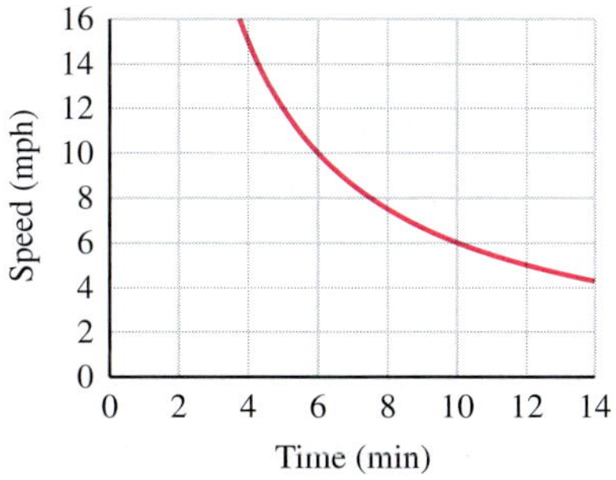

FIGURE 3

If t decreases, then s will increase, and if t increases, then s will decrease. The variable s is *inversely proportional* to the variable t. In this case, the *constant of proportionality* is 60.

Photography If you are familiar with the terminology and mechanics associated with photography, you know that the *f*-stop for a particular lens will increase as the aperture (the maximum diameter of the opening of the lens) decreases. In mathematics we say that *f*-stop and aperture vary inversely with each other. The diagram illustrates this relationship.

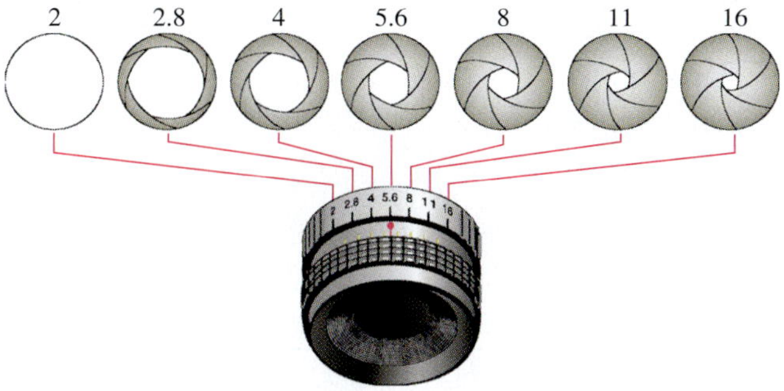

Answer
2. 100 feet

If f is the f-stop and d is the aperture, then their relationship can be written

$$f = \frac{K}{d}$$

In this case, K is the constant of proportionality. (Those of you familiar with photography know that K is also the focal length of the camera lens.)

In General We generalize this discussion of inverse variation as follows: If y varies inversely with x, then

$$y = K\frac{1}{x} \quad \text{or} \quad y = \frac{K}{x}$$

We can also say y is inversely proportional to x. The constant K is again called the constant of variation or proportionality constant.

ENGLISH PHRASE	ALGEBRAIC EQUATION
y is inversely proportional to x	$y = \frac{K}{x}$
s varies inversely with the square of t	$s = \frac{K}{t^2}$
y is inversely proportional to x^4	$y = \frac{K}{x^4}$
z varies inversely with the cube root of t	$z = \frac{K}{\sqrt[3]{t}}$

3. In Example 3, what pressure is needed to produce a volume of 150 cubic feet?

NOTE

The relationship between pressure and volume as given in this example is known as Boyle's law and applies to situations such as those encountered in a piston-cylinder arrangement. It was Robert Boyle (1627–1691) who, in 1662, published the results of some of his experiments that showed, among other things, that the volume of a gas decrease as the pressure increases. This is an example of inverse variation.

EXAMPLE 3

The volume of a gas is inversely proportional to the pressure of the gas on its container. If a pressure of 48 pounds per square inch corresponds to a volume of 50 cubic feet, what pressure is needed to produce a volume of 100 cubic feet?

Solution We can represent volume with V and pressure with P:

$$V = \frac{K}{P}$$

Using $P = 48$ and $V = 50$, we have

$$50 = \frac{K}{48}$$

$$K = 50(48)$$

$$K = 2{,}400$$

The equation that describes the relationship between P and V is

$$V = \frac{2{,}400}{P}$$

Here is a graph of this relationship.

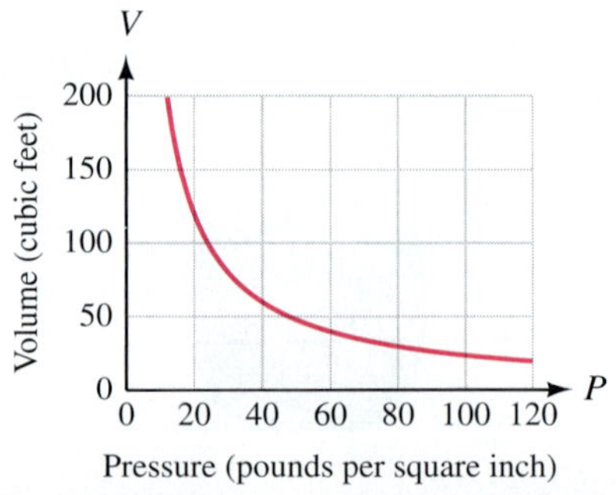

Substituting $V = 100$ into our last equation, we get

$$100 = \frac{2{,}400}{P}$$

$$100P = 2{,}400$$

$$P = \frac{2{,}400}{100}$$

$$P = 24$$

A volume of 100 cubic feet is produced by a pressure of 24 pounds per square inch.

Joint Variation and Other Variation Combinations

Many times relationships among different quantities are described in terms of more than two variables. If the variable y varies directly with *two* other variables, say x and z, then we say y varies *jointly* with x and z. In addition to joint variation, there are many other combinations of direct and inverse variation involving more than two variables. The following table is a list of some variation statements and their equivalent mathematical forms:

ENGLISH PHRASE	ALGEBRAIC EQUATION
y varies jointly with x and z	$y = Kxz$
z varies jointly with r and the square of s	$z = Krs^2$
V is directly proportional to T and inversely proportional to P	$V = \frac{KT}{P}$
F varies jointly with m_1 and m_2 and inversely with the square of r	$F = \frac{Km_1m_2}{r^2}$

EXAMPLE 4 y varies jointly with x and the square of z. When x is 5 and z is 3, y is 180. Find y when x is 2 and z is 4.

Solution The general equation is given by

$$y = Kxz^2$$

Substituting $x = 5$, $z = 3$, and $y = 180$, we have

$$180 = K(5)(3)^2$$

$$180 = 45K$$

$$K = 4$$

The specific equation is

$$y = 4xz^2$$

When $x = 2$ and $z = 4$, the last equation becomes

$$y = 4(2)(4)^2$$

$$y = 128$$

4. y varies jointly with x and the square of z. If y is 81 when x is 2 and z is 9, find y when x is 4 and z is 4.

Answers

3. 16 pounds per square inch
4. 32

5. Use the information in Example 5 to find the resistance of a 300-foot cable made of the same material that is 0.25 inch in diameter.

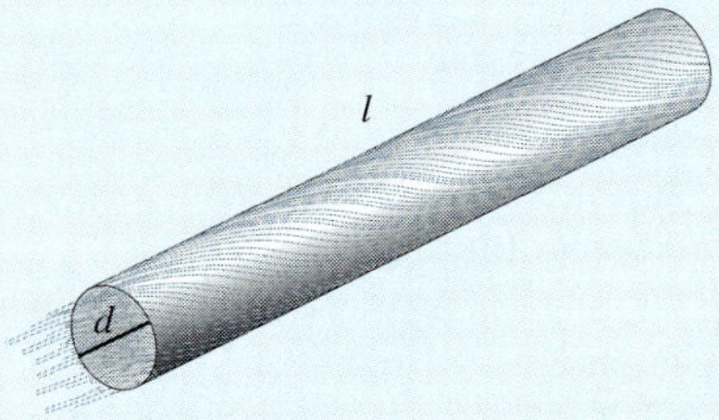

EXAMPLE 5 In electricity, the resistance of a cable is directly proportional to its length and inversely proportional to the square of the diameter. If a 100-foot cable 0.5 inch in diameter has a resistance of 0.2 ohm, what will be the resistance of a cable made from the same material if it is 200 feet long with a diameter of 0.25 inch?

Solution Let R = resistance, l = length, and d = diameter. The equation is

$$R = \frac{Kl}{d^2}$$

When $R = 0.2$, $l = 100$, and $d = 0.5$, the equation becomes

$$0.2 = \frac{K(100)}{(0.5)^2}$$

or

$$K = 0.0005$$

Using this value of K in our original equation, the result is

$$R = \frac{0.0005l}{d^2}$$

When $l = 200$ and $d = 0.25$, the equation becomes

$$R = \frac{0.0005(200)}{(0.25)^2}$$

$$R = 1.6 \text{ ohms}$$

Linking Objectives and Examples

Next to each **objective**, list the examples that are best described by that objective.

A	

Getting Ready for Class

After reading through the preceding section, respond in your own words and in complete sentences.

1. Give an example of a direct variation statement, and then translate it into symbols.
2. Translate the equation $y = \frac{K}{x}$ into words.
3. For the inverse variation equation $y = \frac{3}{x}$, what happens to the values of y as x gets larger?
4. How are direct variation statements and linear equations in two variables related?

Answer

5. 2.4 ohms

Problem Set 3.7

INTERMEDIATE Algebra $f(x)$ Now™

For the following problems, y varies directly with x.

1. If y is 10 when x is 2, find y when x is 6.

2. If y is 20 when x is 5, find y when x is 3.

3. If y is -32 when x is 4, find x when y is -40.

4. If y is -50 when x is 5, find x when y is -70.

For the following problems, r is inversely proportional to s.

5. If r is -3 when s is 4, find r when s is 2.

6. If r is -10 when s is 6, find r when s is -5.

7. If r is 8 when s is 3, find s when r is 48.

8. If r is 12 when s is 5, find s when r is 30.

For the following problems, d varies directly with the square of r.

9. If $d = 10$ when $r = 5$, find d when $r = 10$.

10. If $d = 12$ when $r = 6$, find d when $r = 9$.

11. If $d = 100$ when $r = 2$, find d when $r = 3$.

12. If $d = 50$ when $r = 5$, find d when $r = 7$.

For the following problems, y varies inversely with the square of x.

13. If $y = 45$ when $x = 3$, find y when x is 5.

14. If $y = 12$ when $x = 2$, find y when x is 6.

15. If $y = 18$ when $x = 3$, find y when x is 2.

16. If $y = 45$ when $x = 4$, find y when x is 5.

For the following problems, z varies jointly with x and the square of y.

17. If z is 54 when x and y are 3, find z when $x = 2$ and $y = 4$.

18. If z is 80 when x is 5 and y is 2, find z when $x = 2$ and $y = 5$.

19. If z is 64 when $x = 1$ and $y = 4$, find x when $z = 32$ and $y = 1$.

20. If z is 27 when $x = 6$ and $y = 3$, find x when $z = 50$ and $y = 4$.

Applying the Concepts

21. **Length of a Spring** The length a spring stretches is directly proportional to the force applied. If a force of 5 pounds stretches a spring 3 inches, how much force is necessary to stretch the same spring 10 inches?

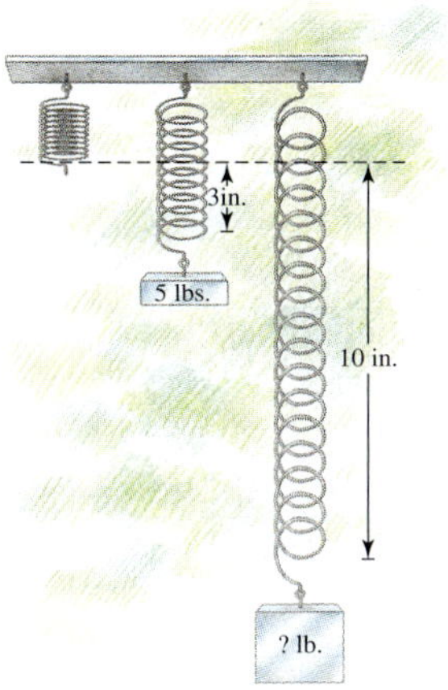

22. **Weight and Surface Area** The weight of a certain material varies directly with the surface area of that material. If 8 square feet weighs half a pound, how much will 10 square feet weigh?

23. **Pressure and Temperature** The temperature of a gas varies directly with its pressure. A temperature of 200 K produces a pressure of 50 pounds per square inch.
   a. Find the equation that relates pressure and temperature.
   b. Graph the equation from part (a) in the first quadrant only.

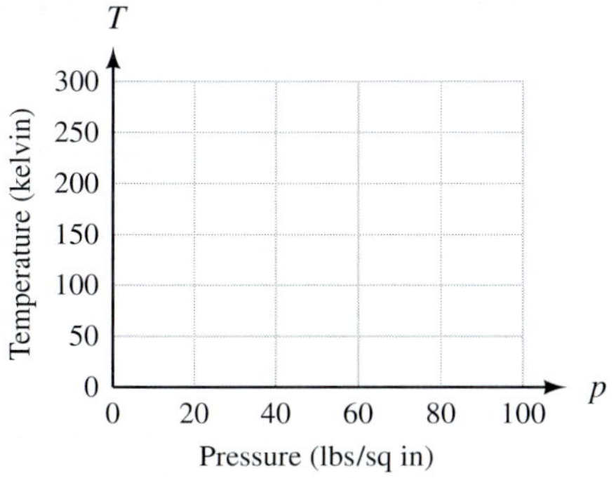

   c. What pressure will the gas have at 280° K?

24. **Circumference and Diameter** The circumference of a wheel is directly proportional to its diameter. A wheel has a circumference of 8.5 feet and a diameter of 2.7 feet.
   a. Find the equation that relates circumference and diameter.
   b. Graph the equation from part (a) in the first quadrant only.

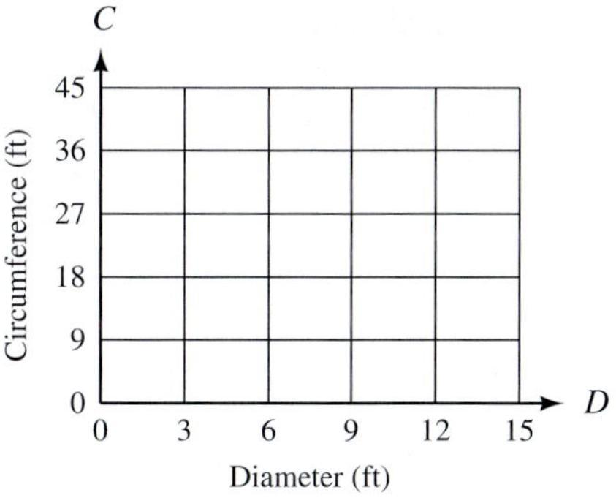

   c. What is the circumference of a wheel that has a diameter of 11.3 feet?

25. **Volume and Pressure** The volume of a gas is inversely proportional to the pressure. If a pressure of 36 pounds per square inch corresponds to a volume of 25 cubic feet, what pressure is needed to produce a volume of 75 cubic feet?

26. **Wave Frequency** The frequency of an electromagnetic wave varies inversely with the wavelength. If a wavelength of 200 meters has a frequency of 800 kilocycles per second, what frequency will be associated with a wavelength of 500 meters?

27. f-Stop and Aperture Diameter The relative aperture or *f*-stop for a camera lens is inversely proportional to the diameter of the aperture. An *f*-stop of 2 corresponds to an aperture diameter of 40 millimeters for the lens on an automatic camera.

a. Find the equation that relates *f*-stop and diameter.

b. Graph the equation from part (a) in the first quadrant only.

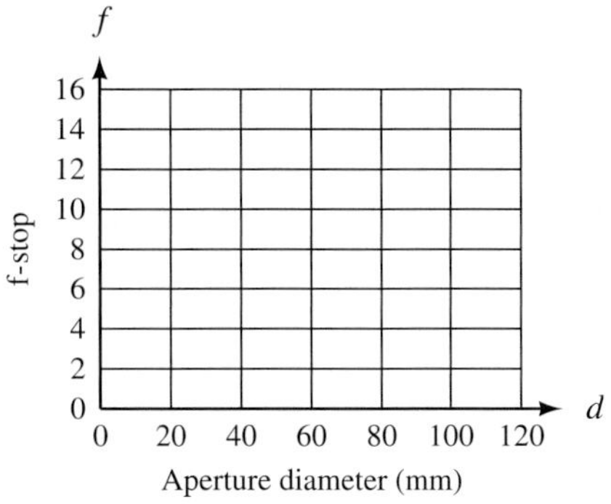

c. What is the *f*-stop of this camera when the aperture diameter is 10 millimeters?

28. f-Stop and Aperture Diameter The relative aperture or *f*-stop for a camera lens is inversely proportional to the diameter of the aperture. An *f*-stop of 2.8 corresponds to an aperture diameter of 75 millimeters for a certain telephoto lens.

a. Find the equation that relates *f*-stop and diameter.

b. Graph the equation from part (a) in the first quadrant only.

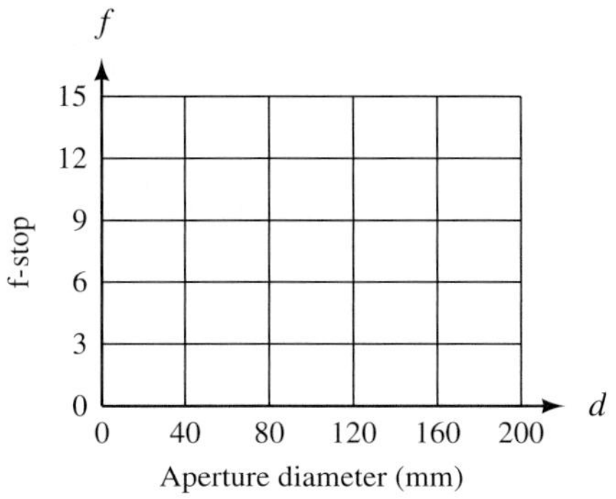

c. What aperture diameter corresponds to an *f*-stop of 5.6?

29. Surface Area of a Cylinder The surface area of a hollow cylinder varies jointly with the height and radius of the cylinder. If a cylinder with radius 3 inches and height 5 inches has a surface area of 94 square inches, what is the surface area of a cylinder with radius 2 inches and height 8 inches?

30. Capacity of a Cylinder The capacity of a cylinder varies jointly with its height and the square of its radius. If a cylinder with a radius of 3 centimeters and a height of 6 centimeters has a capacity of 169.56 cubic centimeters, what will be the capacity of a cylinder with radius 4 centimeters and height 9 centimeters?

31. Electrical Resistance The resistance of a wire varies directly with its length and inversely with the square of its diameter. If 100 feet of wire with diameter 0.01 inch has a resistance of 10 ohms, what is the resistance of 60 feet of the same type of wire if its diameter is 0.02 inch?

32. Volume and Temperature The volume of a gas varies directly with its temperature and inversely with the pressure. If the volume of a certain gas is 30 cubic feet at a temperature of 300 K and a pressure of 20 pounds per square inch, what is the volume of the same gas at 340 K when the pressure is 30 pounds per square inch?

33. Music A musical tone's pitch varies inversely with its wavelength. If one tone has a pitch of 420 vibrations each second and a wavelength of 2.2 meters, find the wavelength of a tone that has a pitch of 720 vibrations each second.

34. Hooke's Law Hooke's law states that the stress (force per unit area) placed on a solid object varies directly with the strain (deformation) produced.

a. Using the variables S_1 for stress and S_2 for strain, state this law in algebraic form.

b. Find the constant, K, if for one type of material $S_1 = 24$ and $S_2 = 72$.

35. Gravity In Book Three of his *Principia,* Isaac Newton (depicted on the postage stamp below) states that there is a single force in the universe that holds everything together, called the force of universal gravity. Newton stated that the force of universal gravity, F, is directly proportional with the product of two masses, m_1 and m_2, and inversely proportional with the square of the distance d between them. Write the equation for Newton's force of universal gravity, using the symbol G as the constant of proportionality.

36. Boyle's Law and Charles's Law Boyle's law states that for low pressures, the pressure of an ideal gas kept at a constant temperature varies inversely with the volume of the gas. Charles's law states that for low pressures, the density of an ideal gas kept at a constant pressure varies inversely with the absolute temperature of the gas.

a. State Boyle's law as an equation using the symbols P, K, and V.

b. State Charles's law as an equation using the symbols D, K, and T.

Maintaining Your Skills

The problems that follow review some of the more important skills you have learned in previous sections and chapters.

Solve the following equations.

37. $x - 5 = 7$

38. $3y = -4$

39. $5 - \frac{4}{7}a = -11$

40. $\frac{1}{5}x - \frac{1}{2} - \frac{1}{10}x + \frac{2}{5} = \frac{3}{10}x + \frac{1}{2}$

41. $5(x - 1) - 2(2x + 3) = 5x - 4$

42. $0.07 - 0.02(3x + 1) = -0.04x + 0.01$

Solve for the indicated variable.

43. $P = 2l + 2w$ for w

44. $A = \frac{1}{2}h(b + B)$ for B

Solve the following inequalities. Write the solution set using interval notation, then graph the solution set.

45. $-5t \le 30$

46. $5 - \frac{3}{2}x > -1$

47. $1.6x - 2 < 0.8x + 2.8$

48. $3(2y + 4) \ge 5(y - 8)$

Solve the following equations.

49. $\left|\frac{1}{4}x - 1\right| = \frac{1}{2}$

50. $\left|\frac{2}{3}a + 4\right| = 6$

51. $|3 - 2x| + 5 = 2$

52. $5 = |3y + 6| - 4$

Review Problems

Solve each inequality and graph the solution set.

53. $\left|\frac{x}{5} + 1\right| \ge \frac{4}{5}$

54. $|2 - 6t| < -5$

55. $|3 - 4t| > -5$

56. $|6y - 1| - 4 \le 2$

Extending the Concepts

57. Human Cannonball A circus company is deciding where to position the net for the human cannonball so that he will land safely during the act. They do this by firing a 100-pound sack of potatoes out of the cannon at different speeds and then measuring how far from the cannon the sack lands. The results are shown in the table.

SPEED IN MILES/HOUR	DISTANCE IN FEET
40	108
50	169
60	243
70	331

The Image Bank/Getty Images

a. Does distance vary directly with the speed, or directly with the square of the speed?

b. Write the equation that describes the relationship between speed and distance.

c. If the cannon will fire a human safely at 55 miles/hour, where should they position the net so the cannonball has a safe landing?

d. How much further will he land if his speed out of the cannon is 56 miles/hour?

Chapter 3 Summary

Linear Equations in Two Variables [3.1, 3.3]

A *linear equation in two variables* is any equation that can be put in *standard form* $ax + by = c$. The graph of every linear equation is a straight line.

EXAMPLES

1. The equation $3x + 2y = 6$ is an example of a linear equation in two variables.

Intercepts [3.1]

The *x-intercept* of an equation is the *x-coordinate* of the point where the graph crosses the x-axis. The *y-intercept* is the y-coordinate of the point where the graph crosses the y-axis. We find the y-intercept by substituting $x = 0$ into the equation and solving for y. The x-intercept is found by letting $y = 0$ and solving for x.

2. To find the x-intercept for $3x + 2y = 6$, we let $y = 0$ and get

$$3x = 6$$
$$x = 2$$

In this case the x-intercept is 2, and the graph crosses the x-axis at (2, 0).

The Slope of a Line [3.2]

The *slope* of the line containing points (x_1, y_1) and (x_2, y_2) is given by

$$\text{Slope} = m = \frac{\text{Rise}}{\text{Run}} = \frac{y_2 - y_1}{x_2 - x_1}$$

Horizontal lines have 0 slope, and vertical lines have no slope.

Parallel lines have equal slopes, and perpendicular lines have slopes that are negative reciprocals.

3. The slope of the line through (6, 9) and (1, −1) is

$$m = \frac{9 - (-1)}{6 - 1} = \frac{10}{5} = 2$$

The Slope-Intercept Form of a Line [3.3]

The equation of a line with slope m and y-intercept b is given by

$$y = mx + b$$

4. The equation of the line with slope 5 and y-intercept 3 is

$$y = 5x + 3$$

The Point-Slope Form of a Line [3.3]

The equation of the line through (x_1, y_1) that has slope m can be written as

$$y - y_1 = m(x - x_1)$$

5. The equation of the line through (3, 2) with slope −4 is

$$y - 2 = -4(x - 3)$$

which can be simplified to

$$y = -4x + 14$$

Linear Inequalities in Two Variables [3.4]

6. The graph of

$$x - y \le 3$$

is

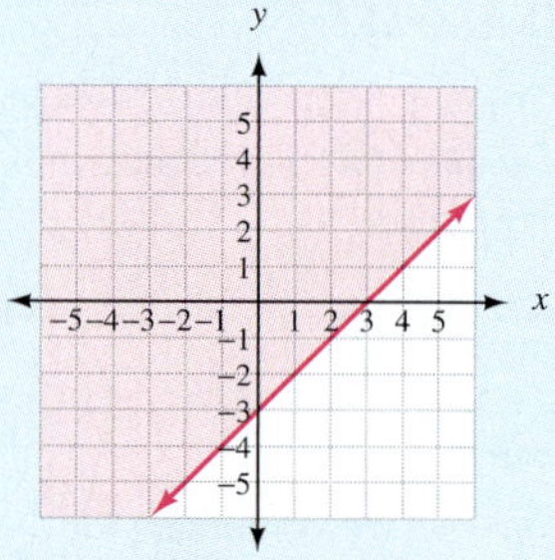

An inequality of the form $ax + by < c$ is a *linear inequality in two variables.* The equation for the boundary of the solution set is given by $ax + by = c$. (This equation is found by simply replacing the inequality symbol with an equal sign.)

To graph a linear inequality, first graph the boundary, using a solid line if the boundary is included in the solution set and a broken line if the boundary is not included in the solution set. Next, choose any point not on the boundary and substitute its coordinates into the original inequality. If the resulting statement is true, the graph lies on the same side of the boundary as the test point. A false statement indicates that the solution set lies on the other side of the boundary.

Relations and Functions [3.5]

7. The relation

$$\{(8, 1), (6, 1), (-3, 0)\}$$

is also a function because no ordered pairs have the same first coordinates. The domain is $\{8, 6, -3\}$ and the range is $\{1, 0\}$.

A *function* is a rule that pairs each element in one set, called the *domain,* with exactly one element from a second set, called the *range.*

A *relation* is any set of ordered pairs. The set of all first coordinates is called the *domain* of the relation, and the set of all second coordinates is the *range* of the relation. A function is a relation in which no two different ordered pairs have the same first coordinates.

Vertical Line Test [3.5]

8. The graph of $x = y^2$ shown in Figure 5 in Section 3.5 fails the vertical line test. It is not the graph of a function.

If a vertical line crosses the graph of a relation in more than one place, the relation cannot be a function. If no vertical line can be found that crosses the graph in more than one place, the relation must be a function.

Function Notation [3.6]

9. If $f(x) = 5x - 3$ then

$$\begin{aligned} f(0) &= 5(0) - 3 \\ &= -3 \\ f(1) &= 5(1) - 3 \\ &= 2 \\ f(-2) &= 5(-2) - 3 \\ &= -13 \\ f(a) &= 5a - 3 \end{aligned}$$

The alternative notation for y is $f(x)$. It is read "f of x" and can be used instead of the variable y when working with functions. The notation y and the notation $f(x)$ are equivalent; that is, $y = f(x)$.

Variation [3.7]

If y *varies directly* with x (y is directly proportional to x), then

$$y = Kx$$

If y *varies inversely* with x (y is inversely proportional to x), then

$$y = \frac{K}{x}$$

If z *varies jointly* with x and y (z is directly proportional to both x and y), then

$$z = Kxy$$

In each case, K is called the *constant of variation.*

10. If y varies directly with x, then

$$y = Kx$$

Then if y is 18 when x is 6,

$$18 = K \cdot 6$$

or

$$K = 3$$

So the equation can be written more specifically as

$$y = 3x$$

If we want to know what y is when x is 4, we simply substitute:

$$y = 3 \cdot 4$$
$$y = 12$$

COMMON MISTAKES

1. When graphing ordered pairs, the most common mistake is to associate the first coordinate with the y-axis and the second with the x-axis. If you make this mistake you would graph (3, 1) by going up 3 and to the right 1, which is just the reverse of what you should do. Remember, the first coordinate is always associated with the horizontal axis, and the second coordinate is always associated with the vertical axis.
2. The two most common mistakes students make when first working with the formula for the slope of a line are the following:
   - **a.** Putting the difference of the x-coordinates over the difference of the y-coordinates.
   - **b.** Subtracting in one order in the numerator and then subtracting in the opposite order in the denominator.
3. When graphing linear inequalities in two variables, remember to graph the boundary with a broken line when the inequality symbol is $<$ or $>$. The only time you use a solid line for the boundary is when the inequality symbol is $\le$ or $\ge$.

Chapter 3 Review/Test

The problems below form a comprehensive review of the material in this chapter. They can be used to study for exams. If you would like to take a practice test on this chapter, you can use the odd-numbered problems. Give yourself an hour and work as many of the odd-numbered problems as possible. When you are finished, or when an hour has passed, check your answers with the answers in the back of the book. You can use the even-numbered problems for a second practice test.

Graph each line. [3.1]

1. $3x + 2y = 6$

2. $y = -\frac{3}{2}x + 1$

3. $x = 3$

Find the slope of the line through the following pairs of points. [3.2]

4. $(5, 2), (3, 6)$

5. $(-4, 2), (3, 2)$

Find x if the line through the two given points has the given slope. [3.2]

6. $(4, x), (1, -3); m = 2$

7. $(-4, 7), (2, x); m = -\frac{1}{3}$

8. Find the slope of any line parallel to the line through $(3, 8)$ and $(5, -2)$. [3.2]

9. The line through $(5, 3y)$ and $(2, y)$ is parallel to a line with slope 4. What is the value of y? [3.2]

Give the equation of the line with the following slope and y-intercept. [3.3]

10. $m = 3, b = 5$

11. $m = -2, b = 0$

Give the slope and y-intercept of each equation. [3.3]

12. $3x - y = 6$

13. $2x - 3y = 9$

Find the equation of the line that contains the given point and has the given slope. [3.3]

14. $(2, 4), m = 2$

15. $(-3, 1), m = -\frac{1}{3}$

Find the equation of the line that contains the given pair of points. [3.3]

16. $(2, 5), (-3, -5)$

17. $(-3, 7), (4, 7)$

18. $(-5, -1), (-3, -4)$

19. Find the equation of the line that is parallel to $2x - y = 4$ and contains the point $(2, -3)$. [3.3]

20. Find the equation of the line perpendicular to $y = -3x + 1$ that has an x-intercept of 2. [3.3]

Graph each linear inequality. [3.4]

21. $y \le 2x - 3$

22. $x \ge -1$

State the domain and range of each relation, and then indicate which relations are also functions. [3.5]

23. $\{(2, 4), (3, 3), (4, 2)\}$

24. $\{(6, 3), (-4, 3), (-2, 0)\}$

If $f = \{(2, -1), (-3, 0), (4, \frac{1}{2}), (\pi, 2)\}$ and $g = \{(2, 2), (-1, 4), (0, 0)\}$, find the following. [3.6]

25. $f(-3)$

26. $f(2) + g(2)$

Let $f(x) = 2x^2 - 4x + 1$ and $g(x) = 3x + 2$, and evaluate each of the following. [3.6]

27. $f(0)$

28. $g(a)$

29. $f[g(0)]$

30. $f[g(1)]$

For the following problems, y varies directly with x. [3.7]

31. If y is 6 when x is 2, find y when x is 8.

32. If y is -3 when x is 5, find y when x is -10.

For the following problems, y varies inversely with the square of x. [3.7]

33. If y is 9 when x is 2, find y when x is 3.

34. If y is 4 when x is 5, find y when x is 2.

Solve each application problem. [3.7]

35. Tension in a Spring The tension t in a spring varies directly with the distance d the spring is stretched. If the tension is 42 pounds when the spring is stretched 2 inches, find the tension when the spring is stretched twice as far.

36. Light Intensity The intensity of a light source varies inversely with the square of the distance from the source. Four feet from the source the intensity is 9 foot-candles. What is the intensity 3 feet from the source?

Chapter 3 Projects

Equations and Inequalities in Two Variables

LIGHT INTENSITY

Group Project

Number of People 2–3

Time Needed 15 minutes

Equipment Paper and pencil

Background I found the following diagram while shopping for some track lighting for my home. I was impressed by the diagram because it displays a lot of useful information in a very efficient manner. As the diagram indicates, the amount of light that falls on a surface depends on how far above the surface the light is placed and how much the light spreads out on the surface. Assume that this light illuminates a circle on a flat surface, and work the following problems.

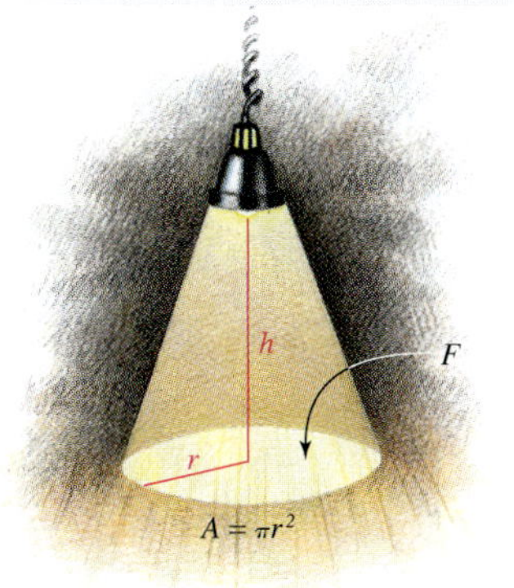

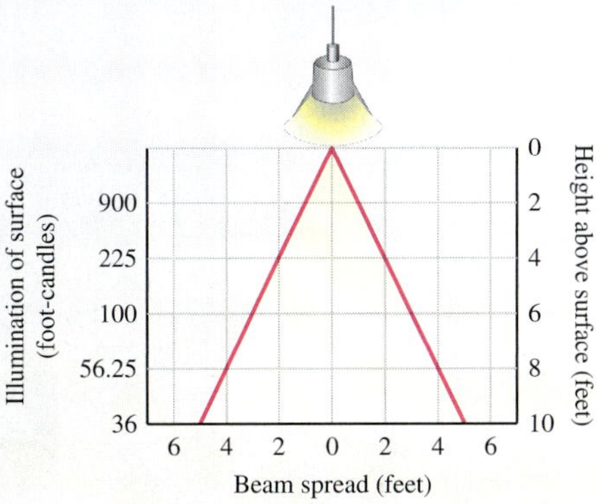

Procedure

a. Fill in each table.

HEIGHT ABOVE SURFACE (FT)	ILLUMINATION (FOOT-CANDLES)
2	
4	
6	
8	
10	

DISTANCE ABOVE SURFACE (FT)	AREA OF ILLUMINATED REGION (FT^2)
2	
4	
6	
8	
10	

b. Use the templates in Figures 1 and 2 to construct line graphs from the data in the tables.

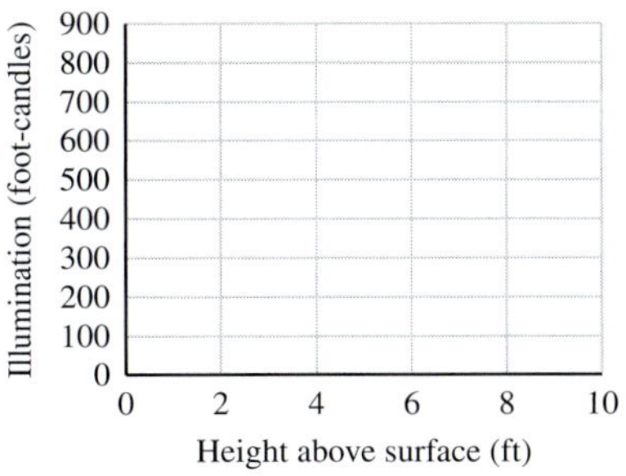

FIGURE 1

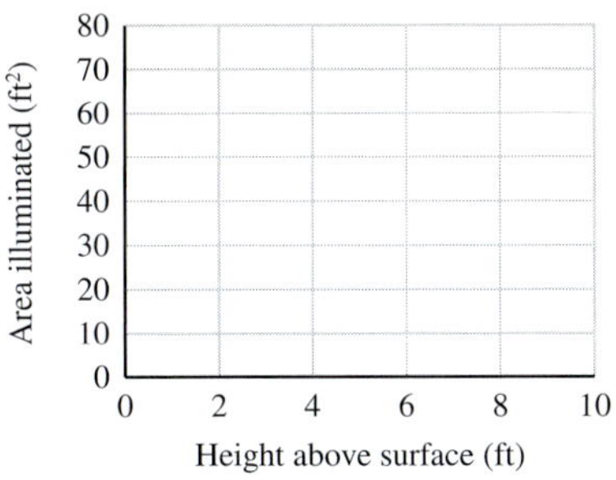

FIGURE 2

c. Which of the relationships is direct variation, and which is inverse variation?

d. Let F represent the number of foot-candles that fall on the surface, h the distance the light source is above the surface, and A the area of the illuminated region. Write an equation that shows the relationship between A and h, then write another equation that gives the relationship between F and h.

DESCARTES AND PASCAL

Research Project

In this chapter, we mentioned that René Descartes, the inventor of the rectangular coordinate system, is the person who made the statement, "I think, therefore, I am." Blaise Pascal, another French philosopher, is responsible for the statement, "The heart has its reasons which reason does not know." Although Pascal and Descartes were contemporaries, the philosophies of the two men differed greatly. Research the philosophy of both Descartes and Pascal, and then write an essay that gives the main points of each man's philosophy. In the essay, show how the quotations given here fit in with the philosophy of the man responsible for the quotation.

René Descartes, 1596–1650

Blaise Pascal, 1623–1662

4

Spencer Grant/PhotoEdit

CHAPTER OUTLINE

INTERMEDIATE Algebra $f(x)$ Now™

Systems of Linear Equations and Inequalities

Suppose you decide to buy a cellular phone and are trying to decide between two rate plans. Plan A is \$18.95 per month plus \$0.48 for each minute, or fraction of a minute, that you use the phone. Plan B is \$34.95 per month plus \$0.36 for each minute, or fraction of a minute. The monthly cost $C(x)$ for each plan can be represented with a linear equation in two variables:

Plan A: $C(x) = 0.48x + 18.95$

Plan B: $C(x) = 0.36x + 34.95$

To compare the two plans, we use the table and graph shown below.

MONTHLY CELLULAR PHONE CHARGES

NUMBER OF MINUTES x	MONTHLY COST PLAN A (\$)	MONTHLY COST PLAN B (\$)
0	18.95	34.95
40	38.15	49.35
80	57.35	63.75
120	76.55	78.15
160	95.75	92.55
200	114.95	106.95
240	134.15	121.35

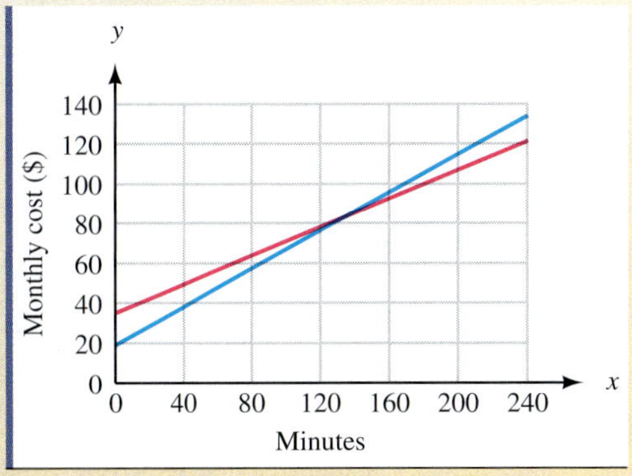

FIGURE 1

The point of intersection of the two lines in Figure 1 is the point at which the monthly costs of the two plans are equal. In this chapter, we will develop methods of finding that point of intersection.

Study Skills

The study skills for this chapter concern the way you approach new situations in mathematics. The first study skill applies to your natural instincts for what does and doesn't work in mathematics. The second study skill gives you a way of testing your instincts.

1 Don't Let Your Intuition Fool You

As you become more experienced and more successful in mathematics, you will be able to trust your mathematical intuition. For now, though, it can get in the way of success. For example, if you ask a beginning algebra student to "subtract 3 from -5" many will answer -2 or 2. Both answers are incorrect, even though they may seem intuitively true.

2 Test Properties About Which You Are Unsure

From time to time you will be in a situation in which you would like to apply a property or rule, but you are not sure if it is true. You can always test a property or statement by substituting numbers for variables. For instance, I always have students that rewrite $(x + 3)^2$ as $x^2 + 9$, thinking that the two expressions are equivalent. The fact that the two expressions are not equivalent becomes obvious when we substitute 10 for x in each one.

When $x = 10$, the expression $(x + 3)^2$ is $(10 + 3)^2 = 13^2 = 169$

When $x = 10$, the expression $x^2 + 9 = 10^2 + 9 = 100 + 9 = 109$

It is not unusual, nor is it wrong, to try occasionally to apply a property that doesn't exist. If you have any doubt about generalizations you are making, test them by replacing variables with numbers and simplifying.

Getting Ready for Chapter 4

Simplify.

1. $0.09(6{,}000)$

2. $1.5(500)$

3. $3(11) - 5(7)$

4. $6 - 12$

5. $3(3) - 5$

6. $4(5) - 3(-3)$

7. $1 + 2(2) - 3(3)$

8. $-4(-1) + 1(2) + 2(6)$

Apply the distributive property, then simplify if possible.

9. $-3(2x - 3y)$

10. $3(4x + 5y)$

11. $6\left(\frac{1}{2}x - \frac{1}{3}y\right)$

12. $12\left(\frac{1}{4}x + \frac{2}{3}y\right)$

13. $10(0.3x + 0.7y)$

14. $100(0.06x + 0.07y)$

Solve.

15. $2(1) + y = 4$

16. $2x - 3(3x - 5) = -6$

17. $2(2y + 6) + 3y = 5$

18. $5\left(\frac{19}{15}\right) + 5y = 9$

19. $4x - 2x = 8$

20. $20x + 9{,}300 > 18{,}000$

Systems of Linear Equations in Two Variables

4.1

OBJECTIVES

A	Solve systems of linear equations in two variables by graphing.
B	Solve systems of linear equations in two variables by the addition method.
C	Solve systems of linear equation in two variables by the substitution method.

Previously we found the graph of an equation of the form $ax + by = c$ to be a straight line. Since the graph is a straight line, the equation is said to be a linear equation. Two linear equations considered together form a *linear system* of equations. For example,

$$3x - 2y = 6$$

$$2x + 4y = 20$$

is a linear system. The solution set to the system is the set of all ordered pairs that satisfy both equations. If we graph each equation on the same set of axes, we can see the solution set (Figure 1).

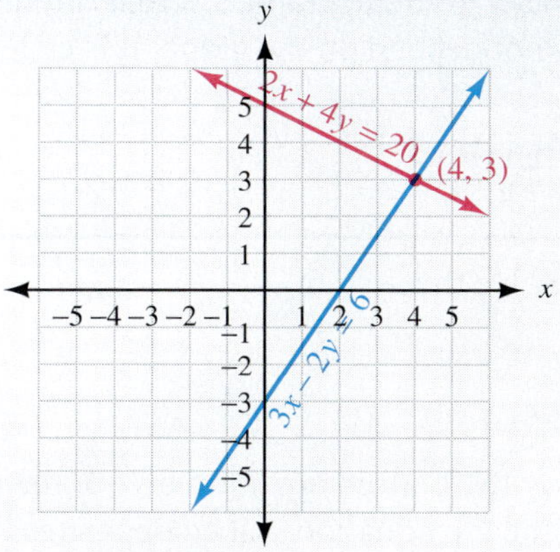

FIGURE 1

The point (4, 3) lies on both lines and therefore must satisfy both equations. It is obvious from the graph that it is the only point that does so. The solution set for the system is {(4, 3)}.

More generally, if $a_1x + b_1y = c_1$ and $a_2x + b_2y = c_2$ are linear equations, then the solution set for the system

$$a_1x + b_1y = c_1$$

$$a_2x + b_2y = c_2$$

can be illustrated through one of the graphs in Figure 2.

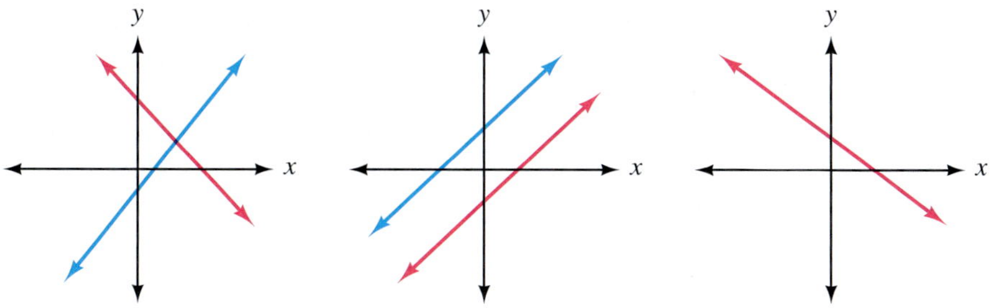

Case I
(One solution)

Case II
(No solutions)

Case III
(Infinite number of solutions)

FIGURE 2

NOTE

It is important that you write solutions to systems of equations in two variables as ordered pairs, like (4, 3); that is, you should always enclose the ordered pairs in parentheses.

Case I The two lines intersect at one and only one point. The coordinates of the point give the solution to the system. This is what usually happens.

Case II The lines are parallel and therefore have no points in common. The solution set to the system is the empty set, ∅. In this case, we say the equations are *inconsistent*.

Case III The lines coincide; that is, their graphs represent the same line. The solution set consists of all ordered pairs that satisfy either equation. In this case, the equations are said to be *dependent*.

In the beginning of this section we found the solution set for the system

$$3x - 2y = 6$$
$$2x + 4y = 20$$

by graphing each equation and then reading the solution set from the graph. Solving a system of linear equations by graphing is the least accurate method. If the coordinates of the point of intersection are not integers, it can be very difficult to read the solution set from the graph. There is another method of solving a linear system that does not depend on the graph. It is called the *addition method*.

The Addition Method

PRACTICE PROBLEMS

1. Solve the system.

$$2x - y = 7$$
$$3x + 4y = -6$$

by first multiplying the top equation by 4 and then adding the result to the bottom equation.

EXAMPLE 1 Solve the system.

$$4x + 3y = 10$$
$$2x + y = 4$$

Solution If we multiply the bottom equation by -3, the coefficients of y in the resulting equation and the top equation will be opposites:

$$4x + 3y = 10 \xrightarrow{\text{No change}} 4x + 3y = 10$$
$$2x + y = 4 \xrightarrow[\text{Multiply by } -3]{} -6x - 3y = -12$$

Adding the left and right sides of the resulting equations, we have

$$\begin{array}{r} 4x + 3y = 10 \\ -6x - 3y = -12 \\ \hline -2x \quad = -2 \end{array}$$

The result is a linear equation in one variable. We have eliminated the variable y from the equations by addition. (It is for this reason we call this method of solving a linear system the *addition method*.) Solving $-2x = -2$ for x, we have

$$x = 1$$

This is the x-coordinate of the solution to our system. To find the y-coordinate, we substitute $x = 1$ into any of the equations containing both the variables x and y. Let's try the second equation in our original system:

$$2(1) + y = 4$$
$$2 + y = 4$$
$$y = 2$$

This is the y-coordinate of the solution to our system. The ordered pair (1, 2) is the solution to the system.

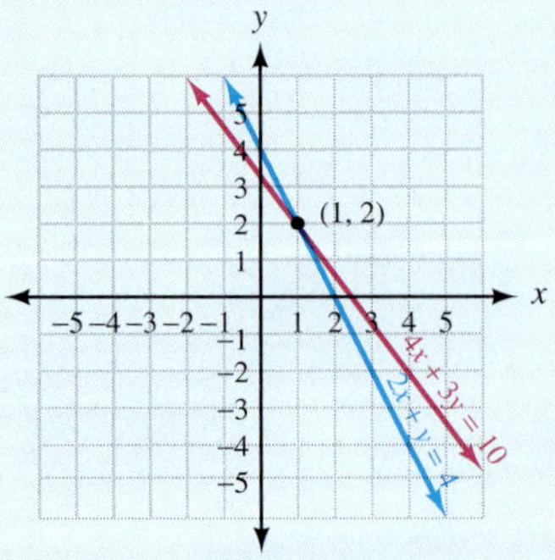

FIGURE 3 **A visual representation of the solution to the system in Example 1**

Checking Solutions We can check our solution by substituting it into both of our equations.

Substituting $x = 1$ and $y = 2$ into $4x + 3y = 10$, we have

$$4(1) + 3(2) \stackrel{?}{=} 10$$
$$4 + 6 \stackrel{?}{=} 10$$
$$10 = 10 \quad \textbf{A true statement}$$

Substituting $x = 1$ and $y = 2$ into $2x + y = 4$, we have

$$2(1) + 2 \stackrel{?}{=} 4$$
$$2 + 2 \stackrel{?}{=} 4$$
$$4 = 4 \quad \textbf{A true statement}$$

Our solution satisfies both equations; therefore, it is a solution to our system of equations.

EXAMPLE 2

Solve the system.

$$3x - 5y = -2$$
$$2x - 3y = 1$$

Solution We can eliminate either variable. Let's decide to eliminate the variable x. We can do so by multiplying the top equation by 2 and the bottom equation by -3, and then adding the left and right sides of the resulting equations:

$$3x - 5y = -2 \xrightarrow{\text{Multiply by 2}} 6x - 10y = -4$$
$$2x - 3y = 1 \xrightarrow[\text{Multiply by } -3]{} -6x + 9y = -3$$
$$-y = -7$$
$$y = 7$$

The y-coordinate of the solution to the system is 7. Substituting this value of y into any of the equations with both x- and y-variables gives $x = 11$. The solution to the system is (11, 7). It is the only ordered pair that satisfies both equations.

Checking Solutions Checking (11, 7) in each equation looks like this

Substituting $x = 11$ and $y = 7$ into $3x - 5y = -2$, we have

$$3(11) - 5(7) \stackrel{?}{=} -2$$
$$33 - 35 \stackrel{?}{=} -2$$
$$-2 = -2 \quad \textbf{A true statement}$$

Substituting $x = 11$ and $y = 7$ into $2x - 3y = 1$, we have

$$2(11) - 3(7) \stackrel{?}{=} 1$$
$$22 - 21 \stackrel{?}{=} 1$$
$$1 = 1 \quad \textbf{A true statement}$$

Our solution satisfies both equations; therefore, (11, 7) is a solution to our system.

EXAMPLE 3

Solve the system.

$$2x - 3y = 4$$
$$4x + 5y = 3$$

Solution We can eliminate x by multiplying the top equation by -2 and adding it to the bottom equation:

$$2x - 3y = 4 \xrightarrow{\text{Multiply by } -2} -4x + 6y = -8$$
$$4x + 5y = 3 \xrightarrow[\text{No change}]{} 4x + 5y = 3$$
$$11y = -5$$
$$y = -\frac{5}{11}$$

2. Solve the system.

$$3x - 2y = -8$$
$$-2x + 3y = 7$$

3. Solve the system.

$$3x - 5y = 2$$
$$2x + 4y = 1$$

Answers

1. (2, −3) **2.** (−2, 1)

The y-coordinate of our solution is $-\frac{5}{11}$. If we were to substitute this value of y back into either of our original equations, we would find the arithmetic necessary to solve for x cumbersome. For this reason, it is probably best to go back to the original system and solve it a second time—for x instead of y. Here is how we do that:

$$\begin{aligned} 2x - 3y = 4 &\xrightarrow{\text{Multiply by 5}} 10x - 15y = 20 \\ 4x + 5y = 3 &\xrightarrow[\text{Multiply by 3}]{} \underline{12x + 15y = 9} \\ &\phantom{\xrightarrow{\text{Multiply by 5}}} 22x = 29 \\ &\phantom{\xrightarrow{\text{Multiply by 5}}} x = \frac{29}{22} \end{aligned}$$

The solution to our system is $\left(\frac{29}{22}, -\frac{5}{11}\right)$. ■

The main idea in solving a system of linear equations by the addition method is to use the multiplication property of equality on one or both of the original equations, if necessary, to make the coefficients of either variable opposites. The following box shows some steps to follow when solving a system of linear equations by the addition method.

Strategy for Solving a System of Linear Equations by the Addition Method

Step 1: Decide which variable to eliminate. (In some cases, one variable will be easier to eliminate than the other. With some practice, you will notice which one it is.)

Step 2: Use the multiplication property of equality on each equation separately to make the coefficients of the variable that is to be eliminated opposites.

Step 3: Add the respective left and right sides of the system together.

Step 4: Solve for the remaining variable.

Step 5: Substitute the value of the variable from step 4 into an equation containing both variables and solve for the other variable. (Or repeat steps 2–4 to eliminate the other variable.)

Step 6: Check your solution in both equations, if necessary.

4. Solve the system.

$$2x + 7y = 3$$
$$4x + 14y = 1$$

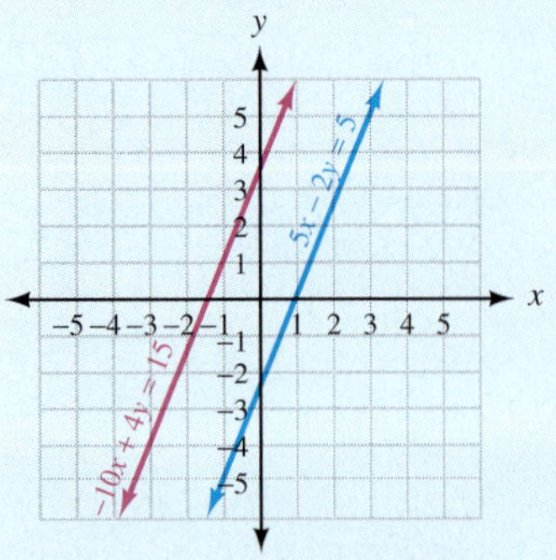

FIGURE 4 **A visual representation of the situation in Example 4—the two lines are parallel**

EXAMPLE 4

Solve the system.

$$5x - 2y = 5$$
$$-10x + 4y = 15$$

Solution We can eliminate y by multiplying the first equation by 2 and adding the result to the second equation:

$$\begin{aligned} 5x - 2y = 5 &\xrightarrow{\text{Multiply by 2}} 10x - 4y = 10 \\ -10x + 4y = 15 &\xrightarrow[\text{No change}]{} \underline{-10x + 4y = 15} \\ &\phantom{\xrightarrow{\text{Multiply by 2}}} 0 = 25 \end{aligned}$$

The result is the false statement $0 = 25$, which indicates there is no solution to the system. If we were to graph the two lines, we would find that they are parallel. In a case like this, we say the system is *inconsistent.* Whenever both variables have been eliminated and the resulting statement is false, the solution set for the system will be the empty set, $\varnothing$. ■

Answers

3. $\left(\frac{13}{22}, -\frac{1}{22}\right)$ **4.** $\varnothing$

EXAMPLE 5 Solve the system.

$$4x + 3y = 2$$
$$8x + 6y = 4$$

Solution Multiplying the top equation by -2 and adding, we can eliminate the variable x:

$$4x + 3y = 2 \xrightarrow{\text{Multiply by } -2} -8x - 6y = -4$$
$$8x + 6y = 4 \xrightarrow[\text{No change}]{} \underline{8x + 6y = 4}$$
$$0 = 0$$

Both variables have been eliminated and the resulting statement $0 = 0$ is true. In this case, the lines coincide and the system is said to be *dependent.* The solution set consists of all ordered pairs that satisfy either equation. We can write the solution set as $\{(x, y) \mid 4x + 3y = 2\}$ or $\{(x, y) \mid 8x + 6y = 4\}$. ■

5. Solve the system.

$$2x + 7y = 3$$
$$4x + 14y = 6$$

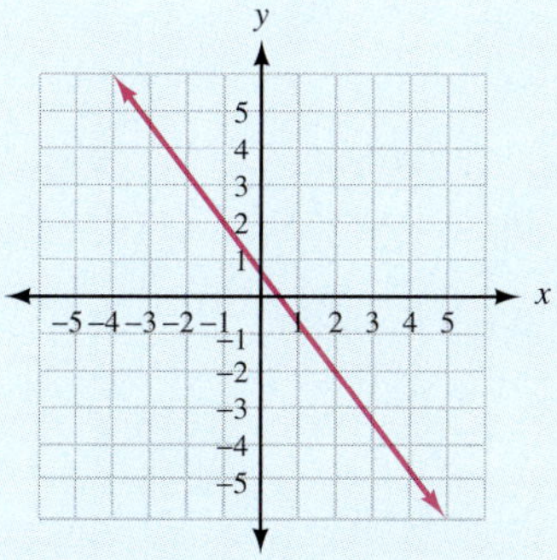

FIGURE 5 **A visual representation of the situation in Example 5—both equations produce the same graph**

Special Cases

The previous two examples illustrate the two special cases in which the graphs of the equations in the system either coincide or are parallel. In both cases the left-hand sides of the equations were multiples of each other. In the case of the dependent equations the right-hand sides were also multiples. We can generalize these observations for the system

$$a_1x + b_1y = c_1$$
$$a_2x + b_2y = c_2$$

Inconsistent System

What Happens	*Geometric Interpretation*	*Algebraic Interpretation*
Both variables are eliminated, and the resulting statement is false.	The lines are parallel, and there is no solution to the system.	$\frac{a_1}{a_2} = \frac{b_1}{b_2} \neq \frac{c_1}{c_2}$

Dependent Equations

What Happens	*Geometric Interpretation*	*Algebraic Interpretation*
Both variables are eliminated, and the resulting statement is true.	The lines coincide, and there are an infinite number of solutions to the system.	$\frac{a_1}{a_2} = \frac{b_1}{b_2} = \frac{c_1}{c_2}$

EXAMPLE 6 Solve the system.

$$\frac{1}{2}x - \frac{1}{3}y = 2$$
$$\frac{1}{4}x + \frac{2}{3}y = 6$$

6. Solve the system.

$$\frac{1}{3}x + \frac{1}{2}y = 4$$
$$\frac{2}{3}x - \frac{1}{4}y = 3$$

Answer
5. Lines coincide

Solution Although we could solve this system without clearing the equations of fractions, there is probably less chance for error if we have only integer coefficients to work with. So let's begin by multiplying both sides of the top equation by 6 and both sides of the bottom equation by 12 to clear each equation of fractions:

$$\frac{1}{2}x - \frac{1}{3}y = 2 \xrightarrow{\text{Times 6}} 3x - 2y = 12$$

$$\frac{1}{4}x + \frac{2}{3}y = 6 \xrightarrow[\text{Times 12}]{} 3x + 8y = 72$$

Now we can eliminate x by multiplying the top equation by -1 and leaving the bottom equation unchanged:

$$\begin{aligned} 3x - 2y = 12 &\xrightarrow{\text{Times } -1} -3x + 2y = -12 \\ 3x + 8y = 72 &\xrightarrow[\text{No change}]{} \underline{3x + 8y = 72} \\ &\qquad\qquad 10y = 60 \\ &\qquad\qquad y = 6 \end{aligned}$$

We can substitute $y = 6$ into any equation that contains both x and y. Let's use $3x - 2y = 12$.

$$\begin{aligned} 3x - 2(6) &= 12 \\ 3x - 12 &= 12 \\ 3x &= 24 \\ x &= 8 \end{aligned}$$

The solution to the system is (8, 6). ■

The Substitution Method

EXAMPLE 7 Solve the system.

$$\begin{aligned} 2x - 3y &= -6 \\ y &= 3x - 5 \end{aligned}$$

Solution The second equation tells us y is $3x - 5$. Substituting the expression $3x - 5$ for y in the first equation, we have

$$2x - 3(3x - 5) = -6$$

The result of the substitution is the elimination of the variable y. Solving the resulting linear equation in x as usual, we have

$$\begin{aligned} 2x - 9x + 15 &= -6 \\ -7x + 15 &= -6 \\ -7x &= -21 \\ x &= 3 \end{aligned}$$

7. Solve the system

$$\begin{aligned} 4x - 2y &= -2 \\ y &= x + 3 \end{aligned}$$

by substituting the expression for y given in the second equation into the first equation.

Answer

6. (6, 4)

Putting $x = 3$ into the second equation in the original system, we have

$$y = 3(3) - 5$$
$$= 9 - 5$$
$$= 4$$

The solution to the system is (3, 4).

Checking Solutions Checking (3, 4) in each equation looks like this

Substituting $x = 3$ and $y = 4$ into $2x - 3y = -6$, we have

$$2(3) - 3(4) \stackrel{?}{=} -6$$
$$6 - 12 \stackrel{?}{=} -6$$
$$-6 = -6 \quad \textbf{A true statement}$$

Substituting $x = 3$ and $y = 4$ into $y = 3x - 5$, we have

$$4 \stackrel{?}{=} 3(3) - 5$$
$$4 \stackrel{?}{=} 9 - 5$$
$$4 = 4 \quad \textbf{A true statement}$$

Our solution satisfies both equations; therefore, (3, 4) is a solution to our system.

Strategy for Solving a System of Equations by the Substitution Method

Step 1: Solve either one of the equations for *x* or *y*. (This step is not necessary if one of the equations is already in the correct form, as in Example 7.)

Step 2: Substitute the expression for the variable obtained in step 1 into the other equation and solve it.

Step 3: Substitute the solution for step 2 into any equation in the system that contains both variables and solve it.

Step 4: Check your results, if necessary.

EXAMPLE 8 Solve by substitution.

$$2x + 3y = 5$$
$$x - 2y = 6$$

Solution To use the substitution method, we must solve one of the two equations for x or y. We can solve for x in the second equation by adding $2y$ to both sides:

$$x - 2y = 6$$
$$x = 2y + 6 \quad \textbf{Add } 2y \textbf{ to both sides.}$$

Substituting the expression $2y + 6$ for x in the first equation of our system, we have

$$2(2y + 6) + 3y = 5$$
$$4y + 12 + 3y = 5$$
$$7y + 12 = 5$$
$$7y = -7$$
$$y = -1$$

Using $y = -1$ in either equation in the original system, we find $x = 4$. The solution is $(4, -1)$.

8. Solve by substitution.

$$5x - 3y = -4$$
$$x + 2y = 7$$

NOTE

Both the substitution method and the addition method can be used to solve any system of linear equations in two variables. Systems like the one in Example 7, however, are easier to solve using the substitution method because one of the variables is already written in terms of the other. A system like the one in Example 2 is easier to solve using the addition method because solving for one of the variables would lead to an expression involving fractions. The system in Example 8 could be solved easily by either method because solving the second equation for x is a one-step process.

Answers

7. (2, 5) **8.** (1, 3)

USING TECHNOLOGY

Graphing Calculators

Solving Systems That Intersect in Exactly One Point

A graphing calculator can be used to solve a system of equations in two variables if the equations intersect in exactly one point. To solve the system shown in Example 3, we first solve each equation for y. Here is the result:

$$2x - 3y = 4 \quad \text{becomes} \quad y = \frac{4 - 2x}{-3}$$

$$4x + 5y = 3 \quad \text{becomes} \quad y = \frac{3 - 4x}{5}$$

Graphing these two functions on the calculator gives a diagram similar to the one in Figure 6.

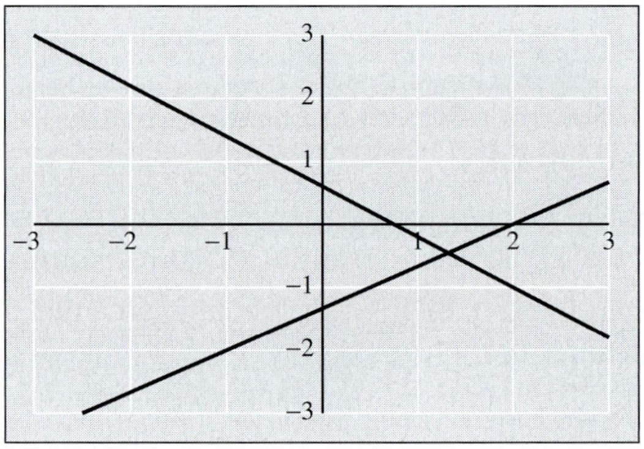

FIGURE 6

Using the Trace and Zoom features, we find that the two lines intersect at $x = 1.32$ and $y = -0.45$, which are the decimal equivalents (accurate to the nearest hundredth) of the fractions found in Example 3.

Linking Objectives and Examples

Next to each **objective**, list the examples that are best described by that objective.

A	
B	
C	

Getting Ready for Class

After reading through the preceding section, respond in your own words and in complete sentences.

1. Two linear equations, each with the same two variables, form a system of equations. How do we define a solution to this system? That is, what form will a solution have, and what properties does a solution possess?
2. When would substitution be more efficient than the addition method in solving two linear equations?
3. Explain what an inconsistent system of linear equations looks like graphically and what would result algebraically when attempting to solve the system.
4. When might the graphing method of solving a system of equations be more desirable than the other techniques, and when might it be less desirable?

Problem Set 4.1

INTERMEDIATE **Algebra** $f(x)$ **Now**™

Solve each system by graphing both equations on the same set of axes and then reading the solution from the graph.

1. $3x - 2y = 6$

$x - y = 1$

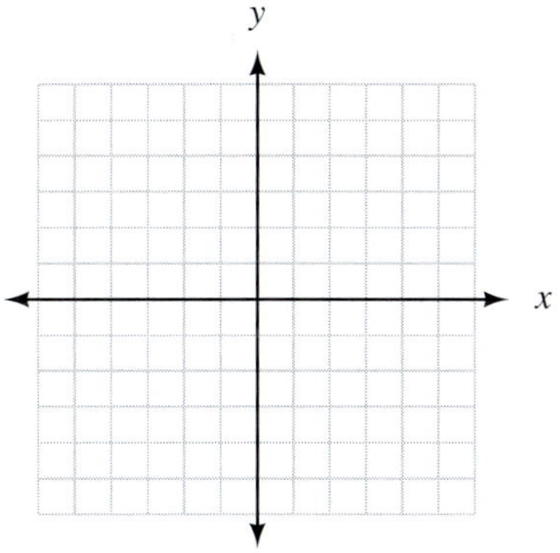

2. $5x - 2y = 10$

$x - y = -1$

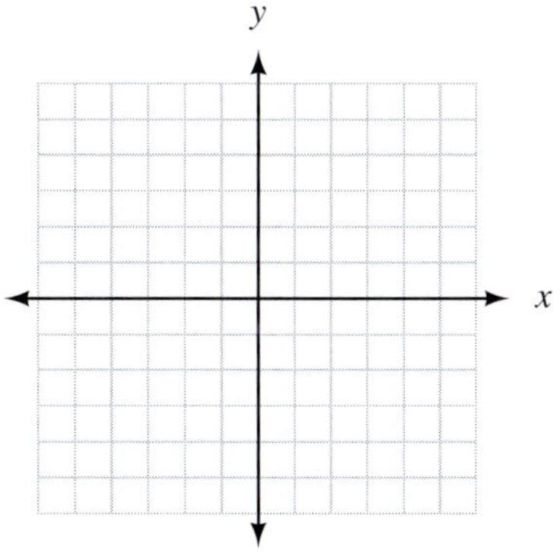

3. $y = \frac{3}{5}x - 3$

$2x - y = -4$

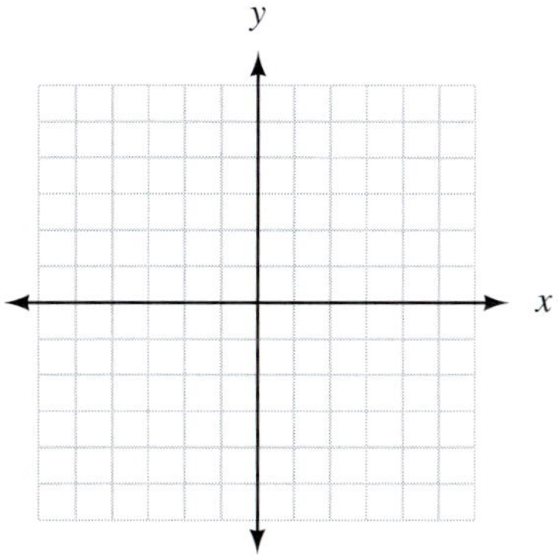

4. $y = \frac{1}{2}x - 2$

$2x - y = -1$

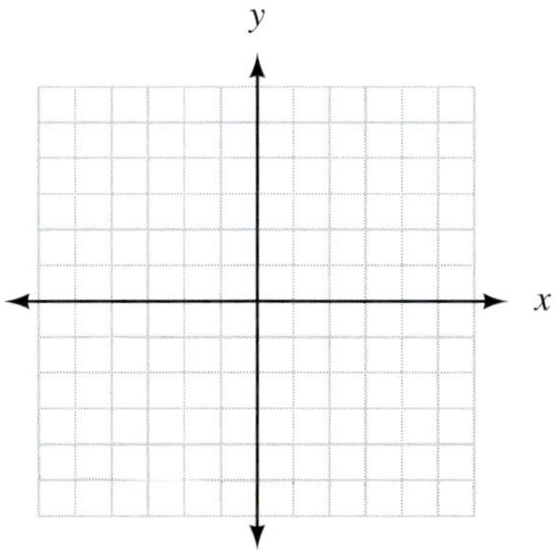

5. $y = \frac{1}{2}x$

$y = -\frac{3}{4}x + 5$

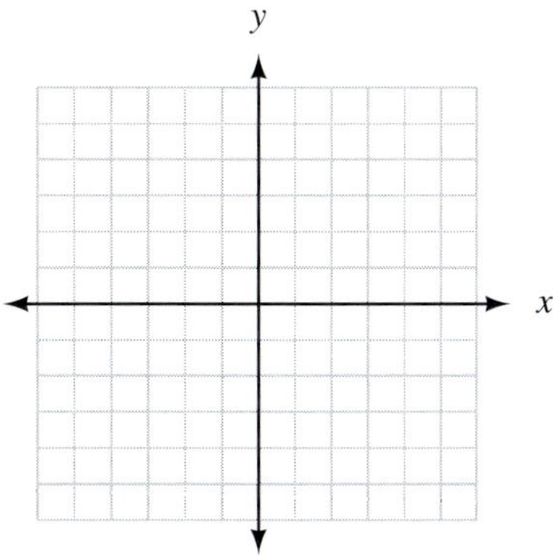

6. $y = \frac{2}{3}x$

$y = -\frac{1}{3}x + 6$

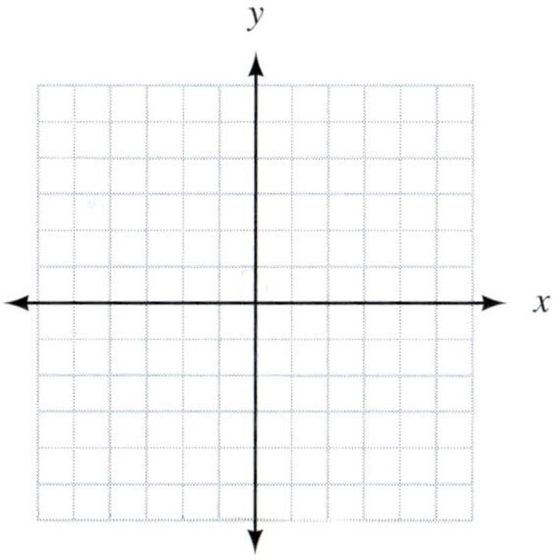

7. $3x + 3y = -2$

$y = -x + 4$

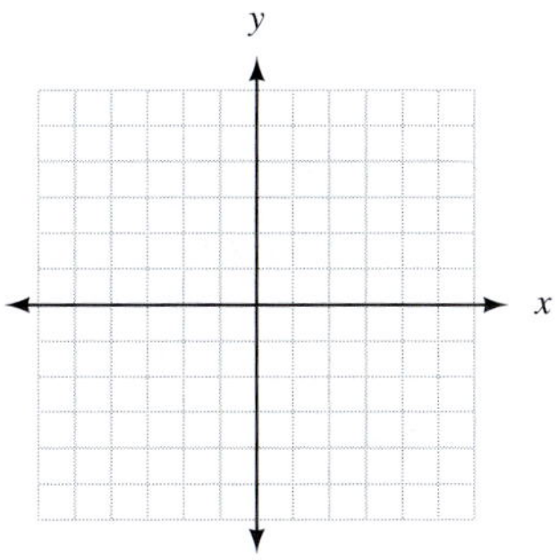

8. $2x - 2y = 6$

$y = x - 3$

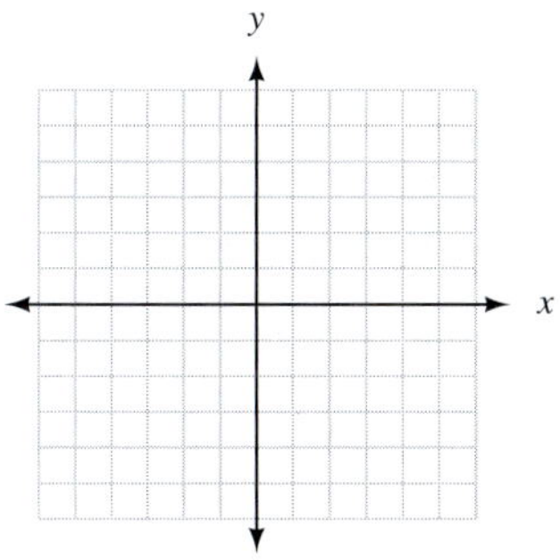

9. $2x - y = 5$

$y = 2x - 5$

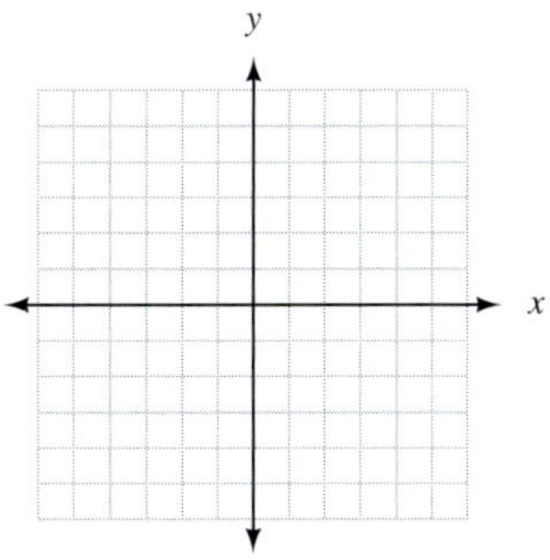

10. $x + 2y = 5$

$y = -\frac{1}{2}x + 3$

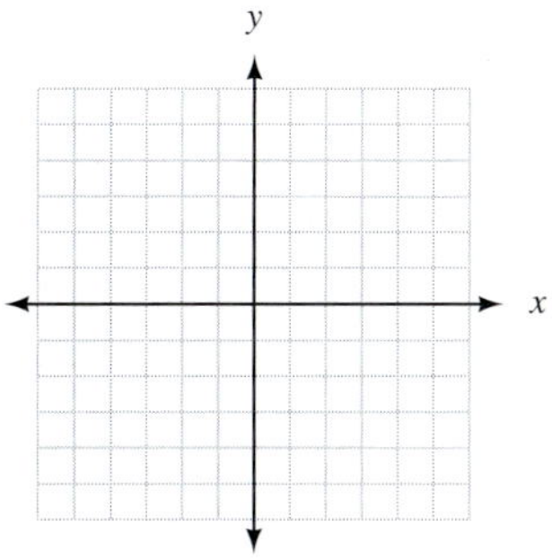

Solve each of the following systems by the addition method.

11. $x + y = 5$
$3x - y = 3$

12. $x - y = 4$
$-x + 2y = -3$

13. $3x + y = 4$
$4x + y = 5$

14. $6x - 2y = -10$
$6x + 3y = -15$

15. $3x - 2y = 6$
$6x - 4y = 12$

16. $4x + 5y = -3$
$-8x - 10y = 3$

17. $x + 2y = 0$
$2x - 6y = 5$

18. $x + 3y = 3$
$2x - 9y = 1$

19. $2x - 5y = 16$
$4x - 3y = 11$

20. $5x - 3y = -11$
$7x + 6y = -12$

21. $6x + 3y = -1$
$9x + 5y = 1$

22. $5x + 4y = -1$
$7x + 6y = -2$

23. $4x + 3y = 14$
$9x - 2y = 14$

24. $7x - 6y = 13$
$6x - 5y = 11$

25. $2x - 5y = 3$
$-4x + 10y = 3$

26. $3x - 2y = 1$
$-6x + 4y = -2$

27. $\frac{1}{4}x - \frac{1}{6}y = -2$
$-\frac{1}{6}x + \frac{1}{5}y = 4$

28. $-\frac{1}{3}x + \frac{1}{4}y = 0$
$\frac{1}{5}x - \frac{1}{10}y = 1$

29. $\frac{1}{2}x + \frac{1}{3}y = 13$
$\frac{2}{5}x + \frac{1}{4}y = 10$

30. $\frac{1}{2}x + \frac{1}{3}y = \frac{2}{3}$
$\frac{2}{3}x + \frac{2}{5}y = \frac{14}{15}$

Solve each of the following systems by the substitution method.

31. $7x - y = 24$
$x = 2y + 9$

32. $3x - y = -8$
$y = 6x + 3$

33. $6x - y = 10$
$y = -\frac{3}{4}x - 1$

34. $2x - y = 6$
$y = -\frac{4}{3}x + 1$

35. $3y + z = 17$
$5y + 20z = 65$

36. $x + y = 850$
$1.5x + y = 1{,}100$

37. $y = 3x - 2$
$y = 4x - 4$

38. $y = 5x - 2$
$y = -2x + 5$

39. $2x - y = 5$
$4x - 2y = 10$

40. $-10x + 8y = -6$
$y = \frac{5}{4}x$

41. $\frac{1}{3}x - \frac{1}{2}y = 0$
$x = \frac{3}{2}y$

42. $\frac{2}{5}x - \frac{2}{3}y = 0$
$y = \frac{3}{5}x$

You may want to read Example 3 again before solving the systems that follow.

43. $4x - 7y = 3$
$5x + 2y = -3$

44. $3x - 4y = 7$
$6x - 3y = 5$

45. $9x - 8y = 4$
$2x + 3y = 6$

46. $4x - 7y = 10$
$-3x + 2y = -9$

47. $3x - 5y = 2$
$7x + 2y = 1$

48. $4x - 3y = -1$
$5x + 8y = 2$

49. Multiply both sides of the second equation in the following system by 100, and then solve as usual.

$$x + y = 10{,}000$$
$$0.06x + 0.05y = 560$$

50. Multiply both sides of the second equation in the following system by 10, and then solve as usual.

$$x + y = 12$$
$$0.20x + 0.50y = 0.30(12)$$

51. What value of c will make the following system a dependent system (one in which the lines coincide)?

$$6x - 9y = 3$$
$$4x - 6y = c$$

52. What value of c will make the following system a dependent system?

$$5x - 7y = c$$
$$-15x + 21y = 9$$

53. Where do the graphs of the lines $x + y = 4$ and $x - 2y = 4$ intersect?

54. Where do the graphs of the line $x = -1$ and $x - 2y = 4$ intersect?

Getting Ready for the Next Section

Simplify.

55. $2 - 2(6)$

56. $2(1) - 2 + 3$

57. $(x + 3y) - 1(x - 2z)$

58. $(x + y + z) + (2x - y + z)$

Solve.

59. $-9y = -9$

60. $30x = 38$

61. $3(1) + 2z = 9$

62. $4\left(\frac{19}{15}\right) - 2y = 4$

Apply the distributive property, then simplify if possible.

63. $2(5x - z)$

64. $-1(x - 2z)$

65. $3(3x + y - 2z)$

66. $2(2x - y + z)$

Maintaining Your Skills

67. Find the slope of the line that contains $(-4, -1)$ and $(-2, 5)$.

68. A line has a slope of $\frac{2}{3}$. Find the slope of any line

a. Parallel to it.

b. Perpendicular to it.

69. Give the slope and y-intercept of the line $2x - 3y = 6$.

70. Give the equation of the line with slope -3 and y-intercept 5.

71. Find the equation of the line with slope $\frac{2}{3}$ that contains the point $(-6, 2)$.

72. Find the equation of the line through $(1, 3)$ and $(-1, -5)$.

73. Find the equation of the line with x-intercept 3 and y-intercept -2.

74. Find the equation of the line through $(-1, 4)$ whose graph is perpendicular to the graph of $y = 2x + 3$.

Extending the Concepts

75. Find a and b so that the line $ax + by = 7$ passes through the points $(1, -2)$ and $(3, 1)$.

76. Find a and b so that the line $ax + by = 2$ passes through the points $(2, 2)$ and $(6, 7)$.

Improving Your Quantitative Literacy

77. The graph here shows the percentage of alcohol- and nonalcohol-related car crashes for the years 1982 through 2004. Although these graphs are not linear, they do intersect, and so we can apply some of the same reasoning we have used in this section to interpret what the intersections mean.

a. How many points of intersection are there on the chart?

b. During what period of years do the graphs intersect?

c. Explain the significance of the points of intersection in terms of car crash fatalities.

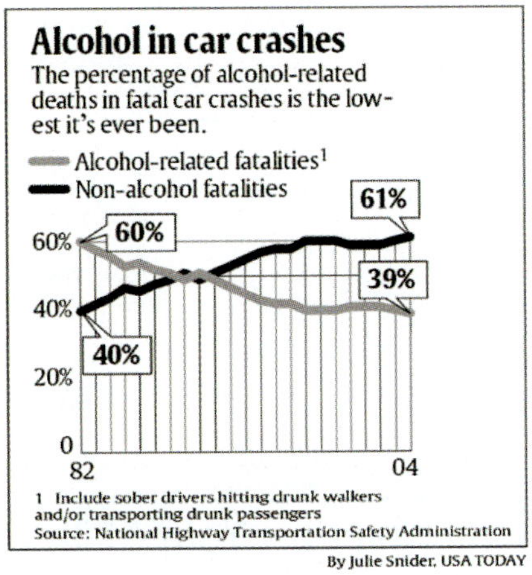

From *USA Today*. Copyright 2005. Reprinted with permission.

Systems of Linear Equations in Three Variables

4.2

A solution to an equation in three variables such as

$$2x + y - 3z = 6$$

Is an ordered triple of numbers (x, y, z). For example, the ordered triples $(0, 0, -2)$, $(2, 2, 0)$, and $(0, 9, 1)$ are solutions to the equation $2x + y - 3z = 6$ since they produce a true statement when their coordinates are substituted for $x, y,$ and z in the equation.

OBJECTIVES

A	Solve systems of linear equations in three variables.

DEFINITION

The **solution set** for a system of three linear equations in three variables is the set of ordered triples that satisfy all three equations.

EXAMPLE 1 Solve the system.

$$\begin{aligned} x + y + z &= 6 && \textbf{(1)} \\ 2x - y + z &= 3 && \textbf{(2)} \\ x + 2y - 3z &= -4 && \textbf{(3)} \end{aligned}$$

Solution We want to find the ordered triple (x, y, z) that satisfies all three equations. We have numbered the equations so it will be easier to keep track of where they are and what we are doing.

There are many ways to proceed. The main idea is to take two different pairs of equations and eliminate the same variable from each pair. We begin by adding equations (1) and (2) to eliminate the y-variable. The resulting equation is numbered (4):

$$\begin{aligned} x + y + z &= 6 && \textbf{(1)} \\ 2x - y + z &= 3 && \textbf{(2)} \\ \hline 3x \quad + 2z &= 9 && \textbf{(4)} \end{aligned}$$

Adding twice equation (2) to equation (3) will also eliminate the variable y. The resulting equation is numbered (5):

$$\begin{aligned} 4x - 2y + 2z &= 6 && \textbf{Twice (2)} \\ x + 2y - 3z &= -4 && \textbf{(3)} \\ \hline 5x \quad - z &= 2 && \textbf{(5)} \end{aligned}$$

Equations (4) and (5) form a linear system in two variables. By multiplying equation (5) by 2 and adding the result to equation (4), we succeed in eliminating the variable z from the new pair of equations:

$$\begin{aligned} 3x + 2z &= 9 && \textbf{(4)} \\ 10x - 2z &= 4 && \textbf{Twice (5)} \\ \hline 13x \quad &= 13 \\ x &= 1 \end{aligned}$$

Substituting $x = 1$ into equation (4), we have

$$\begin{aligned} 3(1) + 2z &= 9 \\ 2z &= 6 \\ z &= 3 \end{aligned}$$

PRACTICE PROBLEMS

1. Solve the system.

$$\begin{aligned} x + 2y + z &= 2 \\ x + y - z &= 6 \\ x - y + 2z &= -7 \end{aligned}$$

Using $x = 1$ and $z = 3$ in equation (1) gives us

$$1 + y + 3 = 6$$
$$y + 4 = 6$$
$$y = 2$$

The solution is the ordered triple (1, 2, 3).

2. Solve the system.
$$3x - 2y + z = 2$$
$$3x + y + 3z = 7$$
$$x + 4y - z = 4$$

EXAMPLE 2

Solve the system.

$$2x + y - z = 3 \quad \textbf{(1)}$$
$$3x + 4y + z = 6 \quad \textbf{(2)}$$
$$2x - 3y + z = 1 \quad \textbf{(3)}$$

Solution It is easiest to eliminate z from the equations. The equation produced by adding (1) and (2) is

$$5x + 5y = 9 \quad \textbf{(4)}$$

The equation that results from adding (1) and (3) is

$$4x - 2y = 4 \quad \textbf{(5)}$$

Equations (4) and (5) form a linear system in two variables. We can eliminate the variable y from this system as follows:

$$5x + 5y = 9 \xrightarrow{\text{Multiply by 2}} 10x + 10y = 18$$
$$4x - 2y = 4 \xrightarrow[\text{Multiply by 5}]{} \underline{20x - 10y = 20}$$
$$30x = 38$$
$$x = \frac{38}{30}$$
$$= \frac{19}{15}$$

Substituting $x = \frac{19}{15}$ into equation (5) or equation (4) and solving for y gives

$$y = \frac{8}{15}$$

Using $x = \frac{19}{15}$ and $y = \frac{8}{15}$ in equation (1), (2), or (3) and solving for z results in

$$z = \frac{1}{15}$$

The ordered triple that satisfies all three equations is $(\frac{19}{15}, \frac{8}{15}, \frac{1}{15})$.

3. Solve the system.
$$3x + 5y - 2z = 1$$
$$6x + 10y - 4z = 2$$
$$x - 8y + z = 4$$

EXAMPLE 3

Solve the system.

$$2x + 3y - z = 5 \quad \textbf{(1)}$$
$$4x + 6y - 2z = 10 \quad \textbf{(2)}$$
$$x - 4y + 3z = 5 \quad \textbf{(3)}$$

Solution Multiplying equation (1) by -2 and adding the result to equation (2) looks like this:

$$-4x - 6y + 2z = -10 \quad \textbf{−2 times (1)}$$
$$\underline{4x + 6y - 2z = 10} \quad \textbf{(2)}$$
$$0 = 0$$

Answers

1. (1, 2, −3) **2.** (1, 1, 1)

All three variables have been eliminated, and we are left with a true statement. As was the case in the previous section, this implies that the two equations are dependent. With a system of three equations in three variables, however, a dependent system can have no solution or an infinite number of solutions.

EXAMPLE 4

Solve the system.

$$\begin{aligned} x - 5y + 4z &= 8 && \textbf{(1)} \\ 3x + y - 2z &= 7 && \textbf{(2)} \\ -9x - 3y + 6z &= 5 && \textbf{(3)} \end{aligned}$$

Solution Multiplying equation (2) by 3 and adding the result to equation (3) produces

$$\begin{aligned} 9x + 3y - 6z &= 21 && \textbf{3 times (2)} \\ -9x - 3y + 6z &= 5 && \textbf{(3)} \\ \hline 0 &= 26 \end{aligned}$$

In this case all three variables have been eliminated, and we are left with a false statement. The two equations are inconsistent: there are no ordered triples that satisfy both equations. The solution set for the system is the empty set, $\varnothing$. If equations (2) and (3) have no ordered triples in common, then certainly (1), (2), and (3) do not either.

EXAMPLE 5

Solve the system.

$$\begin{aligned} x + 3y &= 5 && \textbf{(1)} \\ 6y + z &= 12 && \textbf{(2)} \\ x - 2z &= -10 && \textbf{(3)} \end{aligned}$$

Solution It may be helpful to rewrite the system as

$$\begin{aligned} x + 3y &= 5 && \textbf{(1)} \\ 6y + z &= 12 && \textbf{(2)} \\ x - 2z &= -10 && \textbf{(3)} \end{aligned}$$

Equation (2) does not contain the variable x. If we multiply equation (3) by -1 and add the result to equation (1), we will be left with another equation that does not contain the variable x:

$$\begin{aligned} x + 3y &= 5 && \textbf{(1)} \\ -x + 2z &= 10 && \textbf{−1 times (3)} \\ \hline 3y + 2z &= 15 && \textbf{(4)} \end{aligned}$$

Equations (2) and (4) form a linear system in two variables. Multiplying equation (2) by -2 and adding the result to equation (4) eliminates the variable z:

$$\begin{aligned} 6y + z = 12 &\xrightarrow{\text{Multiply by } -2} & -12y - 2z &= -24 \\ 3y + 2z = 15 &\xrightarrow[\text{No change}]{} & 3y + 2z &= 15 \\ && \hline -9y &= -9 \\ && y &= 1 \end{aligned}$$

Using $y = 1$ in equation (4) and solving for z, we have

$$z = 6$$

Substituting $y = 1$ into equation (1) gives

$$x = 2$$

The ordered triple that satisfies all three equations is (2, 1, 6).

NOTE

The last page of this section contains a discussion of the geometric interpretations associated with systems of equations in three variables.

4. Solve the system.

$$\begin{aligned} 3x - y + 2z &= 4 \\ 6x - 2y + 4z &= 2 \\ 5x - 3y + 7z &= 5 \end{aligned}$$

5. Solve the system.

$$\begin{aligned} x + 2y &= 0 \\ 3y + z &= -3 \\ 2x - z &= 5 \end{aligned}$$

Answers

3. No unique solution; it is a dependent system.
4. No solution, it is an inconsistent system.
5. (4, −2, 3)

The Geometry Behind Equations in Three Variables

We can graph an ordered triple on a coordinate system with three axes. The graph will be a point in space. The coordinate system is drawn in perspective; you have to imagine that the x-axis comes out of the paper and is perpendicular to both the y-axis and the z-axis. To graph the point (3, 4, 5), we move 3 units in the x-direction, 4 units in the y-direction, and then 5 units in the z-direction, as shown in Figure 1.

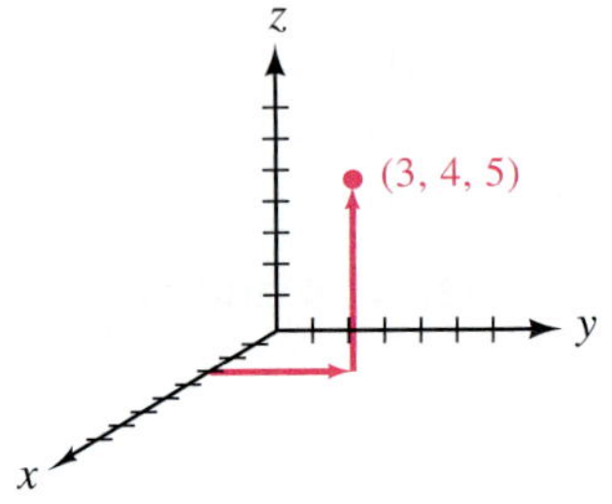

FIGURE 1

Although in actual practice it is sometimes difficult to graph equations in three variables, if we have to graph a linear equation in three variables, we would find that the graph was a plane in space. A system of three equations in three variables is represented by three planes in space.

There are a number of possible ways in which these three planes can intersect, some of which are shown in the margin on this page. There are still other possibilities that are not among those shown in the margin.

In Example 3 we found that equation 1 and 2 were dependent equations. They represent the same plane; that is, they have all their points in common. But the system of equations that they came from has either no solution or an infinite number of solutions. It all depends on the third plane. If the third plane coincides with the first two, then the solution to the system is a plane. If the third plane is parallel to the first two, then there is no solution to the system. And, finally, if the third plane intersects the first two, but does not coincide with them, then the solution to the system is that line of intersection.

In Example 4 we found that trying to eliminate a variable from the second and third equations resulted in a false statement. This means that the two planes represented by these equations are parallel. It makes no difference where the third plane is; there is no solution to the system in Example 4. (If we were to graph the three planes from Example 4, we would obtain a diagram similar to Case 4 in the margin.)

If, in the process of solving a system of linear equations in three variables, we eliminate all the variables from a pair of equations and are left with a false statement, we will say the system is inconsistent. If we eliminate all the variables and are left with a true statement, then we will say the system is a dependent one.

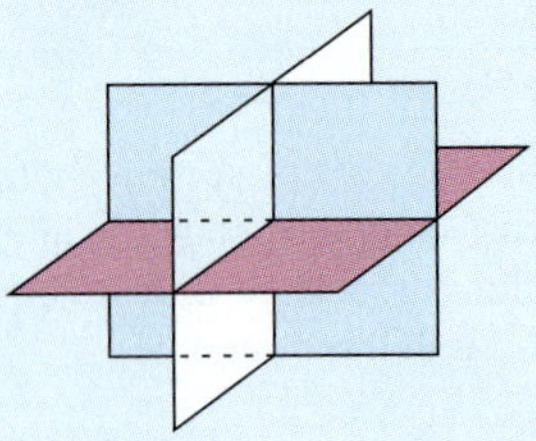

Case 1 The three planes have exactly one point in common. In this case we get one solution to our system, as in Examples 1, 2, and 5.

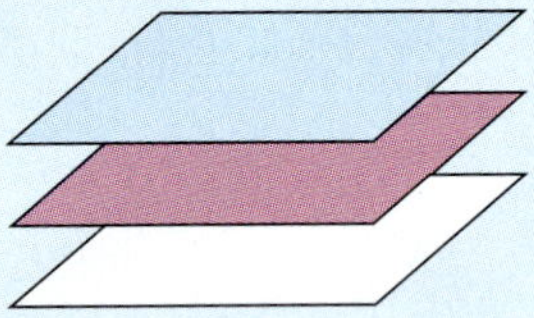

Case 2 The three planes have no points in common because they are all parallel to each other. The system they represent is an inconsistent system.

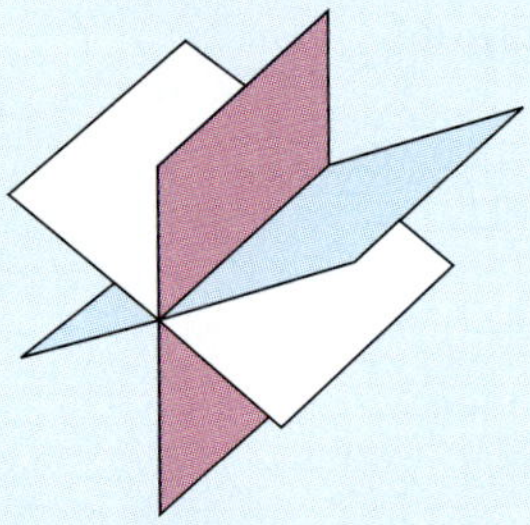

Case 3 The three planes intersect in a line. Any point on the line is a solution to the system of equations is a solution to the system, which is dependent.

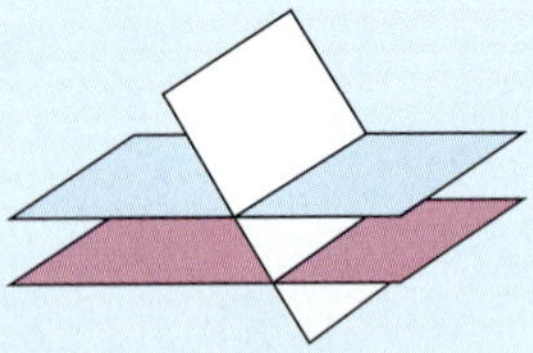

Case 4 In this case the three planes have no points in common. There is no solution to the system; it is an inconsistent system.

Getting Ready for Class

After reading through the preceding section, respond in your own words and in complete sentences.

1. What is an ordered triple of numbers?
2. Explain what it means for (1, 2, 3) to be a solution to a system of linear equations in three variables.
3. Explain in a general way the procedure you would use to solve a system of three linear equations in three variables.
4. How do you know when a system of linear equations in three variables has no solution?

Linking Objectives and Examples

Next to each **objective**, list the examples that are best described by that objective.

A	

Problem Set 4.2

INTERMEDIATE Algebra $f(x)$ Now™

Solve the following systems.

1. $$\begin{aligned} x + y + z &= 4 \\ x - y + 2z &= 1 \\ x - y - 3z &= -4 \end{aligned}$$

2. $$\begin{aligned} x - y - 2z &= -1 \\ x + y + z &= 6 \\ x + y - z &= 4 \end{aligned}$$

3. $$\begin{aligned} x + y + z &= 6 \\ x - y + 2z &= 7 \\ 2x - y - 4z &= -9 \end{aligned}$$

4. $$\begin{aligned} x + y + z &= 0 \\ x + y - z &= 6 \\ x - y + 2z &= -7 \end{aligned}$$

5. $$\begin{aligned} x + 2y + z &= 3 \\ 2x - y + 2z &= 6 \\ 3x + y - z &= 5 \end{aligned}$$

6. $$\begin{aligned} 2x + y - 3z &= -14 \\ x - 3y + 4z &= 22 \\ 3x + 2y + z &= 0 \end{aligned}$$

7. $$\begin{aligned} 2x + 3y - 2z &= 4 \\ x + 3y - 3z &= 4 \\ 3x - 6y + z &= -3 \end{aligned}$$

8. $$\begin{aligned} 4x + y - 2z &= 0 \\ 2x - 3y + 3z &= 9 \\ -6x - 2y + z &= 0 \end{aligned}$$

9. $$\begin{aligned} -x + 4y - 3z &= 2 \\ 2x - 8y + 6z &= 1 \\ 3x - y + z &= 0 \end{aligned}$$

10. $$\begin{aligned} 4x + 6y - 8z &= 1 \\ -6x - 9y + 12z &= 0 \\ x - 2y - 2z &= 3 \end{aligned}$$

11. $$\begin{aligned} \frac{1}{2}x - y + z &= 0 \\ 2x + \frac{1}{3}y + z &= 2 \\ x + y + z &= -4 \end{aligned}$$

12. $$\begin{aligned} \frac{1}{3}x + \frac{1}{2}y + z &= -1 \\ x - y + \frac{1}{5}z &= 1 \\ x + y + z &= 5 \end{aligned}$$

13. $$\begin{aligned} 2x - y - 3z &= 1 \\ x + 2y + 4z &= 3 \\ 4x - 2y - 6z &= 2 \end{aligned}$$

14. $$\begin{aligned} 3x + 2y + z &= 3 \\ x - 3y + z &= 4 \\ -6x - 4y - 2z &= 1 \end{aligned}$$

15. $$\begin{aligned} 2x - y + 3z &= 4 \\ x + 2y - z &= -3 \\ 4x + 3y + 2z &= -5 \end{aligned}$$

16. $$\begin{aligned} 6x - 2y + z &= 5 \\ 3x + y + 3z &= 7 \\ x + 4y - z &= 4 \end{aligned}$$

17. $$\begin{aligned} x + y &= 9 \\ y + z &= 7 \\ x - z &= 2 \end{aligned}$$

18. $$\begin{aligned} x - y &= -3 \\ x + z &= 2 \\ y - z &= 7 \end{aligned}$$

19. $$\begin{aligned} 2x + y &= 2 \\ y + z &= 3 \\ 4x - z &= 0 \end{aligned}$$

20. $$\begin{aligned} 2x + y &= 6 \\ 3y - 2z &= -8 \\ x + z &= 5 \end{aligned}$$

21. $$\begin{aligned} 2x - 3y &= 0 \\ 6y - 4z &= 1 \\ x + 2z &= 1 \end{aligned}$$

22. $$\begin{aligned} 3x + 2y &= 3 \\ y + 2z &= 2 \\ 6x - 4z &= 1 \end{aligned}$$

23. $$\begin{aligned} \frac{1}{2}x + \frac{2}{3}y &= \frac{5}{2} \\ \frac{1}{5}x - \frac{1}{2}z &= -\frac{3}{10} \\ \frac{1}{3}y - \frac{1}{4}z &= \frac{3}{4} \end{aligned}$$

24. $$\begin{aligned} \frac{1}{2}x - \frac{1}{3}y &= \frac{1}{6} \\ \frac{1}{3}y - \frac{1}{3}z &= 1 \\ \frac{1}{5}x - \frac{1}{2}z &= -\frac{4}{5} \end{aligned}$$

25. $$\begin{aligned} \frac{1}{2}x - \frac{1}{4}y + \frac{1}{2}z &= -2 \\ \frac{1}{4}x - \frac{1}{12}y - \frac{1}{3}z &= \frac{1}{4} \\ \frac{1}{6}x + \frac{1}{3}y - \frac{1}{2}z &= \frac{3}{2} \end{aligned}$$

26. $$\begin{aligned} \frac{1}{2}x + \frac{1}{2}y + z &= \frac{1}{2} \\ \frac{1}{2}x - \frac{1}{4}y - \frac{1}{4}z &= 0 \\ \frac{1}{4}x + \frac{1}{12}y + \frac{1}{6}z &= \frac{1}{6} \end{aligned}$$

Applying the Concepts

27. Electric Current In the following diagram of an electrical circuit, x, y, and z represent the amount of current (in amperes) flowing across the 5-ohm, 20-ohm, and 10-ohm resistors, respectively. (In circuit diagrams resistors are represented by -\/\/\/- and potential differences by ⊣⊢.)

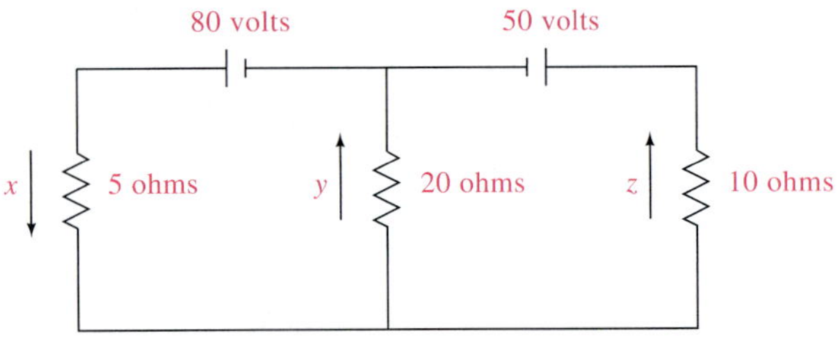

The system of equations used to find the three currents x, y, and z is

$$\begin{aligned} x - y - z &= 0 \\ 5x + 20y &= 80 \\ 20y - 10z &= 50 \end{aligned}$$

Solve the system for all variables.

28. Cost of a Rental Car If a car rental company charges \$10 a day and 8¢ a mile to rent one of its cars, then the cost z, in dollars, to rent a car for x days and drive y miles can be found from the equation

$$z = 10x + 0.08y$$

a. How much does it cost to rent a car for 2 days and drive it 200 miles under these conditions?

b. A second company charges \$12 a day and 6¢ a mile for the same car. Write an equation that gives the cost z, in dollars, to rent a car from this company for x days and drive it y miles.

c. A car is rented from each of the companies mentioned in (a) and (b) for 2 days. To find the mileage at which the cost of renting the cars from each of the two companies will be equal, solve the following system for y:

$$\begin{aligned} z &= 10x + 0.08y \\ z &= 12x + 0.06y \\ x &= 2 \end{aligned}$$

Getting Ready for the Next Section

Simplify.

29. $1(4) - 3(2)$

30. $3(7) - (-2)(5)$

31. $1(1) - 3(-2) + (-2)(-2)$

32. $-4(0)(-2) - (-1)(1)(1) - 1(2)(3)$

33. $-3(-1 - 1) + 4(-2 + 2) - 5[2 - (-2)]$

34. $12 + 4 - (-1) - 6$

Solve.

35. $-5x = 20$

36. $4x - 2x = 8$

Maintaining Your Skills

37. If y varies directly with square of x, and y is 75 when x is 5, find y when x is 7.

38. Suppose y varies directly with the cube of x. If y is 16 when x is 2, find y when x is 3.

39. Suppose y varies inversely with x. If y is 10 when x is 25, find x when y is 5.

40. If y varies inversely with the cube of x, and y is 2 when x is 2, find y when x is 4.

41. Suppose z varies jointly with x and the square of y. If z is 40 when x is 5 and y is 2, find z when x is 2 and y is 5.

42. Suppose z varies jointly with x and the cube of y. If z is 48 when x is 3 and y is 2, find z when x is 4 and y is $\frac{1}{2}$.

Extending the Concepts

Solve each system for the solution (x, y, z, w).

43.
$$\begin{aligned} x + y + z + w &= 10 \\ x + 2y - z + w &= 6 \\ x - y - z + 2w &= 4 \\ x - 2y + z - 3w &= -12 \end{aligned}$$

44.
$$\begin{aligned} x + y + z + w &= 16 \\ x - y + 2z - w &= 1 \\ x + 3y - z - w &= -2 \\ x - 3y - 2z + 2w &= -4 \end{aligned}$$

Introduction to Determinants

4.3

In this section we will expand and evaluate *determinants.* The purpose of this section is simply to be able to find the value of a given determinant. As we will see in the next section, determinants are very useful in solving systems of linear equations. Before we apply determinants to systems of linear equations, however, we must practice calculating the value of some determinants.

OBJECTIVES	
A	Find the value of a 2×2 determinant.
B	Find the value of a 3×3 determinant.

DEFINITION

The value of the **2×2 (2 by 2) determinant**

$$\begin{vmatrix} a & c \\ b & d \end{vmatrix}$$

is given by

$$\begin{vmatrix} a & c \\ b & d \end{vmatrix} = ad - bc$$

From the preceding definition we see that a determinant is simply a square array of numbers with two vertical lines enclosing it. The value of a 2×2 determinant is found by cross-multiplying on the diagonals and then subtracting, a diagram of which looks like

$$\begin{vmatrix} a & c \\ b & d \end{vmatrix} = ad - bc$$

EXAMPLE 1 Find the value of the following 2×2 determinants:

a. $\begin{vmatrix} 1 & 2 \\ 3 & 4 \end{vmatrix} = 1(4) - 3(2) = 4 - 6 = -2$

b. $\begin{vmatrix} 3 & 5 \\ -2 & 7 \end{vmatrix} = 3(7) - (-2)5 = 21 + 10 = 31$

EXAMPLE 2 Solve for x if

$$\begin{vmatrix} x & 2 \\ x & 4 \end{vmatrix} = 8$$

Solution We expand the determinant on the left side to get

$$x(4) - x(2) = 8$$

$$4x - 2x = 8$$

$$2x = 8$$

$$x = 4$$

We now turn our attention to 3×3 determinants. A 3×3 determinant is also a square array of numbers, the value of which is given by the following definition.

PRACTICE PROBLEMS

1. Find the value of each determinant.

a. $\begin{vmatrix} 2 & 1 \\ 4 & 3 \end{vmatrix}$

b. $\begin{vmatrix} 4 & -2 \\ 0 & 3 \end{vmatrix}$

2. Solve for x if

$$\begin{vmatrix} -3 & x \\ 2 & x \end{vmatrix} = 20$$

Answers

1. a. 2 **b.** 12 **2.** −4

DEFINITION

The value of the 3×3 determinant

$$\begin{vmatrix} a_1 & b_1 & c_1 \\ a_2 & b_2 & c_2 \\ a_3 & b_3 & c_3 \end{vmatrix}$$

is given by

$$\begin{vmatrix} a_1 & b_1 & c_1 \\ a_2 & b_2 & c_2 \\ a_3 & b_3 & c_3 \end{vmatrix} = a_1b_2c_3 + a_3b_1c_2 + a_2b_3c_1 - a_3b_2c_1 - a_1b_3c_2 - a_2b_1c_3$$

At first glance, the expansion of a 3×3 determinant looks a little complicated. There are actually two different methods used to find the six products in the preceding definition that simplify matters somewhat.

Method I We begin by writing the determinant with the first two columns repeated on the right:

$$\left|\begin{matrix} a_1 & b_1 & c_1 \\ a_2 & b_2 & c_2 \\ a_3 & b_3 & c_3 \end{matrix}\right| \begin{matrix} a_1 & b_1 \\ a_2 & b_2 \\ a_3 & b_3 \end{matrix}$$

The positive products in the definition come from multiplying down the three full diagonals:

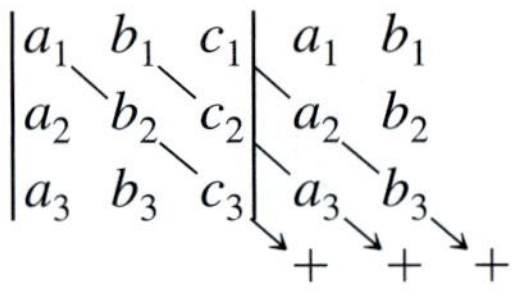

The negative products come from multiplying up the three full diagonals:

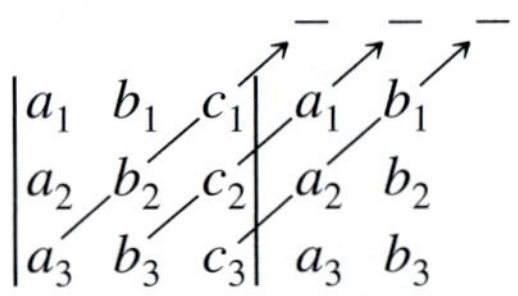

NOTE

Check the products found by multiplying up and down the diagonals given here with the products given in the definition of a 3×3 determinant to see that they match.

EXAMPLE 3 Find the value of

$$\begin{vmatrix} 1 & 3 & -2 \\ 2 & 0 & 1 \\ 4 & -1 & 1 \end{vmatrix}$$

Solution Repeating the first two columns and then finding the products down the diagonals and the products up the diagonals as given in Method 1, we have

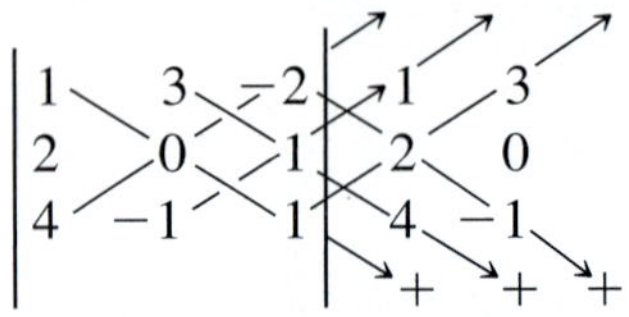

$$= 1(0)(1) + 3(1)(4) + (-2)(2)(-1)$$
$$-4(0)(-2) - (-1)(1)(1) - 1(2)(3)$$
$$= 0 + 12 + 4 - 0 - (-1) - 6$$
$$= 11$$

3. Find the value of

$$\begin{vmatrix} 2 & 0 & -1 \\ 3 & 1 & 2 \\ 5 & -2 & 1 \end{vmatrix}$$

Answer

3. 21

Method 2 The second method of evaluating a 3×3 determinant is called *expansion by minors.*

DEFINITION

The **minor** for an element in a 3×3 determinant is the determinant consisting of the elements remaining when the row and column to which the element belongs are deleted. For example, in the determinant

$$\begin{vmatrix} a_1 & b_1 & c_1 \\ a_2 & b_2 & c_2 \\ a_3 & b_3 & c_3 \end{vmatrix}$$

$$\text{Minor for element } a_1 = \begin{vmatrix} b_2 & c_2 \\ b_3 & c_3 \end{vmatrix}$$

$$\text{Minor for element } b_2 = \begin{vmatrix} a_1 & c_1 \\ a_3 & c_3 \end{vmatrix}$$

$$\text{Minor for element } c_3 = \begin{vmatrix} a_1 & b_1 \\ a_2 & b_2 \end{vmatrix}$$

Before we can evaluate a 3×3 determinant by Method 2, we must first define what is known as the sign array for a 3×3 determinant.

DEFINITION

The **sign array** for a 3×3 determinant is a 3×3 array of signs in the following pattern:

$$\begin{vmatrix} + & - & + \\ - & + & - \\ + & - & + \end{vmatrix}$$

The sign array begins with a plus sign in the upper left-hand corner. The signs then alternate between plus and minus across every row and down every column.

NOTE

If you have read this far and are confused, hang on. After you have done a couple of examples you will find expansion by minors to be a fairly simple process. It just takes a lot of writing to explain it.

To Evaluate a 3 × 3 Determinant by Expansion of Minors

We can evaluate a 3×3 determinant by expanding across any row or down any column as follows:

Step 1: Choose a row or column to expand.

Step 2: Write the product of each element in the row or column chosen in step 1 with its minor.

Step 3: Connect the three products in step 2 with the signs in the corresponding row or column in the sign array.

To illustrate the procedure, we will use the same determinant we used in Example 3.

EXAMPLE 4 Expand across the first row:

$$\begin{vmatrix} 1 & 3 & -2 \\ 2 & 0 & 1 \\ 4 & -1 & 1 \end{vmatrix}$$

4. Expand across the first row:

$$\begin{vmatrix} 2 & 0 & -1 \\ 3 & 1 & 2 \\ 5 & -2 & 1 \end{vmatrix}$$

Solution The products of the three elements in row 1 with their minors are

$$1\begin{vmatrix} 0 & 1 \\ -1 & 1 \end{vmatrix} \qquad 3\begin{vmatrix} 2 & 1 \\ 4 & 1 \end{vmatrix} \qquad (-2)\begin{vmatrix} 2 & 0 \\ 4 & -1 \end{vmatrix}$$

Connecting these three products with the signs from the first row of the sign array, we have

$$+1\begin{vmatrix} 0 & 1 \\ -1 & 1 \end{vmatrix} - 3\begin{vmatrix} 2 & 1 \\ 4 & 1 \end{vmatrix} + (-2)\begin{vmatrix} 2 & 0 \\ 4 & -1 \end{vmatrix}$$

We complete the problem by evaluating each of the three 2×2 determinants and then simplifying the resulting expression:

$$\begin{aligned} &+1[0 - (-1)] - 3(2 - 4) + (-2)(-2 - 0) \\ &\quad = 1(1) - 3(-2) + (-2)(-2) \\ &\quad = 1 + 6 + 4 \\ &\quad = 11 \end{aligned}$$

The results of Examples 3 and 4 match. It makes no difference which method we use—the value of a 3×3 determinant is unique.

NOTE

This method of evaluating a determinant is actually more valuable than our first method because it works with any size determinant from 3×3 to 4×4 to any higher order determinant. Method 1 works only on 3×3 determinants. It cannot be used on a 4×4 determinant.

EXAMPLE 5

Expand down column 2:

$$\begin{vmatrix} 2 & 3 & -2 \\ 1 & 4 & 1 \\ 1 & 5 & -1 \end{vmatrix}$$

Solution We connect the products of elements in column 2 and their minors with the signs from the second column in the sign array:

$$\begin{aligned} \begin{vmatrix} 2 & 3 & -2 \\ 1 & 4 & 1 \\ 1 & 5 & -1 \end{vmatrix} &= -3\begin{vmatrix} 1 & 1 \\ 1 & -1 \end{vmatrix} + 4\begin{vmatrix} 2 & -2 \\ 1 & -1 \end{vmatrix} - 5\begin{vmatrix} 2 & -2 \\ 1 & 1 \end{vmatrix} \\ &= -3(-1 - 1) + 4[-2 - (-2)] - 5[2 - (-2)] \\ &= -3(-2) + 4(0) - 5(4) \\ &= 6 + 0 - 20 \\ &= -14 \end{aligned}$$

5. Expand down column 2:

$$\begin{vmatrix} 0 & 4 & -2 \\ 3 & 1 & 1 \\ 1 & -2 & 0 \end{vmatrix}$$

A Note on the History of Determinants Determinants were originally known as resultants, a name given to them by Pierre Simon Laplace; however, the work of Gottfried Wilhelm Leibniz contains the germ of the original idea of resultants, or determinants.

Linking Objectives and Examples

Next to each **objective**, list the examples that are best described by that objective.

A	
B	

Answers
4. 21 5. 18

Getting Ready for Class

After reading through the preceding section, respond in your own words and in complete sentences.

1. Describe how you evaluate a 2×2 determinant.
2. Name the row and column that the number 3 is in: $\begin{vmatrix} 1 & 3 \\ 5 & 7 \end{vmatrix}$
3. What is the sign array for a 3×3 determinant?
4. If the value of a determinant is 0, does one of the elements have to be 0? Explain.

Problem Set 4.3

INTERMEDIATE Algebra $f(x)$ Now™

Find the value of the following 2×2 determinants.

1. $\begin{vmatrix} 1 & 0 \\ 2 & 3 \end{vmatrix}$

2. $\begin{vmatrix} 5 & 4 \\ 3 & 2 \end{vmatrix}$

3. $\begin{vmatrix} 2 & 1 \\ 3 & 4 \end{vmatrix}$

4. $\begin{vmatrix} 4 & 1 \\ 5 & 2 \end{vmatrix}$

5. $\begin{vmatrix} 0 & 1 \\ 1 & 0 \end{vmatrix}$

6. $\begin{vmatrix} 1 & 0 \\ 0 & 1 \end{vmatrix}$

7. $\begin{vmatrix} -3 & 2 \\ 6 & -4 \end{vmatrix}$

8. $\begin{vmatrix} 8 & -3 \\ -2 & 5 \end{vmatrix}$

Solve each of the following for x.

9. $\begin{vmatrix} 2x & 1 \\ x & 3 \end{vmatrix} = 10$

10. $\begin{vmatrix} 3x & -2 \\ 2x & 3 \end{vmatrix} = 26$

11. $\begin{vmatrix} 1 & 2x \\ 2 & -3x \end{vmatrix} = 21$

12. $\begin{vmatrix} -5 & 4x \\ 1 & -x \end{vmatrix} = 27$

13. $\begin{vmatrix} 2x & -4 \\ x & 2 \end{vmatrix} = -16$

14. $\begin{vmatrix} 3x & -2 \\ x & 4 \end{vmatrix} = -28$

15. $\begin{vmatrix} 11x & -7x \\ 3 & -2 \end{vmatrix} = 3$

16. $\begin{vmatrix} -3x & -5x \\ 4 & 6 \end{vmatrix} = -14$

Find the value of each of the following 3×3 determinants by using Method 1 of this section.

17. $\begin{vmatrix} 1 & 2 & 0 \\ 0 & 2 & 1 \\ 1 & 1 & 1 \end{vmatrix}$

18. $\begin{vmatrix} -1 & 0 & 2 \\ 3 & 0 & 1 \\ 0 & 1 & 3 \end{vmatrix}$

19. $\begin{vmatrix} 1 & 2 & 3 \\ 3 & 2 & 1 \\ 1 & 1 & 1 \end{vmatrix}$

20. $\begin{vmatrix} -1 & 2 & 0 \\ 3 & -2 & 1 \\ 0 & 5 & 4 \end{vmatrix}$

Find the value of each determinant by using Method 2 and expanding across the first row.

21. $\begin{vmatrix} 0 & 1 & 2 \\ 1 & 0 & 1 \\ -1 & 2 & 0 \end{vmatrix}$

22. $\begin{vmatrix} 3 & -2 & 1 \\ 0 & -1 & 0 \\ 2 & 0 & 1 \end{vmatrix}$

23. $\begin{vmatrix} 3 & 0 & 2 \\ 0 & -1 & -1 \\ 4 & 0 & 0 \end{vmatrix}$

24. $\begin{vmatrix} 1 & 1 & 1 \\ 1 & -1 & 1 \\ 1 & 1 & -1 \end{vmatrix}$

Find the value of each of the following determinants.

25. $\begin{vmatrix} 2 & -1 & 0 \\ 1 & 0 & -2 \\ 0 & 1 & 2 \end{vmatrix}$

26. $\begin{vmatrix} 5 & 0 & -4 \\ 0 & 1 & 3 \\ -1 & 2 & -1 \end{vmatrix}$

27. $\begin{vmatrix} 1 & 3 & 7 \\ -2 & 6 & 4 \\ 3 & 7 & -1 \end{vmatrix}$

28. $\begin{vmatrix} 2 & 1 & 5 \\ 6 & -3 & 4 \\ 8 & 9 & -2 \end{vmatrix}$

Applying the Concepts

29. Slope-Intercept Form Show that the following determinant equation is another way to write the slope-intercept form of the equation of a line.

$$\begin{vmatrix} y & x \\ m & 1 \end{vmatrix} = b$$

30. Temperature Conversion Show that the following determinant equation is another way to write the equation $F = \frac{9}{5}C + 32$.

$$\begin{vmatrix} C & F & 1 \\ 5 & 41 & 1 \\ -10 & 14 & 1 \end{vmatrix} = 0$$

31. Amusement Park Income From 1986 to 1990, the annual income of amusement parks was linearly increasing, after which time it remained fairly constant. The annual income y, in billions of dollars, may be found for one of these years by evaluating the following determinant equation, in which x represents the number of years past January 1, 1986.

$$\begin{vmatrix} x & -1.7 \\ 2 & 0.3 \end{vmatrix} = y$$

a. Write the determinant equation in slope-intercept form.

b. Use the equation from part (a) to find the approximate income for amusement parks in the year 1988.

32. Women in Armed Forces From 1981, the enrollment of women in the United States armed forces was linearly increasing until 1990, after which it declined. The approximate number of women, w, enrolled in the armed forces from 1981 to 1990 may be found by evaluating the following determinant equation, in which x represents the number of years past January 1, 1981.

$$\begin{vmatrix} 6{,}509 & -2 \\ 85{,}709 & x \end{vmatrix} = w$$

Use this equation to determine the number of women enrolled in the armed forces on January 1, 1985.

Getting Ready for the Next Section

Simplify.

33. $2(3) - 4(4)$

34. $2(5) - 4(-3)$

35. $1(-1)(-3)$

36. $-(-3)(2)(1)$

37. $-\frac{10}{22}$

38. $\frac{39}{13}$

39. $6(1) - 1(-5) + 1(2)$

40. $-2(-14) + 3(-4) - 1(-10)$

Find the value of each determinant.

41. $\begin{vmatrix} 3 & -5 \\ 2 & 4 \end{vmatrix}$

42. $\begin{vmatrix} 3 & 2 \\ 2 & 1 \end{vmatrix}$

43. $\begin{vmatrix} 6 & 1 & 1 \\ 3 & -1 & 1 \\ -4 & 2 & -3 \end{vmatrix}$

44. $\begin{vmatrix} 1 & 1 & 0 \\ 2 & 0 & 1 \\ 0 & 1 & 2 \end{vmatrix}$

Maintaining Your Skills

For each relation that follows, state the domain and the range, and indicate which are also functions.

45. $\{(1, 2), (3, 4), (4, 2)\}$

46. $\{(0, 0), (1, 1), (0, 1)\}$

47. $\{(3, 1), (2, 3), (1, 2)\}$

48. $\{(-1, 1), (2, -2), (-3, -3)\}$

State whether each of the following graphs is the graph of a function

49.

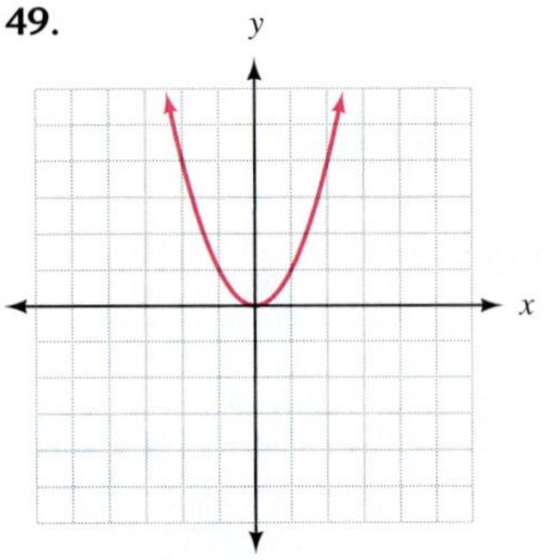

50.

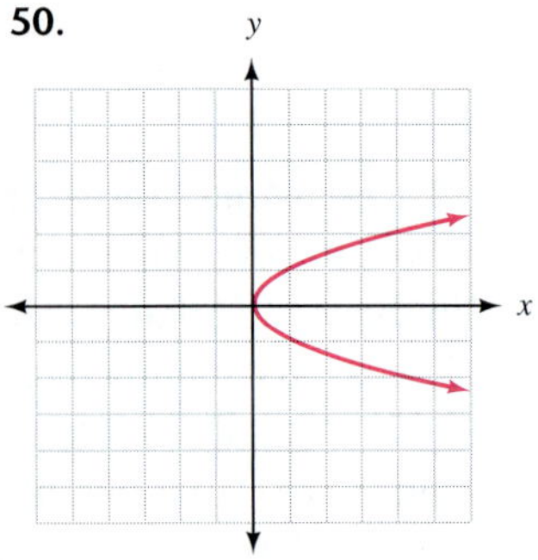

51.

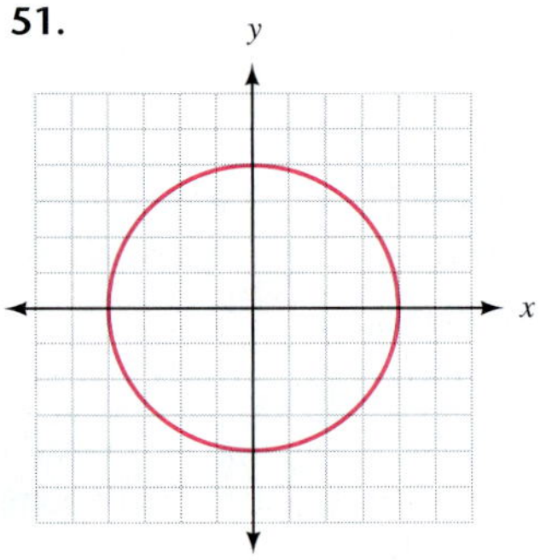

52.

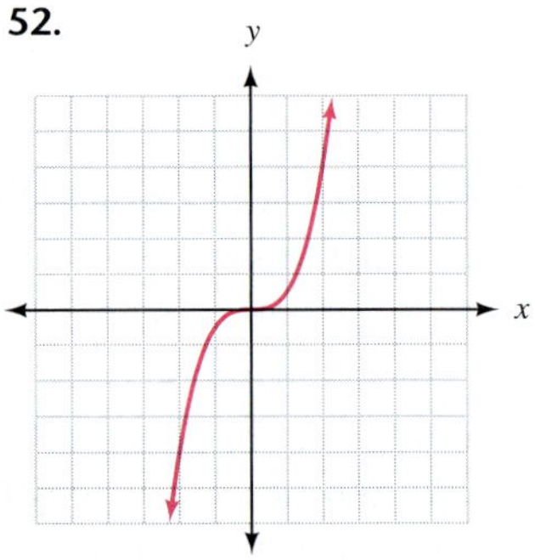

Extending the Concepts

A 4×4 determinant can be evaluated only by using Method 2, expansion by minors; Method 1 will not work. Below is a 4×4 determinant and its associated sign array.

$$\begin{vmatrix} 2 & 0 & 1 & -3 \\ -1 & 2 & 0 & 1 \\ -3 & 0 & 1 & 0 \\ 1 & 1 & 0 & 0 \end{vmatrix} \qquad \begin{vmatrix} + & - & + & - \\ - & + & - & + \\ + & - & + & - \\ - & + & - & + \end{vmatrix}$$

4×4 determinant $\qquad$ 4×4 sign array

53. Use expansion by minors to evaluate the preceding 4×4 determinant by expanding it across row 1.

54. Evaluate the preceding determinant by expanding it down column 4.

55. Use expansion by minors down column 3 to evaluate the preceding determinant.

56. Evaluate the preceding determinant by expanding it across row 4.

Cramer's Rule

4.4

We begin this section with a look at how determinants can be used to solve a system of linear equations in two variables. The method we use is called *Cramer's rule.* We state it here as a theorem without proof.

OBJECTIVES

A	Solve a system of linear equations in two variables using Cramer's rule.
B	Solve a system of linear equations in three variables using Cramer's rule.

Cramer's Rule I

The solution to the system

$$a_1x + b_1y = c_1$$
$$a_2x + b_2y = c_2$$

is given by

$$x = \frac{D_x}{D}, \qquad y = \frac{D_y}{D}$$

where

$$D = \begin{vmatrix} a_1 & b_1 \\ a_2 & b_2 \end{vmatrix} \qquad D_x = \begin{vmatrix} c_1 & b_1 \\ c_2 & b_2 \end{vmatrix} \qquad D_y = \begin{vmatrix} a_1 & c_1 \\ a_2 & c_2 \end{vmatrix} \qquad (D \neq 0)$$

The determinant D is made up of the coefficients of x and y in the original system. The determinants D_x and D_y are found by replacing the coefficients of x or y by the constant terms in the original system. Notice also that Cramer's rule does not apply if $D = 0$. In this case the equations are either inconsistent or dependent.

EXAMPLE 1 Use Cramer's rule to solve

$$2x - 3y = 4$$
$$4x + 5y = 3$$

Solution We begin by calculating the determinants D, D_x, and D_y.

$$D = \begin{vmatrix} 2 & -3 \\ 4 & 5 \end{vmatrix} = 2(5) - 4(-3) = 22$$

$$D_x = \begin{vmatrix} 4 & -3 \\ 3 & 5 \end{vmatrix} = 4(5) - 3(-3) = 29$$

$$D_y = \begin{vmatrix} 2 & 4 \\ 4 & 3 \end{vmatrix} = 2(3) - 4(4) = -10$$

$$x = \frac{D_x}{D} = \frac{29}{22} \quad \text{and} \quad y = \frac{D_y}{D} = \frac{-10}{22} = -\frac{5}{11}$$

The solution set for the system is $\{(\frac{29}{22}, -\frac{5}{11})\}$.

Cramer's rule can also be applied to systems of linear equations in three variables.

PRACTICE PROBLEMS

1. Use Cramer's rule to solve

$$3x - 5y = 2$$
$$2x + 4y = 1$$

Answer

1. $(\frac{13}{22}, -\frac{1}{22})$

9. $\begin{aligned} x + y + z &= 4 \\ x - y - z &= 2 \\ 2x + 2y - z &= 2 \end{aligned}$

10. $\begin{aligned} -x + y + 3z &= 6 \\ x + y + 2z &= 7 \\ 2x + 3y + z &= 4 \end{aligned}$

11. $\begin{aligned} x + y - z &= 2 \\ -x + y + z &= 3 \\ x + y + z &= 4 \end{aligned}$

12. $\begin{aligned} -x - y + z &= 1 \\ x - y + z &= 3 \\ x + y - z &= 4 \end{aligned}$

13. $\begin{aligned} 3x - y + 2z &= 4 \\ 6x - 2y + 4z &= 8 \\ x - 5y + 2z &= 1 \end{aligned}$

14. $\begin{aligned} 2x - 3y + z &= 1 \\ 3x - y - z &= 4 \\ 4x - 6y + 2z &= 3 \end{aligned}$

15. $\begin{aligned} 2x - y + 3z &= 4 \\ x - 5y - 2z &= 1 \\ -4x - 2y + z &= 3 \end{aligned}$

16. $\begin{aligned} 4x - y + 5z &= 1 \\ 2x + 3y + 4z &= 5 \\ x + y + 3z &= 2 \end{aligned}$

17. $\begin{aligned} -x - 7y &= 1 \\ x + 3z &= 11 \\ 2y + z &= 0 \end{aligned}$

18. $\begin{aligned} x + y &= 2 \\ -x + 3z &= 0 \\ 2y + z &= 3 \end{aligned}$

19. $\begin{aligned} x - y &= 2 \\ 3x + z &= 11 \\ y - 2z &= -3 \end{aligned}$

20. $\begin{aligned} 4x + 5y &= -1 \\ 2y + 3z &= -5 \\ x + 2z &= -1 \end{aligned}$

Applying the Concepts

21. Break-Even Point If a company has fixed costs of \$100 per week and each item it produces costs \$10 to manufacture, then the total cost y per week to produce x items is

$$y = 10x + 100$$

If the company sells each item it manufactures for \$12, then the total amount of money y the company brings in for selling x items is

$$y = 12x$$

Use Cramer's rule to solve the system

$$y = 10x + 100$$
$$y = 12x$$

for x to find the number of items the company must sell per week to break even.

22. Break-Even Point Suppose a company has fixed costs of \$200 per week and each item it produces costs \$20 to manufacture.

a. Write an equation that gives the total cost per week y to manufacture x items.

b. If each item sells for \$25, write an equation that gives the total amount of money y the company brings in for selling x items.

c. Use Cramer's rule to find the number of items the company must sell each week to break even.

Getting Ready for the Next Section

Translate into symbols.

23. Two more than 3 times a number

24. One less than twice a number

Simplify.

25. $25 - \frac{385}{9}$

26. $0.30(12)$

27. $0.08(4{,}000)$

28. $500(1.5)$

Apply the distributive property, then simplify.

29. $10(0.2x + 0.5y)$

30. $100(0.09x + 0.08y)$

Solve.

31. $x + (3x + 2) = 26$

32. $5x = 2{,}500$

Solve each system.

33. $3y + z = 17$
$5y + 20z = 65$

34. $x + y = 850$
$1.5x + y = 1{,}100$

Maintaining Your Skills

Let $f(x) = \frac{1}{2}x + 3$ and $g(x) = x^2 - 4$, and find

35. $f(0)$

36. $g(0)$

37. $g(2)$

38. $f(2)$

39. $f(-4)$

40. $g(-6)$

41. $f[g(2)]$

42. $g[f(2)]$

Extending the Concepts

Solve for x and y using Cramer's rule. Your answers will contain the constants a and b.

43. $ax + by = -1$
$bx + ay = 1$

44. $ax + y = b$
$bx + y = a$

45. $a^2x + by = 1$
$b^2x + ay = 1$

46. $ax + by = a$
$bx + ay = a$

47. Name the system of equations for which Cramer's rule yields the following determinants.

$$D = \begin{vmatrix} 1 & 2 \\ 3 & 4 \end{vmatrix} \qquad D_x = \begin{vmatrix} 1 & 2 \\ 0 & 4 \end{vmatrix}$$

48. Name the system of equations for which Cramer's rule yields the following determinants.

$$D = \begin{vmatrix} 1 & 3 & 2 \\ -1 & 0 & 4 \\ 2 & 5 & -1 \end{vmatrix} \qquad D_y = \begin{vmatrix} 1 & 1 & 2 \\ -1 & 3 & 4 \\ 2 & 5 & -1 \end{vmatrix}$$

Applications 4.5

Many times word problems involve more than one unknown quantity. If a problem is stated in terms of two unknowns and we represent each unknown quantity with a different variable, then we must write the relationships between the variables with two equations. The two equations written in terms of the two variables form a system of linear equations that we solve using the methods developed in this chapter. If we find a problem that relates three unknown quantities, then we need three equations to form a linear system we can solve.

Here is our Blueprint for Problem Solving, modified to fit the application problems that you will find in this section.

OBJECTIVES	
A	Solve application problems whose solutions are found through systems of linear equations.

Blueprint for Problem Solving Using a System of Equations

Step 1: ***Read*** the problem, and then mentally ***list*** the items that are known and the items that are unknown.

Step 2: ***Assign variables*** to each of the unknown items; that is, let x = one of the unknown items and y = the other unknown item (and z = the third unknown item, if there is a third one). Then ***translate*** the other ***information*** in the problem to expressions involving the two (or three) variables.

Step 3: ***Reread*** the problem, and then ***write a system of equations,*** using the items and variables listed in steps 1 and 2, that describes the situation.

Step 4: ***Solve the system*** found in step 3.

Step 5: ***Write your answers*** using complete sentences.

Step 6: ***Reread*** the problem, and ***check*** your solution with the original words in the problem.

EXAMPLE 1 One number is 2 more than 3 times another. Their sum is 26. Find the two numbers.

Solution Applying the steps from our Blueprint, we have:

Step 1: ***Read and list.***
We know that we have two numbers, whose sum is 26. One of them is 2 more than 3 times the other. The unknown quantities are the two numbers.

Step 2: ***Assign variables and translate information.***
Let x = one of the numbers and y = the other number.

Step 3: ***Write a system of equations.***
The first sentence in the problem translates into $y = 3x + 2$. The second sentence gives us a second equation: $x + y = 26$. Together, these two equations give us the following system of equations:

$$x + y = 26$$

$$y = 3x + 2$$

PRACTICE PROBLEMS

1. A number is 1 less than twice another. Their sum is 14. Find the two numbers.

Step 4: ***Solve the system.***
Substituting the expression for y from the second equation into the first and solving for x yields

$$x + (3x + 2) = 26$$
$$4x + 2 = 26$$
$$4x = 24$$
$$x = 6$$

Using $x = 6$ in $y = 3x + 2$ gives the second number:

$$y = 3(6) + 2$$
$$y = 20$$

Step 5: ***Write answers.***
The two numbers are 6 and 20.

Step 6: ***Reread and check.***
The sum of 6 and 20 is 26, and 20 is 2 more than 3 times 6.

2. There were 750 tickets sold for a basketball game for a total of \$1,090. If adult tickets cost \$2.00 and children's tickets cost \$1.00, how many of each kind were sold?

EXAMPLE 2

Suppose 850 tickets were sold for a game for a total of \$1,100. If adult tickets cost \$1.50 and children's tickets cost \$1.00, how many of each kind of ticket were sold?

Solution

Step 1: ***Read and list.***
The total number of tickets sold is 850. The total income from tickets is \$1,100. Adult tickets are \$1.50 each. Children's tickets are \$1.00 each. We don't know how many of each type of ticket have been sold.

Step 2: ***Assign variables and translate information.***
We let x = the number of adult tickets and y = the number of children's tickets.

Step 3: ***Write a system of equations.***
The total number of tickets sold is 850, giving us our first equation.

$$x + y = 850$$

Because each adult ticket costs \$1.50 and each children's ticket costs \$1.00 and the total amount of money paid for tickets was \$1,100, a second equation is

$$1.50x + 1.00y = 1{,}100$$

The same information can also be obtained by summarizing the problem with a table. One such table follows. Notice that the two equations we obtained previously are given by the two rows of the table.

	ADULT TICKETS	CHILDREN'S TICKETS	TOTAL
Number	x	y	850
Value	$1.50x$	$1.00y$	1,100

Answer
1. 5, 9

Whether we use a table to summarize the information in the problem or just talk our way through the problem, the system of equations that describes the situation is

$$x + y = 850$$
$$1.50x + 1.00y = 1{,}100$$

Step 4: ***Solve the system.***

If we multiply the second equation by 10 to clear it of decimals, we have the system

$$x + y = 850$$
$$15x + 10y = 11{,}000$$

Multiplying the first equation by −10 and adding the result to the second equation eliminates the variable y from the system:

$$\begin{aligned} -10x - 10y &= -8{,}500 \\ 15x + 10y &= 11{,}000 \\ \hline 5x \qquad &= 2{,}500 \\ x &= 500 \end{aligned}$$

The number of adult tickets sold was 500. To find the number of children's tickets, we substitute $x = 500$ into $x + y = 850$ to get

$$500 + y = 850$$
$$y = 350$$

Step 5: ***Write answers.***

The number of children's tickets is 350, and the number of adult tickets is 500.

Step 6: ***Reread and check.***

The total number of tickets is $350 + 500 = 850$. The amount of money from selling the two types of tickets is

350 children's tickets at \$1.00 each is 350(1.00) = \$350
500 adult tickets at \$1.50 each is 500(1.50) = \$750
The total income from ticket sales is \$1,100

EXAMPLE 3

Suppose a person invests a total of \$10,000 in two accounts. One account earns 8% annually, and the other earns 9% annually. If the total interest earned from both accounts in a year is \$860, how much was invested in each account?

3. A person invests \$12,000 in two accounts. One account earns 6% annually and the other earns 7%. If the total interest in a year is \$790, how much was invested in each account?

Solution

Step 1: ***Read and list.***

The total investment is \$10,000 split between two accounts. One account earns 8% annually, and the other earns 9% annually. The interest from both accounts is \$860 in 1 year. We don't know how much is in each account.

Step 2: ***Assign variables and translate information.***

We let x equal the amount invested at 9% and y be the amount invested at 8%.

Answer

2. 340 adult tickets; 410 children's tickets

Step 3: Write a system of equations.
Because the total investment is \$10,000, one relationship between x and y can be written as

$$x + y = 10{,}000$$

The total interest earned from both accounts is \$860. The amount of interest earned on x dollars at 9% is $0.09x$, whereas the amount of interest earned on y dollars at 8% is $0.08y$. This relationship is represented by the equation

$$0.09x + 0.08y = 860$$

The two equations we have just written can also be found by first summarizing the information from the problem in a table. Again, the two rows of the table yield the two equations just written. Here is the table.

	DOLLARS AT 9%	DOLLARS AT 8%	TOTAL
Number	x	y	10,000
Interest	$0.09x$	$0.08y$	860

The system of equations that describes this situation is given by

$$\begin{aligned} x + \quad y &= 10{,}000 \\ 0.09x + 0.08y &= \quad 860 \end{aligned}$$

Step 4: Solve the system.
Multiplying the second equation by 100 will clear it of decimals. The system that results after doing so is

$$\begin{aligned} x + y &= 10{,}000 \\ 9x + 8y &= 86{,}000 \end{aligned}$$

We can eliminate y from this system by multiplying the first equation by -8 and adding the result to the second equation.

$$\begin{array}{rcr} -8x - 8y &=& -80{,}000 \\ 9x + 8y &=& 86{,}000 \\ \hline x \qquad &=& 6{,}000 \end{array}$$

The amount of money invested at 9% is \$6,000. Because the total investment was \$10,000, the amount invested at 8% must be \$4,000.

Step 5: Write answers.
The amount invested at 8% is \$4,000, and the amount invested at 9% is \$6,000.

Step 6: Reread and check.
The total investment is \$4,000 + \$6,000 = \$10,000. The amount of interest earned from the two accounts is

In 1 year, \$4,000 invested at 8% earns 0.08(4,000) = \$320
In 1 year, \$6,000 invested at 9% earns 0.09(6,000) = \$540
The total interest from the two accounts is \$860

Answer
3. \$5,000 at 6%; \$7,000 at 7%

EXAMPLE 4

How much 20% alcohol solution and 50% alcohol solution must be mixed to get 12 gallons of 30% alcohol solution?

Solution To solve this problem, we must first understand that a 20% alcohol solution is 20% alcohol and 80% water.

Step 1: ***Read and list.***
We will mix two solutions to obtain 12 gallons of solution that is 30% alcohol. One of the solutions is 20% alcohol and the other is 50% alcohol. We don't know how much of each solution we need.

Step 2: ***Assign variables and translate information.***
Let x = the number of gallons of 20% alcohol solution needed and y = the number of gallons of 50% alcohol solution needed.

Step 3: ***Write a system of equations.***
Because we must end up with a total of 12 gallons of solution, one equation for the system is

$$x + y = 12$$

The amount of alcohol in the x gallons of 20% solution is $0.20x$, whereas the amount of alcohol in the y gallons of 50% solution is $0.50y$. Because the total amount of alcohol in the 20% and 50% solutions must add up to the amount of alcohol in the 12 gallons of 30% solution, the second equation in our system can be written as

$$0.20x + 0.50y = 0.30(12)$$

Again, let's make a table that summarizes the information we have to this point in the problem.

	20% SOLUTION	50% SOLUTION	FINAL SOLUTION
Total number of gallons	x	y	12
Gallons of alcohol	$0.20x$	$0.50y$	$0.30(12)$

Our system of equations is

$$\begin{aligned} x + \quad y &= 12 \\ 0.20x + 0.50y &= 0.30(12) = 3.6 \end{aligned}$$

Step 4: ***Solve the system.***
Multiplying the second equation by 10 gives us an equivalent system:

$$\begin{aligned} x + y &= 12 \\ 2x + 5y &= 36 \end{aligned}$$

Multiplying the top equation by -2 to eliminate the x-variable, we have

$$\begin{aligned} -2x - 2y &= -24 \\ 2x + 5y &= 36 \\ \hline 3y &= 12 \\ y &= 4 \end{aligned}$$

4. How much 30% alcohol solution and 70% alcohol solution must be mixed to get 16 gallons of 60% solution?

Substituting $y = 4$ into $x + y = 12$, we solve for x:

$$x + 4 = 12$$

$$x = 8$$

Step 5: ***Write answers.***
It takes 8 gallons of 20% alcohol solution and 4 gallons of 50% alcohol solution to produce 12 gallons of 30% alcohol solution.

Step 6: ***Reread and check.***
If we mix 8 gallons of 20% solution and 4 gallons of 50% solution, we end up with a total of 12 gallons of solution. To check the percentages, we look for the total amount of alcohol in the two initial solutions and in the final solution.

In the initial solutions

The amount of alcohol in 8 gallons of 20% solution is 0.20(8) = 1.6 gallons
The amount of alcohol in 4 gallons of 50% solution is 0.50(4) = 2.0 gallons
The total amount of alcohol in the initial solutions is 3.6 gallons

In the final solution

The amount of alcohol in 12 gallons of 30% solution is 0.30(12) = 3.6 gallons.

5. A boat can travel 20 miles downstream in 2 hours. The same boat can travel 18 miles upstream in 3 hours. What is the speed of the boat in still water, and what is the speed of the current?

EXAMPLE 5 It takes 2 hours for a boat to travel 28 miles downstream (with the current). The same boat can travel 18 miles upstream (against the current) in 3 hours. What is the speed of the boat in still water, and what is the speed of the current of the river?

Solution

Step 1: ***Read and list.***
A boat travels 18 miles upstream and 28 miles downstream. The trip upstream takes 3 hours. The trip downstream takes 2 hours. We don't know the speed of the boat or the speed of the current.

Step 2: ***Assign variables and translate information.***
Let x = the speed of the boat in still water and let y = the speed of the current. The average speed (rate) of the boat upstream is $x - y$ because it is traveling against the current. The rate of the boat downstream is $x + y$ because the boat is traveling with the current.

Step 3: ***Write a system of equations.***
Putting the information into a table, we have

	D (DISTANCE, MILES)	*R* (RATE, MPH)	*T* (TIME, H)
Upstream	18	$x - y$	3
Downstream	28	$x + y$	2

Answer
4. 4 gallons 30%; 12 gallons 70%

The formula for the relationship between distance d, rate r, and time t is $d = rt$ (the rate equation). Because $d = r \cdot t$, the system we need to solve the problem is

$$18 = (x - y) \cdot 3$$
$$28 = (x + y) \cdot 2$$

which is equivalent to

$$6 = x - y$$
$$14 = x + y$$

Step 4: ***Solve the system.***
Adding the two equations, we have

$$20 = 2x$$
$$x = 10$$

Substituting $x = 10$ into $14 = x + y$, we see that

$$y = 4$$

Step 5: ***Write answers.***
The speed of the boat in still water is 10 miles per hour; the speed of the current is 4 miles per hour.

Step 6: ***Reread and check.***
The boat travels at $10 + 4 = 14$ miles per hour downstream, so in 2 hours it will travel $14 \cdot 2 = 28$ miles. The boat travels at $10 - 4 = 6$ miles per hour upstream, so in 3 hours it will travel $6 \cdot 3 = 18$ miles. ■

EXAMPLE 6 A coin collection consists of 14 coins with a total value of \$1.35. If the coins are nickels, dimes, and quarters, and the number of nickels is 3 less than twice the number of dimes, how many of each coin is there in the collection?

Solution This problem will require three variables and three equations.

Step 1: ***Read and list.***
We have 14 coins with a total value of \$1.35. The coins are nickels, dimes, and quarters. The number of nickels is 3 less than twice the number of dimes. We do not know how many of each coin we have.

Step 2: ***Assign variables and translate information.***
Because we have three types of coins, we will have to use three variables. Let's let x = the number of nickels, y = the number of dimes, and z = the number of quarters.

Step 3: ***Write a system of equations.***
Because the total number of coins is 14, our first equation is

$$x + y + z = 14$$

Because the number of nickels is 3 less than twice the number of dimes, a second equation is

$$x = 2y - 3 \quad \text{which is equivalent to} \quad x - 2y = -3$$

6. A collection of nickels, dimes, and quarters consists of 15 coins with a total value of \$1.10. If the number of nickels is one less than 4 times the number of dimes, how many of each coin is contained in the collection?

Answer
5. Boat 8 miles per hour; current 2 miles per hour

Our last equation is obtained by considering the value of each coin and the total value of the collection. Let's write the equation in terms of cents, so we won't have to clear it of decimals later.

$$5x + 10y + 25z = 135$$

Here is our system, with the equations numbered for reference:

$$\begin{aligned} x + y + z &= 14 & \textbf{(1)} \\ x - 2y &= -3 & \textbf{(2)} \\ 5x + 10y + 25z &= 135 & \textbf{(3)} \end{aligned}$$

Step 4: ***Solve the system.***
Let's begin by eliminating x from the first and second equations and the first and third equations. Adding -1 times the second equation to the first equation gives us an equation in only y and z. We call this equation (4).

$$3y + z = 17 \qquad \textbf{(4)}$$

Adding -5 times equation (1) to equation (3) gives us

$$5y + 20z = 65 \qquad \textbf{(5)}$$

We can eliminate z from equations (4) and (5) by adding -20 times (4) to (5). Here is the result:

$$\begin{aligned} -55y &= -275 \\ y &= 5 \end{aligned}$$

Substituting $y = 5$ into equation (4) gives us $z = 2$. Substituting $y = 5$ and $z = 2$ into equation (1) gives us $x = 7$.

Step 5: ***Write answers.***
The collection consists of 7 nickels, 5 dimes, and 2 quarters.

Step 6: ***Reread and check.***
The total number of coins is $7 + 5 + 2 = 14$. The number of nickels, 7, is 3 less than twice the number of dimes, 5. To find the total value of the collection, we have

The value of the 7 nickels is	$7(0.05) = \$0.35$
The value of the 5 dimes is	$5(0.10) = \$0.50$
The value of the 2 quarters is	$2(0.25) = \$0.50$
The total value of the collection is	$\$1.35$

If you go on to take a chemistry class, you may see the next example (or one much like it).

Answer

6. 11 nickels, 3 dimes, 1 quarter

EXAMPLE 7 In a chemistry lab, students record the temperature of water at room temperature and find that it is 77° on the Fahrenheit temperature scale and 25° on the Celsius temperature scale. The water is then heated until it boils. The temperature of the boiling water is 212°F and 100°C. Assume that the relationship between the two temperature scales is a linear one, then use the preceding data to find the formula that gives the Celsius temperature C in terms of the Fahrenheit temperature F.

Solution The data is summarized in Table 1.

TABLE 1

CORRESPONDING TEMPERATURES

IN DEGREES FAHRENHEIT	IN DEGREES CELSIUS
77	25
212	100

If we assume the relationship is linear, then the formula that relates the two temperature scales can be written in slope-intercept form as

$$C = mF + b$$

Substituting $C = 25$ and $F = 77$ into this formula gives us

$$25 = 77m + b$$

Substituting $C = 100$ and $F = 212$ into the formula yields

$$100 = 212m + b$$

Together, the two equations form a system of equations, which we can solve using the addition method.

$$25 = 77m + b \xrightarrow{\text{Multiply by } -1} -25 = -77m - b$$

$$100 = 212m + b \xrightarrow[\text{No change}]{} \underline{100 = 212m + b}$$

$$75 = 135m$$

$$m = \frac{75}{135} = \frac{5}{9}$$

To find the value of b, we substitute $m = \frac{5}{9}$ into $25 = 77m + b$ and solve for b.

$$25 = 77\left(\frac{5}{9}\right) + b$$

$$25 = \frac{385}{9} + b$$

$$b = 25 - \frac{385}{9} = \frac{225}{9} - \frac{385}{9} = -\frac{160}{9}$$

The equation that gives C in terms of F is

$$C = \frac{5}{9}F - \frac{160}{9}$$

7. Suppose the students mentioned in Example 7 take two additional temperature measurements while they are heating the water and find that when $F = 95°$, $C = 35°$, and when $F = 167°$, $C = 75°$. Assume that the relationship between the two temperature scales is linear, then derive the formula that gives C in terms of F using these new data points.

Answer

7. $C = \frac{5}{9}F - \frac{160}{9}$

Linking Objectives and Examples

Next to each **objective**, list the examples that are best described by that objective.

A	

Getting Ready for Class

After reading through the preceding section, respond in your own words and in complete sentences.

1. To apply the Blueprint for Problem Solving to the examples in this section, what is the first step?
2. To apply the Blueprint for Problem Solving to the examples in this section, what is the last step?
3. When working application problems involving boats moving in rivers, how does the current of the river affect the speed of the boat?
4. Write an application problem for which the solution depends on solving the system

$$x + y = 1{,}000$$

$$0.05x + 0.06y = 55$$

Problem Set 4.5

INTERMEDIATE Algebra Now™

Number Problems

1. One number is 3 more than twice another. The sum of the numbers is 18. Find the two numbers.

2. The sum of two numbers is 32. One of the numbers is 4 less than 5 times the other. Find the two numbers.

3. The difference of two numbers is 6. Twice the smaller is 4 more than the larger. Find the two numbers.

4. The larger of two numbers is 5 more than twice the smaller. If the smaller is subtracted from the larger, the result is 12. Find the two numbers.

5. The sum of three numbers is 8. Twice the smallest is 2 less than the largest, and the sum of the largest and smallest is 5. Use a linear system in three variables to find the three numbers.

6. The sum of three numbers is 14. The largest is 4 times the smallest, the sum of the smallest and twice the largest is 18. Use a linear system in three variables to find the three numbers.

Ticket and Interest Problems

7. A total of 925 tickets were sold for a game for a total of $1,150. If adult tickets sold for $2.00 and children's tickets sold for $1.00, how many of each kind of ticket were sold?

8. If tickets for a show cost $2.00 for adults and $1.50 for children, how many of each kind of ticket were sold if a total of 300 tickets were sold for $525?

9. Mr. Jones has $20,000 to invest. He invests part at 6% and the rest at 7%. If he earns $1,280 in interest after 1 year, how much did he invest at each rate?

10. A man invests $17,000 in two accounts. One account earns 5% interest per year and the other earns 6.5%. If his total interest after one year is $970, how much did he invest at each rate?

11. Susan invests twice as much money at 7.5% as she does at 6%. If her total interest after a year is $840, how much does she have invested at each rate?

12. A woman earns $1,350 in interest from two accounts in a year. If she has three times as much invested at 7% as she does at 6%, how much does she have in each account?

13. A man invests \$2,200 in three accounts that pay 6%, 8%, and 9% in annual interest, respectively. He has three times as much invested at 9% as he does at 6%. If his total interest for the year is \$178, how much is invested at each rate?

14. A student has money in three accounts that pay 5%, 7%, and 8% in annual interest. She has three times as much invested at 8% as she does at 5%. If the total amount she has invested is \$1,600 and her interest for the year comes to \$115, how much money does she have in each account?

Mixture Problems

15. How many gallons of 20% alcohol solution and 50% alcohol solution must be mixed to get 9 gallons of 30% alcohol solution?

16. How many ounces of 30% hydrochloric acid solution and 80% hydrochloric acid solution must be mixed to get 10 ounces of 50% hydrochloric acid solution?

17. A mixture of 16% disinfectant solution is to be made from 20% and 14% disinfectant solutions. How much of each solution should be used if 15 gallons of the 16% solution are needed?

18. Paul mixes nuts worth \$1.55 per pound with oats worth \$1.35 per pound to get 25 pounds of trail mix worth \$1.45 per pound. How many pounds of nuts and how many pounds of oats did he use?

19. Metal Alloys Metal workers solve systems of equations when forming metal alloys. If a certain metal alloy is 40% copper and another alloy is 60% copper, then a system of equations may be written to determine the amount of each alloy necessary to make 50 pounds of a metal alloy that is 55% copper. Write the system and determine this amount.

20. A chemist has three different acid solutions. The first acid solution contains 20% acid, the second contains 40%, and the third contains 60%. He wants to use all three solutions to obtain a mixture of 60 liters containing 50% acid, using twice as much of the 60% solution as the 40% solution. How many liters of each solution should be used?

Rate Problems

21. It takes about 2 hours to travel 24 miles downstream and 3 hours to travel 18 miles upstream. What is the speed of the boat in still water? What is the speed of the current of the river?

22. A boat on a river travels 20 miles downstream in only 2 hours. It takes the same boat 6 hours to travel 12 miles upstream. What are the speed of the boat and the speed of the current?

23. An airplane flying with the wind can cover a certain distance in 2 hours. The return trip against the wind takes $2\frac{1}{2}$ hours. How fast is the plane and what is the speed of the air, if the distance is 600 miles?

24. An airplane covers a distance of 1,500 miles in 3 hours when it flies with the wind and $3\frac{1}{3}$ hours when it flies against the wind. What is the speed of the plane in still air?

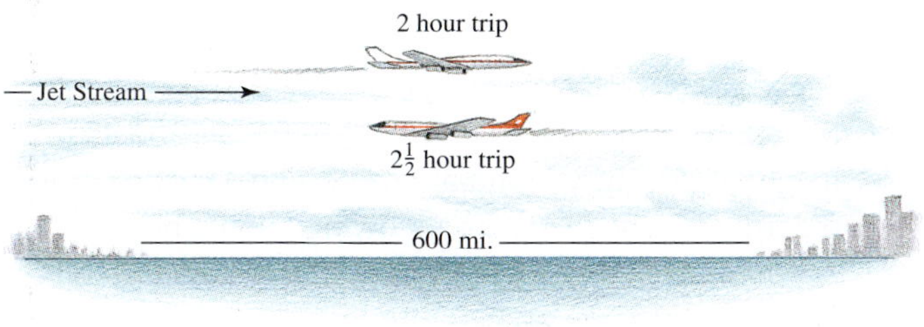

Coin Problems

25. Bob has 20 coins totaling \$1.40. If he has only dimes and nickels, how many of each coin does he have?

26. If Amy has 15 coins totaling \$2.70, and the coins are quarters and dimes, how many of each coin does she have?

27. A collection of nickels, dimes, and quarters consists of 9 coins with a total value of \$1.20. If the number of dimes is equal to the number of nickels, find the number of each type of coin.

28. A coin collection consists of 12 coins with a total value of \$1.20. If the collection consists only of nickels, dimes, and quarters, and the number of dimes is two more than twice the number of nickels, how many of each type of coin are in the collection?

29. A collection of nickels, dimes, and quarters amounts to \$10.00. If there are 140 coins in all and there are twice as many dimes as there are quarters, find the number of nickels.

30. A cash register contains a total of 95 coins consisting of pennies, nickels, dimes, and quarters. There are only 5 pennies and the total value of the coins is \$12.05. Also, there are 5 more quarters than dimes. How many of each coin is in the cash register?

Additional Problems

31. Price and Demand A manufacturing company finds that they can sell 300 items if the price per item is \$2.00, and 400 items if the price is \$1.50 per item. If the relationship between the number of items sold x and the price per item p is a linear one, find a formula that gives x in terms of p. Then use the formula to find the number of items they will sell if the price per item is \$3.00.

32. Price and Demand A company manufactures and sells bracelets. They have found from experience that they can sell 300 bracelets each week if the price per bracelet is \$2.00, but only 150 bracelets are sold if the price is \$2.50 per bracelet. If the relationship between the number of bracelets sold x and the price per bracelet p is a linear one, find a formula that gives x in terms of p. Then use the formula to find the number of bracelets they will sell at \$3.00 each.

33. Height of a Ball A ball is tossed into the air so that the height after 1, 3, and 5 seconds is as given in the following table. If the relationship between the height of the ball h and the time t is quadratic, then the relationship can be written as

$$h = at^2 + bt + c$$

Use the information in the table to write a system of three equations in three variables a, b, and c. Solve the system to find the exact relationship between h and t.

t (SEC)	h (FT)
1	128
3	128
5	0

34. Height of a Ball A ball is tossed into the air and its height above the ground after 1, 3, and 4 seconds is recorded as shown in the following table. The relationship between the height of the ball h and the time t is quadratic and can be written as

$$h = at^2 + bt + c$$

Use the information in the table to write a system of three equations in three variables a, b, and c. Solve the system to find the exact relationship between the variables h and t.

t (SEC)	h (FT)
1	96
3	64
4	0

Getting Ready for the Next Section

35. Does the graph of $x + y < 4$ include the boundary line?

36. Does the graph of $-x + y \le 3$ include the boundary line?

37. Where do the graphs of the lines $x + y = 4$ and $x - 2y = 4$ intersect?

38. Where do the graphs of the line $x = -1$ and $x - 2y = 4$ intersect?

Solve.

39. $20x + 9{,}300 > 18{,}000$

40. $20x + 4{,}800 > 18{,}000$

Maintaining Your Skills

Graph each inequality.

41. $2x + 3y < 6$

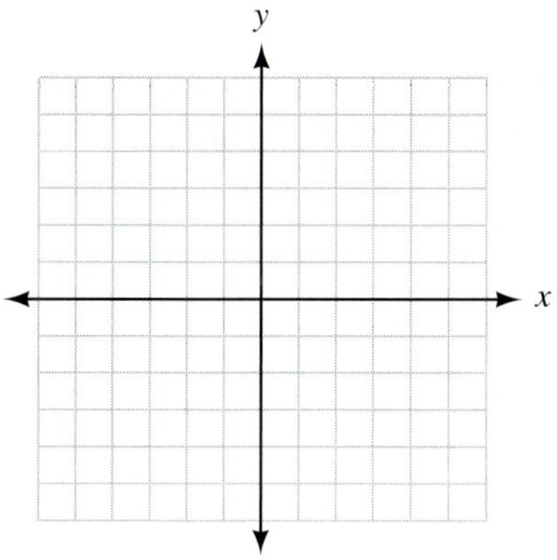

42. $2x + y < -5$

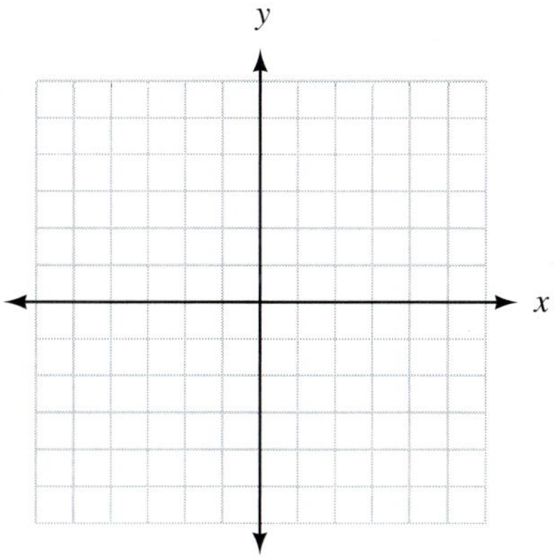

43. $y \ge -3x - 4$

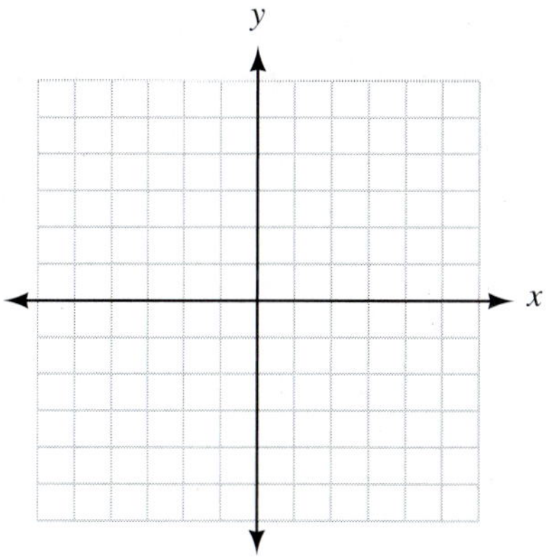

44. $y \geq 2x - 1$

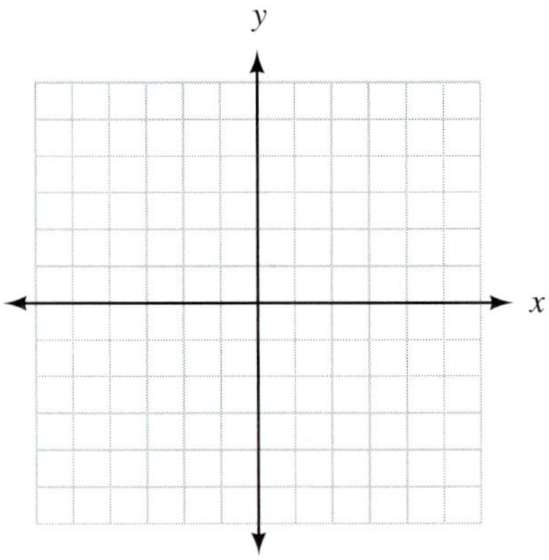

45. $x \geq 3$

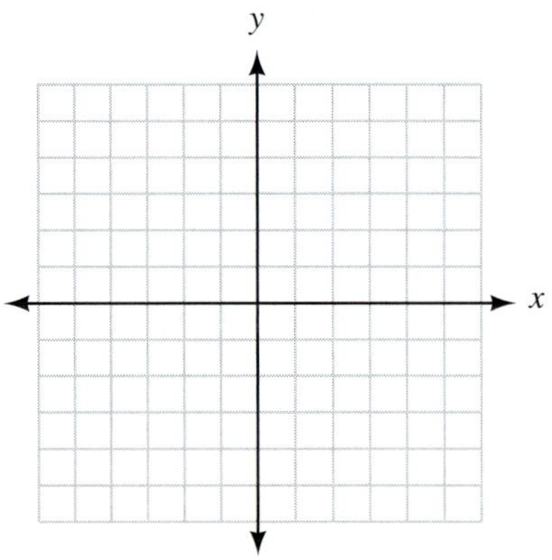

46. $y > -5$

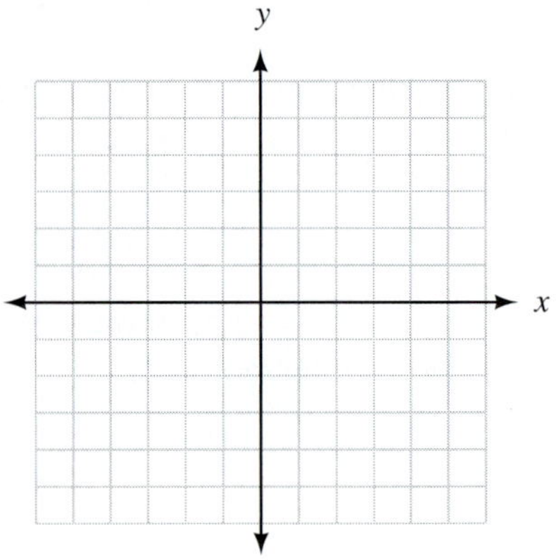

Extending the Concepts

47. **High School Dropout Rate** The high school dropout rates for males and females over the years 1965 to 1990 are shown in the following table.

YEAR	FEMALE DROPOUT RATE (%)	MALE DROPOUT RATE (%)
1965	18	16
1970	16	14
1975	15	13
1980	13	16
1985	12	15
1990	13	14

Plotting the years along the horizontal axis and the dropout rates along the vertical axis, draw a line graph for these data. Draw the female line dashed and the male line solid for easier reading and comparison.

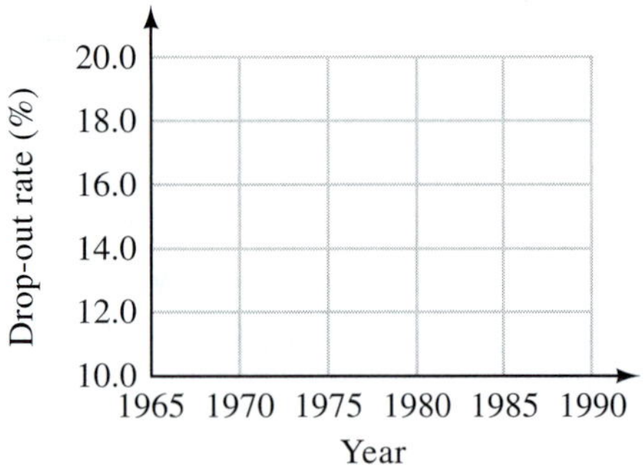

48. **High School Dropout Rate** Refer to the data in the preceding exercise to answer the following questions.

a. Using the slope, determine the time interval when the decline in the female dropout rate was the steepest.

b. Using the slope, determine the time interval when the increase in the male dropout rate was the steepest.

c. What appears unusual about the time period from 1975 to 1980?

d. Are the dropout rates for males and females generally increasing or generally decreasing?

49. **High School Dropout Rate** Refer to the data about high school dropout rates for males and females.

a. Write a linear equation for the dropout rate of males, M, based on the year, x, for the years 1975 and 1980.

b. Write a linear equation for the dropout rate of females, F, based on the year, x, for the years 1975 and 1980.

c. Using your results from parts (a) and (b), determine when the dropout rates for males and females were the same.

Systems of Linear Inequalities

4.6

OBJECTIVES

A	Graph the solution to a system of linear inequalities in two variables.

In the previous chapter, we graphed linear inequalities in two variables. To review, we graph the boundary line, using a solid line if the boundary is part of the solution set and a broken line if the boundary is not part of the solution set. Then we test any point that is not on the boundary line in the original inequality. A true statement tells us that the point lies in the solution set; a false statement tells us the solution set is the other region.

Figure 1 shows the graph of the inequality $x + y < 4$. Note that the boundary is not included in the solution set and is therefore drawn with a broken line. Figure 2 shows the graph of $-x + y \leq 3$. Note that the boundary is drawn with a solid line because it is part of the solution set.

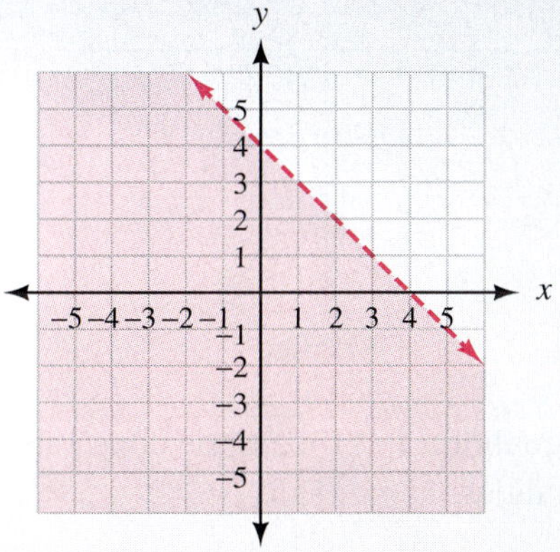

FIGURE 1

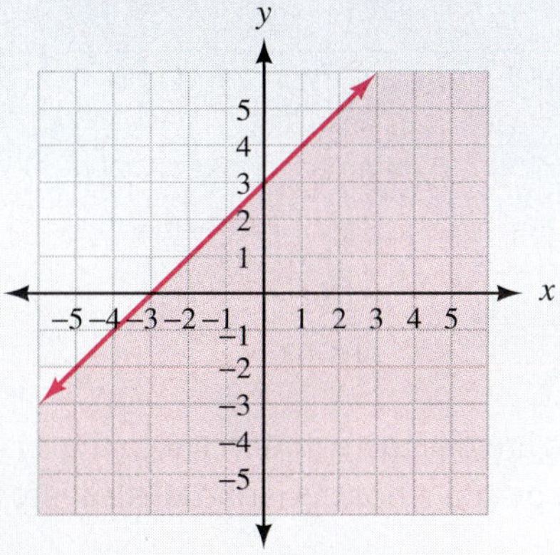

FIGURE 2

If we form a system of inequalities with the two inequalities, the solution set will be all the points common to both solution sets shown in the two figures above: It is the intersection of the two solution sets. Therefore, the solution set for the system of inequalities.

$$x + y < 4$$
$$-x + y \leq 3$$

is all the ordered pairs that satisfy both inequalities. It is the set of points that are below the line $x + y = 4$ and also below (and including) the line $-x + y = 3$. The graph of the solution set to this system is shown in Figure 3. We have written the system in Figure 3 with the word *and* just to remind you that the solution set to a system of equations or inequalities is all the points that satisfy both equations or inequalities.

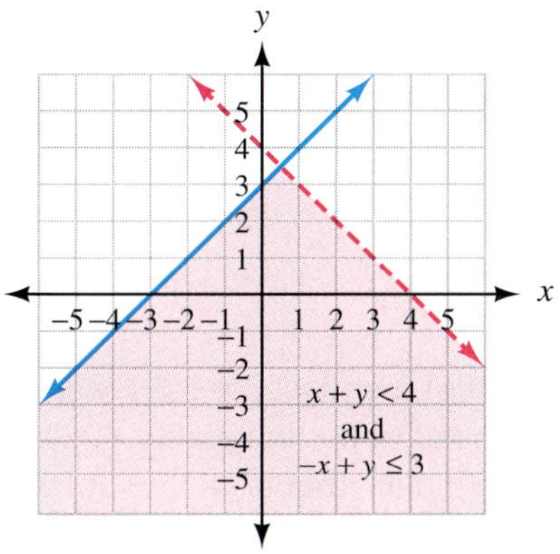

FIGURE 3

PRACTICE PROBLEMS

1. Graph the solution set to the system

$$y \le \frac{1}{3}x + 1$$
$$y > \frac{1}{3}x - 3$$

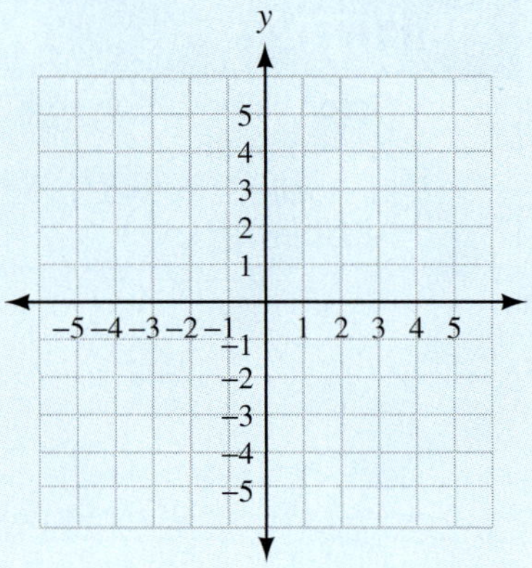

2. Graph the solution set to the system

$$x + y < 3$$
$$x \ge 0$$
$$y \ge 0$$

Answers

1.–2. See Solutions to Selected Practice Problems.

EXAMPLE 1

Graph the solution to the system of linear inequalities.

$$y < \frac{1}{2}x + 3$$

$$y \ge \frac{1}{2}x - 2$$

Solution Figures 4 and 5 show the solution set for each of the inequalities separately.

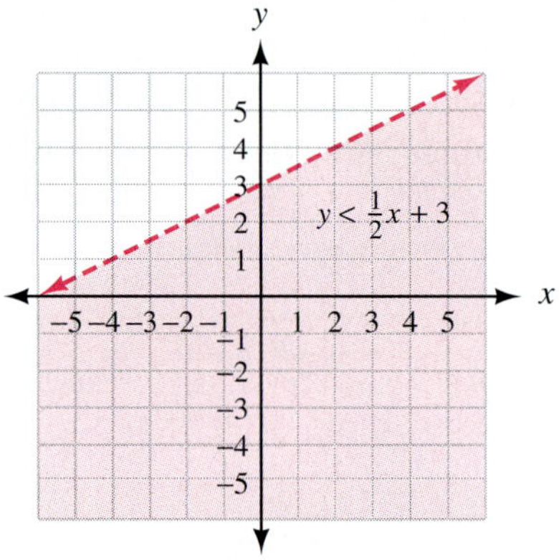

FIGURE 4

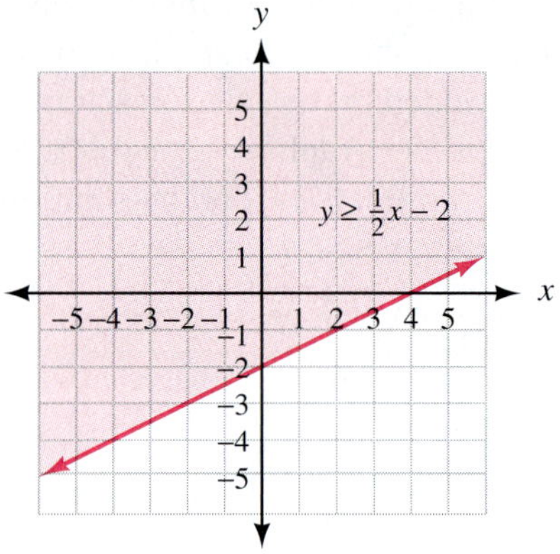

FIGURE 5

Figure 6 is the solution set to the system of inequalities. It is the region consisting of points whose coordinates satisfy both inequalities.

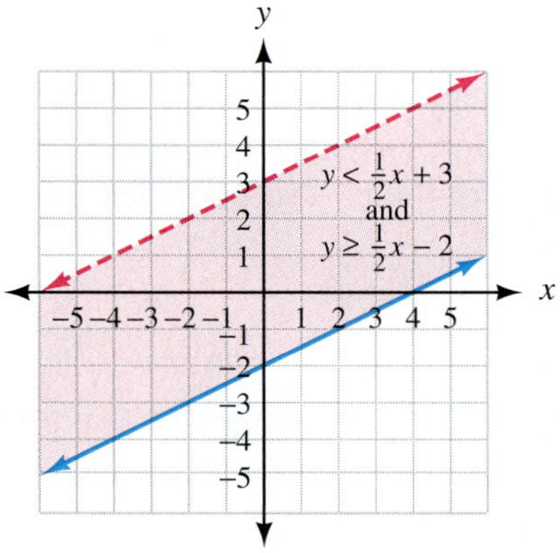

FIGURE 6

EXAMPLE 2

Graph the solution to the system of linear inequalities.

$$x + y < 4$$

$$x \ge 0$$

$$y \ge 0$$

Solution We graphed the first inequality, $x + y < 4$, in Figure 1 at the beginning of this section. The solution set to the inequality $x \ge 0$, shown in Figure 7, is all the points to the right of the y-axis; that is, all the points with x-coordinates that are greater than or equal to 0. Figure 8 shows the graph of $y \ge 0$. It consists of all points with y-coordinates greater than or equal to 0; that is, all points from the x-axis up.

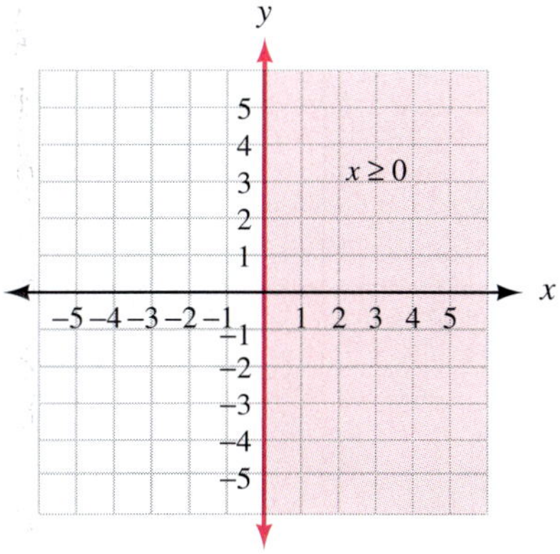

FIGURE 7

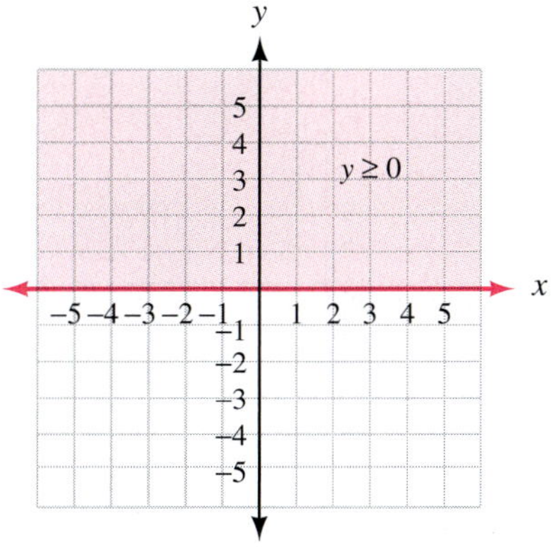

FIGURE 8

The regions shown in Figures 7 and 8 overlap in the first quadrant. Therefore, putting all three regions together we have the points in the first quadrant that are below the line $x + y = 4$. This region is shown in Figure 9, and it is the solution to our system of inequalities.

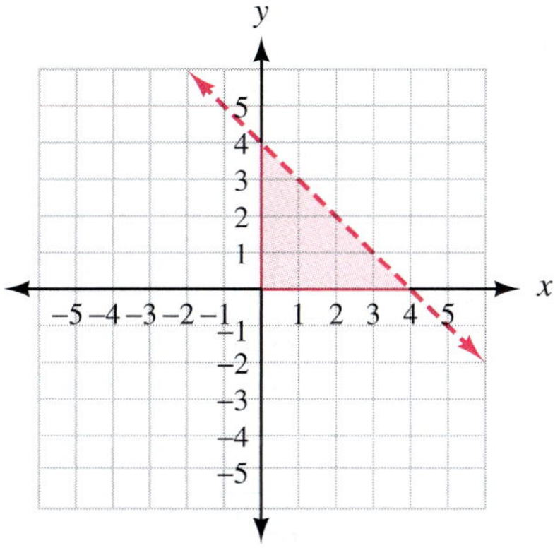

FIGURE 9

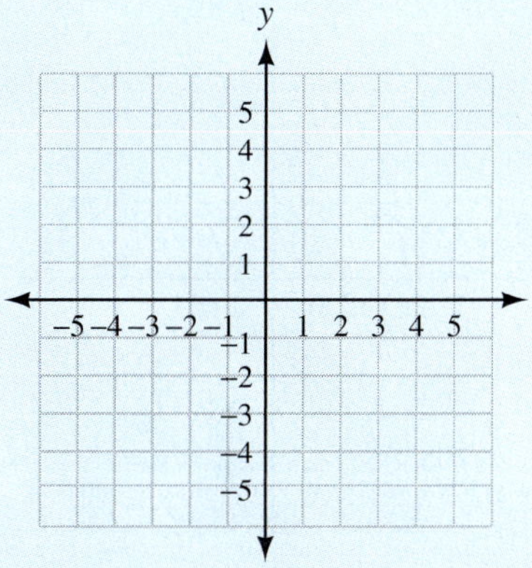

Extending the discussion in Example 2 we can name the points in each of the four quadrants using systems of inequalities.

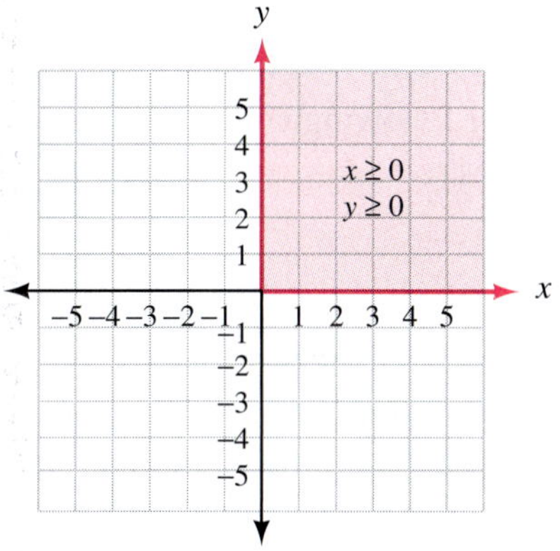

FIGURE 10

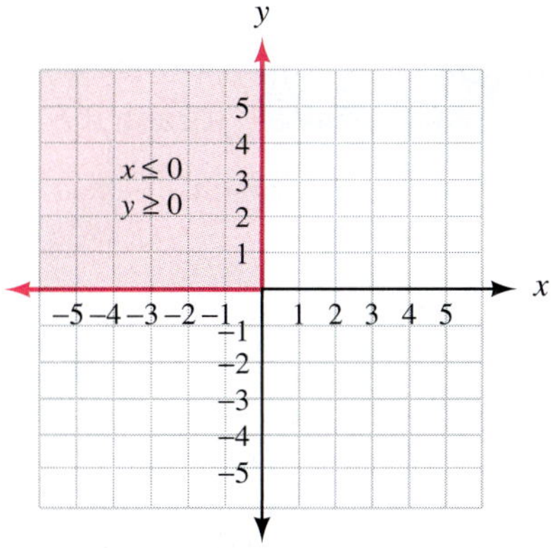

FIGURE 11

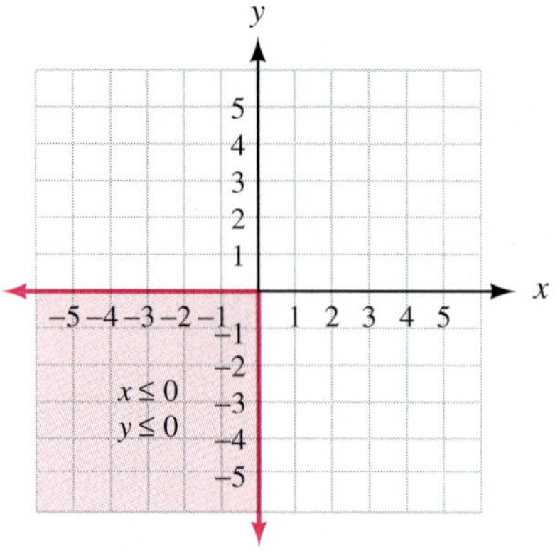

FIGURE 12

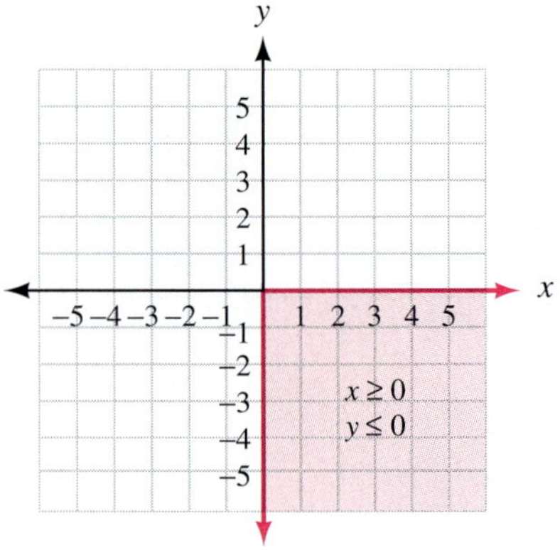

FIGURE 13

3. Graph the solution to the system

$$x > -3$$
$$y < 4$$

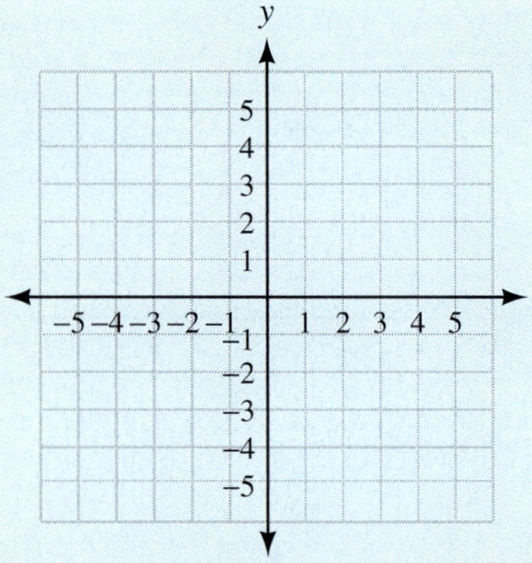

EXAMPLE 3

Graph the solution to the system of linear inequalities.

$$x \le 4$$
$$y \ge -3$$

Solution The solution to this system will consist of all points to the left of and including the vertical line $x = 4$ that intersect with all points above and including the horizontal line $y = -3$. The solution set is shown in Figure 14.

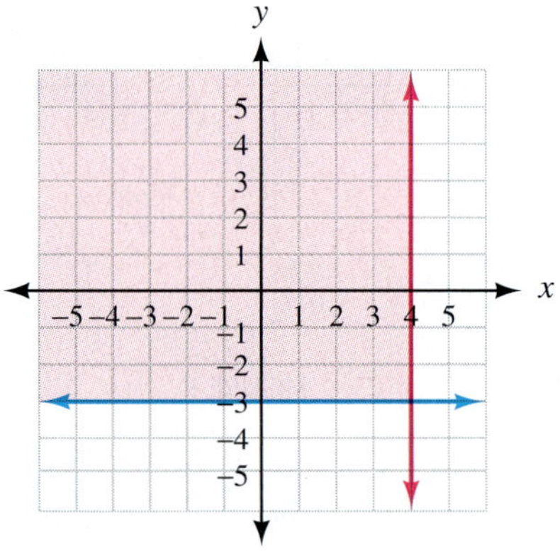

FIGURE 14

4. Graph the solution set for the following system.

$$x - y < 5$$
$$x + y < 5$$
$$x > 1$$

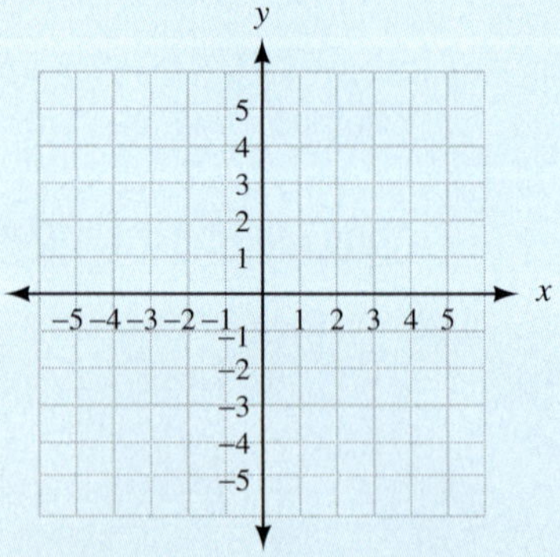

EXAMPLE 4

Graph the solution set for the following system.

$$x - 2y \le 4$$
$$x + y \le 4$$
$$x \ge -1$$

Solution We have three linear inequalities, representing three sections of the coordinate plane. The graph of the solution set for this system will be the intersection of these three sections. The graph of $x - 2y \le 4$ is the section above and including the boundary $x - 2y = 4$. The graph of $x + y \le 4$ is the section below and including the boundary line $x + y = 4$. The graph of $x \ge -1$ is all the points to the right of, and including, the vertical line $x = -1$. The intersection of these three graphs is shown in Figure 15.

Answers

3.–4. See Solutions to Selected Practice Problems.

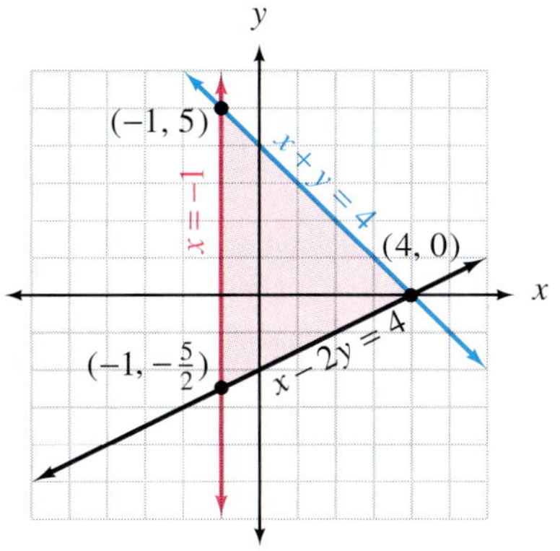

FIGURE 15

EXAMPLE 5

A college basketball arena plans on charging \$20 for certain seats and \$15 for others. They want to bring in more than \$18,000 from all ticket sales and have reserved at least 500 tickets at the \$15 rate. Find a system of inequalities describing all possibilities and sketch the graph. If 620 tickets are sold for \$15, at least how many tickets are sold for \$20?

© Peter Jones/Reuters New Media/Corbis

Solution Let x = the number of \$20 tickets and y = the number of \$15 tickets. We need to write a list of inequalities that describe this situation. That list will form our system of inequalities. First of all, we note that we cannot use negative numbers for either x or y. So, we have our first inequalities:

$$x \geq 0$$
$$y \geq 0$$

Next, we note that they are selling at least 500 tickets for \$15, so we can replace our second inequality with $y \geq 500$. Now our system is

$$x \geq 0$$
$$y \geq 500$$

Now the amount of money brought in by selling \$20 tickets is $20x$, and the amount of money brought in by selling \$15 tickets is $15y$. If the total income from ticket sales is to be more than \$18,000, then $20x + 15y$ must be greater than 18,000. This gives us our last inequality and completes our system.

$$20x + 15y > 18{,}000$$
$$x \geq 0$$
$$y \geq 500$$

5. How would the graph in Figure 16 change if they had reserved only 300 tickets at the \$15 rate?

We have used all the information in the problem to arrive at this system of inequalities. The solution set contains all the values of x and y that satisfy all the conditions given in the problem. Here is the graph of the solution set.

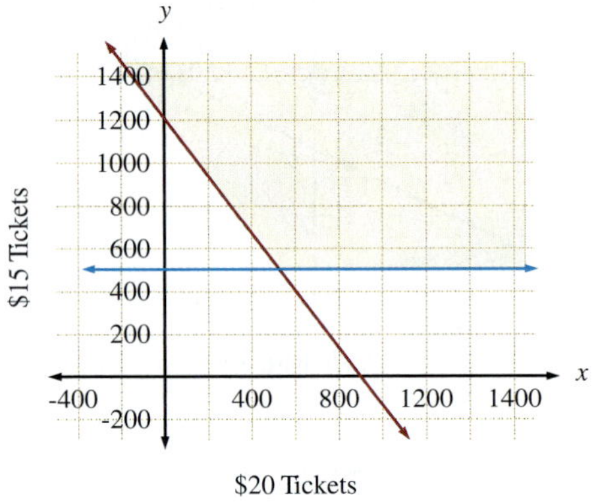

FIGURE 16

If 620 tickets are sold for $15, then we substitute 620 for y in our first inequality to obtain

$20x + 15(620) > 18{,}000$	**Substitute 620 for y**
$20x + 9{,}300 > 18{,}000$	**Multiply**
$20x > 8{,}700$	**Add −9,300 to each side**
$x > 435$	**Divide each side by 20**

If they sell 620 tickets for $15 each, then they need to sell more than 435 tickets at $20 each to bring in more than $18,000. ■

Linking Objectives and Examples

Next to each **objective**, list the examples that are best described by that objective.

A	

Getting Ready for Class

After reading through the preceding section, respond in your own words and in complete sentences.

1. What is the solution set to a system of inequalities?
2. When graphing a system of linear inequalities, how do you find the equations of the boundary lines?
3. For the boundary lines of a system of linear inequalities, when do you use a dotted line rather than a solid line?
4. Once you have graphed the solution set for each inequality in a system, how do you determine the region to shade for the solution to the system of inequalities?

Answer

5. The horizontal line would move down to $y = 300$ from $y = 500$.

Problem Set 4.6

INTERMEDIATE Algebra f(x) Now™

Graph the solution set for each system of linear inequalities.

1. $x + y < 5$
$2x - y > 4$

2. $x + y < 5$
$2x - y < 4$

3. $y < \frac{1}{3}x + 4$
$y \geq \frac{1}{3}x - 3$

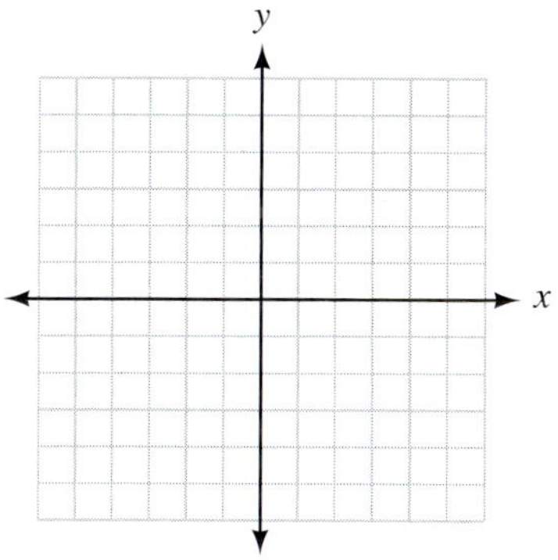

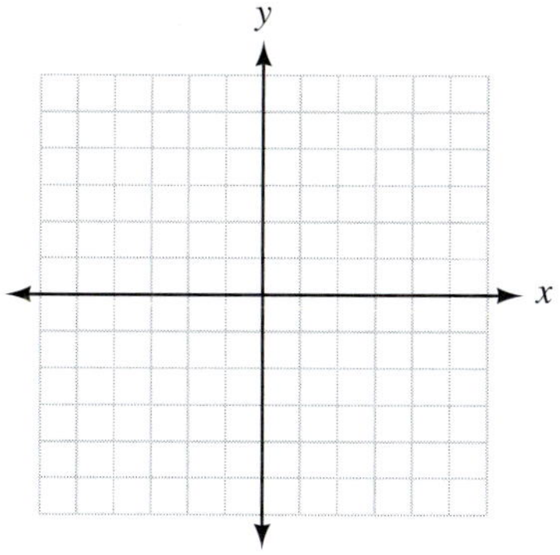

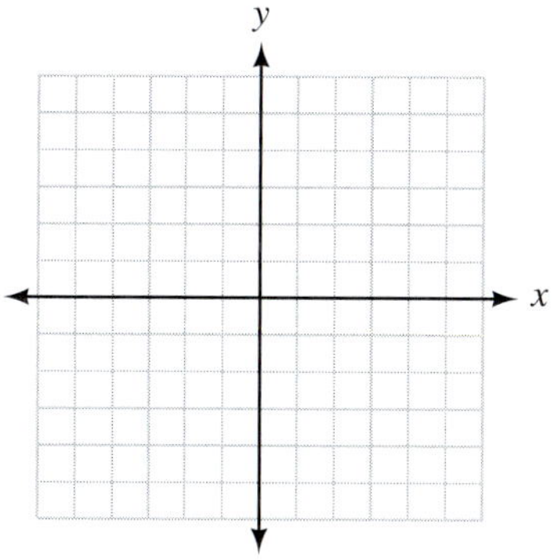

4. $y < 2x + 4$
$y \geq 2x - 3$

5. $x \geq -3$
$y < 2$

6. $x \leq 4$
$y \geq -2$

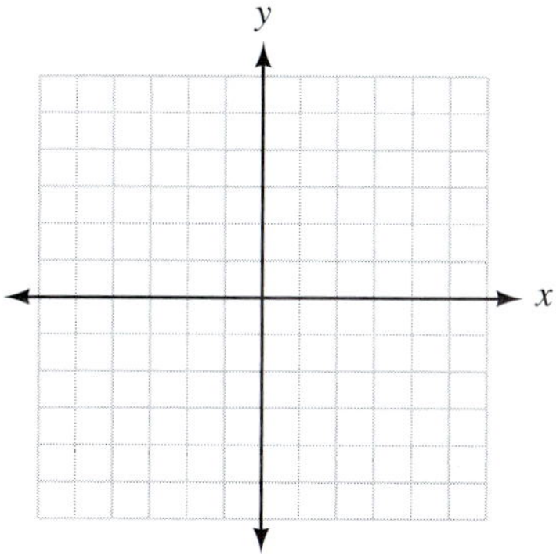

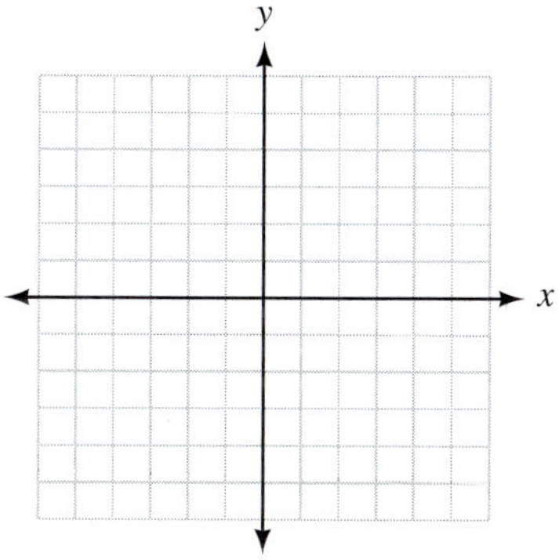

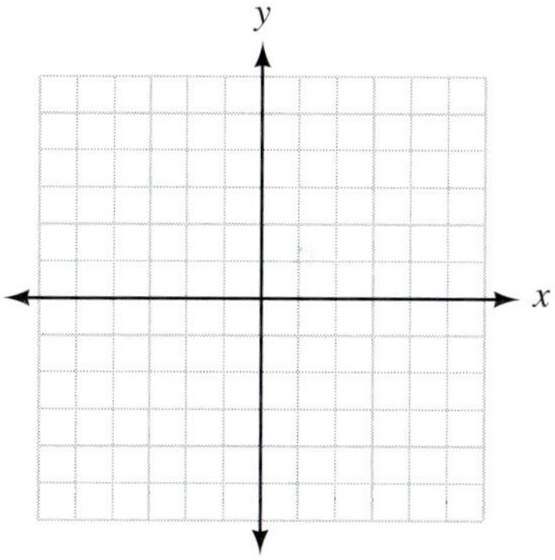

7. $1 \leq x \leq 3$
$2 \leq y \leq 4$

8. $-4 \leq x \leq -2$
$1 \leq y \leq 3$

9. $x + y \leq 4$
$x \geq 0$
$y \geq 0$

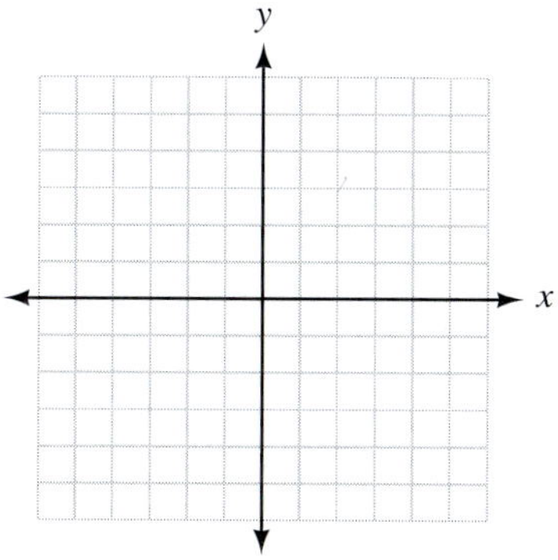

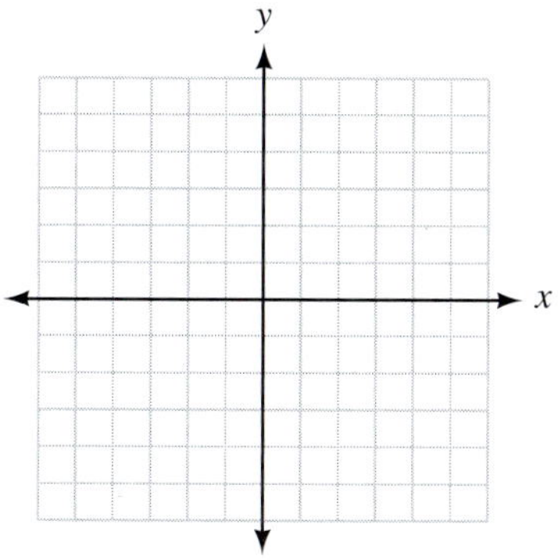

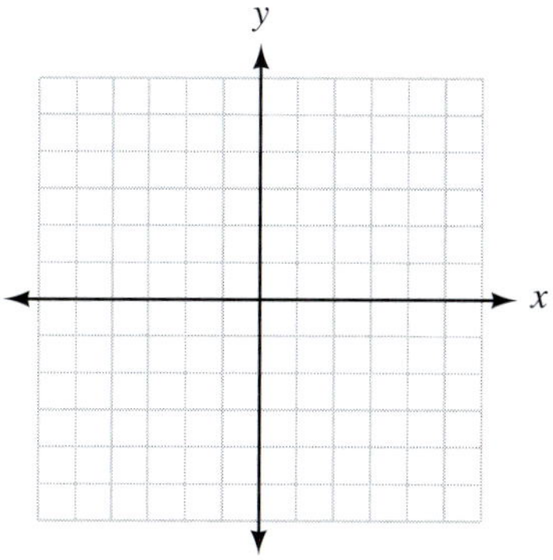

10. $x - y \leq 2$
$x \geq 0$
$y \leq 0$

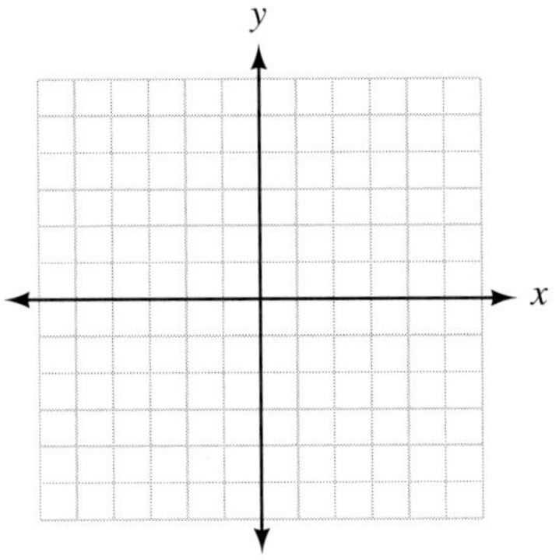

11. $x + y \leq 3$
$x - 3y \leq 3$
$x \geq -2$

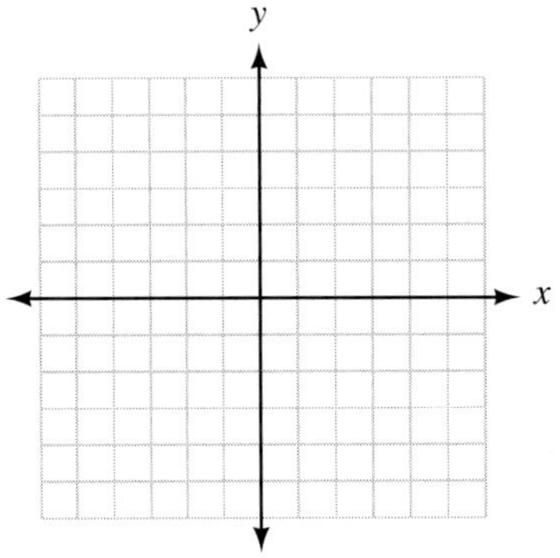

12. $x - y \leq 4$
$x + 2y \leq 4$
$x \geq -1$

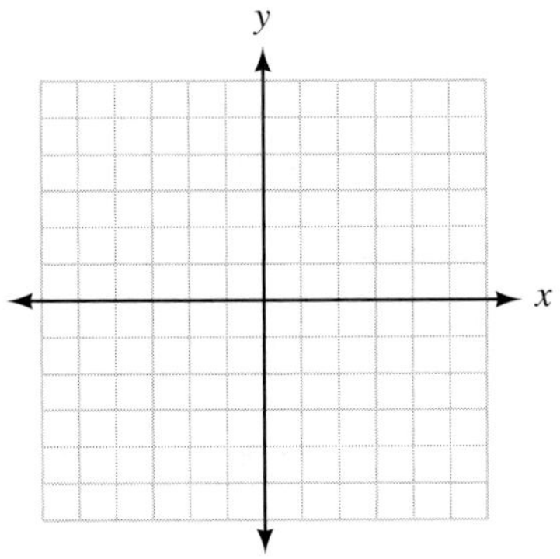

13. $x + y \leq 2$
$-x + y \leq 2$
$y \geq -2$

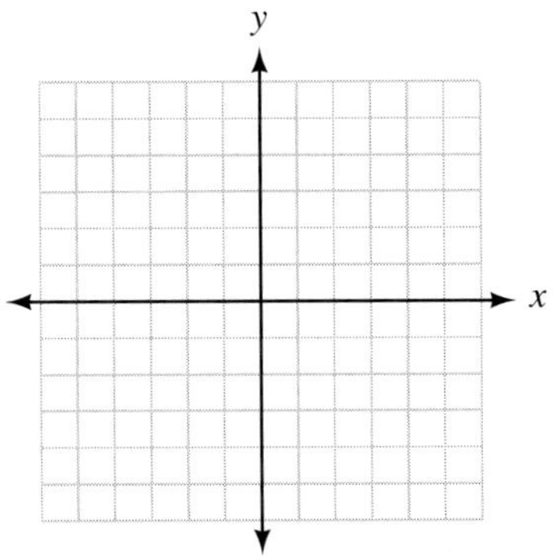

14. $x - y \leq 3$
$-x - y \leq 3$
$y \leq -1$

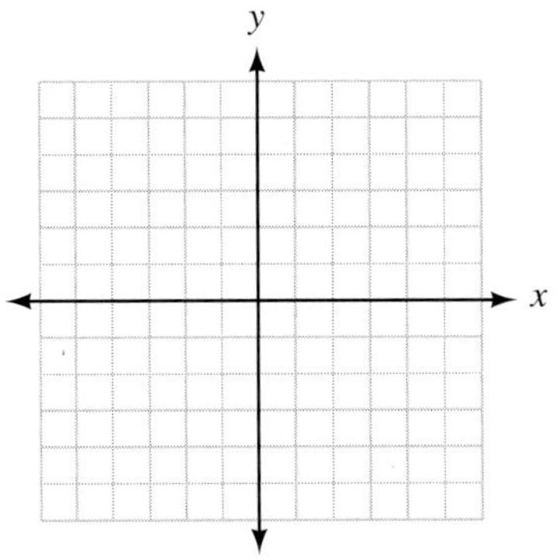

15. $x + y < 5$
$y > x$
$y \geq 0$

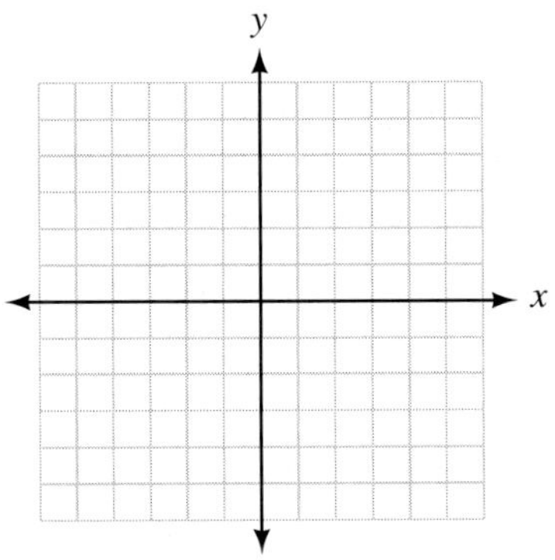

16. $x + y < 5$
$y > x$
$x \geq 0$

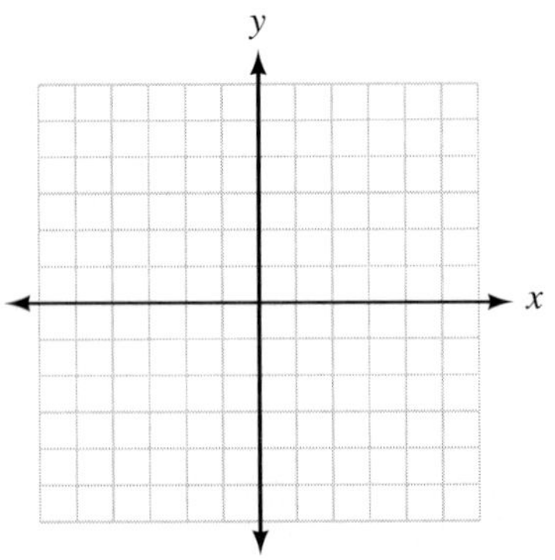

17. $2x + 3y \leq 6$
$x \geq 0$
$y \geq 0$

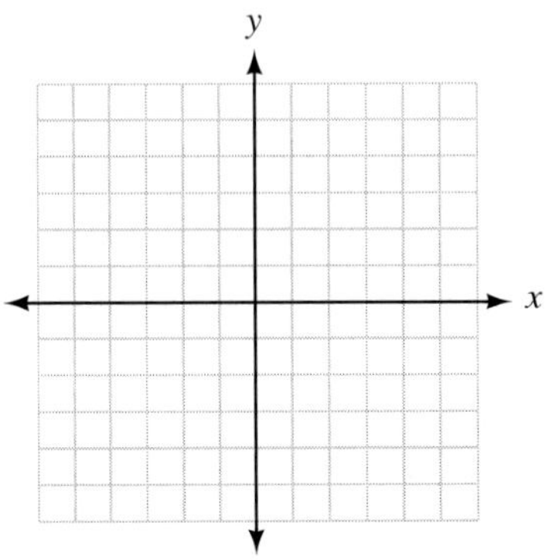

18. $x + 2y \leq 10$
$3x + y \leq 12$
$x \geq 0$
$y \geq 0$

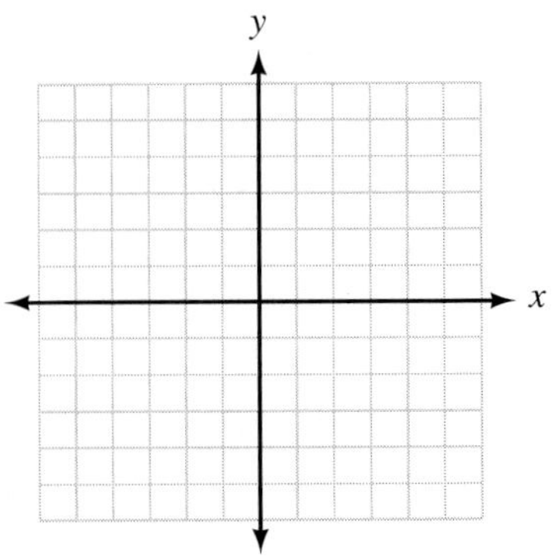

For each figure below, find a system of inequalities that describes the shaded region.

19.

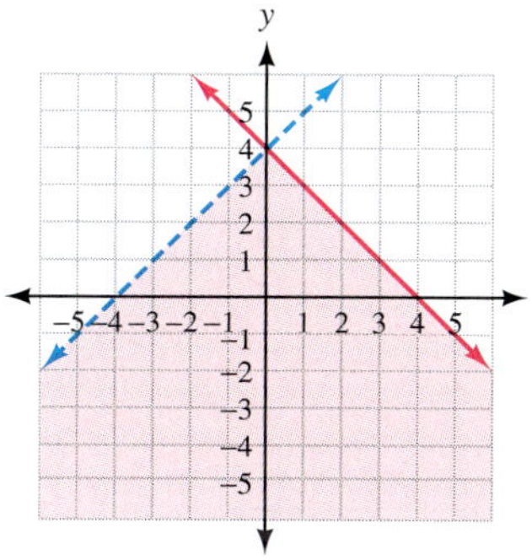

FIGURE 17

20.

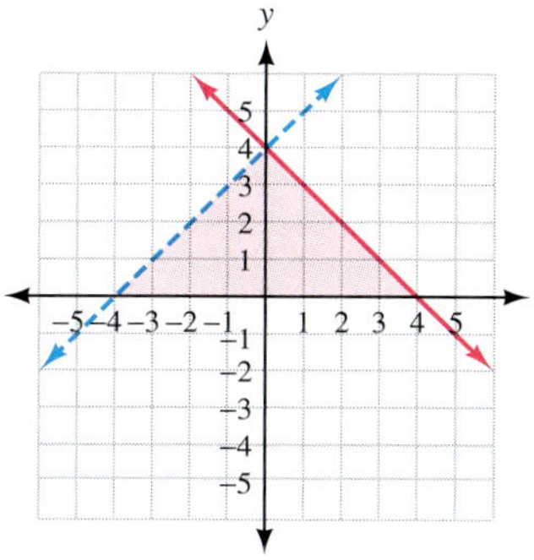

FIGURE 18

21.

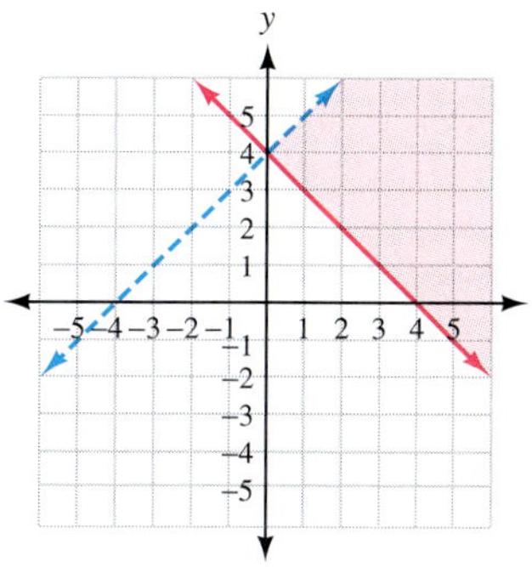

FIGURE 19

22.

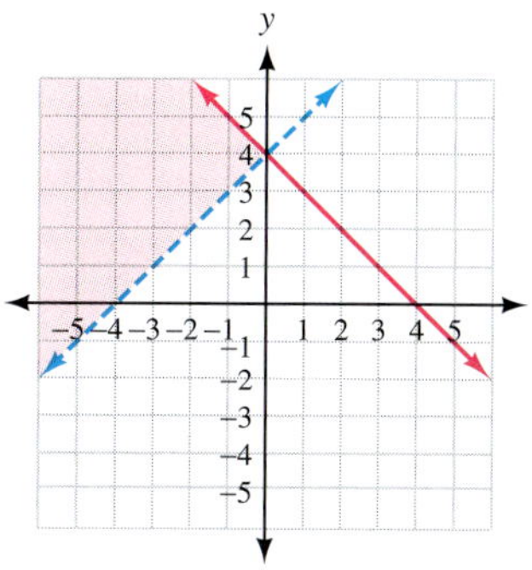

FIGURE 20

Applying the Concepts

23. Office Supplies An office worker wants to purchase some \$0.55 postage stamps and also some \$0.65 postage stamps totaling no more than \$40. It also is desired to have at least twice as many \$0.55 stamps and more than 15 \$0.55 stamps.

a. Find a system of inequalities describing all the possibilities and sketch the graph.

b. If he purchases 20 \$0.55 stamps, what is the maximum number of \$0.65 stamps he can purchase?

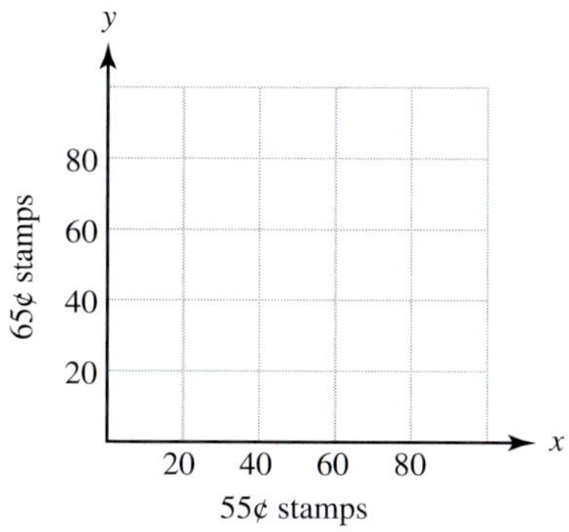

24. Inventory A store sells two brands of VCRs. Customer demand indicates that it is necessary to stock at least twice as many VCRs of brand A as of brand B. At least 30 of brand A and 15 of brand B must be on hand. There is room for not more than 100 VCRs in the store.

a. Find a system of inequalities describing all possibilities, then sketch the graph.

b. If there are 35 VCRs of brand A, what is the maximum number of brand B VCRs on hand?

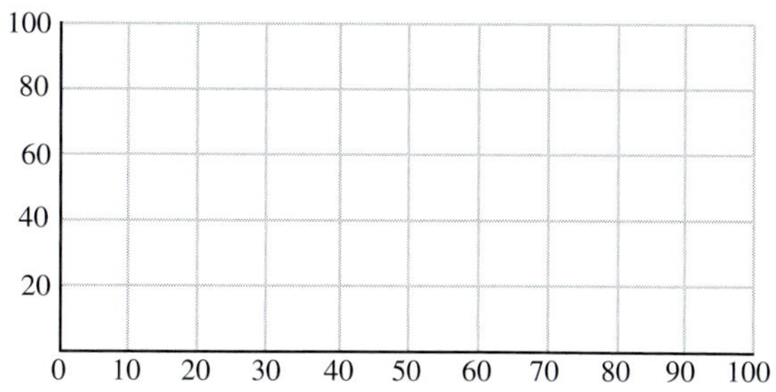

Maintaining Your Skills

For each of the following straight lines, identify the x-intercept, y-intercept, and slope, and sketch the graph.

25. $2x + y = 6$

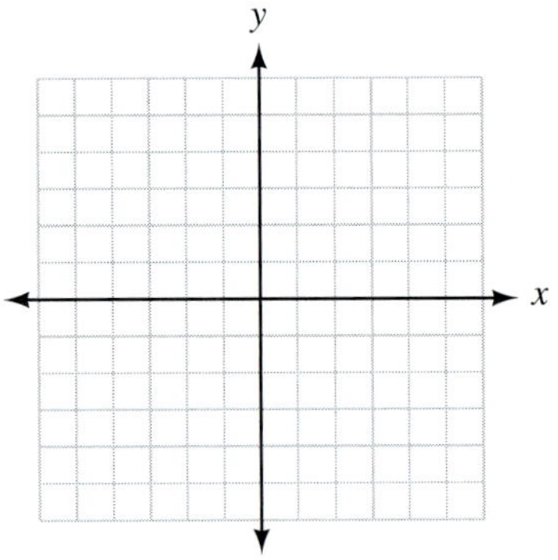

26. $y = \frac{3}{2}x + 4$

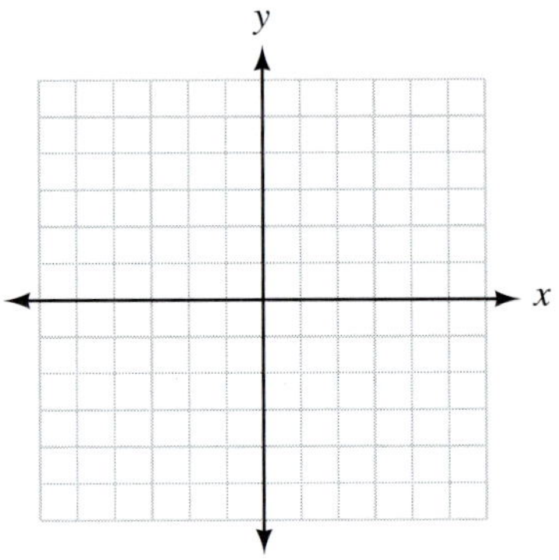

27. $x = -2$

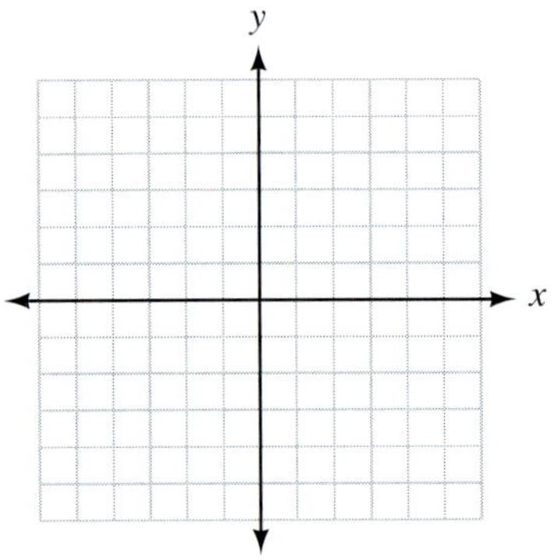

Find the equation for each line.

28. Give the equation of the line through $(-1, 3)$ that has slope $m = 2$.

29. Give the equation of the line through $(-3, 2)$ and $(4, -1)$.

30. Line l contains the point $(5, -3)$ and has a graph parallel to the graph of $2x - 5y = 10$. Find the equation for l.

31. Give the equation of the vertical line through $(4, -7)$.

State the domain and range for the following relations, and indicate which relations are also functions.

32. $\{(-2, 0), (-3, 0), (-2, 1)\}$

33. $y = x^2 - 9$

Let $f(x) = x - 2$, $g(x) = 3x + 4$ and $h(x) = 3x^2 - 2x - 8$, and find the following.

34. $f(3) + g(2)$

35. $h(0) + g(0)$

36. $f[g(2)]$

37. $g[f(2)]$

Solve the following variation problems.

38. Direct Variation Quantity y varies directly with the square of x. If y is 50 when x is 5, find y when x is 3.

39. Joint Variation Quantity z varies jointly with x and the cube of y. If z is 15 when x is 5 and y is 2, find z when x is 2 and y is 3.

Chapter 4 Summary

Systems of Linear Equations [4.1, 4.2]

A system of linear equations consists of two or more linear equations considered simultaneously. The solution set to a linear system in two variables is the set of ordered pairs that satisfy both equations. The solution set to a linear system in three variables consists of all the ordered triples that satisfy each equation in the system.

1. The solution to the system
$x + 2y = 4$
$x - y = 1$
is the ordered pair (2, 1). It is the only ordered pair that satisfies both equations.

To Solve a System by the Addition Method [4.1]

Step 1: Look the system over to decide which variable will be easier to eliminate.

Step 2: Use the multiplication property of equality on each equation separately, if necessary, to ensure that the coefficients of the variable to be eliminated are opposites.

Step 3: Add the left and right sides of the system produced in step 2, and solve the resulting equation.

Step 4: Substitute the solution from step 3 back into any equation with both x- and y-variables, and solve.

Step 5: Check your solution in both equations if necessary.

2. We can eliminate the y-variable from the system in Example 1 by multiplying both sides of the second equation by 2 and adding the result to the first equation:

$$x + 2y = 4 \xrightarrow{\text{No change}} x + 2y = 4$$
$$x - y = 1 \xrightarrow[\text{Times 2}]{} 2x - 2y = 2$$
$$3x = 6$$
$$x = 2$$

Substituting $x = 2$ into either of the original two equations gives $y = 1$. The solution is (2, 1).

To Solve a System by the Substitution Method [4.1]

Step 1: Solve either of the equations for one of the variables (this step is not necessary if one of the equations has the correct form already).

Step 2: Substitute the results of step 1 into the other equation, and solve.

Step 3: Substitute the results of step 2 into an equation with both x- and y-variables, and solve. (The equation produced in step 1 is usually a good one to use.)

Step 4: Check your solution if necessary.

3. We can apply the substitution method to the system in Example 1 by first solving the second equation for x to get

$$x = y + 1$$

Substituting this expression for x into the first equation we have

$$y + 1 + 2y = 4$$
$$3y + 1 = 4$$
$$3y = 3$$
$$y = 1$$

Using $y = 1$ in either of the original equations gives $x = 2$.

Inconsistent and Dependent Equations [4.1, 4.2]

Two linear equations that have no solutions in common are said to be *inconsistent,* whereas two linear equations that have all their solutions in common are said to be *dependent.*

4. If the two lines are parallel, then the system will be inconsistent and the solution is $\varnothing$. If the two lines coincide, then the system is dependent.

2 × 2 Determinants [4.3]

The value of a 2 × 2 determinant is as follows:

$$\begin{vmatrix} a & c \\ b & d \end{vmatrix} = ad - bc$$

5. $\begin{vmatrix} 3 & 4 \\ -2 & 5 \end{vmatrix} = 15 - (-8) = 23$

3 × 3 Determinants [4.3]

6. Expanding $\begin{vmatrix} 1 & 3 & -2 \\ 2 & 0 & 1 \\ 4 & -1 & 1 \end{vmatrix}$ across the first row gives us

$$1\begin{vmatrix} 0 & 1 \\ -1 & 1 \end{vmatrix} - 3\begin{vmatrix} 2 & 1 \\ 4 & 1 \end{vmatrix} - 2\begin{vmatrix} 2 & 0 \\ 4 & -1 \end{vmatrix}$$
$$= 1(1) - 3(-2) - 2(-2)$$
$$= 11$$

The value of a 3 × 3 determinant is given by

$$\begin{vmatrix} a_1 & b_1 & c_1 \\ a_2 & b_2 & c_2 \\ a_3 & b_3 & c_3 \end{vmatrix} = a_1b_2c_3 + a_3b_1c_2 + a_2b_3c_1 - a_3b_2c_1 - a_1b_3c_2 - a_2b_1c_3$$

There are two methods of finding the six products in the expansion of a 3 × 3 determinant. One method involves a cross-multiplication scheme. The other method involves expanding the determinant by minors.

Cramer's Rule for a Linear System in Two Variables [4.4]

7. For the system $\begin{matrix} x + y = 6 \\ 3x - 2y = -2 \end{matrix}$

we have

$$D = \begin{vmatrix} 1 & 1 \\ 3 & -2 \end{vmatrix} = -5$$

$$D_x = \begin{vmatrix} 6 & 1 \\ -2 & -2 \end{vmatrix} = -10$$

$$x = \frac{-10}{-5} = 2$$

$$D_y = \begin{vmatrix} 1 & 6 \\ 3 & -2 \end{vmatrix} = -20$$

$$y = \frac{-20}{-5} = 4$$

The solution to the system

$$a_1x + b_1y = c_1$$
$$a_2x + b_2y = c_2$$

is given by

$$x = \frac{D_x}{D} \quad \text{and} \quad y = \frac{D_y}{D} \quad (D \neq 0)$$

where

$$D = \begin{vmatrix} a_1 & b_1 \\ a_2 & b_2 \end{vmatrix}, \quad D_x = \begin{vmatrix} c_1 & b_1 \\ c_2 & b_2 \end{vmatrix}, \quad D_y = \begin{vmatrix} a_1 & c_1 \\ a_2 & c_2 \end{vmatrix}$$

Cramer's Rule for a Linear System in Three Variables [4.4]

8. For the system

$$x + y = -1$$
$$2x - z = 3$$
$$y + 2z = -1$$

$$D = \begin{vmatrix} 1 & 1 & 0 \\ 2 & 0 & -1 \\ 0 & 1 & 2 \end{vmatrix} = -3$$

$$D_x = \begin{vmatrix} -1 & 1 & 0 \\ 3 & 0 & -1 \\ -1 & 1 & 2 \end{vmatrix} = -6$$

$$x = \frac{-6}{-3} = 2$$

$$D_y = \begin{vmatrix} 1 & -1 & 0 \\ 2 & 3 & -1 \\ 0 & -1 & 2 \end{vmatrix} = 9$$

$$y = \frac{9}{-3} = -3$$

$$D_z = \begin{vmatrix} 1 & 1 & -1 \\ 2 & 0 & 3 \\ 0 & 1 & -1 \end{vmatrix} = -3$$

$$z = \frac{-3}{-3} = 1$$

The solution to the system

$$a_1x + b_1y + c_1z = d_1$$
$$a_2x + b_2y + c_2z = d_2$$
$$a_3x + b_3y + c_3z = d_3$$

is given by

$$x = \frac{D_x}{D}, \quad y = \frac{D_y}{D}, \quad \text{and} \quad z = \frac{D_z}{D} \quad (D \neq 0)$$

where

$$D = \begin{vmatrix} a_1 & b_1 & c_1 \\ a_2 & b_2 & c_2 \\ a_3 & b_3 & c_3 \end{vmatrix} \quad D_x = \begin{vmatrix} d_1 & b_1 & c_1 \\ d_2 & b_2 & c_2 \\ d_3 & b_3 & c_3 \end{vmatrix}$$

$$D_y = \begin{vmatrix} a_1 & d_1 & c_1 \\ a_2 & d_2 & c_2 \\ a_3 & d_3 & c_3 \end{vmatrix} \quad D_z = \begin{vmatrix} a_1 & b_1 & d_1 \\ a_2 & b_2 & d_2 \\ a_3 & b_3 & d_3 \end{vmatrix}$$

Blueprint for Problem Solving Using a System of Equations [4.5]

Step 1: ***Read*** the problem, and then mentally ***list*** the items that are known and the items that are unknown.

Step 2: ***Assign variables*** to each of the unknown items; that is, let x = one of the unknown items and y = the other unknown item (and z = the third unknown item, if there is a third one). Then ***translate*** the other ***information*** in the problem to expressions involving the two (or three) variables.

Step 3: ***Reread*** the problem, and then ***write a system of equations,*** using the items and variables listed in steps 1 and 2, that describes the situation.

Step 4: ***Solve the system*** found in step 3.

Step 5: ***Write your answers*** using complete sentences.

Step 6: ***Reread*** the problem, and ***check*** your solution with the original words in the problem.

9. One number is 2 more than 3 times another. Their sum is 26. Find the numbers.

Step 1: Read and List.
Known items: two numbers, whose sum is 26. One is 2 more than 3 times the other. Unknown items are the two numbers.

Step 2: Assign variables and translate information.
Let x = one of the numbers. Then y = the other number.

Step 3: Write a system of equations.

$$x + y = 26$$
$$y = 3x + 2$$

Step 4: Solve the system.
Substituting the expression for y from the second equation into the first and solving for x yields

$$x + (3x + 2) = 26$$
$$4x + 2 = 26$$
$$4x = 24$$
$$x = 6$$

Using $x = 6$ in $y = 3x + 2$ gives the second number: $y = 20$

Step 5: Write answers.
The two numbers are 6 and 20.

Step 6: Reread and check.
The sum of 6 and 20 is 26, and 20 is 2 more than 3 times 6.

Systems of Linear Inequalities [4.6]

A system of linear inequalities is two or more linear inequalities considered at the same time. To find the solution set to the system, we graph each of the inequalities on the same coordinate system. The solution set is the region that is common to all the regions graphed.

10. The solution set for the system

$$x + y < 4$$
$$-x + y \le 3$$

is shown below.

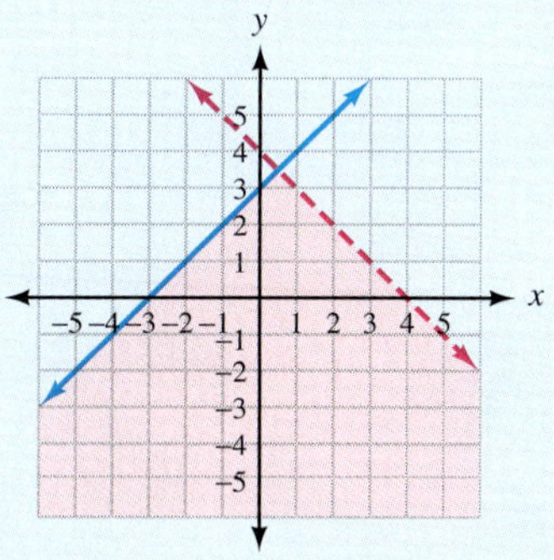

Chapter 4 Review/Test

The problems below form a comprehensive review of the material in this chapter. They can be used to study for exams. If you would like to take a practice test on this chapter, you can use the odd-numbered problems. Give yourself an hour and work as many of the odd-numbered problems as possible. When you are finished, or when an hour has passed, check your answers with the answers in the back of the book. You can use the even-numbered problems for a second practice test.

Solve each system using the addition method. [4.1]

1. $\begin{aligned} x + y &= 4 \\ 2x - y &= 14 \end{aligned}$

2. $\begin{aligned} 3x + y &= 2 \\ 2x + y &= 0 \end{aligned}$

3. $\begin{aligned} 2x - 4y &= 5 \\ -x + 2y &= 3 \end{aligned}$

4. $\begin{aligned} 5x - 2y &= 7 \\ 3x + y &= 2 \end{aligned}$

5. $\begin{aligned} 6x - 5y &= -5 \\ 3x + y &= 1 \end{aligned}$

6. $\begin{aligned} 6x + 4y &= 8 \\ 9x + 6y &= 12 \end{aligned}$

7. $\begin{aligned} 3x - 7y &= 2 \\ -4x + 6y &= -6 \end{aligned}$

8. $\begin{aligned} 6x + 5y &= 9 \\ 4x + 3y &= 6 \end{aligned}$

9. $\begin{aligned} -7x + 4y &= -1 \\ 5x - 3y &= 0 \end{aligned}$

10. $\begin{aligned} \frac{1}{2}x - \frac{3}{4}y &= -4 \\ \frac{1}{4}x + \frac{3}{2}y &= 13 \end{aligned}$

11. $\begin{aligned} \frac{2}{3}x - \frac{1}{6}y &= 0 \\ \frac{4}{3}x + \frac{5}{6}y &= 14 \end{aligned}$

12. $\begin{aligned} -\frac{1}{2}x + \frac{1}{3}y &= -\frac{13}{6} \\ \frac{4}{5}x + \frac{3}{4}y &= \frac{9}{10} \end{aligned}$

Solve each system by the substitution method. [4.1]

13. $\begin{aligned} x + y &= 2 \\ y &= x - 1 \end{aligned}$

14. $\begin{aligned} 2x - 3y &= 5 \\ y &= 2x - 7 \end{aligned}$

15. $\begin{aligned} x + y &= 4 \\ 2x + 5y &= 2 \end{aligned}$

16. $\begin{aligned} x + y &= 3 \\ 2x + 5y &= -6 \end{aligned}$

17. $\begin{aligned} 3x + 7y &= 6 \\ x &= -3y + 4 \end{aligned}$

18. $\begin{aligned} 5x - y &= 4 \\ y &= 5x - 3 \end{aligned}$

Solve each system. [4.2]

19. $\begin{aligned} x + y + z &= 6 \\ x - y - 3z &= -8 \\ x + y - 2z &= -6 \end{aligned}$

20. $\begin{aligned} 3x + 2y + z &= 4 \\ 2x - 4y + z &= -1 \\ x + 6y + 3z &= -4 \end{aligned}$

21. $\begin{aligned} 5x + 8y - 4z &= -7 \\ 7x + 4y + 2z &= -2 \\ 3x - 2y + 8z &= 8 \end{aligned}$

22. $\begin{aligned} 5x - 3y - 6z &= 5 \\ 4x - 6y - 3z &= 4 \\ -x + 9y + 9z &= 7 \end{aligned}$

23. $\begin{aligned} 5x - 2y + z &= 6 \\ -3x + 4y - z &= 2 \\ 6x - 8y + 2z &= -4 \end{aligned}$

24. $\begin{aligned} 4x - 6y + 8z &= 4 \\ 5x + y - 2z &= 4 \\ 6x - 9y + 12z &= 6 \end{aligned}$

25. $\begin{aligned} 2x - y &= 5 \\ 3x - 2z &= -2 \\ 5y + z &= -1 \end{aligned}$

26. $\begin{aligned} x - y &= 2 \\ y - z &= -3 \\ x - z &= -1 \end{aligned}$

Evaluate each determinant. [4.3]

27. $\begin{vmatrix} 2 & 3 \\ -5 & 4 \end{vmatrix}$

28. $\begin{vmatrix} 3 & 0 \\ 5 & -1 \end{vmatrix}$

29. $\begin{vmatrix} 1 & 0 \\ -7 & -3 \end{vmatrix}$

30. $\begin{vmatrix} 3 & 0 & 2 \\ -1 & 4 & 0 \\ 2 & 0 & 0 \end{vmatrix}$

31. $\begin{vmatrix} 3 & -1 & 0 \\ 0 & 2 & -4 \\ 6 & 0 & 2 \end{vmatrix}$

32. $\begin{vmatrix} -3 & -2 & 0 \\ 0 & -4 & 2 \\ 5 & 1 & 1 \end{vmatrix}$

Solve for x. [4.3]

33. $\begin{vmatrix} 2 & 3x \\ -1 & 2x \end{vmatrix} = 4$

34. $\begin{vmatrix} 4x & x \\ 3 & 1 \end{vmatrix} = -4$

Use Cramer's rule to solve each system. [4.4]

35. $3x - 5y = 4$
$7x - 2y = 3$

36. $7x - 5y = 8$
$4x + 3y = 2$

37. $3x - 6y = 9$
$2x - 4y = 6$

38. $6x - 9y = 5$
$7x + 3y = 4$

39. $-6x + 3y = 7$
$5x - 8y = -2$

40. $2x - y + 3z = 4$
$5x + 2y - z = 3$
$-x - 3y + 2z = 1$

41. $4x - 5y = -3$
$2x + 3z = 4$
$3y - z = 8$

42. $2x - 4y = 2$
$4x - 2z = 3$
$4y - z = 2$

Use systems of equations to solve each application problem. In each case, be sure to show the system used. [4.5]

43. Ticket Prices Tickets for the show cost \$2.00 for adults and \$1.50 for children. How many adult tickets and how many children's tickets were sold if a total of 127 tickets were sold for \$214?

44. Coin Collection John has 20 coins totaling \$3.20. If he has only dimes and quarters, how many of each coin does he have?

45. Investments Ms. Jones invests money in two accounts, one of which pays 12% per year, and the other pays 15% per year. If her total investment is \$12,000 and the interest after 1 year is \$1,650, how much is invested in each account?

46. Speed It takes a boat on a river 2 hours to travel 28 miles downstream and 3 hours to travel 30 miles upstream. What is the speed of the boat and the current of the river?

Graph the solution set for each system of linear inequalities. [4.6]

47. $3x + 4y < 12$
$-3x + 2y \le 6$

48. $3x + 4y < 12$
$-3x + 2y \le 6$
$y \ge 0$

49. $3x + 4y < 12$
$x \ge 0$
$y \ge 0$

50. $x > -3$
$y > -2$

Chapter 4 Projects

Equations and Inequalities in Two Variables

SYSTEMS OF LINEAR EQUATIONS AND INEQUALITIES

Group Project

Number of People 2 or 3

Time Needed 10–15 minutes

Equipment Pencil and paper

Background The break-even point for a company occurs when the revenue from sales of a product equals the cost of producing the product. This group project is designed to give you more insight into revenue, cost, and break-even point.

Procedure A company is planning to open a factory to manufacture calculators.

1. It costs them $120,000 to open the factory, and it will cost $10 for each calculator they make. What is the expression for $C(x)$, the cost of making x calculators?

2. They can sell the calculators for $50 each. What is the expression for $R(x)$, their revenue from selling x calculators? Remember that $R = px$, where p is the price per calculator.

3. Graph both the cost equation $C(x)$ and the revenue equation $R(x)$ on a coordinate system like the one below.

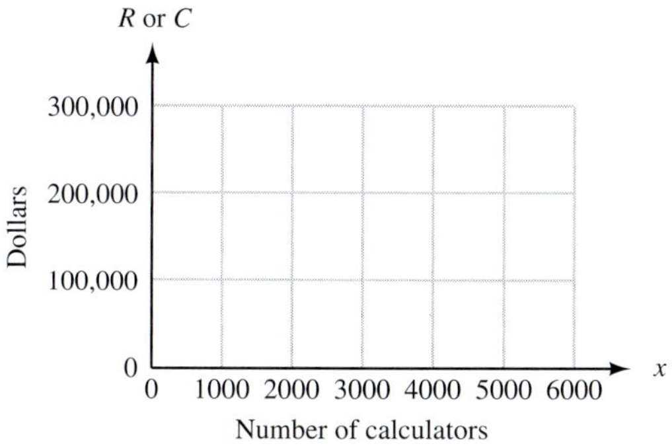

4. The break-even point is the value of x (the number of calculators) for which the revenue is equal to the cost. Where is the break-even point on the graph you produced in Part 3? Estimate the break-even point from the graph.

5. Set the cost equal to the revenue and solve to find x to find the exact value of the break-even point. How many calculators do they need to make and sell to exactly break even? What will be their revenue and their cost for that many calculators?

6. Write an inequality that gives the values of x that will produce a profit for the company. (A profit occurs when the revenue is larger than the cost.)

7. Write an inequality that gives the values of x that will produce a loss for the company. (A loss occurs when the cost is larger than the revenue.)

8. Profit is the difference between revenue and cost, or $P(x) = R(x) - C(x)$. Write the equation for profit and then graph it on a coordinate system like the one below.

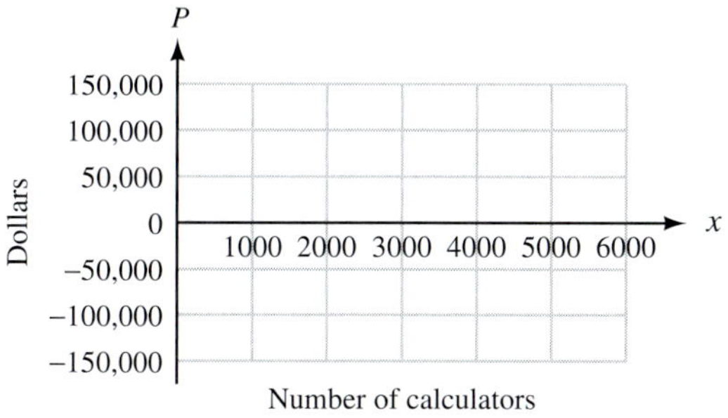

9. How do you recognize the break-even point and the regions of loss and profit on the graph you produced above?

ZENO'S PARADOXES

Research Project

Zeno of Elea was born at about the same time that Pythagoras died. He is responsible for three paradoxes that have come to be known as Zeno's paradoxes. One of the three has to do with a race between Achilles and a tortoise. Achilles is much faster than the tortoise, but the tortoise has a head start. According to Zeno's method of reasoning, Achilles can never pass the tortoise because each time he reaches the place where the tortoise was, the tortoise is gone. Research Zeno's paradox concerning Achilles and the tortoise. Put your findings into essay form that begins with a definition for the word "paradox." Then use Zeno's method of reasoning to describe a race between Achilles and the tortoise—if Achilles runs at 10 miles per hour, the tortoise runs at 1 mile per hour, and the tortoise has a 1-mile head start. Next, use the methods shown in this chapter to find the distance at which Achilles reaches the tortoise and the time at which Achilles reaches the tortoise. Conclude your essay by summarizing what you have done and showing how the two results you have obtained form a paradox.

5

Jack Kurtz/The Image Works

Exponents and Polynomials

CHAPTER OUTLINE

INTERMEDIATE Algebra $f(x)$ Now™

If you go on to take a business course or an economics course, you will find yourself spending lots of time with the three expressions that form the mathematical foundation of business: profit, revenue, and cost. Many times these expressions are given as polynomials, the topic of this chapter. The relationship between the three expressions is known as the profit equation:

$$\text{Profit} = \text{Revenue} - \text{Cost}$$

$$P(x) = R(x) - C(x)$$

The table and graphs shown here were produced on a graphing calculator. They give numerical and graphical descriptions of revenue, profit, and cost for a company that manufactures and sells prerecorded videotapes according to the equations

$$R(x) = 11.5x - 0.05x^2 \quad \text{and} \quad C(x) = 200 + 2x$$

REVENUE, COST, AND PROFIT ON A GRAPHING CALCULATOR

NUMBER OF VIDEOTAPES X	REVENUE Y_1	COST Y_2	PROFIT Y_3
0	0	200	−200
50	450	300	150
100	650	400	250
150	600	500	100
200	300	600	−300

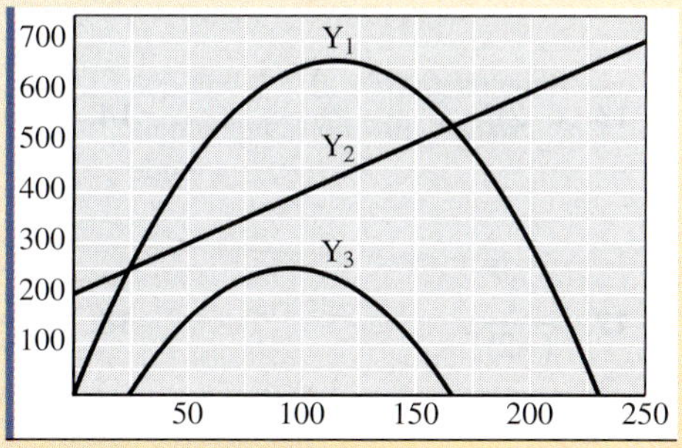

FIGURE 1

By studying the material in this chapter, you will get a head start on learning the equations and relationships that are emphasized in business and economics.

Visit our web site at http://www.brookscole.com

NUMBER WRITTEN IN EXPANDED NOTATION		NUMBER WRITTEN AGAIN IN SCIENTIFIC NOTATION
376,000	=	3.76×10^5
49,500	=	4.95×10^4
3,200	=	3.2×10^3
591	=	5.91×10^2
46	=	4.6×10^1
8	=	8×10^0
0.47	=	4.7×10^{-1}
0.093	=	9.3×10^{-2}
0.00688	=	6.88×10^{-3}
0.0002	=	2×10^{-4}
0.000098	=	9.8×10^{-5}

Calculator Note Many calculators have a key that allows you to enter numbers in scientific notation. The key is labeled

[EXP] or [EE] or [SCI]

To enter the number 3.45×10^6, you first enter the decimal number, then press the scientific notation key, and finally enter the exponent.

3.45 [EXP] 6

We can use our properties of exponents to do arithmetic with numbers written in scientific notation. Here are some examples.

EXAMPLES

Simplify each expression, and write all answers in scientific notation.

21. $(2 \times 10^8)(3 \times 10^{-3}) = (2)(3) \times (10^8)(10^{-3})$

$$= 6 \times 10^5$$

22. $\dfrac{4.8 \times 10^9}{2.4 \times 10^{-3}} = \dfrac{4.8}{2.4} \times \dfrac{10^9}{10^{-3}}$

$$= 2 \times 10^{9-(-3)}$$

$$= 2 \times 10^{12}$$

23. $\dfrac{(6.8 \times 10^5)(3.9 \times 10^{-7})}{7.8 \times 10^{-4}} = \dfrac{(6.8)(3.9)}{7.8} \times \dfrac{(10^5)(10^{-7})}{10^{-4}}$

$$= 3.4 \times 10^2$$

Calculator Note On a scientific calculator with a scientific notation key, you would use the following sequence of keys to do Example 22:

4.8 [EXP] 9 [÷] 2.4 [EXP] 3 [+/−] [=]

Simplify and write all answers in scientific notation.

21. $(3 \times 10^7)(2 \times 10^{-4})$

22. $\dfrac{3.9 \times 10^6}{1.3 \times 10^{-4}}$

23. $\dfrac{(2.4 \times 10^6)(1.8 \times 10^{-4})}{1.2 \times 10^{-3}}$

NOTE

Remember, on some calculators the scientific notation key may be labeled [EE] or [SCI].

Linking Objectives and Examples

Next to each **objective**, list the examples that are best described by that objective.

A	
B	
C	

Answers

21. 6×10^3 **22.** 3×10^{10}

23. 3.6×10^5

Getting Ready for Class

After reading through the preceding section, respond in your own words and in complete sentences.

1. Explain the difference between -2^4 and $(-2)^4$.
2. Explain the difference between 2^5 and 2^{-5}.
3. If a positive base is raised to a negative exponent, can the result be a negative number?
4. State Property 1 for exponents in your own words.

Problem Set 5.1

INTERMEDIATE Algebra $f(x)$ Now™

Evaluate each of the following.

1. 4^2
2. $(-4)^2$
3. -4^2
4. $-(-4)^2$
5. -0.3^3
6. $(-0.3)^3$
7. 2^5
8. 2^4
9. $\left(\frac{1}{2}\right)^3$
10. $\left(\frac{3}{4}\right)^2$
11. $\left(-\frac{5}{6}\right)^2$
12. $\left(-\frac{7}{8}\right)^2$

Use the properties of exponents to simplify each of the following as much as possible.

13. $x^5 \cdot x^4$
14. $x^6 \cdot x^3$
15. $(2^3)^2$
16. $(3^2)^2$
17. $\left(-\frac{2}{3}x^2\right)^3$
18. $\left(-\frac{3}{5}x^4\right)^3$
19. $-3a^2(2a^4)$
20. $5a^7(-4a^6)$

Write each of the following with positive exponents. Then simplify as much as possible.

21. 3^{-2}
22. $(-5)^{-2}$
23. $(-2)^{-5}$
24. 2^{-5}
25. $\left(\frac{3}{4}\right)^{-2}$
26. $\left(\frac{3}{5}\right)^{-2}$
27. $\left(\frac{1}{3}\right)^{-2} + \left(\frac{1}{2}\right)^{-3}$
28. $\left(\frac{1}{2}\right)^{-2} + \left(\frac{1}{3}\right)^{-3}$

Simplify each expression. Write all answers with positive exponents only. (Assume all variables are nonzero.)

29. $x^{-4}x^7$
30. $x^{-3}x^8$
31. $(a^2b^{-5})^3$
32. $(a^4b^{-3})^3$
33. $(5y^4)^{-3}(2y^{-2})^3$
34. $(3y^5)^{-2}(2y^{-4})^3$
35. $\left(\frac{1}{2}x^3\right)\left(\frac{2}{3}x^4\right)\left(\frac{3}{5}x^{-7}\right)$
36. $\left(\frac{1}{7}x^{-3}\right)\left(\frac{7}{8}x^{-5}\right)\left(\frac{8}{9}x^8\right)$
37. $(4a^5b^2)(2b^{-5}c^2)(3a^7c^4)$
38. $(3a^{-2}c^3)(5b^{-6}c^5)(4a^6b^{-2})$
39. $(2x^2y^{-5})^3(3x^{-4}y^2)^{-4}$
40. $(4x^{-4}y^9)^{-2}(5x^4y^{-3})^2$

Use the properties of exponents to simplify each expression. Write all answers with positive exponents only. (Assume all variables are nonzero.)

41. $\dfrac{x^{-1}}{x^9}$

42. $\dfrac{x^{-3}}{x^5}$

43. $\dfrac{a^4}{a^{-6}}$

44. $\dfrac{a^5}{a^{-2}}$

45. $\dfrac{t^{-10}}{t^{-4}}$

46. $\dfrac{t^{-8}}{t^{-5}}$

47. $\left(\dfrac{x^5}{x^3}\right)^6$

48. $\left(\dfrac{x^7}{x^4}\right)^5$

49. $\dfrac{(x^5)^6}{(x^3)^4}$

50. $\dfrac{(x^7)^3}{(x^4)^5}$

51. $\dfrac{(x^{-2})^3(x^3)^{-2}}{x^{10}}$

52. $\dfrac{(x^{-4})^3(x^3)^{-4}}{x^{10}}$

53. $\dfrac{5a^8b^3}{20a^5b^{-4}}$

54. $\dfrac{7a^6b^{-2}}{21a^2b^{-5}}$

55. $\dfrac{(3x^{-2}y^8)^4}{(9x^4y^{-3})^2}$

56. $\dfrac{(6x^{-3}y^{-5})^2}{(3x^{-4}y^{-3})^4}$

57. $\left(\dfrac{8x^2y}{4x^4y^{-3}}\right)^4$

58. $\left(\dfrac{5x^4y^5}{10xy^{-2}}\right)^3$

59. $\left(\dfrac{x^{-5}y^2}{x^{-3}y^5}\right)^{-2}$

60. $\left(\dfrac{x^{-8}y^{-3}}{x^{-5}y^6}\right)^{-1}$

Write each number in scientific notation.

61. 378,000

62. 3,780,000

63. 4,900

64. 490

65. 0.00037

66. 0.000037

67. 0.00495

68. 0.0495

Write each number in expanded form.

69. 5.34×10^3 **70.** 5.34×10^2 **71.** 7.8×10^6 **72.** 7.8×10^4

73. 3.44×10^{-3} **74.** 3.44×10^{-5} **75.** 4.9×10^{-1} **76.** 4.9×10^{-2}

Use the properties of exponents to simplify each of the following expressions. Write all answers in scientific notation.

77. $(4 \times 10^{10})(2 \times 10^{-6})$ **78.** $(3 \times 10^{-12})(3 \times 10^4)$ **79.** $\dfrac{8 \times 10^{14}}{4 \times 10^5}$

80. $\dfrac{6 \times 10^8}{2 \times 10^3}$ **81.** $\dfrac{(5 \times 10^6)(4 \times 10^{-8})}{8 \times 10^4}$ **82.** $\dfrac{(6 \times 10^{-7})(3 \times 10^9)}{5 \times 10^6}$

Problems 83–104 are problems you will see later in the book.

Multiply.

83. $8x^3 \cdot 10y^6$ **84.** $5y^2 \cdot 4x^2$ **85.** $8x^3 \cdot 9y^3$ **86.** $4y^3 \cdot 3x^2$

87. $3x \cdot 5y$ **88.** $3xy \cdot 5z$ **89.** $4x^6y^6 \cdot 3x$ **90.** $16x^4y^4 \cdot 3y$

91. $27a^6c^3 \cdot 2b^2c$ **92.** $8a^3b^3 \cdot 5a^2b$

Divide.

93. $\dfrac{10x^5}{5x^2}$ **94.** $\dfrac{-15x^4}{5x^2}$ **95.** $\dfrac{20x^3}{5x^2}$ **96.** $\dfrac{25x^7}{-5x^2}$

97. $\dfrac{8x^3y^5}{-2x^2y}$ **98.** $\dfrac{-16x^2y^2}{-2x^2y}$ **99.** $\dfrac{4x^4y^3}{-2x^2y}$ **100.** $\dfrac{10a^4b^2}{4a^2b^2}$

Simplify. Write answers without using scientific notation, rounding to the nearest whole number.

101. $\dfrac{2.00 \times 10^8}{3.98 \times 10^6}$ **102.** $\dfrac{2.00 \times 10^8}{3.16 \times 10^5}$

Use a calculator to find each of the following. Write your answer in scientific notation with the first number in each answer rounded to the nearest tenth.

103. $10^{-4.1}$ **104.** $10^{-5.6}$

Applying the Concepts

105. Large Numbers If you are 20 years old, you have been alive for more than 630,000,000 seconds. Write this last number in scientific notation.

106. Fingerprints The FBI has been collecting fingerprint cards since 1924. Their collection has grown to over 200 million cards. They are digitizing the fingerprints. Each fingerprint card turns into about 10 MB of data. (A megabyte [MB] is $2^{20} \approx$ one million bytes.)

a. How many bytes of storage will they need?

b. A compression routine called the WSQ method will compress the bytes by ratio of 12.9 to 1. Approximately how many bytes of storage will the FBI need for the compressed data? (Hint: Divide by 12.9.)

107. Search Engines The chart shows the number of people using different search engines in January of 2005. For each of the following search engines, write the number of visitors in scientific notation.

a. Ask Jeeves

b. AOL

c. Google

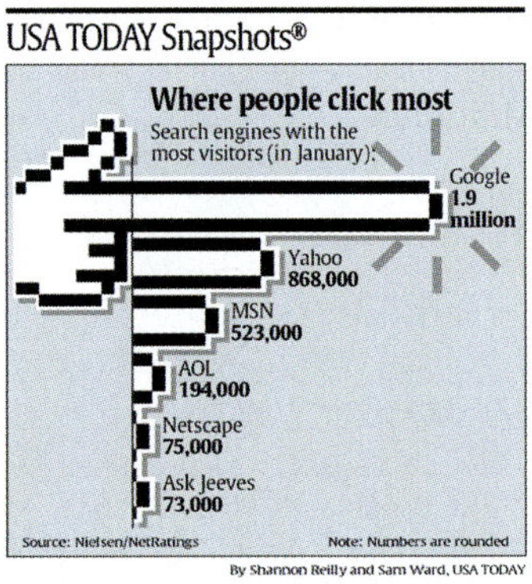

From *USA Today.* Copyright 2005. Reprinted with permission.

108. Our Galaxy The galaxy the Earth resides in is called the Milky Way galaxy. It is a spiral galaxy that contains approximately 200,000,000,000 stars (our Sun is one of them). Write this number in words and in scientific notation.

NASA

109. Light Year A light year, the distance light travels in 1 year, is approximately 5.9×10^{12} miles. The Andromeda galaxy is approximately 1.7×10^{6} light years from our galaxy. Find the distance in miles between our galaxy and the Andromeda galaxy.

110. Distance to the Sun The distance from the Earth to the sun is approximately 9.3×10^{7} miles. If light travels 1.2×10^{7} miles in 1 minute, how many minutes does it take the light from the sun to reach the Earth?

111. Credit Card Debt Outstanding credit-card debt in the United States is over $422 billion.

a. Write the number 422 billion in scientific notation.

b. If there are approximately 60 million households with at least one credit card, find the average credit-card debt per household, to the nearest dollar.

Stockbyte/SuperStock

112. Cone Nebula The photograph was taken by the Hubble telescope in April 2002. The object in the photograph is called the *Cone Nebula.* The distance across the photograph is about 2.5 light-years, which is 14,664,240,000,000 miles. Round this number to the nearest trillion and then write the result in scientific notation.

NASA

Getting Ready for the Next Section

Simplify.

113. $-4x + 9x$

114. $-6x - 2x$

115. $5x^2 + 3x^2$

116. $7x^2 + 3x^2$

117. $-8x^3 + 10x^3$

118. $4x^3 - 7x^3$

119. $2x + 3 - 2x - 8$

120. $9x - 4 - 9x - 10$

121. $-1(2x - 3)$

122. $-1(-3x + 1)$

123. $-3(-3x - 2)$

124. $-4(-5x + 3)$

125. $-500 + 27(100) - 0.1(100)^2$

126. $-500 + 27(170) - 0.1(170)^2$

Maintaining Your Skills

Solve each system by the addition method.

127. $\begin{aligned} 4x + 3y &= 10 \\ 2x + y &= 4 \end{aligned}$

128. $\begin{aligned} 3x - 5y &= -2 \\ 2x - 3y &= 1 \end{aligned}$

129. $\begin{aligned} 4x + 5y &= 5 \\ \tfrac{6}{5}x + y &= 2 \end{aligned}$

130. $\begin{aligned} 4x + 2y &= -2 \\ \tfrac{1}{2}x + y &= 0 \end{aligned}$

Solve each system by the substitution method.

131. $\begin{aligned} x + y &= 3 \\ y &= x + 3 \end{aligned}$

132. $\begin{aligned} x + y &= 6 \\ y &= x - 4 \end{aligned}$

133. $\begin{aligned} 2x - 3y &= -6 \\ y &= 3x - 5 \end{aligned}$

134. $\begin{aligned} 7x - y &= 24 \\ x &= 2y + 9 \end{aligned}$

Extending the Concepts

Assume all variable exponents represent positive integers and simplify each expression.

135. $x^{m+2} \cdot x^{-2m} \cdot x^{m-5}$

136. $x^{m-4}x^{m+9}x^{-2m}$

137. $(y^m)^2(y^{-3})^m(y^{m+3})$

138. $(y^m)^{-4}(y^3)^m(y^{m+6})$

139. $\dfrac{x^{n+2}}{x^{n-3}}$

140. $\dfrac{x^{n-3}}{x^{n-7}}$

Improving Your Quantitative Literacy

141. Computer Science We all use the language of computers to indicate how much memory our computers hold or how much information we can put on a storage device such as a keychain drive. Scientific notation gives us a way to compare the actual numbers associated with the words we use to describe data storage in computers. The smallest amount of data that a computer can hold is measured in bits. A byte is the next largest unit and is equal to 8, or 2^3, bits. Fill in the table below.

	NUMBER OF BYTES	
UNIT	EXPONENTIAL FORM	SCIENTIFIC NOTATION
Kilobyte	$2^{10} = 1{,}024$	
Megabyte	$2^{20} \approx 1{,}048{,}000$	
Gigabyte	$2^{30} \approx 1{,}074{,}000{,}000$	
Terabyte	$2^{40} \approx 1{,}099{,}500{,}000{,}000$	

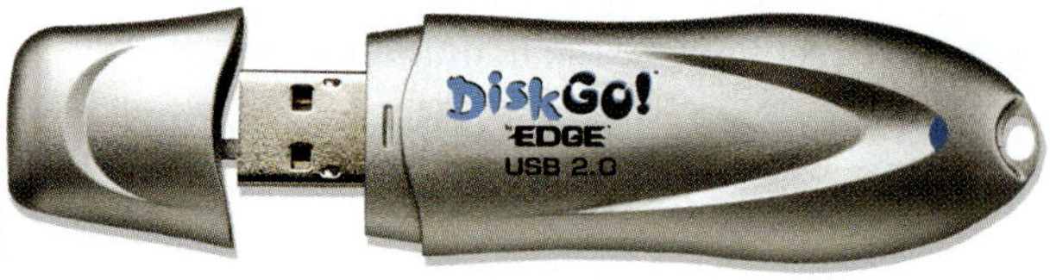

EDGE Tech Corp

Polynomials, Sums, and Differences

5.2

The chart is from a company that duplicates videotapes. It shows the revenue and cost to duplicate a 30-minute video. From the chart you can see that 300 copies will bring in $900 in revenue, with a cost of $600. The profit is the difference between revenue and cost, or $300.

The relationship between profit, revenue, and cost is one application of the polynomials we will study in this section. Let's begin with a definition that we will use to build polynomials.

OBJECTIVES

A	Give the degree of a polynomial.
B	Add and subtract polynomials.
C	Evaluate a polynomial for a given value of its variable.

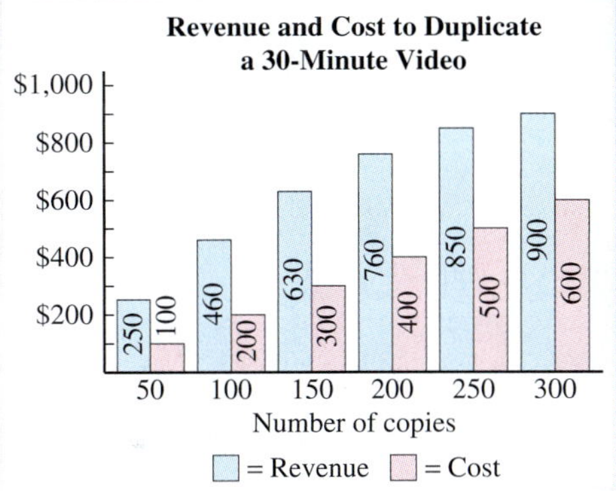

Polynomials in General

DEFINITION

A **term,** or **monomial,** is a constant or the product of a constant and one or more variables raised to whole-number exponents.

The following are monomials, or terms:

$$-16 \qquad 3x^2y \qquad -\frac{2}{5}a^3b^2c \qquad xy^2z$$

The numerical part of each monomial is called the *numerical coefficient,* or just *coefficient* for short. For the preceding terms, the coefficients are -16, 3, $-\frac{2}{5}$, and 1. Notice that the coefficient for xy^2z is understood to be 1.

DEFINITION

A **polynomial** is any finite sum of terms. Because subtraction can be written in terms of addition, finite differences are also included in this definition.

The following are polynomials:

$$2x^2 - 6x + 3 \qquad -5x^2y + 2xy^2 \qquad 4a - 5b + 6c + 7d$$

Polynomials can be classified further according to the number of terms present. If a polynomial consists of two terms, it is said to be a *binomial.* If it has three terms, it is called a *trinomial.* And, as stated, a polynomial with only one term is said to be a *monomial.*

DEFINITION

The **degree** of a polynomial with one variable is the highest power to which the variable is raised in any one term.

EXAMPLES

1. $6x^2 + 2x - 1$ — **A trinomial of degree 2**
2. $5x - 3$ — **A binomial of degree 1**
3. $7x^6 - 5x^3 + 2x - 4$ — **A polynomial of degree 6**
4. $-7x^4$ — **A monomial of degree 4**
5. 15 — **A monomial of degree 0**

Polynomials in one variable are usually written in decreasing powers of the variable. When this is the case, the coefficient of the first term is called the *leading coefficient.* In Example 1, the leading coefficient is 6. In Example 2, it is 5. The leading coefficient in Example 3 is 7.

PRACTICE PROBLEMS

Identify each expression as monomial, binomial, or trinomial, and give the degree of each.

1. $3x + 1$
2. $4x^2 + 2x + 5$
3. -17
4. $4x^5 - 7x^3$
5. $4x^3 - 5x^2 + 2x$

Answers

1. Binomial, 1
2. Trinomial, 2
3. Monomial, 0
4. Binomial, 5
5. Trinomial, 3

DEFINITION

Two or more terms that differ only in their numerical coefficients are called **similar,** or **like,** terms. Since similar terms differ only in their coefficients, they have identical variable parts.

Addition and Subtraction of Polynomials

To add two polynomials, we simply apply the commutative and associative properties to group similar terms together and then use the distributive property as we have in the following example.

6. Add $3x^2 + 2x - 5$ and $2x^2 - 7x + 3$.

EXAMPLE 6 Add $5x^2 - 4x + 2$ and $3x^2 + 9x - 6$.

Solution

$$\begin{aligned}
&(5x^2 - 4x + 2) + (3x^2 + 9x - 6) \\
&= (5x^2 + 3x^2) + (-4x + 9x) + (2 - 6) && \textbf{Commutative and associative properties} \\
&= (5 + 3)x^2 + (-4 + 9)x + (2 - 6) && \textbf{Distributive property} \\
&= 8x^2 + 5x + (-4) \\
&= 8x^2 + 5x - 4
\end{aligned}$$

7. Find the sum of $x^3 + 7x^2 + 3x + 2$ and $-3x^3 - 2x^2 + 3x - 1$.

EXAMPLE 7 Find the sum of $-8x^3 + 7x^2 - 6x + 5$ and $10x^3 + 3x^2 - 2x - 6$.

Solution We can add the two polynomials using the method of Example 6, or we can arrange similar terms in columns and add vertically. Using the column method, we have

$$\begin{array}{r}
-8x^3 + \ \ 7x^2 - 6x + 5 \\
10x^3 + \ \ 3x^2 - 2x - 6 \\
\hline
2x^3 + 10x^2 - 8x - 1
\end{array}$$

NOTE

In practice it is not necessary to show all the steps shown in Example 7. It is important to understand that addition of polynomials is equivalent to combining similar terms.

To find the difference of two polynomials, we need to use the fact that the opposite of a sum is the sum of the opposites; that is,

$$-(a + b) = -a + (-b)$$

One way to remember this is to observe that $-(a + b)$ is equivalent to $-1(a + b) = (-1)a + (-1)b = -a + (-b)$.

If a negative sign directly precedes the parentheses surrounding a polynomial, we may remove the parentheses and the preceding negative sign by changing the sign of each term within the parentheses. For example:

$$\begin{aligned}
-(3x + 4) &= -3x + (-4) = -3x - 4 \\
-(5x^2 - 6x + 9) &= -5x^2 + 6x - 9 \\
-(-x^2 + 7x - 3) &= x^2 - 7x + 3
\end{aligned}$$

8. Subtract. $(4x^2 - 2x + 7) - (7x^2 - 3x + 1)$

EXAMPLE 8 Subtract $(9x^2 - 3x + 5) - (4x^2 + 2x - 3)$.

Solution We subtract by adding the opposite of each term in the polynomial that follows the subtraction sign:

$$\begin{aligned}
&(9x^2 - 3x + 5) - (4x^2 + 2x - 3) \\
&= 9x^2 - 3x + 5 + (-4x^2) + (-2x) + 3 && \textbf{The opposite of a sum is the sum of the opposites} \\
&= (9x^2 - 4x^2) + (-3x - 2x) + (5 + 3) && \textbf{Commutative and associative properties} \\
&= 5x^2 - 5x + 8 && \textbf{Combine similar terms}
\end{aligned}$$

Answers

6. $5x^2 - 5x - 2$
7. $-2x^3 + 5x^2 + 6x + 1$
8. $-3x^2 + x + 6$

EXAMPLE 9 Subtract $4x^2 - 9x + 1$ from $-3x^2 + 5x - 2$.

Solution Again, to subtract, we add the opposite:

$$\begin{aligned}(-3x^2 + 5x - 2) - (4x^2 - 9x + 1) &\\ &= -3x^2 + 5x - 2 - 4x^2 + 9x - 1\\ &= (-3x^2 - 4x^2) + (5x + 9x) + (-2 - 1)\\ &= -7x^2 + 14x - 3\end{aligned}$$

9. Subtract $3x + 5$ from $7x - 4$.

EXAMPLE 10 Simplify $4x - 3[2 - (3x + 4)]$.

Solution Removing the innermost parentheses first, we have

$$\begin{aligned}4x - 3[2 - (3x + 4)] &= 4x - 3(2 - 3x - 4)\\ &= 4x - 3(-3x - 2)\\ &= 4x + 9x + 6\\ &= 13x + 6\end{aligned}$$

10. Simplify $2x - 4[6 - (5x + 3)]$.

EXAMPLE 11 Simplify $(2x + 3) - [(3x + 1) - (x - 7)]$.

Solution

$$\begin{aligned}(2x + 3) - [(3x + 1) - (x - 7)] &= (2x + 3) - (3x + 1 - x + 7)\\ &= (2x + 3) - (2x + 8)\\ &= 2x + 3 - 2x - 8\\ &= -5\end{aligned}$$

11. Simplify.
$(9x - 4) - [(2x + 5) - (x + 3)]$

In the example that follows we will find the value of a polynomial for a given value of the variable.

EXAMPLE 12 Find the value of $5x^3 - 3x^2 + 4x - 5$ when x is 2.

Solution We begin by substituting 2 for x in the original polynomial:

When $x = 2$

the polynomial $5x^3 - 3x^2 + 4x - 5$

becomes

$$\begin{aligned}5 \cdot 2^3 - 3 \cdot 2^2 + 4 \cdot 2 - 5 &= 5 \cdot 8 - 3 \cdot 4 + 4 \cdot 2 - 5\\ &= 40 - 12 + 8 - 5\\ &= 31\end{aligned}$$

12. Find the value of $2x^3 - 3x^2 + 4x - 8$ when x is -2.

Polynomials and Function Notation

Example 12 can be restated using function notation by calling the polynomial $P(x)$ and asking for $P(2)$. The solution would look like this:

If $P(x) = 5x^3 - 3x^2 + 4x - 5$

then

$$\begin{aligned}P(2) &= 5 \cdot 2^3 - 3 \cdot 2^2 + 4 \cdot 2 - 5\\ &= 31\end{aligned}$$

Our next example is stated in terms of function notation.

As we mentioned in the introduction to this chapter, three functions that occur very frequently in business and economics classes are profit, revenue, and cost functions. If a company manufactures and sells x items, then the revenue $R(x)$ is the total amount of money obtained by selling all x items. The cost $C(x)$ is the total amount of money it costs the company to manufacture the x items. The profit $P(x)$ obtained by selling all x items is the difference between the revenue and the cost and is given by the equation

$$P(x) = R(x) - C(x)$$

13. Use the information in Example 13 to find the profit associated with producing and selling 170 programs per week.

EXAMPLE 13 A company produces and sells copies of an accounting program for home computers. The total weekly cost (in dollars) to produce x copies of the program is $C(x) = 8x + 500$. Find its weekly profit if the total revenue obtained

Answers

9. $4x - 9$ 10. $22x - 12$
11. $8x - 6$ 12. -44

from selling all x programs is $R(x) = 35x - 0.1x^2$. How much profit will the company make if it produces and sells 100 programs a week? That is, find $P(100)$.

Solution Using the equation $P(x) = R(x) - C(x)$ and the information given in the problem, we have

$$\begin{aligned} P(x) &= R(x) - C(x) \\ &= 35x - 0.1x^2 - (8x + 500) \\ &= 35x - 0.1x^2 - 8x - 500 \\ &= -500 + 27x - 0.1x^2 \end{aligned}$$

If the company produces and sells 100 copies of the program, its weekly profit will be

$$\begin{aligned} P(100) &= -500 + 27(100) - 0.1(100)^2 \\ &= -500 + 27(100) - 0.1(10{,}000) \\ &= -500 + 2{,}700 - 1{,}000 \\ &= 1{,}200 \end{aligned}$$

The weekly profit is \$1,200.

USING TECHNOLOGY

Graphing Calculators

Your graphing calculator allows you access to a function list in which you can input formulas for a number of functions. There are many ways to use the function list. For the problems in this section, we can use the function list to evaluate a polynomial. Let's use the profit function from Example 13 to illustrate. To input $P(x) = -500 + 27x - 0.1x^2$, we assign $P(x)$ to the first function in our function list, which is Y_1. Call up the function list for your calculator and enter the profit function this way:

$$Y_1 = -500 + 27X - 0.1X^2$$

Now that you have an expression for Y_1, you can evaluate it by substituting a value for X. To do so, quit the function list and store the number 100 into the variable X:

100 → [X] [ENTER]   or   100 [STO] [X] [ENTER]

The display shows 100. Next, display the variable Y_1 (on many calculators, individual variables are on the Y-VAR menu), then press ENTER:

Y_1 [ENTER]

The display will show 1,200.

Now that you have the profit function stored in function Y_1, you can evaluate it for any value of X. Find the weekly profit if 50, 100, 150, 200, 250, and 300 programs are sold.

Linking Objectives and Examples

Next to each **objective**, list the examples that are best described by that objective.

A	
B	
C	

Answer
13. \$1,200

Getting Ready for Class

After reading through the preceding section, respond in your own words and in complete sentences.

1. Is $3x^2 + 2x - \frac{1}{x}$ a polynomial? Explain.
2. What are similar terms?
3. Explain in words how you subtract one polynomial from another.
4. What is revenue?

Problem Set 5.2

INTERMEDIATE Algebra $f(x)$ Now™

Identify those of the following that are monomials, binomials, or trinomials. Give the degree of each, and name the leading coefficient.

1. $5x^2 - 3x + 2$

2. $2x^2 + 4x - 1$

3. $3x - 5$

4. $5y + 3$

5. $8a^2 + 3a - 5$

6. $9a^2 - 8a - 4$

7. $4x^3 - 6x^2 + 5x - 3$

8. $9x^4 + 4x^3 - 2x^2 + x$

9. $-\frac{3}{4}$

10. -16

11. $4x - 5 + 6x^3$

12. $9x + 2 + 3x^3$

Simplify each of the following by combining similar terms.

13. $(4x + 2) + (3x - 1)$

14. $(8x - 5) + (-5x + 4)$

15. $2x^2 - 3x + 10x - 15$

16. $6x^2 - 4x - 15x + 10$

17. $12a^2 + 8ab - 15ab - 10b^2$

18. $28a^2 - 8ab + 7ab - 2b^2$

19. $(5x^2 - 6x + 1) - (4x^2 + 7x - 2)$

20. $(11x^2 - 8x) - (4x^2 - 2x - 7)$

21. $(\frac{1}{2}x^2 - \frac{1}{3}x - \frac{1}{6}) - (\frac{1}{4}x^2 + \frac{7}{12}x) + (\frac{1}{3}x - \frac{1}{12})$

22. $(\frac{2}{3}x^2 - \frac{1}{2}x) - (\frac{1}{4}x^2 + \frac{1}{6}x + \frac{1}{12}) - (\frac{1}{2}x^2 + \frac{1}{4})$

23. $(y^3 - 2y^2 - 3y + 4) - (2y^3 - y^2 + y - 3)$

24. $(8y^3 - 3y^2 + 7y + 2) - (-4y^3 + 6y^2 - 5y - 8)$

25. $(5x^3 - 4x^2) - (3x + 4) + (5x^2 - 7) - (3x^3 + 6)$

26. $(x^3 - x) - (x^2 + x) + (x^3 - 1) - (-3x + 2)$

27. $(\frac{4}{7}x^2 - \frac{1}{7}xy + \frac{1}{14}y^2) - (\frac{1}{2}x^2 - \frac{2}{7}xy - \frac{9}{14}y^2)$

28. $(\frac{1}{5}x^2 - \frac{1}{2}xy + \frac{1}{10}y^2) - (-\frac{3}{10}x^2 + \frac{2}{5}xy - \frac{1}{2}y^2)$

29. $(3a^3 + 2a^2b + ab^2 - b^3) - (6a^3 - 4a^2b + 6ab^2 - b^3)$

30. $(a^3 - 3a^2b + 3ab^2 - b^3) - (a^3 + 3a^2b + 3ab^2 + b^3)$

31. Subtract $2x^2 - 4x$ from $2x^2 - 7x$.

32. Subtract $-3x + 6$ from $-3x + 9$.

33. Find the sum of $x^2 - 6xy + y^2$ and $2x^2 - 6xy - y^2$.

34. Find the sum of $9x^3 - 6x^2 + 2$ and $3x^2 - 5x + 4$.

35. Subtract $-8x^5 - 4x^3 + 6$ from $9x^5 - 4x^3 - 6$.

36. Subtract $4x^4 - 3x^3 - 2x^2$ from $2x^4 + 3x^3 + 4x^2$.

37. Find the sum of $11a^2 + 3ab + 2b^2$, $9a^2 - 2ab + b^2$, and $-6a^2 - 3ab + 5b^2$.

38. Find the sum of $a^2 - ab - b^2$, $a^2 + ab - b^2$, and $a^2 + 2ab + b^2$.

Simplify each of the following. Begin by working on the innermost parentheses.

39. $-[2 - (4 - x)]$

40. $-[-3 - (x - 6)]$

41. $-5[-(x - 3) - (x + 2)]$

42. $-6[(2x - 5) - 3(8x - 2)]$

43. $4x - 5[3 - (x - 4)]$

44. $x - 7[3x - (2 - x)]$

45. $-(3x - 4y) - [(4x + 2y) - (3x + 7y)]$

46. $(8x - y) - [-(2x + y) - (-3x - 6y)]$

47. $4a - \{3a + 2[a - 5(a + 1) + 4]\}$

48. $6a - \{-2a - 6[2a + 3(a - 1) - 6]\}$

49. Find the value of $2x^2 - 3x - 4$ when x is 2.

50. Find the value of $4x^2 + 3x - 2$ when x is -1.

51. If $P(x) = \frac{3}{2}x^2 - \frac{3}{4}x + 1$, find $P(12)$.

52. If $P(x) = \frac{2}{5}x^2 - \frac{1}{10}x + 2$, find $P(10)$.

53. If $Q(x) = x^3 - x^2 + x - 1$, find $Q(-2)$.

54. If $Q(x) = x^3 + x^2 + x + 1$, find $Q(-2)$.

Simplify.

55. a. $(3x - 5) - (3a - 5)$
    b. $(2x + 3) - (2a + 3)$

56. a. $(x^2 - 4) - (a^2 - 4)$
    b. $(x^2 - 1) - (a^2 - 1)$

Applying the Concepts

Problems 57–62 may be solved using a graphing calculator.

57. **Height of an Object** If an object is thrown straight up into the air with a velocity of 128 feet/second, then its height $h(t)$ above the ground t seconds later is given by the formula

$$h(t) = -16t^2 + 128t$$

Find the height after 3 seconds and after 5 seconds. [Find $h(3)$ and $h(5)$.]

58. **Height of an Object** The formula for the height of an object that has been thrown straight up with a velocity of 64 feet/second is

$$h(t) = -16t^2 + 64t$$

Find the height after 1 second and after 3 seconds. [Find $h(1)$ and $h(3)$.]

59. **Profits** The total cost (in dollars) for a company to manufacture and sell x items per week is $C(x) = 60x + 300$. If the revenue brought in by selling all x items is $R(x) = 100x - 0.5x^2$, find the weekly profit. How much profit will be made by producing and selling 60 items each week?

60. **Profits** The total cost (in dollars) for a company to produce and sell x items per week is $C(x) = 200x + 1{,}600$. If the revenue brought in by selling all x items is $R(x) = 300x - 0.6x^2$, find the weekly profit. How much profit will be made by producing and selling 50 items each week?

61. **Profits** Suppose it costs a company selling patterns $C(x) = 800 + 6.5x$ dollars to produce and sell x patterns a month. If the revenue obtained by selling x patterns is $R(x) = 10x - 0.002x^2$, what is the profit equation? How much profit will be made if 1,000 patterns are produced and sold in May?

62. **Profits** Suppose a company manufactures and sells x picture frames each month with a total cost of $C(x) = 1{,}200 + 3.5x$ dollars. If the revenue obtained by selling x frames is $R(x) = 9x - 0.003x^2$, find the profit equation. How much profit will be made if 1,000 frames are manufactured and sold in June?

Getting Ready for the Next Section

Simplify.

63. $2x^2 - 3x + 10x - 15$

64. $12a^2 + 8ab - 15ab - 10b^2$

65. $(6x^3 - 2x^2y + 8xy^2) + (-9x^2y + 3xy^2 - 12y^3)$

66. $(3x^3 - 15x^2 + 18x) + (2x^2 - 10x + 12)$

67. $4x^3(-3x)$

68. $5x^2(-4x)$

69. $4x^3(5x^2)$

70. $5x^2(3x^2)$

71. $(a^3)^2$

72. $(a^4)^2$

73. $11.5(130) - 0.05(130)^2$

74. $-0.05(130)^2 + 9.5(130) - 200$

Maintaining Your Skills

Simplify each expression.

75. $-1(5 - x)$

76. $-1(a - b)$

77. $-1(7 - x)$

78. $-1(6 - y)$

79. $5\left(x - \frac{1}{5}\right)$

80. $7\left(x + \frac{1}{7}\right)$

81. $x\left(1 - \frac{1}{x}\right)$

82. $a\left(1 + \frac{1}{a}\right)$

83. $12\left(\frac{1}{4}x + \frac{2}{3}y\right)$

84. $20\left(\frac{2}{5}x + \frac{1}{4}y\right)$

Extending the Concepts

85. The graphs of two polynomial functions are given in Figures 1 and 2. Use the graphs to find the following.

a. $f(-3)$ **b.** $f(0)$ **c.** $f(1)$ **d.** $g(-1)$

e. $g(0)$ **f.** $g(2)$ **g.** $f[g(2)]$ **h.** $g[f(2)]$

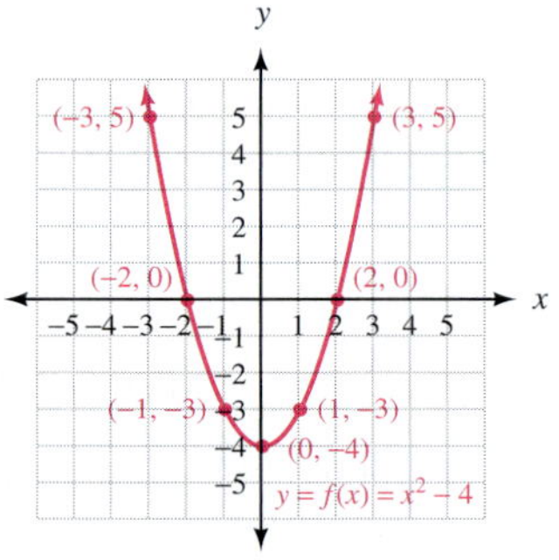

FIGURE 1

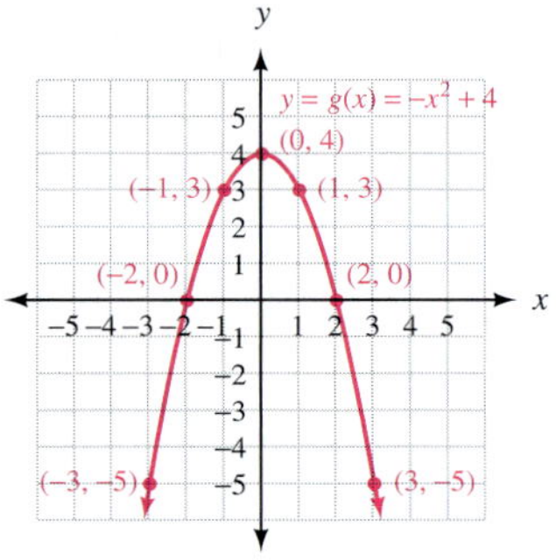

FIGURE 2

Multiplication of Polynomials

5.3

OBJECTIVES

A Multiply polynomials.

In the previous section we found the relationship between profit, revenue, and cost to be

$$P(x) = R(x) - C(x)$$

Revenue itself can be broken down further by another formula common in the business world. The revenue obtained from selling all x items is the product of the number of items sold and the price per item; that is,

$$\text{Revenue} = (\text{number of items sold})(\text{price of each item})$$

For example, if 100 items are sold for \$9 each, the revenue is $100(9) = \$900$. Likewise, if 500 items are sold for \$11 each, then the revenue is $500(11) = \$5{,}500$. In general, if x is the number of items sold and p is the selling price of each item, then we can write

$$R = xp$$

Many times, x and p are polynomials, which means that the expression xp is the product of two polynomials. In this section we learn how to multiply polynomials, and in so doing, increase our understanding of the equations and formulas that describe business applications.

EXAMPLE 1 Find the product of $4x^3$ and $5x^2 - 3x + 1$.

Solution To multiply, we apply the distributive property:

$$\begin{aligned} &4x^3(5x^2 - 3x + 1) \\ &\quad = 4x^3(5x^2) + 4x^3(-3x) + 4x^3(1) && \textbf{Distributive property} \\ &\quad = 20x^5 - 12x^4 + 4x^3 \end{aligned}$$

Notice that we multiply coefficients and add exponents.

EXAMPLE 2 Multiply $2x - 3$ and $x + 5$.

Solution Distributing the $2x - 3$ across the sum $x + 5$ gives us

$$\begin{aligned} &(2x - 3)(x + 5) \\ &\quad = (2x - 3)x + (2x - 3)5 && \textbf{Distributive property} \\ &\quad = 2x(x) + (-3)x + 2x(5) + (-3)5 && \textbf{Distributive property} \\ &\quad = 2x^2 - 3x + 10x - 15 \\ &\quad = 2x^2 + 7x - 15 && \textbf{Combine like terms} \end{aligned}$$

Notice the third line in this example. It consists of all possible products of terms in the first binomial and those of the second binomial. We can generalize this into a rule for multiplying two polynomials.

Rule

To multiply two polynomials, multiply each term in the first polynomial by each term in the second polynomial.

Multiplying polynomials can be accomplished by a method that looks very similar to long multiplication with whole numbers.

PRACTICE PROBLEMS

1. Multiply $5x^2(3x^2 - 4x + 2)$.

2. Multiply $4x - 1$ and $x + 3$.

Answers

1. $15x^4 - 20x^3 + 10x^2$
2. $4x^2 + 11x - 3$

Getting Ready for the Next Section

Simplify.

71. $\dfrac{8a^3}{a}$

72. $\dfrac{-8a^2}{a}$

73. $\dfrac{-48a}{a}$

74. $\dfrac{-32a}{a}$

75. $\dfrac{16a^5b^4}{8a^2b^3}$

76. $\dfrac{12x^4y^5}{3x^3y^3}$

77. $\dfrac{-24a^5b^5}{8a^5b^3}$

78. $\dfrac{-15x^5y^3}{3x^3y^3}$

79. $\dfrac{x^3y^4}{-x^3}$

80. $\dfrac{x^2y^2}{-x^2}$

Maintaining Your Skills

Solve each system.

81. $\begin{aligned} x + y + z &= 6 \\ 2x - y + z &= 3 \\ x + 2y - 3z &= -4 \end{aligned}$

82. $\begin{aligned} x + y + z &= 6 \\ x - y + 2z &= 7 \\ 2x - y - z &= 0 \end{aligned}$

83. $\begin{aligned} 3x + 4y &= 15 \\ 2x - 5z &= -3 \\ 4y - 3z &= 9 \end{aligned}$

84. $\begin{aligned} x + 3y &= 5 \\ 6y + z &= 12 \\ x - 2z &= -10 \end{aligned}$

The Greatest Common Factor and Factoring by Grouping

5.4

In general, factoring is the reverse of multiplication. The diagram to the right illustrates the relationship between factoring and multiplication. Reading from left to right, we say the product of 3 and 7 is 21. Reading in the other direction, from right to left, we say 21 factors into 3 times 7. Or, 3 and 7 are factors of 21.

DEFINITION

The **greatest common factor** for a polynomial is the largest monomial that divides (is a factor of) each term of the polynomial.

The greatest common factor for the polynomial $25x^5 + 20x^4 - 30x^3$ is $5x^3$ since it is the largest monomial that is a factor of each term. We can apply the distributive property and write

$$\begin{aligned} 25x^5 + 20x^4 - 30x^3 &= 5x^3(5x^2) + 5x^3(4x) - 5x^3(6) \\ &= 5x^3(5x^2 + 4x - 6) \end{aligned}$$

The last line is written in factored form.

EXAMPLE 1 Factor the greatest common factor from

$8a^3 - 8a^2 - 48a$

Solution The greatest common factor is $8a$. It is the largest monomial that divides each term of our polynomial. We can write each term in our polynomial as the product of $8a$ and another monomial. Then, we apply the distributive property to factor $8a$ from each term:

$$\begin{aligned} 8a^3 - 8a^2 - 48a &= 8a(a^2) - 8a(a) - 8a(6) \\ &= 8a(a^2 - a - 6) \end{aligned}$$

EXAMPLE 2 Factor the greatest common factor from

$16a^5b^4 - 24a^2b^5 - 8a^3b^3$

Solution The largest monomial that divides each term is $8a^2b^3$. We write each term of the original polynomial in terms of $8a^2b^3$ and apply the distributive property to write the polynomial in factored form:

$$\begin{aligned} 16a^5b^4 - 24a^2b^5 - 8a^3b^3 &= 8a^2b^3(2a^3b) - 8a^2b^3(3b^2) - 8a^2b^3(a) \\ &= 8a^2b^3(2a^3b - 3b^2 - a) \end{aligned}$$

EXAMPLE 3 Factor the greatest common factor from

$5x^2(a + b) - 6x(a + b) - 7(a + b)$

Solution The greatest common factor is $a + b$. Factoring it from each term, we have

$$5x^2(a + b) - 6x(a + b) - 7(a + b) = (a + b)(5x^2 - 6x - 7)$$

EXAMPLE 4 A company manufacturing prerecorded videotapes finds that the total daily revenue for selling x tapes is given by

$$R(x) = 11.5x - 0.05x^2$$

Factor x from each term on the right side of the equation to find the formula that gives the price p in terms of x.

OBJECTIVES

A	Factor by factoring out the greatest common factor.
B	Factor by grouping.

Multiplication

Factors → $3 \cdot 7 = 21$ ← Product

Factoring

PRACTICE PROBLEMS

1. Factor the greatest common factor from $15a^7 - 25a^5 + 30a^3$.

NOTE

The term *largest monomial,* as used here, refers to the monomial with the largest integer exponents whose coefficient has the greatest absolute value.

2. Factor $12x^4y^5 - 9x^3y^4 - 15x^5y^3$.

3. Factor.
$4(a + b)^4 - 6(a + b)^3 + 16(a + b)^2$

4. If the total weekly revenue for the company mentioned in Example 4 is
$R(x) = 80.5x - 0.35x^2$
find the formula that gives the price p in terms of x.

Answers

1. $5a^3(3a^4 - 5a^2 + 6)$
2. $3x^3y^3(4xy^2 - 3y - 5x^2)$
3. $2(a + b)^2[2(a + b)^2 - 3(a + b) + 8]$

Solution We begin by factoring x from the right side of the equation:

$$\text{If} \quad R(x) = 11.5x - 0.05x^2$$
$$\text{then} \quad R(x) = x(11.5 - 0.05x)$$

Because R is always xp, the quantity in parentheses must be p. The price it should charge if it wants to sell x items per day is therefore

$$p = 11.5 - 0.05x$$

Factoring by Grouping

The polynomial $5x + 5y + x^2 + xy$ can be factored by noticing that the first two terms have a 5 in common, whereas the last two have an x in common. Applying the distributive property, we have

$$5x + 5y + x^2 + xy = 5(x + y) + x(x + y)$$

This last expression can be thought of as having two terms, $5(x + y)$ and $x(x + y)$, each of which has a common factor $(x + y)$. We apply the distributive property again to factor $(x + y)$ from each term:

$$5(x + y) + x(x + y)$$
$$= (x + y)(5 + x)$$

5. Factor $ab^3 + b^3 + 6a + 6$.

EXAMPLE 5 Factor $a^2b^2 + b^2 + 8a^2 + 8$.

Solution The first two terms have b^2 in common; the last two have 8 in common:

$$a^2b^2 + b^2 + 8a^2 + 8 = b^2(a^2 + 1) + 8(a^2 + 1)$$
$$= (a^2 + 1)(b^2 + 8)$$

6. Factor $15 - 3y^2 - 5x^2 + x^2y^2$.

EXAMPLE 6 Factor $15 - 5y^4 - 3x^3 + x^3y^4$.

Solution Let's try factoring a 5 from the first two terms and an $-x^3$ from the last two terms:

$$15 - 5y^4 - 3x^3 + x^3y^4 = 5(3 - y^4) - x^3(3 - y^4)$$
$$= (3 - y^4)(5 - x^3)$$

7. Factor by grouping:
$x^3 + 5x^2 + 3x + 15$

EXAMPLE 7 Factor by grouping $x^3 + 2x^2 + 9x + 18$.

Solution We begin by factoring x^2 from the first two terms and 9 from the second two terms:

$$x^3 + 2x^2 + 9x + 18 = x^2(x + 2) + 9(x + 2)$$
$$= (x + 2)(x^2 + 9)$$

Linking Objectives and Examples

Next to each **objective**, list the examples that are best described by that objective.

A	
B	

Answers

4. $p = 80.5 - 0.35x$
5. $(a + 1)(b^3 + 6)$
6. $(5 - y^2)(3 - x^2)$
7. $(x + 5)(x^2 + 3)$

Getting Ready for Class

After reading through the preceding section, respond in your own words and in complete sentences.

1. What is the greatest common factor for a polynomial?
2. After factoring a polynomial, how can you check your result?
3. When would you try to factor by grouping?
4. What is the relationship between multiplication and factoring?

Problem Set 5.4

INTERMEDIATE Algebra f(x) Now™

Factor the greatest common factor from each of the following. (The answers in the back of the book all show greatest common factors whose coefficients are positive.)

1. $10x^3 - 15x^2$

2. $12x^5 + 18x^7$

3. $9y^6 + 18y^3$

4. $24y^4 - 8y^2$

5. $9a^2b - 6ab^2$

6. $30a^3b^4 + 20a^4b^3$

7. $21xy^4 + 7x^2y^2$

8. $14x^6y^3 - 6x^2y^4$

9. $3a^2 - 21a + 30$

10. $3a^2 - 3a - 6$

11. $4x^3 - 16x^2 - 20x$

12. $2x^3 - 14x^2 + 20x$

13. $10x^4y^2 + 20x^3y^3 - 30x^2y^4$

14. $6x^4y^2 + 18x^3y^3 - 24x^2y^4$

15. $-x^2y + xy^2 - x^2y^2$

16. $-x^3y^2 - x^2y^3 - x^2y^2$

17. $4x^3y^2z - 8x^2y^2z^2 + 6xy^2z^3$

18. $7x^4y^3z^2 - 21x^2y^2z^2 - 14x^2y^3z^4$

19. $20a^2b^2c^2 - 30ab^2c + 25a^2bc^2$

20. $8a^3bc^5 - 48a^2b^4c + 16ab^3c^5$

21. $5x(a - 2b) - 3y(a - 2b)$

22. $3a(x - y) - 7b(x - y)$

23. $3x^2(x + y)^2 - 6y^2(x + y)^2$

24. $10x^3(2x - 3y) - 15x^2(2x - 3y)$

25. $2x^2(x + 5) + 7x(x + 5) + 6(x + 5)$

26. $2x^2(x + 2) + 13x(x + 2) + 15(x + 2)$

Factor each of the following by grouping.

27. $3xy + 3y + 2ax + 2a$

28. $5xy^2 + 5y^2 + 3ax + 3a$

29. $x^2y + x + 3xy + 3$

30. $x^3y^3 + 2x^3 + 5x^2y^3 + 10x^2$

31. $3xy^2 - 6y^2 + 4x - 8$

32. $8x^2y - 4x^2 + 6y - 3$

33. $x^2 - ax - bx + ab$

34. $ax - x^2 - bx + ab$

35. $ab + 5a - b - 5$

36. $x^2 - xy - ax + ay$

37. $a^4b^2 + a^4 - 5b^2 - 5$

38. $2a^2 - a^2b - bc^2 + 2c^2$

39. $x^3 + 3x^2 - 4x - 12$

40. $x^3 + 5x^2 - 4x - 20$

41. $x^3 + 2x^2 - 25x - 50$

42. $x^3 + 4x^2 - 9x - 36$

43. $2x^3 + 3x^2 - 8x - 12$

44. $3x^3 + 2x^2 - 27x - 18$

45. $4x^3 + 12x^2 - 9x - 27$

46. $9x^3 + 18x^2 - 4x - 8$

47. The greatest common factor of the binomial $3x - 9$ is 3. The greatest common factor of the binomial $6x - 2$ is 2. What is the greatest common factor of their product, $(3x - 9)(6x - 2)$, when it has been multiplied out?

48. The greatest common factors of the binomials $5x - 10$ and $2x + 4$ are 5 and 2, respectively. What is the greatest common factor of their product, $(5x - 10)(2x + 4)$, when it has been multiplied out?

Applying the Concepts

49. **Investing** If P dollars are placed in a savings account in which the rate of interest r is compounded yearly, then at the end of 1 year the amount of money in the account can be written as $P + Pr$. At the end of 2 years the amount of money in the account is

$$P + Pr + (P + Pr)r$$

Use factoring by grouping to show that this last expression can be written as $P(1 + r)^2$.

50. **Investing** At the end of 3 years, the amount of money in the savings account in Problem 49 will be

$$P(1 + r)^2 + P(1 + r)^2 r$$

Use factoring to show that this last expression can be written as $P(1 + r)^3$.

Use Example 4 as a guide in solving the next four problems.

51. **Price** A company manufacturing prerecorded videotapes finds that the total daily revenue R for selling x tapes at p dollars per tape is given by

$$R(x) = 11.5x - 0.05x^2$$

Factor x from each term on the right side of the equation to find the formula that gives the price p in terms of x. Then, use it to find the price they should charge if they want to sell 125 videotapes per day.

52. **Price** A company producing diskettes for home computers finds that the total daily revenue for selling x diskettes at p dollars per diskette is given by

$$R(x) = 8x - 0.01x^2$$

Use the fact that $R = xp$ and your knowledge of factoring to find a formula that gives the price p in terms of x. Then, use it to find the price they should charge if they want to sell 420 diskettes per day.

53. **Price** The weekly revenue equation for a company selling an inexpensive accounting program for home computers is given by the equation

$$R(x) = 35x - 0.1x^2$$

where x is the number of programs they sell per week. What price p should they charge if they want to sell 65 programs per week?

54. **Price** The weekly revenue equation for a small mail-order company selling boxes of greeting cards is

$$R(x) = 5.9x - 0.004x^2$$

where x is the number of boxes they sell per week. What price p should they charge if they want to sell 200 boxes each week?

Getting Ready for the Next Section

Factor out the greatest common factor.

55. $3x^4 - 9x^3y - 18x^2y^2$

56. $5x^2 + 10x + 30$

57. $2x^2(x - 3) - 4x(x - 3) - 3(x - 3)$

58. $3x^2(x - 2) - 8x(x - 2) + 2(x - 2)$

Multiply.

59. $(x + 2)(3x - 1)$

60. $(x - 2)(3x + 1)$

61. $(x - 1)(3x - 2)$

62. $(x + 1)(3x + 2)$

63. $(x + 2)(x + 3)$

64. $(x - 2)(x - 3)$

65. $(2y + 5)(3y - 7)$

66. $(2y - 5)(3y + 7)$

67. $(4 - 3a)(5 - a)$

68. $(4 - 3a)(5 + a)$

Complete each table.

69.

TWO NUMBERS *a* AND *b*	THEIR PRODUCT *ab*	THEIR SUM *a* + *b*
1, −24		
−1, 24		
2, −12		
−2, 12		
3, −8		
−3, 8		
4, −6		
−4, 6		

70.

TWO NUMBERS *a* AND *b*	THEIR PRODUCT *ab*	THEIR SUM *a* + *b*
1, −54		
−1, 54		
2, −27		
−2, 27		
3, −18		
−3, 18		
6, −9		
−6, 9		

Maintaining Your Skills

Evaluate each determinant.

71. $\begin{vmatrix} 3 & 5 \\ -6 & 2 \end{vmatrix}$

72. $\begin{vmatrix} -2 & 0 \\ 0 & -1 \end{vmatrix}$

73. $\begin{vmatrix} 1 & -2 & 3 \\ 0 & 4 & -1 \\ 2 & -4 & 6 \end{vmatrix}$

74. $\begin{vmatrix} 2 & 0 & 0 \\ 0 & -3 & 0 \\ 0 & 0 & 4 \end{vmatrix}$

Factoring Trinomials 5.5

Factoring Trinomials with a Leading Coefficient of 1

Earlier in this chapter we multiplied binomials:

$$(x - 2)(x + 3) = x^2 + x - 6$$

$$(x + 5)(x + 2) = x^2 + 7x + 10$$

In each case the product of two binomials is a trinomial. The first term in the resulting trinomial is obtained by multiplying the first term in each binomial. The middle term comes from adding the product of the two inside terms with the product of the two outside terms. The last term is the product of the last term in each binomial.

In general,

$$(x + a)(x + b) = x^2 + ax + bx + ab$$

$$= x^2 + (a + b)x + ab$$

Writing this as a factoring problem, we have

$$x^2 + (a + b)x + ab = (x + a)(x + b)$$

To factor a trinomial with a leading coefficient of 1, we simply find the two numbers a and b whose sum is the coefficient of the middle term and whose product is the constant term.

EXAMPLE 1 Factor $x^2 + 2x - 15$.

Solution Again the leading coefficient is 1. We need two integers whose product is -15 and whose sum is $+2$. The integers are $+5$ and -3.

$$x^2 + 2x - 15 = (x + 5)(x - 3)$$

In the preceding example we found factors of $x + 5$ and $x - 3$. These are the only two such factors for $x^2 + 2x - 15$. There is no other pair of binomials $x + a$ and $x + b$ whose product is $x^2 + 2x - 15$.

EXAMPLE 2 Factor $x^2 - xy - 12y^2$.

Solution We need two expressions whose product is $-12y^2$ and whose sum is $-y$. The expressions are $-4y$ and $3y$:

$$x^2 - xy - 12y^2 = (x - 4y)(x + 3y)$$

Checking this result gives

$$(x - 4y)(x + 3y) = x^2 + 3xy - 4xy - 12y^2$$

$$= x^2 - xy - 12y^2$$

EXAMPLE 3 Factor $x^2 - 8x + 6$.

Solution Since there is no pair of integers whose product is 6 and whose sum is -8, the trinomial $x^2 - 8x + 6$ is not factorable. We say it is a *prime polynomial.*

OBJECTIVES

A	Factor trinomials in which the leading coefficient is 1.
B	Factor trinomials in which the leading coefficient is a number other than 1.

PRACTICE PROBLEMS

1. Factor $x^2 - x - 12$.
2. Factor $x^2 + 2xy - 15y^2$.
3. Factor $x^2 + x + 1$.

Answers

1. $(x - 4)(x + 3)$
2. $(x + 5y)(x - 3y)$
3. Does not factor.

4. Factor $5x^2 + 25x + 30$.

NOTE

As a general rule, it is best to factor out the greatest common factor first.

EXAMPLE 4 Factor $3x^4 - 15x^3y - 18x^2y^2$.

Solution The leading coefficient is not 1. Each term is divisible by $3x^2$, however. Factoring this out to begin with we have

$$3x^4 - 15x^3y - 18x^2y^2 = 3x^2(x^2 - 5xy - 6y^2)$$

Factoring the resulting trinomial as in the previous examples gives

$$3x^2(x^2 - 5xy - 6y^2) = 3x^2(x - 6y)(x + y)$$

Factoring Other Trinomials by Trial and Error

We want to turn our attention now to trinomials with leading coefficients other than 1 and with no greatest common factor other than 1.

Suppose we want to factor $3x^2 - x - 2$. The factors will be a pair of binomials. The product of the first terms will be $3x^2$, and the product of the last terms will be -2. We can list all the possible factors along with their products as follows.

POSSIBLE FACTORS	FIRST TERM	MIDDLE TERM	LAST TERM
$(x + 2)(3x - 1)$	$3x^2$	$+5x$	-2
$(x - 2)(3x + 1)$	$3x^2$	$-5x$	-2
$(x + 1)(3x - 2)$	$3x^2$	$+x$	-2
$(x - 1)(3x + 2)$	$3x^2$	$-x$	-2

From the last line we see that the factors of $3x^2 - x - 2$ are $(x - 1)(3x + 2)$. That is,

$$3x^2 - x - 2 = (x - 1)(3x + 2)$$

To factor trinomials with leading coefficients other than 1, when the greatest common factor is 1, we must use trial and error or list all the possible factors. In either case the idea is this: Look only at pairs of binomials whose products give the correct first and last terms, then look for the combination that will give the correct middle term.

5. Factor $3x^2 - x - 2$.

EXAMPLE 5 Factor $2x^2 + 13xy + 15y^2$.

Solution Listing all possible factors the product of whose first terms is $2x^2$ and the product of whose last terms is $+15y^2$ yields

POSSIBLE FACTORS	MIDDLE TERM OF PRODUCT
$(2x - 5y)(x - 3y)$	$-11xy$
$(2x - 3y)(x - 5y)$	$-13xy$
$(2x + 5y)(x + 3y)$	$+11xy$
$(2x + 3y)(x + 5y)$	$+13xy$
$(2x + 15y)(x + y)$	$+17xy$
$(2x - 15y)(x - y)$	$-17xy$
$(2x + y)(x + 15y)$	$+31xy$
$(2x - y)(x - 15y)$	$-31xy$

The fourth line has the correct middle term:

$$2x^2 + 13xy + 15y^2 = (2x + 3y)(x + 5y)$$

Actually, we did not need to check the first two pairs of possible factors in the preceding list. Because all the signs in the trinomial $2x^2 + 13xy + 15y^2$ are positive, the binomial factors must be of the form $(ax + b)(cx + d)$, where a, b, c, and d are all positive.

There are other ways to reduce the number of possible factors to consider. For example, if we were to factor the trinomial $2x^2 - 11x + 12$, we would not have to

Answers
4. $5(x + 3)(x + 2)$
5. $(3x + 2)(x - 1)$

consider the pair of possible factors $(2x - 4)(x - 3)$. If the original trinomial has no greatest common factor other than 1, then neither of its binomial factors will either. The trinomial $2x^2 - 11x + 12$ has a greatest common factor of 1, but the possible factor $2x - 4$ has a greatest common factor of 2: $2x - 4 = 2(x - 2)$. Therefore, we do not need to consider $2x - 4$ as a possible factor.

EXAMPLE 6 Factor $12x^4 + 17x^2 + 6$.

Solution This is a trinomial in x^2:

$$12x^4 + 17x^2 + 6 = (4x^2 + 3)(3x^2 + 2)$$

EXAMPLE 7 Factor $2x^2(x - 3) - 5x(x - 3) - 3(x - 3)$.

Solution We begin by factoring out the greatest common factor $(x - 3)$. Then we factor the trinomial that remains.

$$\begin{aligned} 2x^2(x - 3) - 5x(x - 3) - 3(x - 3) &= (x - 3)(2x^2 - 5x - 3) \\ &= (x - 3)(2x + 1)(x - 3) \\ &= (x - 3)^2(2x + 1) \end{aligned}$$

Another Method of Factoring Trinomials

As an alternative to the trial-and-error method of factoring trinomials, we present the following method. The new method does not require as much trial and error. To use this new method, we must rewrite our original trinomial in such a way that the factoring by grouping method can be applied.

Here are the steps we use to factor $ax^2 + bx + c$.

Step 1: Form the product ac.
Step 2: Find a pair of numbers whose product is ac and whose sum is b.
Step 3: Rewrite the polynomial to be factored so that the middle term bx is written as the sum of two terms whose coefficients are the two numbers found in step 2.
Step 4: Factor by grouping.

EXAMPLE 8 Factor $3x^2 - 10x - 8$ using these steps.

Solution The trinomial $3x^2 - 10x - 8$ has the form $ax^2 + bx + c$, where $a = 3$, $b = -10$, and $c = -8$.

Step 1: The product ac is $3(-8) = -24$.
Step 2: We need to find two numbers whose product is -24 and whose sum is -10. Let's list all the pairs of numbers whose product is -24 to find the pair whose sum is -10.

Product	*Sum*
$1(-24) = -24$	$1 + (-24) = -23$
$-1(24) = -24$	$-1 + 24 = 23$
$2(-12) = -24$	$2 + (-12) = -10$
$-2(12) = -24$	$-2 + 12 = 10$
$3(-8) = -24$	$3 + (-8) = -5$
$-3(8) = -24$	$-3 + 8 = 5$
$4(-6) = -24$	$4 + (-6) = -2$
$-4(6) = -24$	$-4 + 6 = 2$

As you can see, of all the pairs of numbers whose product is -24, only 2 and -12 have a sum of -10.

6. Factor $15x^4 + x^2 - 2$.

7. Factor.
$3x^2(x - 2) - 7x(x - 2) + 2(x - 2)$

8. Factor $8x^2 - 5x - 3$.

Answers
6. $(5x^2 + 2)(3x^2 - 1)$
7. $(x - 2)^2(3x - 1)$

Step 3: We now rewrite our original trinomial so the middle term $-10x$ is written as the sum of $-12x$ and $2x$:

$$3x^2 - 10x - 8 = 3x^2 - 12x + 2x - 8$$

Step 4: Factoring by grouping, we have

$$\begin{aligned} 3x^2 - 12x + 2x - 8 &= 3x(x - 4) + 2(x - 4) \\ &= (x - 4)(3x + 2) \end{aligned}$$

You can see that this method works by multiplying $x - 4$ and $3x + 2$ to get

$$3x^2 - 10x - 8$$

9. Factor $4x^2 + 11x + 6$.

EXAMPLE 9

Factor $9x^2 + 15x + 4$.

Solution In this case $a = 9$, $b = 15$, and $c = 4$. The product ac is $9 \cdot 4 = 36$. Listing all the pairs of numbers whose product is 36 with their corresponding sums, we have

Product	*Sum*
$1(36) = 36$	$1 + 36 = 37$
$2(18) = 36$	$2 + 18 = 20$
$3(12) = 36$	$3 + 12 = 15$
$4(9) = 36$	$4 + 9 = 13$
$6(6) = 36$	$6 + 6 = 12$

Notice we list only positive numbers since both the product and sum we are looking for are positive. The numbers 3 and 12 are the numbers we are looking for. Their product is 36, and their sum is 15. We now rewrite the original polynomial $9x^2 + 15x + 4$ with the middle term written as $3x + 12x$. We then factor by grouping:

$$\begin{aligned} 9x^2 + 15x + 4 &= 9x^2 + 3x + 12x + 4 \\ &= 3x(3x + 1) + 4(3x + 1) \\ &= (3x + 1)(3x + 4) \end{aligned}$$

The polynomial $9x^2 + 15x + 4$ factors into the product

$$(3x + 1)(3x + 4)$$

10. Factor $4x^2 - 4x - 15$.

EXAMPLE 10

Factor $8x^2 - 2x - 15$.

Solution The product ac is $8(-15) = -120$. There are many pairs of numbers whose product is -120. We are looking for the pair whose sum is also -2. The numbers are -12 and 10. Writing $-2x$ as $-12x + 10x$ and then factoring by grouping, we have

$$\begin{aligned} 8x^2 - 2x - 15 &= 8x^2 - 12x + 10x - 15 \\ &= 4x(2x - 3) + 5(2x - 3) \\ &= (2x - 3)(4x + 5) \end{aligned}$$

Linking Objectives and Examples

Next to each **objective**, list the examples that are best described by that objective.

A	
B	

Answers

8. $(x - 1)(8x + 3)$
9. $(x + 2)(4x + 3)$
10. $(2x - 5)(2x + 3)$

Getting Ready for Class

After reading through the preceding section, respond in your own words and in complete sentences.

1. What is a prime polynomial?
2. When factoring polynomials, what should you look for first?
3. How can you check to see that you have factored a trinomial correctly?
4. Describe how to determine the binomial factors of $6x^2 + 5x - 25$.

Problem Set 5.5

INTERMEDIATE Algebra Now™

Factor each of the following trinomials.

1. $x^2 + 7x + 12$
2. $x^2 - 7x + 12$
3. $x^2 - x - 12$
4. $x^2 + x - 12$
5. $y^2 + y - 6$
6. $y^2 - y - 6$
7. $16 - 6x - x^2$
8. $3 + 2x - x^2$
9. $12 + 8x + x^2$
10. $15 - 2x - x^2$

Factor completely by first factoring out the greatest common factor and then factoring the trinomial that remains.

11. $3a^2 - 21a + 30$
12. $3a^2 - 3a - 6$
13. $4x^3 - 16x^2 - 20x$
14. $2x^3 - 14x^2 + 20x$

Factor.

15. $x^2 + 3xy + 2y^2$
16. $x^2 - 5xy - 24y^2$
17. $a^2 + 3ab - 18b^2$
18. $a^2 - 8ab - 9b^2$
19. $x^2 - 2xa - 48a^2$
20. $x^2 + 14xa + 48a^2$
21. $x^2 - 12xb + 36b^2$
22. $x^2 + 10xb + 25b^2$

Factor completely. Be sure to factor out the greatest common factor first if it is other than 1.

23. $3x^2 - 6xy - 9y^2$
24. $5x^2 + 25xy + 20y^2$
25. $2a^5 + 4a^4b + 4a^3b^2$
26. $3a^4 - 18a^3b + 27a^2b^2$
27. $10x^4y^2 + 20x^3y^3 - 30x^2y^4$
28. $6x^4y^2 + 18x^3y^3 - 24x^2y^4$

29. $2x^2 + 7x - 15$

30. $2x^2 - 7x - 15$

31. $2x^2 + x - 15$

32. $2x^2 - x - 15$

33. $2x^2 - 13x + 15$

34. $2x^2 + 13x + 15$

35. $2x^2 - 11x + 15$

36. $2x^2 + 11x + 15$

37. $2x^2 + 7x + 15$

38. $2x^2 + x + 15$

39. $2 + 7a + 6a^2$

40. $2 - 7a + 6a^2$

41. $60y^2 - 15y - 45$

42. $72y^2 + 60y - 72$

43. $6x^4 - x^3 - 2x^2$

44. $3x^4 + 2x^3 - 5x^2$

45. $40r^3 - 120r^2 + 90r$

46. $40r^3 + 200r^2 + 250r$

47. $4x^2 - 11xy - 3y^2$

48. $3x^2 + 19xy - 14y^2$

49. $10x^2 - 3xa - 18a^2$

50. $9x^2 + 9xa - 10a^2$

51. $18a^2 + 3ab - 28b^2$

52. $6a^2 - 7ab - 5b^2$

53. $600 + 800t - 800t^2$

54. $200 - 600t - 350t^2$

55. $9y^4 + 9y^3 - 10y^2$

56. $4y^5 + 7y^4 - 2y^3$

57. $24a^2 - 2a^3 - 12a^4$

58. $60a^2 + 65a^3 - 20a^4$

59. $8x^4y^2 - 2x^3y^3 - 6x^2y^4$

60. $8x^4y^2 - 47x^3y^3 - 6x^2y^4$

61. $300x^4 + 1{,}000x^2 + 300$

62. $600x^4 - 100x^2 - 200$

63. $20a^4 + 37a^2 + 15$

64. $20a^4 + 13a^2 - 15$

65. $9 + 3r^2 - 12r^4$

66. $2 - 4r^2 - 30r^4$

Factor each of the following by first factoring out the greatest common factor and then factoring the trinomial that remains.

67. $2x^2(x + 5) + 7x(x + 5) + 6(x + 5)$

68. $2x^2(x + 2) + 13x(x + 2) + 15(x + 2)$

69. $x^2(2x + 3) + 7x(2x + 3) + 10(2x + 3)$

70. $2x^2(x + 1) + 7x(x + 1) + 6(x + 1)$

71. What polynomial, when factored, gives $(3x + 5y)(3x - 5y)$?

72. What polynomial, when factored, gives $(7x + 2y)(7x - 2y)$?

73. One factor of the trinomial $a^2 + 260a + 2{,}500$ is $a + 10$. What is the other factor?

74. One factor of the trinomial $a^2 - 75a - 2{,}500$ is $a + 25$. What is the other factor?

75. Factor the right side of the equation $y = 4x^2 + 18x - 10$, and then use the result to find y when x is $\frac{1}{2}$, when x is -5, and when x is 2.

76. Factor the right side of the equation $y = 9x^2 + 33x - 12$, and use the result to find y when x is $\frac{1}{3}$, when x is -4, and when x is 3.

Getting Ready for the Next Section

For each problem below, place a number or expression inside the parentheses so that the resulting statement is true.

77. $\frac{25}{64} = (\quad)^2$ **78.** $\frac{4}{9} = (\quad)^2$ **79.** $x^6 = (\quad)^2$ **80.** $x^8 = (\quad)^2$ **81.** $16x^4 = (\quad)^2$ **82.** $81y^4 = (\quad)^2$

Write as a perfect cube.

83. $\frac{1}{8} = (\quad)^3$ **84.** $\frac{1}{27} = (\quad)^3$ **85.** $x^6 = (\quad)^3$ **86.** $x^{12} = (\quad)^3$

87. $27x^3 = (\quad)^3$ **88.** $125y^3 = (\quad)^3$ **89.** $8y^3 = (\quad)^3$ **90.** $1000x^3 = (\quad)^3$

Maintaining Your Skills

Multiply.

91. $(2x - 3)(2x + 3)$ **92.** $(4 - 5x)(4 + 5x)$ **93.** $(2x - 3)^2$ **94.** $(4 - 5x)^2$

95. $(2x - 3)(4x^2 + 6x + 9)$ **96.** $(2x + 3)(4x^2 - 6x + 9)$

Extending the Concepts

Factor completely.

97. $8x^6 + 26x^3y^2 + 15y^4$ **98.** $24x^4 + 6x^2y^3 - 45y^6$

99. $3x^2 + 295x - 500$ **100.** $3x^2 + 594x - 1{,}200$

101. $\frac{1}{8}x^2 + x + 2$ **102.** $\frac{1}{9}x^2 + x + 2$

103. $2x^2 + 1.5x + 0.25$ **104.** $6x^2 + 2x + 0.16$

Special Factoring 5.6

To find the area of the large square in the margin, we can square the length of its side, giving us $(a + b)^2$. However, we can add the areas of the four smaller figures to arrive at the same result.

Since the area of the large square is the same whether we find it by squaring a side or by adding the four smaller areas, we can write the following relationship:

$$(a + b)^2 = a^2 + 2ab + b^2$$

This is the formula for the square of a binomial. The figure gives us a geometric interpretation for one of the special multiplication formulas. We begin this section by looking at the special multiplication formulas from a factoring perspective.

OBJECTIVES	
A	Factor perfect square trinomials.
B	Factor the difference of two squares.
C	Factor the sum or difference of two cubes.

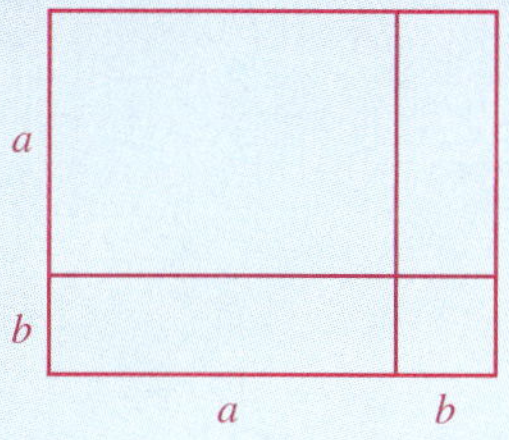

Perfect Square Trinomials

We previously listed some special products found in multiplying polynomials. Two of the formulas looked like this:

$$(a + b)^2 = a^2 + 2ab + b^2$$
$$(a - b)^2 = a^2 - 2ab + b^2$$

If we exchange the left and right sides of each formula, we have two special formulas for factoring:

$$a^2 + 2ab + b^2 = (a + b)^2$$
$$a^2 - 2ab + b^2 = (a - b)^2$$

The left side of each formula is called a *perfect square trinomial.* The right sides are binomial squares. Perfect square trinomials can always be factored using the usual methods for factoring trinomials. However, if we notice that the first and last terms of a trinomial are perfect squares, it is wise to see whether the trinomial factors as a binomial square before attempting to factor by the usual method.

EXAMPLE 1 Factor $x^2 - 6x + 9$.

Solution Since the first and last terms are perfect squares, we attempt to factor according to the preceding formulas:

$$x^2 - 6x + 9 = (x - 3)^2$$

If we expand $(x - 3)^2$, we have $x^2 - 6x + 9$, indicating we have factored correctly. ■

EXAMPLES Factor each of the following perfect square trinomials.

2. $16a^2 + 40ab + 25b^2 = (4a + 5b)^2$

3. $49 - 14t + t^2 = (7 - t)^2$

4. $9x^4 - 12x^2 + 4 = (3x^2 - 2)^2$

5. $(y + 3)^2 + 10(y + 3) + 25 = [(y + 3) + 5]^2 = (y + 8)^2$ ■

EXAMPLE 6 Factor $8x^2 - 24xy + 18y^2$.

Solution We begin by factoring the greatest common factor 2 from each term:

$$8x^2 - 24xy + 18y^2 = 2(4x^2 - 12xy + 9y^2)$$
$$= 2(2x - 3y)^2$$ ■

PRACTICE PROBLEMS

1. Factor $x^2 - 10x + 25$.

Factor.

2. $9a^2 + 42ab + 49b^2$

3. $25 - 10t + t^2$

4. $16x^4 - 24x^2 + 9$

5. $(y + 2)^2 + 8(y + 2) + 16$

6. Factor $27x^2 - 36x + 12$.

Answers

1. $(x - 5)^2$ **2.** $(3a + 7b)^2$
3. $(5 - t)^2$ **4.** $(4x^2 - 3)^2$
5. $(y + 6)^2$ **6.** $3(3x - 2)^2$

The Difference of Two Squares

Recall the formula that results in the difference of two squares: $(a + b)(a - b) = a^2 - b^2$. Writing this as a factoring formula, we have

$$a^2 - b^2 = (a + b)(a - b)$$

Factor.

7. $x^2 - 16$
8. $64 - t^2$
9. $25x^2 - 36y^2$
10. $9x^6 - 1$
11. $x^2 - \frac{25}{64}$

EXAMPLES Each of the following is the difference of two squares. Use the formula $a^2 - b^2 = (a + b)(a - b)$ to factor each one.

7. $x^2 - 25 = x^2 - 5^2 = (x + 5)(x - 5)$
8. $49 - t^2 = 7^2 - t^2 = (7 + t)(7 - t)$
9. $81a^2 - 25b^2 = (9a)^2 - (5b)^2 = (9a + 5b)(9a - 5b)$
10. $4x^6 - 1 = (2x^3)^2 - 1^2 = (2x^3 + 1)(2x^3 - 1)$
11. $x^2 - \frac{4}{9} = x^2 - (\frac{2}{3})^2 = (x + \frac{2}{3})(x - \frac{2}{3})$ ■

As our next example shows, the difference of two fourth powers can be factored as the difference of two squares.

12. Factor $x^4 - 81$.

EXAMPLE 12 Factor $16x^4 - 81y^4$.

Solution The first and last terms are perfect squares. We factor according to the preceding formula:

$$\begin{aligned} 16x^4 - 81y^4 &= (4x^2)^2 - (9y^2)^2 \\ &= (4x^2 + 9y^2)(4x^2 - 9y^2) \end{aligned}$$

Notice that the second factor is also the difference of two squares. Factoring completely, we have

$$16x^4 - 81y^4 = (4x^2 + 9y^2)(2x + 3y)(2x - 3y)$$ ■

NOTE

The sum of two squares never factors into the product of two binomials; that is, if we were to attempt to factor $(4x^2 + 9y^2)$ in the last example, we would be unable to find two binomials (or any other polynomials) whose product is $4x^2 + 9y^2$. The factors do not exist as polynomials.

13. Factor $(x - 4)^2 - 9$.

Here is another example of the difference of two squares.

EXAMPLE 13 Factor $(x - 3)^2 - 25$.

Solution This example has the form $a^2 - b^2$, where a is $x - 3$ and b is 5. We factor it according to the formula for the difference of two squares:

$$\begin{aligned} (x - 3)^2 - 25 &= (x - 3)^2 - 5^2 && \textbf{Write 25 as } 5^2 \\ &= [(x - 3) + 5][(x - 3) - 5] && \textbf{Factor} \\ &= (x + 2)(x - 8) && \textbf{Simplify} \end{aligned}$$

Notice in this example we could have expanded $(x - 3)^2$, subtracted 25, and then factored to obtain the same result:

$$\begin{aligned} (x - 3)^2 - 25 &= x^2 - 6x + 9 - 25 && \textbf{Expand } (x - 3)^2 \\ &= x^2 - 6x - 16 && \textbf{Simplify} \\ &= (x - 8)(x + 2) && \textbf{Factor} \end{aligned}$$ ■

Answers

7. $(x + 4)(x - 4)$ 8. $(8 + t)(8 - t)$
9. $(5x + 6y)(5x - 6y)$
10. $(3x^3 + 1)(3x^3 - 1)$
11. $(x + \frac{5}{8})(x - \frac{5}{8})$
12. $(x^2 + 9)(x + 3)(x - 3)$
13. $(x - 7)(x - 1)$

EXAMPLE 14 Factor $x^2 - 10x + 25 - y^2$.

Solution Notice the first three items form a perfect square trinomial; that is, $x^2 - 10x + 25 = (x - 5)^2$. If we replace the first three terms by $(x - 5)^2$, the expression that results has the form $a^2 - b^2$. We can factor as we did in Example 13:

$x^2 - 10x + 25 - y^2 = (x^2 - 10x + 25) - y^2$	**Group first three terms together**
$= (x - 5)^2 - y^2$	**This has the form $a^2 - b^2$**
$= [(x - 5) + y][(x - 5) - y]$	**Factor according to the formula $a^2 - b^2 = (a + b)(a - b)$**
$= (x - 5 + y)(x - 5 - y)$	**Simplify**

We could check this result by multiplying the two factors together. (You may want to do that to convince yourself that we have the correct result.)

EXAMPLE 15 Factor completely $x^3 + 2x^2 - 9x - 18$.

Solution We use factoring by grouping to begin and then factor the difference of two squares:

$$\begin{aligned} x^3 + 2x^2 - 9x - 18 &= x^2(x + 2) - 9(x + 2) \\ &= (x + 2)(x^2 - 9) \\ &= (x + 2)(x + 3)(x - 3) \end{aligned}$$

The Sum and Difference of Two Cubes

Here are the formulas for factoring the sum and difference of two cubes:

$$a^3 + b^3 = (a + b)(a^2 - ab + b^2)$$

$$a^3 - b^3 = (a - b)(a^2 + ab + b^2)$$

Since these formulas are unfamiliar, it is important that we verify them.

EXAMPLE 16 Verify the two formulas.

Solution We verify the formulas by multiplying the right sides and comparing the results with the left sides:

$$\begin{array}{r} a^2 - ab + b^2 \\ a + b \\ \hline a^3 - a^2b + ab^2 \\ a^2b - ab^2 + b^3 \\ \hline a^3 \qquad\qquad + b^3 \end{array}$$

The first formula is correct.

$$\begin{array}{r} a^2 + ab + b^2 \\ a - b \\ \hline a^3 + a^2b + ab^2 \\ -a^2b - ab^2 - b^3 \\ \hline a^3 \qquad\qquad - b^3 \end{array}$$

The second formula is correct.

14. Factor $x^2 - 6x + 9 - y^2$.

15. Factor completely. $x^3 + 5x^2 - 4x - 20$

16. Multiply $(x - 3)(x^2 + 3x + 9)$.

Answers

14. $(x - 3 + y)(x - 3 - y)$
15. $(x + 5)(x + 2)(x - 2)$
16. $x^3 - 27$

Here are some examples using the formulas for factoring the sum and difference of two cubes.

Factor.

17. $27 + x^3$

18. $8x^3 + y^3$

19. $a^3 - \frac{1}{27}$

20. $x^6 - 1$

EXAMPLE 17

Factor $64 + t^3$.

Solution The first term is the cube of 4 and the second term is the cube of t. Therefore,

$$64 + t^3 = 4^3 + t^3$$
$$= (4 + t)(16 - 4t + t^2)$$

EXAMPLE 18

Factor $27x^3 + 125y^3$.

Solution Writing both terms as perfect cubes, we have

$$27x^3 + 125y^3 = (3x)^3 + (5y)^3$$
$$= (3x + 5y)(9x^2 - 15xy + 25y^2)$$

EXAMPLE 19

Factor $a^3 - \frac{1}{8}$.

Solution The first term is the cube of a, whereas the second term is the cube of $\frac{1}{2}$:

$$a^3 - \frac{1}{8} = a^3 - \left(\frac{1}{2}\right)^3$$
$$= \left(a - \frac{1}{2}\right)\left(a^2 + \frac{1}{2}a + \frac{1}{4}\right)$$

EXAMPLE 20

Factor $x^6 - y^6$.

Solution We have a choice of how we want to write the two terms to begin. We can write the expression as the difference of two squares, $(x^3)^2 - (y^3)^2$, or as the difference of two cubes, $(x^2)^3 - (y^2)^3$. It is better to use the difference of two squares if we have a choice:

$$x^6 - y^6 = (x^3)^2 - (y^3)^2$$
$$= (x^3 - y^3)(x^3 + y^3)$$
$$= (x - y)(x^2 + xy + y^2)(x + y)(x^2 - xy + y^2)$$

Try this example again writing the first line as the difference of two cubes instead of the difference of two squares. It will become apparent why it is better to use the difference of two squares.

Linking Objectives and Examples

Next to each **objective**, list the examples that are best described by that objective.

A	
B	
C	

Answers

17. $(3 + x)(9 - 3x + x^2)$
18. $(2x + y)(4x^2 - 2xy + y^2)$
19. $\left(a - \frac{1}{3}\right)\left(a^2 + \frac{1}{3}a + \frac{1}{9}\right)$
20. $(x + 1)(x^2 - x + 1)(x - 1)(x^2 + x + 1)$

Getting Ready for Class

After reading through the preceding section, respond in your own words and in complete sentences.

1. In what cases can you factor a binomial?
2. What is a perfect square trinomial?
3. Is it possible to factor the sum of two squares?
4. Write the formula you use to factor the sum of two cubes.

Problem Set 5.6

INTERMEDIATE Algebra $f(x)$ Now™

Factor each perfect square trinomial.

1. $x^2 - 6x + 9$

2. $x^2 + 10x + 25$

3. $a^2 - 12a + 36$

4. $36 - 12a + a^2$

5. $25 - 10t + t^2$

6. $64 + 16t + t^2$

7. $\frac{1}{9}x^2 + 2x + 9$

8. $\frac{1}{4}x^2 - 2x + 4$

9. $4y^4 - 12y^2 + 9$

10. $9y^4 + 12y^2 + 4$

11. $16a^2 + 40ab + 25b^2$

12. $25a^2 - 40ab + 16b^2$

13. $\frac{1}{25} + \frac{1}{10}t^2 + \frac{1}{16}t^4$

14. $\frac{1}{9} - \frac{1}{3}t^3 + \frac{1}{4}t^6$

15. $16x^2 - 48x + 36$

16. $36x^2 + 48x + 16$

17. $75a^3 + 30a^2 + 3a$

18. $45a^4 - 30a^3 + 5a^2$

19. $(x + 2)^2 + 6(x + 2) + 9$

20. $(x + 5)^2 + 4(x + 5) + 4$

Factor each as the difference of two squares. Be sure to factor completely.

21. $x^2 - 9$

22. $x^2 - 16$

23. $49x^2 - 64y^2$

24. $81x^2 - 49y^2$

25. $4a^2 - \frac{1}{4}$

26. $25a^2 - \frac{1}{25}$

27. $x^2 - \frac{9}{25}$

28. $x^2 - \frac{25}{36}$

29. $9x^2 - 16y^2$

30. $25x^2 - 49y^2$

31. $250 - 10t^2$

32. $640 - 10t^2$

Factor each as the difference of two squares. Be sure to factor completely.

33. $x^4 - 81$

34. $x^4 - 16$

35. $9x^6 - 1$

36. $25x^6 - 1$

37. $16a^4 - 81$

38. $81a^4 - 16b^4$

39. $\frac{1}{81} - \frac{y^4}{16}$

40. $\frac{1}{25} - \frac{y^4}{64}$

Factor completely.

41. $x^6 - y^6$

42. $x^6 - 1$

43. $2a^7 - 128a$

44. $128a^8 - 2a^2$

45. $(x - 2)^2 - 9$

46. $(x + 2)^2 - 9$

47. $(y + 4)^2 - 16$

48. $(y - 4)^2 - 16$

49. $x^2 - 10x + 25 - y^2$

50. $x^2 - 6x + 9 - y^2$

51. $a^2 + 8a + 16 - b^2$

52. $a^2 + 12a + 36 - b^2$

53. $x^2 + 2xy + y^2 - a^2$

54. $a^2 + 2ab + b^2 - y^2$

55. $x^3 + 3x^2 - 4x - 12$

56. $x^3 + 5x^2 - 4x - 20$

57. $x^3 + 2x^2 - 25x - 50$

58. $x^3 + 4x^2 - 9x - 36$

59. $2x^3 + 3x^2 - 8x - 12$

60. $3x^3 + 2x^2 - 27x - 18$

61. $4x^3 + 12x^2 - 9x - 27$

62. $9x^3 + 18x^2 - 4x - 8$

Factor each of the following as the sum or difference of two cubes.

63. $x^3 - y^3$

64. $x^3 + y^3$

65. $a^3 + 8$

66. $a^3 - 8$

67. $27 + x^3$

68. $27 - x^3$

69. $y^3 - 1$

70. $y^3 + 1$

71. $10r^3 - 1{,}250$

72. $10r^3 + 1{,}250$

73. $64 + 27a^3$

74. $27 - 64a^3$

75. $8x^3 - 27y^3$

76. $27x^3 - 8y^3$

77. $t^3 + \frac{1}{27}$

78. $t^3 - \frac{1}{27}$

79. $27x^3 - \frac{1}{27}$

80. $8x^3 + \frac{1}{8}$

81. $64a^3 + 125b^3$

82. $125a^3 - 27b^3$

83. Find two values of b that will make $9x^2 + bx + 25$ a perfect square trinomial.

84. Find a value of c that will make $49x^2 - 42x + c$ a perfect square trinomial.

Getting Ready for the Next Section

Factor out the greatest common factor.

85. $y^3 + 25y$

86. $y^4 + 36y^2$

87. $2ab^5 + 8ab^4 + 2ab^3$

88. $3a^2b^3 + 6a^2b^2 - 3a^2b$

Factor by grouping.

89. $4x^2 - 6x + 2ax - 3a$

90. $6x^2 - 4x + 3ax - 2a$

Factor the difference of squares.

91. $x^2 - 4$

92. $x^2 - 9$

Factor the perfect square trinomial.

93. $x^2 - 6x + 9$

94. $x^2 - 10x + 25$

Factor.

95. $6a^2 - 11a + 4$

96. $6x^2 - x - 15$

Factor the sum or difference of cubes.

97. $x^3 + 8$

98. $x^3 - 27$

Maintaining Your Skills

Solve each system by using Cramer's rule.

99. $$\begin{aligned} 4x - 7y &= 3 \\ 5x + 2y &= -3 \end{aligned}$$

100. $$\begin{aligned} 9x - 8y &= 4 \\ 2x + 3y &= 6 \end{aligned}$$

101. $$\begin{aligned} 3x + 4y &= 15 \\ 2x - 5z &= -3 \\ 4y - 3z &= 9 \end{aligned}$$

102. $$\begin{aligned} x + 3y &= 5 \\ 6y + z &= 12 \\ x - 2z &= -10 \end{aligned}$$

Factoring: A General Review

5.7

In this section we will review the different methods of factoring that we have presented in the previous sections of this chapter. This section is important because it will give you an opportunity to factor a variety of polynomials.

We begin this section by listing the steps that can be used to factor polynomials of any type.

OBJECTIVES

A	Factor a variety of polynomials.

To Factor a Polynomial

Step 1: If the polynomial has a greatest common factor other than 1, then factor out the greatest common factor.

Step 2: If the polynomial has two terms (it is a binomial), then see if it is the difference of two squares or the sum or difference of two cubes, and then factor accordingly. Remember, if it is the sum of two squares it will not factor.

Step 3: If the polynomial has three terms (a trinomial), then it is either a perfect square trinomial, which will factor into the square of a binomial, or it is not a perfect square trinomial, in which case we try to write it as the product of two binomials using the methods developed in this chapter.

Step 4: If the polynomial has more than three terms, then try to factor it by grouping.

Step 5: As a final check, see if any of the factors you have written can be factored further. If you have overlooked a common factor, you can catch it here.

Here are some examples illustrating how we use the steps in our list. There are no new factoring problems in this section. The problems here are all similar to the problems you have seen before. What is different is that they are not all of the same type.

EXAMPLE 1

Factor $2x^5 - 8x^3$.

Solution First we check to see if the greatest common factor is other than 1. Since the greatest common factor is $2x^3$, we begin by factoring it out. Once we have done so, we notice that the binomial that remains is the difference of two squares, which we factor according to the formula $a^2 - b^2 = (a + b)(a - b)$.

$2x^5 - 8x^3 = 2x^3(x^2 - 4)$ **Factor out the greatest common factor, $2x^3$**

$= 2x^3(x + 2)(x - 2)$ **Factor the difference of two squares**

EXAMPLE 2

Factor $3x^4 - 18x^3 + 27x^2$.

Solution Step 1 is to factor out the greatest common factor $3x^2$. After we have done so, we notice that the trinomial that remains is a perfect square trinomial, which will factor as the square of a binomial.

$3x^4 - 18x^3 + 27x^2 = 3x^2(x^2 - 6x + 9)$ **Factor out $3x^2$**

$= 3x^2(x - 3)^2$ **$x^2 - 6x + 9$ is the square of $x - 3$**

PRACTICE PROBLEMS

1. Factor $3x^8 - 27x^6$.

2. Factor $4x^4 + 40x^3 + 100x^2$.

Answers

1. $3x^6(x + 3)(x - 3)$ 2. $4x^2(x + 5)^2$

3. Factor $y^4 + 36y^2$.

EXAMPLE 3

Factor $y^3 + 25y$.

Solution We begin by factoring out the y that is common to both terms. The binomial that remains after we have done so is the sum of two squares, which does not factor, so after the first step, we are finished.

$$y^3 + 25y = y(y^2 + 25)$$

4. Factor $6x^2 - x - 15$.

EXAMPLE 4

Factor $6a^2 - 11a + 4$.

Solution Here we have a trinomial that does not have a greatest common factor other than 1. Since it is not a perfect square trinomial, we factor it by trial and error. Without showing all the different possibilities, here is the answer.

$$6a^2 - 11a + 4 = (3a - 4)(2a - 1)$$

5. Factor $3x^5 - 81x^2$.

EXAMPLE 5

Factor $2x^4 + 16x$.

Solution This binomial has a greatest common factor of $2x$. The binomial that remains after the $2x$ has been factored from each term is the sum of two cubes, which we factor according to the formula $a^3 + b^3 = (a + b)(a^2 - ab + b^2)$.

$$2x^4 + 16x = 2x(x^3 + 8) \quad \text{Factor } 2x \text{ from each term}$$

$$= 2x(x + 2)(x^2 - 2x + 4) \quad \text{The sum of two cubes}$$

6. Factor $3a^2b^3 + 6a^2b^2 - 3a^2b$.

EXAMPLE 6

Factor $2ab^5 + 8ab^4 + 2ab^3$.

Solution The greatest common factor is $2ab^3$. We begin by factoring it from each term. After that we find that the trinomial that remains cannot be factored further.

$$2ab^5 + 8ab^4 + 2ab^3 = 2ab^3(b^2 + 4b + 1)$$

7. Factor $x^2 - 10x + 25 - b^2$. (*Hint:* Group the first three terms together.)

EXAMPLE 7

Factor $4x^2 - 6x + 2ax - 3a$.

Solution Our polynomial has four terms, so we factor by grouping.

$$4x^2 - 6x + 2ax - 3a = 2x(2x - 3) + a(2x - 3)$$

$$= (2x - 3)(2x + a)$$

Linking Objectives and Examples

Next to each **objective**, list the examples that are best described by that objective.

A	

Answers

3. $y^2(y^2 + 36)$ 4. $(3x - 5)(2x + 3)$
5. $3x^2(x - 3)(x^2 + 3x + 9)$
6. $3a^2b(b^2 + 2b - 1)$
7. $(x - 5 + b)(x - 5 - b)$

Getting Ready for Class

After reading through the preceding section, respond in your own words and in complete sentences.

1. How do you know when you've factored completely?
2. If a polynomial has four terms, what method of factoring should you try?
3. What is the first step in factoring a polynomial?
4. What do we call a polynomial that does not factor?

Problem Set 5.7

INTERMEDIATE Algebra $f(x)$ Now™

Factor each of the following polynomials completely. Once you are finished factoring, none of the factors you obtain should be factorable. Also, note that the even-numbered problems are not necessarily similar to the odd-numbered problems that precede them in this problem set.

1. $x^2 - 81$

2. $x^2 - 18x + 81$

3. $x^2 + 2x - 15$

4. $15x^2 + 13x - 6$

5. $x^2(x + 2) + 6x(x + 2) + 9(x + 2)$

6. $12x^2 - 11x + 2$

7. $x^2y^2 + 2y^2 + x^2 + 2$

8. $21y^2 - 25y - 4$

9. $2a^3b + 6a^2b + 2ab$

10. $6a^2 - ab - 15b^2$

11. $x^2 + x + 1$

12. $x^2y + 3y + 2x^2 + 6$

13. $12a^2 - 75$

14. $18a^2 - 50$

15. $9x^2 - 12xy + 4y^2$

16. $x^3 - x^2$

17. $25 - 10t + t^2$

18. $t^2 + 4t + 4 - y^2$

19. $4x^3 + 16xy^2$

20. $16x^2 + 49y^2$

21. $2y^3 + 20y^2 + 50y$

22. $x^2 + 5bx - 2ax - 10ab$

23. $a^7 + 8a^4b^3$

24. $5a^2 - 45b^2$

25. $t^2 + 6t + 9 - x^2$

26. $36 + 12t + t^2$

Factor completely.

27. $x^3 + 5x^2 - 9x - 45$

28. $x^3 + 5x^2 - 16x - 80$

29. $5a^2 + 10ab + 5b^2$

30. $3a^3b^2 + 15a^2b^2 + 3ab^2$

31. $x^2 + 49$

32. $16 - x^4$

33. $3x^2 + 15xy + 18y^2$

34. $3x^2 + 27xy + 54y^2$

35. $9a^2 + 2a + \frac{1}{9}$

36. $18 - 2a^2$

37. $x^2(x - 3) - 14x(x - 3) + 49(x - 3)$

38. $x^2 + 3ax - 2bx - 6ab$

39. $x^2 - 64$

40. $9x^2 - 4$

41. $8 - 14x - 15x^2$

42. $5x^4 + 14x^2 - 3$

43. $49a^7 - 9a^5$

44. $a^6 - b^6$

45. $r^2 - \frac{1}{25}$

46. $27 - r^3$

47. $49x^2 + 9y^2$

48. $12x^4 - 62x^3 + 70x^2$

49. $100x^2 - 100x - 600$

50. $100x^2 - 100x - 1,200$

51. $25a^3 + 20a^2 + 3a$

52. $16a^5 - 54a^2$

53. $3x^4 - 14x^2 - 5$

54. $8 - 2x - 15x^2$

55. $24a^5b - 3a^2b$

56. $18a^4b^2 - 24a^3b^3 + 8a^2b^4$

57. $64 - r^3$

58. $r^2 - \frac{1}{9}$

59. $20x^4 - 45x^2$

60. $16x^3 + 16x^2 + 3x$

61. $400t^2 - 900$

62. $900 - 400t^2$

63. $16x^5 - 44x^4 + 30x^3$

64. $16x^2 + 16x - 1$

65. $y^6 - 1$

66. $25y^7 - 16y^5$

67. $50 - 2a^2$

68. $4a^2 + 2a + \frac{1}{4}$

69. $12x^4y^2 + 36x^3y^3 + 27x^2y^4$

70. $16x^3y^2 - 4xy^2$

71. $x^2 - 4x + 4 - y^2$

72. $x^2 - 12x + 36 - b^2$

Getting Ready for the Next Section

Simplify.

73. $x^2 + (x + 1)^2$

74. $x^2 + (x + 3)^2$

75. $\dfrac{16t^2 - 64t + 48}{16}$

76. $\dfrac{100p^2 - 1{,}300p + 4{,}000}{100}$

Factor each of the following.

77. $x^2 - 2x - 24$

78. $x^2 - x - 6$

79. $2x^3 - 5x^2 - 3x$

80. $3x^3 - 5x^2 - 2x$

81. $x^3 + 2x^2 - 9x - 18$

82. $x^3 + 5x^2 - 4x - 20$

Solve.

83. $x - 6 = 0$

84. $x + 4 = 0$

85. $2x + 1 = 0$

86. $3x + 1 = 0$

Maintaining Your Skills

The following problems are taken from the book *Algebra for the Practical Man,* written by J. E. Thompson and published by D. Van Nostrand Company in 1931.

87. A man spent \$112.80 for 108 geese and ducks, each goose costing 14 dimes and each duck 6 dimes. How many of each did he buy?

88. If 15 pounds of tea and 10 pounds of coffee together cost \$15.50, while 25 pounds of tea and 13 pounds of coffee at the same prices cost \$24.55, find the price per pound of each.

89. A number of oranges at the rate of three for \$0.10 and apples at \$0.15 a dozen cost, together, \$6.80. Five times as many oranges and one-fourth as many apples at the same rates would have cost \$25.45. How many of each were bought?

90. An estate is divided among three persons: A, B, and C. A's share is three times that of B and B's share is twice that of C. If A receives \$9,000 more than C, how much does each receive?

5.8 Solving Equations by Factoring

In this section we will use our knowledge of factoring to solve equations. Most of the equations we will solve in this section are *quadratic equations.* Here is the definition of a quadratic equation.

DEFINITION

Any equation that can be written in the form

$$ax^2 + bx + c = 0$$

where a, b, and c are constants and a is not 0 ($a \neq 0$) is called a **quadratic equation.** The form $ax^2 + bx + c = 0$ is called **standard form** for quadratic equations.

Each of the following is a quadratic equation:

$$2x^2 = 5x + 3 \qquad 5x^2 = 75 \qquad 4x^2 - 3x + 2 = 0$$

Notation For a quadratic equation written in standard form, the first term ax^2 is called the *quadratic term;* the second term bx is the *linear term;* and the last term c is called the *constant term.*

In the past we have noticed that the number 0 is a special number. There is another property of 0 that is the key to solving quadratic equations. It is called the *zero-factor property.*

Zero-Factor Property

For all real numbers r and s,

$$r \cdot s = 0 \quad \text{if and only if} \quad r = 0 \quad \text{or} \quad s = 0 \quad \text{(or both)}$$

EXAMPLE 1 Solve $x^2 - 2x - 24 = 0$.

Solution We begin by factoring the left side as $(x - 6)(x + 4)$ and get

$$(x - 6)(x + 4) = 0$$

Now both $(x - 6)$ and $(x + 4)$ represent real numbers. We notice that their product is 0. By the zero-factor property, one or both of them must be 0:

$$x - 6 = 0 \quad \text{or} \quad x + 4 = 0$$

We have used factoring and the zero-factor property to rewrite our original second-degree equation as two first-degree equations connected by the word *or.* Completing the solution, we solve the two first-degree equations:

$$x - 6 = 0 \quad \text{or} \quad x + 4 = 0$$

$$x = 6 \quad \text{or} \quad x = -4$$

We check our solutions in the original equation as follows:

Check $x = 6$	Check $x = -4$
$6^2 - 2(6) - 24 \stackrel{?}{=} 0$	$(-4)^2 - 2(-4) - 24 \stackrel{?}{=} 0$
$36 - 12 - 24 \stackrel{?}{=} 0$	$16 + 8 - 24 \stackrel{?}{=} 0$
$0 = 0$	$0 = 0$

In both cases the result is a true statement, which means that both 6 and −4 are solutions to the original equation.

OBJECTIVES

A	Solve equations by factoring.
B	Apply the Blueprint for Problem Solving to solve application problems whose solutions involve quadratic equations.
C	Solve problems that contain formulas that are quadratic.

NOTE

The third equation is clearly a quadratic equation since it is in standard form. (Notice that a is 4, b is −3, and c is 2.) The first two equations are also quadratic because they could be put in the form $ax^2 + bx + c = 0$ by using the addition property of equality.

PRACTICE PROBLEMS

1. Solve $x^2 - x - 6 = 0$.

NOTE

What the zero-factor property says in words is that we can't multiply and get 0 without multiplying by 0; that is, if we multiply two numbers and get 0, then one or both of the original two numbers we multiplied must have been 0.

NOTE

We are placing a question mark over the equal sign because we don't know yet if the expression on the left will be equal to the expression on the right.

Answer

1. −2, 3

Although the next equation is not quadratic, the method we use is similar.

2. Solve $\frac{1}{2}x^3 = \frac{5}{6}x^2 + \frac{1}{3}x$.

EXAMPLE 2 Solve $\frac{1}{3}x^3 = \frac{5}{6}x^2 + \frac{1}{2}x$.

Solution We can simplify our work if we clear the equation of fractions. Multiplying both sides by the LCD, 6, we have

$$\mathbf{6} \cdot \frac{1}{3}x^3 = \mathbf{6} \cdot \frac{5}{6}x^2 + \mathbf{6} \cdot \frac{1}{2}x$$

$$2x^3 = 5x^2 + 3x$$

Next we add $-5x^2$ and $-3x$ to each side so that the right side will become 0.

$$2x^3 - 5x^2 - 3x = 0 \qquad \text{Standard form}$$

We factor the left side and then use the zero-factor property to set each factor to 0.

$$x(2x^2 - 5x - 3) = 0 \qquad \text{Factor out the greatest common factor}$$

$$x(2x + 1)(x - 3) = 0 \qquad \text{Continue factoring}$$

$$x = 0 \quad \text{or} \quad 2x + 1 = 0 \quad \text{or} \quad x - 3 = 0 \qquad \text{Zero-factor property}$$

Solving each of the resulting equations, we have

$$x = 0 \quad \text{or} \quad x = -\frac{1}{2} \quad \text{or} \quad x = 3$$

To generalize the preceding example, here are the steps used in solving a quadratic equation by factoring.

To Solve an Equation by Factoring

Step 1: Write the equation in standard form.

Step 2: Factor the left side.

Step 3: Use the zero-factor property to set each factor equal to 0.

Step 4: Solve the resulting linear equations.

3. Solve $100x^2 = 500x$.

EXAMPLE 3 Solve $100x^2 = 300x$.

Solution We begin by writing the equation in standard form and factoring:

$$100x^2 = 300x$$

$$100x^2 - 300x = 0 \qquad \text{Standard form}$$

$$100x(x - 3) = 0 \qquad \text{Factor}$$

Using the zero-factor property to set each factor to 0, we have

$$100x = 0 \quad \text{or} \quad x - 3 = 0$$

$$x = 0 \quad \text{or} \quad x = 3$$

The two solutions are 0 and 3.

4. Solve $(x + 1)(x + 2) = 12$.

EXAMPLE 4 Solve $(x - 2)(x + 1) = 4$.

Solution We begin by multiplying the two factors on the left side. (Notice that it would be incorrect to set each of the factors on the left side equal to 4. The fact that the product is 4 does not imply that either of the factors must be 4.)

Answers
2. $0, -\frac{1}{3}, 2$ **3.** 0, 5

$$(x - 2)(x + 1) = 4$$

$$x^2 - x - 2 = 4 \quad \textbf{Multiply the left side}$$

$$x^2 - x - 6 = 0 \quad \textbf{Standard form}$$

$$(x - 3)(x + 2) = 0 \quad \textbf{Factor}$$

$$x - 3 = 0 \quad \text{or} \quad x + 2 = 0 \quad \textbf{Zero-factor property}$$

$$x = 3 \quad \text{or} \quad x = -2$$

EXAMPLE 5

Solve for x: $x^3 + 2x^2 - 9x - 18 = 0$.

5. Solve $x^3 + 5x^2 - 4x - 20 = 0$.

Solution We start with factoring by grouping.

$$x^3 + 2x^2 - 9x - 18 = 0$$

$$x^2(x + 2) - 9(x + 2) = 0$$

$$(x + 2)(x^2 - 9) = 0$$

$$(x + 2)(x - 3)(x + 3) = 0 \quad \textbf{The difference of two squares}$$

$$x + 2 = 0 \quad \text{or} \quad x - 3 = 0 \quad \text{or} \quad x + 3 = 0 \quad \textbf{Set factors to 0}$$

$$x = -2 \quad \text{or} \quad x = 3 \quad \text{or} \quad x = -3$$

We have three solutions: −2, 3, and −3.

EXAMPLE 6

The sum of the squares of two consecutive integers is 25. Find the two integers.

6. One integer is 3 more than another. The sum of their squares is 29. Find the two integers.

Solution We apply the Blueprint for Problem Solving to solve this application problem. Remember, step 1 in the blueprint is done mentally.

Step 1: Read and list.
Known items: Two consecutive integers. If we add their squares, the result is 25.
Unknown items: The two integers

Step 2: Assign a variable and translate information.
Let x = the first integer; then $x + 1$ = the next consecutive integer.

Step 3: Reread and write an equation.
Since the sum of the squares of the two integers is 25, the equation that describes the situation is

$$x^2 + (x + 1)^2 = 25$$

Step 4: Solve the equation.

$$x^2 + (x + 1)^2 = 25$$

$$x^2 + (x^2 + 2x + 1) = 25$$

$$2x^2 + 2x - 24 = 0$$

$$x^2 + x - 12 = 0$$

$$(x + 4)(x - 3) = 0$$

$$x = -4 \quad \text{or} \quad x = 3$$

Step 5: Write the answer.
If $x = -4$, then $x + 1 = -3$. If $x = 3$, then $x + 1 = 4$. The two integers are −4 and −3, or the two integers are 3 and 4.

Step 6: Reread and check.
The two integers in each pair are consecutive integers, and the sum of the squares of either pair is 25.

Answers
4. −5, 2 **5.** −5, −2, 2
6. −5, −2, or 2, 5

Another application of quadratic equations involves the Pythagorean theorem, an important theorem from geometry. The theorem gives the relationship between the sides of any right triangle (a triangle with a 90-degree angle). We state it here without proof.

Pythagorean Theorem

In any right triangle, the square of the longest side (hypotenuse) is equal to the sum of the squares of the other two sides (legs).

$$c^2 = a^2 + b^2$$

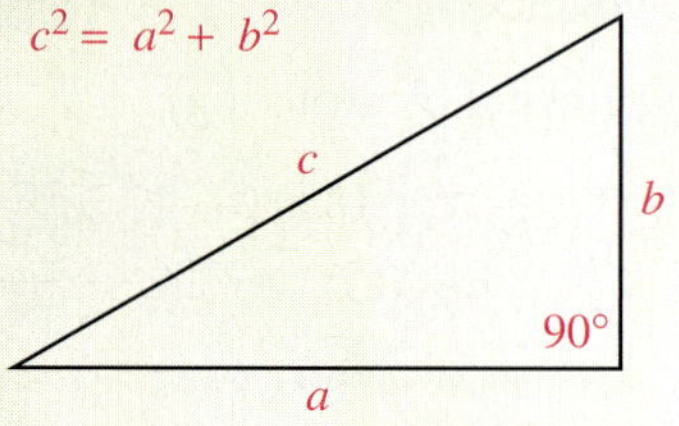

7. The longest side of a right triangle is 4 more than the shortest side. The third side is 2 more than the shortest side. Find the length of each side.

EXAMPLE 7

The lengths of the three sides of a right triangle are given by three consecutive integers. Find the lengths of the three sides.

Solution

Step 1: ***Read and list.***

Known items: A right triangle. The three sides are three consecutive integers.

Unknown items: The three sides

Step 2: ***Assign a variable and translate information.***

Let x = first integer (shortest side).

Then $x + 1$ = next consecutive integer

$x + 2$ = last consecutive integer (longest side)

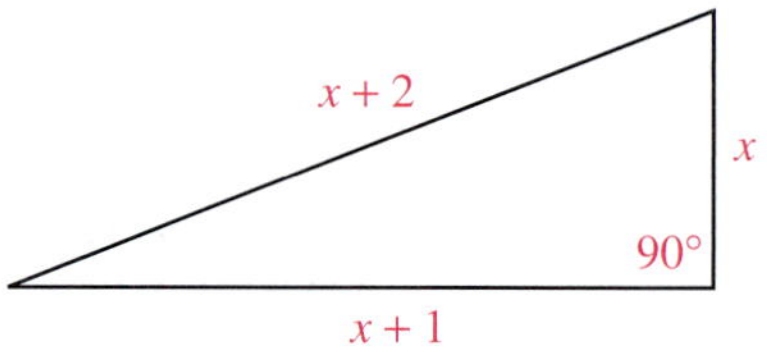

Step 3: ***Reread and write an equation.***

By the Pythagorean theorem, we have

$$(x + 2)^2 = (x + 1)^2 + x^2$$

Step 4: ***Solve the equation.***

$$x^2 + 4x + 4 = x^2 + 2x + 1 + x^2$$

$$x^2 - 2x - 3 = 0$$

$$(x - 3)(x + 1) = 0$$

$$x = 3 \quad \text{or} \quad x = -1$$

Step 5: ***Write the answer.***

Since x is the length of a side in a triangle, it must be a positive number. Therefore, $x = -1$ cannot be used.

The shortest side is 3. The other two sides are 4 and 5.

Step 6: ***Reread and check.***
The three sides are given by consecutive integers. The square of the longest side is equal to the sum of the squares of the two shorter sides. ■

Our next two examples involve formulas that are quadratic.

EXAMPLE 8 If an object is projected into the air with an initial vertical velocity of v feet/second, its height h, in feet, above the ground after t seconds will be given by

$$h = vt - 16t^2$$

Find t if $v = 64$ feet/second and $h = 48$ feet.

Solution Substituting $v = 64$ and $h = 48$ into the preceding formula, we have

$$48 = 64t - 16t^2$$

which is a quadratic equation. We write it in standard form and solve by factoring:

$$16t^2 - 64t + 48 = 0$$

$$t^2 - 4t + 3 = 0 \quad \text{Divide each side by 16}$$

$$(t - 1)(t - 3) = 0$$

$$t - 1 = 0 \quad \text{or} \quad t - 3 = 0$$

$$t = 1 \quad \text{or} \quad t = 3$$

Here is how we interpret our results: If an object is projected upward with an initial vertical velocity of 64 feet/second, it will be 48 feet above the ground after 1 second and after 3 seconds; that is, it passes 48 feet going up and also coming down. ■

8. Use the formula in Example 8 to find t if $v = 64$ feet/second and $h = 64$ feet.

EXAMPLE 9 A manufacturer of small portable radios knows that the number of radios she can sell each week is related to the price of the radios by the equation $x = 1{,}300 - 100p$, where x is the number of radios and p is the price per radio. What price should she charge for each radio if she wants the weekly revenue to be \$4,000?

Solution The formula for total revenue is $R = xp$. Since we want R in terms of p, we substitute $1{,}300 - 100p$ for x in the equation $R = xp$:

If $\quad R = xp$

and $\quad x = 1{,}300 - 100p$

then $\quad R = (1{,}300 - 100p)p$

We want to find p when R is 4,000. Substituting 4,000 for R in the formula gives us

$$4{,}000 = (1{,}300 - 100p)p$$

$$4{,}000 = 1{,}300p - 100p^2$$

9. Use the information in Example 9 to find the price that should be charged if the weekly revenue is to be \$3,600.

Answers
7. 6, 8, 10 8. 2 seconds

which is a quadratic equation. To write it in standard form, we add $100p^2$ and $-1{,}300p$ to each side, giving us

$$100p^2 - 1{,}300p + 4{,}000 = 0$$

$$p^2 - 13p + 40 = 0 \qquad \textbf{Divide each side by 100}$$

$$(p - 5)(p - 8) = 0$$

$$p - 5 = 0 \quad \text{or} \quad p - 8 = 0$$

$$p = 5 \quad \text{or} \quad p = 8$$

If she sells the radios for \$5 each or for \$8 each she will have a weekly revenue of \$4,000.

Linking Objectives and Examples

Next to each **objective**, list the examples that are best described by that objective.

A	
B	
C	

Getting Ready for Class

After reading through the preceding section, respond in your own words and in complete sentences.

1. Explain the Pythagorean theorem in words.
2. What is the first step in solving an equation by factoring?
3. Describe the zero-factor property in your own words.
4. Write an application problem for which the solution depends on solving the equation $x^2 + (x + 1)^2 = 313$.

Answer

9. \$4 or \$9

Problem Set 5.8

INTERMEDIATE Algebra $f(x)$ Now™

Solve each equation.

1. $x^2 - 5x - 6 = 0$

2. $x^2 + 5x - 6 = 0$

3. $x^3 - 5x^2 + 6x = 0$

4. $x^3 + 5x^2 + 6x = 0$

5. $3y^2 + 11y - 4 = 0$

6. $3y^2 - y - 4 = 0$

7. $60x^2 - 130x + 60 = 0$

8. $90x^2 + 60x - 80 = 0$

9. $\frac{1}{10}t^2 - \frac{5}{2} = 0$

10. $\frac{2}{7}t^2 - \frac{7}{2} = 0$

11. $100x^4 = 400x^3 + 2{,}100x^2$

12. $100x^4 = -400x^3 + 2{,}100x^2$

13. $\frac{1}{5}y^2 - 2 = -\frac{3}{10}y$

14. $\frac{1}{2}y^2 + \frac{5}{3} = \frac{17}{6}y$

15. $9x^2 - 12x = 0$

16. $4x^2 + 4x = 0$

17. $0.02r + 0.01 = 0.15r^2$

18. $0.02r - 0.01 = -0.08r^2$

19. $9a^3 = 16a$

20. $16a^3 = 25a$

21. $-100x = 10x^2$

22. $800x = 100x^2$

23. $(x + 6)(x - 2) = -7$

24. $(x - 7)(x + 5) = -20$

25. $(y - 4)(y + 1) = -6$

26. $(y - 6)(y + 1) = -12$

27. $(x + 1)^2 = 3x + 7$

28. $(x + 2)^2 = 9x$

29. $(2r + 3)(2r - 1) = -(3r + 1)$

30. $(3r + 2)(r - 1) = -(7r - 7)$

31. $x^3 + 3x^2 - 4x - 12 = 0$

32. $x^3 + 5x^2 - 4x - 20 = 0$

33. $x^3 + 2x^2 - 25x - 50 = 0$

34. $x^3 + 4x^2 - 9x - 36 = 0$

35. $2x^3 + 3x^2 - 8x - 12 = 0$

36. $3x^3 + 2x^2 - 27x - 18 = 0$

37. $4x^3 + 12x^2 - 9x - 27 = 0$

38. $9x^3 + 18x^2 - 4x - 8 = 0$

Problems 39–48 are problems you will see later in the book. Solve each equation.

39. $3x^2 + x = 10$

40. $y^2 + y - 20 = 2y$

41. $12(x + 3) + 12(x - 3) = 3(x^2 - 9)$

42. $8(x + 2) + 8(x - 2) = 3(x^2 - 4)$

43. $(y + 3)^2 + y^2 = 9$

44. $(2y + 4)^2 + y^2 = 4$

45. $(x + 3)^2 + 1^2 = 2$

46. $(x - 3)^2 + (-1)^2 = 10$

47. $(x + 2)(x) = 2^3$

48. $(x + 3)(x) = 2^2$

Applying the Concepts

49. Consecutive Integers The sum of the squares of two consecutive odd integers is 34. Find the two integers.

50. Consecutive Integers The sum of the squares of two consecutive even integers is 100. Find the two integers.

51. Consecutive Integers The square of the sum of two consecutive integers is 81. Find the two integers.

52. Consecutive Integers Find two consecutive even integers whose sum squared is 100.

53. Right Triangle A 25-foot ladder is leaning against a building. The base of the ladder is 7 feet from the side of the building. How high does the ladder reach along the side of the building?

54. Right Triangle Noreen wants to place her 13-foot ramp against the side of her house so that the top of the ramp rests on a ledge that is 5 feet above the ground. How far will the base of the ramp be from the house.

55. **Right Triangle** The lengths of the three sides of a right triangle are given by three consecutive even integers. Find the lengths of the three sides.

56. **Right Triangle** The longest side of a right triangle is 3 less than twice the shortest side. The third side measures 12 inches. Find the length of the shortest side.

57. **Geometry** The length of a rectangle is 2 feet more than 3 times the width. If the area is 16 square feet, find the width and the length.

58. **Geometry** The length of a rectangle is 4 yards more than twice the width. If the area is 70 square yards, find the width and the length.

59. **Geometry** The base of a triangle is 2 inches more than 4 times the height. If the area is 36 square inches, find the base and the height.

60. **Geometry** The height of a triangle is 4 feet less than twice the base. If the area is 48 square feet, find the base and the height.

61. **Projectile Motion** If an object is thrown straight up into the air with an initial velocity of 32 feet per second, then its height above the ground at any time t is given by the formula $h = 32t - 16t^2$. Find the times at which the object is on the ground by letting $h = 0$ in the equation and solving for t.

62. **Projectile Motion** An object is projected into the air with an initial velocity of 64 feet per second. Its height at any time t is given by the formula $h = 64t - 16t^2$. Find the times at which the object is on the ground.

The formula $h = vt - 16t^2$ gives the height h, in feet, of an object projected into the air with an initial vertical velocity v, in feet per second, after t seconds.

63. **Projectile Motion** If an object is projected upward with an initial velocity of 48 feet per second, at what times will it reach a height of 32 feet above the ground?

64. **Projectile Motion** If an object is projected upward into the air with an initial velocity of 80 feet per second, at what times will it reach a height of 64 feet above the ground?

65. **Projectile Motion** An object is projected into the air with a vertical velocity of 24 feet per second. At what times will the object be on the ground? (It is on the ground when h is 0.)

66. **Projectile Motion** An object is projected into the air with a vertical velocity of 20 feet per second. At what times will the object be on the ground?

67. **Height of a Bullet** A bullet is fired into the air with an initial upward velocity of 80 feet per second from the top of a building 96 feet high. The equation that gives the height of the bullet at any time t is $h = 96 + 80t - 16t^2$. At what times will the bullet be 192 feet in the air?

68. **Height of an Arrow** An arrow is shot into the air with an upward velocity of 48 feet per second from a hill 32 feet high. The equation that gives the height of the arrow at any time t is $h = 32 + 48t - 16t^2$. Find the times at which the arrow will be 64 feet above the ground.

69. Price and Revenue A company that manufactures typewriter ribbons knows that the number of ribbons x it can sell each week is related to the price per ribbon p by the equation $x = 1{,}200 - 100p$. At what price should it sell the ribbons if it wants the weekly revenue to be \$3,200? (*Remember:* The equation for revenue is $R = xp$.)

70. Price and Revenue A company manufactures diskettes for home computers. It knows from past experience that the number of diskettes x it can sell each day is related to the price per diskette p by the equation $x = 800 - 100p$. At what price should it sell its diskettes if it wants the daily revenue to be \$1,200?

71. Price and Revenue The relationship between the number of calculators x a company sells per day and the price of each calculator p is given by the equation $x = 1{,}700 - 100p$. At what price should the calculators be sold if the daily revenue is to be \$7,000?

72. Price and Revenue The relationship between the number of pencil sharpeners x a company can sell each week and the price of each sharpener p is given by the equation $x = 1{,}800 - 100p$. At what price should the sharpeners be sold if the weekly revenue is to be \$7,200?

Maintaining Your Skills

Solve each system.

73. $2x - 5y = -8$
$3x + y = 5$

74. $4x - 7y = -2$
$-5x + 6y = -3$

75. $\frac{1}{3}x - \frac{1}{6}y = 3$
$-\frac{1}{5}x + \frac{1}{4}y = 0$

76. $2x - 5y = 14$
$y = 3x + 8$

77. $2x - y + z = 9$
$x + y - 3z = -2$
$3x + y - z = 6$

78. Number Problem A number is 1 less than twice another. Their sum is 14. Find the two numbers.

79. Investing John invests twice as much money at 6% as he does at 5%. If his investments earn a total of \$680 in 1 year, how much does he have invested at each rate?

80. Speed of a Boat A boat can travel 20 miles downstream in 2 hours. The same boat can travel 18 miles upstream in 3 hours. What is the speed of the boat in still water, and what is the speed of the current?

Graph the solution set for each system.

81. $3x + 2y < 6$
$-2x + 3y < 6$

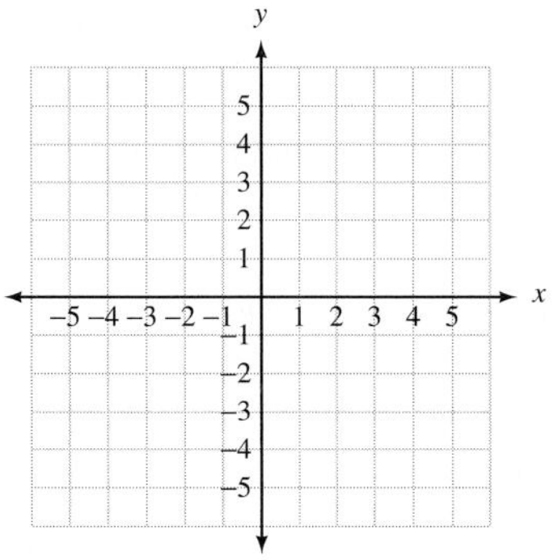

82. $y \le x + 3$
$y > x - 4$

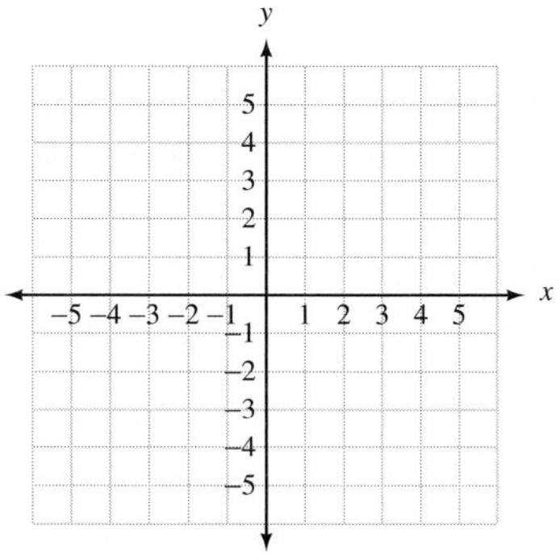

83. $x \le 4$
$y < 2$

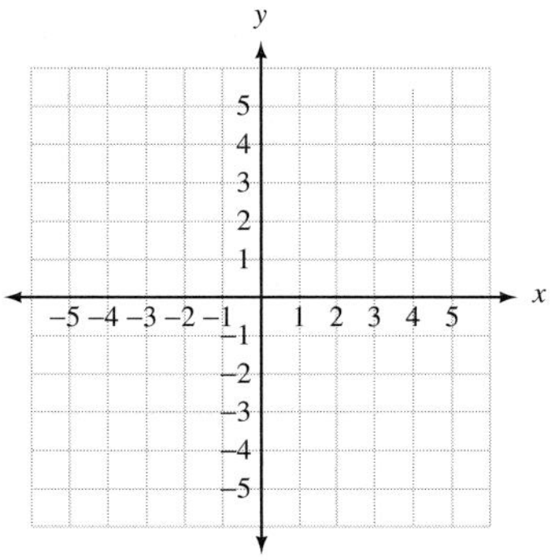

84. $2x + y < 4$
$x \ge 0$
$y \ge 0$

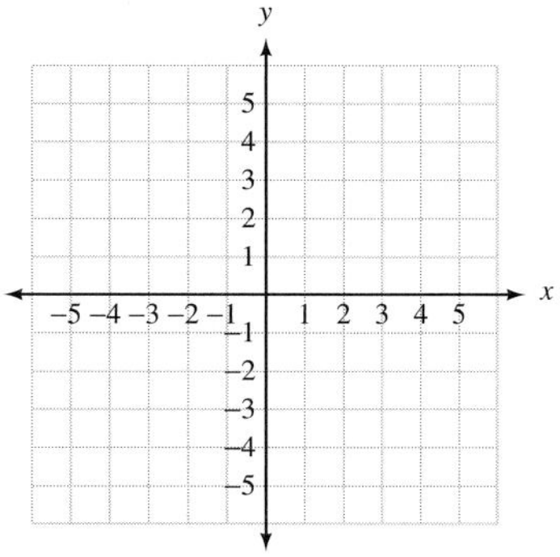

Chapter 5 Summary

Properties of Exponents [5.1]

If a and b represent real numbers and r and s represent integers, then

1. $a^r \cdot a^s = a^{r+s}$
2. $(a^r)^s = a^{r \cdot s}$
3. $(ab)^r = a^r \cdot b^r$
4. $a^{-r} = \dfrac{1}{a^r} \qquad (a \neq 0)$
5. $\left(\dfrac{a}{b}\right)^r = \dfrac{a^r}{b^r} \qquad (b \neq 0)$
6. $\dfrac{a^r}{a^s} = a^{r-s} \qquad (a \neq 0)$
7. $a^1 = a$
   $a^0 = 1 \qquad (a \neq 0)$

EXAMPLES

1. These expressions illustrate the properties of exponents.
   a. $x^2 \cdot x^3 = x^{2+3} = x^5$
   b. $(x^2)^3 = x^{2 \cdot 3} = x^6$
   c. $(3x)^2 = 3^2 \cdot x^2 = 9x^2$
   d. $2^{-3} = \frac{1}{2^3} = \frac{1}{8}$
   e. $\left(\frac{x}{5}\right)^2 = \frac{x^2}{5^2} = \frac{x^2}{25}$
   f. $\frac{x^7}{x^5} = x^{7-5} = x^2$
   g. $3^1 = 3$
      $3^0 = 1$

Scientific Notation [5.1]

A number is written in scientific notation when it is written as the product of a number between 1 and 10 and an integer power of 10; that is, when it has the form

$$n \times 10^r$$

where $1 \leq n < 10$ and $r =$ an integer.

2. $49{,}800{,}000 = 4.98 \times 10^7$
   $0.00462 = 4.62 \times 10^{-3}$

Addition of Polynomials [5.2]

To add two polynomials, simply combine the coefficients of similar terms.

3. $(3x^2 + 2x - 5) + (4x^2 - 7x + 2)$
   $= 7x^2 - 5x - 3$

Negative Signs Preceding Parentheses [5.2]

If there is a negative sign directly preceding the parentheses surrounding a polynomial, we may remove the parentheses and preceding negative sign by changing the sign of each term within the parentheses. (This procedure is actually just another application of the distributive property.)

4. $-(2x^2 - 8x - 9)$
   $= -2x^2 + 8x + 9$

Multiplication of Polynomials [5.3]

To multiply two polynomials, multiply each term in the first by each term in the second.

5. $(3x - 5)(x + 2)$
   $= 3x^2 + 6x - 5x - 10$
   $= 3x^2 + x - 10$

Special Products [5.3]

$$(a + b)^2 = a^2 + 2ab + b^2$$
$$(a - b)^2 = a^2 - 2ab + b^2$$
$$(a + b)(a - b) = a^2 - b^2$$

6. The following are examples of the three special products:

$$(x + 3)^2 = x^2 + 6x + 9$$
$$(5 - x)^2 = 25 - 10x + x^2$$
$$(x + 7)(x - 7) = x^2 - 49$$

Business Applications [5.2, 5.3, 5.4]

If a company manufacturers and sells x items at p dollars per item, then the revenue R is given by the formula

$$R = xp$$

If the total cost to manufacture all x items is C, then the profit obtained from selling all x items is

$$P = R - C$$

7. A company makes x items each week and sells them for p dollars each, according to the equation $p = 35 - 0.1x$. Then, the revenue is

$$R = x(35 - 0.1x) = 35x - 0.1x^2$$

If the total cost to make all x items is $C = 8x + 500$, then the profit gained by selling the x items is

$$P = 35x - 0.1x^2 - (8x + 500)$$
$$= -500 + 27x - 0.1x^2$$

Greatest Common Factor [5.4]

The greatest common factor of a polynomial is the largest monomial (the monomial with the largest coefficient and highest exponent) that divides each term of the polynomial. The first step in factoring a polynomial is to factor the greatest common factor (if it is other than 1) out of each term.

8. The greatest common factor of $10x^5 - 15x^4 + 30x^3$ is $5x^3$. Factoring it out of each term, we have

$$5x^3(2x^2 - 3x + 6)$$

Factoring Trinomials [5.5]

We factor a trinomial by writing it as the product of two binomials. (This refers to trinomials whose greatest common factor is 1.) Each factorable trinomial has a unique set of factors. Finding the factors is sometimes a matter of trial and error.

9. $x^2 + 5x + 6 = (x + 2)(x + 3)$

$x^2 - 5x + 6 = (x - 2)(x - 3)$

$x^2 + x - 6 = (x - 2)(x + 3)$

$x^2 - x - 6 = (x + 2)(x - 3)$

Special Factoring [5.6]

$a^2 + 2ab + b^2 = (a + b)^2$ **Perfect square trinomials**

$a^2 - 2ab + b^2 = (a - b)^2$

$a^2 - b^2 = (a - b)(a + b)$ **Difference of two squares**

$a^3 - b^3 = (a - b)(a^2 + ab + b^2)$ **Difference of two cubes**

$a^3 + b^3 = (a + b)(a^2 - ab + b^2)$ **Sum of two cubes**

10. Here are some binomials that have been factored this way.

$$x^2 + 6x + 9 = (x + 3)^2$$
$$x^2 - 6x + 9 = (x - 3)^2$$
$$x^2 - 9 = (x + 3)(x - 3)$$
$$x^3 - 27 = (x - 3)(x^2 + 3x + 9)$$
$$x^3 + 27 = (x + 3)(x^2 - 3x + 9)$$

To Factor Polynomials in General [5.7]

Step 1: If the polynomial has a greatest common factor other than 1, then factor out the greatest common factor.

Step 2: If the polynomial has two terms (it is a binomial), then see if it is the difference of two squares, or the sum or difference of two cubes, and then factor accordingly. Remember, if it is the sum of two squares it will not factor.

Step 3: If the polynomial has three terms (a trinomial), then it is either a perfect square trinomial, which will factor into the square of a binomial, or it is not a perfect square trinomial, in which case you use one of the methods developed in Section 5.5.

Step 4: If the polynomial has more than three terms, then try to factor it by grouping.

Step 5: As a final check, see if any of the factors you have written can be factored further. If you have overlooked a common factor, you can catch it here.

11. Factor completely.

a. $3x^3 - 6x^2 = 3x^2(x - 2)$

b. $x^2 - 9 = (x + 3)(x - 3)$
$x^3 - 8 = (x - 2)(x^2 + 2x + 4)$
$x^3 + 27 = (x + 3)(x^2 - 3x + 9)$

c. $x^2 - 6x + 9 = (x - 3)^2$
$6x^2 - 7x - 5 = (2x + 1)(3x - 5)$

d. $x^2 + ax + bx + ab$
$= x(x + a) + b(x + a)$
$= (x + a)(x + b)$

To Solve an Equation by Factoring [5.8]

Step 1: Write the equation in standard form.

Step 2: Factor the left side.

Step 3: Use the zero-factor property to set each factor equal to zero.

Step 4: Solve the resulting linear equations.

12. Solve $x^2 - 5x = -6$.

$x^2 - 5x + 6 = 0$

$(x - 3)(x - 2) = 0$

$x - 3 = 0$ or $x - 2 = 0$

$x = 3$ or $x = 2$

COMMON MISTAKES

When we subtract one polynomial from another, it is common to forget to add the opposite of each term in the second polynomial. For example

$$(6x - 5) - (3x + 4) = 6x - 5 - 3x + 4 \quad \textbf{Mistake}$$

This mistake occurs if the negative sign outside the second set of parentheses is not distributed over all terms inside the parentheses. To avoid this mistake, remember: The opposite of a sum is the sum of the opposites, or,

$$-(3x + 4) = -3x + (-4)$$

Chapter 5 Review/Test

The problems below form a comprehensive review of the material in this chapter. They can be used to study for exams. If you would like to take a practice test on this chapter, you can use the odd-numbered problems. Give yourself an hour and work as many of the odd-numbered problems as possible. When you are finished, or when an hour has passed, check your answers with the answers in the back of the book. You can use the even-numbered problems for a second practice test.

Simplify each of the following. [5.1]

1. $x^3 \cdot x^7$

2. $(5x^3)^2$

3. $(2x^3y)^2(-2x^4y^2)^3$

Write with positive exponents, and then simplify. [5.1]

4. 2^{-3}

5. $(\frac{2}{3})^{-2}$

6. $2^{-2} + 4^{-1}$

Write in scientific notation. [5.1]

7. 34,500,000

8. 0.00357

Write in expanded form. [5.1]

9. 4.45×10^4

10. 4.45×10^{-4}

Simplify each expression. All answers should contain positive exponents only. (Assume all variables are nonnegative.) [5.1]

11. $\dfrac{a^{-4}}{a^5}$

12. $\dfrac{(4x^2)(-3x^3)^2}{(12x^{-2})^2}$

13. $\dfrac{x^n x^{3n}}{x^{4n-2}}$

Simplify each expression as much as possible. Write all answers in scientific notation. [5.1]

14. $(2 \times 10^3)(4 \times 10^{-5})$

15. $\dfrac{(600{,}000)(0.000008)}{(4{,}000)(3{,}000{,}000)}$

Simplify by combining similar terms. [5.2]

16. $(6x^2 - 3x + 2) - (4x^2 + 2x - 5)$

17. $(x^3 - x) - (x^2 + x) + (x^3 - 3) - (x^2 + 1)$

18. Subtract $2x^2 - 3x + 1$ from $3x^2 - 5x - 2$.

19. Simplify $-3[2x - 4(3x + 1)]$.

20. Find the value of $2x^2 - 3x + 1$ when x is -2.

Multiply. [5.3]

21. $3x(4x^2 - 2x + 1)$

22. $2a^2b^3(a^2 + 2ab + b^2)$

23. $(6 - y)(3 - y)$

24. $(2x^2 - 1)(3x^2 + 4)$

25. $2t(t + 1)(t - 3)$

26. $(x + 3)(x^2 - 3x + 9)$

27. $(2x - 3)(4x^2 + 6x + 9)$

28. $(a^2 - 2)^2$

29. $(3x + 5)^2$

30. $(4x - 3y)^2$

31. $\left(x - \dfrac{1}{3}\right)\left(x + \dfrac{1}{3}\right)$

32. $(2a + b)(2a - b)$

33. $(x - 1)^3$

34. $(x^m + 2)(x^m - 2)$

Factor out the greatest common factor. [5.4]

35. $6x^4y - 9xy^4 + 18x^3y^3$

36. $4x^2(x + y)^2 - 8y^2(x + y)^2$

Factor by grouping. [5.4, 5.6]

37. $8x^2 + 10 - 4x^2y - 5y$

38. $x^3 + 8b^2 - x^3y^2 - 8y^2b^2$

Factor completely. [5.4, 5.5]

39. $x^2 - 5x + 6$

40. $2x^3 + 4x^2 - 30x$

41. $20a^2 - 41ab + 20b^2$

42. $6x^4 - 11x^3 - 10x^2$

43. $24x^2y - 6xy - 45y$

Factor completely. [5.6]

44. $x^4 - 16$

45. $3a^4 + 18a^2 + 27$

46. $a^3 - 8$

47. $5x^3 + 30x^2y + 45xy^2$

48. $3a^3b - 27ab^3$

49. $x^2 - 10x + 25 - y^2$

50. $36 - 25a^2$

51. $x^3 + 4x^2 - 9x - 36$

Solve each equation. [5.8]

52. $x^2 + 5x + 6 = 0$

53. $\frac{5}{6}y^2 = \frac{1}{4}y + \frac{1}{3}$

54. $9x^2 - 25 = 0$

55. $5x^2 = -10x$

56. $(x + 2)(x - 5) = 8$

57. $x^3 + 4x^2 - 9x - 36 = 0$

Solve each application. In each case be sure to show the equation used. [5.8]

58. Consecutive Numbers The product of two consecutive even integers is 80. Find the two integers.

59. Consecutive Numbers The sum of the squares of two consecutive integers is 41. Find the two integers.

60. Geometry The lengths of the three sides of a right triangle are given by three consecutive integers. Find the three sides.

61. Geometry The lengths of three sides of a right triangle are given by three consecutive even integers. Find the three sides.

Chapter 5 Projects

Exponents and Polynomials

DISCOVERING PASCAL'S TRIANGLE

Group Project

Number of People 3

Time Needed 20 minutes

Equipment Paper and pencils

Background The triangular array of numbers shown here is known as Pascal's triangle, after the French philosopher Blaise Pascal (1623–1662).

```
          1
        1   1
      1   2   1
    1   3   3   1
  1   4   6   4   1
1   5  10  10   5   1
```

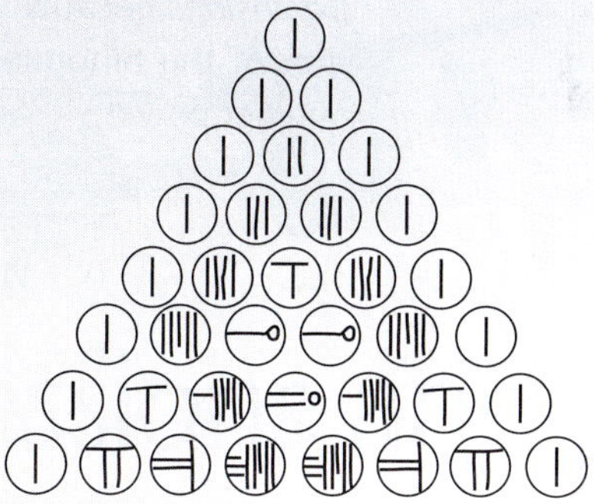

Pascal's triangle in Japanese (1781)

Procedure Look at Pascal's triangle and discover how the numbers in each row of the triangle are obtained from the numbers in the row above it.

1. Once you have discovered how to extend the triangle, write the next two rows.
2. Pascal's triangle can be linked to the Fibonacci sequence by rewriting Pascal's triangle so that the 1s on the left side of the triangle line up under one another and the other columns are equally spaced to the right of the first column. Rewrite Pascal's triangle as indicated and then look along the diagonals of the new array until you discover how the Fibonacci sequence can be obtained from it.
3. The diagram above shows Pascal's triangle as written in Japanese in 1781. Use your knowledge of Pascal's triangle to translate the numbers written in Japanese into our number system. Then write down the Japanese numbers from 1 to 20.

Linking Objectives and Examples

Next to each **objective**, list the examples that are best described by that objective.

A	
B	
C	

Getting Ready for Class

After reading through the preceding section, respond in your own words and in complete sentences.

1. What is a rational expression?
2. Explain how to determine if a rational expression is in "lowest terms."
3. When is a rational expression undefined?
4. Explain the process we use to reduce a rational expression or a fraction to lowest terms.

Problem Set 6.1

INTERMEDIATE Algebra $f(x)$ Now™

1. If $g(x) = \frac{x+3}{x-1}$, find $g(0)$, $g(-3)$, $g(3)$, $g(-1)$, and $g(1)$, if possible.

2. If $g(x) = \frac{x-2}{x-1}$, find $g(0)$, $g(-2)$, $g(2)$, $g(-1)$, and $g(1)$, if possible.

3. If $h(t) = \frac{t-3}{t+1}$, find $h(0)$, $h(-3)$, $h(3)$, $h(-1)$, and $h(1)$, if possible.

4. If $h(t) = \frac{t-2}{t+1}$, find $h(0)$, $h(-2)$, $h(2)$, $h(-1)$, and $h(1)$, if possible.

State the domain for each rational function.

5. $f(x) = \frac{x-3}{x-1}$

6. $g(x) = \frac{x^2-4}{x-2}$

7. $h(t) = \frac{t-4}{t^2-16}$

8. $h(t) = \frac{t-5}{t^2-25}$

Reduce each rational expression to lowest terms.

9. $\frac{x^2-16}{6x+24}$

10. $\frac{5x+25}{x^2-25}$

11. $\frac{12x-9y}{3x^2+3xy}$

12. $\frac{x^3-xy^2}{4x+4y}$

13. $\frac{a^4-81}{a-3}$

14. $\frac{a+4}{a^2-16}$

15. $\frac{a^2-4a-12}{a^2+8a+12}$

16. $\frac{a^2-7a+12}{a^2-9a+20}$

17. $\frac{20y^2-45}{10y^2-5y-15}$

18. $\frac{54y^2-6}{18y^2-60y+18}$

19. $\frac{a^3+b^3}{a^2-b^2}$

20. $\frac{a^2-b^2}{a^3-b^3}$

Reduce to lowest terms.

21. $\dfrac{8x^4 - 8x}{4x^4 + 4x^3 + 4x^2}$

22. $\dfrac{6x^5 - 48x^2}{12x^3 + 24x^2 + 48x}$

23. $\dfrac{6x^2 + 7xy - 3y^2}{6x^2 + xy - y^2}$

24. $\dfrac{4x^2 - y^2}{4x^2 - 8xy - 5y^2}$

25. $\dfrac{ax + 2x + 3a + 6}{ay + 2y - 4a - 8}$

26. $\dfrac{ax - x - 5a + 5}{ax + x - 5a - 5}$

27. $\dfrac{x^2 + bx - 3x - 3b}{x^2 - 2bx - 3x + 6b}$

28. $\dfrac{x^2 - 3ax - 2x + 6a}{x^2 - 3ax + 2x - 6a}$

29. $\dfrac{x^3 + 3x^2 - 4x - 12}{x^2 + x - 6}$

30. $\dfrac{x^3 + 5x^2 - 4x - 20}{x^2 + 7x + 10}$

31. $\dfrac{4x^4 - 25}{6x^3 - 4x^2 + 15x - 10}$

32. $\dfrac{16x^4 - 49}{8x^3 - 12x^2 + 14x - 21}$

Refer to Examples 5 and 6 in this section, and reduce the following to lowest terms.

33. $\dfrac{x - 4}{4 - x}$

34. $\dfrac{6 - x}{x - 6}$

35. $\dfrac{y^2 - 36}{6 - y}$

36. $\dfrac{1 - y}{y^2 - 1}$

37. $\dfrac{1 - 9a^2}{9a^2 - 6a + 1}$

38. $\dfrac{1 - a^2}{a^2 - 2a + 1}$

Simplify each expression.

39. $\dfrac{(3x - 5) - (3a - 5)}{x - a}$

40. $\dfrac{(2x + 3) - (2a + 3)}{x - a}$

41. $\dfrac{(x^2 - 4) - (a^2 - 4)}{x - a}$

42. $\dfrac{(x^2 - 1) - (a^2 - 1)}{x - a}$

Let $f(x) = \dfrac{x^2 - 4}{x - 2}$ and $g(x) = x + 2$, and evaluate the following expressions, if possible.

43. $f(0)$ and $g(0)$

44. $f(1)$ and $g(1)$

45. $f(2)$ and $g(2)$

46. $f(3)$ and $g(3)$

Let $f(x) = \dfrac{x^2 - 1}{x - 1}$ and $g(x) = x + 1$, and evaluate the following expressions, if possible.

47. $f(0)$ and $g(0)$

48. $f(1)$ and $g(1)$

49. $f(2)$ and $g(2)$

50. $f(-1)$ and $g(-1)$

51. Graph the equation $y = \dfrac{x^2 - 4}{x - 2}$. Then explain how this graph is different from the graph of $y = x + 2$.

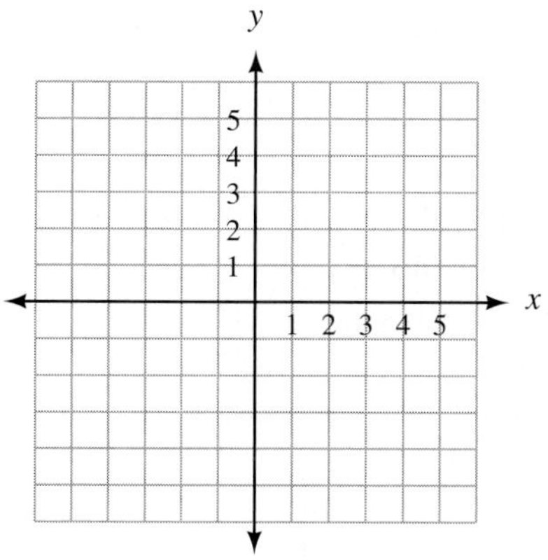

52. Graph the equation $y = \dfrac{x^2 - 1}{x - 1}$. Then explain how this graph is different from the graph of $y = x + 1$.

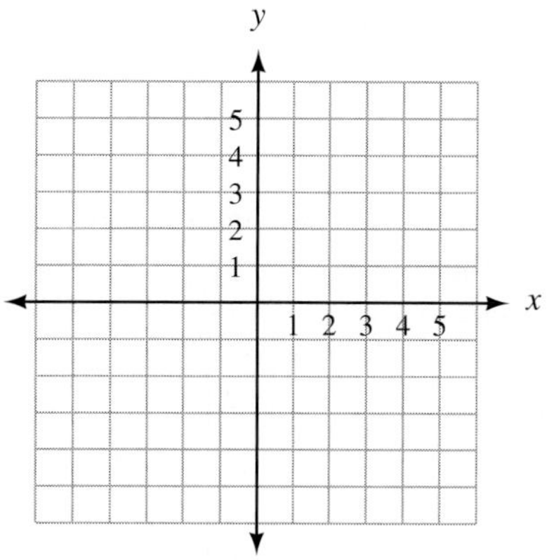

Applying the Concepts

53. Diet The following rational function is the one we mentioned in the introduction to this chapter. The quantity $W(x)$ is the weight (in pounds) of the person after x weeks of dieting. Use the function to fill in the table. Then compare your results with the graph in the chapter introduction.

$$W(x) = \frac{80(2x + 15)}{x + 6}$$

WEEKS x	WEIGHT (POUNDS) $W(x)$
0	
1	
4	
12	
24	

54. Drag Racing The following rational function gives the speed $V(x)$, in miles per hour, of a dragster at each second x during a quarter-mile race. Use the function to fill in the table.

$$V(x) = \frac{340x}{x + 3}$$

TIME (SEC) x	SPEED (MI/HR) $V(x)$
0	
1	
2	
3	
4	
5	
6	

Average Speed For Problems 55 and 56, use 3.14 as an approximation for π. Round answers to the nearest tenth.

55. A person riding a Ferris wheel with a diameter of 65 feet travels once around the wheel in 30 seconds. What is the average speed of the rider in feet per second?

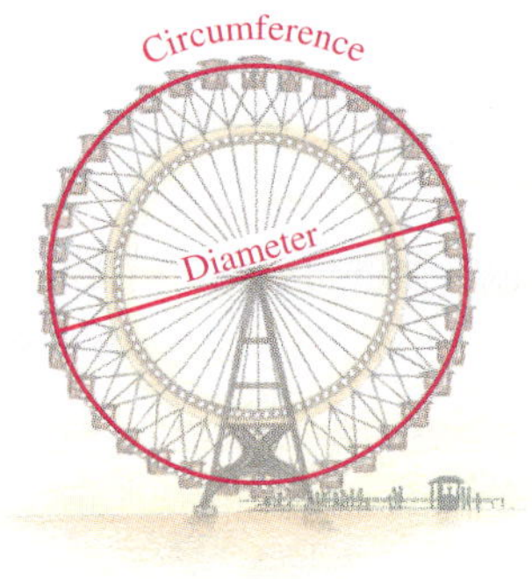

56. A person riding a Ferris wheel with a diameter of 102 feet travels once around the wheel in 3.5 minutes. What is the average speed of the rider in feet per minute?

57. Average Speed The Ferris wheel in Problem 55 has a circumference of 204 feet (to the nearest foot). If a ride on the wheel takes from 20 to 50 seconds, then the relationship between the average speed of a rider and the amount of time it takes to complete one revolution is given by the function

$$r(t) = \frac{204}{t} \qquad 20 \le t \le 50$$

where $r(t)$ is in feet per second and t is in seconds.

a. State the domain for this function.

b. Graph this function.

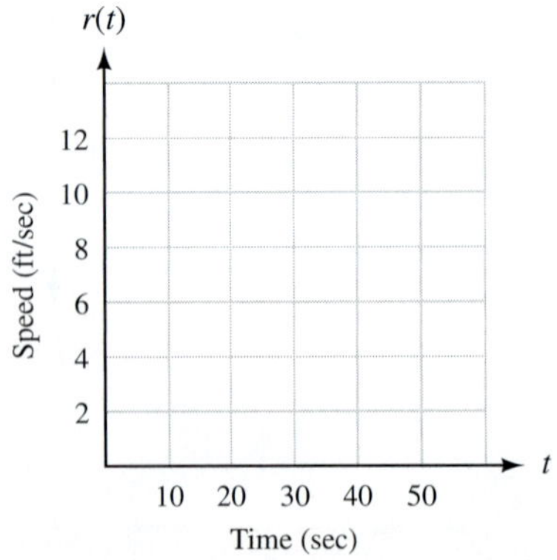

58. Average Speed The Ferris wheel in Problem 56 has a circumference of 320 feet (to the nearest foot). If a ride on the wheel takes from 3 to 5 minutes, then the relationship between the average speed of a rider and the amount of time it takes to complete one revolution is given by the function

$$r(t) = \frac{320}{t} \qquad 3 \le t \le 5$$

where $r(t)$ is in feet per minute and t is in minutes.

a. State the domain for this function.

b. Graph this function.

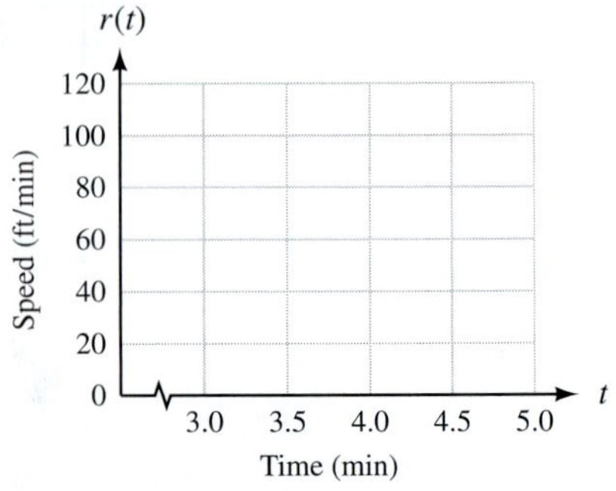

59. Intensity of Light The relationship between the intensity of light that falls on a surface from a 100-watt light bulb and the distance from that surface is given by the rational function

$$I(d) = \frac{120}{d^2} \qquad \text{for } 1 \le d \le 6$$

where $I(d)$ is the intensity of light (in lumens per square foot) and d is the distance (in feet) from the light bulb to the surface.

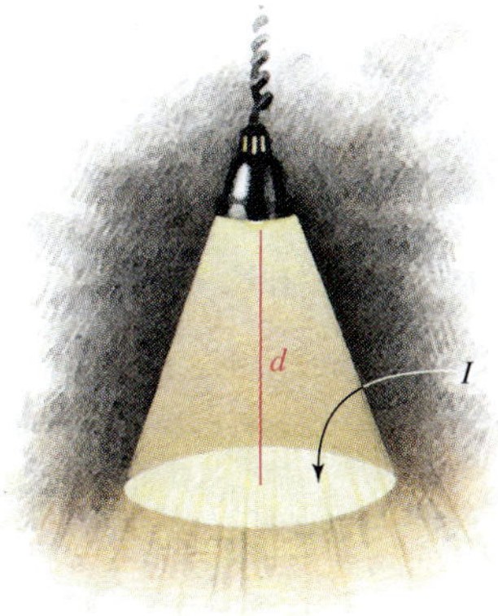

a. State the domain for this function.

b. Graph this function.

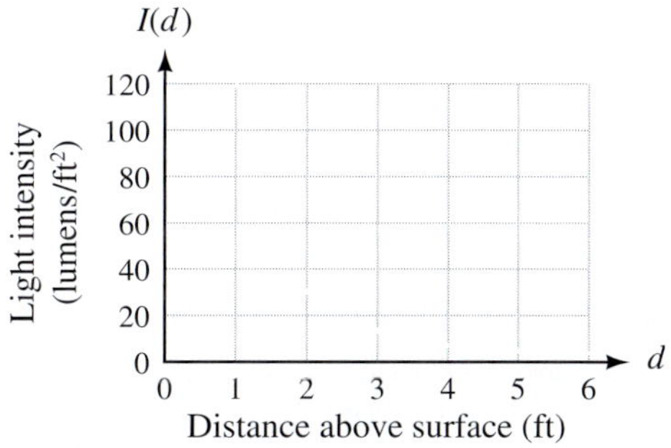

60. Average Speed If it takes Maria t minutes to run a mile, then her average speed $s(t)$ is given by the rational function

$$s(t) = \frac{60}{t} \qquad \text{for } 6 \le t \le 12$$

where $s(t)$ is in miles per hour and t is in minutes.

a. State the domain for this function.

b. Graph this function.

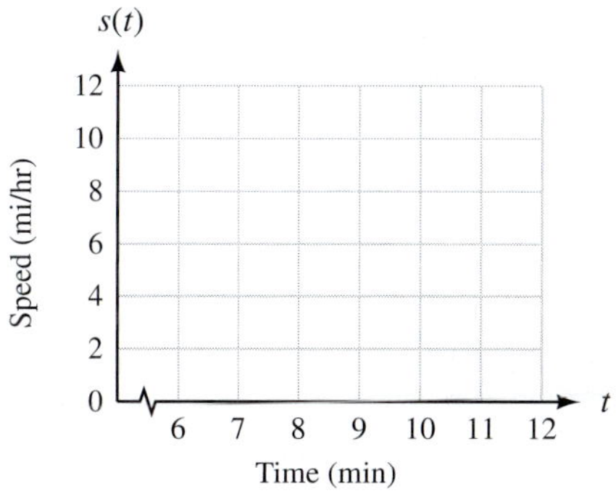

Average Speed The abbreviation "rpm" stands for revolutions per minute. If a point on a circle rotates at 300 rpm, then it rotates through one complete revolution 300 times every minute. The length of time it takes to rotate once around the circle is $\frac{1}{300}$ minute. Use 3.14 as an approximation for π.

61. A $3\frac{1}{2}$-inch diskette, when placed in the disk drive of a computer, rotates at 300 rpm (1 revolution takes $\frac{1}{300}$ minute). Find the average speed of a point 2 inches from the center of the diskette. Then find the average speed of a point 1.5 inches from the center of the diskette.

62. A 5-inch fixed disk in a computer rotates at 3,600 rpm. Find the average speed of a point 2 inches from the center of the disk. Then find the average speed of a point 1.5 inches from the center.

Getting Ready for the Next Section

Divide.

63. $\dfrac{10x^5}{5x^2}$

64. $\dfrac{-15x^4}{5x^2}$

65. $\dfrac{4x^4y^3}{-2x^2y}$

66. $\dfrac{10a^4b^2}{4a^2b^2}$

67. $4{,}628 \div 25$

68. $7{,}546 \div 35$

Multiply.

69. $2x^2(2x - 4)$

70. $3x^2(x - 2)$

71. $(2x - 4)(2x^2 + 4x + 5)$

72. $(x - 2)(3x^2 + 6x + 15)$

Subtract.

73. $(2x^2 - 7x + 9) - (2x^2 - 4x)$

74. $(x^2 - 6xy - 7y^2) - (x^2 + xy)$

Factor.

75. $x^2 - a^2$

76. $x^2 - 1$

77. $x^2 - 6xy - 7y^2$

78. $2x^2 - 5xy + 3y^2$

Maintaining Your Skills

Subtract as indicated.

79. Subtract $x^2 + 2x + 1$ from $4x^2 - 5x + 5$.

80. Subtract $3x^2 - 5x + 2$ from $7x^2 + 6x + 4$.

81. Subtract $10x - 20$ from $10x - 11$.

82. Subtract $-6x - 18$ from $-6x + 5$.

83. Subtract $4x^3 - 8x^2$ from $4x^3$.

84. Subtract $2x^2 + 6x$ from $2x^2$.

Extending the Concepts

85. The graphs of two rational functions are given in Figures 1 and 2. Use the graphs to find the following.
a. $f(2)$ **b.** $f(-1)$ **c.** $f(0)$ **d.** $g(3)$ **e.** $g(6)$ **f.** $g(-1)$ **g.** $f(g(6))$ **h.** $g(f(-2))$

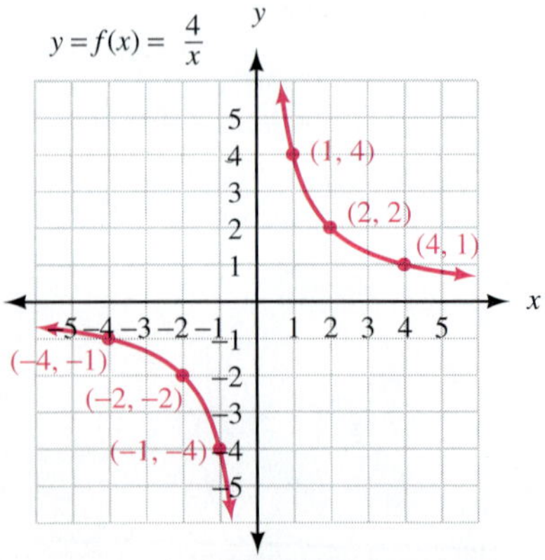

FIGURE 1

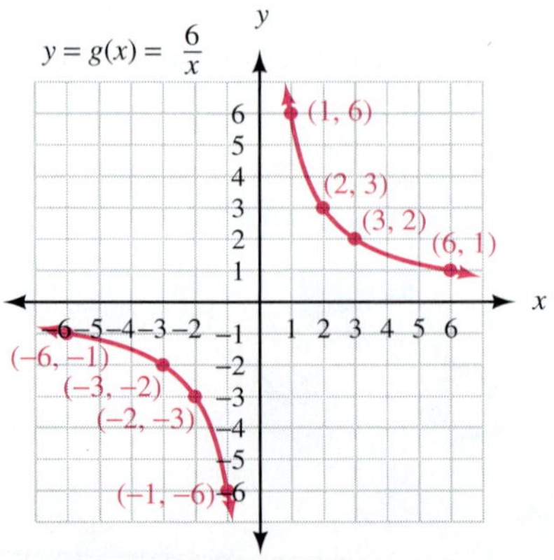

FIGURE 2

6.2 Division of Polynomials

First Bank of San Luis Obispo charges \$2.00 per month and \$0.15 per check for a regular checking account. So, if you write x checks in one month, the total monthly cost of the checking account will be $C(x) = 2.00 + 0.15x$. From this formula, we see that the more checks we write in a month, the more we pay for the account. But it is also true that the more checks we write in a month, the lower the cost per check. To find the cost per check, we use the average cost function. To find the average cost function, we divide the total cost by the number of checks written.

$$\text{Average Cost} = \overline{C}(x) = \frac{C(x)}{x} = \frac{2.00 + 0.15x}{x}$$

This last expression gives us the average cost per check for each of the x checks written. To work with this last expression, we need to know something about division with polynomials, and that is what we will cover in this section.

We begin this section by considering division of a polynomial by a monomial. This is the simplest kind of polynomial division. The rest of the section is devoted to division of a polynomial by a polynomial. This kind of division is similar to long division with whole numbers.

OBJECTIVES	
A	Divide a polynomial by a monomial.
B	Divide a polynomial by a polynomial.

Dividing a Polynomial by a Monomial

To divide a polynomial by a monomial, we use the definition of division and apply the distributive property. The following example illustrates the procedure.

EXAMPLE 1 Divide $\dfrac{10x^5 - 15x^4 + 20x^3}{5x^2}$.

Solution

$$= (10x^5 - 15x^4 + 20x^3) \cdot \frac{1}{5x^2}$$ **Dividing by $5x^2$ is the same as multiplying by $\frac{1}{5x^2}$**

$$= 10x^5 \cdot \frac{1}{5x^2} - 15x^4 \cdot \frac{1}{5x^2} + 20x^3 \cdot \frac{1}{5x^2}$$ **Distributive property**

$$= \frac{10x^5}{5x^2} - \frac{15x^4}{5x^2} + \frac{20x^3}{5x^2}$$ **Multiplying by $\frac{1}{5x^2}$ is the same as dividing by $5x^2$**

$$= 2x^3 - 3x^2 + 4x$$ **Divide coefficients, subtract exponents**

Notice that division of a polynomial by a monomial is accomplished by dividing each term of the polynomial by the monomial. The first two steps are usually not shown in a problem like this. They are part of Example 1 to justify distributing $5x^2$ under all three terms of the polynomial $10x^5 - 15x^4 + 20x^3$.

Here are some more examples of this kind of division.

PRACTICE PROBLEMS

1. Divide $\dfrac{12x^4 - 18x^3 + 24x^2}{6x}$.

Answer

1. $2x^3 - 3x^2 + 4x$

Divide.

2. $\dfrac{27x^4y^7 - 81x^5y^3}{-9x^3y^2}$

3. $\dfrac{12a^5 + 8a^4 + 16a^3 + 4a^2}{8a^4}$

EXAMPLES Divide. Write all results with positive exponents.

2. $$\frac{8x^3y^5 - 16x^2y^2 + 4x^4y^3}{-2x^2y} = \frac{8x^3y^5}{-2x^2y} + \frac{-16x^2y^2}{-2x^2y} + \frac{4x^4y^3}{-2x^2y}$$
$$= -4xy^4 + 8y - 2x^2y^2$$

3. $$\frac{10a^4b^2 + 8ab^3 - 12a^3b + 6ab}{4a^2b^2} = \frac{10a^4b^2}{4a^2b^2} + \frac{8ab^3}{4a^2b^2} - \frac{12a^3b}{4a^2b^2} + \frac{6ab}{4a^2b^2}$$
$$= \frac{5a^2}{2} + \frac{2b}{a} - \frac{3a}{b} + \frac{3}{2ab}$$

Notice in Example 3 that the result is not a polynomial because of the last three terms. If we were to write each as a product, some of the variables would have negative exponents. For example, the second term would be

$$\frac{2b}{a} = 2a^{-1}b$$

The divisor in each of the previous examples was a monomial. We now want to turn our attention to division of polynomials in which the divisor has two or more terms.

Dividing a Polynomial by a Polynomial

4. Divide $\dfrac{2x^2 - 5xy + 3y^2}{x - y}$.

EXAMPLE 4 Divide $\dfrac{x^2 - 6xy - 7y^2}{x + y}$.

Solution In this case, we can factor the numerator and perform our division by simply dividing out common factors, just like we did in the previous section:

$$\frac{x^2 - 6xy - 7y^2}{x + y} = \frac{\cancel{(x + y)}(x - 7y)}{\cancel{x + y}}$$
$$= x - 7y$$

Difference Quotients The diagram in Figure 1 is an important diagram from calculus. Although it may look complicated, the point of it is simple: The slope of the line passing through the points P and Q is given by the formula

$$\text{Slope of line through } PQ = m = \frac{f(x) - f(a)}{x - a}$$

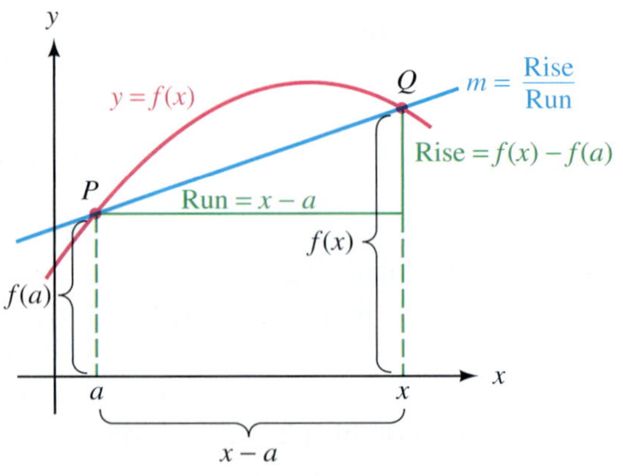

FIGURE 1

The expression $\dfrac{f(x) - f(a)}{x - a}$ is called a difference quotient. When $f(x)$ is a polynomial, the difference quotient will be a rational expression.

Answers

2. $-3xy^5 + 9x^2y$
3. $\frac{3a}{2} + 1 + \frac{2}{a} + \frac{1}{2a^2}$
4. $2x - 3y$

EXAMPLE 5 If $f(x) = 3x - 5$, find $\frac{f(x) - f(a)}{x - a}$.

Solution

$$\frac{f(x) - f(a)}{x - a} = \frac{(3x - 5) - (3a - 5)}{x - a}$$

$$= \frac{3x - 5 - 3a + 5}{x - a}$$

$$= \frac{3x - 3a}{x - a}$$

$$= \frac{3(x - a)}{x - a}$$

$$= 3$$

5. If $f(x) = 2x + 3$, find $\frac{f(x) - f(a)}{x - a}$.

EXAMPLE 6 If $f(x) = x^2 - 4$, find $\frac{f(x) - f(a)}{x - a}$ and simplify.

Solution Because $f(x) = x^2 - 4$ and $f(a) = a^2 - 4$, we have

$$\frac{f(x) - f(a)}{x - a} = \frac{(x^2 - 4) - (a^2 - 4)}{x - a}$$

$$= \frac{x^2 - 4 - a^2 + 4}{x - a}$$

$$= \frac{x^2 - a^2}{x - a}$$

$$= \frac{(x + a)\cancel{(x - a)}}{\cancel{x - a}}$$ **Factor and divide out common factor**

$$= x + a$$

6. If $f(x) = x^2 - 1$, find $\frac{f(x) - f(a)}{x - a}$.

Long Division

For the type of division shown in Examples 4 through 6, the denominator must be a factor of the numerator. When the denominator is not a factor of the numerator, or in the case where we can't factor the numerator, the method used in Examples 4 through 6 won't work. We need to develop a new method for these cases. Since this new method is very similar to long division with whole numbers, we will review it here.

7. Divide $35\overline{)7{,}546}$.

EXAMPLE 7 Divide $25\overline{)4{,}628}$.

Solution

```
      1     ← Estimate 25 into 46
25)4,628
   2 5      ← Multiply 1 × 25 = 25
   2 1      ← Subtract 46 − 25 = 21
```

```
      1
25)4,628
   2 5↓
   2 12     ← Bring down the 2
```

NOTE

You may realize when looking over this example that you don't have a very good idea why you proceed as you do with the steps in long division. What you do know is the process always works. We are going to approach the explanation for division of two polynomials with this in mind; that is, we won't always be sure why the steps we use are important, only that they always produce the correct result.

Answers

5. 2 6. $x + a$

These are the four basic steps in long division: estimate, multiply, subtract, and bring down the next term. To complete the problem, we simply perform the same four steps:

$$\begin{array}{rl}
18 & \leftarrow \textbf{8 is the estimate} \\
25\overline{)4{,}628} & \\
\underline{2\,5} & \\
2\,12 & \\
\underline{2\,00}\downarrow & \leftarrow \textbf{Multiply to get 200} \\
128 & \leftarrow \textbf{Subtract to get 12, then bring down the 8}
\end{array}$$

One more time:

$$\begin{array}{rl}
185 & \leftarrow \textbf{5 is the estimate} \\
25\overline{)4{,}628} & \\
\underline{2\,5} & \\
2\,12 & \\
\underline{2\,00}\downarrow & \\
128 & \\
\underline{125} & \leftarrow \textbf{Multiply to get 125} \\
3 & \leftarrow \textbf{Subtract to get 3}
\end{array}$$

Since 3 is less than 25 and we have no more terms to bring down, we have our answer:

$$\frac{4{,}628}{25} = 185 + \frac{3}{25}$$

To check our answer, we multiply 185 by 25 and then add 3 to the result:

$$25(185) + 3 = 4{,}625 + 3 = 4{,}628$$

Long division with polynomials is very similar to long division with whole numbers. Both use the same four basic steps: estimate, multiply, subtract, and bring down the next term. We use long division with polynomials when the denominator has two or more terms and is not a factor of the numerator. Here is an example.

8. Divide $\frac{3x^2 - 8x - 1}{x - 3}$.

EXAMPLE 8 Divide $\frac{2x^2 - 7x + 9}{x - 2}$.

Solution

$$\begin{array}{rl}
2x\phantom{{}-7x+9} & \leftarrow \textbf{Estimate } 2x^2 \div x = 2x \\
x - 2\overline{)\;2x^2 - 7x + 9} & \\
\underline{\overset{-}{\not+}\,2x^2 \overset{+}{\not-}\,4x}\phantom{{}+9} & \leftarrow \textbf{Multiply } 2x(x-2) = 2x^2 - 4x \\
-3x\phantom{{}+9} & \leftarrow \textbf{Subtract } (2x^2 - 7x) - (2x^2 - 4x) = -3x
\end{array}$$

$$\begin{array}{rl}
2x\phantom{{}-7x+9} & \\
x - 2\overline{)\;2x^2 - 7x + 9} & \\
\underline{\overset{-}{\not+}\,2x^2 \overset{+}{\not-}\,4x}\phantom{{}+9} & \\
-3x + 9 & \leftarrow \textbf{Bring down the 9}
\end{array}$$

Notice we change the signs on $2x^2 - 4x$ and add in the subtraction step. Subtracting a polynomial is equivalent to adding its opposite.

Answer
7. $215 + \frac{3}{5}$

We repeat the four steps.

$$
\begin{array}{rl}
2x - 3 & \leftarrow \textbf{−3 is the estimate: } -3x \div x = -3 \\
x - 2\overline{)\ 2x^2 - 7x + 9} & \\
\overset{-}{\cancel{+}}\, 2x^2 \overset{+}{\cancel{-}}\, 4x & \\
\hline
-3x + 9 & \\
\overset{+}{\cancel{-}}\, 3x \overset{-}{\cancel{+}}\, 6 & \leftarrow \textbf{Multiply } -3(x-2) = -3x + 6 \\
\hline
3 & \leftarrow \textbf{Subtract } (-3x + 9) - (-3x + 6) = 3
\end{array}
$$

Since we have no other term to bring down, we have our answer:

$$\frac{2x^2 - 7x + 9}{x - 2} = 2x - 3 + \frac{3}{x - 2}$$

To check, we multiply $(2x - 3)(x - 2)$ to get $2x^2 - 7x + 6$; then, adding the remainder 3 to this result, we have $2x^2 - 7x + 9$. ■

In setting up a division problem involving two polynomials, we must remember two things: (1) both polynomials should be in decreasing powers of the variable, and (2) neither should skip any powers from the highest power down to the constant term. If there are any missing terms, they can be filled in using a coefficient of 0.

EXAMPLE 9

Divide $2x - 4\overline{)4x^3 - 6x - 11}$.

Solution Since the trinomial is missing a term in x^2, we can fill it in with $0x^2$:

$$4x^3 - 6x - 11 = 4x^3 + 0x^2 - 6x - 11$$

Adding $0x^2$ does not change our original problem.

$$
\begin{array}{r}
2x^2 + 4x + 5 \\
2x - 4\overline{)\ 4x^3 + 0x^2 - 6x - 11} \\
\overset{-}{\cancel{+}}\, 4x^3 \overset{+}{\cancel{-}}\, 8x^2 \qquad\qquad \\
\hline
+ 8x^2 - 6x \qquad \\
\overset{-}{\cancel{+}}\, 8x^2 \overset{+}{\cancel{-}}\, 16x \qquad \\
\hline
+ 10x - 11 \\
\overset{-}{\cancel{+}}\, 10x \overset{+}{\cancel{-}}\, 20 \\
\hline
+9
\end{array}
$$

Notice: **Adding the $0x^2$ term gives us a column in which to write $-8x^2$**

$$\frac{4x^3 - 6x - 11}{2x - 4} = 2x^2 + 4x + 5 + \frac{9}{2x - 4}$$

To check this result, we multiply $2x - 4$ and $2x^2 + 4x + 5$:

$$
\begin{array}{r}
2x^2 + 4x + 5 \\
2x - 4 \\
\hline
4x^3 + 8x^2 + 10x \qquad\quad \\
- 8x^2 - 16x - 20 \\
\hline
4x^3 \qquad\quad - 6x - 20
\end{array}
$$

Adding 9 (the remainder) to this result gives us the polynomial $4x^3 - 6x - 11$. Our answer checks. ■

9. Divide $x - 2\overline{)3x^3 + 3x + 1}$.

Answers

8. $3x + 1 + \dfrac{2}{x - 3}$

9. $3x^2 + 6x + 15 + \dfrac{31}{x - 2}$

10. Divide $\dfrac{2x^2 - 5xy + 3y^2}{x - y}$.

For our last example in this section, let's do Example 4 again, but this time use long division.

EXAMPLE 10 Divide $\dfrac{x^2 - 6xy - 7y^2}{x + y}$.

Solution

$$
\begin{array}{r l}
 & x \quad - 7y \\
x + y \,\big) & \overline{x^2 - 6xy - 7y^2} \\
 & \underline{\cancel{+} x^2 \cancel{+} xy} \\
 & \quad - 7xy - 7y^2 \\
 & \quad \underline{\cancel{-} 7xy \cancel{-} 7y^2} \\
 & \qquad\qquad 0
\end{array}
$$

In this case, the remainder is 0 and we have

$$\frac{x^2 - 6xy - 7y^2}{x + y} = x - 7y$$

which is easy to check since

$$(x + y)(x - 7y) = x^2 - 6xy - 7y^2$$

Linking Objectives and Examples

Next to each **objective**, list the examples that are best described by that objective.

A	
B	

Getting Ready for Class

After reading through the preceding section, respond in your own words and in complete sentences.

1. What are the four steps used in long division with polynomials?
2. What does it mean to have a remainder of 0?
3. When must long division be performed, and when can factoring be used to divide polynomials?
4. What property of real numbers is the key to dividing a polynomial by a monomial?

Answer
10. $2x - 3y$

Problem Set 6.2

INTERMEDIATE Algebra $f(x)$ Now™

Find the following quotients.

1. $\dfrac{4x^3 - 8x^2 + 6x}{2x}$

2. $\dfrac{6x^3 + 12x^2 - 9x}{3x}$

3. $\dfrac{10x^4 + 15x^3 - 20x^2}{-5x^2}$

4. $\dfrac{12x^5 - 18x^4 - 6x^3}{6x^3}$

5. $\dfrac{8y^5 + 10y^3 - 6y}{4y^3}$

6. $\dfrac{6y^4 - 3y^3 + 18y^2}{9y^2}$

7. $\dfrac{5x^3 - 8x^2 - 6x}{-2x^2}$

8. $\dfrac{-9x^5 + 10x^3 - 12x}{-6x^4}$

9. $\dfrac{28a^3b^5 + 42a^4b^3}{7a^2b^2}$

10. $\dfrac{a^2b + ab^2}{ab}$

11. $\dfrac{10x^3y^2 - 20x^2y^3 - 30x^3y^3}{-10x^2y}$

12. $\dfrac{9x^4y^4 + 18x^3y^4 - 27x^2y^4}{-9xy^3}$

Divide by factoring numerators and then dividing out common factors.

13. $\dfrac{x^2 - x - 6}{x - 3}$

14. $\dfrac{x^2 - x - 6}{x + 2}$

15. $\dfrac{2a^2 - 3a - 9}{2a + 3}$

16. $\dfrac{2a^2 + 3a - 9}{2a - 3}$

17. $\dfrac{5x^2 - 14xy - 24y^2}{x - 4y}$

18. $\dfrac{5x^2 - 26xy - 24y^2}{5x + 4y}$

19. $\dfrac{x^3 - y^3}{x - y}$

20. $\dfrac{x^3 + 8}{x + 2}$

21. $\dfrac{y^4 - 16}{y - 2}$

22. $\dfrac{y^4 - 81}{y - 3}$

23. $\dfrac{x^3 + 2x^2 - 25x - 50}{x - 5}$

24. $\dfrac{x^3 + 2x^2 - 25x - 50}{x + 5}$

Divide using the long division method.

25. $\dfrac{x^2 - 5x - 7}{x + 2}$

26. $\dfrac{x^2 + 4x - 8}{x - 3}$

27. $\dfrac{2x^3 - 3x^2 - 4x + 5}{x + 1}$

28. $\dfrac{3x^3 - 5x^2 + 2x - 1}{x - 2}$

29. $\dfrac{2y^3 - 9y^2 - 17y + 39}{2y - 3}$

30. $\dfrac{3y^3 - 19y^2 + 17y + 4}{3y - 4}$

31. $\dfrac{6y^3 - 8y + 5}{2y - 4}$

32. $\dfrac{9y^3 - 6y^2 + 8}{3y - 3}$

33. $\dfrac{a^4 - 2a + 5}{a - 2}$

34. $\dfrac{a^4 + a^3 - 1}{a + 2}$

35. $\dfrac{y^4 - 16}{y - 2}$

36. $\dfrac{y^4 - 81}{y - 3}$

Applying the Concepts

Difference Quotients For each function below, find the difference quotient $\frac{f(x) - f(a)}{x - a}$ and then simplify.

37. $f(x) = 4x$

38. $f(x) = -3x$

39. $f(x) = 5x + 3$

40. $f(x) = 6x - 5$

41. $f(x) = x^2$

42. $f(x) = 3x^2$

43. $f(x) = x^2 + 1$

44. $f(x) = x^2 - 3$

45. The Factor Theorem The factor theorem of algebra states that if $x - a$ is a factor of a polynomial, $P(x)$, then $P(a) = 0$. Verify the following.

a. That $x - 2$ is a factor of $P(x) = x^3 - 3x^2 + 5x - 6$, and that $P(2) = 0$

b. That $x - 5$ is a factor of $P(x) = x^4 - 5x^3 - x^2 + 6x - 5$, and that $P(5) = 0$

46. The Remainder Theorem The remainder theorem of algebra states that if a polynomial, $P(x)$, is divided by $x - a$, then the remainder is $P(a)$. Verify the remainder theorem by showing that when $P(x) = x^2 - x + 3$ is divided by $x - 2$ the remainder is 5 and that $P(2) = 5$.

47. Checking Account First Bank of San Luis Obispo charges \$2.00 per month and \$0.15 per check for a regular checking account. As we mentioned in the introduction to this section, the total monthly cost of this account is $C(x) = 2.00 + 0.15x$. To find the average cost of each of the x checks, we divide the total cost by the number of checks written; that is,

$$\overline{C}(x) = \frac{C(x)}{x}$$

a. Use the total cost function to fill in the following table.

x	1	5	10	15	20
$C(x)$					

b. Find the formula for the average cost function, $\overline{C}(x)$.

c. Use the average cost function to fill in the following table.

x	1	5	10	15	20
$\overline{C}(x)$					

d. What happens to the average cost as more checks are written?

48. Average Cost A company that manufactures computer diskettes uses the function $C(x) = 200 + 2x$ to represent the daily cost of producing x diskettes.

a. Find the average cost function, $\overline{C}(x)$.

b. Use the average cost function to fill in the following table: Round to the nearest dollar.

x	1	5	10	20	50
$\overline{C}(x)$					

c. What happens to the average cost as more items are produced?

Getting Ready for the Next Section

Multiply or divide as indicated.

49. $\frac{6}{7} \cdot \frac{14}{18}$

50. $\frac{6}{8} \div \frac{3}{5}$

51. $5y^2 \cdot 4x^2$

52. $4y^3 \cdot 3x^2$

53. $9x^4 \cdot 8y^5$

54. $6x^4 \cdot 12y^5$

Factor.

55. $x^2 - 4$

56. $x^2 - 6x + 9$

57. $x^3 - x^2y$

58. $a^2 - 5a + 6$

59. $2y^2 - 2$

60. $xa + xb + ya + yb$

Maintaining Your Skills

Divide.

61. $\frac{3}{5} \div \frac{2}{7}$

62. $\frac{2}{7} \div \frac{3}{5}$

63. $\frac{3}{4} \div \frac{6}{11}$

64. $\frac{6}{8} \div \frac{3}{5}$

65. $\frac{4}{9} \div 8$

66. $\frac{3}{7} \div 6$

67. $8 \div \frac{1}{4}$

68. $12 \div \frac{2}{3}$

Write each expression with positive exponents and simplify as much as possible.

69. $\left(\frac{1}{3}\right)^{-2} + \left(\frac{1}{2}\right)^{-3}$

70. $\left(\frac{1}{2}\right)^{-3} - \left(\frac{1}{3}\right)^{-3}$

Simplify, and write your answers with positive exponents only.

71. $(9x^{-4}y^{9})^{-2}(3x^{2}y^{-1})^{4}$

72. $(4x^{4}y^{-3})^{2}(2x^{-6}y^{4})^{-3}$

Extending the Concepts

Divide.

73. $\dfrac{4x^5 - x^4 - 20x^3 + 8x^2 - 15}{x^2 - 5}$

74. $\dfrac{4x^5 + 2x^4 + x^3 - 20x^2 - 10x - 5}{x^3 - 5}$

75. $\dfrac{0.5x^3 - 0.3x^2 + 0.22x + 0.06}{x + 0.2}$

76. $\dfrac{0.6x^3 - 1.1x^2 - 0.1x + 0.6}{0.3x + 0.2}$

77. $\dfrac{3x^2 + x - 9}{2x + 4}$

78. $\dfrac{2x^2 - x - 9}{3x + 6}$

79. $\dfrac{2x^2 + \frac{1}{3}x + \frac{5}{3}}{3x - 1}$

80. $\dfrac{x^2 + \frac{3}{5}x + \frac{8}{5}}{5x - 2}$

Multiplication and Division of Rational Expressions

6.3

If you have ever taken a home videotape to be duplicated, you know the amount you pay for the duplication service depends on the number of copies you have made: The more copies you have made, the lower the charge per copy. The following demand function gives the price (in dollars) per tape $p(x)$ a company charges for making x copies of a 30-minute videotape. As you can see, it is a rational function.

$$p(x) = \frac{2(x + 60)}{x + 5}$$

The graph in Figure 1 shows this function from $x = 0$ to $x = 100$. As you can see, the more copies that are made, the lower the price per copy.

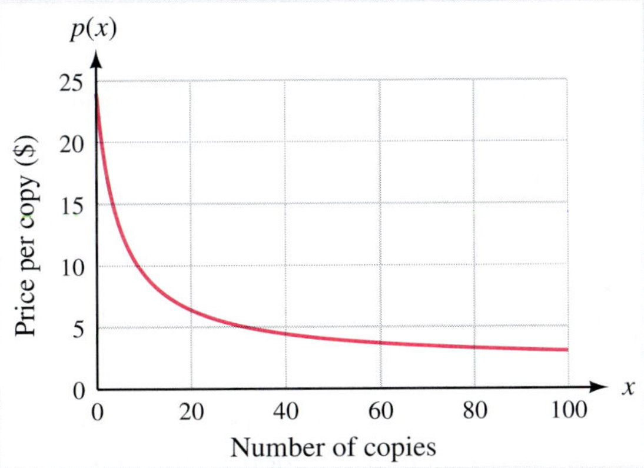

FIGURE 1

If we were interested in finding the revenue function for this situation, we would multiply the number of copies made x by the price per copy $p(x)$. This involves multiplication with a rational expression, which is one of the topics we cover in this section.

In the previous section we found the process of reducing rational expressions to lowest terms to be the same process used in reducing fractions to lowest terms. The similarity also holds for the process of multiplication or division of rational expressions.

Multiplication with fractions is the simplest of the four basic operations. To multiply two fractions we simply multiply numerators and multiply denominators; that is, if a, b, c, and d are real numbers, with $b \neq 0$ and $d \neq 0$, then

$$\frac{a}{b} \cdot \frac{c}{d} = \frac{ac}{bd}$$

EXAMPLE 1 Multiply $\frac{6}{7} \cdot \frac{14}{18}$.

Solution

$$\frac{6}{7} \cdot \frac{14}{18} = \frac{6(14)}{7(18)} \quad \textbf{Multiply numerators and denominators}$$

$$= \frac{\cancel{2} \cdot \cancel{3}(2 \cdot \cancel{7})}{\cancel{7}(\cancel{2} \cdot \cancel{3} \cdot 3)} \quad \textbf{Factor}$$

$$= \frac{2}{3} \quad \textbf{Divide out common factors}$$

OBJECTIVES

A Multiply and divide rational expressions.

PRACTICE PROBLEMS

1. Multiply $\frac{3}{4} \cdot \frac{12}{27}$.

Answer

1. $\frac{1}{3}$

Our next example is similar to some of the problems we worked in an earlier chapter. We multiply fractions whose numerators and denominators are monomials by multiplying numerators and multiplying denominators and then reducing to lowest terms. Here is how it looks.

2. Multiply $\frac{6x^4}{4y^9} \cdot \frac{12y^5}{3x^2}$.

NOTE

Notice how we factor the coefficients just enough so that we can see the factors they have in common. If you want to show this step without showing the factoring, it would look like this:

$$\frac{\overset{2}{\cancel{8}} \cdot \overset{1}{\cancel{9}}x^3y^3}{\underset{3}{\cancel{27}} \cdot \underset{3}{\cancel{12}}x^2y^8}$$

EXAMPLE 2

Multiply $\frac{8x^3}{27y^8} \cdot \frac{9y^3}{12x^2}$.

Solution We multiply numerators and denominators without actually carrying out the multiplication:

$$\frac{8x^3}{27y^8} \cdot \frac{9y^3}{12x^2} = \frac{8 \cdot 9x^3y^3}{27 \cdot 12x^2y^8}$$ **Multiply numerators** **Multiply denominators**

$$= \frac{\cancel{4} \cdot 2 \cdot \cancel{9}x^3y^3}{\cancel{9} \cdot 3 \cdot \cancel{4} \cdot 3x^2y^8}$$ **Factor coefficients**

$$= \frac{2x}{9y^5}$$ **Divide out common factors**

The product of two rational expressions is the product of their numerators over the product of their denominators.

Once again, we should mention that the little slashes we have drawn through the factors are used to denote the factors we have divided out of the numerator and denominator.

3. Multiply $\frac{x+5}{x^2-25} \cdot \frac{x-5}{x^2-10x+25}$.

EXAMPLE 3

Multiply $\frac{x-3}{x^2-4} \cdot \frac{x+2}{x^2-6x+9}$.

Solution We begin by multiplying numerators and denominators. We then factor all polynomials and divide out factors common to the numerator and denominator:

$$\frac{x-3}{x^2-4} \cdot \frac{x+2}{x^2-6x+9} = \frac{(x-3)(x+2)}{(x^2-4)(x^2-6x+9)}$$ **Multiply**

$$= \frac{\cancel{(x-3)}\cancel{(x+2)}}{\cancel{(x+2)}(x-2)\cancel{(x-3)}(x-3)}$$ **Factor**

$$= \frac{1}{(x-2)(x-3)}$$ **Divide out common factors**

The first two steps can be combined to save time. We can perform the multiplication and factoring steps together.

4. Multiply $\frac{3y^2-3y}{3y-12} \cdot \frac{y^2-2y-8}{y^2+3y+2}$.

EXAMPLE 4

Multiply $\frac{2y^2-4y}{2y^2-2} \cdot \frac{y^2-2y-3}{y^2-5y+6}$.

Solution

$$\frac{2y^2-4y}{2y^2-2} \cdot \frac{y^2-2y-3}{y^2-5y+6} = \frac{\cancel{2}y\cancel{(y-2)}\cancel{(y-3)}\cancel{(y+1)}}{\cancel{2}\cancel{(y+1)}(y-1)\cancel{(y-3)}\cancel{(y-2)}}$$

$$= \frac{y}{y-1}$$

Notice in both of the preceding examples that we did not actually multiply the polynomials as we did in the chapter on exponents and polynomials. It would be senseless to do that since we would then have to factor each of the resulting products to reduce them to lowest terms.

The quotient of two rational expressions is the product of the first and the reciprocal of the second; that is, we find the quotient of two rational expressions

Answers

2. $\frac{6x^2}{y^4}$ 3. $\frac{1}{(x-5)^2}$ 4. $\frac{y(y-1)}{y+1}$

the same way we find the quotient of two fractions. Here is an example that reviews division with fractions.

EXAMPLE 5 Divide $\frac{6}{8} \div \frac{3}{5}$.

Solution

$$\frac{6}{8} \div \frac{3}{5} = \frac{6}{8} \cdot \frac{5}{3} \quad \text{Write division in terms of multiplication}$$

$$= \frac{6(5)}{8(3)} \quad \text{Multiply numerators and denominators}$$

$$= \frac{\cancel{2} \cdot \cancel{3}(5)}{\cancel{2} \cdot 2 \cdot 2(\cancel{3})} \quad \text{Factor}$$

$$= \frac{5}{4} \quad \text{Divide out common factors}$$

To divide one rational expression by another, we use the definition of division to multiply by the reciprocal of the expression that follows the division symbol.

EXAMPLE 6 Divide $\frac{8x^3}{5y^2} \div \frac{4x^2}{10y^6}$.

Solution First we rewrite the problem in terms of multiplication. Then we multiply.

$$\frac{8x^3}{5y^2} \div \frac{4x^2}{10y^6} = \frac{8x^3}{5y^2} \cdot \frac{10y^6}{4x^2}$$

$$= \frac{\overset{2}{\cancel{8}} \cdot \overset{2}{\cancel{10}}x^3y^6}{\cancel{4} \cdot \cancel{5}x^2y^2}$$

$$= 4xy^4$$

EXAMPLE 7 Divide $\frac{x^2 - y^2}{x^2 - 2xy + y^2} \div \frac{x^3 + y^3}{x^3 - x^2y}$.

Solution We begin by writing the problem as the product of the first and the reciprocal of the second and then proceed as in the previous two examples:

$$\frac{x^2 - y^2}{x^2 - 2xy + y^2} \div \frac{x^3 + y^3}{x^3 - x^2y} \quad \text{Multiply by the reciprocal of the divisor}$$

$$= \frac{x^2 - y^2}{x^2 - 2xy + y^2} \cdot \frac{x^3 - x^2y}{x^3 + y^3}$$

$$= \frac{\cancel{(x-y)}\cancel{(x+y)}(x^2)\cancel{(x-y)}}{\cancel{(x-y)}\cancel{(x-y)}\cancel{(x+y)}(x^2 - xy + y^2)} \quad \text{Factor and multiply}$$

$$= \frac{x^2}{x^2 - xy + y^2} \quad \text{Divide out common factors}$$

Here are some more examples of multiplication and division with rational expressions.

EXAMPLE 8 Perform the indicated operations.

$$\frac{a^2 - 8a + 15}{a + 4} \cdot \frac{a + 2}{a^2 - 5a + 6} \div \frac{a^2 - 3a - 10}{a^2 + 2a - 8}$$

5. Divide $\frac{5}{9} \div \frac{10}{27}$.

6. Divide $\frac{9x^4}{4y^3} \div \frac{3x^2}{8y^5}$.

7. Divide.

$$\frac{xy^2 - y^3}{x^2 - y^2} \div \frac{x^3 + y^3}{x^2 + 2xy + y^2}$$

8. Perform the indicated operations.

$$\frac{a^2 + 3a - 4}{a - 4} \cdot \frac{a + 3}{a^2 - 4a + 3} \div \frac{a + 1}{a^2 - 2a - 3}$$

Answers

5. $\frac{3}{2}$ 6. $6x^2y^2$ 7. $\frac{y^2}{x^2 - xy + y^2}$

Solution First we rewrite the division as multiplication by the reciprocal. Then we proceed as usual.

$$\frac{a^2-8a+15}{a+4}\cdot\frac{a+2}{a^2-5a+6}\div\frac{a^2-3a-10}{a^2+2a-8}$$

$$=\frac{(a^2-8a+15)(a+2)(a^2+2a-8)}{(a+4)(a^2-5a+6)(a^2-3a-10)}$$ **Change division to multiplication by the reciprocal**

$$=\frac{(a-5)(a-3)(a+2)(a+4)(a-2)}{(a+4)(a-3)(a-2)(a-5)(a+2)}$$ **Factor**

$$=1$$ **Divide out common factors**

Our next example involves factoring by grouping. As you may have noticed, working the problems in this chapter gives you a very detailed review of factoring.

9. Multiply.
$$\frac{xa+xb-ya-yb}{xa+2x+ya+2y}\cdot\frac{xa+2x+ya+2y}{xa+xb+ya+yb}$$

10. Multiply.
$$(5x^2-45)\cdot\frac{3}{5x-15}$$

EXAMPLE 9 Multiply $\frac{xa+xb+ya+yb}{xa-xb-ya+yb}\cdot\frac{xa+xb-ya-yb}{xa-xb+ya-yb}$

Solution We will factor each polynomial by grouping, which takes two steps.

$$\frac{xa+xb+ya+yb}{xa-xb-ya+yb}\cdot\frac{xa+xb-ya-yb}{xa-xb+ya-yb}$$

$$=\frac{x(a+b)+y(a+b)}{x(a-b)-y(a-b)}\cdot\frac{x(a+b)-y(a+b)}{x(a-b)+y(a-b)}$$

$$=\frac{(a+b)(x+y)(a+b)(x-y)}{(a-b)(x-y)(a-b)(x+y)}$$ **Factor by grouping**

$$=\frac{(a+b)^2}{(a-b)^2}$$

EXAMPLE 10 Multiply $(4x^2-36)\cdot\frac{12}{4x+12}$.

Solution We can think of $4x^2-36$ as having a denominator of 1. Thinking of it in this way allows us to proceed as we did in the previous examples.

$$(4x^2-36)\cdot\frac{12}{4x+12}$$

$$=\frac{4x^2-36}{1}\cdot\frac{12}{4x+12}$$ **Write $4x^2-36$ with denominator 1**

$$=\frac{4(x-3)(x+3)12}{4(x+3)}$$ **Factor**

$$=12(x-3)$$ **Divide out common factors**

Linking Objectives and Examples

Next to each **objective**, list the examples that are best described by that objective.

A	

Getting Ready for Class

After reading through the preceding section, respond in your own words and in complete sentences.

1. Summarize the steps used to multiply fractions.
2. What is the first step in multiplying two rational expressions?
3. Why is factoring important when multiplying and dividing rational expressions?
4. How is division with rational expressions different than multiplication of rational expressions?

Answers

8. $\frac{(a+4)(a+3)}{a-4}$ 9. $\frac{x-y}{x+y}$

10. $3(x+3)$

Problem Set 6.3

INTERMEDIATE Algebra $f(x)$ Now™

Perform the indicated operations involving fractions.

1. $\frac{2}{9} \cdot \frac{3}{4}$

2. $\frac{5}{6} \cdot \frac{7}{8}$

3. $\frac{3}{4} \div \frac{1}{3}$

4. $\frac{3}{8} \div \frac{5}{4}$

5. $\frac{3}{7} \cdot \frac{14}{24} \div \frac{1}{2}$

6. $\frac{6}{5} \cdot \frac{10}{36} \div \frac{3}{4}$

7. $\frac{10x^2}{5y^2} \cdot \frac{15y^3}{2x^4}$

8. $\frac{8x^3}{7y^4} \cdot \frac{14y^6}{16x^2}$

9. $\frac{11a^2b}{5ab^2} \div \frac{22a^3b^2}{10ab^4}$

10. $\frac{8ab^3}{9a^2b} \div \frac{16a^2b^2}{18ab^3}$

11. $\frac{6x^2}{5y^3} \cdot \frac{11z^2}{2x^2} \div \frac{33z^5}{10y^8}$

12. $\frac{4x^3}{7y^2} \cdot \frac{6z^5}{5x^6} \div \frac{24z^2}{35x^6}$

Perform the indicated operations. Be sure to write all answers in lowest terms.

13. $\frac{x^2 - 9}{x^2 - 4} \cdot \frac{x - 2}{x - 3}$

14. $\frac{x^2 - 16}{x^2 - 25} \cdot \frac{x - 5}{x - 4}$

15. $\frac{y^2 - 1}{y + 2} \cdot \frac{y^2 + 5y + 6}{y^2 + 2y - 3}$

16. $\frac{y - 1}{y^2 - y - 6} \cdot \frac{y^2 + 5y + 6}{y^2 - 1}$

17. $\frac{3x - 12}{x^2 - 4} \cdot \frac{x^2 + 6x + 8}{x - 4}$

18. $\frac{x^2 + 5x + 1}{4x - 4} \cdot \frac{x - 1}{x^2 + 5x + 1}$

Perform the indicated operations. Be sure to write all answers in lowest terms.

19. $\dfrac{5x + 2y}{25x^2 - 5xy - 6y^2} \cdot \dfrac{20x^2 - 7xy - 3y^2}{4x + y}$

20. $\dfrac{7x + 3y}{42x^2 - 17xy - 15y^2} \cdot \dfrac{12x^2 - 4xy - 5y^2}{2x + y}$

21. $\dfrac{a^2 - 5a + 6}{a^2 - 2a - 3} \div \dfrac{a - 5}{a^2 + 3a + 2}$

22. $\dfrac{a^2 + 7a + 12}{a - 5} \div \dfrac{a^2 + 9a + 18}{a^2 - 7a + 10}$

23. $\dfrac{4t^2 - 1}{6t^2 + t - 2} \div \dfrac{8t^3 + 1}{27t^3 + 8}$

24. $\dfrac{9t^2 - 1}{6t^2 + 7t - 3} \div \dfrac{27t^3 + 1}{8t^3 + 27}$

25. $\dfrac{2x^2 - 5x - 12}{4x^2 + 8x + 3} \div \dfrac{x^2 - 16}{2x^2 + 7x + 3}$

26. $\dfrac{x^2 - 2x + 1}{3x^2 + 7x - 20} \div \dfrac{x^2 + 3x - 4}{3x^2 - 2x - 5}$

27. $\dfrac{6a^2b + 2ab^2 - 20b^3}{4a^2b - 16b^3} \cdot \dfrac{10a^2 - 22ab + 4b^2}{27a^3 - 125b^3}$

28. $\dfrac{12a^2b - 3ab^2 - 42b^3}{9a^2 - 36b^2} \cdot \dfrac{6a^2 - 15ab + 6b^2}{8a^3b - b^4}$

29. $\dfrac{360x^3 - 490x}{36x^2 + 84x + 49} \cdot \dfrac{30x^2 + 83x + 56}{150x^3 + 65x^2 - 280x}$

30. $\dfrac{490x^2 - 640}{49x^2 - 112x + 64} \cdot \dfrac{28x^2 - 95x + 72}{56x^3 - 62x^2 - 144x}$

31. $\dfrac{x^5 - x^2}{5x^5 - 5x} \cdot \dfrac{10x^4 - 10x^2}{2x^4 + 2x^3 + 2x^2}$

32. $\dfrac{2x^4 - 16x}{3x^6 - 48x^2} \cdot \dfrac{6x^5 + 24x^3}{4x^4 + 8x^3 + 16x^2}$

33. $\dfrac{a^2 - 16b^2}{a^2 - 8ab + 16b^2} \cdot \dfrac{a^2 - 9ab + 20b^2}{a^2 - 7ab + 12b^2} \div \dfrac{a^2 - 25b^2}{a^2 - 6ab + 9b^2}$

34. $\dfrac{a^2 - 6ab + 9b^2}{a^2 - 4b^2} \cdot \dfrac{a^2 - 5ab + 6b^2}{(a - 3b)^2} \div \dfrac{a^2 - 9b^2}{a^2 - ab - 6b^2}$

35. $\dfrac{2y^2 - 7y - 15}{42y^2 - 29y - 5} \cdot \dfrac{12y^2 - 16y + 5}{7y^2 - 36y + 5} \div \dfrac{4y^2 - 9}{49y^2 - 1}$

36. $\dfrac{8y^2 + 18y - 5}{21y^2 - 16y + 3} \cdot \dfrac{35y^2 - 22y + 3}{6y^2 + 17y + 5} \div \dfrac{16y^2 - 1}{9y^2 - 1}$

37. $\dfrac{xy - 2x + 3y - 6}{xy + 2x - 4y - 8} \cdot \dfrac{xy + x - 4y - 4}{xy - x + 3y - 3}$

38. $\dfrac{ax + bx + 2a + 2b}{ax - 3a + bx - 3b} \cdot \dfrac{ax - bx - 3a + 3b}{ax - bx - 2a + 2b}$

39. $\dfrac{xy^2 - y^2 + 4xy - 4y}{xy - 3y + 4x - 12} \div \dfrac{xy^3 + 2xy^2 + y^3 + 2y^2}{xy^2 - 3y^2 + 2xy - 6y}$

40. $\dfrac{4xb - 8b + 12x - 24}{xb^2 + 3b^2 + 3xb + 9b} \div \dfrac{4xb - 8b - 8x + 16}{xb^2 + 3b^2 - 2xb - 6b}$

41. $\dfrac{2x^3 + 10x^2 - 8x - 40}{x^3 + 4x^2 - 9x - 36} \cdot \dfrac{x^2 + x - 12}{2x^2 + 14x + 20}$

42. $\dfrac{x^3 + 2x^2 - 9x - 18}{x^4 + 3x^3 - 4x^2 - 12x} \cdot \dfrac{x^3 + 5x^2 + 6x}{x^2 - x - 6}$

Use the method shown in Example 10 to find the following products.

43. $(3x - 6) \cdot \dfrac{x}{x - 2}$

44. $(4x + 8) \cdot \dfrac{x}{x + 2}$

45. $(x^2 - 25) \cdot \dfrac{2}{x - 5}$

46. $(x^2 - 49) \cdot \dfrac{5}{x + 7}$

47. $(x^2 - 3x + 2) \cdot \dfrac{3}{3x - 3}$

48. $(x^2 - 3x + 2) \cdot \dfrac{-1}{x - 2}$

49. $(y - 3)(y - 4)(y + 3) \cdot \dfrac{-1}{y^2 - 9}$

50. $(y + 1)(y + 4)(y - 1) \cdot \dfrac{3}{y^2 - 1}$

51. $a(a + 5)(a - 5) \cdot \dfrac{a + 1}{a^2 + 5a}$

52. $a(a + 3)(a - 3) \cdot \dfrac{a - 1}{a^2 - 3a}$

Applying the Concepts

At the beginning of this section we introduced the demand equation shown here. Use it to work Problems 53–56.

$$p(x) = \frac{2(x + 60)}{x + 5}$$

53. Demand Equation Use the demand equation to fill in the table. Then compare your results with the graph shown in Figure 1 of this section.

NUMBER OF COPIES	PRICE PER COPY ($)
1	
10	
20	
50	
100	

54. Demand Equation To find the revenue for selling 50 copies of a tape, we multiply the price per tape by 50. Find the revenue for selling 50 tapes.

55. Revenue Find the revenue for selling 100 tapes.

56. Revenue Find the revenue equation $R(x)$.

57. Area The following box has a square top. The front face of the box has an area of $A = x^3 - 2x^2 - 2x - 3$. The height of the box is $h = x^2 + x + 1$. Find a formula for the area of the top square in terms of x.

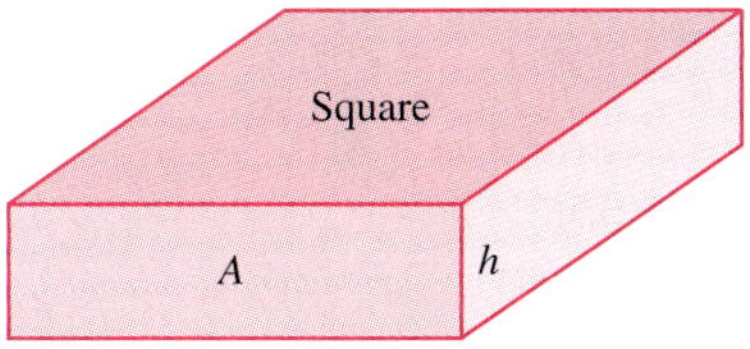

58. Surface Area of a Cylinder The surface area of the cylinder in the figure is defined as the area of its two circular bases and the lateral, or side, area. The surface area may be found by the formula

$$A = 2\pi r^2 + 2\pi rh$$

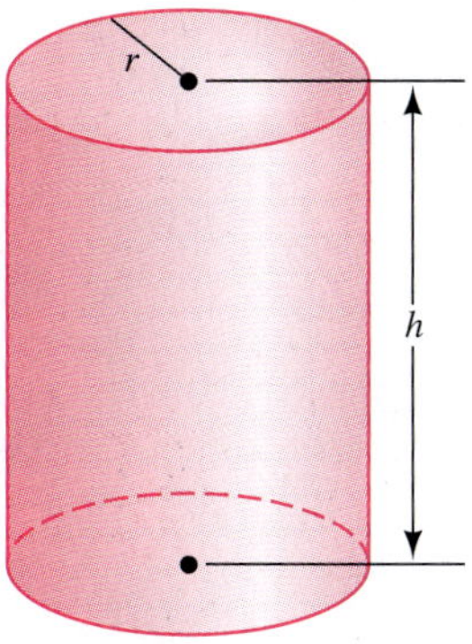

If the surface area is 6π, and $h = 2$, find r.

Getting Ready for the Next Section

Combine.

59. $\frac{4}{9} + \frac{2}{9}$

60. $\frac{3}{8} + \frac{1}{8}$

61. $\frac{3}{14} + \frac{7}{30}$

62. $\frac{3}{10} + \frac{11}{42}$

Multiply.

63. $-1(7 - x)$

64. $-1(3 - x)$

Factor.

65. $x^2 - 1$

66. $x^2 - 2x - 3$

67. $2x + 10$

68. $x^2 + 4x + 3$

69. $a^3 - b^3$

70. $8y^3 - 27$

Maintaining Your Skills

Multiply.

71. $2x^2(5x^3 + 4x - 3)$

72. $3x^3(7x^2 - 4x - 8)$

73. $(3a - 1)(4a + 5)$

74. $(6a - 3)(2a + 1)$

75. $(3x + 7)(4y - 2)$

76. $(x + 2a)(2 - 3b)$

77. $(3 - t^2)^2$

78. $(2 - t^3)^2$

79. $3(x + 1)(x + 2)(x + 3)$

80. $4(x - 1)(x - 2)(x - 3)$

Addition and Subtraction of Rational Expressions

6.4

This section is concerned with addition and subtraction of rational expressions. In the first part of this section we will look at addition of expressions that have the same denominator. In the second part of this section we will look at addition of expressions that have different denominators.

OBJECTIVES	
A	Add and subtract rational expressions with the same denominator.
B	Add and subtract rational expressions with different denominators.

Addition and Subtraction with the Same Denominator

To add two expressions that have the same denominator, we simply add numerators and put the sum over the common denominator. Since the process we use to add and subtract rational expressions is the same process used to add and subtract fractions, we will begin with an example involving fractions.

EXAMPLE 1 Add $\frac{4}{9} + \frac{2}{9}$.

Solution We add fractions with the same denominator by using the distributive property. Here is a detailed look at the steps involved.

$$\begin{aligned} \frac{4}{9} + \frac{2}{9} &= 4\left(\tfrac{1}{9}\right) + 2\left(\tfrac{1}{9}\right) \\ &= (4 + 2)\left(\tfrac{1}{9}\right) && \textbf{Distributive property} \\ &= 6\left(\tfrac{1}{9}\right) \\ &= \frac{6}{9} \\ &= \frac{2}{3} && \textbf{Divide numerator and denominator by common factor 3} \end{aligned}$$

Note that the important thing about the fractions in this example is that they each have a denominator of 9. If they did not have the same denominator, we could not have written them as two terms with a factor of $\frac{1}{9}$ in common. Without the $\frac{1}{9}$ common to each term, we couldn't apply the distributive property. And without the distributive property, we would not have been able to add the two fractions. ■

In the examples that follow, we will not show all the steps we showed in Example 1. The steps are shown in Example 1 so that you will see why both fractions must have the same denominator before we can add them. In practice we simply add numerators and place the result over the common denominator.

We add and subtract rational expressions with the same denominator by combining numerators and writing the result over the common denominator. Then we reduce the result to lowest terms, if possible. Example 2 shows this process in detail. If you see the similarities between operations on rational numbers and operations on rational expressions, this chapter will look like an extension of rational numbers rather than a completely new set of topics.

EXAMPLE 2 Add $\frac{x}{x^2 - 1} + \frac{1}{x^2 - 1}$.

Solution Since the denominators are the same, we simply add numerators:

$$\begin{aligned} \frac{x}{x^2 - 1} + \frac{1}{x^2 - 1} &= \frac{x + 1}{x^2 - 1} && \textbf{Add numerators} \\ &= \frac{\cancel{x + 1}}{(x - 1)\cancel{(x + 1)}} && \textbf{Factor denominator} \\ &= \frac{1}{x - 1} && \textbf{Divide out common factor } x + 1 \end{aligned}$$

■

PRACTICE PROBLEMS

1. Add $\frac{3}{8} + \frac{1}{8}$.

2. Add $\frac{x}{x^2 - 9} + \frac{3}{x^2 - 9}$.

Answers

1. $\frac{1}{2}$ 2. $\frac{1}{x - 3}$

Our next example involves subtraction of rational expressions. Pay careful attention to what happens to the signs of the terms in the numerator of the second expression when we subtract it from the first expression.

3. Subtract $\frac{2x-7}{x-2} - \frac{x-5}{x-2}$.

EXAMPLE 3 Subtract $\frac{2x-5}{x-2} - \frac{x-3}{x-2}$.

Solution Since each expression has the same denominator, we simply subtract the numerator in the second expression from the numerator in the first expression and write the difference over the common denominator $x - 2$. We must be careful, however, that we subtract both terms in the second numerator. To ensure that we do, we will enclose that numerator in parentheses.

$$\frac{2x-5}{x-2} - \frac{x-3}{x-2} = \frac{2x-5-(x-3)}{x-2} \quad \text{Subtract numerators}$$

$$= \frac{2x-5-x+3}{x-2} \quad \text{Remove parentheses}$$

$$= \frac{x-2}{x-2} \quad \text{Combine similar terms in the numerator}$$

$$= 1 \quad \text{Reduce (or divide)}$$

Note the +3 in the numerator of the second step. It is a very common mistake to write that as −3 by forgetting to subtract both terms in the numerator of the second expression. Whenever the expression we are subtracting has two or more terms in its numerator, we have to watch for this mistake.

NOTE

If you got an answer of

$$\frac{x-12}{x-2}$$

for Practice Problem 3, then you made the mistake talked about in Example 3. Look at the first two lines in the solution to Example 3 again and see if you can find your mistake.

Next we consider addition and subtraction of fractions and rational expressions that have different denominators.

Addition and Subtraction with Different Denominators

Before we look at an example of addition of fractions with different denominators, we need to review the definition for the least common denominator.

DEFINITION

The **least common denominator,** abbreviated LCD, for a set of denominators is the smallest expression that is divisible by each of the denominators.

The first step in combining two fractions is to find the LCD. Once we have the common denominator, we rewrite each fraction as an equivalent fraction with the common denominator. After that, we simply add or subtract as we did in our first three examples.

Example 4 is a review of the step-by-step procedure used to add two fractions with different denominators.

4. Add $\frac{3}{10} + \frac{11}{42}$.

EXAMPLE 4 Add $\frac{3}{14} + \frac{7}{30}$.

Solution

Step 1: Find the LCD.
To do this, we first factor both denominators into prime factors.

Factor 14: $14 = 2 \cdot 7$

Factor 30: $30 = 2 \cdot 3 \cdot 5$

Answer
3. 1

Since the LCD must be divisible by 14, it must have factors of $2 \cdot 7$. It must also be divisible by 30 and, therefore, have factors of $2 \cdot 3 \cdot 5$. We do not need to repeat the 2 that appears in both the factors of 14 and those of 30. Therefore,

$$\text{LCD} = 2 \cdot 3 \cdot 5 \cdot 7 = 210$$

Step 2: Change to equivalent fractions.

Since we want each fraction to have a denominator of 210 and at the same time keep its original value, we multiply each by 1 in the appropriate form.

Change $\frac{3}{14}$ to a fraction with denominator 210:

$$\frac{3}{14} \cdot \frac{\mathbf{15}}{\mathbf{15}} = \frac{45}{210}$$

Change $\frac{7}{30}$ to a fraction with denominator 210:

$$\frac{7}{30} \cdot \frac{\mathbf{7}}{\mathbf{7}} = \frac{49}{210}$$

Step 3: Add numerators of equivalent fractions found in step 2:

$$\frac{45}{210} + \frac{49}{210} = \frac{94}{210}$$

Step 4: Reduce to lowest terms if necessary:

$$\frac{94}{210} = \frac{47}{105}$$

The main idea in adding fractions is to write each fraction again with the LCD for a denominator. In doing so, we must be sure not to change the value of either of the original fractions.

NOTE

When we multiply $\frac{3}{14}$ by $\frac{15}{15}$ we obtain a fraction with the same value as $\frac{3}{14}$ (because we multiplied by 1) but with the common denominator 210.

EXAMPLE 5

Add $\dfrac{-2}{x^2 - 2x - 3} + \dfrac{3}{x^2 - 9}$.

Solution

Step 1: Factor each denominator and build the LCD from the factors:

$$\left.\begin{aligned} x^2 - 2x - 3 &= (x - 3)(x + 1) \\ x^2 - 9 &= (x - 3)(x + 3) \end{aligned}\right\} \text{LCD} = (x - 3)(x + 3)(x + 1)$$

Step 2: Change each rational expression to an equivalent expression that has the LCD for a denominator:

$$\frac{-2}{x^2 - 2x - 3} = \frac{-2}{(x - 3)(x + 1)} \cdot \frac{\mathbf{(x + 3)}}{\mathbf{(x + 3)}} = \frac{-2x - 6}{(x - 3)(x + 3)(x + 1)}$$

$$\frac{3}{x^2 - 9} = \frac{3}{(x - 3)(x + 3)} \cdot \frac{\mathbf{(x + 1)}}{\mathbf{(x + 1)}} = \frac{3x + 3}{(x - 3)(x + 3)(x + 1)}$$

Step 3: Add numerators of the rational expressions found in step 2:

$$\frac{-2x - 6}{(x - 3)(x + 3)(x + 1)} + \frac{3x + 3}{(x - 3)(x + 3)(x + 1)} = \frac{x - 3}{(x - 3)(x + 3)(x + 1)}$$

Step 4: Reduce to lowest terms by dividing out the common factor $x - 3$:

$$= \frac{1}{(x + 3)(x + 1)}$$

5. Add $\dfrac{-3}{x^2 - 2x - 8} + \dfrac{4}{x^2 - 16}$.

Answers

4. $\frac{59}{105}$ 5. $\dfrac{1}{(x + 4)(x + 2)}$

6. Add $\frac{x-4}{2x-6}+\frac{3}{x^2-9}$.

EXAMPLE 6

Subtract $\frac{x+4}{2x+10}-\frac{5}{x^2-25}$.

Solution We begin by factoring each denominator:

$$\frac{x+4}{2x+10}-\frac{5}{x^2-25}=\frac{x+4}{2(x+5)}-\frac{5}{(x+5)(x-5)}$$

The LCD is $2(x+5)(x-5)$. Completing the problem we have

$$=\frac{x+4}{2(x+5)}\cdot\frac{\mathbf{(x-5)}}{\mathbf{(x-5)}}-\frac{5}{(x+5)(x-5)}\cdot\frac{\mathbf{2}}{\mathbf{2}}$$

$$=\frac{x^2-x-20}{2(x+5)(x-5)}-\frac{10}{2(x+5)(x-5)}$$

$$=\frac{x^2-x-30}{2(x+5)(x-5)}$$

To see if this expression will reduce, we factor the numerator into $(x-6)(x+5)$.

$$=\frac{(x-6)\cancel{(x+5)}}{2\cancel{(x+5)}(x-5)}$$

$$=\frac{x-6}{2(x-5)}$$

7. Subtract.

$\frac{2x-4}{x^2+5x+4}-\frac{x-4}{x^2+6x+8}$

EXAMPLE 7

Subtract $\frac{2x-2}{x^2+4x+3}-\frac{x-1}{x^2+5x+6}$.

Solution We factor each denominator and build the LCD from those factors:

$$\frac{2x-2}{x^2+4x+3}-\frac{x-1}{x^2+5x+6}$$

$$=\frac{2x-2}{(x+3)(x+1)}-\frac{x-1}{(x+3)(x+2)}$$

$$=\frac{2x-2}{(x+3)(x+1)}\cdot\frac{\mathbf{(x+2)}}{\mathbf{(x+2)}}-\frac{x-1}{(x+3)(x+2)}\cdot\frac{\mathbf{(x+1)}}{\mathbf{(x+1)}}$$ **The LCD is $(x+1)(x+2)(x+3)$**

$$=\frac{2x^2+2x-4}{(x+1)(x+2)(x+3)}-\frac{x^2-1}{(x+1)(x+2)(x+3)}$$ **Multiply out each numerator**

$$=\frac{(2x^2+2x-4)-(x^2-1)}{(x+1)(x+2)(x+3)}$$

$$=\frac{x^2+2x-3}{(x+1)(x+2)(x+3)}$$ **Subtract numerators**

$$=\frac{\cancel{(x+3)}(x-1)}{(x+1)(x+2)\cancel{(x+3)}}$$ **Factor numerator to see if we can reduce**

$$=\frac{x-1}{(x+1)(x+2)}$$ **Reduce**

Answers

6. $\frac{x+2}{2(x+3)}$ 7. $\frac{x-1}{(x+1)(x+2)}$

EXAMPLE 8 Add $\frac{x^2}{x-7} + \frac{6x+7}{7-x}$.

Solution In the first section of this chapter we were able to reverse the terms in a factor such as $7 - x$ by factoring -1 from each term. In a problem like this, the same result can be obtained by multiplying the numerator and denominator by -1:

$$\frac{x^2}{x-7} + \frac{6x+7}{7-x} \cdot \frac{\mathbf{-1}}{\mathbf{-1}} = \frac{x^2}{x-7} + \frac{-6x-7}{x-7}$$

$$= \frac{x^2-6x-7}{x-7} \qquad \textbf{Add numerators}$$

$$= \frac{\cancel{(x-7)}(x+1)}{\cancel{(x-7)}} \qquad \textbf{Factor numerator}$$

$$= x + 1 \qquad \textbf{Divide out } x - 7$$

For our next example we will look at a problem in which we combine a whole number and a rational expression.

EXAMPLE 9 Subtract $2 - \frac{9}{3x+1}$.

Solution To subtract these two expressions, we think of 2 as a rational expression with a denominator of 1.

$$2 - \frac{9}{3x+1} = \frac{2}{1} - \frac{9}{3x+1}$$

The LCD is $3x + 1$. Multiplying the numerator and denominator of the first expression by $3x + 1$ gives us a rational expression equivalent to 2 but with a denominator of $3x + 1$.

$$\frac{2}{1} \cdot \frac{\mathbf{(3x+1)}}{\mathbf{(3x+1)}} - \frac{9}{3x+1} = \frac{6x+2-9}{3x+1}$$

$$= \frac{6x-7}{3x+1}$$

The numerator and denominator of this last expression do not have any factors in common other than 1, so the expression is in lowest terms.

EXAMPLE 10 Write an expression for the sum of a number and twice its reciprocal. Then, simplify that expression.

Solution If x is the number, then its reciprocal is $\frac{1}{x}$. Twice its reciprocal is $\frac{2}{x}$. The sum of the number and twice its reciprocal is

$$x + \frac{2}{x}$$

To combine these two expressions, we think of the first term x as a rational expression with a denominator of 1. The least common denominator is x:

$$x + \frac{2}{x} = \frac{x}{1} + \frac{2}{x}$$

$$= \frac{x}{1} \cdot \frac{\mathbf{x}}{\mathbf{x}} + \frac{2}{x}$$

$$= \frac{x^2+2}{x}$$

8. Add $\frac{x^2}{x-4} + \frac{x+12}{4-x}$

9. Add $2 + \frac{25}{5x-1}$

10. One number is three times another. Write an expression for the sum of the reciprocals of the two numbers. Then simplify that expression.

Answers

8. $x + 3$ **9.** $\frac{10x+23}{5x-1}$ **10.** $\frac{4}{3x}$

Linking Objectives and Examples

Next to each **objective**, list the examples that are best described by that objective.

A	
B	

Getting Ready for Class

After reading through the preceding section, respond in your own words and in complete sentences.

1. Briefly describe how you would add two rational expressions that have the same denominator.
2. Why is factoring important in finding a least common denominator?
3. What is the last step in adding or subtracting two rational expressions?
4. Explain how you would change the fraction $\frac{5}{x-3}$ to an equivalent fraction with denominator $x^2 - 9$.

Problem Set 6.4

INTERMEDIATE Algebra $f(x)$ Now™

Combine the following fractions.

1. $\frac{3}{4} + \frac{1}{2}$

2. $\frac{5}{6} + \frac{1}{3}$

3. $\frac{2}{5} - \frac{1}{15}$

4. $\frac{5}{8} - \frac{1}{4}$

5. $\frac{5}{6} + \frac{7}{8}$

6. $\frac{3}{4} + \frac{2}{3}$

7. $\frac{9}{48} - \frac{3}{54}$

8. $\frac{6}{28} - \frac{5}{42}$

9. $\frac{3}{4} - \frac{1}{8} + \frac{2}{3}$

10. $\frac{1}{3} - \frac{5}{6} + \frac{5}{12}$

Combine the following rational expressions. Reduce all answers to lowest terms.

11. $\frac{x}{x+3} + \frac{3}{x+3}$

12. $\frac{5x}{5x+2} + \frac{2}{5x+2}$

13. $\frac{4}{y-4} - \frac{y}{y-4}$

14. $\frac{8}{y+8} + \frac{y}{y+8}$

15. $\frac{x}{x^2-y^2} - \frac{y}{x^2-y^2}$

16. $\frac{x}{x^2-y^2} + \frac{y}{x^2-y^2}$

17. $\frac{2x-3}{x-2} - \frac{x-1}{x-2}$

18. $\frac{2x-4}{x+2} - \frac{x-6}{x+2}$

19. $\frac{1}{a} + \frac{2}{a^2} - \frac{3}{a^3}$

20. $\frac{3}{a} + \frac{2}{a^2} - \frac{1}{a^3}$

21. $\frac{7x-2}{2x+1} - \frac{5x-3}{2x+1}$

22. $\frac{7x-1}{3x+2} - \frac{4x-3}{3x+2}$

Combine the following rational expressions. Reduce all answers to lowest terms.

23. $\frac{2}{t^2} - \frac{3}{2t}$

24. $\frac{5}{3t} - \frac{4}{t^2}$

25. $\frac{3x + 1}{2x - 6} - \frac{x + 2}{x - 3}$

26. $\frac{x + 1}{x - 2} - \frac{4x + 7}{5x - 10}$

27. $\frac{6x + 5}{5x - 25} - \frac{x + 2}{x - 5}$

28. $\frac{4x + 2}{3x + 12} - \frac{x - 2}{x + 4}$

29. $\frac{x + 1}{2x - 2} - \frac{2}{x^2 - 1}$

30. $\frac{x + 7}{2x + 12} + \frac{6}{x^2 - 36}$

31. $\frac{1}{a - b} - \frac{3ab}{a^3 - b^3}$

32. $\frac{1}{a + b} + \frac{3ab}{a^3 + b^3}$

33. $\frac{1}{2y - 3} - \frac{18y}{8y^3 - 27}$

34. $\frac{1}{3y - 2} - \frac{18y}{27y^3 - 8}$

35. $\frac{x}{x^2 - 5x + 6} - \frac{3}{3 - x}$

36. $\frac{x}{x^2 + 4x + 4} - \frac{2}{2 + x}$

37. $\dfrac{2}{4t-5}+\dfrac{9}{8t^2-38t+35}$

38. $\dfrac{3}{2t-5}+\dfrac{21}{8t^2-14t-15}$

39. $\dfrac{1}{a^2-5a+6}+\dfrac{3}{a^2-a-2}$

40. $\dfrac{-3}{a^2+a-2}+\dfrac{5}{a^2-a-6}$

41. $\dfrac{1}{8x^3-1}-\dfrac{1}{4x^2-1}$

42. $\dfrac{1}{27x^3-1}-\dfrac{1}{9x^2-1}$

43. $\dfrac{4}{4x^2-9}-\dfrac{6}{8x^2-6x-9}$

44. $\dfrac{9}{9x^2+6x-8}-\dfrac{6}{9x^2-4}$

45. $\dfrac{4a}{a^2+6a+5}-\dfrac{3a}{a^2+5a+4}$

46. $\dfrac{3a}{a^2+7a+10}-\dfrac{2a}{a^2+6a+8}$

47. $\dfrac{2x-1}{x^2+x-6}-\dfrac{x+2}{x^2+5x+6}$

48. $\dfrac{4x+1}{x^2+5x+4}-\dfrac{x+3}{x^2+4x+3}$

49. $\dfrac{2x-8}{3x^2+8x+4}+\dfrac{x+3}{3x^2+5x+2}$

50. $\dfrac{5x+3}{2x^2+5x+3}-\dfrac{3x+9}{2x^2+7x+6}$

51. $\dfrac{2}{x^2 + 5x + 6} - \dfrac{4}{x^2 + 4x + 3} + \dfrac{3}{x^2 + 3x + 2}$

52. $\dfrac{-5}{x^2 + 3x - 4} + \dfrac{5}{x^2 + 2x - 3} + \dfrac{1}{x^2 + 7x + 12}$

53. $\dfrac{2x + 8}{x^2 + 5x + 6} - \dfrac{x + 5}{x^2 + 4x + 3} - \dfrac{x - 1}{x^2 + 3x + 2}$

54. $\dfrac{2x + 11}{x^2 + 9x + 20} - \dfrac{x + 1}{x^2 + 7x + 12} - \dfrac{x + 6}{x^2 + 8x + 15}$

55. $2 + \dfrac{3}{2x + 1}$

56. $3 - \dfrac{2}{2x + 3}$

57. $5 + \dfrac{2}{4 - t}$

58. $7 + \dfrac{3}{5 - t}$

59. $x - \dfrac{4}{2x + 3}$

60. $x - \dfrac{5}{3x + 4} + 1$

61. $\dfrac{x}{x + 2} + \dfrac{1}{2x + 4} - \dfrac{3}{x^2 + 2x}$

62. $\dfrac{x}{x + 3} + \dfrac{7}{3x + 9} - \dfrac{2}{x^2 + 3x}$

63. $\dfrac{1}{x} + \dfrac{x}{2x + 4} - \dfrac{2}{x^2 + 2x}$

64. $\dfrac{1}{x} + \dfrac{x}{3x + 9} - \dfrac{3}{x^2 + 3x}$

Applying the Concepts

65. **Number Problem** Write an expression for the sum of a number and 4 times its reciprocal. Then, simplify that expression.

66. **Number Problem** Write an expression for the sum of the reciprocals of two consecutive integers. Then, simplify that expression.

67. **Optometry** The formula

$$P = \frac{1}{a} + \frac{1}{b}$$

is used by optometrists to help determine how strong to make the lenses for a pair of eyeglasses. If a is 10 and b is 0.2, find the corresponding value of P.

68. **Quadratic Formula** Later in the book we will work with the quadratic formula. The derivation of the formula requires that you can add the fractions below. Add the fractions.

$$-\frac{c}{a} + \left(\frac{b}{2a}\right)^2$$

69. **Elliptical Orbits** Consider two objects, A and B, that move in the same direction along an elliptical path at constant but different velocities.

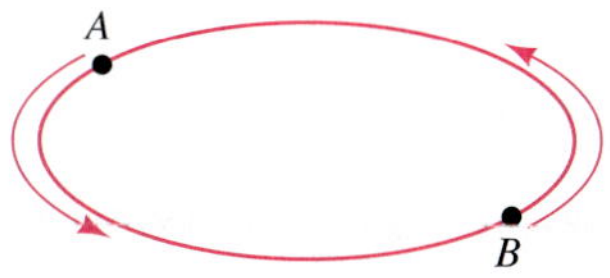

It can be shown that the time, T, it takes for the two objects to meet can be found from the formula

$$\frac{1}{T} = \frac{1}{t_A} - \frac{1}{t_B}$$

where t_A = time required for object A to orbit, and t_B = time required for object B to orbit.

a. If $t_A = 24$ months and $t_B = 30$ months, when will these two objects meet?

b. If $t_A = t_B$ what can one conclude?

70. **Average Velocity** If a car travels at a constant velocity, v_1, for 10 miles and then at a constant, but different velocity, v_2, for the next 10 miles, it can be shown that the cars average velocity, v_{avg}, over these 20 miles satisfies the equation

$$\frac{2}{v_{avg}} = \frac{1}{v_1} + \frac{1}{v_2}$$

Find the average velocity of a car that travels a constant 45 miles per hour for 10 miles and then increases to a constant 60 miles per hour for the next 10 miles. Round to the nearest tenth.

Getting Ready for the Next Section

Divide.

71. $\frac{3}{4} \div \frac{5}{8}$

72. $\frac{2}{3} \div \frac{5}{6}$

Multiply.

73. $x\left(1 + \frac{2}{x}\right)$

74. $3\left(x + \frac{1}{3}\right)$

75. $3x\left(\frac{1}{x} - \frac{1}{3}\right)$

76. $3x\left(\frac{1}{x} + \frac{1}{3}\right)$

Factor.

77. $x^2 - 4$

78. $x^2 - x - 6$

Maintaining Your Skills

Write each number in scientific notation.

79. 54,000

80. 768,000

81. 0.00034

82. 0.0359

Write each number in expanded form.

83. 6.44×10^3

84. 2.5×10^2

85. 6.44×10^{-3}

86. 2.5×10^{-2}

Simplify each expression as much as possible. Write all answers in scientific notation.

87. $(3 \times 10^8)(4 \times 10^{-5})$

88. $\frac{8 \times 10^{-3}}{4 \times 10^{-6}}$

Extending the Concepts

Simplify.

89. $\left(1 - \frac{1}{x}\right)\left(1 - \frac{1}{x+1}\right)\left(1 - \frac{1}{x+2}\right)\left(1 - \frac{1}{x+3}\right)$

90. $\left(1 + \frac{1}{x}\right)\left(1 + \frac{1}{x+1}\right)\left(1 + \frac{1}{x+2}\right)\left(1 + \frac{1}{x+3}\right)$

Complex Fractions

6.5

The quotient of two fractions or two rational expressions is called a *complex fraction.* This section is concerned with the simplification of complex fractions.

OBJECTIVES

A Simplify complex fractions.

EXAMPLE 1 Simplify $\dfrac{\frac{3}{4}}{\frac{5}{8}}$.

Solution There are generally two methods that can be used to simplify complex fractions.

Method 1 We can multiply the numerator and denominator of the complex fraction by the LCD for both of the fractions, which in this case is 8.

$$\frac{\frac{3}{4}}{\frac{5}{8}} = \frac{\frac{3}{4}\cdot\mathbf{8}}{\frac{5}{8}\cdot\mathbf{8}} = \frac{6}{5}$$

Method 2 Instead of dividing by $\frac{5}{8}$ we can multiply by $\frac{8}{5}$.

$$\frac{\frac{3}{4}}{\frac{5}{8}} = \frac{3}{4}\cdot\frac{8}{5} = \frac{24}{20} = \frac{6}{5}$$

Here are some examples of complex fractions involving rational expressions. Most can be solved using either of the two methods shown in Example 1.

EXAMPLE 2 Simplify $\dfrac{\frac{1}{x}+\frac{1}{y}}{\frac{1}{x}-\frac{1}{y}}$.

Solution This problem is most easily solved using Method 1. We begin by multiplying both the numerator and denominator by the quantity xy, which is the LCD for all the fractions:

$$\frac{\frac{1}{x}+\frac{1}{y}}{\frac{1}{x}-\frac{1}{y}} = \frac{\left(\frac{1}{x}+\frac{1}{y}\right)\cdot\mathbf{xy}}{\left(\frac{1}{x}-\frac{1}{y}\right)\cdot\mathbf{xy}}$$

$$= \frac{\frac{1}{x}(xy)+\frac{1}{y}(xy)}{\frac{1}{x}(xy)-\frac{1}{y}(xy)}$$ **Apply the distributive property to distribute xy over both terms in the numerator and denominator**

$$= \frac{y+x}{y-x}$$

PRACTICE PROBLEMS

1. Simplify $\dfrac{\frac{2}{3}}{\frac{5}{6}}$. (Try both methods.)

NOTE

You should become proficient at both methods. Each method can be useful in simplifying complex fractions.

2. Simplify $\dfrac{\frac{1}{x}-\frac{1}{3}}{\frac{1}{x}+\frac{1}{3}}$.

Answers

1. $\frac{4}{5}$ 2. $\frac{3-x}{3+x}$

3. Simplify $\dfrac{\frac{x+5}{x^2-16}}{\frac{x^2-25}{x-4}}$.

NOTE

We could have used method 1 just as easily: we would have multiplied the numerator and denominator of the complex fraction by $(x+3)(x-3)$.

EXAMPLE 3 Simplify $\dfrac{\frac{x-2}{x^2-9}}{\frac{x^2-4}{x+3}}$.

Solution Applying method 2, we have

$$\frac{\frac{x-2}{x^2-9}}{\frac{x^2-4}{x+3}} = \frac{x-2}{x^2-9} \cdot \frac{x+3}{x^2-4}$$

$$= \frac{\cancel{(x-2)}\cancel{(x+3)}}{\cancel{(x+3)}(x-3)(x+2)\cancel{(x-2)}}$$

$$= \frac{1}{(x-3)(x+2)}$$

4. Simplify $\dfrac{1-\frac{9}{x^2}}{1-\frac{1}{x}-\frac{6}{x^2}}$.

EXAMPLE 4 Simplify $\dfrac{1-\frac{4}{x^2}}{1-\frac{1}{x}-\frac{6}{x^2}}$.

Solution The simplest way to simplify this complex fraction is to multiply the numerator and denominator by the LCD, x^2:

$$\frac{1-\frac{4}{x^2}}{1-\frac{1}{x}-\frac{6}{x^2}} = \frac{\mathbf{x^2}\left(1-\frac{4}{x^2}\right)}{\mathbf{x^2}\left(1-\frac{1}{x}-\frac{6}{x^2}\right)} \qquad \textbf{Multiply numerator and denominator by } \boldsymbol{x^2}$$

$$= \frac{x^2 \cdot 1 - x^2 \cdot \frac{4}{x^2}}{x^2 \cdot 1 - x^2 \cdot \frac{1}{x} - x^2 \cdot \frac{6}{x^2}} \qquad \textbf{Distributive property}$$

$$= \frac{x^2-4}{x^2-x-6} \qquad \textbf{Simplify}$$

$$= \frac{(x-2)\cancel{(x+2)}}{(x-3)\cancel{(x+2)}} \qquad \textbf{Factor}$$

$$= \frac{x-2}{x-3} \qquad \textbf{Reduce}$$

5. Simplify $2 + \dfrac{5}{x-\frac{1}{5}}$.

EXAMPLE 5 Simplify $2 - \dfrac{3}{x+\frac{1}{3}}$.

Solution First we simplify the expression that follows the subtraction sign.

$$2 - \frac{3}{x+\frac{1}{3}} = 2 - \frac{\mathbf{3} \cdot 3}{\mathbf{3}\left(x+\frac{1}{3}\right)} = 2 - \frac{9}{3x+1}$$

Now we subtract by rewriting the first term, 2, with the LCD, $3x + 1$.

$$2 - \frac{9}{3x+1} = \frac{2}{1} \cdot \frac{\mathbf{3x+1}}{\mathbf{3x+1}} - \frac{9}{3x+1}$$

$$= \frac{6x+2-9}{3x+1} = \frac{6x-7}{3x+1}$$

Answers

3. $\dfrac{1}{(x+4)(x-5)}$ 4. $\dfrac{x+3}{x+2}$

5. $\dfrac{10x+23}{5x-1}$

Getting Ready for Class

After reading through the preceding section, respond in your own words and in complete sentences.

1. What is a complex fraction?
2. Explain how a least common denominator can be used to simplify a complex fraction.
3. Explain how some complex fractions can be converted to division problems. When is it more efficient to convert a complex fraction to a division problem of rational expressions?
4. Which method of simplifying complex fractions do you prefer? Why?

Linking Objectives and Examples

Next to each **objective**, list the examples that are best described by that objective.

A	

Problem Set 6.5

INTERMEDIATE Algebra Now™

Simplify each of the following as much as possible.

1. $\dfrac{\frac{3}{4}}{\frac{2}{3}}$

2. $\dfrac{\frac{5}{9}}{\frac{7}{12}}$

3. $\dfrac{\frac{1}{3} - \frac{1}{4}}{\frac{1}{2} + \frac{1}{8}}$

4. $\dfrac{\frac{1}{6} - \frac{1}{3}}{\frac{1}{4} - \frac{1}{8}}$

5. $\dfrac{3 + \frac{2}{5}}{1 - \frac{3}{7}}$

6. $\dfrac{2 + \frac{5}{6}}{1 - \frac{7}{8}}$

7. $\dfrac{\frac{1}{x}}{1 + \frac{1}{x}}$

8. $\dfrac{1 - \frac{1}{x}}{\frac{1}{x}}$

9. $\dfrac{1 + \frac{1}{a}}{1 - \frac{1}{a}}$

10. $\dfrac{1 - \frac{2}{a}}{1 - \frac{3}{a}}$

11. $\dfrac{\frac{1}{x} - \frac{1}{y}}{\frac{1}{x} + \frac{1}{y}}$

12. $\dfrac{\frac{1}{x} + \frac{2}{y}}{\frac{2}{x} + \frac{1}{y}}$

13. $\dfrac{\frac{x - 5}{x^2 - 4}}{\frac{x^2 - 25}{x + 2}}$

14. $\dfrac{\frac{3x + 1}{x^2 - 49}}{\frac{9x^2 - 1}{x - 7}}$

15. $\dfrac{\frac{4a}{2a^3 + 2}}{\frac{8a}{4a + 4}}$

16. $\dfrac{\frac{2a}{3a^3 - 3}}{\frac{4a}{6a - 6}}$

17. $\dfrac{1 - \dfrac{9}{x^2}}{1 - \dfrac{1}{x} - \dfrac{6}{x^2}}$

18. $\dfrac{4 - \dfrac{1}{x^2}}{4 + \dfrac{4}{x} + \dfrac{1}{x^2}}$

19. $\dfrac{2 + \dfrac{5}{a} - \dfrac{3}{a^2}}{2 - \dfrac{5}{a} + \dfrac{2}{a^2}}$

20. $\dfrac{3 + \dfrac{5}{a} - \dfrac{2}{a^2}}{3 - \dfrac{10}{a} + \dfrac{3}{a^2}}$

21. $\dfrac{1 + \dfrac{1}{x + 3}}{1 - \dfrac{1}{x + 3}}$

22. $\dfrac{1 + \dfrac{1}{x - 2}}{1 - \dfrac{1}{x - 2}}$

23. $\dfrac{1 - \dfrac{1}{a + 1}}{1 + \dfrac{1}{a - 1}}$

24. $\dfrac{\dfrac{1}{a - 1} + 1}{\dfrac{1}{a + 1} - 1}$

25. $\dfrac{\dfrac{1}{x + 3} + \dfrac{1}{x - 3}}{\dfrac{1}{x + 3} - \dfrac{1}{x - 3}}$

26. $\dfrac{\dfrac{1}{x + a} + \dfrac{1}{x - a}}{\dfrac{1}{x + a} - \dfrac{1}{x - a}}$

27. $\dfrac{\dfrac{y + 1}{y - 1} + \dfrac{y - 1}{y + 1}}{\dfrac{y + 1}{y - 1} - \dfrac{y - 1}{y + 1}}$

28. $\dfrac{\dfrac{y - 1}{y + 1} - \dfrac{y + 1}{y - 1}}{\dfrac{y - 1}{y + 1} + \dfrac{y + 1}{y - 1}}$

29. $1 - \dfrac{x}{1 - \dfrac{1}{x}}$

30. $x - \dfrac{1}{x - \dfrac{1}{2}}$

31. $1 + \dfrac{1}{1 + \dfrac{1}{1 + 1}}$

32. $1 - \dfrac{1}{1 - \dfrac{1}{1 - \frac{1}{2}}}$

Applying the Concepts

33. **Difference Quotient** For each rational function below, find the difference quotient

$$\frac{f(x) - f(a)}{x - a}$$

a. $f(x) = \frac{4}{x}$

b. $f(x) = \frac{1}{x + 1}$

c. $f(x) = \frac{1}{x^2}$

34. **Difference Quotient** For each rational function below, find the difference quotient

$$\frac{f(x) - f(2)}{x - 2}$$

a. $f(x) = \frac{4}{x}$

b. $f(x) = \frac{1}{x + 1}$

c. $f(x) = \frac{1}{x^2}$

35. **Optics** The formula $f = \frac{ab}{a + b}$ is used in optics to find the focal length of a lens. Show that the formula $f = (a^{-1} + b^{-1})^{-1}$ is equivalent to the preceding formula by rewriting it without the negative exponents and then simplifying the results.

36. **Optics** Show that the expression $(a^{-1} - b^{-1})^{-1}$ can be simplified to $\frac{ab}{b - a}$ by first writing it without the negative exponents and then simplifying the result.

37. **Doppler Effect** The change in the pitch of a sound (such as a train whistle) as an object passes is called the Doppler effect, named after C. J. Doppler (1803–1853). A person will *hear* a sound with a frequency, h, according to the formula

$$h = \frac{f}{1 + \frac{v}{s}}$$

where f is the actual frequency of the sound being produced, s is the speed of sound (about 740 miles per hour), and v is the velocity of the moving object.

a. Examine this fraction, and then explain why h and f approach the same value as v becomes smaller and smaller.

b. Solve this formula for v.

38. **Work Problem** A water storage tank has two drains. It can be shown that the time it takes to empty the tank if both drains are open is given by the formula

$$\frac{1}{\frac{1}{a} + \frac{1}{b}}$$

where a = time it takes for the first drain to empty the tank, and b = time for the second drain to empty the tank.

a. Simplify this complex fraction.

b. Find the amount of time needed to empty the tank using both drains if, used alone, the first drain empties the tank in 4 hours and the second drain can empty the tank in 3 hours.

Getting Ready for the Next Section

Multiply.

39. $x(y - 2)$

40. $x(y - 1)$

41. $6\left(\frac{x}{2} - 3\right)$

42. $6\left(\frac{x}{3} + 1\right)$

43. $xab \cdot \frac{1}{x}$

44. $xab\left(\frac{1}{b} + \frac{1}{a}\right)$

Factor.

45. $y^2 - 25$

46. $x^2 - 3x + 2$

47. $xa + xb$

48. $xy - y$

Solve.

49. $5x - 4 = 6$

50. $y^2 + y - 20 = 2y$

Maintaining Your Skills

Solve each equation.

51. $3x + 60 = 15$

52. $3x - 18 = 4$

53. $3(y - 3) = 2(y - 2)$

54. $5(y + 2) = 4(y + 1)$

55. $10 - 2(x + 3) = x + 1$

56. $15 - 3(x - 1) = x - 2$

57. $x^2 - x - 12 = 0$

58. $3x^2 + x - 10 = 0$

59. $(x + 1)(x - 6) = -12$

60. $(x + 1)(x - 4) = -6$

Equations Involving Rational Expressions

6.6

The first step in solving an equation that contains one or more rational expressions is to find the LCD for all denominators in the equation. We then multiply both sides of the equation by the LCD to clear the equation of all fractions—that is, after we have multiplied through by the LCD, each term in the resulting equation will have a denominator of 1.

	OBJECTIVES
A	Solve equations containing rational expressions.
B	Solve formulas containing rational expressions.

EXAMPLE 1 Solve $\frac{x}{2} - 3 = \frac{2}{3}$.

Solution The LCD for 2 and 3 is 6. Multiplying both sides by 6, we have

$$\mathbf{6}\left(\frac{x}{2} - 3\right) = \mathbf{6}\left(\frac{2}{3}\right)$$

$$6\left(\frac{x}{2}\right) - 6(3) = 6\left(\frac{2}{3}\right)$$

$$3x - 18 = 4$$

$$3x = 22$$

$$x = \frac{22}{3}$$

Multiplying both sides of an equation by the LCD clears the equation of fractions because the LCD has the property that all the denominators divide it evenly.

EXAMPLE 2 Solve $\frac{6}{a-4} = \frac{3}{8}$.

Solution The LCD for $a - 4$ and 8 is $8(a - 4)$. Multiplying both sides by this quantity yields

$$\mathbf{8(a-4)} \cdot \frac{6}{a-4} = \mathbf{8(a-4)} \cdot \frac{3}{8}$$

$$48 = (a-4) \cdot 3$$

$$48 = 3a - 12$$

$$60 = 3a$$

$$20 = a$$

The solution set is {20}, which checks in the original equation.

When we multiply both sides of an equation by an expression containing the variable, we must be sure to check our solutions. The multiplication property of equality does not allow multiplication by 0. If the expression we multiply by contains the variable, then it has the possibility of being 0. In the last example we multiplied both sides by $8(a - 4)$. This gives a restriction $a \neq 4$ for any solution we come up with.

PRACTICE PROBLEMS

1. Solve $\frac{x}{3} + 1 = \frac{1}{2}$.

2. $\frac{2}{a+5} = \frac{1}{3}$.

Answers

1. $-\frac{3}{2}$ 2. 1

3. Solve $\frac{x}{x+1} - \frac{1}{2} = \frac{-1}{x+1}$.

EXAMPLE 3

Solve $\frac{x}{x-2} + \frac{2}{3} = \frac{2}{x-2}$.

Solution The LCD is $3(x-2)$. We are assuming $x \neq 2$ when we multiply both sides of the equation by $3(x-2)$:

$$\mathbf{3(x-2)} \cdot \left[\frac{x}{x-2} + \frac{2}{3}\right] = \mathbf{3(x-2)} \cdot \frac{2}{x-2}$$

$$3x + (x-2) \cdot 2 = 3 \cdot 2$$

$$3x + 2x - 4 = 6$$

$$5x - 4 = 6$$

$$5x = 10$$

$$x = 2$$

The only possible solution is $x = 2$. Checking this value back in the original equation gives

$$\frac{2}{2-2} + \frac{2}{3} \stackrel{?}{=} \frac{2}{2-2}$$

$$\frac{2}{0} + \frac{2}{3} \stackrel{?}{=} \frac{2}{0}$$

The first and last terms are undefined. The proposed solution, $x = 2$, does not check in the original equation. The solution set is the empty set. There is no solution to the original equation. ■

NOTE

In the process of solving the equation, we multiplied both sides by $3(x-2)$, solved for x, and got $x = 2$ for our solution. But when x is 2, the quantity $3(x-2) = 3(2-2) = 3(0) = 0$, which means we multiplied both sides of our equation by 0, which is not allowed under the multiplication property of equality.

When the proposed solution to an equation is not actually a solution, it is called an *extraneous* solution. In the last example, $x = 2$ is an extraneous solution.

4. Solve.
$\frac{x}{x^2-9} - \frac{1}{x+3} = \frac{1}{4x-12}$

EXAMPLE 4

Solve $\frac{5}{x^2-3x+2} - \frac{1}{x-2} = \frac{1}{3x-3}$.

Solution Writing the equation again with the denominators in factored form, we have

$$\frac{5}{(x-2)(x-1)} - \frac{1}{x-2} = \frac{1}{3(x-1)}$$

The LCD is $3(x-2)(x-1)$. Multiplying through by the LCD, we have

$$\mathbf{3(x-2)(x-1)} \cdot \frac{5}{(x-2)(x-1)} - \mathbf{3(x-2)(x-1)} \cdot \frac{1}{(x-2)}$$

$$= \mathbf{3(x-2)(x-1)} \cdot \frac{1}{3(x-1)}$$

$$3 \cdot 5 - 3(x-1) \cdot 1 = (x-2) \cdot 1$$

$$15 - 3x + 3 = x - 2$$

$$-3x + 18 = x - 2$$

$$-4x + 18 = -2$$

$$-4x = -20$$

$$x = 5$$

Checking the proposed solution $x = 5$ in the original equation yields a true statement. Try it and see. ■

NOTE

We can check the proposed solution in any of the equations obtained before multiplying through by the LCD. We cannot check the proposed solution in an equation obtained *after* multiplying through by the LCD since, if we have multiplied by 0, the resulting equations will not be equivalent to the original one.

Answers
3. No solution 4. 9

EXAMPLE 5

Solve $3 + \frac{1}{x} = \frac{10}{x^2}$.

Solution To clear the equation of denominators, we multiply both sides by x^2:

$$\mathbf{x^2}\left(3 + \frac{1}{x}\right) = \mathbf{x^2}\left(\frac{10}{x^2}\right)$$

$$3(x^2) + \left(\frac{1}{x}\right)(x^2) = \left(\frac{10}{x^2}\right)(x^2)$$

$$3x^2 + x = 10$$

Rewrite in standard form, and solve:

$$3x^2 + x - 10 = 0$$

$$(3x - 5)(x + 2) = 0$$

$$3x - 5 = 0 \quad \text{or} \quad x + 2 = 0$$

$$x = \frac{5}{3} \quad \text{or} \quad x = -2$$

The solution set is $\{-2, \frac{5}{3}\}$. Both solutions check in the original equation. Remember: We have to check *all solutions* any time we multiply both sides of the equation by an expression that contains the variable, just to be sure we haven't multiplied by 0.

5. Solve $1 - \frac{2}{x} = \frac{8}{x^2}$.

EXAMPLE 6

Solve $\frac{y - 4}{y^2 - 5y} = \frac{2}{y^2 - 25}$.

Solution Factoring each denominator, we find the LCD is $y(y - 5)(y + 5)$. Multiplying each side of the equation by the LCD clears the equation of denominators and leads us to our possible solutions:

$$\mathbf{y(y - 5)(y + 5)} \cdot \frac{y - 4}{y(y - 5)} = \frac{2}{(y - 5)(y + 5)} \cdot \mathbf{y(y - 5)(y + 5)}$$

$$(y + 5)(y - 4) = 2y$$

$$y^2 + y - 20 = 2y \qquad \text{Multiply out the left side}$$

$$y^2 - y - 20 = 0 \qquad \text{Add } -2y \text{ to each side}$$

$$(y - 5)(y + 4) = 0$$

$$y - 5 = 0 \quad \text{or} \quad y + 4 = 0$$

$$y = 5 \quad \text{or} \quad y = -4$$

The two possible solutions are 5 and -4. If we substitute -4 for y in the original equation, we find that it leads to a true statement. It is, therefore, a solution. However, on the other hand, if we substitute 5 for y in the original equation, we find that both sides of the equation are undefined. The only solution to our original equation is $y = -4$. The other possible solution $y = 5$ is extraneous.

6. Solve $\frac{y + 1}{3(y + 4)} = \frac{8}{(y + 4)(y - 4)}$.

Answers

5. $-2, 4$ 6. 7

7. Solve for y: $x = \frac{y+2}{y-1}$.

8. Solve the formula $\frac{1}{a} = \frac{1}{x} + \frac{1}{b}$ for x.

EXAMPLE 7

Solve for y: $x = \dfrac{y-4}{y-2}$.

Solution To solve for y, we first multiply each side by $y - 2$ to obtain

$$x(y-2) = y - 4$$

$$xy - 2x = y - 4 \quad \text{Distributive property}$$

$$xy - y = 2x - 4 \quad \text{Collect all terms containing } y \text{ on the left side}$$

$$y(x-1) = 2x - 4 \quad \text{Factor } y \text{ from each term on the left side}$$

$$y = \frac{2x-4}{x-1} \quad \text{Divide each side by } x - 1$$

EXAMPLE 8

Solve the formula $\frac{1}{x} = \frac{1}{b} + \frac{1}{a}$ for x.

Solution We begin by multiplying both sides by the least common denominator xab. As you can see from our previous examples, multiplying both sides of an equation by the LCD is equivalent to multiplying each term of both sides by the LCD:

$$\mathbf{xab} \cdot \frac{1}{x} = \frac{1}{b} \cdot \mathbf{xab} + \frac{1}{a} \cdot \mathbf{xab}$$

$$ab = xa + xb$$

$$ab = (a+b)x \quad \text{Factor } x \text{ from the right side}$$

$$\frac{ab}{a+b} = x$$

We know we are finished because the variable we were solving for is alone on one side of the equation and does not appear on the other side.

Linking Objectives and Examples

Next to each **objective**, list the examples that are best described by that objective.

A	
B	

Getting Ready for Class

After reading through the preceding section, respond in your own words and in complete sentences.

1. Explain how a least common denominator can be used to simplify an equation.
2. What is an extraneous solution?
3. Is it possible for an equation containing rational expressions to have no solutions?
4. What is the last step in solving an equation that contains rational expressions?

Answers

7. $y = \frac{x+2}{x-1}$ 8. $x = \frac{ab}{b-a}$

Problem Set 6.6

INTERMEDIATE Algebra $f(x)$ Now™

Solve each of the following equations:

1. $\frac{x}{5} + 4 = \frac{5}{3}$

2. $\frac{x}{5} = \frac{x}{2} - 9$

3. $\frac{a}{3} + 2 = \frac{4}{5}$

4. $\frac{a}{4} + \frac{1}{2} = \frac{2}{3}$

5. $\frac{y}{2} + \frac{y}{4} + \frac{y}{6} = 3$

6. $\frac{y}{3} - \frac{y}{6} + \frac{y}{2} = 1$

7. $\frac{5}{2x} = \frac{1}{x} + \frac{3}{4}$

8. $\frac{1}{2a} = \frac{2}{a} - \frac{3}{8}$

9. $\frac{1}{x} = \frac{1}{3} - \frac{2}{3x}$

10. $\frac{5}{2x} = \frac{2}{x} - \frac{1}{12}$

11. $\frac{2x}{x-3} + 2 = \frac{2}{x-3}$

12. $\frac{2}{x+5} = \frac{2}{5} - \frac{x}{x+5}$

13. $1 - \frac{1}{x} = \frac{12}{x^2}$

14. $2 + \frac{5}{x} = \frac{3}{x^2}$

15. $y - \frac{4}{3y} = -\frac{1}{3}$

16. $\frac{y}{2} - \frac{4}{y} = -\frac{7}{2}$

Getting Ready for the Next Section

Multiply.

59. $39.3 \cdot 60$

60. $1{,}100 \cdot 60 \cdot 60$

Divide. Round to the nearest tenth, if necessary.

61. $65{,}000 \div 5{,}280$

62. $3{,}960{,}000 \div 5{,}280$

Multiply.

63. $2x\left(\frac{1}{x} + \frac{1}{2x}\right)$

64. $3x\left(\frac{1}{x} + \frac{1}{3x}\right)$

Solve.

65. $12(x + 3) + 12(x - 3) = 3(x^2 - 9)$

66. $40 + 2x = 60 - 3x$

67. $\frac{1}{10} - \frac{1}{12} = \frac{1}{x}$

68. $\frac{1}{x} + \frac{1}{2x} = 2$

Maintaining Your Skills

69. Number Problem Twice the sum of a number and 3 is 16. Find the number.

70. Number Problem The sum of two consecutive odd integers is 48. Find the two integers.

71. Geometry The length of a rectangle is 3 less than twice the width. The perimeter is 42 meters. Find the length and width.

72. Geometry The smaller angle in a triangle is one fourth as large as the largest angle. The third angle is 9 degrees more than the smallest angle. Find the measure of all three angles.

73. Consecutive Integers The sum of the squares of two consecutive integers is 61. Find the integers.

74. Consecutive Integers The square of the sum of two consecutive integers is 121. Find the two integers.

75. Geometry The lengths of the sides of a right triangle are given by three consecutive integers. Find the lengths of the three sides.

76. Geometry The longest side of a right triangle is 8 inches more than the shortest side. The other side is 7 inches more than the shortest side. Find the lengths of the three sides.

Extending the Concepts

Solve each equation.

77. $\frac{12}{x} + \frac{8}{x^2} - \frac{75}{x^3} - \frac{50}{x^4} = 0$

78. $\frac{45}{x} + \frac{18}{x^2} - \frac{80}{x^3} - \frac{32}{x^4} = 0$

79. $\frac{1}{x^3} - \frac{1}{3x^2} - \frac{1}{4x} + \frac{1}{12} = 0$

80. $\frac{1}{x^3} - \frac{1}{2x^2} - \frac{1}{9x} + \frac{1}{18} = 0$

Applications

6.7

We begin this section with some application problems, the solutions to which involve equations that contain rational expressions. As you will see, the solutions to the examples show only the essential steps from our Blueprint for Problem Solving. Recall that step 1 was done mentally; we read the problem and mentally list the items that are known and the items that are unknown. This is an essential part of problem solving. Now that you have had experience with application problems, however, you are doing step 1 automatically.

Also in this section we will look at a method of solving conversion problems that is called *unit analysis.* With unit analysis, we can convert expressions with units of feet per minute to equivalent expressions in miles per hour. This method of converting between different units of measure is used often in chemistry, physics, and engineering classes.

OBJECTIVES

A	Solve application problems using equations containing rational expressions.
B	Solve conversion problems using unit analysis.

EXAMPLE 1

One number is twice another. The sum of their reciprocals is 2. Find the numbers.

Solution Let x = the smaller number. The larger number is $2x$. Their reciprocals are $\frac{1}{x}$ and $\frac{1}{2x}$. The equation that describes the situation is

$$\frac{1}{x} + \frac{1}{2x} = 2$$

Multiplying both sides by the LCD $2x$, we have

$$\mathbf{2x} \cdot \frac{1}{x} + \mathbf{2x} \cdot \frac{1}{2x} = \mathbf{2x}(2)$$

$$2 + 1 = 4x$$

$$3 = 4x$$

$$x = \frac{3}{4}$$

The smaller number is $\frac{3}{4}$. The larger is $2\left(\frac{3}{4}\right) = \frac{6}{4} = \frac{3}{2}$. Adding their reciprocals, we have

$$\frac{4}{3} + \frac{2}{3} = \frac{6}{3} = 2$$

The sum of the reciprocals of $\frac{3}{4}$ and $\frac{3}{2}$ is 2. ■

PRACTICE PROBLEMS

1. One number is 3 times another. The sum of their reciprocals is $\frac{4}{3}$. Find the numbers.

EXAMPLE 2

The speed of a boat in still water is 20 miles per hour. It takes the same amount of time for the boat to travel 3 miles downstream (with the current) as it does to travel 2 miles upstream (against the current). Find the speed of the current.

Solution The following table will be helpful in finding the equation necessary to solve this problem.

	d (DISTANCE)	*r* (RATE)	*t* (TIME)
Upstream			
Downstream			

2. A boat can travel at 15 miles per hour in still water. If it takes the same amount of time for the boat to travel 2 miles downstream as it does to travel 1 mile upstream, find the speed of the current.

Answer

1. 1, 3

If we let x = the speed of the current, the speed (rate) of the boat upstream is $(20 - x)$ since it is traveling against the current. The rate downstream is $(20 + x)$ since the boat is then traveling with the current. The distance traveled upstream is 2 miles, and the distance traveled downstream is 3 miles. Putting the information given here into the table, we have

	d	r	t
Upstream	2	$20 - x$	
Downstream	3	$20 + x$	

To fill in the last two spaces in the table we must use the relationship $d = r \cdot t$. Since we know the spaces to be filled in are in the time column, we solve the equation $d = r \cdot t$ for t and get

$$t = \frac{d}{r}$$

The completed table then is

	d	r	t
Upstream	2	$20 - x$	$\frac{2}{20 - x}$
Downstream	3	$20 + x$	$\frac{3}{20 + x}$

Reading the problem again, we find that the time moving upstream is equal to the time moving downstream, or

$$\frac{2}{20 - x} = \frac{3}{20 + x}$$

Multiplying both sides by the LCD $(20 - x)(20 + x)$ gives

$$(20 + x) \cdot 2 = 3(20 - x)$$

$$40 + 2x = 60 - 3x$$

$$5x = 20$$

$$x = 4$$

The speed of the current is 4 miles per hour.

3. The current of a river is 2 miles per hour. It takes a motorboat a total of 3 hours to travel 8 miles upstream and return 8 miles downstream. What is the speed of the boat in still water?

EXAMPLE 3 The current of a river is 3 miles per hour. It takes a motorboat a total of 3 hours to travel 12 miles upstream and return 12 miles downstream. What is the speed of the boat in still water?

Solution This time we let x = the speed of the boat in still water. Then, we fill in as much of the table as possible using the information given in the problem. For instance, since we let x = the speed of the boat in still water, the rate upstream (against the current) must be $x - 3$. The rate downstream (with the current) is $x + 3$.

	d	r	t
Upstream	12	$x - 3$	
Downstream	12	$x + 3$	

Answer
2. 5 mph

The last two boxes can be filled in using the relationship

$$t = \frac{d}{r}$$

	d	r	t
Upstream	12	$x - 3$	$\frac{12}{x-3}$
Downstream	12	$x + 3$	$\frac{12}{x+3}$

The total time for the trip up and back is 3 hours:

$$\text{Time upstream} + \text{Time downstream} = \text{Total time}$$

$$\frac{12}{x-3} + \frac{12}{x+3} = 3$$

Multiplying both sides by $(x - 3)(x + 3)$, we have

$$12(x + 3) + 12(x - 3) = 3(x^2 - 9)$$

$$12x + 36 + 12x - 36 = 3x^2 - 27$$

$$3x^2 - 24x - 27 = 0$$

$$x^2 - 8x - 9 = 0 \quad \textbf{Divide both sides by 3}$$

$$(x - 9)(x + 1) = 0$$

$$x = 9 \quad \text{or} \quad x = -1$$

The speed of the motorboat in still water is 9 miles per hour. (We don't use $x = -1$ because the speed of the motorboat cannot be a negative number.)

EXAMPLE 4

An inlet pipe can fill a pool in 10 hours, and the drain can empty it in 12 hours. If the pool is empty and both the inlet pipe and drain are open, how long will it take to fill the pool?

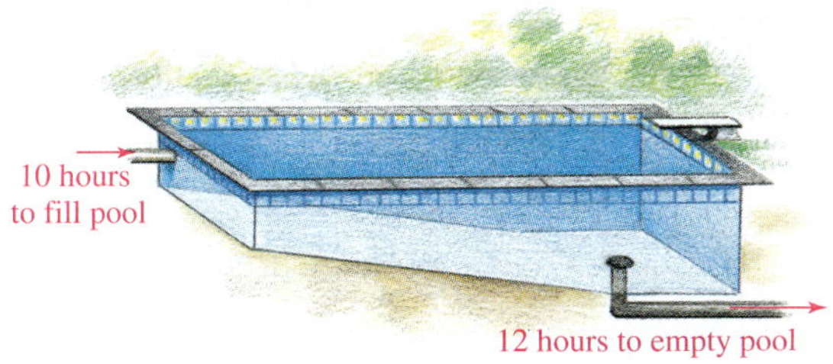

Solution It is helpful to think in terms of how much work is done by each pipe in 1 hour.

Let x = the time it takes to fill the pool with both pipes open.

If the inlet pipe can fill the pool in 10 hours, then in 1 hour it is $\frac{1}{10}$ full. If the outlet pipe empties the pool in 12 hours, then in 1 hour it is $\frac{1}{12}$ empty. If the pool can be filled in x hours with both the inlet pipe and the drain open, then in 1 hour it is $\frac{1}{x}$ full when both pipes are open.

Here is the equation:

In 1 hour

$$\left[\begin{matrix}\text{Amount filled by} \\ \text{inlet pipe}\end{matrix}\right] - \left[\begin{matrix}\text{Amount emptied by} \\ \text{the drain}\end{matrix}\right] = \left[\begin{matrix}\text{Fraction of pool filled} \\ \text{with both pipes open}\end{matrix}\right]$$

$$\frac{1}{10} - \frac{1}{12} = \frac{1}{x}$$

4. The hot-water faucet can fill a sink in 3 minutes. The drain will empty the sink in 4 minutes. If the hot-water faucet is on and the drain is open, how long will it take to fill the sink?

Answer

3. 6 mph

Multiplying through by $60x$, we have

$$\mathbf{60x} \cdot \frac{1}{10} - \mathbf{60x} \cdot \frac{1}{12} = \mathbf{60x} \cdot \frac{1}{x}$$

$$6x - 5x = 60$$

$$x = 60$$

It takes 60 hours to fill the pool if both the inlet pipe and the drain are open.

Unit Analysis

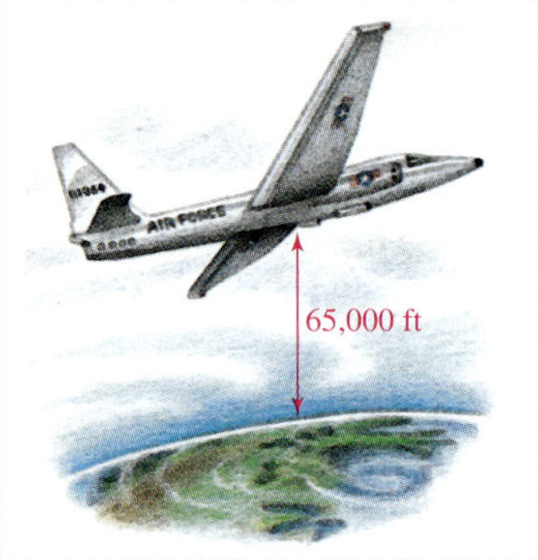

In the 1950s the United States had a spy plane, the U-2, that could fly at an altitude of 65,000 feet. Do you know how many miles are in 65,000 feet?

We can solve problems like this by using a method called *unit analysis.* With unit analysis, we analyze the units we are given and the units for which we are asked, and then multiply by the appropriate *conversion factor.* Since 1 mile is 5,280 feet, the conversion factor we use is

$$\frac{1 \text{ mile}}{5{,}280 \text{ feet}}$$

which is the number 1. Multiplying 65,000 feet by this conversion factor we have the following:

$$65{,}000 \text{ feet} = \frac{65{,}000 \text{ feet}}{1} \cdot \frac{1 \text{ mile}}{5{,}280 \text{ feet}}$$

We treat the units common to the numerator and denominator in the same way we treat factors common to the numerator and denominator: We divide out common units, just as we divide out common factors. In the preceding expression we have feet common to the numerator and denominator. Dividing them out leaves us with miles only. Here is the complete problem.

$$65{,}000 \text{ feet} = \frac{65{,}000 \cancel{\text{feet}}}{1} \cdot \frac{1 \text{ mile}}{5{,}280 \cancel{\text{feet}}}$$

$$= \frac{65{,}000}{5{,}280} \text{ mile}$$

$$= 12.3 \text{ miles, to the nearest tenth of a mile}$$

The key to solving a problem like this one lies in choosing the appropriate conversion factor. The fact that 1 mile = 5,280 feet yields two conversion factors, each of which is equal to the number 1. They are

$$\frac{1 \text{ mile}}{5{,}280 \text{ feet}} \quad \text{and} \quad \frac{5{,}280 \text{ feet}}{1 \text{ mile}}$$

The conversion factor we choose depends on the units we are given and the units with which we want to end up. Multiplying any expression by either of the two conversion factors leaves the value of the original expression unchanged because each of the conversion factors is simply the number 1.

4. 12 minutes

EXAMPLE 5

In the first section of this chapter we found a rider on the first Ferris wheel was traveling at approximately 39.3 feet per minute. Convert 39.3 feet per minute to miles per hour.

Solution We know that 5,280 feet = 1 mile and 60 minutes = 1 hour. Therefore, we have the following conversion factors, each of which is equal to 1.

$$\frac{5{,}280 \text{ feet}}{1 \text{ mile}} \quad \frac{1 \text{ mile}}{5{,}280 \text{ feet}} \quad \frac{60 \text{ minutes}}{1 \text{ hour}} \quad \frac{1 \text{ hour}}{60 \text{ minutes}}$$

The conversion factors we choose to multiply by are the ones that will allow us to divide out the units we are converting from and leave us with the units we are converting to. Specifically, we want to get rid of feet and be left with miles. Likewise, we want to get rid of minutes and be left with hours. Here is the conversion process that will accomplish these goals:

$$\begin{aligned} 39.3 \text{ feet per minute} &= \frac{39.3 \text{ } \cancel{\text{feet}}}{1 \text{ } \cancel{\text{minute}}} \cdot \frac{1 \text{ mile}}{5{,}280 \text{ } \cancel{\text{feet}}} \cdot \frac{60 \text{ } \cancel{\text{minutes}}}{1 \text{ hour}} \\ &= \frac{39.3 \cdot 60 \text{ miles}}{5{,}280 \text{ hours}} \\ &= 0.45 \text{ miles per hour, to the nearest hundredth} \end{aligned}$$

5. In Practice Problem 7 of the first section of this chapter we found a rider on the Ferris wheel called *Colossus* was traveling at approximately 13.0 feet per second. Give the speed in miles per hour. Round to the nearest tenth.

EXAMPLE 6

In 1993, a ski resort in Vermont advertised their new high-speed chair lift as "the world's fastest chair lift, with a speed of 1,100 feet per second." Show why the speed cannot be correct.

Solution To solve this problem, we can convert feet per second into miles per hour, a unit of measure we are more familiar with on an intuitive level.

$$\begin{aligned} 1{,}100 \text{ feet per second} &= \frac{1{,}100 \text{ } \cancel{\text{feet}}}{1 \text{ } \cancel{\text{second}}} \cdot \frac{1 \text{ mile}}{5{,}280 \text{ } \cancel{\text{feet}}} \cdot \frac{60 \text{ } \cancel{\text{seconds}}}{1 \text{ } \cancel{\text{minute}}} \cdot \frac{60 \text{ } \cancel{\text{minutes}}}{1 \text{ hour}} \\ &= \frac{1{,}100 \cdot 60 \cdot 60 \text{ miles}}{5{,}280 \text{ hours}} \\ &= 750 \text{ miles per hour} \end{aligned}$$

Obviously, there is a mistake in the advertisement.

6. Assume that the mistake in the advertisement is that feet per second should read feet per minute. Is 1,100 feet per minute a reasonable speed for a chair lift?

Getting Ready for Class

After reading through the preceding section, respond in your own words and in complete sentences.

1. Briefly list the steps in the Blueprint for Problem Solving that you have used previously to solve application problems.
2. Write an application problem for which the solution depends on solving the equation $\frac{1}{2} + \frac{1}{3} = \frac{1}{x}$.
3. What is a conversion factor in unit analysis?
4. What conversion factors would you use to convert feet per second to miles per hour?

Linking Objectives and Examples

Next to each **objective**, list the examples that are best described by that objective.

A	
B	

Answers

5. 8.9 mph
6. 12.5 mph is a reasonable speed for a chair lift.

Problem Set 6.7

INTERMEDIATE Algebra $f(x)$ Now™

Solve each of the following word problems. Be sure to show the equation in each case.

Number Problems

1. One number is 3 times another. The sum of their reciprocals is $\frac{20}{3}$. Find the numbers.

2. One number is 3 times another. The sum of their reciprocals is $\frac{4}{9}$. Find the numbers.

3. The sum of a number and its reciprocal is $\frac{10}{3}$. Find the number.

4. The sum of a number and twice its reciprocal is $\frac{27}{5}$. Find the number.

5. The sum of the reciprocals of two consecutive integers is $\frac{7}{12}$. Find the two integers.

6. Find two consecutive even integers, the sum of whose reciprocals is $\frac{3}{4}$.

7. If a certain number is added to the numerator and denominator of $\frac{7}{9}$, the result is $\frac{5}{6}$. Find the number.

8. Find the number you would add to both the numerator and denominator of $\frac{8}{11}$ so the result would be $\frac{6}{7}$.

Rate Problems

9. The speed of a boat in still water is 5 miles per hour. If the boat travels 3 miles downstream in the same amount of time it takes to travel 1.5 miles upstream, what is the speed of the current?

10. A boat, which moves at 18 miles per hour in still water, travels 14 miles downstream in the same amount of time it takes to travel 10 miles upstream. Find the speed of the current.

11. The current of a river is 2 miles per hour. A boat travels to a point 8 miles upstream and back again in 3 hours. What is the speed of the boat in still water?

12. A motorboat travels at 4 miles per hour in still water. It goes 12 miles upstream and 12 miles back again in a total of 8 hours. Find the speed of the current of the river.

13. Train A has a speed 15 miles per hour greater than that of train B. If train A travels 150 miles in the same time train B travels 120 miles, what are the speeds of the two trains?

14. A train travels 30 miles per hour faster than a car. If the train covers 120 miles in the same time the car covers 80 miles, what is the speed of each of them?

15. A small airplane flies 810 miles from Los Angeles to Portland, OR, with an average speed of 270 miles per hour. An hour and a half after the plane leaves, a Boeing 747 leaves Los Angeles for Portland. Both planes arrive in Portland at the same time. What was the average speed of the 747?

16. Lou leaves for a cross-country excursion traveling on a bicycle at 20 miles per hour. His friends are driving the trip and will meet him at several rest stops along the way. The first stop is scheduled 30 miles from the original starting point. If the people driving leave 15 minutes after Lou from the same place, how fast will they have to drive to reach the first rest stop at the same time as Lou?

17. A tour bus leaves Sacramento every Friday evening at 5:00 P.M. for a 270-mile trip to Las Vegas. This week, however, the bus leaves at 5:30 P.M. To arrive in Las Vegas on time, the driver drives 6 miles per hour faster than usual. What is the bus's usual speed?

18. A bakery delivery truck leaves the bakery at 5:00 A.M. each morning on its 140-mile route. One day the driver gets a late start and does not leave the bakery until 5:30 A.M. To finish her route on time the driver drives 5 miles per hour faster than usual. At what speed does she usually drive?

Work Problems

19. A water tank can be filled by an inlet pipe in 8 hours. It takes twice that long for the outlet pipe to empty the tank. How long will it take to fill the tank if both pipes are open?

20. It takes 10 hours to fill a pool with the inlet pipe. It can be emptied in 15 hours with the outlet pipe. If the pool is half full to begin with, how long will it take to fill it from there if both pipes are open?

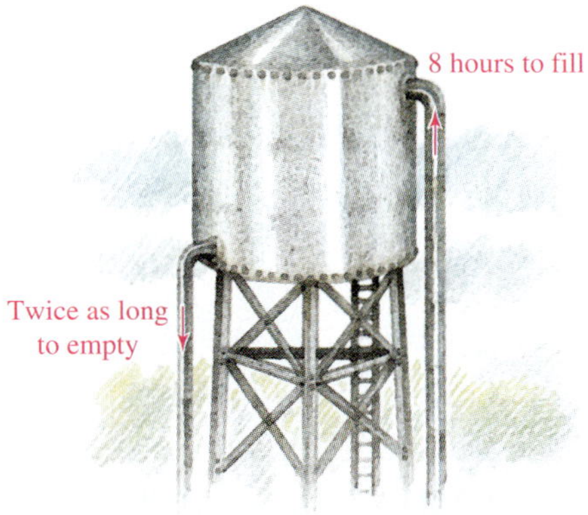

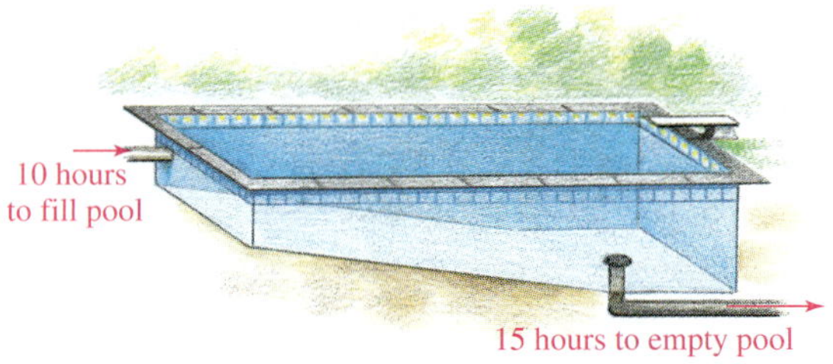

21. A sink can be filled from the faucet in 5 minutes. It takes only 3 minutes to empty the sink when the drain is open. If the sink is full and both the faucet and the drain are open, how long will it take to empty the sink?

22. A sink is one quarter full when both the faucet and the drain are opened. The faucet alone can fill the sink in 6 minutes, whereas it takes 8 minutes to empty it with the drain. How long will it take to fill the remaining three quarters of the sink?

23. A sink has two faucets: one for hot water and one for cold water. The sink can be filled by a cold-water faucet in 3.5 minutes. If both faucets are open, the sink is filled in 2.1 minutes. How long does it take to fill the sink with just the hot-water faucet open?

24. A water tank is being filled by two inlet pipes. Pipe A can fill the tank in $4\frac{1}{2}$ hours, but both pipes together can fill the tank in 2 hours. How long does it take to fill the tank using only pipe B?

Unit Analysis Problems

Give your answers to the following problems to the nearest tenth.

25. The South Coast Shopping Mall in Costa Mesa, California, covers an area of 2,224,750 square feet. If 1 acre = 43,560 square feet, how many acres does the South Coast Shopping Mall cover?

26. The relationship between liters and cubic inches, both of which are measures of volume, is 0.0164 liters = 1 cubic inch. If a Ford Mustang has a motor with a displacement of 4.9 liters, what is the displacement in cubic inches?

27. The Forest chair lift at the Northstar ski resort in Lake Tahoe is 5,750 feet long. If a ride on this chair lift takes 11 minutes, what is the average speed of the lift in miles per hour?

28. The Bear Paw chair lift at the Northstar ski resort in Lake Tahoe is 790 feet long. If a ride on this chair lift takes 2.2 minutes, what is the average speed of the lift in miles per hour?

29. A sprinter runs 100 meters in 10.8 seconds. What is the sprinter's average speed in miles per hour? (1 meter = 3.28 feet)

30. A runner covers 400 meters in 49.8 seconds. What is the average speed of the runner in miles per hour?

31. A person riding a Ferris wheel with a diameter of 65 feet travels once around the wheel in 30 seconds. What is the average speed of the rider in miles per hour?

32. A person riding a Ferris wheel with a diameter of 102 feet travels once around the wheel in 3.5 minutes. What is the average speed of the rider in miles per hour?

33. A $3\frac{1}{2}$-inch diskette, when placed in the disk drive of a computer, rotates at 300 rpm (meaning one revolution takes 1/300 minute). Find the average speed of a point 2 inches from the center of the diskette in miles per hour.

34. **Golf** The length of a golf course is the sum of the lengths of all 18 holes on the course. It is measured in yards. Round each of the lengths in the chart to the nearest hundred yards, then convert the result to miles, and round to the nearest tenth of a mile.

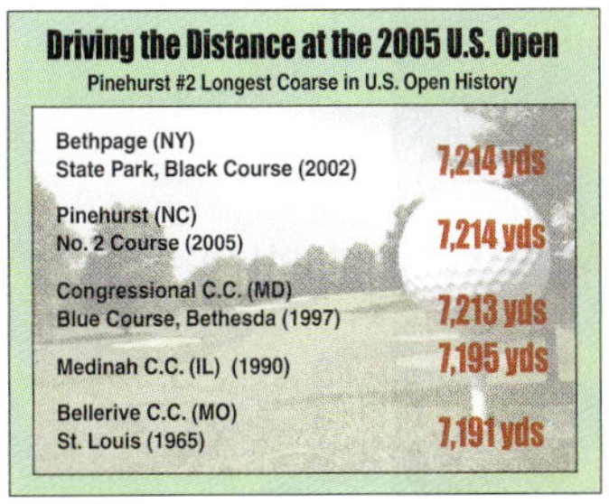

Data from *USA Today*. Copyright 2005.

Multiplication [6.3]

To multiply two rational numbers or rational expressions, multiply numerators and multiply denominators. In symbols,

$$\frac{P}{Q} \cdot \frac{R}{S} = \frac{PR}{QS} \qquad (Q \neq 0 \text{ and } S \neq 0)$$

In practice, we don't really multiply, but rather, we factor and then divide out common factors.

5. $\dfrac{x+1}{x^2-4} \cdot \dfrac{x+2}{3x+3}$

$= \dfrac{\cancel{(x+1)}\,\cancel{(x+2)}}{(x-2)\cancel{(x+2)}\,(3)\cancel{(x+1)}}$

$= \dfrac{1}{3(x-2)}$

Division [6.3]

To divide one rational expression by another, we use the definition of division to rewrite our division problem as an equivalent multiplication problem. Instead of dividing by a rational expression we multiply by its reciprocal. In symbols,

$$\frac{P}{Q} \div \frac{R}{S} = \frac{P}{Q} \cdot \frac{S}{R} = \frac{PS}{QR} \qquad (Q \neq 0, S \neq 0, R \neq 0)$$

6. $\dfrac{x^2-y^2}{x^3+y^3} \div \dfrac{x-y}{x^2-xy+y^2}$

$= \dfrac{x^2-y^2}{x^3+y^3} \cdot \dfrac{x^2-xy+y^2}{x-y}$

$= \dfrac{\cancel{(x+y)}\,\cancel{(x-y)}\,\cancel{(x^2-xy+y^2)}}{\cancel{(x+y)}\,\cancel{(x^2-xy+y^2)}\cancel{(x-y)}}$

$= 1$

Least Common Denominator [6.4]

The *least common denominator,* LCD, for a set of denominators is the smallest quantity divisible by each of the denominators.

7. The LCD for $\dfrac{2}{x-3}$ and $\dfrac{3}{5}$ is $5(x-3)$.

Addition and Subtraction [6.4]

If P, Q, and R represent polynomials, $R \neq 0$, then

$$\frac{P}{R} + \frac{Q}{R} = \frac{P+Q}{R} \quad \text{and} \quad \frac{P}{R} - \frac{Q}{R} = \frac{P-Q}{R}$$

When adding or subtracting rational expressions with different denominators, we must find the LCD for all denominators and change each rational expression to an equivalent expression that has the LCD.

8. $\dfrac{2}{x-3} + \dfrac{3}{5}$

$= \dfrac{2}{x-3} \cdot \dfrac{\mathbf{5}}{\mathbf{5}} + \dfrac{3}{5} \cdot \dfrac{\mathbf{x-3}}{\mathbf{x-3}}$

$= \dfrac{3x+1}{5(x-3)}$

Complex Fractions [6.5]

A rational expression that contains, in its numerator or denominator, other rational expressions is called a *complex fraction.* One method of simplifying a complex fraction is to multiply the numerator and denominator by the LCD for all denominators.

9. $\dfrac{\frac{1}{x}+\frac{1}{y}}{\frac{1}{x}-\frac{1}{y}} = \dfrac{\mathbf{xy}\left(\frac{1}{x}+\frac{1}{y}\right)}{\mathbf{xy}\left(\frac{1}{x}-\frac{1}{y}\right)}$

$= \dfrac{y+x}{y-x}$

Equations Involving Rational Expressions [6.6]

To solve an equation involving rational expressions, we first find the LCD for all denominators appearing on either side of the equation. We then multiply both sides by the LCD to clear the equation of all fractions and solve as usual.

10. Solve $\frac{x}{2} + 3 = \frac{1}{3}$.

$\mathbf{6}(\frac{x}{2}) + \mathbf{6} \cdot 3 = \mathbf{6} \cdot \frac{1}{3}$

$3x + 18 = 2$

$x = -\frac{16}{3}$

COMMON MISTAKES

1. Attempting to divide the numerator and denominator of a rational expression by a quantity that is not a factor of both. Like this:

$$\frac{\cancel{x^2} - \overset{3}{\cancel{9x}} + \overset{2}{\cancel{20}}}{\cancel{x^2} - \underset{1}{\cancel{3x}} - \underset{1}{\cancel{10}}} \qquad \textbf{Mistake}$$

   This makes no sense at all. The numerator and denominator must be factored completely before any factors they have in common can be recognized:

$$\frac{x^2 - 9x + 20}{x^2 - 3x - 10} = \frac{\cancel{(x-5)}(x-4)}{\cancel{(x-5)}(x+2)}$$
$$= \frac{x-4}{x+2}$$

2. Forgetting to check solutions to equations involving rational expressions. When we multiply both sides of an equation by a quantity containing the variable, we must be sure to check for extraneous solutions.

Chapter 6 Review/Test

The problems below form a comprehensive review of the material in this chapter. They can be used to study for exams. If you would like to take a practice test on this chapter, you can use the odd-numbered problems. Give yourself an hour and work as many of the odd-numbered problems as possible. When you are finished, or when an hour has passed, check your answers with the answers in the back of the book. You can use the even-numbered problems for a second practice test.

Reduce to lowest terms. [6.1]

1. $\dfrac{125x^4yz^3}{35x^2y^4z^3}$ **2.** $\dfrac{a^3 - ab^2}{4a + 4b}$ **3.** $\dfrac{x^2 - 25}{x^2 + 10x + 25}$ **4.** $\dfrac{ax + x - 5a - 5}{ax - x - 5a + 5}$

Divide. If the denominator is a factor of the numerator, you may want to factor the numerator and divide out the common factor. [6.2]

5. $\dfrac{12x^3 + 8x^2 + 16x}{4x^2}$ **6.** $\dfrac{27a^2b^3 - 15a^3b^2 + 21a^4b^4}{-3a^2b^2}$

7. $\dfrac{x^{6n} - x^{5n}}{x^{3n}}$ **8.** $\dfrac{x^2 - x - 6}{x - 3}$ **9.** $\dfrac{5x^2 - 14xy - 24y^2}{x - 4y}$ **10.** $\dfrac{y^4 - 16}{y - 2}$

11. $\dfrac{8x^2 - 26x - 9}{2x - 7}$ **12.** $\dfrac{2y^3 - 9y^2 - 17y + 39}{2y - 3}$

Multiply and divide as indicated. [6.3]

13. $\dfrac{3}{4} \cdot \dfrac{12}{15} \div \dfrac{1}{3}$ **14.** $\dfrac{15x^2y}{8xy^2} \div \dfrac{10xy}{4x}$

15. $\dfrac{x^3 - 1}{x^4 - 1} \cdot \dfrac{x^2 - 1}{x^2 + x + 1}$ **16.** $\dfrac{a^2 + 5a + 6}{a + 1} \cdot \dfrac{a + 5}{a^2 + 2a - 3} \div \dfrac{a^2 + 7a + 10}{a^2 - 1}$

17. $\dfrac{ax + bx + 2a + 2b}{ax - 3a + bx - 3b} \div \dfrac{ax - bx - 2a + 2b}{ax - bx - 3a + 3b}$ **18.** $(4x^2 - 9) \cdot \dfrac{x + 3}{2x + 3}$

Add and subtract as indicated. [6.4]

19. $\dfrac{3}{5} - \dfrac{1}{10} + \dfrac{8}{15}$ **20.** $\dfrac{5}{x - 5} - \dfrac{x}{x - 5}$

21. $\dfrac{1}{x} + \dfrac{1}{x^2} + \dfrac{1}{x^3}$ **22.** $\dfrac{8}{y^2 - 16} - \dfrac{7}{y^2 - y - 12}$

23. $\dfrac{x - 2}{x^2 + 5x + 4} - \dfrac{x - 4}{2x^2 + 12x + 16}$ **24.** $3 + \dfrac{4}{5x - 2}$

Simplify each complex fraction. [6.5]

25. $\dfrac{1 + \frac{2}{3}}{1 - \frac{2}{3}}$ **26.** $\dfrac{\frac{4a}{2a^3 + 2}}{\frac{8a}{4a + 4}}$ **27.** $1 + \dfrac{1}{x + \frac{1}{x}}$ **28.** $\dfrac{1 - \frac{9}{x^2}}{1 - \frac{1}{x} - \frac{6}{x^2}}$

Solve each equation. [6.6]

29. $\dfrac{3}{x - 1} = \dfrac{3}{5}$ **30.** $\dfrac{x + 1}{3} + \dfrac{x - 3}{4} = \dfrac{1}{6}$ **31.** $\dfrac{5}{y + 1} = \dfrac{4}{y + 2}$ **32.** $\dfrac{x + 6}{x + 3} - 2 = \dfrac{3}{x + 3}$

33. $\dfrac{4}{x^2 - x - 12} + \dfrac{1}{x^2 - 9} = \dfrac{2}{x^2 - 7x + 12}$ **34.** $\dfrac{a + 4}{a^2 + 5a} = \dfrac{-2}{a^2 - 25}$

35. **Distance, Rate, and Time** A car makes a 120-mile trip 10 miles per hour faster than a truck. The truck takes 2 hours longer to make the trip. What are the speeds of the car and the truck? [6.7]

36. **Average Speed** A jogger covers 3.5 miles in 28 minutes. Find the average speed of the jogger in miles per hour. [6.7]

37. **Unit Analysis** The speed of sound is 1,088 feet per second. Convert the speed of sound to miles per hour. Round your answer to the nearest whole number. [6.7]

Chapter 6 Projects

RATIONAL EXPRESSIONS

Group Project

Number of People 3

Time Needed 10–15 minutes

Equipment Pencil and paper

Procedure The four problems shown here all involve the same rational expressions. Often, students who have worked problems successfully on their homework have trouble when they take a test on rational expressions because the problems are mixed up and do not have similar instructions. Noticing similarities and differences between the types of problems involving rational expressions can help with this situation.

Problem 1: Add: $\frac{-2}{x^2 - 2x - 3} + \frac{3}{x^2 - 9}$

Problem 2: Divide: $\frac{-2}{x^2 - 2x - 3} \div \frac{3}{x^2 - 9}$

Problem 3: Solve: $\frac{-2}{x^2 - 2x - 3} + \frac{3}{x^2 - 9} = -1$

Problem 4: Simplify: $\dfrac{\frac{-2}{x^2 - 2x - 3}}{\frac{3}{x^2 - 9}}$

1. Which problems here do not require the use of a least common denominator?

2. Which two problems involve multiplying by the least common denominator?

3. Which of the problems will have an answer that is one or two numbers but no variables?

4. Work each of the four problems.

FERRIS WHEEL AND *THE THIRD MAN*

Research Project

Among the large Ferris wheels built around the turn of the twentieth century was one built in Vienna in 1897. It is the only one of those large wheels that is still in operation today. Known as the Riesenrad, it has a diameter of 197 feet and can carry a total of 800 people. A brochure that gives some statistics associated with the Riesenrad indicates that passengers riding it travel at 2 feet 6 inches per second. You can check the accuracy of this number by watching the movie *The Third Man.* In the movie, Orson Welles rides the Riesenrad through one complete revolution. Play *The Third Man* on a VCR so you can view the Riesenrad in operation. Use the pause button and the timer on the VCR to time how long it takes Orson Welles to ride once around the wheel. Then calculate his average speed during the ride. Use your results to either prove or disprove the claim that passengers travel at 2 feet 6 inches per second on the Riesenrad. When you have finished, write your procedures and results in essay form.

© Hulton Getty/Stone/Getty Images

© Selznick Films/The Kobal Collection

7

Charles Rotkin/Corbis

CHAPTER OUTLINE

INTERMEDIATE Algebra $f(x)$ Now™

Rational Exponents and Roots

Ecology and conservation are topics that interest most college students. If our rivers and oceans are to be preserved for future generations, we need to work to eliminate pollution from our waters. If a river is flowing at 1 meter per second and a pollutant is entering the river at a constant rate, the shape of the pollution plume can often be modeled by the simple equation

$$y = \sqrt{x}$$

The following table and graph were produced from the equation.

WIDTH OF A POLLUTANT PLUME

DISTANCE FROM SOURCE (METERS) x	WIDTH OF PLUME (METERS) y
0	0
1	1
4	2
9	3
16	4

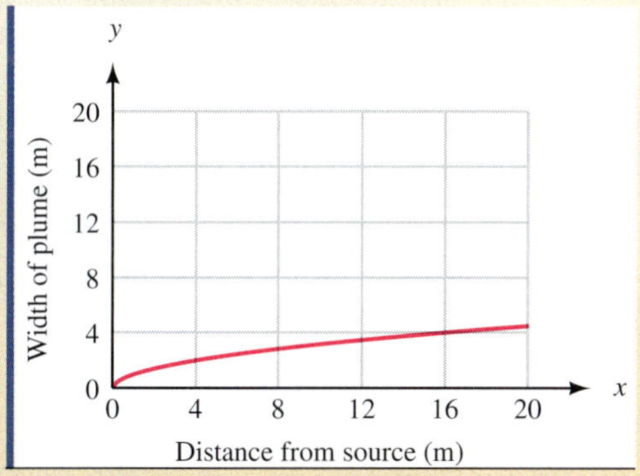

FIGURE 1

To visualize how Figure 1 models the pollutant plume, imagine that the river is flowing from left to right, parallel to the x-axis, with the x-axis as one of its banks. The pollutant is entering the river from the bank at (0, 0).

By modeling pollution with mathematics, we can use our knowledge of mathematics to help control and eliminate pollution.

Study Skills

This is the last chapter in which we will mention study skills. You know by now what works best for you and what you have to do to achieve your goals for this course. From now on, it is simply a matter of sticking with the things that work for you and avoiding the things that do not work. It seems simple, but as with anything that takes effort, it is up to you to see that you maintain the skills that get you where you want to be in the course.

If you intend to take more classes in mathematics, and you want to ensure your success in those classes, then you can work toward this goal: *Become a student who can learn mathematics on your own.* Most people who have degrees in mathematics were students who could learn mathematics on their own. This doesn't mean that you have to learn it all on your own; it simply means that if you have to, you can learn it on your own. Attaining this goal gives you independence and puts you in control of your success in any math class you take.

Getting Ready for Chapter 7

Evaluate the following.

1. 8^2

2. $(-8)^2$

3. -8^2

4. $-(-8)^2$

5. $(-2)^3$

6. -2^3

7. $12\left(\frac{1}{3}\right)$

8. $8\left(\frac{1}{4}\right)$

9. $\frac{1}{3} + \frac{5}{6}$

10. $\frac{3}{5} \cdot \frac{5}{9}$

11. $-12 - 10$

12. $-7 - (-11)$

13. $2 - \frac{3}{2}$

14. $3 + \frac{5}{4}$

Use rules for exponents to simplify the following.

15. $x^8 \cdot x^{-3}$

16. $(y^2)^3$

17. $(3r^8s^{20})^5$

18. $\frac{x^{-3}}{x^{-4}}$

19. 3^{-3}

20. $\left(\frac{3}{2}\right)^{-4}$

Rational Exponents

7.1

OBJECTIVES

A	Simplify radical expressions using the definition for roots.
B	Simplify expressions with rational exponents.

Figure 1 shows a square in which each of the four sides is 1 inch long. To find the square of the length of the diagonal c, we apply the Pythagorean theorem:

$$c^2 = 1^2 + 1^2$$
$$c^2 = 2$$

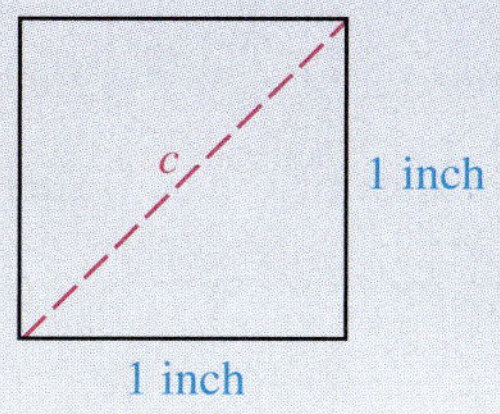

FIGURE 1

Because we know that c is positive and that its square is 2, we call c the *positive square root* of 2, and we write $c = \sqrt{2}$. Associating numbers, such as $\sqrt{2}$, with the diagonal of a square or rectangle allows us to analyze some interesting items from geometry. One particularly interesting geometric object that we will study in this section is shown in Figure 2. It is constructed from a right triangle, and the length of the diagonal is found from the Pythagorean theorem. We will come back to this figure at the end of this section.

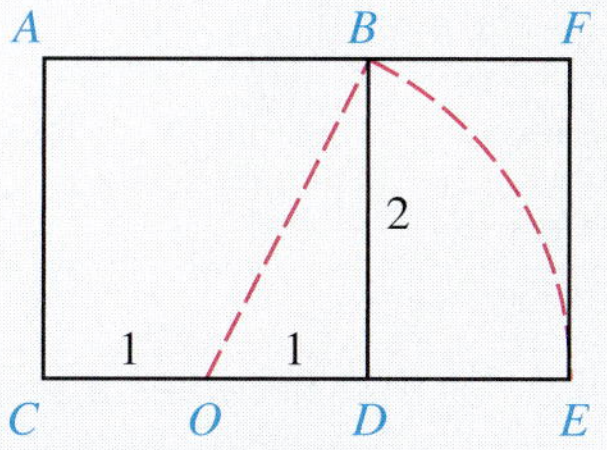

The Golden Rectangle

FIGURE 2

Previously, we developed notation (exponents) to give us the square, cube, or any other power of a number. For instance, if we wanted the square of 3, we wrote $3^2 = 9$. If we wanted the cube of 3, we wrote $3^3 = 27$. In this section, we will develop notation that will take us in the reverse direction—that is, from the square of a number, say 25, back to the original number, 5.

DEFINITION

If x is a nonnegative real number, then the expression $\sqrt{x}$ is called the **positive square root** of x and is such that

$$(\sqrt{x})^2 = x$$

In words: $\sqrt{x}$ is the nonnegative number we square to get x.

The negative square root of x, $-\sqrt{x}$, is defined in a similar manner.

EXAMPLE 1 The positive square root of 64 is 8 because 8 is the positive number with the property $8^2 = 64$. The negative square root of 64 is -8 since -8 is the negative number whose square is 64. We can summarize both of these facts by saying

$$\sqrt{64} = 8 \quad \text{and} \quad -\sqrt{64} = -8$$

NOTE

It is a common mistake to assume that an expression like $\sqrt{25}$ indicates both square roots, 5 and -5. The expression $\sqrt{25}$ indicates only the positive square root of 25, which is 5. If we want the negative square root, we must use a negative sign: $-\sqrt{25} = -5$.

PRACTICE PROBLEMS

1. Give the two square roots of 36.

Answer

1. 6, -6

The higher roots, cube roots, fourth roots, and so on are defined by definitions similar to that of square roots.

DEFINITION

If x is a real number and n is a positive integer, then

Positive square root of x, $\sqrt{x}$, is such that $(\sqrt{x})^2 = x \quad x \geq 0$

Cube root of x, $\sqrt[3]{x}$, is such that $(\sqrt[3]{x})^3 = x$

Positive fourth root of x, $\sqrt[4]{x}$, is such that $(\sqrt[4]{x})^4 = x \quad x \geq 0$

Fifth root of x, $\sqrt[5]{x}$, is such that $(\sqrt[5]{x})^5 = x$

⋮

The ***n*th root of *x***, $\sqrt[n]{x}$, is such that $(\sqrt[n]{x})^n = x \quad x \geq 0$ if n is even

The following is a table of the most common roots used in this book. Any of the roots that are unfamiliar should be memorized.

SQUARE ROOTS		CUBE ROOTS	FOURTH ROOTS
$\sqrt{0} = 0$	$\sqrt{49} = 7$	$\sqrt[3]{0} = 0$	$\sqrt[4]{0} = 0$
$\sqrt{1} = 1$	$\sqrt{64} = 8$	$\sqrt[3]{1} = 1$	$\sqrt[4]{1} = 1$
$\sqrt{4} = 2$	$\sqrt{81} = 9$	$\sqrt[3]{8} = 2$	$\sqrt[4]{16} = 2$
$\sqrt{9} = 3$	$\sqrt{100} = 10$	$\sqrt[3]{27} = 3$	$\sqrt[4]{81} = 3$
$\sqrt{16} = 4$	$\sqrt{121} = 11$	$\sqrt[3]{64} = 4$	
$\sqrt{25} = 5$	$\sqrt{144} = 12$	$\sqrt[3]{125} = 5$	
$\sqrt{36} = 6$	$\sqrt{169} = 13$		

Notation An expression like $\sqrt[3]{8}$ that involves a root is called a *radical expression.* In the expression $\sqrt[3]{8}$, the ^{3} is called the *index,* the $\sqrt{}$ is the *radical sign,* and 8 is called the *radicand.* The index of a radical must be a positive integer greater than 1. If no index is written, it is assumed to be 2.

Roots and Negative Numbers

When dealing with negative numbers and radicals, the only restriction concerns negative numbers under even roots. We can have negative signs in front of radicals and negative numbers under odd roots and still obtain real numbers. Here are some examples to help clarify this. In the last section of this chapter we will see how to deal with even roots of negative numbers.

EXAMPLES Simplify each expression, if possible.

2. $\sqrt[3]{-8} = -2$ because $(-2)^3 = -8$.
3. $\sqrt{-4}$ is not a real number since there is no real number whose square is -4.
4. $-\sqrt{25} = -5$ is the negative square root of 25.
5. $\sqrt[5]{-32} = -2$ because $(-2)^5 = -32$.
6. $\sqrt[4]{-81}$ is not a real number since there is no real number we can raise to the fourth power and obtain -81.

Variables Under a Radical

From the preceding examples it is clear that we must be careful that we do not try to take an even root of a negative number. For this reason, we will assume that all variables appearing under a radical sign represent nonnegative numbers.

NOTE

We have restricted the even roots in this definition to nonnegative numbers. Even roots of negative numbers exist but are not represented by real numbers; that is, $\sqrt{-4}$ is not a real number since there is no real number whose square is -4.

Simplify, if possible.

2. $\sqrt[3]{-64}$
3. $\sqrt{-25}$
4. $-\sqrt{4}$
5. $\sqrt[5]{-1}$
6. $\sqrt[4]{-16}$

Answers

2. -4 3. Not a real number
4. -2 5. -1
6. Not a real number

EXAMPLES Assume all variables represent nonnegative numbers and simplify each expression as much as possible.

7. $\sqrt{25a^4b^6} = 5a^2b^3$ because $(5a^2b^3)^2 = 25a^4b^6$.

8. $\sqrt[3]{x^6y^{12}} = x^2y^4$ because $(x^2y^4)^3 = x^6y^{12}$.

9. $\sqrt[4]{81r^8s^{20}} = 3r^2s^5$ because $(3r^2s^5)^4 = 81r^8s^{20}$.

Simplify.

7. $\sqrt{81a^4b^8}$

8. $\sqrt[3]{8x^3y^9}$

9. $\sqrt[4]{81a^4b^8}$

Rational Numbers as Exponents

Next we develop a second kind of notation involving exponents that will allow us to designate square roots, cube roots, and so on in another way.

Consider the equation $x = 8^{1/3}$. Although we have not encountered fractional exponents before, let's assume that all the properties of exponents hold in this case. Cubing both sides of the equation, we have

$$x^3 = (8^{1/3})^3$$

$$x^3 = 8^{(1/3)(3)}$$

$$x^3 = 8^1$$

$$x^3 = 8$$

The last line tells us that x is the number whose cube is 8. It must be true, then, that x is the cube root of 8, $x = \sqrt[3]{8}$. Since we started with $x = 8^{1/3}$, it follows that

$$8^{1/3} = \sqrt[3]{8}$$

It seems reasonable, then, to define fractional exponents as indicating roots. Here is the formal definition.

DEFINITION

If x is a real number and n is a positive integer greater than 1, then

$$x^{1/n} = \sqrt[n]{x} \qquad (x \geq 0 \text{ when } n \text{ is even})$$

In words: The quantity $x^{1/n}$ is the nth root of x.

With this definition we have a way of representing roots with exponents. Here are some examples.

EXAMPLES Write each expression as a root and then simplify, if possible.

10. $8^{1/3} = \sqrt[3]{8} = 2$

11. $36^{1/2} = \sqrt{36} = 6$

12. $-25^{1/2} = -\sqrt{25} = -5$

13. $(-25)^{1/2} = \sqrt{-25}$, which is not a real number

14. $\left(\frac{4}{9}\right)^{1/2} = \sqrt{\frac{4}{9}} = \frac{2}{3}$

Simplify.

10. $9^{1/2}$

11. $27^{1/3}$

12. $-49^{1/2}$

13. $(-49)^{1/2}$

14. $\left(\frac{16}{25}\right)^{1/2}$

The properties of exponents developed in a previous chapter were applied to integer exponents only. We will now extend these properties to include rational exponents also. We do so without proof.

Answers

7. $9a^2b^4$ **8.** $2xy^3$ **9.** $3ab^2$
10. 3 **11.** 3 **12.** -7
13. Not a real number **14.** $\frac{4}{5}$

Properties of Exponents

If a and b are real numbers and r and s are rational numbers, and a and b are nonnegative whenever r and s indicate even roots, then

1. $a^r \cdot a^s = a^{r+s}$
2. $(a^r)^s = a^{rs}$
3. $(ab)^r = a^r b^r$
4. $a^{-r} = \dfrac{1}{a^r} \quad (a \neq 0)$
5. $\left(\dfrac{a}{b}\right)^r = \dfrac{a^r}{b^r} \quad (b \neq 0)$
6. $\dfrac{a^r}{a^s} = a^{r-s} \quad (a \neq 0)$

NOTE
Unless we state otherwise, assume all variables throughout this section and problem set represent nonnegative numbers.

There are times when rational exponents can simplify our work with radicals. Here are Examples 8 and 9 again, but this time we will work them using rational exponents.

EXAMPLES Write each radical with a rational exponent and then simplify.

Simplify.

15. $\sqrt[3]{8x^3y^9}$

16. $\sqrt[4]{81a^4b^8}$

15. $\sqrt[3]{x^6y^{12}} = (x^6y^{12})^{1/3}$
$= (x^6)^{1/3}(y^{12})^{1/3}$
$= x^2y^4$

16. $\sqrt[4]{81r^8s^{20}} = (81r^8s^{20})^{1/4}$
$= 81^{1/4}(r^8)^{1/4}(s^{20})^{1/4}$
$= 3r^2s^5$

So far, the numerators of all the rational exponents we have encountered have been 1. The next theorem extends the work we can do with rational exponents to rational exponents with numerators other than 1.

We can extend our properties of exponents with the following theorem.

The Rational Exponent Theorem

If a is a nonnegative real number, m is an integer, and n is a positive integer, then

$$a^{m/n} = (a^{1/n})^m = (a^m)^{1/n}$$

Proof We can prove this theorem using the properties of exponents. Since $\frac{m}{n} = m\left(\frac{1}{n}\right)$ we have

$$a^{m/n} = a^{m(1/n)} = (a^m)^{1/n} \qquad\qquad a^{m/n} = a^{(1/n)(m)} = (a^{1/n})^m$$

NOTE
On a scientific calculator, Example 17 would look like this:

8 [y^x] [(] 2 [÷] 3 [)] [=]

Here are some examples that illustrate how we use this theorem.

EXAMPLES Simplify as much as possible.

Simplify.

17. $9^{3/2}$

18. $16^{3/4}$

Answers
15. $2xy^3$ **16.** $3ab^2$ **17.** 27 **18.** 8

17. $8^{2/3} = (8^{1/3})^2$	**Rational exponent theorem**
$= 2^2$	**Definition of fractional exponents**
$= 4$	**The square of 2 is 4**
18. $25^{3/2} = (25^{1/2})^3$	**Rational exponent theorem**
$= 5^3$	**Definition of fractional exponents**
$= 125$	**The cube of 5 is 125**

19. $9^{-3/2} = (9^{1/2})^{-3}$ Rational exponent theorem
$= 3^{-3}$ Definition of fractional exponents
$= \frac{1}{3^3}$ Property 4 for exponents
$= \frac{1}{27}$ The cube of 3 is 27

20. $\left(\frac{27}{8}\right)^{-4/3} = \left[\left(\frac{27}{8}\right)^{1/3}\right]^{-4}$ Rational exponent theorem
$= \left(\frac{3}{2}\right)^{-4}$ Definition of fractional exponents
$= \left(\frac{2}{3}\right)^{4}$ Property 4 for exponents
$= \frac{16}{81}$ The fourth power of $\frac{2}{3}$ is $\frac{16}{81}$

19. $8^{-2/3}$

20. $\left(\frac{16}{81}\right)^{-3/4}$

EXAMPLES Assume all variables represent positive quantities and simplify as much as possible.

21. $x^{1/3} \cdot x^{5/6} = x^{1/3+5/6}$ Property 1
$= x^{2/6+5/6}$ LCD is 6
$= x^{7/6}$ Add fractions

22. $(y^{2/3})^{3/4} = y^{(2/3)(3/4)}$ Property 2
$= y^{1/2}$ Multiply fractions: $\frac{2}{3} \cdot \frac{3}{4} = \frac{6}{12} = \frac{1}{2}$

23. $\frac{z^{1/3}}{z^{1/4}} = z^{1/3-1/4}$ Property 6
$= z^{4/12-3/12}$ LCD is 12
$= z^{1/12}$ Subtract fractions

24. $\frac{(x^{-3}y^{1/2})^4}{x^{10}y^{3/2}} = \frac{(x^{-3})^4(y^{1/2})^4}{x^{10}y^{3/2}}$ Property 3
$= \frac{x^{-12}y^2}{x^{10}y^{3/2}}$ Property 2
$= x^{-22}y^{1/2}$ Property 6
$= \frac{y^{1/2}}{x^{22}}$ Property 4

Simplify.

21. $x^{1/2} \cdot x^{1/4}$

22. $(y^{3/5})^{5/6}$

23. $\frac{z^{3/4}}{z^{2/3}}$

24. $\frac{(x^{1/3}y^{-3})^6}{x^4y^{10}}$

Facts From Geometry: The Pythagorean Theorem (Again) and the Golden Rectangle

Now that we have had some experience working with square roots, we can rewrite the Pythagorean theorem using a square root. If triangle *ABC* is a right triangle with $C = 90°$, then the length of the longest side is the *positive square root* of the sum of the squares of the other two sides (see Figure 3).

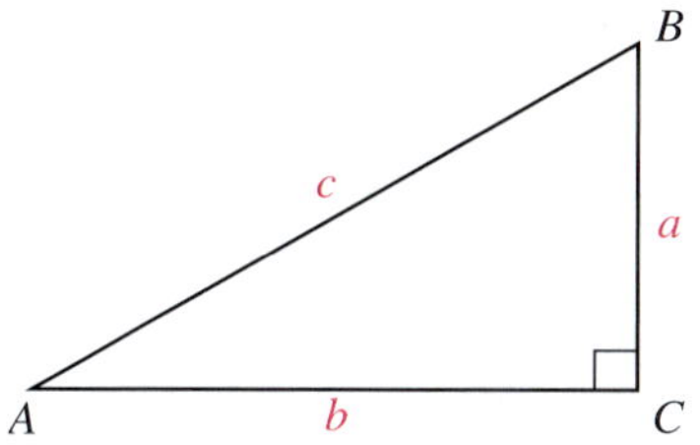

$$c = \sqrt{a^2 + b^2}$$

FIGURE 3

Answers

19. $\frac{1}{4}$ 20. $\frac{27}{8}$ 21. $x^{3/4}$ 22. $y^{1/2}$
23. $z^{1/12}$ 24. $\frac{1}{x^2y^{28}}$

NOTE

In the introduction to this section we mentioned the golden rectangle. Its origins can be traced back more than 2,000 years to the Greek civilization that produced Pythagoras, Socrates, Plato, Aristotle, and Euclid. The most important mathematical work to come from that Greek civilization was Euclid's *Elements,* an elegantly written summary of all that was known about geometry at that time in history. Euclid's *Elements,* according to Howard Eves, an authority on the history of mathematics, exercised a greater influence on scientific thinking than any other work. Here is how we construct a golden rectangle from a square of side 2, using the same method that Euclid used in his *Elements.*

Constructing a Golden Rectangle from a Square of Side 2

Step 1: Draw a square with a side of length two. Connect the midpoint of side *CD* to corner *B*. (Note that we have labeled the midpoint of segment *CD* with the letter *O*.)

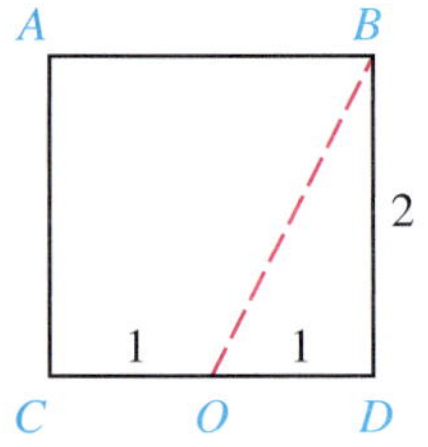

Step 2: Drop the diagonal from step 1 down so it aligns with side *CD*.

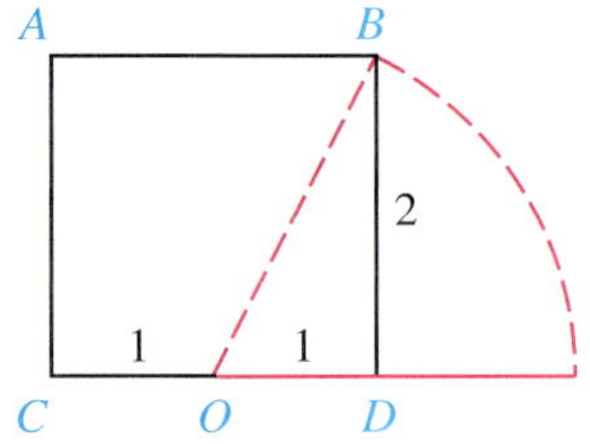

Step 3: Form rectangle *ACEF*. This is a golden rectangle.

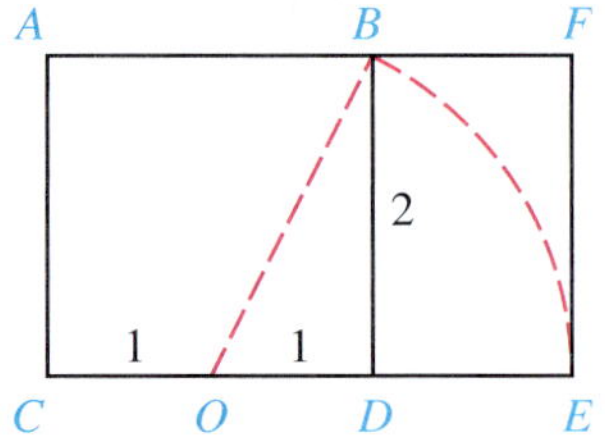

All golden rectangles are constructed from squares. Every golden rectangle, no matter how large or small it is, will have the same shape. To associate a number with the shape of the golden rectangle, we use the ratio of its length to its width. This ratio is called the *golden ratio.* To calculate the golden ratio, we must first find the length of the diagonal we used to construct the golden rectangle. Figure 4 shows the golden rectangle that we constructed from a square of side 2. The length of the diagonal *OB* is found by applying the Pythagorean theorem to triangle *OBD*.

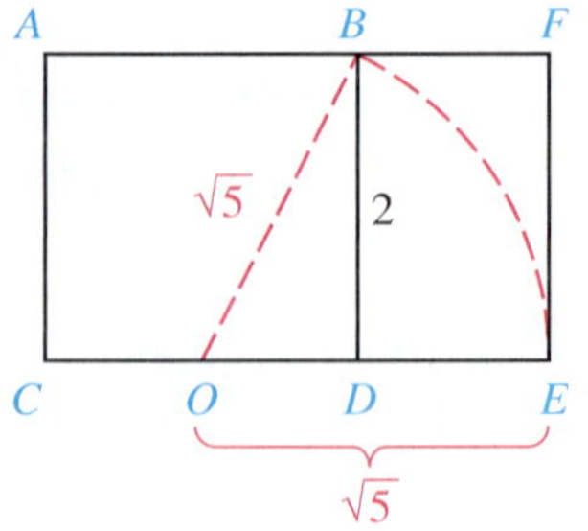

FIGURE 4

The length of segment *OE* is equal to the length of diagonal *OB*; both are $\sqrt{5}$. Since the distance from *C* to *O* is 1, the length *CE* of the golden rectangle is $1 + \sqrt{5}$. Now we can find the golden ratio:

$$\text{Golden ratio} = \frac{\text{length}}{\text{width}} = \frac{CE}{EF} = \frac{1 + \sqrt{5}}{2}$$

USING TECHNOLOGY

Graphing Calculators—A Word of Caution

Some graphing calculators give surprising results when evaluating expressions such as $(-8)^{2/3}$. As you know from reading this section, the expression $(-8)^{2/3}$ simplifies to 4, either by taking the cube root first and then squaring the result, or by squaring the base first and then taking the cube root of the result. Here are three different ways to evaluate this expression on your calculator:

1. (−8)^(2/3) **To evaluate $(-8)^{2/3}$**
2. ((−8)^2)^(1/3) **To evaluate $((-8)^2)^{1/3}$**
3. ((−8)^(1/3))^2 **To evaluate $((-8)^{1/3})^2$**

Note any difference in the results.

Next, graph each of the following functions, one at a time.

1. $Y_1 = X^{2/3}$ **2.** $Y_2 = (X^2)^{1/3}$ **3.** $Y_3 = (X^{1/3})^2$

The correct graph is shown in Figure 5. Note which of your graphs match the correct graph.

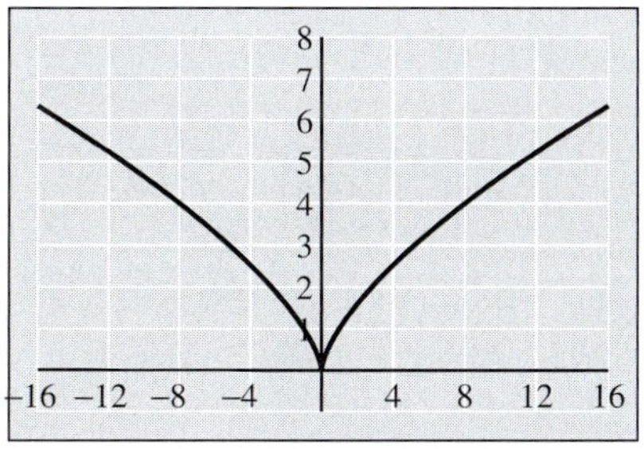

FIGURE 5

Different calculators evaluate exponential expressions in different ways. You should use the method (or methods) that gave you the correct graph.

Getting Ready for Class

After reading through the preceding section, respond in your own words and in complete sentences.

1. Every real number has two square roots. Explain the notation we use to tell them apart. Use the square roots of 3 for examples.
2. Explain why a square root of −4 is not a real number.
3. We use the notation $\sqrt{2}$ to represent the positive square root of 2. Explain why there isn't a simpler way to express the positive square root of 2.
4. For the expression $a^{m/n}$, explain the significance of the numerator m and the significance of the denominator n in the exponent.

Linking Objectives and Examples

Next to each **objective**, list the examples that are best described by that objective.

A	
B	

Problem Set 7.1 INTERMEDIATE Algebra f(x) Now™

Find each of the following roots, if possible.

1. $\sqrt{144}$
2. $-\sqrt{144}$
3. $\sqrt{-144}$
4. $\sqrt{-49}$
5. $-\sqrt{49}$
6. $\sqrt{49}$
7. $\sqrt[3]{-27}$
8. $-\sqrt[3]{27}$
9. $\sqrt[4]{16}$
10. $-\sqrt[4]{16}$
11. $\sqrt[4]{-16}$
12. $-\sqrt[4]{-16}$
13. $\sqrt{0.04}$
14. $\sqrt{0.81}$
15. $\sqrt[3]{0.008}$
16. $\sqrt[3]{0.125}$

Simplify each expression. Assume all variables represent nonnegative numbers.

17. $\sqrt{36a^8}$
18. $\sqrt{49a^{10}}$
19. $\sqrt[3]{27a^{12}}$
20. $\sqrt[3]{8a^{15}}$
21. $\sqrt[5]{32x^{10}y^5}$
22. $\sqrt[5]{32x^5y^{10}}$
23. $\sqrt[4]{16a^{12}b^{20}}$
24. $\sqrt[4]{81a^{24}b^8}$

Use the definition of rational exponents to write each of the following with the appropriate root. Then simplify.

25. $36^{1/2}$
26. $49^{1/2}$
27. $-9^{1/2}$
28. $-16^{1/2}$
29. $8^{1/3}$
30. $-8^{1/3}$
31. $(-8)^{1/3}$
32. $-27^{1/3}$
33. $32^{1/5}$
34. $81^{1/4}$
35. $\left(\frac{81}{25}\right)^{1/2}$
36. $\left(\frac{64}{125}\right)^{1/3}$

Use the rational exponent theorem to simplify each of the following as much as possible.

37. $27^{2/3}$ **38.** $8^{4/3}$ **39.** $25^{3/2}$ **40.** $81^{3/4}$

Simplify each expression. Remember, negative exponents give reciprocals.

41. $27^{-1/3}$ **42.** $9^{-1/2}$ **43.** $81^{-3/4}$ **44.** $4^{-3/2}$

45. $\left(\frac{25}{36}\right)^{-1/2}$ **46.** $\left(\frac{16}{49}\right)^{-1/2}$ **47.** $\left(\frac{81}{16}\right)^{-3/4}$ **48.** $\left(\frac{27}{8}\right)^{-2/3}$

49. $16^{1/2} + 27^{1/3}$ **50.** $25^{1/2} + 100^{1/2}$ **51.** $8^{-2/3} + 4^{-1/2}$ **52.** $49^{-1/2} + 25^{-1/2}$

Use the properties of exponents to simplify each of the following as much as possible. Assume all bases are positive.

53. $x^{3/5} \cdot x^{1/5}$ **54.** $x^{3/4} \cdot x^{5/4}$ **55.** $(a^{3/4})^{4/3}$ **56.** $(a^{2/3})^{3/4}$

57. $\frac{x^{1/5}}{x^{3/5}}$ **58.** $\frac{x^{2/7}}{x^{5/7}}$ **59.** $\frac{x^{5/6}}{x^{2/3}}$ **60.** $\frac{x^{7/8}}{x^{8/7}}$

61. $(x^{3/5}y^{5/6}z^{1/3})^{3/5}$ **62.** $(x^{3/4}y^{1/8}z^{5/6})^{4/5}$ **63.** $\frac{a^{3/4}b^{2}}{a^{7/8}b^{1/4}}$ **64.** $\frac{a^{1/3}b^{4}}{a^{3/5}b^{1/3}}$

65. $\frac{(y^{2/3})^{3/4}}{(y^{1/3})^{3/5}}$ **66.** $\frac{(y^{5/4})^{2/5}}{(y^{1/4})^{4/3}}$ **67.** $\left(\frac{a^{-1/4}}{b^{1/2}}\right)^{8}$ **68.** $\left(\frac{a^{-1/5}}{b^{1/3}}\right)^{15}$

69. $\frac{(r^{-2}s^{1/3})^{6}}{r^{8}s^{3/2}}$ **70.** $\frac{(r^{-5}s^{1/2})^{4}}{r^{12}s^{5/2}}$ **71.** $\frac{(25a^{6}b^{4})^{1/2}}{(8a^{-9}b^{3})^{-1/3}}$ **72.** $\frac{(27a^{3}b^{6})^{1/3}}{(81a^{8}b^{-4})^{1/4}}$

Applying the Concepts

73. **Maximum Speed** The maximum speed (v) that an automobile can travel around a curve of radius r without skidding is given by the equation

$$v = \left(\frac{5r}{2}\right)^{1/2}$$

where v is in miles per hour and r is measured in feet. What is the maximum speed a car can travel around a curve with a radius of 250 feet without skidding?

74. **Golden Ratio** The golden ratio is the ratio of the length to the width in any golden rectangle. The exact value of this number is $\frac{1+\sqrt{5}}{2}$. Use a calculator to find a decimal approximation to this number and round it to the nearest thousandth.

75. **Chemistry** Figure 6 shows part of a model of a magnesium oxide (MgO) crystal. Each corner of the square is at the center of one oxygen ion (O^{2-}), and the center of the middle ion is at the center of the square. The radius for each oxygen ion is 126 picometers (pm), and the radius for each magnesium ion (Mg^{2+}) is 86 picometers.
   a. Find the length of the side of the square. Write your answer in picometers.
   b. Find the length of the diagonal of the square. Write your answer in picometers.
   c. If 1 meter is 10^{12} picometers, give the length of the diagonal of the square in meters.

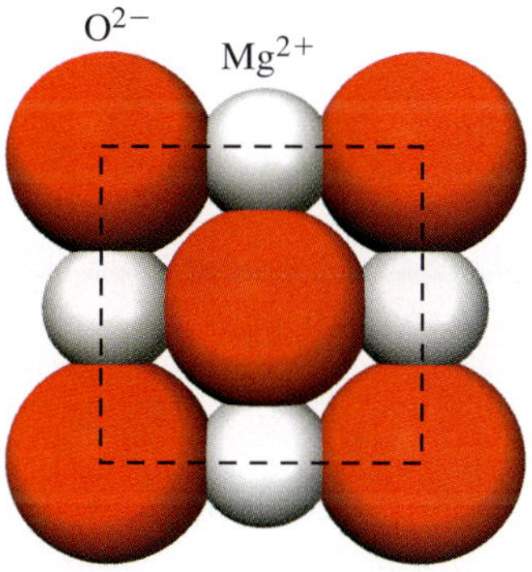

FIGURE 6 (Susan M. Young)

76. **Geometry** The length of each side of the cube shown in Figure 7 is 1 inch.
   a. Find the length of the diagonal CH.
   b. Find the length of the diagonal CF.

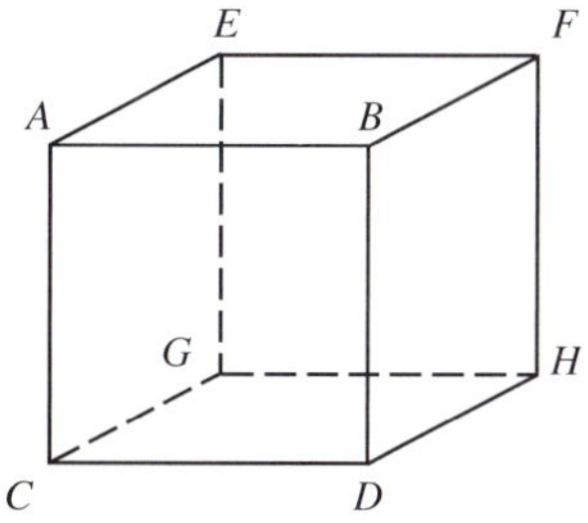

FIGURE 7

77. **Comparing Graphs** Identify the graph with the correct equation.
   a. $y = x$
   b. $y = x^2$
   c. $y = x^{2/3}$
   d. What are the two points of intersection of all three graphs?

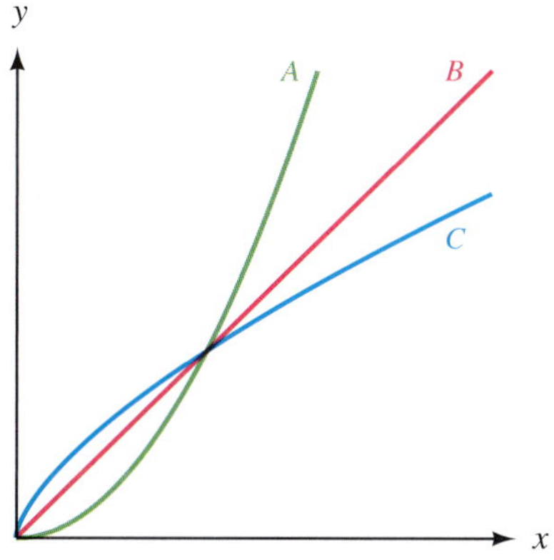

78. **Falling Objects** The time t in seconds it takes an object to fall d feet is given by the equation

$$t = \frac{1}{4}\sqrt{d}$$

   a. The Sears Tower in Chicago is 1,450 feet tall. How long would it take a penny to fall to the ground from the top of the Sears Tower?
   b. An object took 30 seconds to fall to the ground. From what distance must it have been dropped?

Getting Ready for the Next Section

Simplify.

79. $x^2(x^4 - x^3)$

80. $(x^2 - 3)(x^2 + 5)$

81. $(3a - 2b)(4a - b)$

82. $(x^2 + 3)^2$

83. $(x^3 - 2)(x^3 + 2)$

84. $(a - b)(a^2 + ab + b^2)$

85. $\dfrac{15x^2y - 20x^4y^2}{5xy}$

86. $\dfrac{12a^3b^2 - 24a^2b^4}{3ab}$

Factor.

87. $x^2 - 3x - 10$

88. $x^2 + x - 12$

89. $6x^2 + 11x - 10$

90. $10x^2 - x - 3$

Use rules of exponents to simplify.

91. $x^{2/3} \cdot x^{4/3}$

92. $x^{1/4} \cdot x^{3/4}$

93. $(t^{1/2})^2$

94. $(x^{3/2})^2$

95. $\dfrac{x^{2/3}}{x^{1/3}}$

96. $\dfrac{x^{1/2}}{x^{1/2}}$

Maintaining Your Skills

Multiply.

97. $x^2(x^4 - x)$

98. $5x^2(2x^3 - x)$

99. $(x - 3)(x + 5)$

100. $(x - 2)(x + 2)$

101. $(x^2 - 5)^2$

102. $(x^2 + 5)^2$

103. $(x - 3)(x^2 + 3x + 9)$

104. $(x + 3)(x^2 - 3x + 9)$

Extending the Concepts

105. Show that the expression $(a^{1/2} + b^{1/2})^2$ is not equal to $a + b$ by replacing a with 9 and b with 4 in both expressions and then simplifying each.

106. Show that the statement $(a^2 + b^2)^{1/2} = a + b$ is not, in general, true by replacing a with 3 and b with 4 and then simplifying both sides.

107. You may have noticed, if you have been using a calculator to find roots, that you can find the fourth root of a number by pressing the square root button twice. Written in symbols, this fact looks like this:

$$\sqrt{\sqrt{a}} = \sqrt[4]{a} \qquad (a \ge 0)$$

Show that this statement is true by rewriting each side with exponents instead of radical notation and then simplifying the left side.

108. Show that the following statement is true by rewriting each side with exponents instead of radical notation and then simplifying the left side.

$$\sqrt[3]{\sqrt{a}} = \sqrt[6]{a} \qquad (a \ge 0)$$

More Expressions Involving Rational Exponents

7.2

Suppose you purchased 10 silver proof coin sets in 1997 for \$21 each, for a total investment of \$210. Three years later, in 2000, you find that each set is worth \$30, which means that your 10 sets have a total value of \$300.

You can calculate the annual rate of return on this investment using a formula that involves rational exponents. The annual rate of return will tell you at what interest rate you would have to invest your original \$210 for it to be worth \$300 three years later. As you will see at the end of this section, the annual rate of return on this investment is 12.6%, which is a good return on your money.

In this section we will look at multiplication, division, factoring, and simplification of some expressions that resemble polynomials but contain rational exponents. The problems in this section will be of particular interest to you if you are planning to take either an engineering calculus class or a business calculus class. As was the case in the previous section, we will assume all variables represent nonnegative real numbers. That way, we will not have to worry about the possibility of introducing undefined terms—even roots of negative numbers—into any of our examples. Let's begin this section with a look at multiplication of expressions containing rational exponents.

OBJECTIVES

A	Multiply expressions with rational exponents.
B	Divide expressions with rational exponents.
C	Factor expressions with rational exponents.
D	Add and subtract expressions with rational exponents.

EXAMPLE 1

Multiply $x^{2/3}(x^{4/3} - x^{1/3})$.

Solution Applying the distributive property and then simplifying the resulting terms, we have:

$$x^{2/3}(x^{4/3} - x^{1/3}) = x^{2/3}x^{4/3} - x^{2/3}x^{1/3} \quad \text{Distributive property}$$
$$= x^{6/3} - x^{3/3} \quad \text{Add exponents}$$
$$= x^2 - x \quad \text{Simplify}$$

EXAMPLE 2

Multiply $(x^{2/3} - 3)(x^{2/3} + 5)$.

Solution Applying the FOIL method, we multiply as if we were multiplying two binomials:

$$(x^{2/3} - 3)(x^{2/3} + 5) = x^{2/3}x^{2/3} + 5x^{2/3} - 3x^{2/3} - 15$$
$$= x^{4/3} + 2x^{2/3} - 15$$

PRACTICE PROBLEMS

1. Multiply $x^{1/3}(x^{2/3} - x^{1/3})$.

2. Multiply $(x^{3/5} + 2)(x^{3/5} - 7)$.

Answers

1. $x - x^{2/3}$ 2. $x^{6/5} - 5x^{3/5} - 14$

3. Multiply.
$(5a^{1/2} - 4b^{1/2})(3a^{1/2} - b^{1/2})$

EXAMPLE 3 Multiply $(3a^{1/3} - 2b^{1/3})(4a^{1/3} - b^{1/3})$.

Solution Again, we use the FOIL method to multiply:

$$\begin{aligned}(3a^{1/3} - 2b^{1/3})(4a^{1/3} - b^{1/3}) \\ &= 3a^{1/3}4a^{1/3} - 3a^{1/3}b^{1/3} - 2b^{1/3}4a^{1/3} + 2b^{1/3}b^{1/3} \\ &= 12a^{2/3} - 11a^{1/3}b^{1/3} + 2b^{2/3}\end{aligned}$$

4. Expand $(t^{1/3} + 3)^2$.

EXAMPLE 4 Expand $(t^{1/2} - 5)^2$.

Solution We can use the definition of exponents and the FOIL method:

$$\begin{aligned}(t^{1/2} - 5)^2 &= (t^{1/2} - 5)(t^{1/2} - 5) \\ &= t^{1/2}t^{1/2} - 5t^{1/2} - 5t^{1/2} + 25 \\ &= t - 10t^{1/2} + 25\end{aligned}$$

We can obtain the same result by using the formula for the square of a binomial, $(a - b)^2 = a^2 - 2ab + b^2$.

$$\begin{aligned}(t^{1/2} - 5)^2 &= (t^{1/2})^2 - 2t^{1/2} \cdot 5 + 5^2 \\ &= t - 10t^{1/2} + 25\end{aligned}$$

5. Multiply.
$(x^{5/2} - 3^{1/2})(x^{5/2} + 3^{1/2})$

EXAMPLE 5 Multiply $(x^{3/2} - 2^{3/2})(x^{3/2} + 2^{3/2})$.

Solution This product has the form $(a - b)(a + b)$, which will result in the difference of two squares, $a^2 - b^2$:

$$\begin{aligned}(x^{3/2} - 2^{3/2})(x^{3/2} + 2^{3/2}) &= (x^{3/2})^2 - (2^{3/2})^2 \\ &= x^3 - 2^3 \\ &= x^3 - 8\end{aligned}$$

6. Multiply.
$(a^{1/3} + b^{1/3})(a^{2/3} - a^{1/3}b^{1/3} + b^{2/3})$

EXAMPLE 6 Multiply $(a^{1/3} - b^{1/3})(a^{2/3} + a^{1/3}b^{1/3} + b^{2/3})$.

Solution We can find this product by multiplying in columns:

$$\begin{array}{lllll} & a^{2/3} & + a^{1/3}b^{1/3} & + b^{2/3} & \\ & & a^{1/3} & - b^{1/3} & \\ \hline a & + a^{2/3}b^{1/3} & + a^{1/3}b^{2/3} & & \\ & - a^{2/3}b^{1/3} & - a^{1/3}b^{2/3} & & - b \\ \hline a & & & & - b \end{array}$$

The product is $a - b$.

NOTE
For Example 7 and Practice Problem 7, assume x and y are positive numbers.

Our next example involves division with expressions that contain rational exponents. As you will see, this kind of division is very similar to division of a polynomial by a monomial.

7. Divide $\dfrac{36x^{3/4}y^{1/4} - 18x^{5/4}y^{1/4}}{6x^{1/4}y^{1/4}}$.

EXAMPLE 7 Divide $\dfrac{15x^{2/3}y^{1/3} - 20x^{4/3}y^{2/3}}{5x^{1/3}y^{1/3}}$.

Solution We can approach this problem in the same way we approached division by a monomial. We simply divide each term in the numerator by the term in the denominator:

$$\begin{aligned}\frac{15x^{2/3}y^{1/3} - 20x^{4/3}y^{2/3}}{5x^{1/3}y^{1/3}} &= \frac{15x^{2/3}y^{1/3}}{5x^{1/3}y^{1/3}} - \frac{20x^{4/3}y^{2/3}}{5x^{1/3}y^{1/3}} \\ &= 3x^{1/3} - 4xy^{1/3}\end{aligned}$$

Answers
3. $15a - 17a^{1/2}b^{1/2} + 4b$
4. $t^{2/3} + 6t^{1/3} + 9$
5. $x^5 - 3$ 6. $a + b$
7. $6x^{1/2} - 3x$

The next three examples involve factoring. In the first example, we are told what to factor from each term of an expression.

EXAMPLE 8 Factor $3(x-2)^{1/3}$ from $12(x-2)^{4/3} - 9(x-2)^{1/3}$, and then simplify, if possible.

Solution This solution is similar to factoring out the greatest common factor:

$$12(x-2)^{4/3} - 9(x-2)^{1/3} = 3(x-2)^{1/3}[4(x-2) - 3]$$
$$= 3(x-2)^{1/3}(4x - 11)$$

Although an expression containing rational exponents is not a polynomial—remember, a polynomial must have exponents that are whole numbers—we are going to treat the expressions that follow as if they were polynomials.

EXAMPLE 9 Factor $x^{2/3} - 3x^{1/3} - 10$ as if it were a trinomial.

Solution We can think of $x^{2/3} - 3x^{1/3} - 10$ as if it is a trinomial in which the variable is $x^{1/3}$. To see this, replace $x^{1/3}$ with y to get

$$y^2 - 3y - 10$$

Since this trinomial in y factors as $(y-5)(y+2)$, we can factor our original expression similarly:

$$x^{2/3} - 3x^{1/3} - 10 = (x^{1/3} - 5)(x^{1/3} + 2)$$

Remember, with factoring, we can always multiply our factors to check that we have factored correctly.

EXAMPLE 10 Factor $6x^{2/5} + 11x^{1/5} - 10$ as if it were a trinomial.

Solution We can think of the expression in question as a trinomial in $x^{1/5}$.

$$6x^{2/5} + 11x^{1/5} - 10 = (3x^{1/5} - 2)(2x^{1/5} + 5)$$

In our next example, we combine two expressions by applying the methods we used to add and subtract fractions or rational expressions.

EXAMPLE 11 Subtract $(x^2+4)^{1/2} - \dfrac{x^2}{(x^2+4)^{1/2}}$.

Solution To combine these two expressions, we need to find a least common denominator, change to equivalent fractions, and subtract numerators. The least common denominator is $(x^2+4)^{1/2}$.

$$(x^2+4)^{1/2} - \frac{x^2}{(x^2+4)^{1/2}} = \frac{(x^2+4)^{1/2}}{1} \cdot \frac{\mathbf{(x^2+4)^{1/2}}}{\mathbf{(x^2+4)^{1/2}}} - \frac{x^2}{(x^2+4)^{1/2}}$$
$$= \frac{x^2 + 4 - x^2}{(x^2+4)^{1/2}}$$
$$= \frac{4}{(x^2+4)^{1/2}}$$

8. Factor $2(x-3)^{1/2}$ from $8(x-3)^{3/2} - 6(x-3)^{1/2}$.

9. Factor $x^{2/3} - 4x^{1/3} - 21$.

10. Factor $6x^{2/3} + 19x^{1/3} + 10$.

11. Subtract.

$$(x^2-3)^{1/2} - \frac{x^2}{(x^2-3)^{1/2}}$$

Answers

8. $2(x-3)^{1/2}(4x - 15)$
9. $(x^{1/3} - 7)(x^{1/3} + 3)$
10. $(3x^{1/3} + 2)(2x^{1/3} + 5)$
11. $\dfrac{-3}{(x^2-3)^{1/2}}$

12. Find the annual rate of return on a coin collection that was purchased for \$600 and sold 3 years later for \$800.

EXAMPLE 12

If you purchase an investment for P dollars and t years later it is worth A dollars, then the annual rate of return r on that investment is given by the formula

$$r = \left(\frac{A}{P}\right)^{1/t} - 1$$

Find the annual rate of return on a coin collection that was purchased for \$210 and sold 3 years later for \$300.

Solution Using $A = 300$, $P = 210$, and $t = 3$ in the formula, we have

$$r = \left(\frac{300}{210}\right)^{1/3} - 1$$

The easiest way to simplify this expression is with a calculator.

[(] 300 [÷] 210 [)] [^] [(] 1 [÷] 3 [)] [−] 1 [=]

Allowing three decimal places, the result is 0.126. The annual return on the coin collection is approximately 12.6%. To do as well with a savings account, we would have to invest the original \$210 in an account that paid 12.6%, compounded annually.

Linking Objectives and Examples

Next to each **objective**, list the examples that are best described by that objective.

A	
B	
C	
D	

Getting Ready for Class

After reading through the preceding section, respond in your own words and in complete sentences.

1. When multiplying expressions with fractional exponents, when do we add the fractional exponents?
2. Is it possible to multiply two expressions with fractional exponents and end up with an expression containing only integer exponents? Support your answer with examples.
3. Write an application modeled by the equation $r = \left(\frac{1{,}000}{600}\right)^{1/8} - 1$.
4. When can you use the FOIL method with expressions that contain rational exponents?

Answer
12. 10.1%

Problem Set 7.2 INTERMEDIATE Algebra f(x) Now™

Multiply. (Assume all variables in this problem set represent nonnegative real numbers.)

1. $x^{2/3}(x^{1/3} + x^{4/3})$

2. $x^{2/5}(x^{3/5} - x^{8/5})$

3. $a^{1/2}(a^{3/2} - a^{1/2})$

4. $a^{1/4}(a^{3/4} + a^{7/4})$

5. $2x^{1/3}(3x^{8/3} - 4x^{5/3} + 5x^{2/3})$

6. $5x^{1/2}(4x^{5/2} + 3x^{3/2} + 2x^{1/2})$

7. $4x^{1/2}y^{3/5}(3x^{3/2}y^{-3/5} - 9x^{-1/2}y^{7/5})$

8. $3x^{4/5}y^{1/3}(4x^{6/5}y^{-1/3} - 12x^{-4/5}y^{5/3})$

9. $(x^{2/3} - 4)(x^{2/3} + 2)$

10. $(x^{2/3} - 5)(x^{2/3} + 2)$

11. $(a^{1/2} - 3)(a^{1/2} - 7)$

12. $(a^{1/2} - 6)(a^{1/2} - 2)$

13. $(4y^{1/3} - 3)(5y^{1/3} + 2)$

14. $(5y^{1/3} - 2)(4y^{1/3} + 3)$

15. $(5x^{2/3} + 3y^{1/2})(2x^{2/3} + 3y^{1/2})$

16. $(4x^{2/3} - 2y^{1/2})(5x^{2/3} - 3y^{1/2})$

17. $(t^{1/2} + 5)^2$

18. $(t^{1/2} - 3)^2$

19. $(x^{3/2} + 4)^2$

20. $(x^{3/2} - 6)^2$

21. $(a^{1/2} - b^{1/2})^2$

22. $(a^{1/2} + b^{1/2})^2$

23. $(2x^{1/2} - 3y^{1/2})^2$

24. $(5x^{1/2} + 4y^{1/2})^2$

25. $(a^{1/2} - 3^{1/2})(a^{1/2} + 3^{1/2})$

26. $(a^{1/2} - 5^{1/2})(a^{1/2} + 5^{1/2})$

27. $(x^{3/2} + y^{3/2})(x^{3/2} - y^{3/2})$

28. $(x^{5/2} + y^{5/2})(x^{5/2} - y^{5/2})$

29. $(t^{1/2} - 2^{3/2})(t^{1/2} + 2^{3/2})$

30. $(t^{1/2} - 5^{3/2})(t^{1/2} + 5^{3/2})$

31. $(2x^{3/2} + 3^{1/2})(2x^{3/2} - 3^{1/2})$

32. $(3x^{1/2} + 2^{3/2})(3x^{1/2} - 2^{3/2})$

33. $(x^{1/3} + y^{1/3})(x^{2/3} - x^{1/3}y^{1/3} + y^{2/3})$

34. $(x^{1/3} - y^{1/3})(x^{2/3} + x^{1/3}y^{1/3} + y^{2/3})$

35. $(a^{1/3} - 2)(a^{2/3} + 2a^{1/3} + 4)$

36. $(a^{1/3} + 3)(a^{2/3} - 3a^{1/3} + 9)$

37. $(2x^{1/3} + 1)(4x^{2/3} - 2x^{1/3} + 1)$

38. $(3x^{1/3} - 1)(9x^{2/3} + 3x^{1/3} + 1)$

39. $(t^{1/4} - 1)(t^{1/4} + 1)(t^{1/2} + 1)$

40. $(t^{1/4} - 2)(t^{1/4} + 2)(t^{1/2} + 4)$

Divide. (Assume all variables represent positive numbers.)

41. $\dfrac{18x^{3/4} + 27x^{1/4}}{9x^{1/4}}$

42. $\dfrac{25x^{1/4} + 30x^{3/4}}{5x^{1/4}}$

43. $\dfrac{12x^{2/3}y^{1/3} - 16x^{1/3}y^{2/3}}{4x^{1/3}y^{1/3}}$

44. $\dfrac{12x^{4/3}y^{1/3} - 18x^{1/3}y^{4/3}}{6x^{1/3}y^{1/3}}$

45. $\dfrac{21a^{7/5}b^{3/5} - 14a^{2/5}b^{8/5}}{7a^{2/5}b^{3/5}}$

46. $\dfrac{24a^{9/5}b^{3/5} - 16a^{4/5}b^{8/5}}{8a^{4/5}b^{3/5}}$

47. Factor $3(x-2)^{1/2}$ from $12(x-2)^{3/2} - 9(x-2)^{1/2}$.

48. Factor $4(x+1)^{1/3}$ from $4(x+1)^{4/3} + 8(x+1)^{1/3}$.

49. Factor $5(x-3)^{7/5}$ from $5(x-3)^{12/5} - 15(x-3)^{7/5}$.

50. Factor $6(x+3)^{8/7}$ from $6(x+3)^{15/7} - 12(x+3)^{8/7}$.

51. Factor $3(x+1)^{1/2}$ from $9x(x+1)^{3/2} + 6(x+1)^{1/2}$.

52. Factor $4x(x+1)^{1/2}$ from $4x^2(x+1)^{1/2} + 8x(x+1)^{3/2}$.

Factor each of the following as if it were a trinomial.

53. $x^{2/3} - 5x^{1/3} + 6$

54. $x^{2/3} - x^{1/3} - 6$

55. $a^{2/5} - 2a^{1/5} - 8$

56. $a^{2/5} + 2a^{1/5} - 8$

57. $2y^{2/3} - 5y^{1/3} - 3$

58. $3y^{2/3} + 5y^{1/3} - 2$

59. $9t^{2/5} - 25$

60. $16t^{2/5} - 49$

61. $4x^{2/7} + 20x^{1/7} + 25$

62. $25x^{2/7} - 20x^{1/7} + 4$

Simplify each of the following to a single fraction. (Assume all variables represent positive numbers.)

63. $\dfrac{3}{x^{1/2}} + x^{1/2}$

64. $\dfrac{2}{x^{1/2}} - x^{1/2}$

65. $x^{2/3} + \dfrac{5}{x^{1/3}}$

66. $x^{3/4} - \dfrac{7}{x^{1/4}}$

67. $\dfrac{3x^2}{(x^3+1)^{1/2}} + (x^3+1)^{1/2}$

68. $\dfrac{x^3}{(x^2-1)^{1/2}} + 2x(x^2-1)^{1/2}$

69. $\dfrac{x^2}{(x^2+4)^{1/2}} - (x^2+4)^{1/2}$

70. $\dfrac{x^5}{(x^2-2)^{1/2}} + 4x^3(x^2-2)^{1/2}$

Applying the Concepts

71. Investing A coin collection is purchased as an investment for \$500 and sold 4 years later for \$900. Find the annual rate of return on the investment.

72. Investing An investor buys stock in a company for \$800. Five years later, the same stock is worth \$1,600. Find the annual rate of return on the stocks.

Getting Ready for the Next Section

Simplify. Assume all variable are positive real numbers.

73. $\sqrt{25}$

74. $\sqrt{4}$

75. $\sqrt{6^2}$

76. $\sqrt{3^2}$

77. $\sqrt{16x^4y^2}$

78. $\sqrt{4x^6y^8}$

79. $\sqrt{(5y)^2}$

80. $\sqrt{(8x^3)^2}$

81. $\sqrt[3]{27}$

82. $\sqrt[3]{-8}$

83. $\sqrt[3]{2^3}$

84. $\sqrt[3]{(-5)^3}$

85. $\sqrt[3]{8a^3b^3}$

86. $\sqrt[3]{64a^6b^3}$

Fill in the blank.

87. $50 =$ ______ $\cdot 2$

88. $12 =$ ______ $\cdot 3$

89. $48x^4y^3 =$ ______ $\cdot y$

90. $40a^5b^4 =$ ______ $\cdot 5a^2b$

91. $12x^7y^6 =$ ______ $\cdot 3x$

92. $54a^6b^2c^4 =$ ______ $\cdot 2b^2c$

Maintaining Your Skills

Reduce to lowest terms.

93. $\dfrac{x^2 - 9}{x^4 - 81}$

94. $\dfrac{6 - a - a^2}{3 - 2a - a^2}$

Divide.

95. $\dfrac{15x^2y - 20x^4y^2}{5xy}$

96. $\dfrac{12x^3y^2 - 24x^2y^3}{6xy}$

Divide using long division.

97. $\dfrac{10x^2 + 7x - 12}{2x + 3}$

98. $\dfrac{6x^2 - x - 35}{2x - 5}$

99. $\dfrac{x^3 - 125}{x - 5}$

100. $\dfrac{x^3 + 64}{x + 4}$

Simplified Form for Radicals

7.3

OBJECTIVES

A	Write radical expressions in simplified form.
B	Rationalize a denominator that contains only one term.

Earlier in this chapter, we showed how the Pythagorean theorem can be used to construct a golden rectangle. In a similar manner, the Pythagorean theorem can be used to construct the attractive spiral shown here.

The Spiral of Roots

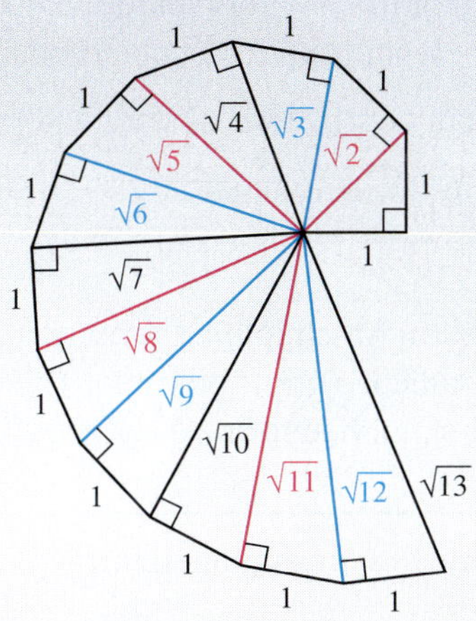

This spiral is called the Spiral of Roots because each of the diagonals is the positive square root of one of the positive integers. At the end of this section, we will use the Pythagorean theorem and some of the material in this section to construct this spiral.

In this section we will use radical notation instead of rational exponents. We will begin by stating two properties of radicals. Following this, we will give a definition for simplified form for radical expressions. The examples in this section show how we use the properties of radicals to write radical expressions in simplified form.

There are two properties of radicals. For these two properties, we will assume a and b are nonnegative real numbers whenever n is an even number.

Property 1 for Radicals

$$\sqrt[n]{ab} = \sqrt[n]{a}\,\sqrt[n]{b}$$

In words: The nth root of a product is the product of the nth roots.

Proof of Property 1

$\sqrt[n]{ab} = (ab)^{1/n}$	**Definition of fractional exponents**
$= a^{1/n}b^{1/n}$	**Exponents distribute over products**
$= \sqrt[n]{a}\,\sqrt[n]{b}$	**Definition of fractional exponents**

Property 2 for Radicals

$$\sqrt[n]{\frac{a}{b}} = \frac{\sqrt[n]{a}}{\sqrt[n]{b}} \qquad (b \neq 0)$$

In words: The nth root of a quotient is the quotient of the nth roots.

NOTE

There is no property for radicals that says the nth root of a sum is the sum of the nth roots; that is,

$$\sqrt[n]{a+b} \neq \sqrt[n]{a} + \sqrt[n]{b}$$

The proof of property 2 is similar to the proof of property 1.

The two properties of radicals allow us to change the form of and simplify radical expressions without changing their value.

NOTE

Writing a radical expression in simplified form does not always result in a simpler-looking expression. Simplified form for radicals is a way of writing radicals so they are easiest to work with.

Simplified Form for Radical Expressions

A radical expression is in *simplified form* if

1. None of the factors of the radicand (the quantity under the radical sign) can be written as powers greater than or equal to the index—that is, no perfect squares can be factors of the quantity under a square root sign, no perfect cubes can be factors of what is under a cube root sign, and so forth;
2. There are no fractions under the radical sign; and
3. There are no radicals in the denominator.

Satisfying the first condition for simplified form actually amounts to taking as much out from under the radical sign as possible. The following examples illustrate the first condition for simplified form.

PRACTICE PROBLEMS

1. Write $\sqrt{18}$ in simplified form.

EXAMPLE 1

Write $\sqrt{50}$ in simplified form.

Solution The largest perfect square that divides 50 is 25. We write 50 as $25 \cdot 2$ and apply property 1 for radicals:

$$\begin{aligned} \sqrt{50} &= \sqrt{25 \cdot 2} && \mathbf{50 = 25 \cdot 2} \\ &= \sqrt{25}\,\sqrt{2} && \textbf{Property 1} \\ &= 5\sqrt{2} && \sqrt{\mathbf{25}} = \mathbf{5} \end{aligned}$$

We have taken as much as possible out from under the radical sign—in this case, factoring 25 from 50 and then writing $\sqrt{25}$ as 5.

2. Write $\sqrt{50x^2y^3}$ in simplified form. Assume $x, y \ge 0$.

NOTE

Unless we state otherwise, assume all variables throughout this section represent nonnegative numbers. Further, if a variable appears in a denominator, assume the variable cannot be 0.

EXAMPLE 2

Write in simplified form: $\sqrt{48x^4y^3}$, where $x, y \ge 0$.

Solution The largest perfect square that is a factor of the radicand is $16x^4y^2$. Applying property 1 again, we have

$$\begin{aligned} \sqrt{48x^4y^3} &= \sqrt{16x^4y^2 \cdot 3y} \\ &= \sqrt{16x^4y^2}\,\sqrt{3y} \\ &= 4x^2y\sqrt{3y} \end{aligned}$$

3. Write $\sqrt[3]{54a^4b^3}$ in simplified form.

EXAMPLE 3

Write $\sqrt[3]{40a^5b^4}$ in simplified form.

Solution We now want to factor the largest perfect cube from the radicand. We write $40a^5b^4$ as $8a^3b^3 \cdot 5a^2b$ and proceed as we did in Examples 1 and 2.

$$\begin{aligned} \sqrt[3]{40a^5b^4} &= \sqrt[3]{8a^3b^3 \cdot 5a^2b} \\ &= \sqrt[3]{8a^3b^3}\,\sqrt[3]{5a^2b} \\ &= 2ab\sqrt[3]{5a^2b} \end{aligned}$$

Here are some further examples concerning the first condition for simplified form.

Answers

1. $3\sqrt{2}$ 2. $5xy\sqrt{2y}$
3. $3ab\sqrt[3]{2a}$

EXAMPLES Write each expression in simplified form.

4. $\sqrt{12x^7y^6} = \sqrt{4x^6y^6 \cdot 3x}$

$= \sqrt{4x^6y^6}\sqrt{3x}$

$= 2x^3y^3\sqrt{3x}$

5. $\sqrt[3]{54a^6b^2c^4} = \sqrt[3]{27a^6c^3 \cdot 2b^2c}$

$= \sqrt[3]{27a^6c^3}\sqrt[3]{2b^2c}$

$= 3a^2c\sqrt[3]{2b^2c}$

The second property of radicals is used to simplify a radical that contains a fraction.

EXAMPLE 6 Simplify $\sqrt{\dfrac{3}{4}}$.

Solution Applying property 2 for radicals, we have

$$\sqrt{\frac{3}{4}} = \frac{\sqrt{3}}{\sqrt{4}} \quad \textbf{Property 2}$$

$$= \frac{\sqrt{3}}{2} \quad \sqrt{4} = 2$$

The last expression is in simplified form because it satisfies all three conditions for simplified form.

EXAMPLE 7 Write $\sqrt{\dfrac{5}{6}}$ in simplified form.

Solution Proceeding as in Example 6, we have

$$\sqrt{\frac{5}{6}} = \frac{\sqrt{5}}{\sqrt{6}}$$

The resulting expression satisfies the second condition for simplified form since neither radical contains a fraction. It does, however, violate condition 3 since it has a radical in the denominator. Getting rid of the radical in the denominator is called *rationalizing the denominator* and is accomplished, in this case, by multiplying the numerator and denominator by $\sqrt{6}$:

$$\frac{\sqrt{5}}{\sqrt{6}} = \frac{\sqrt{5}}{\sqrt{6}} \cdot \frac{\sqrt{6}}{\sqrt{6}}$$

$$= \frac{\sqrt{30}}{\sqrt{6^2}}$$

$$= \frac{\sqrt{30}}{6}$$

EXAMPLES Rationalize the denominator.

8. $\dfrac{4}{\sqrt{3}} = \dfrac{4}{\sqrt{3}} \cdot \dfrac{\sqrt{3}}{\sqrt{3}}$

$= \dfrac{4\sqrt{3}}{\sqrt{3^2}}$

$= \dfrac{4\sqrt{3}}{3}$

Write each expression in simplified form.

4. $\sqrt{75x^5y^8}$

5. $\sqrt[4]{48a^8b^5c^4}$

6. Simplify $\sqrt{\dfrac{5}{9}}$.

7. Write $\sqrt{\dfrac{2}{3}}$ in simplified form.

NOTE

The idea behind rationalizing the denominator is to produce a perfect square under the square root sign in the denominator. This is accomplished by multiplying both the numerator and denominator by the appropriate radical.

8. Rationalize the denominator.

$$\frac{5}{\sqrt{2}}$$

Answers

4. $5x^2y^4\sqrt{3x}$ **5.** $2a^2bc\sqrt[4]{3b}$
6. $\dfrac{\sqrt{5}}{3}$ **7.** $\dfrac{\sqrt{6}}{3}$ **8.** $\dfrac{5\sqrt{2}}{2}$

9. Rationalize the denominator.

$$\frac{3\sqrt{5x}}{\sqrt{2y}}$$

9. $$\frac{2\sqrt{3x}}{\sqrt{5y}} = \frac{2\sqrt{3x}}{\sqrt{5y}} \cdot \frac{\sqrt{\mathbf{5y}}}{\sqrt{\mathbf{5y}}}$$

$$= \frac{2\sqrt{15xy}}{\sqrt{(5y)^2}}$$

$$= \frac{2\sqrt{15xy}}{5y}$$

When the denominator involves a cube root, we must multiply by a radical that will produce a perfect cube under the cube root sign in the denominator, as our next example illustrates.

10. Rationalize the denominator.

$$\frac{5}{\sqrt[3]{9}}$$

EXAMPLE 10 Rationalize the denominator in $\frac{7}{\sqrt[3]{4}}$.

Solution Since $4 = 2^2$, we can multiply both numerator and denominator by $\sqrt[3]{2}$ and obtain $\sqrt[3]{2^3}$ in the denominator.

$$\frac{7}{\sqrt[3]{4}} = \frac{7}{\sqrt[3]{2^2}}$$

$$= \frac{7}{\sqrt[3]{2^2}} \cdot \frac{\sqrt[3]{\mathbf{2}}}{\sqrt[3]{\mathbf{2}}}$$

$$= \frac{7\sqrt[3]{2}}{\sqrt[3]{2^3}}$$

$$= \frac{7\sqrt[3]{2}}{2}$$

11. Simplify $\sqrt{\dfrac{48x^3y^4}{7z}}$.

EXAMPLE 11 Simplify $\sqrt{\dfrac{12x^5y^3}{5z}}$.

Solution We use property 2 to write the numerator and denominator as two separate radicals:

$$\sqrt{\frac{12x^5y^3}{5z}} = \frac{\sqrt{12x^5y^3}}{\sqrt{5z}}$$

Simplifying the numerator, we have

$$\frac{\sqrt{12x^5y^3}}{\sqrt{5z}} = \frac{\sqrt{4x^4y^2}\sqrt{3xy}}{\sqrt{5z}}$$

$$= \frac{2x^2y\sqrt{3xy}}{\sqrt{5z}}$$

To rationalize the denominator, we multiply the numerator and denominator by $\sqrt{5z}$:

$$\frac{2x^2y\sqrt{3xy}}{\sqrt{5z}} \cdot \frac{\sqrt{\mathbf{5z}}}{\sqrt{\mathbf{5z}}} = \frac{2x^2y\sqrt{15xyz}}{\sqrt{(5z)^2}}$$

$$= \frac{2x^2y\sqrt{15xyz}}{5z}$$

The Square Root of a Perfect Square

So far in this chapter we have assumed that all our variables are nonnegative when they appear under a square root symbol. There are times, however, when this is not the case.

Answers

9. $\frac{3\sqrt{10xy}}{2y}$ **10.** $\frac{5\sqrt[3]{3}}{3}$

11. $\frac{4xy^2\sqrt{21xz}}{7z}$

Consider the following two statements:

$$\sqrt{3^2} = \sqrt{9} = 3 \quad \text{and} \quad \sqrt{(-3)^2} = \sqrt{9} = 3$$

Whether we operate on 3 or -3, the result is the same: Both expressions simplify to 3. The other operation we have worked with in the past that produces the same result is absolute value; that is,

$$|3| = 3 \quad \text{and} \quad |-3| = 3$$

This leads us to the next property of radicals.

Property 3 for Radicals

If a is a real number, then $\sqrt{a^2} = |a|$.

The result of this discussion and property 3 is simply this:

If we know a is positive, then $\sqrt{a^2} = a$.

If we know a is negative, then $\sqrt{a^2} = |a|$.

If we don't know if a is positive or negative, then $\sqrt{a^2} = |a|$.

EXAMPLES Simplify each expression. Do *not* assume the variables represent positive numbers.

12. $\sqrt{9x^2} = 3|x|$

13. $\sqrt{x^3} = |x|\sqrt{x}$

14. $\sqrt{x^2 - 6x + 9} = \sqrt{(x-3)^2} = |x-3|$

15. $\sqrt{x^3 - 5x^2} = \sqrt{x^2(x-5)} = |x|\sqrt{x-5}$

As you can see, we must use absolute value symbols when we take a square root of a perfect square, unless we know the base of the perfect square is a positive number. The same idea holds for higher even roots, but not for odd roots. With odd roots, no absolute value symbols are necessary.

EXAMPLES Simplify each expression.

16. $\sqrt[3]{(-2)^3} = \sqrt[3]{-8} = -2$

17. $\sqrt[3]{(-5)^3} = \sqrt[3]{-125} = -5$

We can extend this discussion to all roots as follows:

Extending Property 3 for Radicals

If a is a real number, then

$$\sqrt[n]{a^n} = |a| \quad \text{if} \quad n \text{ is even}$$

$$\sqrt[n]{a^n} = a \quad \text{if} \quad n \text{ is odd}$$

Simplify each expression. Do not assume the variables represent nonnegative numbers.

12. $\sqrt{16x^2}$

13. $\sqrt{25x^3}$

14. $\sqrt{x^2 + 10x + 25}$

15. $\sqrt{2x^3 + 7x^2}$

Simplify each expression.

16. $\sqrt[3]{(-3)^3}$

17. $\sqrt[3]{(-1)^3}$

Answers

12. $4|x|$ **13.** $5|x|\sqrt{x}$

14. $|x+5|$ **15.** $|x|\sqrt{2x+7}$

16. -3 **17.** -1

Linking Objectives and Examples

Next to each **objective**, list the examples that are best described by that objective.

A	
B	

Getting Ready for Class

After reading through the preceding section, respond in your own words and in complete sentences.

1. Explain why this statement is false: "The square root of a sum is the sum of the square roots."
2. What is simplified form for an expression that contains a square root?
3. Why is it not necessarily true that $\sqrt{a^2} = a$?
4. What does it mean to rationalize the denominator in an expression?

Problem Set 7.3

INTERMEDIATE **Algebra** $f(x)$ **Now**™

Use property 1 for radicals to write each of the following expressions in simplified form. (Assume all variables are nonnegative through Problem 70.)

1. $\sqrt{8}$ **2.** $\sqrt{32}$ **3.** $\sqrt{98}$ **4.** $\sqrt{75}$ **5.** $\sqrt{288}$ **6.** $\sqrt{128}$

7. $\sqrt{80}$ **8.** $\sqrt{200}$ **9.** $\sqrt{48}$ **10.** $\sqrt{27}$ **11.** $\sqrt{675}$ **12.** $\sqrt{972}$

13. $\sqrt[3]{54}$ **14.** $\sqrt[3]{24}$ **15.** $\sqrt[3]{128}$ **16.** $\sqrt[3]{162}$ **17.** $\sqrt[3]{432}$ **18.** $\sqrt[3]{1,536}$

19. $\sqrt[5]{64}$ **20.** $\sqrt[4]{48}$ **21.** $\sqrt{18x^3}$ **22.** $\sqrt{27x^5}$ **23.** $\sqrt[4]{32y^7}$ **24.** $\sqrt[5]{32y^7}$

25. $\sqrt[3]{40x^4y^7}$ **26.** $\sqrt[3]{128x^6y^2}$ **27.** $\sqrt{48a^2b^3c^4}$ **28.** $\sqrt{72a^4b^3c^2}$ **29.** $\sqrt[3]{48a^2b^3c^4}$ **30.** $\sqrt[3]{72a^4b^3c^2}$

31. $\sqrt[5]{64x^8y^{12}}$ **32.** $\sqrt[4]{32x^9y^{10}}$ **33.** $\sqrt[5]{243x^7y^{10}z^5}$ **34.** $\sqrt[5]{64x^8y^4z^{11}}$

Substitute the given number into the expression $\sqrt{b^2 - 4ac}$, and then simplify.

35. $a = 2, b = -6, c = 3$ **36.** $a = 6, b = 7, c = -5$

37. $a = 1, b = 2, c = 6$ **38.** $a = 2, b = 5, c = 3$

39. $a = \frac{1}{2}, b = -\frac{1}{2}, c = -\frac{5}{4}$ **40.** $a = \frac{7}{4}, b = -\frac{3}{4}, c = -2$

Rationalize the denominator in each of the following expressions.

41. $\dfrac{2}{\sqrt{3}}$

42. $\dfrac{3}{\sqrt{2}}$

43. $\dfrac{5}{\sqrt{6}}$

44. $\dfrac{7}{\sqrt{5}}$

45. $\sqrt{\dfrac{1}{2}}$

46. $\sqrt{\dfrac{1}{3}}$

47. $\sqrt{\dfrac{1}{5}}$

48. $\sqrt{\dfrac{1}{6}}$

49. $\dfrac{4}{\sqrt[3]{2}}$

50. $\dfrac{5}{\sqrt[3]{3}}$

51. $\dfrac{2}{\sqrt[3]{9}}$

52. $\dfrac{3}{\sqrt[3]{4}}$

53. $\sqrt[4]{\dfrac{3}{2x^2}}$

54. $\sqrt[4]{\dfrac{5}{3x^2}}$

55. $\sqrt[4]{\dfrac{8}{y}}$

56. $\sqrt[4]{\dfrac{27}{y}}$

57. $\sqrt[3]{\dfrac{4x}{3y}}$

58. $\sqrt[3]{\dfrac{7x}{6y}}$

59. $\sqrt[3]{\dfrac{2x}{9y}}$

60. $\sqrt[3]{\dfrac{5x}{4y}}$

61. $\sqrt[4]{\dfrac{1}{8x^3}}$

62. $\sqrt[4]{\dfrac{8}{9x^3}}$

Write each of the following in simplified form.

63. $\sqrt{\dfrac{27x^3}{5y}}$

64. $\sqrt{\dfrac{12x^5}{7y}}$

65. $\sqrt{\dfrac{75x^3y^2}{2z}}$

66. $\sqrt{\dfrac{50x^2y^3}{3z}}$

67. $\sqrt[3]{\dfrac{16a^4b^3}{9c}}$

68. $\sqrt[3]{\dfrac{54a^5b^4}{25c^2}}$

69. $\sqrt[3]{\dfrac{8x^3y^6}{9z}}$

70. $\sqrt[3]{\dfrac{27x^6y^3}{2z^2}}$

Simplify each expression. Do *not* assume the variables represent positive numbers.

71. $\sqrt{25x^2}$

72. $\sqrt{49x^2}$

73. $\sqrt{27x^3y^2}$

74. $\sqrt{40x^3y^2}$

75. $\sqrt{x^2 - 10x + 25}$

76. $\sqrt{x^2 - 16x + 64}$

77. $\sqrt{4x^2 + 12x + 9}$

78. $\sqrt{16x^2 + 40x + 25}$

79. $\sqrt{4a^4 + 16a^3 + 16a^2}$

80. $\sqrt{9a^4 + 18a^3 + 9a^2}$

81. $\sqrt{4x^3 - 8x^2}$

82. $\sqrt{18x^3 - 9x^2}$

83. Show that the statement $\sqrt{a + b} = \sqrt{a} + \sqrt{b}$ is not true by replacing a with 9 and b with 16 and simplifying both sides.

84. Find a pair of values for a and b that will make the statement $\sqrt{a + b} = \sqrt{a} + \sqrt{b}$ true.

Extending the Concepts

Factor each radicand into the product of prime factors. Then simplify each radical.

109. $\sqrt[3]{8{,}640}$

110. $\sqrt{8{,}640}$

111. $\sqrt[3]{10{,}584}$

112. $\sqrt{10{,}584}$

Assume a is a positive number, and rationalize each denominator.

113. $\dfrac{1}{\sqrt[10]{a^3}}$

114. $\dfrac{1}{\sqrt[12]{a^7}}$

115. $\dfrac{1}{\sqrt[20]{a^{11}}}$

116. $\dfrac{1}{\sqrt[15]{a^{13}}}$

117. Show that the two expressions $\sqrt{x^2 + 1}$ and $x + 1$ are not, in general, equal to each other by graphing $y = \sqrt{x^2 + 1}$ and $y = x + 1$ in the same viewing window.

118. Show that the two expressions $\sqrt{x^2 + 9}$ and $x + 3$ are not, in general, equal to each other by graphing $y = \sqrt{x^2 + 9}$ and $y = x + 3$ in the same viewing window.

119. Approximately how far apart are the graphs in Problem 117 when $x = 2$?

120. Approximately how far apart are the graphs in Problem 118 when $x = 2$?

121. For what value of x are the expressions $\sqrt{x^2 + 1}$ and $x + 1$ equal?

122. For what value of x are the expressions $\sqrt{x^2 + 9}$ and $x + 3$ equal?

Addition and Subtraction of Radical Expressions

7.4

We have been able to add and subtract polynomials by combining similar terms. The same idea applies to addition and subtraction of radical expressions.

OBJECTIVES

A	Add and subtract radicals.
B	Construct golden rectangles from squares.

DEFINITION

Two radicals are said to be **similar radicals** if they have the same index and the same radicand.

The expressions $5\sqrt[3]{7}$ and $-8\sqrt[3]{7}$ are similar since the index is ^{3} in both cases and the radicands are 7. The expressions $3\sqrt[4]{5}$ and $7\sqrt[3]{5}$ are not similar since they have different indices, and the expressions $2\sqrt[5]{8}$ and $3\sqrt[5]{9}$ are not similar because the radicands are not the same.

EXAMPLE 1

Combine $5\sqrt{3} - 4\sqrt{3} + 6\sqrt{3}$.

Solution All three radicals are similar. We apply the distributive property to get

$$\begin{aligned} 5\sqrt{3} - 4\sqrt{3} + 6\sqrt{3} &= (5 - 4 + 6)\sqrt{3} \\ &= 7\sqrt{3} \end{aligned}$$

EXAMPLE 2

Combine $3\sqrt{8} + 5\sqrt{18}$.

Solution The two radicals do not seem to be similar. We must write each in simplified form before applying the distributive property.

$$\begin{aligned} 3\sqrt{8} + 5\sqrt{18} &= 3\sqrt{4 \cdot 2} + 5\sqrt{9 \cdot 2} \\ &= 3\sqrt{4}\,\sqrt{2} + 5\sqrt{9}\,\sqrt{2} \\ &= 3 \cdot 2\sqrt{2} + 5 \cdot 3\sqrt{2} \\ &= 6\sqrt{2} + 15\sqrt{2} \\ &= (6 + 15)\sqrt{2} \\ &= 21\sqrt{2} \end{aligned}$$

The result of Example 2 can be generalized to the following rule for sums and differences of radical expressions.

Rule

To add or subtract radical expressions, put each in simplified form and apply the distributive property if possible. We can add only similar radicals. We must write each expression in simplified form for radicals before we can tell if the radicals are similar.

EXAMPLE 3

Combine $7\sqrt{75xy^3} - 4y\sqrt{12xy}$, where $x, y \geq 0$.

Solution We write each expression in simplified form and combine similar radicals:

$$\begin{aligned} 7\sqrt{75xy^3} - 4y\sqrt{12xy} &= 7\sqrt{25y^2}\,\sqrt{3xy} - 4y\sqrt{4}\,\sqrt{3xy} \\ &= 35y\sqrt{3xy} - 8y\sqrt{3xy} \\ &= (35y - 8y)\sqrt{3xy} \\ &= 27y\sqrt{3xy} \end{aligned}$$

PRACTICE PROBLEMS

1. Combine $3\sqrt{5} - 2\sqrt{5} + 4\sqrt{5}$.

2. Combine $4\sqrt{50} + 3\sqrt{8}$.

3. Assume $x, y \geq 0$ and combine: $4\sqrt{18x^2y} - 3x\sqrt{50y}$

Answers

1. $5\sqrt{5}$ 2. $26\sqrt{2}$ 3. $-3x\sqrt{2y}$

4. Combine $2\sqrt[3]{27a^2b^4} + 3b\sqrt[3]{125a^2b}$.

EXAMPLE 4 Combine $10\sqrt[3]{8a^4b^2} + 11a\sqrt[3]{27ab^2}$.

Solution Writing each radical in simplified form and combining similar terms, we have

$$\begin{aligned} 10\sqrt[3]{8a^4b^2} + 11a\sqrt[3]{27ab^2} &= 10\sqrt[3]{8a^3}\sqrt[3]{ab^2} + 11a\sqrt[3]{27}\sqrt[3]{ab^2} \\ &= 20a\sqrt[3]{ab^2} + 33a\sqrt[3]{ab^2} \\ &= 53a\sqrt[3]{ab^2} \end{aligned}$$

5. Combine $\frac{\sqrt{5}}{3} + \frac{1}{\sqrt{5}}$.

EXAMPLE 5 Combine $\frac{\sqrt{3}}{2} + \frac{1}{\sqrt{3}}$.

Solution We begin by writing the second term in simplified form.

$$\begin{aligned} \frac{\sqrt{3}}{2} + \frac{1}{\sqrt{3}} &= \frac{\sqrt{3}}{2} + \frac{1}{\sqrt{3}} \cdot \frac{\mathbf{\sqrt{3}}}{\mathbf{\sqrt{3}}} \\ &= \frac{\sqrt{3}}{2} + \frac{\sqrt{3}}{3} \\ &= \frac{1}{2}\sqrt{3} + \frac{1}{3}\sqrt{3} \\ &= \left(\frac{1}{2} + \frac{1}{3}\right)\sqrt{3} \\ &= \frac{5}{6}\sqrt{3} = \frac{5\sqrt{3}}{6} \end{aligned}$$

6. Construct a golden rectangle from a square of side 6. Then show that the ratio of the length to the width is the golden ratio.

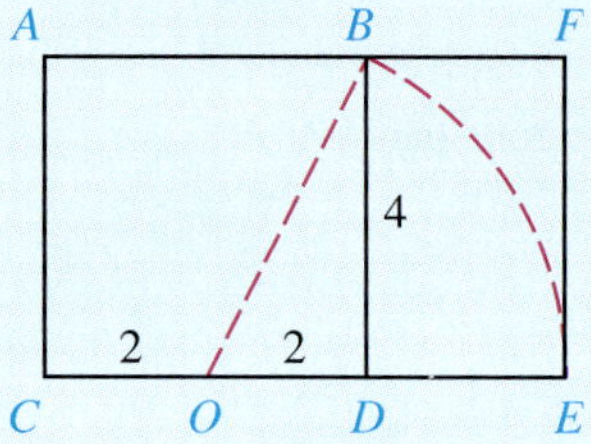

FIGURE 1

As you can see, showing that the ratio of length to width in this rectangle is the golden ratio depends on our ability to write $\sqrt{20}$ as $2\sqrt{5}$ and our ability to reduce to lowest terms by factoring and then dividing out the common factor 2 from the numerator and denominator.

EXAMPLE 6 Construct a golden rectangle from a square of side 4. Then show that the ratio of the length to the width is the golden ratio $\frac{1 + \sqrt{5}}{2}$.

Solution Figure 1 shows the golden rectangle constructed from a square of side 4.

The length of the diagonal OB is found from the Pythagorean theorem.

$$OB = \sqrt{2^2 + 4^2} = \sqrt{4 + 16} = \sqrt{20} = 2\sqrt{5}$$

The ratio of the length to the width for the rectangle is the golden ratio.

$$\text{Golden ratio} = \frac{CE}{EF} = \frac{2 + 2\sqrt{5}}{4} = \frac{\not{2}(1 + \sqrt{5})}{\not{2} \cdot 2} = \frac{1 + \sqrt{5}}{2}$$

Linking Objectives and Examples

Next to each **objective**, list the examples that are best described by that objective.

A	
B	

Getting Ready for Class

After reading through the preceding section, respond in your own words and in complete sentences.

1. What are similar radicals?
2. When can we add two radical expressions?
3. What is the first step when adding or subtracting expressions containing radicals?
4. What is the golden ratio, and where does it come from?

Answers

4. $21b\sqrt[3]{a^2b}$ 5. $\frac{8\sqrt{5}}{15}$

6. See Solutions to Selected Practice Problems.

Problem Set 7.4

INTERMEDIATE Algebra Now™

Combine the following expressions. (Assume any variables under an even root are nonnegative.)

1. $3\sqrt{5} + 4\sqrt{5}$

2. $6\sqrt{3} - 5\sqrt{3}$

3. $3x\sqrt{7} - 4x\sqrt{7}$

4. $6y\sqrt{a} + 7y\sqrt{a}$

5. $5\sqrt[3]{10} - 4\sqrt[3]{10}$

6. $6\sqrt[4]{2} + 9\sqrt[4]{2}$

7. $8\sqrt[5]{6} - 2\sqrt[5]{6} + 3\sqrt[5]{6}$

8. $7\sqrt[6]{7} - \sqrt[6]{7} + 4\sqrt[6]{7}$

9. $3x\sqrt{2} - 4x\sqrt{2} + x\sqrt{2}$

10. $5x\sqrt{6} - 3x\sqrt{6} - 2x\sqrt{6}$

11. $\sqrt{20} - \sqrt{80} + \sqrt{45}$

12. $\sqrt{8} - \sqrt{32} - \sqrt{18}$

13. $4\sqrt{8} - 2\sqrt{50} - 5\sqrt{72}$

14. $\sqrt{48} - 3\sqrt{27} + 2\sqrt{75}$

15. $5x\sqrt{8} + 3\sqrt{32x^2} - 5\sqrt{50x^2}$

16. $2\sqrt{50x^2} - 8x\sqrt{18} - 3\sqrt{72x^2}$

17. $5\sqrt[3]{16} - 4\sqrt[3]{54}$

18. $\sqrt[3]{81} + 3\sqrt[3]{24}$

19. $\sqrt[3]{x^4y^2} + 7x\sqrt[3]{xy^2}$

20. $2\sqrt[3]{x^8y^6} - 3y^2\sqrt[3]{8x^8}$

21. $5a^2\sqrt{27ab^3} - 6b\sqrt{12a^5b}$

22. $9a\sqrt{20a^3b^2} + 7b\sqrt{45a^5}$

23. $b\sqrt[3]{24a^5b} + 3a\sqrt[3]{81a^2b^4}$

24. $7\sqrt[3]{a^4b^3c^2} - 6ab\sqrt[3]{ac^2}$

25. $5x\sqrt[4]{3y^5} + y\sqrt[4]{243x^4y} + \sqrt[4]{48x^4y^5}$

26. $x\sqrt[4]{5xy^8} + y\sqrt[4]{405x^5y^4} + y^2\sqrt[4]{80x^5}$

27. $\dfrac{\sqrt{2}}{2} + \dfrac{1}{\sqrt{2}}$

28. $\dfrac{\sqrt{3}}{3} + \dfrac{1}{\sqrt{3}}$

29. $\dfrac{\sqrt{5}}{3} + \dfrac{1}{\sqrt{5}}$

30. $\dfrac{\sqrt{6}}{2} + \dfrac{1}{\sqrt{6}}$

31. $\sqrt{x} - \dfrac{1}{\sqrt{x}}$

32. $\sqrt{x} + \dfrac{1}{\sqrt{x}}$

33. $\dfrac{\sqrt{18}}{6} + \sqrt{\dfrac{1}{2}} + \dfrac{\sqrt{2}}{2}$

34. $\dfrac{\sqrt{12}}{6} + \sqrt{\dfrac{1}{3}} + \dfrac{\sqrt{3}}{3}$

35. $\sqrt{6} - \sqrt{\dfrac{2}{3}} + \sqrt{\dfrac{1}{6}}$

36. $\sqrt{15} - \sqrt{\dfrac{3}{5}} + \sqrt{\dfrac{5}{3}}$

37. $\sqrt[3]{25} + \dfrac{3}{\sqrt[3]{5}}$

38. $\sqrt[4]{8} + \dfrac{1}{\sqrt[4]{2}}$

39. Use a calculator to find a decimal approximation for $\sqrt{12}$ and for $2\sqrt{3}$.

40. Use a calculator to find decimal approximations for $\sqrt{50}$ and $5\sqrt{2}$.

41. Use a calculator to find a decimal approximation for $\sqrt{8} + \sqrt{18}$. Is it equal to the decimal approximation for $\sqrt{26}$ or $\sqrt{50}$?

42. Use a calculator to find a decimal approximation for $\sqrt{3} + \sqrt{12}$. Is it equal to the decimal approximation for $\sqrt{15}$ or $\sqrt{27}$?

Applying the Concepts

43. Golden Rectangle Construct a golden rectangle from a square of side 8. Then show that the ratio of the length to the width is the golden ratio $\dfrac{1 + \sqrt{5}}{2}$.

44. Golden Rectangle Construct a golden rectangle from a square of side 10. Then show that the ratio of the length to the width is the golden ratio $\dfrac{1 + \sqrt{5}}{2}$.

45. Golden Rectangle To show that all golden rectangles have the same ratio of length to width, construct a golden rectangle from a square of side $2x$. Then show that the ratio of the length to the width is the golden ratio.

46. Golden Rectangle To show that all golden rectangles have the same ratio of length to width, construct a golden rectangle from a square of side x. Then show that the ratio of the length to the width is the golden ratio.

47. Equilateral Triangles A triangle is equilateral if it has three equal sides. The triangle in the figure is equilateral with each side of length $2x$. Find the ratio of the height to a side.

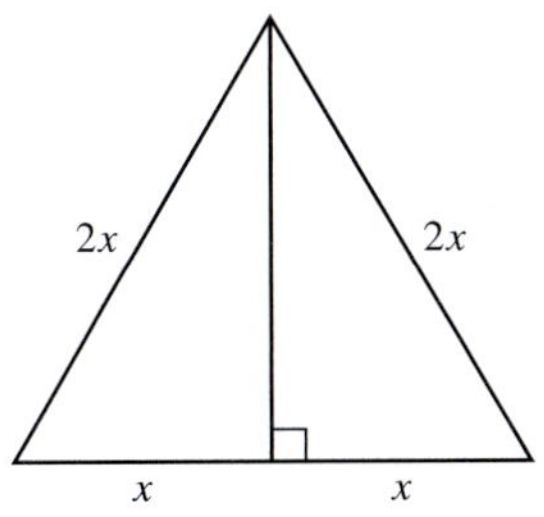

48. Pyramids Refer to the diagram of a square pyramid below. Find the ratio of the height h of the pyramid to the altitude a.

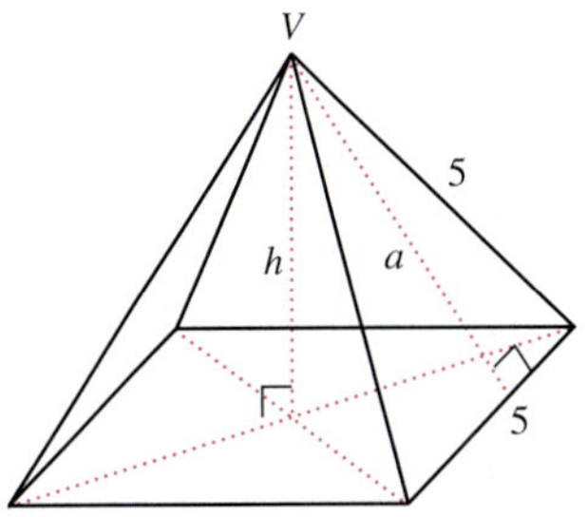

Getting Ready for the Next Section

Simplify the following.

49. $3 \cdot 2$

50. $5 \cdot 7$

51. $(x + y)(4x - y)$

52. $(2x + y)(x - y)$

53. $(x + 3)^2$

54. $(3x - 2y)^2$

55. $(x - 2)(x + 2)$

56. $(2x + 5)(2x - 5)$

Simplify the following expressions.

57. $2\sqrt{18}$

58. $5\sqrt{36}$

59. $(\sqrt{6})^2$

60. $(\sqrt{2})^2$

61. $(3\sqrt{x})^2$

62. $(2\sqrt{y})^2$

Rationalize the denominator.

63. $\dfrac{\sqrt{3}}{\sqrt{2}}$

64. $\dfrac{\sqrt{5}}{\sqrt{6}}$

Maintaining Your Skills

Add and subtract as indicated.

65. $\dfrac{2a - 4}{a + 2} - \dfrac{a - 6}{a + 2}$

66. $\dfrac{2a - 3}{a - 2} - \dfrac{a - 1}{a - 2}$

67. $3 + \dfrac{4}{3 - t}$

68. $6 + \dfrac{2}{5 - t}$

69. $\dfrac{3}{2x - 5} - \dfrac{39}{8x^2 - 14x - 15}$

70. $\dfrac{2}{4x - 5} + \dfrac{9}{8x^2 - 38x + 35}$

71. $\dfrac{1}{x - y} - \dfrac{3xy}{x^3 - y^3}$

72. $\dfrac{1}{x + y} + \dfrac{3xy}{x^3 + y^3}$

7.5 Multiplication and Division of Radical Expressions

We have worked with the golden rectangle more than once in this chapter. The following is one such golden rectangle.

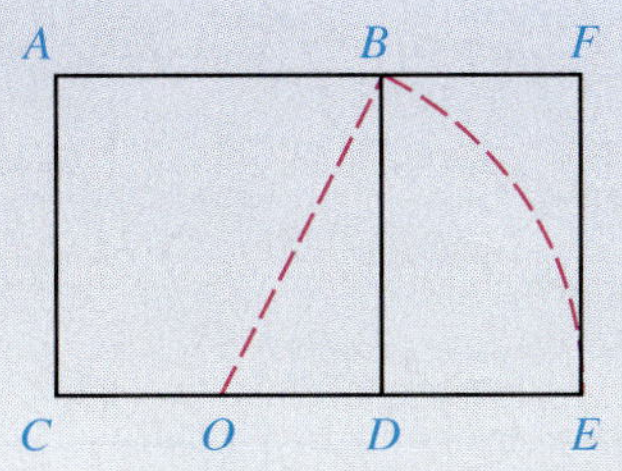

By now you know that in any golden rectangle constructed from a square (of any size) the ratio of the length to the width will be

$$\frac{1+\sqrt{5}}{2}$$

which we call the golden ratio. What is interesting is that the smaller rectangle on the right, *BFED*, is also a golden rectangle. We will use the mathematics developed in this section to confirm this fact.

In this section we will look at multiplication and division of expressions that contain radicals. As you will see, multiplication of expressions that contain radicals is very similar to multiplication of polynomials. The division problems in this section are just an extension of the work we did previously when we rationalized denominators.

OBJECTIVES

A	Multiply expressions containing radicals.
B	Rationalize a denominator containing two terms.

EXAMPLE 1

Multiply $(3\sqrt{5})(2\sqrt{7})$.

Solution We can rearrange the order and grouping of the numbers in this product by applying the commutative and associative properties. Following this, we apply property 1 for radicals and multiply:

$$(3\sqrt{5})(2\sqrt{7}) = (3\cdot 2)(\sqrt{5}\sqrt{7}) \quad \textbf{Commutative and associative properties}$$
$$= (3\cdot 2)(\sqrt{5\cdot 7}) \quad \textbf{Property 1 for radicals}$$
$$= 6\sqrt{35} \quad \textbf{Multiplication}$$

In practice, it is not necessary to show the first two steps.

EXAMPLE 2

Multiply $\sqrt{3}(2\sqrt{6} - 5\sqrt{12})$.

Solution Applying the distributive property, we have

$$\sqrt{3}(2\sqrt{6} - 5\sqrt{12}) = \sqrt{3}\cdot 2\sqrt{6} - \sqrt{3}\cdot 5\sqrt{12}$$
$$= 2\sqrt{18} - 5\sqrt{36}$$

Writing each radical in simplified form gives

$$2\sqrt{18} - 5\sqrt{36} = 2\sqrt{9}\sqrt{2} - 5\sqrt{36}$$
$$= 6\sqrt{2} - 30$$

PRACTICE PROBLEMS

1. Multiply $(7\sqrt{3})(5\sqrt{11})$.

2. Multiply $\sqrt{2}(3\sqrt{5} - 4\sqrt{2})$.

Answers

1. $35\sqrt{33}$ **2.** $3\sqrt{10} - 8$

3. Multiply: $(\sqrt{2} + \sqrt{7})(\sqrt{2} - 3\sqrt{7})$

EXAMPLE 3

Multiply $(\sqrt{3} + \sqrt{5})(4\sqrt{3} - \sqrt{5})$.

Solution The same principle that applies when multiplying two binomials applies to this product. We must multiply each term in the first expression by each term in the second one. Any convenient method can be used. Let's use the FOIL method.

$$\begin{aligned}(\sqrt{3} + \sqrt{5})(4\sqrt{3} - \sqrt{5}) &= \overset{F}{\sqrt{3} \cdot 4\sqrt{3}} - \overset{O}{\sqrt{3}\sqrt{5}} + \overset{I}{\sqrt{5} \cdot 4\sqrt{3}} - \overset{L}{\sqrt{5}\sqrt{5}} \\ &= 4 \cdot 3 - \sqrt{15} + 4\sqrt{15} - 5 \\ &= 12 + 3\sqrt{15} - 5 \\ &= 7 + 3\sqrt{15}\end{aligned}$$

4. Expand and simplify. $(\sqrt{x} + 5)^2$

EXAMPLE 4

Expand and simplify $(\sqrt{x} + 3)^2$.

Solution 1 We can write this problem as a multiplication problem and proceed as we did in Example 3:

$$\begin{aligned}(\sqrt{x} + 3)^2 &= (\sqrt{x} + 3)(\sqrt{x} + 3) \\ &= \overset{F}{\sqrt{x} \cdot \sqrt{x}} + \overset{O}{3\sqrt{x}} + \overset{I}{3\sqrt{x}} + \overset{L}{3 \cdot 3} \\ &= x + 3\sqrt{x} + 3\sqrt{x} + 9 \\ &= x + 6\sqrt{x} + 9\end{aligned}$$

Solution 2 We can obtain the same result by applying the formula for the square of a sum: $(a + b)^2 = a^2 + 2ab + b^2$.

$$\begin{aligned}(\sqrt{x} + 3)^2 &= (\sqrt{x})^2 + 2(\sqrt{x})(3) + 3^2 \\ &= x + 6\sqrt{x} + 9\end{aligned}$$

5. Expand $(5\sqrt{a} - 3\sqrt{b})^2$ and simplify the result.

EXAMPLE 5

Expand $(3\sqrt{x} - 2\sqrt{y})^2$ and simplify the result.

Solution Let's apply the formula for the square of a difference, $(a - b)^2 = a^2 - 2ab + b^2$.

$$\begin{aligned}(3\sqrt{x} - 2\sqrt{y})^2 &= (3\sqrt{x})^2 - 2(3\sqrt{x})(2\sqrt{y}) + (2\sqrt{y})^2 \\ &= 9x - 12\sqrt{xy} + 4y\end{aligned}$$

6. Expand and simplify. $(\sqrt{x+3} - 1)^2$

EXAMPLE 6

Expand and simplify $(\sqrt{x+2} - 1)^2$.

Solution Applying the formula $(a - b)^2 = a^2 - 2ab + b^2$, we have

$$\begin{aligned}(\sqrt{x+2} - 1)^2 &= (\sqrt{x+2})^2 - 2\sqrt{x+2}(1) + 1^2 \\ &= x + 2 - 2\sqrt{x+2} + 1 \\ &= x + 3 - 2\sqrt{x+2}\end{aligned}$$

7. Multiply. $(\sqrt{5} + \sqrt{3})(\sqrt{5} - \sqrt{3})$

EXAMPLE 7

Multiply $(\sqrt{6} + \sqrt{2})(\sqrt{6} - \sqrt{2})$.

Solution We notice the product is of the form $(a + b)(a - b)$, which always gives the difference of two squares, $a^2 - b^2$:

$$\begin{aligned}(\sqrt{6} + \sqrt{2})(\sqrt{6} - \sqrt{2}) &= (\sqrt{6})^2 - (\sqrt{2})^2 \\ &= 6 - 2 \\ &= 4\end{aligned}$$

Answers

3. $-19 - 2\sqrt{14}$ 4. $x + 10\sqrt{x} + 25$
5. $25a - 30\sqrt{ab} + 9b$
6. $x + 4 - 2\sqrt{x+3}$ 7. 2

The two expressions $(\sqrt{6} + \sqrt{2})$ and $(\sqrt{6} - \sqrt{2})$ are called *conjugates.* In general, the conjugate of $\sqrt{a} + \sqrt{b}$ is $\sqrt{a} - \sqrt{b}$. If a and b are integers, multiplying conjugates of this form always produces a rational number.

Division with radical expressions is the same as rationalizing the denominator. In a previous section we were able to divide $\sqrt{3}$ by $\sqrt{2}$ by rationalizing the denominator:

$$\frac{\sqrt{3}}{\sqrt{2}} = \frac{\sqrt{3}}{\sqrt{2}} \cdot \frac{\sqrt{\mathbf{2}}}{\sqrt{\mathbf{2}}} = \frac{\sqrt{6}}{2}$$

We can accomplish the same result with expressions such as

$$\frac{6}{\sqrt{5} - \sqrt{3}}$$

by multiplying the numerator and denominator by the conjugate of the denominator.

NOTE

We can prove that conjugates always multiply to yield a rational number as follows: If a and b are positive integers, then

$(\sqrt{a} + \sqrt{b})(\sqrt{a} - \sqrt{b})$
$= \sqrt{a}\sqrt{a} - \sqrt{a}\sqrt{b} + \sqrt{a}\sqrt{b} - \sqrt{b}\sqrt{b}$
$= a - \sqrt{ab} + \sqrt{ab} - b$
$= a - b$

which is rational if a and b are rational.

EXAMPLE 8 Divide $\dfrac{6}{\sqrt{5} - \sqrt{3}}$. (Rationalize the denominator.)

Solution Since the product of two conjugates is a rational number, we multiply the numerator and denominator by the conjugate of the denominator.

$$\frac{6}{\sqrt{5} - \sqrt{3}} = \frac{6}{\sqrt{5} - \sqrt{3}} \cdot \frac{(\sqrt{\mathbf{5}} + \sqrt{\mathbf{3}})}{(\sqrt{\mathbf{5}} + \sqrt{\mathbf{3}})}$$

$$= \frac{6\sqrt{5} + 6\sqrt{3}}{(\sqrt{5})^2 - (\sqrt{3})^2}$$

$$= \frac{6\sqrt{5} + 6\sqrt{3}}{5 - 3}$$

$$= \frac{6\sqrt{5} + 6\sqrt{3}}{2}$$

The numerator and denominator of this last expression have a factor of 2 in common. We can reduce to lowest terms by factoring 2 from the numerator and then dividing both the numerator and denominator by 2:

$$= \frac{\not{2}(3\sqrt{5} + 3\sqrt{3})}{\not{2}}$$

$$= 3\sqrt{5} + 3\sqrt{3}$$

8. Divide $\dfrac{3}{\sqrt{7} - \sqrt{3}}$.

EXAMPLE 9 Rationalize the denominator $\dfrac{\sqrt{5} - 2}{\sqrt{5} + 2}$.

Solution To rationalize the denominator, we multiply the numerator and denominator by the conjugate of the denominator:

$$\frac{\sqrt{5} - 2}{\sqrt{5} + 2} = \frac{\sqrt{5} - 2}{\sqrt{5} + 2} \cdot \frac{(\sqrt{\mathbf{5}} - \mathbf{2})}{(\sqrt{\mathbf{5}} - \mathbf{2})}$$

$$= \frac{5 - 2\sqrt{5} - 2\sqrt{5} + 4}{(\sqrt{5})^2 - 2^2}$$

$$= \frac{9 - 4\sqrt{5}}{5 - 4}$$

$$= \frac{9 - 4\sqrt{5}}{1}$$

$$= 9 - 4\sqrt{5}$$

9. Rationalize the denominator:

$$\frac{\sqrt{10} - 3}{\sqrt{10} + 3}$$

Answers

8. $\dfrac{3\sqrt{7} + 3\sqrt{3}}{4}$ 9. $19 - 6\sqrt{10}$

10. If side AC in Figure 1 were 4 instead of 2, show that the smaller rectangle $BDEF$ is also a golden rectangle by finding the ratio of its length to its width.

EXAMPLE 10 A golden rectangle constructed from a square of side 2 is shown in Figure 1. Show that the smaller rectangle $BDEF$ is also a golden rectangle by finding the ratio of its length to its width.

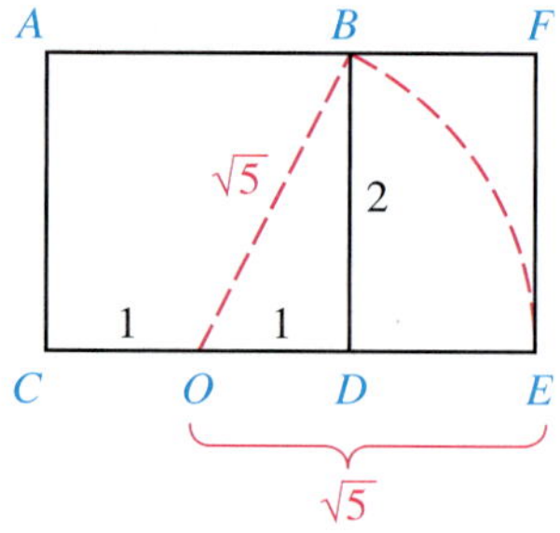

FIGURE 1

Solution First we find expressions for the length and width of the smaller rectangle.

$$\text{Length} = EF = 2$$

$$\text{Width} = DE = \sqrt{5} - 1$$

Next, we find the ratio of length to width.

$$\text{Ratio of length to width} = \frac{EF}{DE} = \frac{2}{\sqrt{5} - 1}$$

To show that the small rectangle is a golden rectangle, we must show that the ratio of length to width is the golden ratio. We do so by rationalizing the denominator.

$$\begin{aligned}
\frac{2}{\sqrt{5} - 1} &= \frac{2}{\sqrt{5} - 1} \cdot \frac{\mathbf{\sqrt{5} + 1}}{\mathbf{\sqrt{5} + 1}} \\
&= \frac{2(\sqrt{5} + 1)}{5 - 1} \\
&= \frac{2(\sqrt{5} + 1)}{4} \\
&= \frac{\sqrt{5} + 1}{2} \qquad \textbf{Divide out common factor 2}
\end{aligned}$$

Since addition is commutative, this last expression is the golden ratio. Therefore, the small rectangle in Figure 1 is a golden rectangle.

Linking Objectives and Examples

Next to each **objective**, list the examples that are best described by that objective.

A	
B	

Answer

10. See Solutions to Selected Practice Problems.

Getting Ready for Class

After reading through the preceding section, respond in your own words and in complete sentences.

1. Explain why $(\sqrt{5} + \sqrt{2})^2 \neq 5 + 2$.
2. Explain in words how you would rationalize the denominator in the expression $\dfrac{\sqrt{3}}{\sqrt{5} - \sqrt{2}}$.
3. What are conjugates?
4. What result is guaranteed when multiplying radical expressions that are conjugates?

Problem Set 7.5

INTERMEDIATE Algebra $f(x)$ Now™

Multiply. (Assume all expressions appearing under a square root symbol represent nonnegative numbers throughout this problem set.)

1. $\sqrt{6}\sqrt{3}$

2. $\sqrt{6}\sqrt{2}$

3. $(2\sqrt{3})(5\sqrt{7})$

4. $(3\sqrt{5})(2\sqrt{7})$

5. $(4\sqrt{6})(2\sqrt{15})(3\sqrt{10})$

6. $(4\sqrt{35})(2\sqrt{21})(5\sqrt{15})$

7. $(3\sqrt[3]{3})(6\sqrt[3]{9})$

8. $(2\sqrt[3]{2})(6\sqrt[3]{4})$

9. $\sqrt{3}(\sqrt{2} - 3\sqrt{3})$

10. $\sqrt{2}(5\sqrt{3} + 4\sqrt{2})$

11. $6\sqrt[3]{4}(2\sqrt[3]{2} + 1)$

12. $7\sqrt[3]{5}(3\sqrt[3]{25} - 2)$

13. $(\sqrt{3} + \sqrt{2})(3\sqrt{3} - \sqrt{2})$

14. $(\sqrt{5} - \sqrt{2})(3\sqrt{5} + 2\sqrt{2})$

15. $(\sqrt{x} + 5)(\sqrt{x} - 3)$

16. $(\sqrt{x} + 4)(\sqrt{x} + 2)$

17. $(3\sqrt{6} + 4\sqrt{2})(\sqrt{6} + 2\sqrt{2})$

18. $(\sqrt{7} - 3\sqrt{3})(2\sqrt{7} - 4\sqrt{3})$

19. $(\sqrt{3} + 4)^2$

20. $(\sqrt{5} - 2)^2$

21. $(\sqrt{x} - 3)^2$

22. $(\sqrt{x} + 4)^2$

23. $(2\sqrt{a} - 3\sqrt{b})^2$

24. $(5\sqrt{a} - 2\sqrt{b})^2$

25. $(\sqrt{x - 4} + 2)^2$

26. $(\sqrt{x - 3} + 2)^2$

27. $(\sqrt{x - 5} - 3)^2$

28. $(\sqrt{x - 3} - 4)^2$

29. $(\sqrt{3} - \sqrt{2})(\sqrt{3} + \sqrt{2})$

30. $(\sqrt{5} - \sqrt{2})(\sqrt{5} + \sqrt{2})$

31. $(\sqrt{a} + 7)(\sqrt{a} - 7)$

32. $(\sqrt{a} + 5)(\sqrt{a} - 5)$

33. $(5 - \sqrt{x})(5 + \sqrt{x})$

34. $(3 - \sqrt{x})(3 + \sqrt{x})$

35. $(\sqrt{x - 4} + 2)(\sqrt{x - 4} - 2)$

36. $(\sqrt{x + 3} + 5)(\sqrt{x + 3} - 5)$

37. $(\sqrt{3} + 1)^3$

38. $(\sqrt{5} - 2)^3$

Rationalize the denominator in each of the following.

39. $\dfrac{\sqrt{2}}{\sqrt{6}-\sqrt{2}}$

40. $\dfrac{\sqrt{5}}{\sqrt{5}+\sqrt{3}}$

41. $\dfrac{\sqrt{5}}{\sqrt{5}+1}$

42. $\dfrac{\sqrt{7}}{\sqrt{7}-1}$

43. $\dfrac{\sqrt{x}}{\sqrt{x}-3}$

44. $\dfrac{\sqrt{x}}{\sqrt{x}+2}$

45. $\dfrac{\sqrt{5}}{2\sqrt{5}-3}$

46. $\dfrac{\sqrt{7}}{3\sqrt{7}-2}$

47. $\dfrac{3}{\sqrt{x}-\sqrt{y}}$

48. $\dfrac{2}{\sqrt{x}+\sqrt{y}}$

49. $\dfrac{\sqrt{6}+\sqrt{2}}{\sqrt{6}-\sqrt{2}}$

50. $\dfrac{\sqrt{5}-\sqrt{3}}{\sqrt{5}+\sqrt{3}}$

51. $\dfrac{\sqrt{7}-2}{\sqrt{7}+2}$

52. $\dfrac{\sqrt{11}+3}{\sqrt{11}-3}$

53. $\dfrac{\sqrt{a}+\sqrt{b}}{\sqrt{a}-\sqrt{b}}$

54. $\dfrac{\sqrt{a}-\sqrt{b}}{\sqrt{a}+\sqrt{b}}$

55. $\dfrac{\sqrt{x}+2}{\sqrt{x}-2}$

56. $\dfrac{\sqrt{x}-3}{\sqrt{x}+3}$

57. $\dfrac{2\sqrt{3}-\sqrt{7}}{3\sqrt{3}+\sqrt{7}}$

58. $\dfrac{5\sqrt{6}+2\sqrt{2}}{\sqrt{6}-\sqrt{2}}$

59. $\dfrac{3\sqrt{x}+2}{1+\sqrt{x}}$

60. $\dfrac{5\sqrt{x}-1}{2+\sqrt{x}}$

61. Show that the product $(\sqrt[3]{2}+\sqrt[3]{3})(\sqrt[3]{4}-\sqrt[3]{6}+\sqrt[3]{9})$ is 5.

62. Show that the product $(\sqrt[3]{x}+2)(\sqrt[3]{x^2}-2\sqrt[3]{x}+4)$ is $x+8$.

Each statement below is false. Correct the right side of each one.

63. $5(2\sqrt{3}) = 10\sqrt{15}$

64. $3(2\sqrt{x}) = 6\sqrt{3x}$

65. $(\sqrt{x}+3)^2 = x+9$

66. $(\sqrt{x}-7)^2 = x-49$

67. $(5\sqrt{3})^2 = 15$

68. $(3\sqrt{5})^2 = 15$

Applying the Concepts

69. Gravity If an object is dropped from the top of a 100-foot building, the amount of time t (in seconds) that it takes for the object to be h feet from the ground is given by the formula

$$t = \frac{\sqrt{100-h}}{4}$$

How long does it take before the object is 50 feet from the ground? How long does it take to reach the ground? (When it is on the ground, h is 0.)

70. Gravity Use the formula given in Problem 69 to determine h if t is 1.25 seconds.

71. Golden Rectangle Rectangle $ACEF$ in Figure 2 is a golden rectangle. If side AC is 6 inches, show that the smaller rectangle $BDEF$ is also a golden rectangle.

72. Golden Rectangle Rectangle $ACEF$ in Figure 2 is a golden rectangle. If side AC is 1 inch, show that the smaller rectangle $BDEF$ is also a golden rectangle.

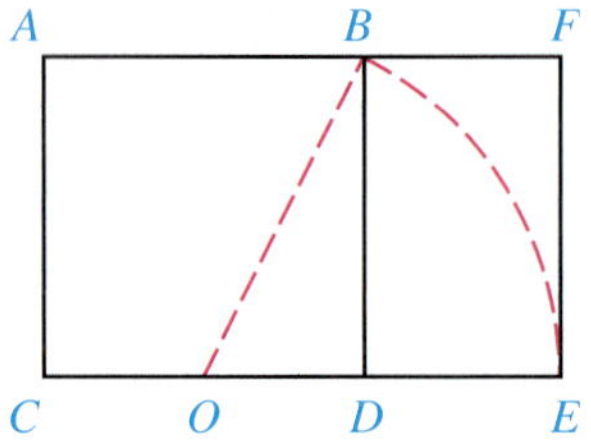

FIGURE 2

73. Golden Rectangle If side AC in Figure 2 is $2x$, show the rectangle $BDEF$ is a golden rectangle.

74. Golden Rectangle If side AC in Figure 2 is x, show that rectangle $BDEF$ is a golden rectangle.

Getting Ready for the Next Section

Simplify.

75. $(t + 5)^2$

76. $(x - 4)^2$

77. $\sqrt{x} \cdot \sqrt{x}$

78. $\sqrt{3x} \cdot \sqrt{3x}$

Solve.

79. $3x + 4 = 5^2$

80. $4x - 7 = 3^2$

81. $t^2 + 7t + 12 = 0$

82. $x^2 - 3x - 10 = 0$

83. $t^2 + 10t + 25 = t + 7$

84. $x^2 - 4x + 4 = x - 2$

85. $(x + 4)^2 = x + 6$

86. $(x - 6)^2 = x - 4$

87. Is $x = 7$ a solution to $\sqrt{3x + 4} = 5$?

88. Is $x = 4$ a solution to $\sqrt{4x - 7} = -3$?

89. Is $t = -6$ a solution to $t + 5 = \sqrt{t + 7}$?

90. Is $t = -3$ a solution to $t + 5 = \sqrt{t + 7}$?

Maintaining Your Skills

Simplify each complex fraction.

91. $\dfrac{\frac{1}{4} - \frac{1}{3}}{\frac{1}{2} + \frac{1}{6}}$

92. $\dfrac{\frac{1}{8} - \frac{1}{3}}{\frac{1}{4} - \frac{1}{3}}$

93. $\dfrac{1 - \dfrac{2}{y}}{1 + \dfrac{2}{y}}$

94. $\dfrac{1 + \dfrac{3}{y}}{1 - \dfrac{3}{y}}$

95. $\dfrac{4 + \dfrac{4}{x} + \dfrac{1}{x^2}}{4 - \dfrac{1}{x^2}}$

96. $\dfrac{1 - \dfrac{1}{x} - \dfrac{6}{x^2}}{1 - \dfrac{9}{x^2}}$

Equations with Radicals

7.6

This section is concerned with solving equations that involve one or more radicals. The first step in solving an equation that contains a radical is to eliminate the radical from the equation. To do so, we need an additional property.

OBJECTIVES

A	Solve equations containing radicals.
B	Graph simple square root and cube root equations in two variables.

Squaring Property of Equality

If both sides of an equation are squared, the solutions to the original equation are solutions to the resulting equation.

We will never lose solutions to our equations by squaring both sides. We may, however, introduce *extraneous solutions.* Extraneous solutions satisfy the equation obtained by squaring both sides of the original equation, but they do not satisfy the original equation.

We know that if two real numbers a and b are equal, then so are their squares:

$$\text{If} \quad a = b$$
$$\text{then} \quad a^2 = b^2$$

However, extraneous solutions are introduced when we square opposites; that is, even though opposites are not equal, their squares are. For example,

$$5 = -5 \quad \textbf{A false statement}$$
$$(5)^2 = (-5)^2 \quad \textbf{Square both sides}$$
$$25 = 25 \quad \textbf{A true statement}$$

We are free to square both sides of an equation any time it is convenient. We must be aware, however, that doing so may introduce extraneous solutions. We must, therefore, check all our solutions in the original equation if at any time we square both sides of the original equation.

EXAMPLE 1 Solve for x: $\sqrt{3x+4} = 5$.

Solution We square both sides and proceed as usual:

$$\sqrt{3x+4} = 5$$
$$(\sqrt{3x+4})^2 = 5^2$$
$$3x + 4 = 25$$
$$3x = 21$$
$$x = 7$$

Checking $x = 7$ in the original equation, we have

$$\sqrt{3(7)+4} \stackrel{?}{=} 5$$
$$\sqrt{21+4} \stackrel{?}{=} 5$$
$$\sqrt{25} \stackrel{?}{=} 5$$
$$5 = 5$$

The solution $x = 7$ satisfies the original equation.

PRACTICE PROBLEMS

1. Solve for x: $\sqrt{2x+4} = 4$

Answer

1. 6

2. Solve $\sqrt{7x - 3} = -5$.

EXAMPLE 2

Solve $\sqrt{4x - 7} = -3$.

Solution Squaring both sides, we have

$$\sqrt{4x - 7} = -3$$
$$(\sqrt{4x - 7})^2 = (-3)^2$$
$$4x - 7 = 9$$
$$4x = 16$$
$$x = 4$$

Checking $x = 4$ in the original equation gives

$$\sqrt{4(4) - 7} \stackrel{?}{=} -3$$
$$\sqrt{16 - 7} \stackrel{?}{=} -3$$
$$\sqrt{9} \stackrel{?}{=} -3$$
$$3 = -3$$

The solution $x = 4$ produces a false statement when checked in the original equation. Since $x = 4$ was the only possible solution, there is no solution to the original equation. The possible solution $x = 4$ is an extraneous solution. It satisfies the equation obtained by squaring both sides of the original equation, but it does not satisfy the original equation.

NOTE

The fact that there is no solution to the equation in Example 2 was obvious to begin with. Notice that the left side of the equation is the *positive* square root of $4x - 7$, which must be a positive number or 0. The right side of the equation is -3. Since we cannot have a number that is either positive or zero equal to a negative number, there is no solution to the equation.

3. Solve $\sqrt{4x + 5} + 2 = 7$.

EXAMPLE 3

Solve $\sqrt{5x - 1} + 3 = 7$.

Solution We must isolate the radical on the left side of the equation. If we attempt to square both sides without doing so, the resulting equation will also contain a radical. Adding -3 to both sides, we have

$$\sqrt{5x - 1} + 3 = 7$$
$$\sqrt{5x - 1} = 4$$

We can now square both sides and proceed as usual:

$$(\sqrt{5x - 1})^2 = 4^2$$
$$5x - 1 = 16$$
$$5x = 17$$
$$x = \frac{17}{5}$$

Checking $x = \frac{17}{5}$, we have

$$\sqrt{5\left(\frac{17}{5}\right) - 1} + 3 \stackrel{?}{=} 7$$
$$\sqrt{17 - 1} + 3 \stackrel{?}{=} 7$$
$$\sqrt{16} + 3 \stackrel{?}{=} 7$$
$$4 + 3 \stackrel{?}{=} 7$$
$$7 = 7$$

Answers
2. No solution 3. 5

EXAMPLE 4

Solve $t + 5 = \sqrt{t + 7}$.

Solution This time, squaring both sides of the equation results in a quadratic equation:

$$(t + 5)^2 = (\sqrt{t + 7})^2 \quad \textbf{Square both sides}$$

$$t^2 + 10t + 25 = t + 7$$

$$t^2 + 9t + 18 = 0 \quad \textbf{Standard form}$$

$$(t + 3)(t + 6) = 0 \quad \textbf{Factor the left side}$$

$$t + 3 = 0 \quad \text{or} \quad t + 6 = 0 \quad \textbf{Set factors equal to 0}$$

$$t = -3 \quad \text{or} \quad t = -6$$

We must check each solution in the original equation:

Check $t = -3$	Check $t = -6$
$-3 + 5 \stackrel{?}{=} \sqrt{-3 + 7}$	$-6 + 5 \stackrel{?}{=} \sqrt{-6 + 7}$
$2 \stackrel{?}{=} \sqrt{4}$	$-1 \stackrel{?}{=} \sqrt{1}$
$2 = 2$	$-1 = 1$
A true statement	A false statement

Since $t = -6$ does not check, our only solution is $t = -3$.

4. Solve $t - 6 = \sqrt{t - 4}$.

EXAMPLE 5

Solve $\sqrt{x - 3} = \sqrt{x} - 3$.

Solution We begin by squaring both sides. Note what happens when we square the right side of the equation, and compare the square of the right side with the square of the left side. You must convince yourself that these results are correct. (The note in the margin will help if you are having trouble convincing yourself that what is written below is true.)

$$(\sqrt{x - 3})^2 = (\sqrt{x} - 3)^2$$

$$x - 3 = x - 6\sqrt{x} + 9$$

Now we still have a radical in our equation, so we will have to square both sides again. Before we do, though, let's isolate the remaining radical.

$$x - 3 = x - 6\sqrt{x} + 9$$

$$-3 = -6\sqrt{x} + 9 \quad \textbf{Add } -x \textbf{ to each side}$$

$$-12 = -6\sqrt{x} \quad \textbf{Add } -9 \textbf{ to each side}$$

$$2 = \sqrt{x} \quad \textbf{Divide each side by } -6$$

$$4 = x \quad \textbf{Square each side}$$

Our only possible solution is $x = 4$, which we check in our original equation as follows:

$$\sqrt{4 - 3} \stackrel{?}{=} \sqrt{4} - 3$$

$$\sqrt{1} \stackrel{?}{=} 2 - 3$$

$$1 = -1 \quad \textbf{A false statement}$$

Substituting 4 for x in the original equation yields a false statement. Since 4 was our only possible solution, there is no solution to our equation.

5. Solve $\sqrt{x - 9} = \sqrt{x} - 3$.

NOTE

It is very important that you realize that the square of $(\sqrt{x} - 3)$ is not $x + 9$. Remember, when we square a difference with two terms, we use the formula

$$(a - b)^2 = a^2 - 2ab + b^2$$

Applying this formula to $(\sqrt{x} - 3)^2$, we have

$$(\sqrt{x} - 3)^2 = (\sqrt{x})^2 - 2(\sqrt{x})(3) + 3^2$$
$$= x - 6\sqrt{x} + 9$$

Answers

4. 8 5. 9

Here is another example of an equation for which we must apply our squaring property twice before all radicals are eliminated.

6. Solve $\sqrt{x+4} = 2 - \sqrt{3x}$.

EXAMPLE 6 Solve $\sqrt{x+1} = 1 - \sqrt{2x}$.

Solution This equation has two separate terms involving radical signs. Squaring both sides gives

$$x + 1 = 1 - 2\sqrt{2x} + 2x$$

$$-x = -2\sqrt{2x} \quad \text{Add } -2x \text{ and } -1 \text{ to both sides}$$

$$x^2 = 4(2x) \quad \text{Square both sides}$$

$$x^2 - 8x = 0 \quad \text{Standard form}$$

Our equation is a quadratic equation in standard form. To solve for x, we factor the left side and set each factor equal to 0.:

$$x(x - 8) = 0 \quad \text{Factor left side}$$

$$x = 0 \quad \text{or} \quad x - 8 = 0 \quad \text{Set factors equal to 0}$$

$$x = 8$$

Since we squared both sides of our equation, we have the possibility that one or both of the solutions are extraneous. We must check each one in the original equation:

Check $x = 8$	Check $x = 0$
$\sqrt{8+1} \stackrel{?}{=} 1 - \sqrt{2 \cdot 8}$	$\sqrt{0+1} \stackrel{?}{=} 1 - \sqrt{2 \cdot 0}$
$\sqrt{9} \stackrel{?}{=} 1 - \sqrt{16}$	$\sqrt{1} \stackrel{?}{=} 1 - \sqrt{0}$
$3 \stackrel{?}{=} 1 - 4$	$1 \stackrel{?}{=} 1 - 0$
$3 = -3$	$1 = 1$
A false statement	A true statement

Since $x = 8$ does not check, it is an extraneous solution. Our only solution is $x = 0$.

7. Solve $\sqrt{x+2} = \sqrt{x+3} - 1$.

EXAMPLE 7 Solve $\sqrt{x+1} = \sqrt{x+2} - 1$.

Solution Squaring both sides we have

$$(\sqrt{x+1})^2 = (\sqrt{x+2} - 1)^2$$

$$x + 1 = x + 2 - 2\sqrt{x+2} + 1$$

Once again we are left with a radical in our equation. Before we square each side again, we must isolate the radical on the right side of the equation.

$$x + 1 = x + 3 - 2\sqrt{x+2} \quad \text{Simplify the right side}$$

$$1 = 3 - 2\sqrt{x+2} \quad \text{Add } -x \text{ to each side}$$

$$-2 = -2\sqrt{x+2} \quad \text{Add } -3 \text{ to each side}$$

$$1 = \sqrt{x+2} \quad \text{Divide each side by } -2$$

$$1 = x + 2 \quad \text{Square both sides}$$

$$-1 = x \quad \text{Add } -2 \text{ to each side}$$

Answer
6. 0

Checking our only possible solution, $x = -1$, in our original equation, we have

$$\sqrt{-1+1} \stackrel{?}{=} \sqrt{-1+2} - 1$$
$$\sqrt{0} \stackrel{?}{=} \sqrt{1} - 1$$
$$0 \stackrel{?}{=} 1 - 1$$
$$0 = 0 \qquad \textbf{A true statement}$$

Our solution checks.

It is also possible to raise both sides of an equation to powers greater than 2. We only need to check for extraneous solutions when we raise both sides of an equation to an even power. Raising both sides of an equation to an odd power will not produce extraneous solutions.

EXAMPLE 8 Solve $\sqrt[3]{4x+5} = 3$.

Solution Cubing both sides we have

$$(\sqrt[3]{4x+5})^3 = 3^3$$
$$4x + 5 = 27$$
$$4x = 22$$
$$x = \frac{22}{4}$$
$$x = \frac{11}{2}$$

We do not need to check $x = \frac{11}{2}$ since we raised both sides to an odd power.

We end this section by looking at graphs of some equations that contain radicals.

EXAMPLE 9 Graph $y = \sqrt{x}$ and $y = \sqrt[3]{x}$.

Solution The graphs are shown in Figures 1 and 2. Notice that the graph of $y = \sqrt{x}$ appears in the first quadrant only because in the equation $y = \sqrt{x}$, x and y cannot be negative.

The graph of $y = \sqrt[3]{x}$ appears in quadrants 1 and 3 since the cube root of a positive number is also a positive number and the cube root of a negative number is a negative number; that is, when x is positive, y will be positive and when x is negative, y will be negative.

The graphs of both equations will contain the origin since $y = 0$ when $x = 0$ in both equations.

8. Solve $\sqrt[3]{3x-7} = 2$.

9. Graph each equation:
$y = \sqrt{x} + 3$
$y = \sqrt{x+3}$

Answers
7. -2 **8.** 5

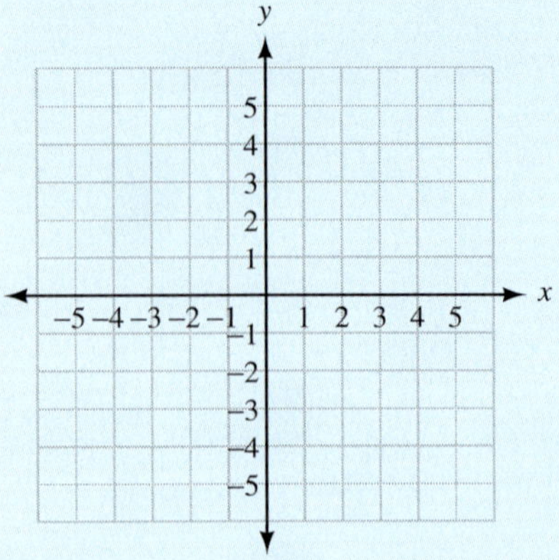

x	y
−4	undefined
−1	undefined
0	0
1	1
4	2
9	3
16	4

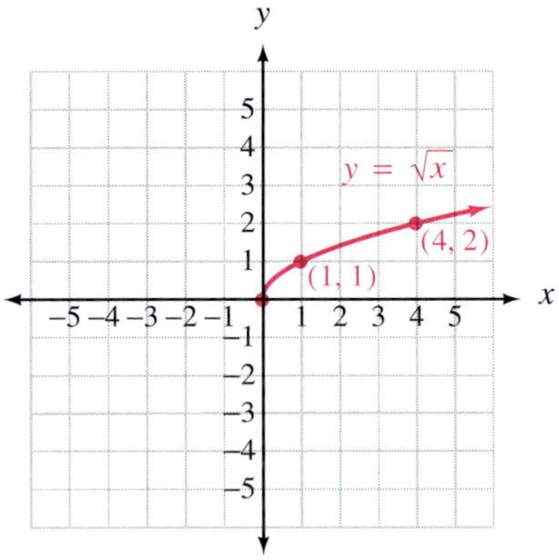

FIGURE 1

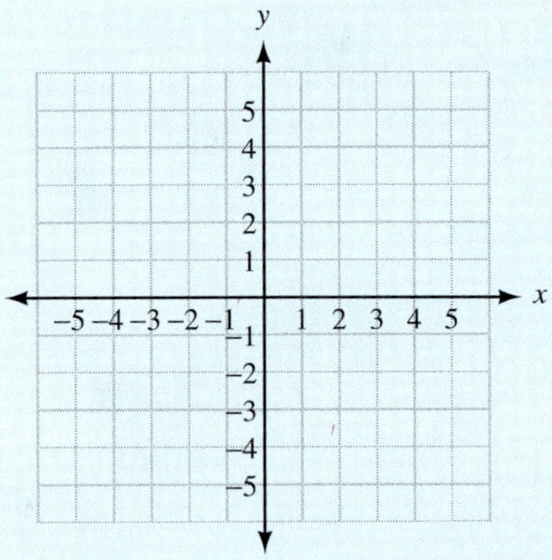

x	y
−27	−3
−8	−2
−1	−1
0	0
1	1
8	2
27	3

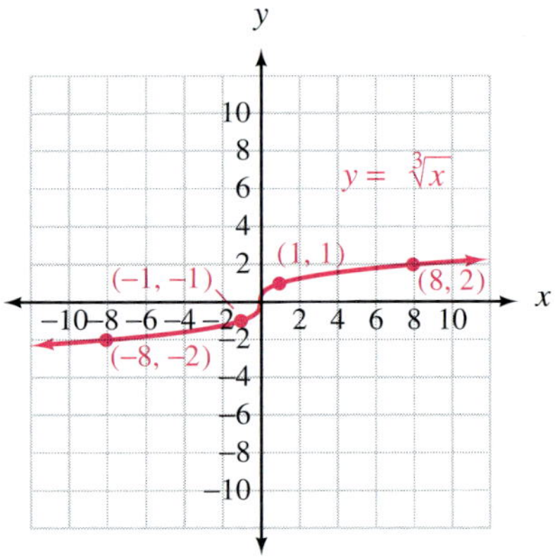

FIGURE 2

Linking Objectives and Examples

Next to each **objective**, list the examples that are best described by that objective.

A	
B	

Getting Ready for Class

After reading through the preceding section, respond in your own words and in complete sentences.

1. What is the squaring property of equality?
2. Under what conditions do we obtain extraneous solutions to equations that contain radical expressions?
3. If we have raised both sides of an equation to a power, when is it not necessary to check for extraneous solutions?
4. When will you need to apply the squaring property of equality twice in the process of solving an equation containing radicals?

Answer

9. See Solutions to Selected Practice Problems.

Problem Set 7.6

INTERMEDIATE Algebra $f(x)$ Now™

Solve each of the following equations.

1. $\sqrt{2x+1} = 3$
2. $\sqrt{3x+1} = 4$
3. $\sqrt{4x+1} = -5$
4. $\sqrt{6x+1} = -5$
5. $\sqrt{2y-1} = 3$
6. $\sqrt{3y-1} = 2$
7. $\sqrt{5x-7} = -1$
8. $\sqrt{8x+3} = -6$
9. $\sqrt{2x-3} - 2 = 4$
10. $\sqrt{3x+1} - 4 = 1$
11. $\sqrt{4x+1} + 3 = 2$
12. $\sqrt{5a-3} + 6 = 2$
13. $\sqrt[4]{3x+1} = 2$
14. $\sqrt[4]{4x+1} = 3$
15. $\sqrt[3]{2x-5} = 1$
16. $\sqrt[3]{5x+7} = 2$
17. $\sqrt[3]{3a+5} = -3$
18. $\sqrt[3]{2a+7} = -2$
19. $\sqrt{y-3} = y-3$
20. $\sqrt{y+3} = y-3$

21. $\sqrt{a+2} = a+2$

22. $\sqrt{a+10} = a-2$

23. $\sqrt{2x+4} = \sqrt{1-x}$

24. $\sqrt{3x+4} = -\sqrt{2x+3}$

25. $\sqrt{4a+7} = -\sqrt{a+2}$

26. $\sqrt{7a-1} = \sqrt{2a+4}$

27. $\sqrt[4]{5x-8} = \sqrt[4]{4x-1}$

28. $\sqrt[4]{6x+7} = \sqrt[4]{x+2}$

29. $x+1 = \sqrt{5x+1}$

30. $x-1 = \sqrt{6x+1}$

31. $t+5 = \sqrt{2t+9}$

32. $t+7 = \sqrt{2t+13}$

33. $\sqrt{y-8} = \sqrt{8-y}$

34. $\sqrt{2y+5} = \sqrt{5y+2}$

35. $\sqrt[3]{3x+5} = \sqrt[3]{5-2x}$

36. $\sqrt[3]{4x+9} = \sqrt[3]{3-2x}$

The following equations will require that you square both sides twice before all the radicals are eliminated. Solve each equation using the methods shown in Examples 5, 6, and 7.

37. $\sqrt{x-8} = \sqrt{x} - 2$

38. $\sqrt{x+3} = \sqrt{x} - 3$

39. $\sqrt{x+1} = \sqrt{x} + 1$

40. $\sqrt{x-1} = \sqrt{x} - 1$

41. $\sqrt{x+8} = \sqrt{x-4} + 2$

42. $\sqrt{x+5} = \sqrt{x-3} + 2$

43. $\sqrt{x-5} - 3 = \sqrt{x-8}$

44. $\sqrt{x-3} - 4 = \sqrt{x-3}$

45. $\sqrt{x+4} = 2 - \sqrt{2x}$

46. $\sqrt{5x+1} = 1 + \sqrt{5x}$

47. $\sqrt{2x+4} = \sqrt{x+3} + 1$

48. $\sqrt{2x-1} = \sqrt{x-4} + 2$

Graph each equation.

49. $y = 2\sqrt{x}$

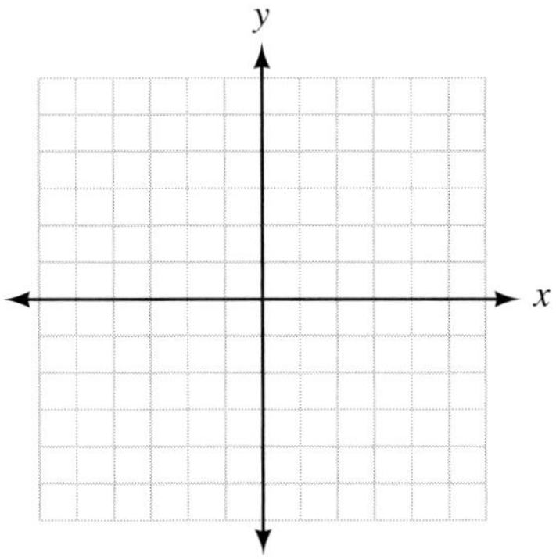

50. $y = -2\sqrt{x}$

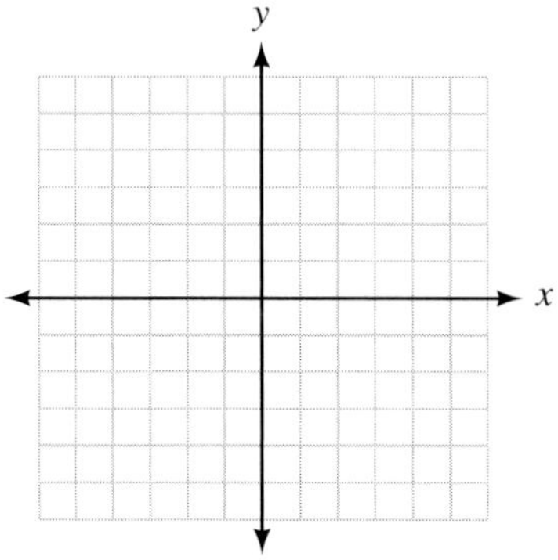

51. $y = \sqrt{x} - 2$

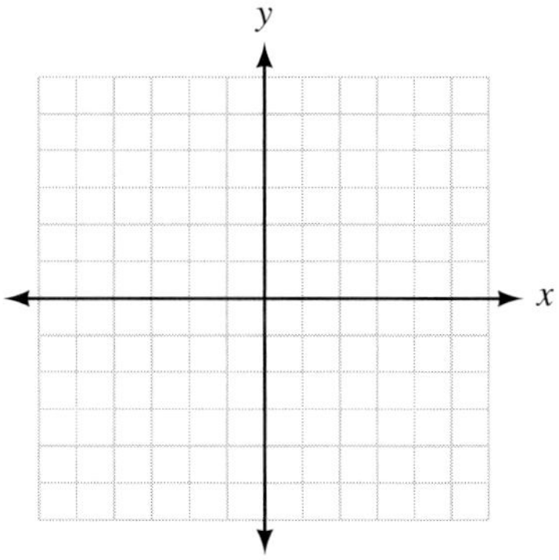

52. $y = \sqrt{x} + 2$

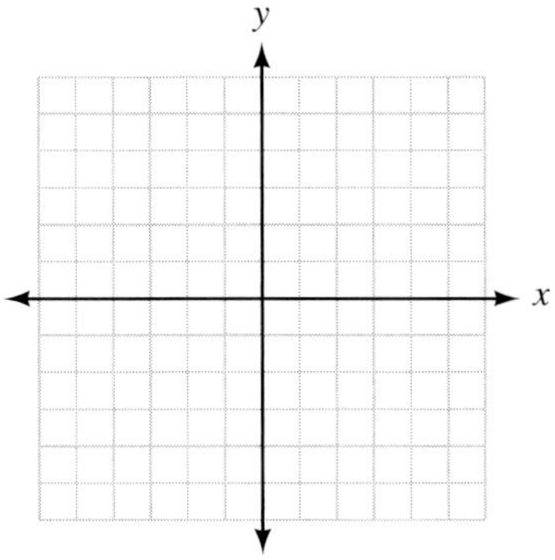

53. $y = \sqrt{x - 2}$

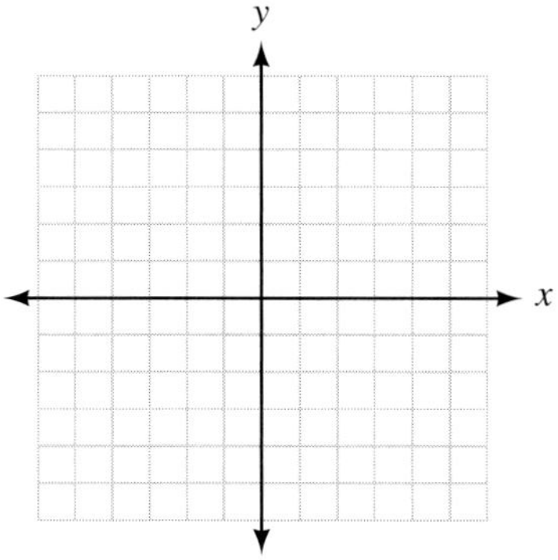

54. $y = \sqrt{x + 2}$

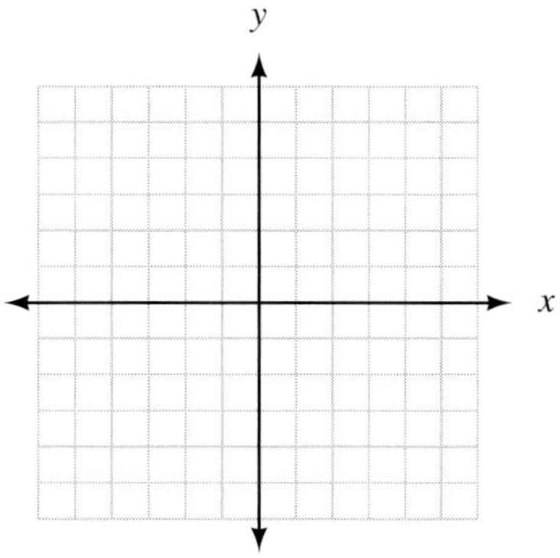

55. $y = 3\sqrt[3]{x}$

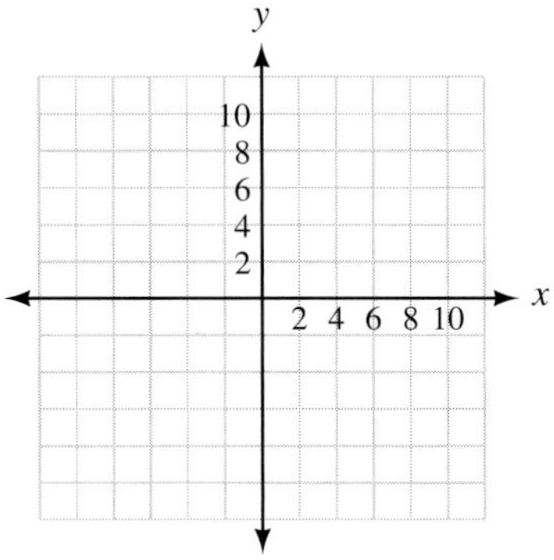

56. $y = -3\sqrt[3]{x}$

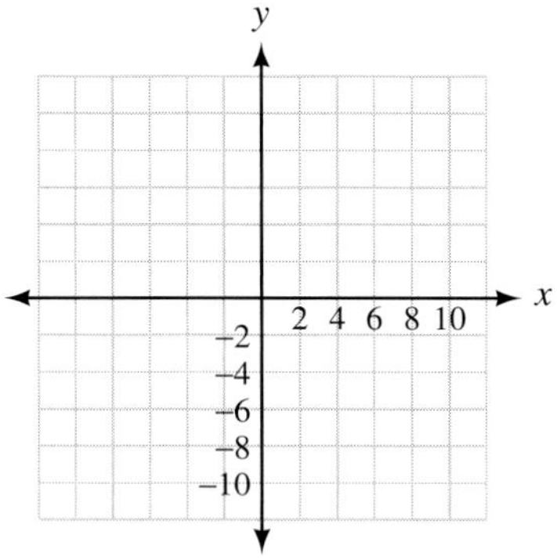

57. $y = \sqrt[3]{x} + 3$

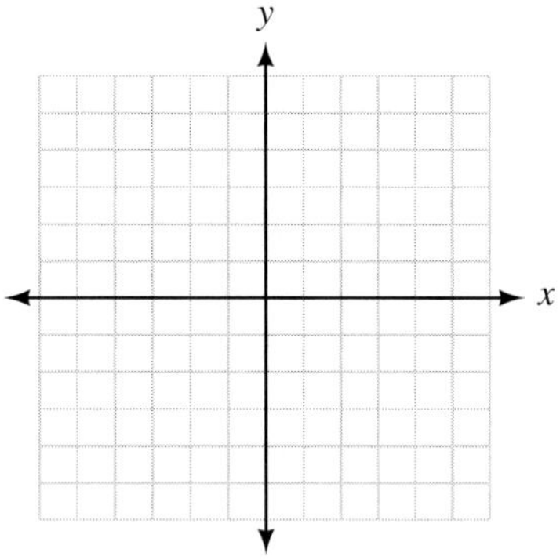

58. $y = \sqrt[3]{x} - 3$

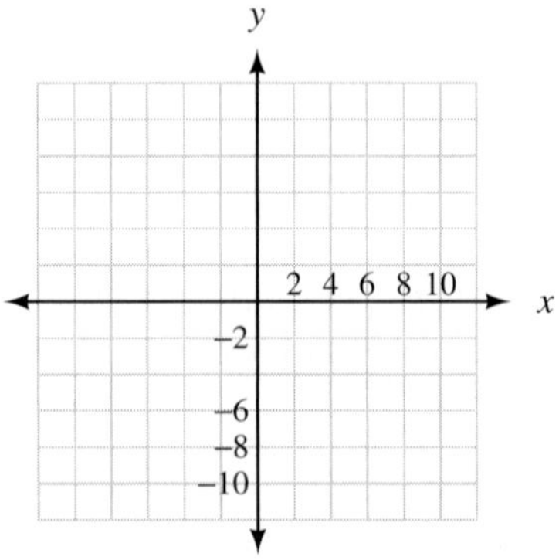

59. $y = \sqrt[3]{x + 3}$

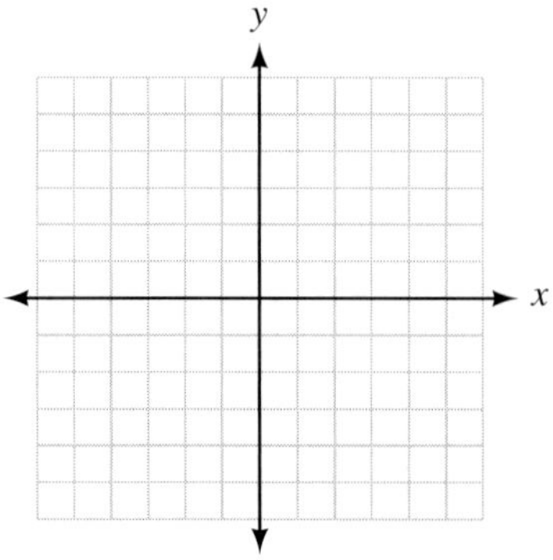

60. $y = \sqrt[3]{x - 3}$

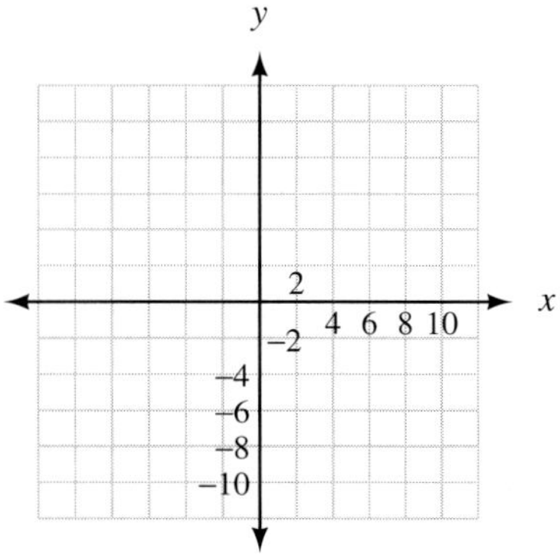

Applying the Concepts

61. **Solving a Formula** Solve the following formula for h:

$$t = \frac{\sqrt{100 - h}}{4}$$

62. **Solving a Formula** Solve the following formula for h:

$$t = \sqrt{\frac{2h - 40t}{g}}$$

63. **Pendulum Clock** The length of time (T) in seconds it takes the pendulum of a grandfather clock to swing through one complete cycle is given by the formula

$$T = 2\pi\sqrt{\frac{L}{32}}$$

where L is the length, in feet, of the pendulum, and π is approximately $\frac{22}{7}$. How long must the pendulum be if one complete cycle takes 2 seconds?

64. **Pendulum Clock** Solve the formula in Problem 63 for L.

Pollution A long straight river, 100 meters wide, is flowing at 1 meter per second. A pollutant is entering the river at a constant rate from one of its banks. As the pollutant disperses in the water, it forms a plume that is modeled by the equation $y = \sqrt{x}$. Use this information to answer the following questions.

65. How wide is the plume 25 meters down river from the source of the pollution?

66. How wide is the plume 100 meters down river from the source of the pollution?

67. How far down river from the source of the pollution does the plume reach halfway across the river?

68. How far down river from the source of the pollution does the plume reach the other side of the river?

Getting Ready for the Next Section

Simplify.

69. $\sqrt{25}$ **70.** $\sqrt{49}$ **71.** $\sqrt{12}$ **72.** $\sqrt{50}$

73. $(-1)^{15}$ **74.** $(-1)^{20}$ **75.** $(-1)^{50}$ **76.** $(-1)^{5}$

Solve.

77. $3x = 12$ **78.** $4 = 8y$ **79.** $4x - 3 = 5$ **80.** $7 = 2y - 1$

Perform the indicated operation.

81. $(3 + 4x) + (7 - 6x)$ **82.** $(2 - 5x) + (-1 + 7x)$ **83.** $(7 + 3x) - (5 + 6x)$

84. $(5 - 2x) - (9 - 4x)$ **85.** $(3 - 4x)(2 + 5x)$ **86.** $(8 + x)(7 - 3x)$

87. $2x(4 - 6x)$ **88.** $3x(7 + 2x)$ **89.** $(2 + 3x)^2$

90. $(3 + 5x)^2$ **91.** $(2 - 3x)(2 + 3x)$ **92.** $(4 - 5x)(4 + 5x)$

Maintaining Your Skills

Multiply.

93. $\sqrt{2}(\sqrt{3} - \sqrt{2})$ **94.** $(\sqrt{x} - 4)(\sqrt{x} + 5)$ **95.** $(\sqrt{x} + 5)^2$ **96.** $(\sqrt{5} + \sqrt{3})(\sqrt{5} - \sqrt{3})$

Rationalize the denominator.

97. $\dfrac{\sqrt{x}}{\sqrt{x} + 3}$ **98.** $\dfrac{\sqrt{5} - \sqrt{3}}{\sqrt{5} + \sqrt{3}}$

Complex Numbers

7.7

The stamp shown here was issued by Germany in 1977 to commemorate the 200th anniversary of the birth of mathematician Carl Gauss. The number $-5+6i$ shown on the stamp is a complex number, as are the other numbers on the stamp. Working with complex numbers gives us a way to solve a wider variety of equations. For example, the equation $x^2 = -9$ has no real number solutions since the square of a real number is always positive. We have been unable to work with square roots of negative numbers like $\sqrt{-25}$ and $\sqrt{-16}$ for the same reason. Complex numbers allow us to expand our work with radicals to include square roots of negative numbers and solve equations like $x^2 = -9$ and $x^2 = -64$. Our work with complex numbers is based on the following definition.

DEFINITION

The number *i* is such that $i = \sqrt{-1}$ (which is the same as saying $i^2 = -1$).

The number i, as we have defined it here, is not a real number. Because of the way we have defined i, we can use it to simplify square roots of negative numbers.

Square Roots of Negative Numbers

If a is a positive number, then $\sqrt{-a}$ can always be written as $i\sqrt{a}$; that is,

$$\sqrt{-a} = i\sqrt{a} \quad \text{if } a \text{ is a positive number}$$

To justify our rule, we simply square the quantity $i\sqrt{a}$ to obtain $-a$. Here is what it looks like when we do so:

$$(i\sqrt{a})^2 = i^2 \cdot (\sqrt{a})^2$$
$$= -1 \cdot a$$
$$= -a$$

Here are some examples that illustrate the use of our new rule.

EXAMPLES Write each square root in terms of the number i.

1. $\sqrt{-25} = i\sqrt{25} = i \cdot 5 = 5i$
2. $\sqrt{-49} = i\sqrt{49} = i \cdot 7 = 7i$
3. $\sqrt{-12} = i\sqrt{12} = i \cdot 2\sqrt{3} = 2i\sqrt{3}$
4. $\sqrt{-17} = i\sqrt{17}$

If we assume all the properties of exponents hold when the base is i, we can write any power of i as i, -1, $-i$, or 1. Using the fact that $i^2 = -1$, we have

$$i^1 = i$$
$$i^2 = -1$$
$$i^3 = i^2 \cdot i = -1(i) = -i$$
$$i^4 = i^2 \cdot i^2 = -1(-1) = 1$$

OBJECTIVES

A	Simplify square roots of negative numbers.
B	Simplify powers of i.
C	Solve for unknown variables by equating real parts and equating imaginary parts of two complex numbers.
D	Add and subtract complex numbers.
E	Multiply complex numbers.
F	Divide complex numbers.

PRACTICE PROBLEMS

1. $\sqrt{-36}$
2. $-\sqrt{-64}$
3. $\sqrt{-18}$
4. $-\sqrt{-19}$

NOTE

In Examples 3 and 4 we wrote i before the radical simply to avoid confusion. If we were to write the answer to 3 as $2\sqrt{3}i$, some people would think the i was under the radical sign and it is not.

Answers

1. $6i$ 2. $-8i$ 3. $3i\sqrt{2}$ 4. $-i\sqrt{19}$

Since $i^4 = 1$, i^5 will simplify to i, and we will begin repeating the sequence $i, -1, -i, 1$ as we simplify higher powers of i: Any power of i simplifies to $i, -1, -i$, or 1. The easiest way to simplify higher powers of i is to write them in terms of i^2. For instance, to simplify i^{21}, we would write it as

$$(i^2)^{10} \cdot i \qquad \text{because } 2 \cdot 10 + 1 = 21$$

Then, since $i^2 = -1$, we have

$$(-1)^{10} \cdot i = 1 \cdot i = i$$

EXAMPLES

Simplify as much as possible.

5. $i^{30} = (i^2)^{15} = (-1)^{15} = -1$

6. $i^{11} = (i^2)^5 \cdot i = (-1)^5 \cdot i = (-1)i = -i$

7. $i^{40} = (i^2)^{20} = (-1)^{20} = 1$

Simplify.

5. i^{20}

6. i^{23}

7. i^{50}

DEFINITION

A **complex number** is any number that can be put in the form

$$a + bi$$

where a and b are real numbers and $i = \sqrt{-1}$. The form $a + bi$ is called **standard form** for complex numbers. The number a is called the **real part** of the complex number. The number b is called the **imaginary part** of the complex number.

Every real number is a complex number. For example, 8 can be written as $8 + 0i$. Likewise, $-\frac{1}{2}$, π, $\sqrt{3}$, and -9 are complex numbers because they can all be written in the form $a + bi$:

$$-\frac{1}{2} = -\frac{1}{2} + 0i \qquad \pi = \pi + 0i$$

$$\sqrt{3} = \sqrt{3} + 0i \qquad -9 = -9 + 0i$$

The real numbers occur when $b = 0$. When $b \neq 0$, we have complex numbers that contain i, such as $2 + 5i$, $6 - i$, $4i$, and $\frac{1}{2}i$. These numbers are called *imaginary numbers.* The diagram explains this further.

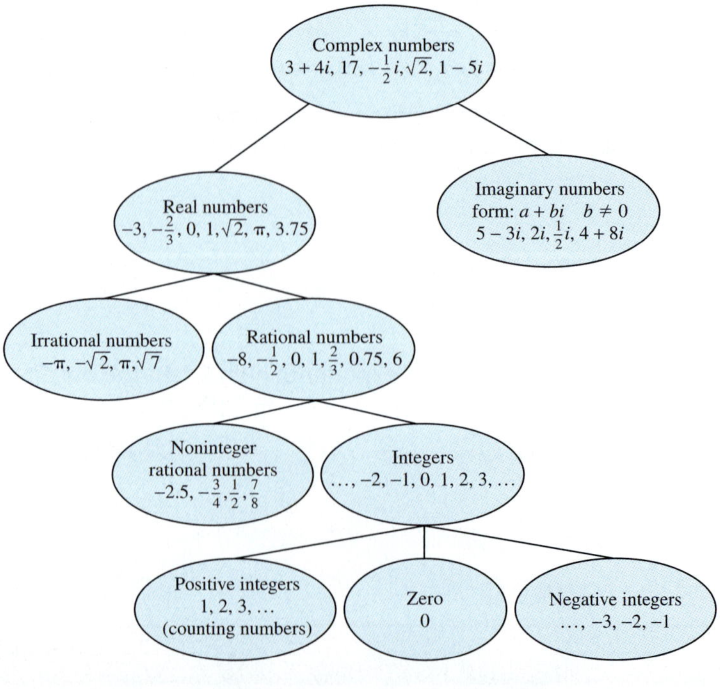

Answers

5. 1 **6.** $-i$ **7.** -1

Equality for Complex Numbers

Two complex numbers are equal if and only if their real parts are equal and their imaginary parts are equal; that is, for real numbers a, b, c, and d,

$$a + bi = c + di \quad \text{if and only if} \quad a = c \quad \text{and} \quad b = d$$

EXAMPLE 8 Find x and y if $3x + 4i = 12 - 8yi$.

Solution Since the two complex numbers are equal, their real parts are equal and their imaginary parts are equal:

$$3x = 12 \qquad \text{and} \qquad 4 = -8y$$

$$x = 4 \qquad\qquad y = -\frac{1}{2}$$

EXAMPLE 9 Find x and y if $(4x - 3) + 7i = 5 + (2y - 1)i$.

Solution The real parts are $4x - 3$ and 5. The imaginary parts are 7 and $2y - 1$:

$$4x - 3 = 5 \qquad \text{and} \qquad 7 = 2y - 1$$

$$4x = 8 \qquad\qquad 8 = 2y$$

$$x = 2 \qquad\qquad y = 4$$

Addition and Subtraction of Complex Numbers

To add two complex numbers, add their real parts and add their imaginary parts; that is, if a, b, c, and d are real numbers, then

$$(a + bi) + (c + di) = (a + c) + (b + d)i$$

If we assume that the commutative, associative, and distributive properties hold for the number i, then the definition of addition is simply an extension of these properties.

We define subtraction in a similar manner. If a, b, c, and d are real numbers, then

$$(a + bi) - (c + di) = (a - c) + (b - d)i$$

EXAMPLES Add or subtract as indicated.

10. $(3 + 4i) + (7 - 6i) = (3 + 7) + (4 - 6)i = 10 - 2i$

11. $(7 + 3i) - (5 + 6i) = (7 - 5) + (3 - 6)i = 2 - 3i$

12. $(5 - 2i) - (9 - 4i) = (5 - 9) + (-2 + 4)i = -4 + 2i$

Multiplication of Complex Numbers

Since complex numbers have the same form as binomials, we find the product of two complex numbers the same way we find the product of two binomials.

EXAMPLE 13 Multiply $(3 - 4i)(2 + 5i)$.

Solution Multiplying each term in the second complex number by each term in the first, we have

$$(3 - 4i)(2 + 5i) = 3 \cdot 2 + 3 \cdot 5i - 2 \cdot 4i - 5i(4i)$$

$$= 6 + 15i - 8i - 20i^2$$

8. Find x and y if
$4x + 7i = 8 - 14yi$

9. Find x and y if
$(2x - 1) + 9i = 5 + (4y + 1)i$

Add or subtract as indicated.

10. $(2 + 6i) + (3 - 4i)$

11. $(6 + 5i) - (4 + 3i)$

12. $(7 - i) - (8 - 2i)$

13. Multiply $(2 + 3i)(1 - 4i)$.

Answers
8. $x = 2, y = -\frac{1}{2}$ **9.** $x = 3, y = 2$
10. $5 + 2i$ **11.** $2 + 2i$ **12.** $-1 + i$

Combining similar terms and using the fact that $i^2 = -1$, we can simplify as follows:

$$6 + 15i - 8i - 20i^2 = 6 + 7i - 20(-1)$$
$$= 6 + 7i + 20$$
$$= 26 + 7i$$

The product of the complex numbers $3 - 4i$ and $2 + 5i$ is the complex number $26 + 7i$.

14. Multiply $-3i(2 + 3i)$.

EXAMPLE 14

Multiply $2i(4 - 6i)$.

Solution Applying the distributive property gives us

$$2i(4 - 6i) = 2i \cdot 4 - 2i \cdot 6i$$
$$= 8i - 12i^2$$
$$= 12 + 8i$$

15. Expand $(2 + 4i)^2$.

EXAMPLE 15

Expand $(3 + 5i)^2$.

Solution We treat this like the square of a binomial. Remember, $(a + b)^2 = a^2 + 2ab + b^2$.

$$(3 + 5i)^2 = 3^2 + 2(3)(5i) + (5i)^2$$
$$= 9 + 30i + 25i^2$$
$$= 9 + 30i - 25$$
$$= -16 + 30i$$

NOTE

We can obtain the same result by writing $(3 + 5i)^2$ as $(3 + 5i)$ times $(3 + 5i)$ and applying the FOIL method as we did with the problem in Example 13.

16. Multiply $(3 - 5i)(3 + 5i)$.

EXAMPLE 16

Multiply $(2 - 3i)(2 + 3i)$.

Solution This product has the form $(a - b)(a + b)$, which we know results in the difference of two squares, $a^2 - b^2$:

$$(2 - 3i)(2 + 3i) = 2^2 - (3i)^2$$
$$= 4 - 9i^2$$
$$= 4 + 9$$
$$= 13$$

The product of the two complex numbers $2 - 3i$ and $2 + 3i$ is the real number 13. The two complex numbers $2 - 3i$ and $2 + 3i$ are called complex conjugates. The fact that their product is a real number is very useful.

DEFINITION

The complex numbers $a + bi$ and $a - bi$ are called **complex conjugates.** One important property they have is that their product is the real number $a^2 + b^2$. Here's why:

$$(a + bi)(a - bi) = a^2 - (bi)^2$$
$$= a^2 - b^2i^2$$
$$= a^2 - b^2(-1)$$
$$= a^2 + b^2$$

Answers

13. $14 - 5i$ **14.** $9 - 6i$
15. $-12 + 16i$ **16.** 34

Division with Complex Numbers

The fact that the product of two complex conjugates is a real number is the key to division with complex numbers.

EXAMPLE 17 Divide $\dfrac{2+i}{3-2i}$.

Solution We want a complex number in standard form that is equivalent to the quotient $\dfrac{2+i}{3-2i}$. We need to eliminate i from the denominator. Multiplying the numerator and denominator by $3 + 2i$ will give us what we want:

$$\begin{aligned}\frac{2+i}{3-2i} &= \frac{2+i}{3-2i}\cdot\frac{\mathbf{(3+2i)}}{\mathbf{(3+2i)}}\\ &= \frac{6+4i+3i+2i^2}{9-4i^2}\\ &= \frac{6+7i-2}{9+4}\\ &= \frac{4+7i}{13}\\ &= \frac{4}{13}+\frac{7}{13}i\end{aligned}$$

Dividing the complex number $2 + i$ by $3 - 2i$ gives the complex number $\frac{4}{13} + \frac{7}{13}i$.

17. Divide $\dfrac{3+2i}{2-5i}$.

EXAMPLE 18 Divide $\dfrac{7-4i}{i}$.

Solution The conjugate of the denominator is $-i$. Multiplying numerator and denominator by this number, we have

$$\begin{aligned}\frac{7-4i}{i} &= \frac{7-4i}{i}\cdot\frac{\mathbf{-i}}{\mathbf{-i}}\\ &= \frac{-7i+4i^2}{-i^2}\\ &= \frac{-7i+4(-1)}{-(-1)}\\ &= -4-7i\end{aligned}$$

18. Divide $\dfrac{3+2i}{i}$.

Getting Ready for Class

After reading through the preceding section, respond in your own words and in complete sentences.

1. What is the number i?
2. What is a complex number?
3. What kind of number results when we multiply complex conjugates?
4. Explain how to divide complex numbers.

Linking Objectives and Examples

Next to each **objective**, list the examples that are best described by that objective.

A	
B	
C	
D	
E	
F	

Answers

17. $-\frac{4}{29} + \frac{19}{29}i$ **18.** $2 - 3i$

Problem Set 7.7

INTERMEDIATE Algebra Now™

Write the following in terms of i, and simplify as much as possible.

1. $\sqrt{-36}$

2. $\sqrt{-49}$

3. $-\sqrt{-25}$

4. $-\sqrt{-81}$

5. $\sqrt{-72}$

6. $\sqrt{-48}$

7. $-\sqrt{-12}$

8. $-\sqrt{-75}$

Write each of the following as i, -1, $-i$, or 1.

9. i^{28}

10. i^{31}

11. i^{26}

12. i^{37}

13. i^{75}

14. i^{42}

Find x and y so each of the following equations is true.

15. $2x + 3yi = 6 - 3i$

16. $4x - 2yi = 4 + 8i$

17. $2 - 5i = -x + 10yi$

18. $4 + 7i = 6x - 14yi$

19. $2x + 10i = -16 - 2yi$

20. $4x - 5i = -2 + 3yi$

21. $(2x - 4) - 3i = 10 - 6yi$

22. $(4x - 3) - 2i = 8 + yi$

23. $(7x - 1) + 4i = 2 + (5y + 2)i$

24. $(5x + 2) - 7i = 4 + (2y + 1)i$

Combine the following complex numbers.

25. $(2 + 3i) + (3 + 6i)$

26. $(4 + i) + (3 + 2i)$

27. $(3 - 5i) + (2 + 4i)$

28. $(7 + 2i) + (3 - 4i)$

29. $(5 + 2i) - (3 + 6i)$

30. $(6 + 7i) - (4 + i)$

31. $(3 - 5i) - (2 + i)$

32. $(7 - 3i) - (4 + 10i)$

33. $[(3 + 2i) - (6 + i)] + (5 + i)$

34. $[(4 - 5i) - (2 + i)] + (2 + 5i)$

35. $[(7 - i) - (2 + 4i)] - (6 + 2i)$

36. $[(3 - i) - (4 + 7i)] - (3 - 4i)$

37. $(3 + 2i) - [(3 - 4i) - (6 + 2i)]$

38. $(7 - 4i) - [(-2 + i) - (3 + 7i)]$

39. $(4 - 9i) + [(2 - 7i) - (4 + 8i)]$

40. $(10 - 2i) - [(2 + i) - (3 - i)]$

Find the following products.

41. $3i(4 + 5i)$

42. $2i(3 + 4i)$

43. $6i(4 - 3i)$

44. $11i(2 - i)$

45. $(3 + 2i)(4 + i)$

46. $(2 - 4i)(3 + i)$

47. $(4 + 9i)(3 - i)$

48. $(5 - 2i)(1 + i)$

49. $(1 + i)^3$

50. $(1 - i)^3$

51. $(2 - i)^3$

52. $(2 + i)^3$

53. $(2 + 5i)^2$

54. $(3 + 2i)^2$

55. $(1 - i)^2$

56. $(1 + i)^2$

57. $(3 - 4i)^2$

58. $(6 - 5i)^2$

59. $(2 + i)(2 - i)$

60. $(3 + i)(3 - i)$

61. $(6 - 2i)(6 + 2i)$

62. $(5 + 4i)(5 - 4i)$

63. $(2 + 3i)(2 - 3i)$

64. $(2 - 7i)(2 + 7i)$

65. $(10 + 8i)(10 - 8i)$

66. $(11 - 7i)(11 + 7i)$

Find the following quotients. Write all answers in standard form for complex numbers.

67. $\dfrac{2 - 3i}{i}$

68. $\dfrac{3 + 4i}{i}$

69. $\dfrac{5 + 2i}{-i}$

70. $\dfrac{4 - 3i}{-i}$

71. $\dfrac{4}{2 - 3i}$

72. $\dfrac{3}{4 - 5i}$

73. $\dfrac{6}{-3 + 2i}$

74. $\dfrac{-1}{-2 - 5i}$

75. $\dfrac{2 + 3i}{2 - 3i}$

76. $\dfrac{4 - 7i}{4 + 7i}$

77. $\dfrac{5 + 4i}{3 + 6i}$

78. $\dfrac{2 + i}{5 - 6i}$

Applying the Concepts

79. Electric Circuits Complex numbers may be applied to electrical circuits. Electrical engineers use the fact that resistance R to electrical flow of the electrical current I and the voltage V are related by the formula $V = RI$. (Voltage is measured in volts, resistance in ohms, and current in amperes.) Find the resistance to electrical flow in a circuit that has a voltage $V = (80 + 20i)$ volts and current $I = (-6 + 2i)$ amps.

80. Electric Circuits Refer to the information about electrical circuits in Problem 79, and find the current in a circuit that has a resistance of $(4 + 10i)$ ohms and a voltage of $(5 - 7i)$ volts.

Maintaining Your Skills

Solve each equation.

81. $\dfrac{t}{3} - \dfrac{1}{2} = -1$

82. $\dfrac{x}{x-2} + \dfrac{2}{3} = \dfrac{2}{x-2}$

83. $2 + \dfrac{5}{y} = \dfrac{3}{y^2}$

84. $1 - \dfrac{1}{y} = \dfrac{12}{y^2}$

Solve each application problem.

85. The sum of a number and its reciprocal is $\frac{41}{20}$. Find the number.

86. It takes an inlet pipe 8 hours to fill a tank. The drain can empty the tank in 6 hours. If the tank is full and both the inlet pipe and drain are open, how long will it take to drain the tank?

Extending the Concepts

87. Show that $-i$ and $\dfrac{1}{i}$ (the opposite and the reciprocal of i) are the same number.

88. Show that i^{2n+1} is the same as i for all positive even integers n.

89. Show that $x = 1 + i$ is a solution to the equation $x^2 - 2x + 2 = 0$.

90. Show that $x = 1 - i$ is a solution to the equation $x^2 - 2x + 2 = 0$.

91. Show that $x = 2 + i$ is a solution to the equation $x^3 - 11x + 20 = 0$.

92. Show that $x = 2 - i$ is a solution to the equation $x^3 - 11x + 20 = 0$.

Chapter 7 Summary

Square Roots [7.1]

Every positive real number x has two square roots. The *positive square root* of x is written $\sqrt{x}$, and the *negative square root* of x is written $-\sqrt{x}$. Both the positive and the negative square roots of x are numbers we square to get x; that is,

$$\left.\begin{aligned} (\sqrt{x})^2 &= x \\ \text{and} \quad (-\sqrt{x})^2 &= x \end{aligned}\right\} \quad \text{for } x \geq 0$$

EXAMPLES

1. The number 49 has two square roots, 7 and -7. They are written like this:

$$\sqrt{49} = 7 \qquad -\sqrt{49} = -7$$

Higher Roots [7.1]

In the expression $\sqrt[n]{a}$, n is the *index*, a is the *radicand*, and $\sqrt{}$ is the *radical sign*. The expression $\sqrt[n]{a}$ is such that

$$(\sqrt[n]{a})^n = a \qquad a \geq 0 \text{ when } n \text{ is even}$$

2. $\sqrt[3]{8} = 2$
   $\sqrt[3]{-27} = -3$

Rational Exponents [7.1, 7.2]

Rational exponents are used to indicate roots. The relationship between rational exponents and roots is as follows:

$$a^{1/n} = \sqrt[n]{a} \qquad \text{and} \qquad a^{m/n} = (a^{1/n})^m = (a^m)^{1/n}$$

$$a \geq 0 \text{ when } n \text{ is even}$$

3. $25^{1/2} = \sqrt{25} = 5$
   $8^{2/3} = (\sqrt[3]{8})^2 = 2^2 = 4$
   $9^{3/2} = (\sqrt{9})^3 = 3^3 = 27$

Properties of Radicals [7.3]

If a and b are nonnegative real numbers whenever n is even, then

1. $\sqrt[n]{ab} = \sqrt[n]{a}\sqrt[n]{b}$
2. $\sqrt[n]{\dfrac{a}{b}} = \dfrac{\sqrt[n]{a}}{\sqrt[n]{b}} \qquad (b \neq 0)$

4. $\sqrt{4 \cdot 5} = \sqrt{4}\sqrt{5} = 2\sqrt{5}$
   $\sqrt{\dfrac{7}{9}} = \dfrac{\sqrt{7}}{\sqrt{9}} = \dfrac{\sqrt{7}}{3}$

Simplified Form for Radicals [7.3]

A radical expression is said to be in *simplified form*

1. If there is no factor of the radicand that can be written as a power greater than or equal to the index;
2. If there are no fractions under the radical sign; and
3. If there are no radicals in the denominator.

5. $\sqrt{\dfrac{4}{5}} = \dfrac{\sqrt{4}}{\sqrt{5}}$
   $= \dfrac{2}{\sqrt{5}} \cdot \dfrac{\sqrt{5}}{\sqrt{5}}$
   $= \dfrac{2\sqrt{5}}{5}$

Addition and Subtraction of Radical Expressions [7.4]

We add and subtract radical expressions by using the distributive property to combine similar radicals. Similar radicals are radicals with the same index and the same radicand.

6. $5\sqrt{3} - 7\sqrt{3} = (5 - 7)\sqrt{3}$
   $= -2\sqrt{3}$
   $\sqrt{20} + \sqrt{45} = 2\sqrt{5} + 3\sqrt{5}$
   $= (2 + 3)\sqrt{5}$
   $= 5\sqrt{5}$

Multiplication of Radical Expressions [7.5]

7. $(\sqrt{x}+2)(\sqrt{x}+3)$
$= \sqrt{x}\,\sqrt{x} + 3\sqrt{x} + 2\sqrt{x} + 2\cdot 3$
$= x + 5\sqrt{x} + 6$

We multiply radical expressions in the same way that we multiply polynomials. We can use the distributive property and the FOIL method.

Rationalizing the Denominator [7.3, 7.5]

8. $$\frac{3}{\sqrt{2}} = \frac{3}{\sqrt{2}} \cdot \frac{\mathbf{\sqrt{2}}}{\mathbf{\sqrt{2}}} = \frac{3\sqrt{2}}{2}$$

$$\frac{3}{\sqrt{5}-\sqrt{3}} = \frac{3}{\sqrt{5}-\sqrt{3}} \cdot \frac{\mathbf{\sqrt{5}+\sqrt{3}}}{\mathbf{\sqrt{5}+\sqrt{3}}}$$
$$= \frac{3\sqrt{5}+3\sqrt{3}}{5-3}$$
$$= \frac{3\sqrt{5}+3\sqrt{3}}{2}$$

When a fraction contains a square root in the denominator, we rationalize the denominator by multiplying numerator and denominator by

1. The square root itself if there is only one term in the denominator, or
2. The conjugate of the denominator if there are two terms in the denominator.

Rationalizing the denominator is also called division of radical expressions.

Squaring Property of Equality [7.6]

9. $\sqrt{2x+1} = 3$
$(\sqrt{2x+1})^2 = 3^2$
$2x + 1 = 9$
$x = 4$

We may square both sides of an equation any time it is convenient to do so, as long as we check all resulting solutions in the original equation.

Complex Numbers [7.7]

10. $3 + 4i$ is a complex number.

Addition

$(3+4i) + (2-5i) = 5 - i$

Multiplication

$(3+4i)(2-5i)$
$= 6 - 15i + 8i - 20i^2$
$= 6 - 7i + 20$
$= 26 - 7i$

Division

$$\frac{2}{3+4i} = \frac{2}{3+4i} \cdot \frac{\mathbf{3-4i}}{\mathbf{3-4i}}$$
$$= \frac{6-8i}{9+16}$$
$$= \frac{6}{25} - \frac{8}{25}i$$

A *complex number* is any number that can be put in the form

$$a + bi$$

where a and b are real numbers and $i = \sqrt{-1}$. The *real part* of the complex number is a, and b is the *imaginary part.*

If a, b, c, and d are real numbers, then we have the following definitions associated with complex numbers:

1. Equality

$$a + bi = c + di \quad \text{if and only if} \quad a = c \text{ and } b = d$$

2. Addition and subtraction

$$(a+bi) + (c+di) = (a+c) + (b+d)i$$
$$(a+bi) - (c+di) = (a-c) + (b-d)i$$

3. Multiplication

$$(a+bi)(c+di) = (ac-bd) + (ad+bc)i$$

4. Division is similar to rationalizing the denominator.

Chapter 7 Review/Test

The problems below form a comprehensive review of the material in this chapter. They can be used to study for exams. If you would like to take a practice test on this chapter, you can use the odd-numbered problems. Give yourself an hour and work as many of the odd-numbered problems as possible. When you are finished, or when an hour has passed, check your answers with the answers in the back of the book. You can use the even-numbered problems for a second practice test.

Simplify each expression as much as possible. [7.1]

1. $\sqrt{49}$ **2.** $(-27)^{1/3}$ **3.** $16^{1/4}$ **4.** $9^{3/2}$ **5.** $\sqrt[5]{32x^{15}y^{10}}$ **6.** $8^{-4/3}$

Use the properties of exponents to simplify each expression. Assume all bases represent positive numbers. [7.1]

7. $x^{2/3} \cdot x^{4/3}$ **8.** $(a^{2/3}b^{4/3})^3$ **9.** $\dfrac{a^{3/5}}{a^{1/4}}$ **10.** $\dfrac{a^{2/3}b^3}{a^{1/4}b^{1/3}}$

Multiply. [7.2]

11. $(3x^{1/2} + 5y^{1/2})(4x^{1/2} - 3y^{1/2})$ **12.** $(a^{1/3} - 5)^2$

13. Divide: $\dfrac{28x^{5/6} + 14x^{7/6}}{7x^{1/3}}$. (Assume $x > 0$.) [7.2]

14. Factor $2(x-3)^{1/4}$ from $8(x-3)^{5/4} - 2(x-3)^{1/4}$. [7.2]

15. Simplify $x^{3/4} + \dfrac{5}{x^{1/4}}$ into a single fraction. (Assume $x > 0$.) [7.2]

Write each expression in simplified form for radicals. (Assume all variables represent nonnegative numbers.) [7.3]

16. $\sqrt{12}$ **17.** $\sqrt{50}$ **18.** $\sqrt[3]{16}$ **19.** $\sqrt{18x^2}$ **20.** $\sqrt{80a^3b^4c^2}$ **21.** $\sqrt[4]{32a^4b^5c^6}$

Rationalize the denominator in each expression. [7.3]

22. $\dfrac{3}{\sqrt{2}}$ **23.** $\dfrac{6}{\sqrt[3]{2}}$

Write each expression in simplified form. (Assume all variables represent positive numbers.) [7.3]

24. $\sqrt{\dfrac{48x^3}{7y}}$ **25.** $\sqrt[3]{\dfrac{40x^2y^3}{3z}}$

Combine the following expressions. (Assume all variables represent positive numbers.) [7.4]

26. $5x\sqrt{6} + 2x\sqrt{6} - 9x\sqrt{6}$ **27.** $\sqrt{12} + \sqrt{3}$ **28.** $\dfrac{3}{\sqrt{5}} + \sqrt{5}$

29. $3\sqrt{8} - 4\sqrt{72} + 5\sqrt{50}$ **30.** $3b\sqrt{27a^5b} + 2a\sqrt{3a^3b^3}$ **31.** $2x\sqrt[3]{xy^3z^2} - 6y\sqrt[3]{x^4z^2}$

Multiply. [7.5]

32. $\sqrt{2}(\sqrt{3} - 2\sqrt{2})$ **33.** $(\sqrt{x} - 2)(\sqrt{x} - 3)$

Rationalize the denominator. (Assume $x, y > 0$.) [7.5]

34. $\dfrac{3}{\sqrt{5} - 2}$ **35.** $\dfrac{\sqrt{7} + \sqrt{5}}{\sqrt{7} - \sqrt{5}}$ **36.** $\dfrac{3\sqrt{7}}{3\sqrt{7} - 4}$

Solve each equation. [7.6]

37. $\sqrt{4a + 1} = 1$ **38.** $\sqrt[3]{3x - 8} = 1$ **39.** $\sqrt{3x + 1} - 3 = 1$ **40.** $\sqrt{x + 4} = \sqrt{x} - 2$

Graph each equation. [7.6]

41. $y = 3\sqrt{x}$ **42.** $y = \sqrt[3]{x} + 2$

Write each of the following as i, -1, $-i$, or 1. [7.7]

43. i^{24}

44. i^{27}

Find x and y so that each of the following equations is true. [7.7]

45. $3 - 4i = -2x + 8yi$

46. $(3x + 2) - 8i = -4 + 2yi$

Combine the following complex numbers. [7.7]

47. $(3 + 5i) + (6 - 2i)$

48. $(2 + 5i) - [(3 + 2i) + (6 - i)]$

Multiply. [7.7]

49. $3i(4 + 2i)$

50. $(2 + 3i)(4 + i)$

51. $(4 + 2i)^2$

52. $(4 + 3i)(4 - 3i)$

Divide. Write all answers in standard form for complex numbers. [7.7]

53. $\frac{3 + i}{i}$

54. $\frac{-3}{2 + i}$

55. Construction The roof of the house shown in Figure 1 is to extend up 13.5 feet above the ceiling, which is 36 feet across. Find the length of one side of the roof.

FIGURE 1

56. Surveying A surveyor is attempting to find the distance across a pond. From a point on one side of the pond he walks 25 yards to the end of the pond and then makes a 90-degree turn and walks another 60 yards before coming to a point directly across the pond from the point at which he started. What is the distance across the pond? (See Figure 2.)

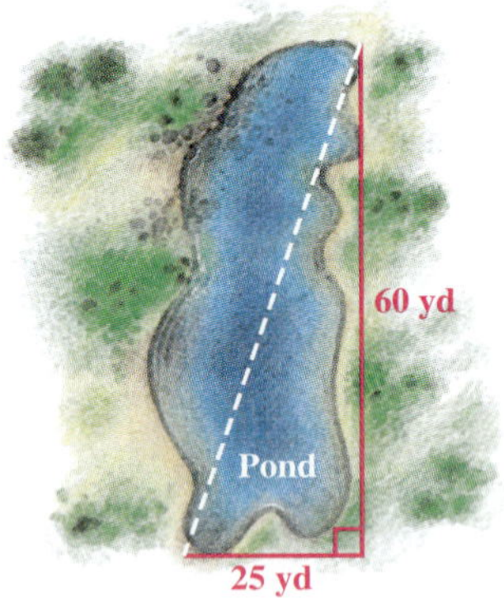

FIGURE 2

Chapter 7 Projects

Rational Exponents and Roots

CONSTRUCTING THE SPIRAL OF ROOTS

Group Project

Number of People 3

Time Needed 20 minutes

Equipment Two sheets of graph paper (4 or 5 squares per inch) and pencils.

Background The spiral of roots gives us a way to visualize the positive square roots of the counting numbers, and in so doing, we see many line segments whose lengths are irrational numbers.

Procedure You are to construct a spiral of roots from a line segment 1 inch long. The graph paper you have contains either 4 or 5 squares per inch, allowing you to accurately draw 1-inch line segments. Because the lines on the graph paper are perpendicular to one another, if you are careful, you can use the graph paper to connect one line segment to another so that they form a right angle.

1. Fold one of the pieces of graph paper so it can be used as a ruler.
2. Use the folded paper to draw a line segment 1-inch long, just to the right of the middle of the unfolded paper. On the end of this segment, attach another segment of 1-inch length at a right angle to the first one. Connect the end points of the segments to form a right triangle. Label each side of this triangle. When you are finished, your work should resemble Figure 1.
3. On the end of the hypotenuse of the triangle, attach a 1-inch line segment so that the two segments form a right angle. (Use the folded paper to do this.) Draw the hypotenuse of this triangle. Label all the sides of this second triangle. Your work should resemble Figure 2.
4. Continue to draw a new right triangle by attaching 1-inch line segments at right angles to the previous hypotenuse. Label all the sides of each triangle.
5. Stop when you have drawn a hypotenuse $\sqrt{8}$ inches long.

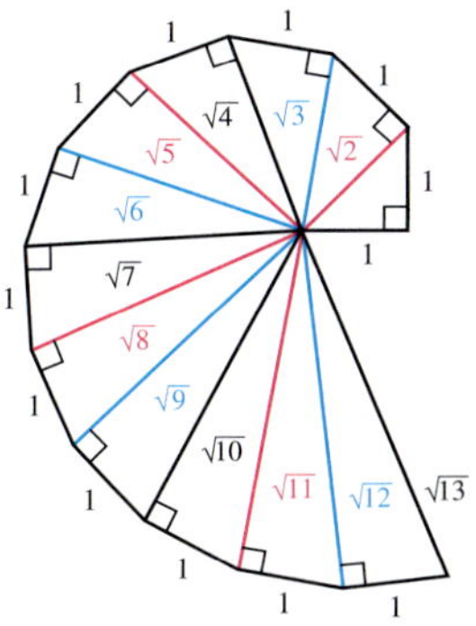

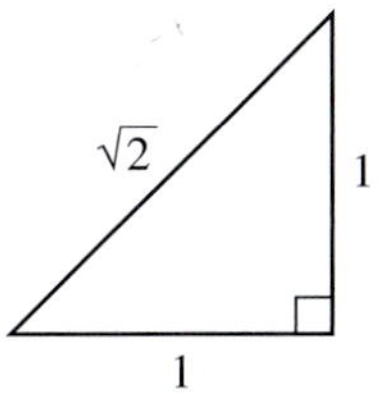

FIGURE 1

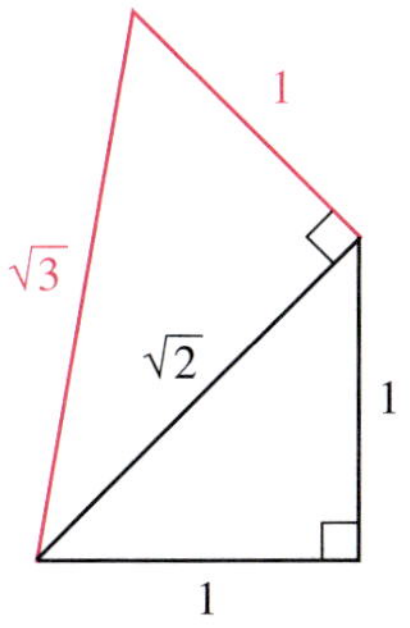

FIGURE 2

MARIA GAETANA AGNESI (1718–1799)

Research Project

© Culver Pictures

January 9, 1999, was the 200th anniversary of the death of Maria Agnesi, the author of *Instituzioni analitiche ad uso della gioventu italiana* (1748), a calculus textbook considered to be the best book of its time and the first surviving mathematical work written by a woman. Maria Agnesi is also famous for a curve that is named for her. The curve is called the Witch of Agnesi in English, but its actual translation is the Locus of Agnesi. The foundation of the curve is shown in the figure. Research the Witch of Agnesi and then explain, in essay form, how the diagram in the figure is used to produce the Witch of Agnesi. Include a rough sketch of the curve, starting with a circle of diameter a as shown in the figure. Then, for comparison, sketch the curve again, starting with a circle of diameter $2a$.

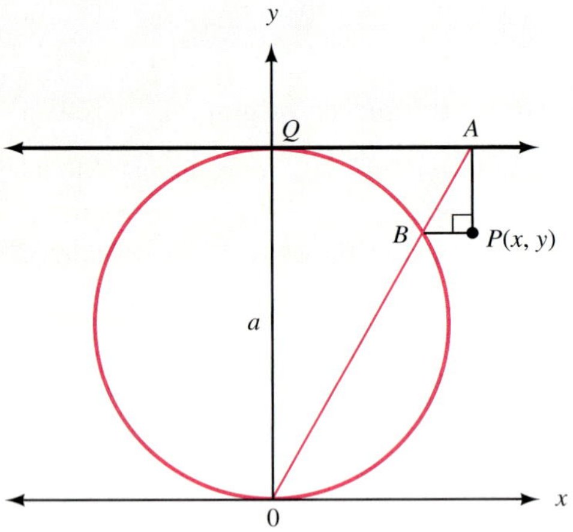

8

The Image Bank/Getty Images

CHAPTER OUTLINE

Quadratic Functions

If you have been to the circus or the county fair, you may have witnessed one of the more spectacular acts, the human cannonball. The human cannonball shown in the photograph will reach a height of 70 feet and travel a distance of 160 feet before landing in a safety net. In this chapter, we use this information to derive the equation

$$f(x) = -\frac{7}{640}(x - 80)^2 + 70 \quad \text{for } 0 \le x \le 160$$

which describes the path flown by this particular cannonball. The table and graph below were constructed from this equation.

PATH OF A HUMAN CANNONBALL	
x (FEET)	$f(x)$ (NEAREST FOOT)
0	0
40	53
80	70
120	53
160	0

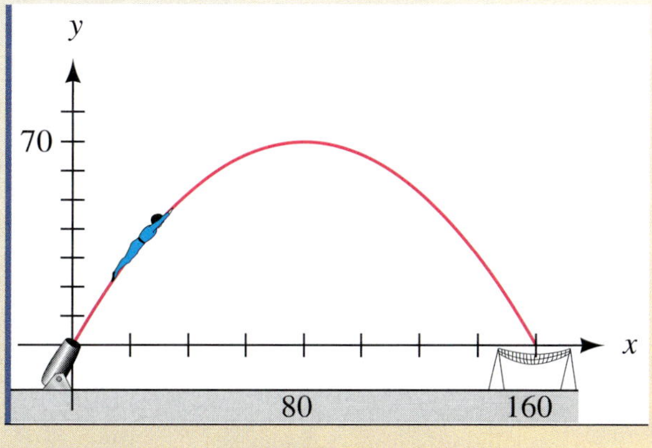

FIGURE 1

All objects that are projected into the air, whether they are basketballs, bullets, arrows, or coins, follow parabolic paths like the one shown in Figure 1. Studying the material in this chapter will give you a more mathematical hold on the world around you.

Getting Ready for Chapter 8

Simplify.

1. $\dfrac{3+5}{2}$

2. $\dfrac{3-5}{2}$

3. $2+\dfrac{25}{4}$

4. $-\dfrac{7}{3}+\dfrac{16}{9}$

Solve.

5. $x^2 = 25$

6. $y^2 = 9$

7. $x^2 = 12$

8. $y^2 = 18$

Find $\left(\dfrac{b}{2}\right)^2$ when

9. $b = 6$

10. $b = 4$

11. $b = 5$

12. $b = 7$

13. $b = \dfrac{8}{3}$

14. $b = \dfrac{2}{5}$

Factor.

15. $x^2 - 6x + 9$

16. $x^2 - 8x + 16$

17. $x^2 + 12x + 36$

18. $125t^3 + 1$

Simplify.

19. $-\dfrac{c}{a}+\left(\dfrac{b}{2a}\right)^2$

20. $(3a-2)^2 + 2(3a-2) - 3$

8.1 Completing the Square

Table 1 is taken from the trail map given to skiers at the Northstar at Tahoe Ski Resort in Lake Tahoe, California. The table gives the length of each chair lift at Northstar, along with the change in elevation from the beginning of the lift to the end of the lift.

Right triangles are good mathematical models for chair lifts. In this section, we will use our knowledge of right triangles, along with the new material developed in the section, to solve problems involving chair lifts and a variety of other examples.

OBJECTIVES	
A	Solve quadratic equations by taking the square root of both sides.
B	Solve quadratic equations by completing the square.
C	Use quadratic equations to solve for missing parts of right triangles.

TABLE 1

FROM THE TRAIL MAP FOR THE NORTHSTAR AT TAHOE SKI RESORT

LIFT INFORMATION

LIFT	VERTICAL RISE (FEET)	LENGTH (FEET)
Big Springs Gondola	480	4,100
Bear Paw Double	120	790
Echo Triple	710	4,890
Aspen Express Quad	900	5,100
Forest Double	1,170	5,750
Lookout Double	960	4,330
Comstock Express Quad	1,250	5,900
Rendezvous Triple	650	2,900
Schaffer Camp Triple	1,860	6,150
Chipmunk Tow Lift	28	280
Bear Cub Tow Lift	120	750

In this section, we will develop the first of our new methods of solving quadratic equations. The new method is called *completing the square.* Completing the square on a quadratic equation allows us to obtain solutions, regardless of whether the equation can be factored. Before we solve equations by completing the square, we need to learn how to solve equations by taking square roots of both sides.

Consider the equation

$$x^2 = 16$$

We could solve it by writing it in standard form, factoring the left side, and proceeding as we have done previously. We can shorten our work considerably, however, if we simply notice that x must be either the positive square root of 16 or the negative square root of 16; that is,

$$\text{If} \quad x^2 = 16$$

$$\text{then} \quad x = \sqrt{16} \quad \text{or} \quad x = -\sqrt{16}$$

$$x = 4 \quad \text{or} \quad x = -4$$

We can generalize this result into a theorem as follows.

Theorem 1

If $a^2 = b$ where b is a real number, then $a = \sqrt{b}$ or $a = -\sqrt{b}$.

Notation The expression $a = \sqrt{b}$ or $a = -\sqrt{b}$ can be written in shorthand form as $a = \pm\sqrt{b}$. The symbol $\pm$ is read "plus or minus."

We can apply Theorem 1 to some fairly complicated quadratic equations.

PRACTICE PROBLEMS

1. Solve $(3x + 2)^2 = 16$.

2. Solve $(4x - 3)^2 = -50$.

EXAMPLE 1

Solve $(2x - 3)^2 = 25$.

Solution

$(2x - 3)^2 = 25$	
$2x - 3 = \pm\sqrt{25}$	**Theorem 1**
$2x - 3 = \pm 5$	$\sqrt{25} = 5$
$2x = 3 \pm 5$	**Add 3 to both sides**
$x = \dfrac{3 \pm 5}{2}$	**Divide both sides by 2**

The last equation can be written as two separate statements:

$$x = \frac{3 + 5}{2} \quad \text{or} \quad x = \frac{3 - 5}{2}$$

$$= \frac{8}{2} \qquad\qquad = \frac{-2}{2}$$

$$= 4 \quad \text{or} \quad = -1$$

The solution set is $\{4, -1\}$.

Notice that we could have solved the equation in Example 1 by expanding the left side, writing the resulting equation in standard form, and then factoring. The problem would look like this:

$(2x - 3)^2 = 25$	**Original equation**
$4x^2 - 12x + 9 = 25$	**Expand the left side**
$4x^2 - 12x - 16 = 0$	**Add −25 to each side**
$4(x^2 - 3x - 4) = 0$	**Begin factoring**
$4(x - 4)(x + 1) = 0$	**Factor completely**
$x - 4 = 0$ or $x + 1 = 0$	**Set variable factors equal to 0**
$x = 4$ or $x = -1$	

As you can see, solving the equation by factoring leads to the same two solutions.

NOTE

We cannot solve the equation in Example 2 by factoring. If we expand the left side and write the resulting equation in standard form, we are left with a quadratic equation that does not factor:

$(3x - 1)^2 = -12$ Equation from Example 2

$9x^2 - 6x + 1 = -12$ Expand the left side

$9x^2 - 6x + 13 = 0$ Standard form, but not factorable

EXAMPLE 2

Solve for x: $(3x - 1)^2 = -12$.

Solution

$(3x - 1)^2 = -12$	
$3x - 1 = \pm\sqrt{-12}$	**Theorem 1**
$3x - 1 = \pm 2i\sqrt{3}$	$\sqrt{-12} = 2i\sqrt{3}$
$3x = 1 \pm 2i\sqrt{3}$	**Add 1 to both sides**
$x = \dfrac{1 \pm 2i\sqrt{3}}{3}$	**Divide both sides by 3**

The solution set is $\left\{\dfrac{1 + 2i\sqrt{3}}{3}, \dfrac{1 - 2i\sqrt{3}}{3}\right\}$

Answer

1. $\frac{2}{3}, -2$

Both solutions are complex. Here is a check of the first solution:

When $x = \frac{1 + 2i\sqrt{3}}{3}$

the equation $(3x - 1)^2 = -12$

becomes $\left(3 \cdot \frac{1 + 2i\sqrt{3}}{3} - 1\right)^2 \stackrel{?}{=} -12$

or $(1 + 2i\sqrt{3} - 1)^2 \stackrel{?}{=} -12$

$$(2i\sqrt{3})^2 \stackrel{?}{=} -12$$

$$4 \cdot i^2 \cdot 3 \stackrel{?}{=} -12$$

$$12(-1) \stackrel{?}{=} -12$$

$$-12 = -12$$

EXAMPLE 3 Solve $x^2 + 6x + 9 = 12$.

Solution We can solve this equation as we have the equations in Examples 1 and 2 if we first write the left side as $(x + 3)^2$.

$x^2 + 6x + 9 = 12$	**Original equation**
$(x + 3)^2 = 12$	**Write $x^2 + 6x + 9$ as $(x + 3)^2$**
$x + 3 = \pm 2\sqrt{3}$	**Theorem 1**
$x = -3 \pm 2\sqrt{3}$	**Add −3 to each side**

We have two irrational solutions: $-3 + 2\sqrt{3}$ and $-3 - 2\sqrt{3}$. What is important about this problem, however, is the fact that the equation was easy to solve because the left side was a perfect square trinomial.

3. Solve $x^2 + 10x + 25 = 20$.

Completing the Square

The method of completing the square is simply a way of transforming any quadratic equation into an equation of the form found in the preceding three examples.

The key to understanding the method of completing the square lies in recognizing the relationship between the last two terms of any perfect square trinomial whose leading coefficient is 1.

Consider the following list of perfect square trinomials and their corresponding binomial squares:

$$x^2 - 6x + 9 = (x - 3)^3$$

$$x^2 + 8x + 16 = (x + 4)^2$$

$$x^2 - 10x + 25 = (x - 5)^2$$

$$x^2 + 12x + 36 = (x + 6)^2$$

In each case the leading coefficient is 1. A more important observation comes from noticing the relationship between the linear and constant terms (middle and last terms) in each trinomial. Observe that the constant term in each case is the square of half the coefficient of x in the middle term. For example, in the last expression, the constant term 36 is the square of half of 12, where 12 is the coefficient of x in the middle term. (Notice also that the second terms in all the binomials on the right side are half the coefficients of the middle terms of the trinomials

Answers

2. $\frac{3 \pm 5i\sqrt{2}}{4}$ 3. $-5 \pm 2\sqrt{5}$

on the left side.) We can use these observations to build our own perfect square trinomials and, in doing so, solve some quadratic equations.

Consider the following equation:

$$x^2 + 6x = 3$$

We can think of the left side as having the first two terms of a perfect square trinomial. We need only add the correct constant term. If we take half the coefficient of x, we get 3. If we then square this quantity, we have 9. Adding the 9 to both sides, the equation becomes

NOTE

This is the step in which we actually complete the square.

$$x^2 + 6x + \mathbf{9} = 3 + \mathbf{9}$$

The left side is the perfect square $(x + 3)^2$; the right side is 12:

$$(x + 3)^2 = 12$$

The equation is now in the correct form. We can apply Theorem 1 and finish the solution:

$$(x + 3)^2 = 12$$

$$x + 3 = \pm\sqrt{12} \qquad \textbf{Theorem 1}$$

$$x + 3 = \pm 2\sqrt{3}$$

$$x = -3 \pm 2\sqrt{3}$$

The solution set is $\{-3 + 2\sqrt{3}, -3 - 2\sqrt{3}\}$. The method just used is called *completing the square* since we complete the square on the left side of the original equation by adding the appropriate constant term.

4. Solve $x^2 + 3x - 4 = 0$ by completing the square.

EXAMPLE 4 Solve by completing the square: $x^2 + 5x - 2 = 0$.

Solution We must begin by adding 2 to both sides. (The left side of the equation, as it is, is not a perfect square because it does not have the correct constant term. We will simply "move" that term to the other side and use our own constant term.)

$$x^2 + 5x = 2 \qquad \textbf{Add 2 to each side}$$

We complete the square by adding the square of half the coefficient of the linear term to both sides:

$$x^2 + 5x + \mathbf{\frac{25}{4}} = 2 + \mathbf{\frac{25}{4}} \qquad \textbf{Half of 5 is } \tfrac{5}{2}\textbf{, the square of which is } \tfrac{25}{4}$$

$$\left(x + \frac{5}{2}\right)^2 = \frac{33}{4} \qquad 2 + \tfrac{25}{4} = \tfrac{8}{4} + \tfrac{25}{4} = \tfrac{33}{4}$$

$$x + \frac{5}{2} = \pm\sqrt{\frac{33}{4}} \qquad \textbf{Theorem 1}$$

$$= \pm\frac{\sqrt{33}}{2} \qquad \textbf{Simplify the radical}$$

$$x = -\frac{5}{2} \pm \frac{\sqrt{33}}{2} \qquad \textbf{Add } -\tfrac{5}{2} \textbf{ to both sides}$$

$$= \frac{-5 \pm \sqrt{33}}{2}$$

NOTE

We can use a calculator to get decimal approximations to these solutions. If $\sqrt{33} \approx 5.74$, then

$$\frac{-5 + 5.74}{2} = 0.37$$

$$\frac{-5 - 5.74}{2} = -5.37$$

The solution set is $\left\{\frac{-5 + \sqrt{33}}{2}, \frac{-5 - \sqrt{33}}{2}\right\}$.

Answer

4. 1, −4

EXAMPLE 5 Solve for x: $3x^2 - 8x + 7 = 0$.

5. Solve for x: $5x^2 - 3x + 2 = 0$.

Solution

$$3x^2 - 8x + 7 = 0$$

$$3x^2 - 8x = -7 \qquad \textbf{Add } \mathbf{-7} \textbf{ to both sides}$$

We cannot complete the square on the left side because the leading coefficient is not 1. We take an extra step and divide both sides by 3:

$$\frac{3x^2}{3} - \frac{8x}{3} = -\frac{7}{3}$$

$$x^2 - \frac{8}{3}x = -\frac{7}{3}$$

Half of $\frac{8}{3}$ is $\frac{4}{3}$, the square of which is $\frac{16}{9}$.

$$x^2 - \frac{8}{3}x + \mathbf{\frac{16}{9}} = -\frac{7}{3} + \mathbf{\frac{16}{9}} \qquad \textbf{Add } \tfrac{16}{9} \textbf{ to both sides}$$

$$\left(x - \frac{4}{3}\right)^2 = -\frac{5}{9} \qquad \textbf{Simplify right side}$$

$$x - \frac{4}{3} = \pm\sqrt{-\frac{5}{9}} \qquad \textbf{Theorem 1}$$

$$x - \frac{4}{3} = \pm\frac{i\sqrt{5}}{3} \qquad \sqrt{-\frac{5}{9}} = \frac{\sqrt{-5}}{3} = \frac{i\sqrt{5}}{3}$$

$$x = \frac{4}{3} \pm \frac{i\sqrt{5}}{3} \qquad \textbf{Add } \tfrac{4}{3} \textbf{ to both sides}$$

$$x = \frac{4 \pm i\sqrt{5}}{3}$$

The solution set is $\left\{\frac{4 + i\sqrt{5}}{3}, \frac{4 - i\sqrt{5}}{3}\right\}$.

To Solve a Quadratic Equation by Completing the Square

To summarize the method used in the preceding two examples, we list the following steps:

Step 1: Write the equation in the form $ax^2 + bx = c$.

Step 2: If the leading coefficient is not 1, divide both sides by the coefficient so that the resulting equation has a leading coefficient of 1; that is, if $a \neq 1$, then divide both sides by a.

Step 3: Add the square of half the coefficient of the linear term to both sides of the equation.

Step 4: Write the left side of the equation as the square of a binomial, and simplify the right side if possible.

Step 5: Apply Theorem 1, and solve as usual.

Answer

5. $\frac{3 \pm i\sqrt{31}}{10}$

Facts from Geometry: More Special Triangles

The triangles shown in Figures 1 and 2 occur frequently in mathematics.

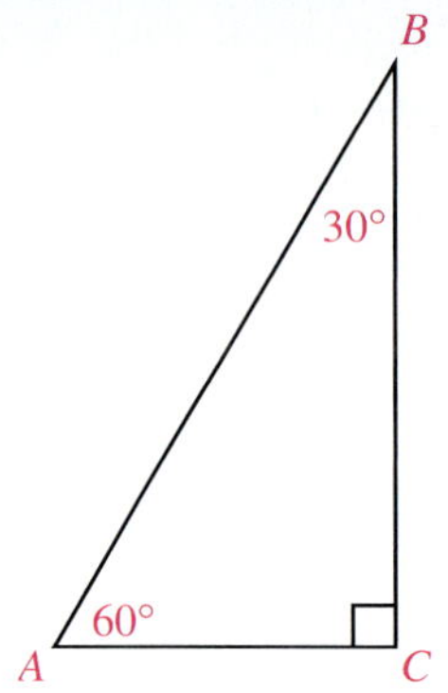

FIGURE 1

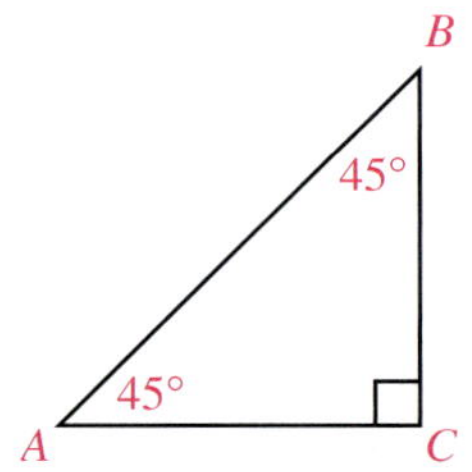

FIGURE 2

Note that both of the triangles are right triangles. We refer to the triangle in Figure 1 as a 30°–60°–90° triangle and the triangle in Figure 2 as a 45°–45°–90° triangle.

6. If the shortest side in a 30°–60°–90° triangle is 2 inches long, find the lengths of the other two sides.

EXAMPLE 6 If the shortest side in a 30°–60°–90° triangle is 1 inch, find the lengths of the other two sides.

Solution In Figure 3 triangle *ABC* is a 30°–60°–90° triangle in which the shortest side *AC* is 1 inch long. Triangle *DBC* is also a 30°–60°–90° triangle in which the shortest side *DC* is 1 inch long.

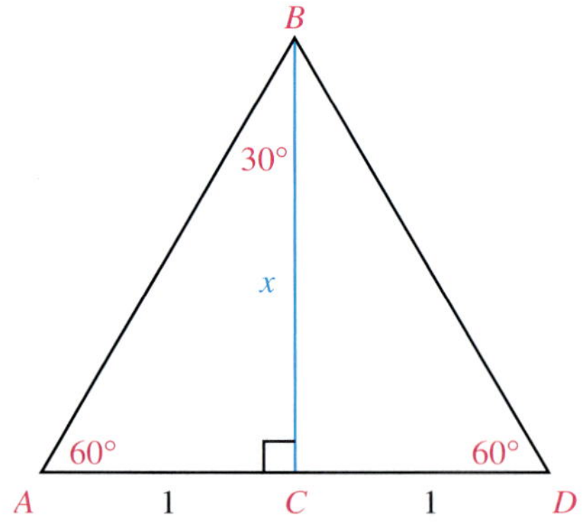

FIGURE 3

Notice that the large triangle *ABD* is an equilateral triangle because each of its interior angles is 60°. Each side of triangle *ABD* is 2 inches long. Side *AB* in triangle *ABC* is therefore 2 inches. To find the length of side *BC*, we use the Pythagorean theorem.

$$BC^2 + AC^2 = AB^2$$

$$x^2 + 1^2 = 2^2$$

$$x^2 + 1 = 4$$

$$x^2 = 3$$

$$x = \sqrt{3} \text{ inches}$$

Note that we write only the positive square root because x is the length of a side in a triangle and is therefore a positive number. ■

Answer

6. 4 inches and $2\sqrt{3}$ inches

EXAMPLE 7 Table 1 in the introduction to this section gives the vertical rise of the Forest Double chair lift as 1,170 feet and the length of the chair lift as 5,750 feet. To the nearest foot, find the horizontal distance covered by a person riding this lift.

Solution Figure 4 is a model of the Forest Double chair lift. A rider gets on the lift at point A and exits at point B. The length of the lift is AB.

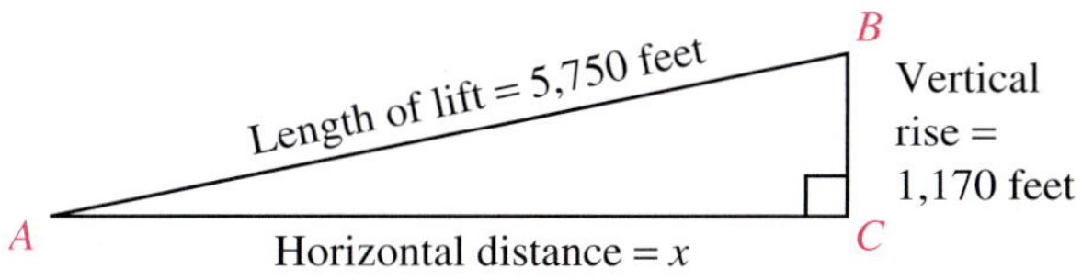

FIGURE 4

To find the horizontal distance covered by a person riding the chair lift, we use the Pythagorean theorem.

$$5{,}750^2 = x^2 + 1{,}170^2 \quad \textbf{Pythagorean theorem}$$

$$33{,}062{,}500 = x^2 + 1{,}368{,}900 \quad \textbf{Simplify squares}$$

$$x^2 = 33{,}062{,}500 - 1{,}368{,}900 \quad \textbf{Solve for } x^2$$

$$x^2 = 31{,}693{,}600 \quad \textbf{Simplify the right side}$$

$$x = \sqrt{31{,}693{,}600} \quad \textbf{Theorem 1}$$

$$= 5{,}630 \text{ feet (to the nearest foot)}$$

A rider getting on the lift at point A and riding to point B will cover a horizontal distance of approximately 5,630 feet.

Getting Ready for Class

After reading through the preceding section, respond in your own words and in complete sentences.

1. What kind of equation do we solve using the method of completing the square?
2. Explain in words how you would complete the square on $x^2 - 16x = 4$.
3. What is the relationship between the shortest side and the longest side in a 30°–60°–90° triangle?
4. What two expressions together are equivalent to $x = \pm 4$?

7. Table 1 in the introduction to this chapter gives the vertical rise of the Lookout Double chair lift as 960 feet and the length of the chair lift as 4,330 feet. To the nearest foot, find the horizontal distance covered by a person riding this lift.

Linking Objectives and Examples

Next to each **objective**, list the examples that are best described by that objective.

A	
B	
C	

Answer

7. 4,222 feet

Problem Set 8.1

INTERMEDIATE Algebra $f(x)$ Now™

Solve the following equations.

1. $x^2 = 25$

2. $x^2 = 16$

3. $y^2 = \frac{3}{4}$

4. $y^2 = \frac{5}{9}$

5. $x^2 + 12 = 0$

6. $x^2 + 8 = 0$

7. $4a^2 - 45 = 0$

8. $9a^2 - 20 = 0$

9. $(2y - 1)^2 = 25$

10. $(3y + 7)^2 = 1$

11. $(2a + 3)^2 = -9$

12. $(3a - 5)^2 = -49$

13. $x^2 + 8x + 16 = -27$

14. $x^2 - 12x + 36 = -8$

15. $4a^2 - 12a + 9 = -4$

16. $9a^2 - 12a + 4 = -9$

17. $(x + 5)^2 + (x - 5)^2 = 52$

18. $(2x + 1)^2 + (2x - 1)^2 = 10$

Copy each of the following, and fill in the blanks so that the left side of each is a perfect square trinomial; that is, complete the square.

19. $x^2 + 12x +$ ______ $= (x +$ ______$)^2$

20. $x^2 + 6x +$ ______ $= (x +$ ______$)^2$

21. $x^2 - 4x +$ ______ $= (x -$ ______$)^2$

22. $x^2 - 2x +$ ______ $= (x -$ ______$)^2$

23. $a^2 - 10a +$ ______ $= (a -$ ______$)^2$

24. $a^2 - 8a +$ ______ $= (a -$ ______$)^2$

25. $x^2 + 5x +$ ______ $= (x +$ ______$)^2$

26. $x^2 + 3x +$ ______ $= (x +$ ______$)^2$

27. $y^2 - 7y +$ ______ $= (y -$ ______$)^2$

28. $y^2 - y +$ ______ $= (y -$ ______$)^2$

Solve each of the following quadratic equations by completing the square.

29. $x^2 + 4x = 12$

30. $x^2 - 2x = 8$

31. $x^2 + 12x = -27$

32. $x^2 - 6x = 16$

33. $a^2 - 2a + 5 = 0$

34. $a^2 + 10a + 22 = 0$

35. $y^2 - 8y + 1 = 0$

36. $y^2 + 6y - 1 = 0$

37. $x^2 - 5x - 3 = 0$

38. $x^2 - 5x - 2 = 0$

39. $2x^2 - 4x - 8 = 0$

40. $3x^2 - 9x - 12 = 0$

41. $3t^2 - 8t + 1 = 0$

42. $5t^2 + 12t - 1 = 0$

43. $4x^2 - 3x + 5 = 0$

44. $7x^2 - 5x + 2 = 0$

Applying the Concepts

45. Geometry If the shortest side in a 30°–60°–90° triangle is $\frac{1}{2}$ inch long, find the lengths of the other two sides.

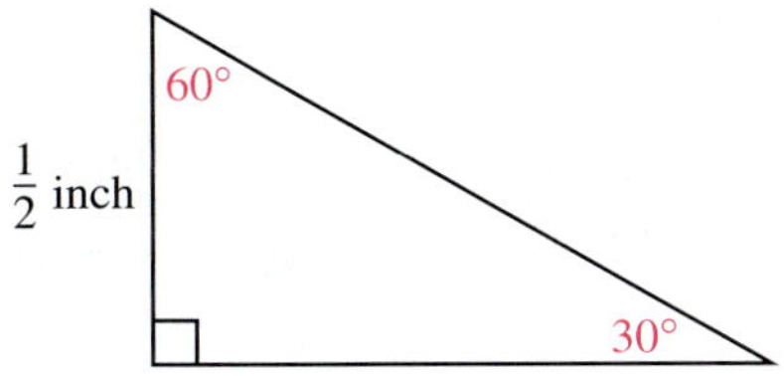

46. Geometry If the length of the longest side of a 30°–60°–90° triangle is x, find the length of the other two sides in terms of x.

47. Geometry If the length of the shorter sides of a 45°–45°–90° triangle is 1 inch, find the length of the hypotenuse.

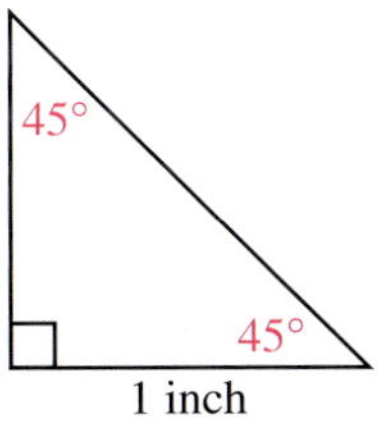

48. Geometry If the length of the shorter sides of a 45°–45°–90° triangle is x, find the length of the hypotenuse, in terms of x.

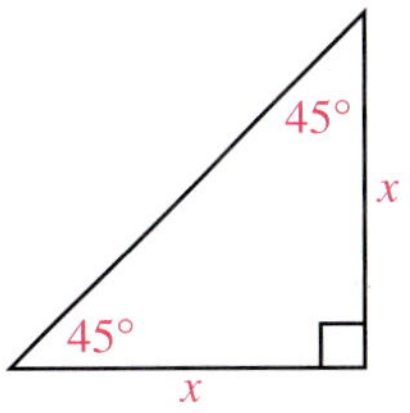

49. Chair Lift Use Table 1 from the introduction to this chapter to find the horizontal distance covered by a person riding the Bear Paw Double chair lift. Round your answer to the nearest foot.

50. Chair Lift Use Table 1 from the introduction to this chapter to find the horizontal distance covered by a person riding the Big Springs Gondola lift. Round your answer to the nearest foot.

51. Interest Rate Suppose a deposit of \$3,000 in a savings account that paid an annual interest rate r (compounded yearly) is worth \$3,456 after 2 years. Using the formula $A = P(1 + r)^t$, we have

$$3{,}456 = 3{,}000(1 + r)^2$$

Solve for r to find the annual interest rate.

52. Special Triangles In Figure 5, triangle ABC has angles 45° and 30° and height x. Find the lengths of sides AB, BC, and AC, in terms of x.

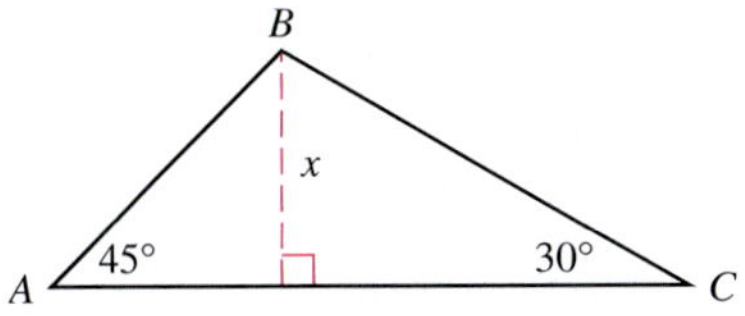

FIGURE 5

53. **Length of an Escalator** An escalator in a department store is to carry people a vertical distance of 20 feet between floors. How long is the escalator if it makes an angle of 45° with the ground? (See Figure 6.)

FIGURE 6

54. **Dimensions of a Tent** A two-person tent is to be made so the height at the center is 4 feet. If the sides of the tent are to meet the ground at an angle of 60° and the tent is to be 6 feet in length, how many square feet of material will be needed to make the tent? (Figure 7, assume that the tent has a floor and is closed at both ends.) Give your answer to the nearest tenth of a square foot.

FIGURE 7

Getting Ready for the Next Section

Simplify.

55. $\sqrt{49 - 4(6)(-5)}$

56. $\sqrt{49 - 4(6)2}$

57. $\sqrt{(-27)^2 - 4(0.1)(1{,}700)}$

58. $\sqrt{25 - 4(4)(-10)}$

59. $\dfrac{-7 + \sqrt{169}}{12}$

60. $\dfrac{-7 - \sqrt{169}}{12}$

61. Simplify $\sqrt{b^2 - 4ac}$ when $a = 6$, $b = 7$, and $c = -5$.

62. Simplify $\sqrt{b^2 - 4ac}$ when $a = 2$, $b = -6$, and $c = 3$.

Factor.

63. $27t^3 - 8$

64. $125t^3 + 1$

Maintaining Your Skills

Write each of the following in simplified form for radicals.

65. $\sqrt{45}$

66. $\sqrt{24}$

67. $\sqrt{27y^5}$

68. $\sqrt{8y^3}$

69. $\sqrt[3]{54x^6y^5}$

70. $\sqrt[3]{16x^9y^7}$

Rationalize the denominator.

71. $\dfrac{3}{\sqrt{2}}$

72. $\dfrac{5}{\sqrt{3}}$

73. $\dfrac{2}{\sqrt[3]{4}}$

74. $\dfrac{3}{\sqrt[3]{2}}$

75. **Fermat's Last Theorem** As we mentioned in a previous chapter, the postage stamp shows Fermat's last theorem, which states that if n is an integer greater than 2, then there are no positive integers x, y, and z that will make the formula $x^n + y^n = z^n$ true.

Use the formula $x^n + y^n = z^n$ to

a. find z if $n = 2$, $x = 6$, and $y = 8$.

b. find y if $n = 2$, $x = 5$, and $z = 13$.

8.2 The Quadratic Formula

OBJECTIVES	
A	Solve quadratic equations by the quadratic formula.
B	Solve application problems using quadratic equations.

In this section we will use the method of completing the square from the preceding section to derive the quadratic formula. The *quadratic formula* is a very useful tool in mathematics. It allows us to solve all types of quadratic equations.

The Quadratic Theorem

For any quadratic equation in the form $ax^2 + bx + c = 0$, where $a \neq 0$, the two solutions are

$$x = \frac{-b + \sqrt{b^2 - 4ac}}{2a} \quad \text{and} \quad x = \frac{-b - \sqrt{b^2 - 4ac}}{2a}$$

Proof We will prove the quadratic theorem by completing the square on $ax^2 + bx + c = 0$;

$$ax^2 + bx + c = 0$$

$$ax^2 + bx = -c \qquad \textbf{Add } -c \textbf{ to both sides}$$

$$x^2 + \frac{b}{a}x = -\frac{c}{a} \qquad \textbf{Divide both sides by } a$$

To complete the square on the left side, we add the square of $\frac{1}{2}$ of $\frac{b}{a}$ to both sides. $\left(\frac{1}{2} \text{ of } \frac{b}{a} \text{ is } \frac{b}{2a}.\right)$

$$x^2 + \frac{b}{a}x + \left(\frac{b}{2a}\right)^2 = -\frac{c}{a} + \left(\frac{b}{2a}\right)^2$$

We now simplify the right side as a separate step. We square the second term and combine the two terms by writing each with the least common denominator $4a^2$.

$$-\frac{c}{a} + \left(\frac{b}{2a}\right)^2 = -\frac{c}{a} + \frac{b^2}{4a^2} = \frac{\mathbf{4a}}{\mathbf{4a}}\left(\frac{-c}{a}\right) + \frac{b^2}{4a^2} = \frac{-4ac + b^2}{4a^2}$$

It is convenient to write this last expression as

$$\frac{b^2 - 4ac}{4a^2}$$

Continuing with the proof, we have

$$x^2 + \frac{b}{a}x + \left(\frac{b}{2a}\right)^2 = \frac{b^2 - 4ac}{4a^2}$$

$$\left(x + \frac{b}{2a}\right)^2 = \frac{b^2 - 4ac}{4a^2} \qquad \textbf{Write left side as a binomial square}$$

$$x + \frac{b}{2a} = \pm\frac{\sqrt{b^2 - 4ac}}{2a} \qquad \textbf{Theorem 1}$$

$$x = -\frac{b}{2a} \pm \frac{\sqrt{b^2 - 4ac}}{2a} \qquad \textbf{Add } -\frac{b}{2a} \textbf{ to both sides}$$

$$= \frac{-b \pm \sqrt{b^2 - 4ac}}{2a}$$

Our proof is now complete. What we have is this: If our equation is in the form $ax^2 + bx + c = 0$ (standard form), where $a \neq 0$, the two solutions are always given by the formula

$$x = \frac{-b \pm \sqrt{b^2 - 4ac}}{2a}$$

This formula is known as the *quadratic formula.* If we substitute the coefficients a, b, and c of any quadratic equation in standard form into the formula, we need only perform some basic arithmetic to arrive at the solution set.

PRACTICE PROBLEMS

1. Solve $6x^2 + 7x + 2 = 0$ using the quadratic formula.

EXAMPLE 1 Use the quadratic formula to solve $6x^2 + 7x - 5 = 0$.

Solution Using $a = 6$, $b = 7$, and $c = -5$ in the formula

$$x = \frac{-b \pm \sqrt{b^2 - 4ac}}{2a}$$

we have $$x = \frac{-7 \pm \sqrt{49 - 4(6)(-5)}}{2(6)}$$

or $$x = \frac{-7 \pm \sqrt{49 + 120}}{12}$$

$$= \frac{-7 \pm \sqrt{169}}{12}$$

$$= \frac{-7 \pm 13}{12}$$

We separate the last equation into the two statements

$$x = \frac{-7 + 13}{12} \quad \text{or} \quad x = \frac{-7 - 13}{12}$$

$$x = \frac{1}{2} \quad \text{or} \quad x = -\frac{5}{3}$$

The solution set is $\{\frac{1}{2}, -\frac{5}{3}\}$.

Whenever the solutions to a quadratic equation are rational numbers, as they are in Example 1, it means that the original equation was solvable by factoring. To illustrate, let's solve the equation from Example 1 again but this time by factoring:

$$6x^2 + 7x - 5 = 0 \quad \textbf{Equation in standard form}$$

$$(3x + 5)(2x - 1) = 0 \quad \textbf{Factor the left side}$$

$$3x + 5 = 0 \quad \text{or} \quad 2x - 1 = 0 \quad \textbf{Set factors equal to 0}$$

$$x = -\frac{5}{3} \quad \text{or} \quad x = \frac{1}{2}$$

When an equation can be solved by factoring, then factoring is usually the faster method of solution. It is best to try to factor first, and then if you have trouble factoring, go to the quadratic formula. It always works.

2. Solve $\frac{x^2}{2} + x = \frac{1}{3}$.

EXAMPLE 2 Solve $\frac{x^2}{3} - x = -\frac{1}{2}$.

Solution Multiplying through by 6 and writing the result in standard form, we have

$$2x^2 - 6x + 3 = 0$$

Answer

1. $-\frac{1}{2}, -\frac{2}{3}$

the left side of which is not factorable. Therefore, we use the quadratic formula with $a = 2$, $b = -6$, and $c = 3$. The two solutions are given by

$$x = \frac{-(-6) \pm \sqrt{36 - 4(2)(3)}}{2(2)}$$

$$= \frac{6 \pm \sqrt{12}}{4}$$

$$= \frac{6 \pm 2\sqrt{3}}{4} \qquad \sqrt{12} = \sqrt{4 \cdot 3} = \sqrt{4}\,\sqrt{3} = 2\sqrt{3}$$

We can reduce this last expression to lowest terms by factoring 2 from the numerator and denominator and then dividing the numerator and denominator by 2:

$$x = \frac{\not{2}(3 \pm \sqrt{3})}{\not{2} \cdot 2} = \frac{3 \pm \sqrt{3}}{2}$$ ■

EXAMPLE 3 Solve $\dfrac{1}{x+2} - \dfrac{1}{x} = \dfrac{1}{3}$.

Solution To solve this equation, we must first put it in standard form. To do so, we must clear the equation of fractions by multiplying each side by the LCD for all the denominators, which is $3x(x + 2)$. Multiplying both sides by the LCD, we have

$$\mathbf{3x(x+2)}\left(\frac{1}{x+2} - \frac{1}{x}\right) = \frac{1}{3} \cdot \mathbf{3x(x+2)} \qquad \textbf{Multiply each by the LCD}$$

$$3x(\cancel{x+2}) \cdot \frac{1}{\cancel{x+2}} - 3\cancel{x}(x+2) \cdot \frac{1}{\cancel{x}} = \frac{1}{\cancel{3}} \cdot \cancel{3}x(x+2)$$

$$3x - 3(x+2) = x(x+2)$$

$$3x - 3x - 6 = x^2 + 2x \qquad \textbf{Multiplication}$$

$$-6 = x^2 + 2x \qquad \textbf{Simplify left side}$$

$$0 = x^2 + 2x + 6 \qquad \textbf{Add 6 to each side}$$

Since the right side of our last equation is not factorable, we use the quadratic formula. From our last equation, we have $a = 1$, $b = 2$, and $c = 6$. Using these numbers for a, b, and c in the quadratic formula gives us

$$x = \frac{-2 \pm \sqrt{4 - 4(1)(6)}}{2(1)}$$

$$= \frac{-2 \pm \sqrt{4 - 24}}{2} \qquad \textbf{Simplify inside the radical}$$

$$= \frac{-2 \pm \sqrt{-20}}{2} \qquad \mathbf{4 - 24 = -20}$$

$$= \frac{-2 \pm 2i\sqrt{5}}{2} \qquad \mathbf{-20 = i\sqrt{20} = i\sqrt{4}\,\sqrt{5} = 2i\sqrt{5}}$$

$$= \frac{\not{2}(-1 \pm i\sqrt{5})}{\not{2})} \qquad \textbf{Factor 2 from the numerator}$$

$$= -1 \pm i\sqrt{5} \qquad \textbf{Divide numerator and denominator by 2}$$

Since neither of the two solutions, $-1 + i\sqrt{5}$ nor $-1 - i\sqrt{5}$, will make any of the denominators in our original equation 0, they are both solutions. ■

Although the equation in our next example is not a quadratic equation, we solve it by using both factoring and the quadratic formula.

3. Solve $\dfrac{1}{x+4} - \dfrac{1}{x} = \dfrac{1}{2}$.

Answers

2. $\dfrac{-3 \pm \sqrt{15}}{3}$ **3.** $-2 \pm 2i$

4. Solve $8t^3 - 27 = 0$.

EXAMPLE 4

Solve $27t^3 - 8 = 0$.

Solution It would be a mistake to add 8 to each side of this equation and then take the cube root of each side because we would lose two of our solutions. Instead, we factor the left side, and then set the factors equal to 0:

$$27t^3 - 8 = 0 \quad \text{Equation in standard form}$$

$$(3t - 2)(9t^2 + 6t + 4) = 0 \quad \text{Factor as the difference of two cubes}$$

$$3t - 2 = 0 \quad \text{or} \quad 9t^2 + 6t + 4 = 0 \quad \text{Set each factor equal to 0}$$

The first equation leads to a solution of $t = \frac{2}{3}$. The second equation does not factor, so we use the quadratic formula with $a = 9$, $b = 6$, and $c = 4$:

$$t = \frac{-6 \pm \sqrt{36 - 4(9)(4)}}{2(9)}$$

$$= \frac{-6 \pm \sqrt{36 - 144}}{18}$$

$$= \frac{-6 \pm \sqrt{-108}}{18}$$

$$= \frac{-6 \pm 6i\sqrt{3}}{18} \quad \sqrt{-108} = i\sqrt{36 \cdot 3} = 6i\sqrt{3}$$

$$= \frac{\not{6}(-1 \pm i\sqrt{3})}{\not{6} \cdot 3} \quad \text{Factor 6 from the numerator and denominator}$$

$$= \frac{-1 \pm i\sqrt{3}}{3} \quad \text{Divide out common factor 6}$$

The three solutions to our original equation are

$$\frac{2}{3}, \quad \frac{-1 + i\sqrt{3}}{3}, \quad \text{and} \quad \frac{-1 - i\sqrt{3}}{3}$$

5. An object thrown upward with an initial velocity of 32 feet per second rises and falls according to the equation

$$s = 32t - 16t^2$$

where s is the height of the object above the ground at any time t. At what times t will the object be 12 feet above the ground?

EXAMPLE 5

If an object is thrown downward with an initial velocity of 20 feet per second, the distance $s(t)$, in feet, it travels in t seconds is given by the function $s(t) = 20t + 16t^2$. How long does it take the object to fall 40 feet?

Solution We let $s(t) = 40$, and solve for t:

When $s(t) = 40$

the function $s(t) = 20t + 16t^2$

becomes $40 = 20t + 16t^2$

or $16t^2 + 20t - 40 = 0$

$4t^2 + 5t - 10 = 0$ **Divide by 4**

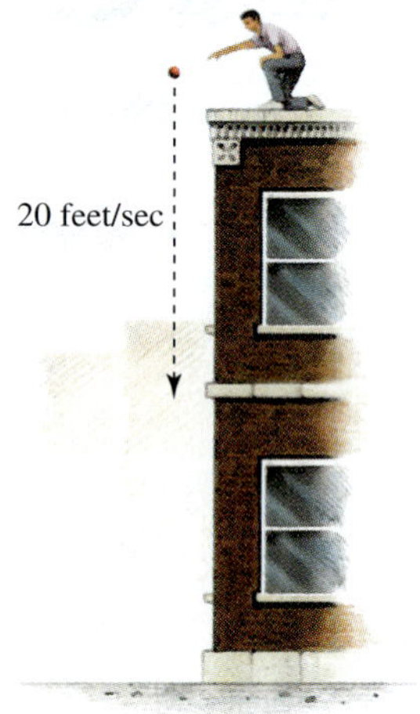

Using the quadratic formula, we have

$$t = \frac{-5 \pm \sqrt{25 - 4(4)(-10)}}{2(4)}$$

$$= \frac{-5 \pm \sqrt{185}}{8}$$

$$= \frac{-5 + \sqrt{185}}{8} \quad \text{or} \quad \frac{-5 - \sqrt{185}}{8}$$

The second solution is impossible since it is a negative number and time t must be positive. It takes

$$t = \frac{-5 + \sqrt{185}}{8} \quad \text{or approximately} \quad \frac{-5 + 13.60}{8} \approx 1.08 \text{ seconds}$$

for the object to fall 40 feet.

Answers

4. $\frac{-3 \pm 3i\sqrt{3}}{4}, \frac{3}{2}$

5. $\frac{3}{2}$ second, $\frac{1}{2}$ second

Recall that the relationship between profit, revenue, and cost is given by the formula

$$P(x) = R(x) - C(x)$$

where $P(x)$ is the profit, $R(x)$ is the total revenue, and $C(x)$ is the total cost of producing and selling x items.

EXAMPLE 6 A company produces and sells copies of an accounting program for home computers. The total weekly cost (in dollars) to produce x copies of the program is $C(x) = 8x + 500$, and the weekly revenue for selling all x copies of the program is $R(x) = 35x - 0.1x^2$. How many programs must be sold each week for the weekly profit to be \$1,200?

Solution Substituting the given expressions for $R(x)$ and $C(x)$ in the equation $P(x) = R(x) - C(x)$, we have a polynomial in x that represents the weekly profit $P(x)$:

$$\begin{aligned} P(x) &= R(x) - C(x) \\ &= 35x - 0.1x^2 - (8x + 500) \\ &= 35x - 0.1x^2 - 8x - 500 \\ &= -500 + 27x - 0.1x^2 \end{aligned}$$

Setting this expression equal to 1,200, we have a quadratic equation to solve that gives us the number of programs x that need to be sold each week to bring in a profit of \$1,200:

$$1{,}200 = -500 + 27x - 0.1x^2$$

We can write this equation in standard form by adding the opposite of each term on the right side of the equation to both sides of the equation. Doing so produces the following equation:

$$0.1x^2 - 27x + 1{,}700 = 0$$

Applying the quadratic formula to this equation with $a = 0.1$, $b = -27$, and $c = 1{,}700$, we have

$$\begin{aligned} x &= \frac{27 \pm \sqrt{(-27)^2 - 4(0.1)(1{,}700)}}{2(0.1)} \\ &= \frac{27 \pm \sqrt{729 - 680}}{0.2} \\ &= \frac{27 \pm \sqrt{49}}{0.2} \\ &= \frac{27 \pm 7}{0.2} \end{aligned}$$

Writing this last expression as two separate expressions, we have our two solutions:

$$x = \frac{27 + 7}{0.2} \quad \text{or} \quad x = \frac{27 - 7}{0.2}$$

$$= \frac{34}{0.2} \qquad\qquad = \frac{20}{0.2}$$

$$= 170 \qquad\qquad = 100$$

The weekly profit will be \$1,200 if the company produces and sells 100 programs or 170 programs. ■

What is interesting about the equation we solved in Example 6 is that it has rational solutions, meaning it could have been solved by factoring. But looking

6. Use the information in Example 6 to find the number of programs the company must sell each week for its weekly profit to be \$1,320.

Answer

6. 130 or 140 programs

back at the equation, factoring does not seem like a reasonable method of solution because the coefficients are either very large or very small. So, there are times when using the quadratic formula is a faster method of solution, even though the equation you are solving is factorable.

USING TECHNOLOGY

Graphing Calculators

More About Example 5

We can solve the problem discussed in Example 5 by graphing the function $Y_1 = 20X + 16X^2$ in a window with X from 0 to 2 (because X is taking the place of t and we know t is a positive quantity) and Y from 0 to 50 (because we are looking for X when Y_1 is 40). Graphing Y_1 gives a graph similar to the graph in Figure 1. Using the Zoom and Trace features at $Y_1 = 40$ gives us $X = 1.08$ to the nearest hundredth, matching the results we obtained by solving the original equation algebraically.

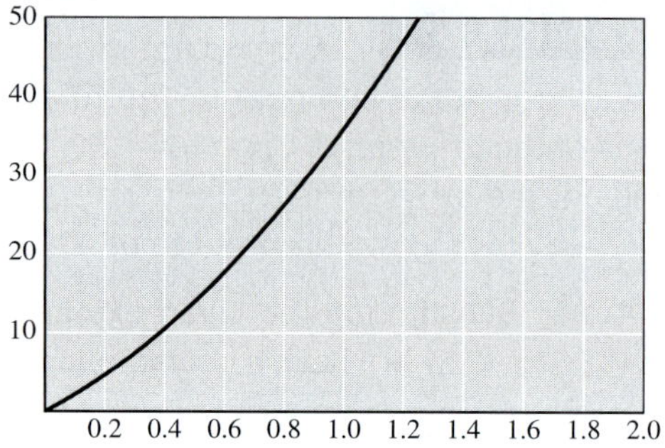

FIGURE 1

More About Example 6

To visualize the functions in Example 6, we set up our calculator this way:

$Y_1 = 35X - .1X^2$ **Revenue function**

$Y_2 = 8X + 500$ **Cost function**

$Y_3 = Y_1 - Y_2$ **Profit function**

Window: X from 0 to 350, Y from 0 to 3500

Graphing these functions produces graphs similar to the ones shown in Figure 2. The lower graph is the graph of the profit function. Using the Zoom and Trace features on the lower graph at $Y_3 = 1{,}200$ produces two corresponding values of X, 170 and 100, which match the results in Example 6.

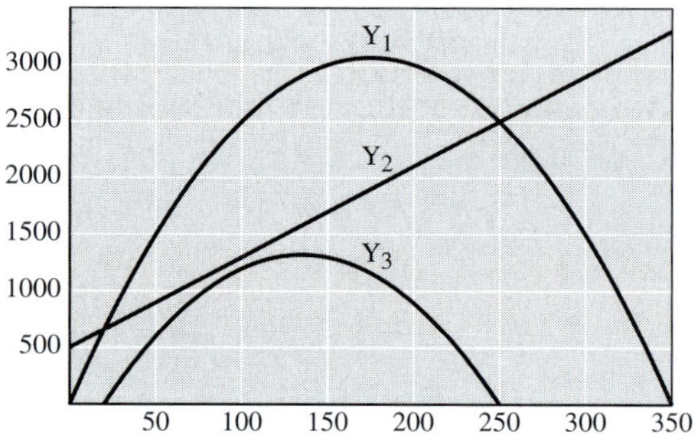

FIGURE 2

We will continue this discussion of the relationship between graphs of functions and solutions to equations in the Using Technology material in the next section.

Linking Objectives and Examples

Next to each **objective**, list the examples that are best described by that objective.

A	
B	

Getting Ready for Class

After reading through the preceding section, respond in your own words and in complete sentences.

1. What is the quadratic formula?
2. Under what circumstances should the quadratic formula be applied?
3. When would the quadratic formula result in complex solutions?
4. When will the quadratic formula result in only one solution?

Problem Set 8.2

INTERMEDIATE Algebra $f(x)$ Now™

Solve each equation in each problem using the quadratic formula.

1. a. $3x^2 + 4x - 2 = 0$

b. $3x^2 - 4x - 2 = 0$

c. $3x^2 + 4x + 2 = 0$

d. $2x^2 + 4x - 3 = 0$

e. $2x^2 - 4x + 3 = 0$

2. a. $3x^2 + 6x - 2 = 0$

b. $3x^2 - 6x - 2 = 0$

c. $3x^2 + 6x + 2 = 0$

d. $2x^2 + 6x + 3 = 0$

e. $2x^2 + 6x - 3 = 0$

3. a. $x^2 - 2x + 2 = 0$

b. $x^2 - 2x + 5 = 0$

c. $x^2 + 2x + 2 = 0$

4. a. $x^2 - 4x + 5 = 0$

b. $x^2 + 4x + 5 = 0$

c. $a^2 + 4a + 1 = 0$

Solve each equation. Use factoring or the quadratic formula, whichever is appropriate. (Try factoring first. If you have any difficulty factoring, then go right to the quadratic formula.)

5. $\frac{1}{6}x^2 - \frac{1}{2}x + \frac{1}{3} = 0$

6. $\frac{1}{4}x^2 + \frac{1}{4}x - \frac{1}{2} = 0$

7. $\frac{x^2}{2} + 1 = \frac{2x}{3}$

8. $\frac{x^2}{2} + \frac{2}{3} = -\frac{2x}{3}$

9. $y^2 - 5y = 0$

10. $2y^2 + 10y = 0$

11. $30x^2 + 40x = 0$

12. $50x^2 - 20x = 0$

13. $\frac{2t^2}{3} - t = -\frac{1}{6}$

14. $\frac{t^2}{3} - \frac{t}{2} = -\frac{3}{2}$

15. $0.01x^2 + 0.06x - 0.08 = 0$

16. $0.02x^2 - 0.03x + 0.05 = 0$

17. $2x + 3 = -2x^2$

18. $2x - 3 = 3x^2$

19. $100x^2 - 200x + 100 = 0$

20. $100x^2 - 600x + 900 = 0$

21. $\frac{1}{2}r^2 = \frac{1}{6}r - \frac{2}{3}$

22. $\frac{1}{4}r^2 = \frac{2}{5}r + \frac{1}{10}$

23. $(x - 3)(x - 5) = 1$

24. $(x - 3)(x + 1) = -6$

25. $(x + 3)^2 + (x - 8)(x - 1) = 16$

26. $(x - 4)^2 + (x + 2)(x + 1) = 9$

27. $\frac{x^2}{3} - \frac{5x}{6} = \frac{1}{2}$

28. $\frac{x^2}{6} + \frac{5}{6} = -\frac{x}{3}$

Multiply both sides of each equation by its LCD. Then solve the resulting equation.

29. $\frac{1}{x+1} - \frac{1}{x} = \frac{1}{2}$

30. $\frac{1}{x+1} + \frac{1}{x} = \frac{1}{3}$

31. $\frac{1}{y-1} + \frac{1}{y+1} = 1$

32. $\frac{2}{y+2} + \frac{3}{y-2} = 1$

33. $\frac{1}{x+2} + \frac{1}{x+3} = 1$

34. $\frac{1}{x+3} + \frac{1}{x+4} = 1$

35. $\frac{6}{r^2-1} - \frac{1}{2} = \frac{1}{r+1}$

36. $2 + \frac{5}{r-1} = \frac{12}{(r-1)^2}$

Solve each equation. In each case you will have three solutions.

37. $x^3 - 8 = 0$

38. $x^3 - 27 = 0$

39. $8a^3 + 27 = 0$

40. $27a^3 + 8 = 0$

41. $125t^3 - 1 = 0$

42. $64t^3 + 1 = 0$

Each of the following equations has three solutions. Look for the greatest common factor, then use the quadratic formula to find all solutions.

43. $2x^3 + 2x^2 + 3x = 0$

44. $6x^3 - 4x^2 + 6x = 0$

45. $3y^4 = 6y^3 - 6y^2$

46. $4y^4 = 16y^3 - 20y^2$

47. $6t^5 + 4t^4 = -2t^3$

48. $8t^5 + 2t^4 = -10t^3$

Applying the Concepts

49. Falling Object An object is thrown downward with an initial velocity of 5 feet per second. The relationship between the distance s it travels and time t is given by $s = 5t + 16t^2$. How long does it take the object to fall 74 feet?

50. Coin Toss A coin is tossed upward with an initial velocity of 32 feet per second from a height of 16 feet above the ground. The equation giving the object's height h at any time t is $h = 16 + 32t - 16t^2$. Does the object ever reach a height of 32 feet?

51. Profit The total cost (in dollars) for a company to manufacture and sell x items per week is $C = 60x + 300$, whereas the revenue brought in by selling all x items is $R = 100x - 0.5x^2$. How many items must be sold to obtain a weekly profit of \$300?

52. Profit Suppose a company manufactures and sells x picture frames each month with a total cost of $C = 1{,}200 + 3.5x$ dollars. If the revenue obtained by selling x frames is $R = 9x - 0.002x^2$, find the number of frames it must sell each month if its monthly profit is to be \$2,300.

Kathleen Olson

53. Photograph Cropping The following figure shows a photographic image on a 10.5-centimeter by 8.2-centimeter background. The overall area of the background is to be reduced to 80% of its original area by cutting off (cropping) equal strips on all four sides. What is the width of the strip that is cut from each side?

Pixtal/Superstock

54. Area of a Garden A garden measures 20.3 meters by 16.4 meters. To double the area of the garden, strips of equal width are added to all four sides.

a. Draw a diagram that illustrates these conditions.

b. What are the new overall dimensions of the garden?

55. **Area and Perimeter** A rectangle has a perimeter of 20 yards and an area of 15 square yards.
  a. Write two equations that state these facts in terms of the rectangle's length, l, and its width, w.
  b. Solve the two equations from part (a) to determine the actual length and width of the rectangle.
  c. Explain why two answers are possible to part (b).

56. **Population Size** Writing in 1829, former President James Madison made some predictions about the growth of the population of the United States. The populations he predicted fit the equation

$$y = 0.029x^2 - 1.39x + 42$$

where y is the population in millions of people x years from 1829.

Library of Congress

  a. Use the equation to determine the approximate year President Madison would have predicted that the U.S. population would reach 100,000,000.
  b. If the U.S. population in 2006 was approximately 300 million, were President Madison's predictions accurate in the long term? Explain why or why not.

Getting Ready for the Next Section

Find the value of $b^2 - 4ac$ when

57. $a = 1, b = -3, c = -40$

58. $a = 2, b = 3, c = 4$

59. $a = 4, b = 12, c = 9$

60. $a = -3, b = 8, c = -1$

Solve.

61. $k^2 - 144 = 0$

62. $36 - 20k = 0$

Multiply.

63. $(x - 3)(x + 2)$

64. $(t - 5)(t + 5)$

65. $(x - 3)(x - 3)(x + 2)$

66. $(t - 5)(t + 5)(t - 3)$

Maintaining Your Skills

Divide, using long division.

67. $\dfrac{8y^2 - 26y - 9}{2y - 7}$

68. $\dfrac{6y^2 + 7y - 18}{3y - 4}$

69. $\dfrac{x^3 + 9x^2 + 26x + 24}{x + 2}$

70. $\dfrac{x^3 + 6x^2 + 11x + 6}{x + 3}$

Simplify each expression. (Assume $x, y > 0$.)

71. $25^{1/2}$

72. $8^{1/3}$

73. $\left(\dfrac{9}{25}\right)^{3/2}$

74. $\left(\dfrac{16}{81}\right)^{3/4}$

75. $8^{-2/3}$

76. $4^{-3/2}$

77. $\dfrac{(49x^8y^{-4})^{1/2}}{(27x^{-3}y^9)^{-1/3}}$

78. $\dfrac{(x^{-2}y^{1/3})^6}{x^{-10}y^{3/2}}$

Extending the Concepts

So far, all the equations we have solved have had coefficients that were rational numbers. Here are some equations that have irrational coefficients and some that have complex coefficients. Solve each equation. (Remember, $i^2 = -1$.)

79. $x^2 + \sqrt{3}x - 6 = 0$

80. $x^2 - \sqrt{5}x - 5 = 0$

81. $\sqrt{2}x^2 + 2x - \sqrt{2} = 0$

82. $\sqrt{7}x^2 + 2\sqrt{2}x - \sqrt{7} = 0$

83. $x^2 + ix + 2 = 0$

84. $x^2 + 3ix - 2 = 0$

Additional Items Involving Solutions to Equations

8.3

In this section we will do two things. First, we will define the discriminant and use it to find the kind of solutions a quadratic equation has without solving the equation. Second, we will use the zero-factor property to build equations from their solutions.

OBJECTIVES	
A	Find the number and kind of solutions to a quadratic equation by using the discriminant.
B	Find an unknown constant in quadratic equation so that there is exactly one solution.
C	Find an equation from its solutions.

The Discriminant

The quadratic formula

$$x = \frac{-b \pm \sqrt{b^2 - 4ac}}{2a}$$

gives the solutions to any quadratic equation in standard form. When working with quadratic equations, there are times when it is important only to know what kind of solutions the equation has.

DEFINITION

The expression under the radical in the quadratic formula is called the **discriminant:**

$$\text{Discriminant} = D = b^2 - 4ac$$

The discriminant indicates the number and type of solutions to a quadratic equation, when the original equation has integer coefficients. For example, if we were to use the quadratic formula to solve the equation $2x^2 + 2x + 3 = 0$, we would find the discriminant to be

$$b^2 - 4ac = 2^2 - 4(2)(3) = -20$$

Since the discriminant appears under a square root symbol, we have the square root of a negative number in the quadratic formula. Our solutions therefore would be complex numbers. Similarly, if the discriminant were 0, the quadratic formula would yield

$$x = \frac{-b \pm \sqrt{0}}{2a} = \frac{-b \pm 0}{2a} = \frac{-b}{2a}$$

and the equation would have one rational solution, the number $\frac{-b}{2a}$.

The following table gives the relationship between the discriminant and the type of solutions to the equation.

For the equation $ax^2 + bx + c = 0$ where a, b, and c are integers and $a \neq 0$:

IF THE DISCRIMINANT $b^2 - 4ac$ IS	THEN THE EQUATION WILL HAVE
Negative	Two complex solutions containing i
Zero	One rational solution
A positive number that is also a perfect square	Two rational solutions
A positive number that is not a perfect square	Two irrational solutions

In the second and third cases, when the discriminant is 0 or a positive perfect square, the solutions are rational numbers. The quadratic equations in these two cases are the ones that can be factored.

EXAMPLES

For each equation, give the number and kind of solutions.

1. $x^2 - 3x - 40 = 0$

Solution Using $a = 1$, $b = -3$, and $c = -40$ in $b^2 - 4ac$, we have $(-3)^2 - 4(1)(-40) = 9 + 160 = 169$.

The discriminant is a perfect square. The equation therefore has two rational solutions.

2. $2x^2 - 3x + 4 = 0$

Solution Using $a = 2$, $b = -3$, and $c = 4$, we have

$$b^2 - 4ac = (-3)^2 - 4(2)(4) = 9 - 32 = -23$$

The discriminant is negative, implying the equation has two complex solutions that contain i.

3. $4x^2 - 12x + 9 = 0$

Solution Using $a = 4$, $b = -12$, and $c = 9$, the discriminant is

$$b^2 - 4ac = (-12)^2 - 4(4)(9) = 144 - 144 = 0$$

Since the discriminant is 0, the equation will have one rational solution.

4. $x^2 + 6x = 8$

Solution We first must put the equation in standard form by adding -8 to each side. If we do so, the resulting equation is

$$x^2 + 6x - 8 = 0$$

Now we identify a, b, and c as 1, 6, and -8, respectively:

$$b^2 - 4ac = 6^2 - 4(1)(-8) = 36 + 32 = 68$$

The discriminant is a positive number but not a perfect square. Therefore, the equation will have two irrational solutions.

EXAMPLE 5

Find an appropriate k so that the equation $4x^2 - kx = -9$ has exactly one rational solution.

Solution We begin by writing the equation in standard form:

$$4x^2 - kx + 9 = 0$$

Using $a = 4$, $b = -k$, and $c = 9$, we have

$$b^2 - 4ac = (-k)^2 - 4(4)(9)$$
$$= k^2 - 144$$

An equation has exactly one rational solution when the discriminant is 0. We set the discriminant equal to 0 and solve:

$$k^2 - 144 = 0$$
$$k^2 = 144$$
$$k = \pm 12$$

Choosing k to be 12 or -12 will result in an equation with one rational solution.

PRACTICE PROBLEMS

Give the number and kind of solution to each equation.

1. $x^2 - 3x - 28 = 0$
2. $x^2 - 6x + 9 = 0$
3. $3x^2 - 2x + 4 = 0$
4. $x^2 + 1 = 4x$
5. Find k so that the equation $9x^2 + kx = -4$ has exactly one rational solution.

Answers

1. Two rational
2. One rational
3. Two complex
4. Two irrational
5. ± 12

Building Equations From Their Solutions

Suppose we know that the solutions to an equation are $x = 3$ and $x = -2$. We can find equations with these solutions by using the zero-factor property. First, let's write our solutions as equations with 0 on the right side:

If	$x = 3$	**First solution**
then	$x - 3 = 0$	**Add −3 to each side**
and if	$x = -2$	**Second solution**
then	$x + 2 = 0$	**Add 2 to each side**

Now, since both $x - 3$ and $x + 2$ are 0, their product must be 0 also. Therefore, we can write

$$(x - 3)(x + 2) = 0 \quad \textbf{Zero-factor property}$$

$$x^2 - x - 6 = 0 \quad \textbf{Multiply out the left side}$$

Many other equations have 3 and −2 as solutions. For example, any constant multiple of $x^2 - x - 6 = 0$, such as $5x^2 - 5x - 30 = 0$, also has 3 and −2 as solutions. Similarly, any equation built from positive integer powers of the factors $x - 3$ and $x + 2$ will also have 3 and −2 as solutions. One such equation is

$$(x - 3)^2(x + 2) = 0$$

$$(x^2 - 6x + 9)(x + 2) = 0$$

$$x^3 - 4x^2 - 3x + 18 = 0$$

In mathematics we distinguish between the solutions to this last equation and those to the equation $x^2 - x - 6 = 0$ by saying $x = 3$ is a solution of *multiplicity 2* in the equation $x^3 - 4x^2 - 3x + 18 = 0$ and a solution of *multiplicity 1* in the equation $x^2 - x - 6 = 0$.

EXAMPLE 6 Find an equation that has solutions $t = 5$, $t = -5$, and $t = 3$.

Solution First, we use the given solutions to write equations that have 0 on their right sides:

If	$t = 5$	$t = -5$	$t = 3$
then	$t - 5 = 0$	$t + 5 = 0$	$t - 3 = 0$

Since $t - 5$, $t + 5$, and $t - 3$ are all 0, their product is also 0 by the zero-factor property. An equation with solutions of 5, −5, and 3 is

$$(t - 5)(t + 5)(t - 3) = 0 \quad \textbf{Zero-factor property}$$

$$(t^2 - 25)(t - 3) = 0 \quad \textbf{Multiply first two binomials}$$

$$t^3 - 3t^2 - 25t + 75 = 0 \quad \textbf{Complete the multiplication}$$

The last line gives us an equation with solutions of 5, −5, and 3. Remember, many other equations have these same solutions. ■

6. Find an equation that has solutions $t = -2$, $t = 2$, and $t = 3$.

Answer

6. $t^3 - 3t^2 - 4t + 12 = 0$

7. Find an equation with solutions $x = -\frac{3}{4}$ and $x = \frac{1}{5}$.

EXAMPLE 7 Find an equation with solutions $x = -\frac{2}{3}$ and $x = \frac{4}{5}$.

Solution The solution $x = -\frac{2}{3}$ can be rewritten as $3x + 2 = 0$ as follows:

$x = -\dfrac{2}{3}$	**The first solution**
$3x = -2$	**Multiply each side by 3**
$3x + 2 = 0$	**Add 2 to each side**

Similarly, the solution $x = \frac{4}{5}$ can be rewritten as $5x - 4 = 0$:

$x = \dfrac{4}{5}$	**The second solution**
$5x = 4$	**Multiply each side by 5**
$5x - 4 = 0$	**Add −4 to each side**

Since both $3x + 2$ and $5x - 4$ are 0, their product is 0 also, giving us the equation we are looking for:

$(3x + 2)(5x - 4) = 0$	**Zero-factor property**
$15x^2 - 2x - 8 = 0$	**Multiplication**

USING TECHNOLOGY

Graphing Calculators
Solving Equations

Now that we have explored the relationship between equations and their solutions, we can look at how a graphing calculator can be used in the solution process. To begin, let's solve the equation $x^2 = x + 2$ using techniques from algebra: writing it in standard form, factoring, and then setting each factor equal to 0.

		$x^2 - x - 2 = 0$	**Standard form**
		$(x - 2)(x + 1) = 0$	**Factor**
$x - 2 = 0$	or	$x + 1 = 0$	**Set each factor equal to 0**
$x = 2$	or	$x = -1$	**Solve**

Our original equation, $x^2 = x + 2$, has two solutions: $x = 2$ and $x = -1$. To solve the equation using a graphing calculator, we need to associate it with an equation (or equations) in two variables. One way to do this is to associate the left side with the equation $y = x^2$ and the right side of the equation with $y = x + 2$. To do so, we set up the functions list in our calculator this way:

$Y_1 = X^2$

$Y_2 = X + 2$

Window: X from −5 to 5, Y from −5 to 5

Graphing these functions in this window will produce a graph similar to the one shown in Figure 1.

Answer

7. $20x^2 + 11x - 3 = 0$

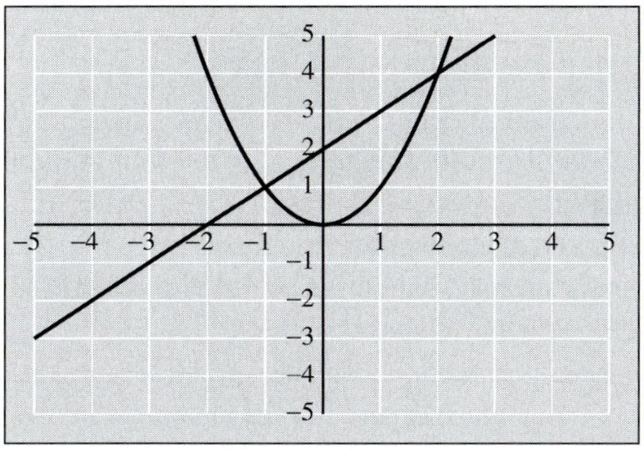

Figure 1

If we use the Trace feature to find the coordinates of the points of intersection, we find that the two curves intersect at $(-1, 1)$ and $(2, 4)$. We note that the x-coordinates of these two points match the solutions to the equation $x^2 = x + 2$, which we found using algebraic techniques. This makes sense because if two graphs intersect at a point (x, y), then the coordinates of that point satisfy both equations. If a point (x, y) satisfies both $y = x^2$ and $y = x + 2$, then, for that particular point, $x^2 = x + 2$. From this we conclude that the x-coordinates of the points of intersection are solutions to our original equation. Here is a summary of what we have discovered:

Conclusion 1 If the graph of two functions $y = f(x)$ and $y = g(x)$ intersect in the coordinate plane, then the x-coordinates of the points of intersection are solutions to the equation $f(x) = g(x)$.

A second method of solving our original equation $x^2 = x + 2$ graphically requires the use of one function instead of two. To begin, we write the equation in standard form as $x^2 - x - 2 = 0$. Next, we graph the function $y = x^2 - x - 2$. The x-intercepts of the graph are the points with y-coordinates of 0. They therefore satisfy the equation $0 = x^2 - x - 2$, which is equivalent to our original equation. The graph in Figure 2 shows $Y_1 = X^2 - X - 2$ in a window with X from -5 to 5 and Y from -5 to 5.

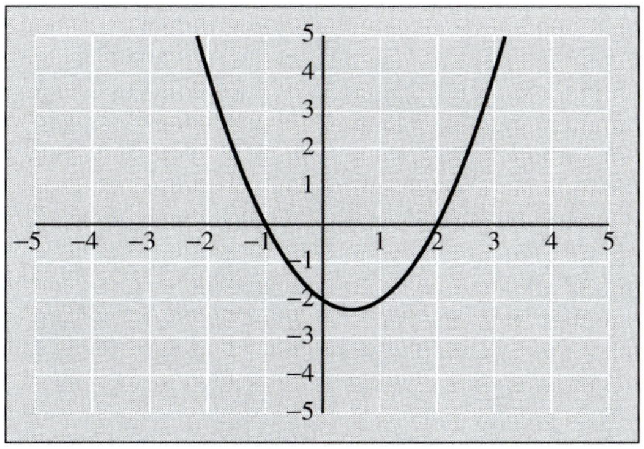

Figure 2

Using the Trace feature, we find that the x-intercepts of the graph are $x = -1$ and $x = 2$, which match the solutions to our original equation $x^2 = x + 2$. We can summarize the relationship between solutions to an equation and the intercepts of its associated graph this way:

Conclusion 2 If $y = f(x)$ is a function, then any x-intercept on the graph of $y = f(x)$ is a solution to the equation $f(x) = 0$.

Linking Objectives and Examples

Next to each **objective**, list the examples that are best described by that objective.

A	
B	
C	

Getting Ready for Class

After reading through the preceding section, respond in your own words and in complete sentences.

1. What is the discriminant?
2. What kind of solutions do we get to a quadratic equation when the discriminant is negative?
3. What does it mean for a solution to have multiplicity 3?
4. When will a quadratic equation have two rational solutions?

Problem Set 8.3

INTERMEDIATE Algebra f(x) Now™

Use the discriminant to find the number and kind of solutions for each of the following equations.

1. $x^2 - 6x + 5 = 0$
2. $x^2 - x - 12 = 0$
3. $4x^2 - 4x = -1$
4. $9x^2 + 12x = -4$
5. $x^2 + x - 1 = 0$
6. $x^2 - 2x + 3 = 0$
7. $2y^2 = 3y + 1$
8. $3y^2 = 4y - 2$
9. $x^2 - 9 = 0$
10. $4x^2 - 81 = 0$
11. $5a^2 - 4a = 5$
12. $3a = 4a^2 - 5$

Determine k so that each of the following has exactly one real solution.

13. $x^2 - kx + 25 = 0$
14. $x^2 + kx + 25 = 0$
15. $x^2 = kx - 36$
16. $x^2 = kx - 49$
17. $4x^2 - 12x + k = 0$
18. $9x^2 + 30x + k = 0$
19. $kx^2 - 40x = 25$
20. $kx^2 - 2x = -1$
21. $3x^2 - kx + 2 = 0$
22. $5x^2 + kx + 1 = 0$

For each of the following problems, find an equation that has the given solutions.

23. $x = 5, x = 2$

24. $x = -5, x = -2$

25. $t = -3, t = 6$

26. $t = -4, t = 2$

27. $y = 2, y = -2, y = 4$

28. $y = 1, y = -1, y = 3$

29. $x = \frac{1}{2}, x = 3$

30. $x = \frac{1}{3}, x = 5$

31. $t = -\frac{3}{4}, t = 3$

32. $t = -\frac{4}{5}, t = 2$

33. $x = 3, x = -3, x = \frac{5}{6}$

34. $x = 5, x = -5, x = \frac{2}{3}$

35. $a = -\frac{1}{2}, a = \frac{3}{5}$

36. $a = -\frac{1}{3}, a = \frac{4}{7}$

37. $x = -\frac{2}{3}, x = \frac{2}{3}, x = 1$

38. $x = -\frac{4}{5}, x = \frac{4}{5}, x = -1$

39. $x = 2, x = -2, x = 3, x = -3$

40. $x = 1, x = -1, x = 5, x = -5$

Getting Ready for the Next Section

Simplify.

41. $(x + 3)^2 - 2(x + 3) - 8$

42. $(x - 2)^2 - 3(x - 2) - 10$

43. $(2a - 3)^2 - 9(2a - 3) + 20$

44. $(3a - 2)^2 + 2(3a - 2) - 3$

45. $2(4a + 2)^2 - 3(4a + 2) - 20$

46. $6(2a + 4)^2 - (2a + 4) - 2$

Solve.

47. $x^2 = \frac{1}{4}$

48. $x^2 = -2$

49. $\sqrt{x} = -3$

50. $\sqrt{x} = 2$

51. $x + 3 = 4$

52. $x + 3 = -2$

53. $y^2 - 2y - 8 = 0$

54. $y^2 + y - 6 = 0$

55. $4y^2 + 7y - 2 = 0$

56. $6x^2 - 13x - 5 = 0$

Maintaining Your Skills

Multiply. (Assume all variables represent positive numbers for the rest of the problems in this section.)

57. $a^4(a^{3/2} - a^{1/2})$

58. $(a^{1/2} - 5)(a^{1/2} + 3)$

59. $(x^{3/2} - 3)^2$

60. $(x^{1/2} - 8)(x^{1/2} + 8)$

Divide.

61. $\dfrac{30x^{3/4} - 25x^{5/4}}{5x^{1/4}}$

62. $\dfrac{45x^{5/3}y^{7/3} - 36x^{8/3}y^{4/3}}{9x^{2/3}y^{1/3}}$

63. Factor $5(x - 3)^{1/2}$ from $10(x - 3)^{3/2} - 15(x - 3)^{1/2}$.

64. Factor $2(x + 1)^{1/3}$ from $8(x + 1)^{4/3} - 2(x + 1)^{1/3}$.

Factor each of the following.

65. $2x^{2/3} - 11x^{1/3} + 12$

66. $9x^{2/3} + 12x^{1/3} + 4$

Extending the Concepts

Find all solutions to the following equations. Solve using algebra and by graphing. If rounding is necessary, round to the nearest hundredth. A calculator can be used in these problems.

67. $x^2 = 4x + 5$

68. $4x^2 = 8x + 5$

69. $x^2 - 1 = 2x$

70. $4x^2 - 1 = 4x$

Find all solutions to each equation. If rounding is necessary, round to the nearest hundredth.

71. $2x^3 - x^2 - 2x + 1 = 0$

72. $3x^3 - 2x^2 - 3x + 2 = 0$

73. $2x^3 + 2 = x^2 + 4x$

74. $3x^3 - 9x = 2x^2 - 6$

Equations Quadratic in Form

8.4

We are now in a position to put our knowledge of quadratic equations to work to solve a variety of equations.

OBJECTIVES

A	Solve equations that are reducible to a quadratic equation.
B	Solve application problems using equations quadratic in form.

EXAMPLE 1

Solve $(x + 3)^2 - 2(x + 3) - 8 = 0$.

Solution We can see that this equation is quadratic in form by replacing $x + 3$ with another variable, say y. Replacing $x + 3$ with y we have

$$y^2 - 2y - 8 = 0$$

We can solve this equation by factoring the left side and then setting each factor equal to 0.

$$y^2 - 2y - 8 = 0$$
$$(y - 4)(y + 2) = 0 \quad \textbf{Factor}$$
$$y - 4 = 0 \quad \text{or} \quad y + 2 = 0 \quad \textbf{Set factors to 0}$$
$$y = 4 \quad \text{or} \quad y = -2$$

Since our original equation was written in terms of the variable x, we would like our solutions in terms of x also. Replacing y with $x + 3$ and then solving for x we have

$$x + 3 = 4 \quad \text{or} \quad x + 3 = -2$$
$$x = 1 \quad \text{or} \quad x = -5$$

The solutions to our original equation are 1 and -5.

The method we have just shown lends itself well to other types of equations that are quadratic in form, as we will see. In this example, however, there is another method that works just as well. Let's solve our original equation again, but this time, let's begin by expanding $(x + 3)^2$ and $2(x + 3)$.

$$(x + 3)^2 - 2(x + 3) - 8 = 0$$
$$x^2 + 6x + 9 - 2x - 6 - 8 = 0 \quad \textbf{Multiply}$$
$$x^2 + 4x - 5 = 0 \quad \textbf{Combine similar terms}$$
$$(x - 1)(x + 5) = 0 \quad \textbf{Factor}$$
$$x - 1 = 0 \quad \text{or} \quad x + 5 = 0 \quad \textbf{Set factors to 0}$$
$$x = 1 \quad \text{or} \quad x = -5$$

As you can see, either method produces the same result.

EXAMPLE 2

Solve $4x^4 + 7x^2 = 2$.

Solution This equation is quadratic in x^2. We can make it easier to look at by using the substitution $y = x^2$. (The choice of the letter y is arbitrary. We could just as easily use the substitution $m = x^2$.) Making the substitution $y = x^2$ and then solving the resulting equation we have

$$4y^2 + 7y = 2$$
$$4y^2 + 7y - 2 = 0 \quad \textbf{Standard form}$$
$$(4y - 1)(y + 2) = 0 \quad \textbf{Factor}$$
$$4y - 1 = 0 \quad \text{or} \quad y + 2 = 0 \quad \textbf{Set factors to 0}$$
$$y = \frac{1}{4} \quad \text{or} \quad y = -2$$

PRACTICE PROBLEMS

1. Solve $(x - 2)^2 - 3(x - 2) - 10 = 0$.

2. Solve $6x^4 - 13x^2 = 5$.

Answer

1. 7, 0

Now we replace y with x^2 to solve for x:

$$x^2 = \frac{1}{4} \quad \text{or} \quad x^2 = -2$$

$$x = \pm\sqrt{\frac{1}{4}} \quad \text{or} \quad x = \pm\sqrt{-2} \qquad \textbf{Theorem 1}$$

$$x = \pm\frac{1}{2} \quad \text{or} \quad x = \pm i\sqrt{2}$$

The solution set is $\{\frac{1}{2}, -\frac{1}{2}, i\sqrt{2}, -i\sqrt{2}\}$. ■

3. Solve for x: $x - \sqrt{x} - 12 = 0$.

EXAMPLE 3

Solve for x: $x + \sqrt{x} - 6 = 0$.

Solution To see that this equation is quadratic in form, we have to notice that $(\sqrt{x})^2 = x$; that is, the equation can be rewritten as

$$(\sqrt{x})^2 + \sqrt{x} - 6 = 0$$

Replacing $\sqrt{x}$ with y and solving as usual, we have

$$y^2 + y - 6 = 0$$

$$(y + 3)(y - 2) = 0$$

$$y + 3 = 0 \quad \text{or} \quad y - 2 = 0$$

$$y = -3 \quad \text{or} \quad y = 2$$

Again, to find x, we replace with $\sqrt{x}$ and solve:

$$\sqrt{x} = -3 \quad \text{or} \quad \sqrt{x} = 2$$

$$x = 9 \qquad x = 4 \qquad \textbf{Square both sides of each equation}$$

Since we squared both sides of each equation, we have the possibility of obtaining extraneous solutions. We have to check both solutions in our original equation.

When	$x = 9$	When	$x = 4$
the equation	$x + \sqrt{x} - 6 = 0$	the equation	$x + \sqrt{x} - 6 = 0$
becomes	$9 + \sqrt{9} - 6 \stackrel{?}{=} 0$	becomes	$4 + \sqrt{4} - 6 \stackrel{?}{=} 0$
	$9 + 3 - 6 \stackrel{?}{=} 0$		$4 + 2 - 6 \stackrel{?}{=} 0$
	$6 \neq 0$		$0 = 0$
	This means 9 is extraneous.		This means 4 is a solution.

The only solution to the equation $x + \sqrt{x} - 6 = 0$ is $x = 4$. ■

We should note here that the two possible solutions, 9 and 4, to the equation in Example 3 can be obtained by another method. Instead of substituting for $\sqrt{x}$, we can isolate it on one side of the equation and then square both sides to clear the equation of radicals.

$$x + \sqrt{x} - 6 = 0$$

$$\sqrt{x} = -x + 6 \qquad \textbf{Isolate } x$$

$$x = x^2 - 12x + 36 \qquad \textbf{Square both sides}$$

$$0 = x^2 - 13x + 36 \qquad \textbf{Add } -x \textbf{ to both sides}$$

$$0 = (x - 4)(x - 9) \qquad \textbf{Factor}$$

$$x - 4 = 0 \quad \text{or} \quad x - 9 = 0$$

$$x = 4 \qquad x = 9$$

Answers

2. $\pm\frac{i\sqrt{3}}{3}, \pm\frac{\sqrt{10}}{2}$ 3. 16

We obtain the same two possible solutions. Since we squared both sides of the equation to find them, we would have to check each one in the original equation. As was the case in Example 3, only $x = 4$ is a solution; $x = 9$ is extraneous.

EXAMPLE 4 If an object is tossed into the air with an upward velocity of 12 feet per second from the top of a building h feet high, the time it takes for the object to hit the ground below is given by the formula

$$16t^2 - 12t - h = 0$$

Solve this formula for t.

4. Solve $16t^2 - 10t - h = 0$ for t.

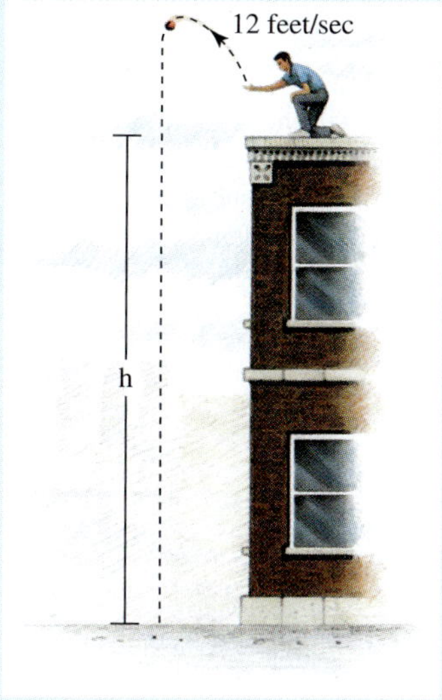

Solution The formula is in standard form and is quadratic in t. The coefficients a, b, and c that we need to apply to the quadratic formula are $a = 16$, $b = -12$, and $c = -h$. Substituting these quantities into the quadratic formula, we have

$$t = \frac{12 \pm \sqrt{144 - 4(16)(-h)}}{2(16)}$$

$$= \frac{12 \pm \sqrt{144 + 64h}}{32}$$

We can factor the perfect square 16 from the two terms under the radical and simplify our radical somewhat:

$$t = \frac{12 \pm \sqrt{16(9 + 4h)}}{32}$$

$$= \frac{12 \pm 4\sqrt{9 + 4h}}{32}$$

Now we can reduce to lowest terms by factoring a 4 from the numerator and denominator.

$$t = \frac{\cancel{4}(3 \pm \sqrt{9 + 4h})}{\cancel{4} \cdot 8}$$

$$= \frac{3 \pm \sqrt{9 + 4h}}{8}$$

If we were given a value of h, we would find that one of the solutions to this last formula would be a negative number. Since time is always measured in positive units, we wouldn't use that solution.

More About the Golden Ratio

Previously, we derived the golden ratio $\frac{1 + \sqrt{5}}{2}$ by finding the ratio of length to width for a golden rectangle. The golden ratio was actually discovered before the golden rectangle by the Greeks who lived before Euclid. The early Greeks found the golden ratio by dividing a line segment into two parts so that the ratio of the shorter part to the longer part was the same as the ratio of the longer part to the whole segment. When they divided a line segment in this manner, they said it was divided in "extreme and mean ratio." Figure 1 illustrates a line segment divided this way.

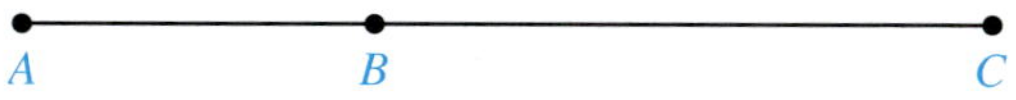

FIGURE 1

Answer

4. $t = \frac{5 \pm \sqrt{25 + 16h}}{16}$

5. If the length of segment AB in Figure 1 is 2 inches, and B divides AC in "extreme and mean ratio," find BC, and then show that BC is twice the golden ratio.

If point B divides segment AC in "extreme and mean ratio," then

$$\frac{\text{Length of shorter segment}}{\text{Length of longer segment}} = \frac{\text{Length of longer segment}}{\text{Length of whole segment}}$$

$$\frac{AB}{BC} = \frac{BC}{AC}$$

EXAMPLE 5 If the length of segment AB in Figure 1 is 1 inch, find the length of BC so that the whole segment AC is divided in "extreme and mean ratio."

Solution Using Figure 1 as a guide, if we let x = the length of segment BC, then the length of AC is $x + 1$. If B divides AC into "extreme and mean ratio," then the ratio of AB to BC must equal the ratio of BC to AC. Writing this relationship using the variable x, we have

$$\frac{1}{x} = \frac{x}{x+1}$$

If we multiply both sides of this equation by the LCD $x(x + 1)$ we have

$$x + 1 = x^2$$

$$0 = x^2 - x - 1 \quad \textbf{Write equation in standard form}$$

Since this last equation is not factorable, we apply the quadratic formula.

$$x = \frac{1 \pm \sqrt{(-1)^2 - 4(1)(-1)}}{2}$$

$$= \frac{1 \pm \sqrt{5}}{2}$$

Our equation has two solutions, which we approximate using decimals:

$$\frac{1+\sqrt{5}}{2} \approx 1.618 \qquad \frac{1-\sqrt{5}}{2} \approx -0.618$$

Since we originally let x equal the length of segment BC, we use only the positive solution to our equation. As you can see, the positive solution is the golden ratio.

USING TECHNOLOGY

Graphing Calculators
More About Example 1

As we mentioned earlier, algebraic expressions entered into a graphing calculator do not have to be simplified to be evaluated. This fact applies to equations as well. We can graph the equation $y = (x + 3)^2 - 2(x + 3) - 8$ to assist us in solving the equation in Example 1. The graph is shown in Figure 2. Using the Zoom and Trace features at the x-intercepts gives us $x = 1$ and $x = -5$ as the solutions to the equation $0 = (x + 3)^2 - 2(x + 3) - 8$.

Answer

5. $BC = 1 + \sqrt{5}$, which is twice the golden ratio

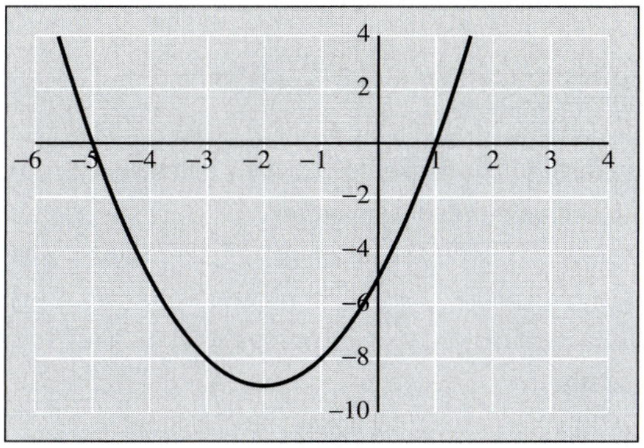

FIGURE 2

More About Example 2

Figure 3 shows the graph of $y = 4x^4 + 7x^2 - 2$. As we expect, the x-intercepts give the real number solutions to the equation $0 = 4x^4 + 7x^2 - 2$. The complex solutions do not appear on the graph.

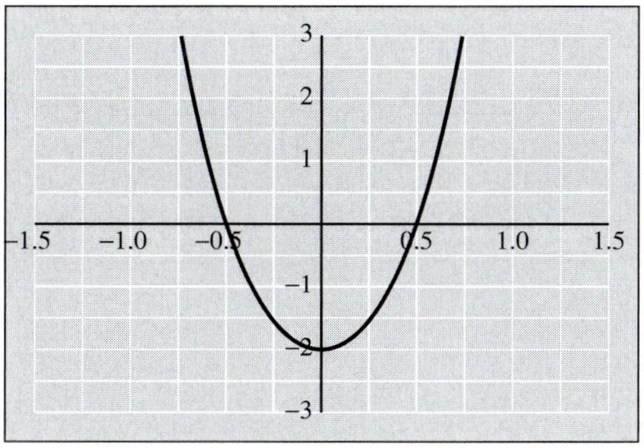

FIGURE 3

More About Example 3

In solving the equation in Example 3, we found that one of the possible solutions was an extraneous solution. If we solve the equation $x + \sqrt{x} - 6 = 0$ by graphing the function $y = x + \sqrt{x} - 6$, we find that the extraneous solution, 9, is not an x-intercept. Figure 4 shows that the only solution to the equation occurs at the x-intercept 4.

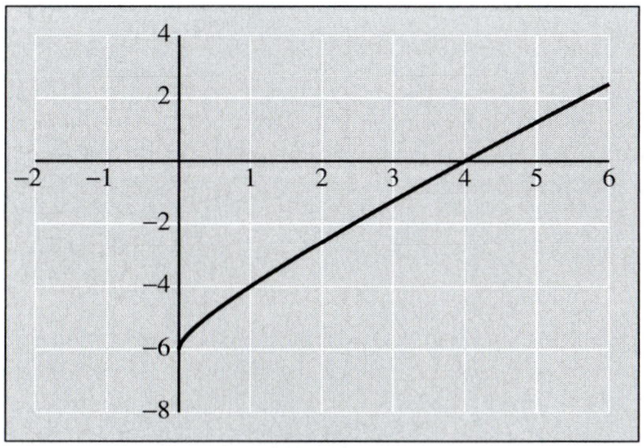

FIGURE 4

Linking Objectives and Examples

Next to each **objective**, list the examples that are best described by that objective.

A	
B	

Getting Ready for Class

After reading through the preceding section, respond in your own words and in complete sentences.

1. What does it mean for an equation to be quadratic in form?
2. What are all the circumstances in solving equations (that we have studied) in which it is necessary to check for extraneous solutions?
3. How would you start to solve the equation $x + \sqrt{x} - 6 = 0$?
4. What does it mean for a line segment to be divided in "extreme and mean ratio"?

Problem Set 8.4

INTERMEDIATE Algebra $f(x)$ Now™

Solve each equation.

1. $(x - 3)^2 + 3(x - 3) + 2 = 0$

2. $(x + 4)^2 - (x + 4) - 6 = 0$

3. $2(x + 4)^2 + 5(x + 4) - 12 = 0$

4. $3(x - 5)^2 + 14(x - 5) - 5 = 0$

5. $x^4 - 6x^2 - 27 = 0$

6. $x^4 + 2x^2 - 8 = 0$

7. $x^4 + 9x^2 = -20$

8. $x^4 - 11x^2 = -30$

9. $(2a - 3)^2 - 9(2a - 3) = -20$

10. $(3a - 2)^2 + 2(3a - 2) = 3$

11. $2(4a + 2)^2 = 3(4a + 2) + 20$

12. $6(2a + 4)^2 = (2a + 4) + 2$

13. $6t^4 = -t^2 + 5$

14. $3t^4 = -2t^2 + 8$

15. $9x^4 - 49 = 0$

16. $25x^4 - 9 = 0$

Solve each of the following equations. Remember, if you square both sides of an equation in the process of solving it, you have to check all solutions in the original equation.

17. $x - 7\sqrt{x} + 10 = 0$

18. $x - 6\sqrt{x} + 8 = 0$

19. $t - 2\sqrt{t} - 15 = 0$

20. $t - 3\sqrt{t} - 10 = 0$

21. $6x + 11\sqrt{x} = 35$

22. $2x + \sqrt{x} = 15$

23. $(a - 2) - 11\sqrt{a - 2} + 30 = 0$

24. $(a - 3) - 9\sqrt{a - 3} + 20 = 0$

25. $(2x + 1) - 8\sqrt{2x + 1} + 15 = 0$

26. $(2x - 3) - 7\sqrt{2x - 3} + 12 = 0$

27. Solve the formula $16t^2 - vt - h = 0$ for t.

28. Solve the formula $16t^2 + vt + h = 0$ for t.

29. Solve the formula $kx^2 + 8x + 4 = 0$ for x.

30. Solve the formula $k^2x^2 + kx + 4 = 0$ for x.

31. Solve $x^2 + 2xy + y^2 = 0$ for x by using the quadratic formula with $a = 1$, $b = 2y$, and $c = y^2$.

32. Solve $x^2 - 2xy + y^2 = 0$ for x by using the quadratic formula, with $a = 1$, $b = -2y$, and $c = y^2$.

Applying the Concepts

For Problems 33–34, t is in seconds.

33. Falling Object An object is tossed into the air with an upward velocity of 8 feet per second from the top of a building h feet high. The time it takes for the object to hit the ground below is given by the formula $16t^2 - 8t - h = 0$. Solve this formula for t.

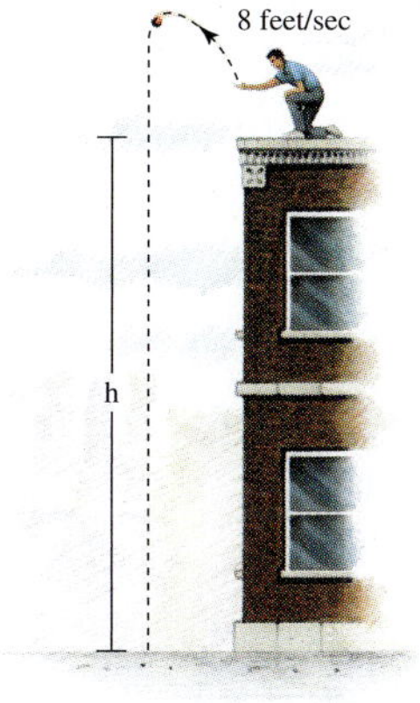

34. Falling Object An object is tossed into the air with an upward velocity of 6 feet per second from the top of a building h feet high. The time it takes for the object to hit the ground below is given by the formula $16t^2 - 6t - h = 0$. Solve this formula for t.

35. Saint Louis Arch The shape of the famous "Gateway to the West" arch in Saint Louis can be modeled by a parabola. The equation for one such parabola is:

$$y = -\frac{1}{150}x^2 + \frac{21}{5}x$$

where x and y are in feet.

a. Sketch the graph of the arch's equation on a coordinate axis.
b. Approximately how far do you have to walk to get from one side of the arch to the other?

Royalty-Free/Corbis

36. Area and Perimeter A total of 160 yards of fencing is to be used to enclose part of a lot that borders on a river. This situation is shown in the following diagram.

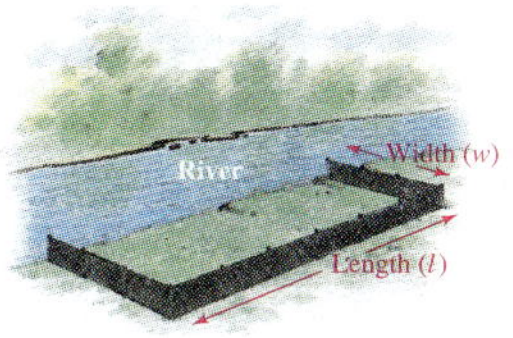

a. Write an equation that gives the relationship between the length and width and the 160 yards of fencing.
b. The formula for the area that is enclosed by the fencing and the river is $A = lw$. Solve the equation in part (a) for l, and then use the result to write the area in terms of w only.
c. Make a table that gives at least five possible values of w and associated area A.
d. From the pattern in your table shown in part (c), what is the largest area that can be enclosed by the 160 yards of fencing? (Try some other table values if necessary.)

Golden Ratio Use Figure 1 from this section as a guide to working Problems 37–38.

37. If AB in Figure 1 is 4 inches, and B divides AC in "extreme and mean ratio," find BC, and then show that BC is 4 times the golden ratio.

38. If AB in Figure 1 is $\frac{1}{2}$ inch, and B divides AC in "extreme and mean ratio," find BC, and then show that the ratio of BC to AB is the golden ratio.

Getting Ready for the Next Section

39. Evaluate $y = 3x^2 - 6x + 1$ for $x = 1$.

40. Evaluate $y = -2x^2 + 6x - 5$ for $x = \frac{3}{2}$.

41. Let $P(x) = -0.1x^2 + 27x - 500$ and find $P(135)$.

42. Let $P(x) = -0.1x^2 + 12x - 400$ and find $P(600)$.

Solve.

43. $0 = a(80)^2 + 70$

44. $0 = a(80)^2 + 90$

45. $x^2 - 6x + 5 = 0$

46. $x^2 - 3x - 4 = 0$

47. $-x^2 - 2x + 3 = 0$

48. $-x^2 + 4x + 12 = 0$

49. $2x^2 - 6x + 5 = 0$

50. $x^2 - 4x + 5 = 0$

Fill in the blanks to complete the square.

51. $x^2 - 6x + \square = (x - \square)^2$

52. $x^2 + 10x + \square = (x + \square)^2$

53. $y^2 + 2y + \square = (y + \square)^2$

54. $y^2 - 12y + \square = (y - \square)^2$

Maintaining Your Skills

Combine, if possible.

55. $5\sqrt{7} - 2\sqrt{7}$

56. $6\sqrt{2} - 9\sqrt{2}$

57. $\sqrt{18} - \sqrt{8} + \sqrt{32}$

58. $\sqrt{50} + \sqrt{72} - \sqrt{8}$

59. $9x\sqrt{20x^3y^2} + 7y\sqrt{45x^5}$

60. $5x^2\sqrt{27xy^3} - 6y\sqrt{12x^5y}$

Multiply.

61. $(\sqrt{5} - 2)(\sqrt{5} + 8)$

62. $(2\sqrt{3} - 7)(2\sqrt{3} + 7)$

63. $(\sqrt{x} + 2)^2$

64. $(3 - \sqrt{x})(3 + \sqrt{x})$

Rationalize the denominator.

65. $\dfrac{\sqrt{7}}{\sqrt{7} - 2}$

66. $\dfrac{\sqrt{5} - \sqrt{2}}{\sqrt{5} + \sqrt{2}}$

Extending the Concepts

Find the x- and y-intercepts.

67. $y = x^3 - 4x$

68. $y = x^4 - 10x^2 + 9$

69. $y = 3x^3 + x^2 - 27x - 9$

70. $y = 2x^3 + x^2 - 8x - 4$

71. The graph of $y = 2x^3 - 7x^2 - 5x + 4$ crosses the x-axis at $x = 4$. Where else does it cross the x-axis?

72. The graph of $y = 6x^3 + x^2 - 12x + 5$ crosses the x-axis at $x = 1$. Where else does it cross the x-axis?

Graphing Parabolas

8.5

OBJECTIVES	
A	Graph a parabola.
B	Solve application problems using information from a graph.
C	Find an equation from its graph.

The solution set to the equation

$$y = x^2 - 3$$

consists of ordered pairs. One method of graphing the solution set is to find a number of ordered pairs that satisfy the equation and to graph them. We can obtain some ordered pairs that are solutions to $y = x^2 - 3$ by use of a table as follows:

x	$y = x^2 - 3$	y	SOLUTIONS
−3	$y = (-3)^2 - 3 = 9 - 3 = 6$	6	(−3, 6)
−2	$y = (-2)^2 - 3 = 4 - 3 = 1$	1	(−2, 1)
−1	$y = (-1)^2 - 3 = 1 - 3 = -2$	−2	(−1, −2)
0	$y = 0^2 - 3 = 0 - 3 = -3$	−3	(0, −3)
1	$y = 1^2 - 3 = 1 - 3 = -2$	−2	(1, −2)
2	$y = 2^2 - 3 = 4 - 3 = 1$	1	(2, 1)
3	$y = 3^2 - 3 = 9 - 3 = 6$	6	(3, 6)

Graphing these solutions and then connecting them with a smooth curve, we have the graph of $y = x^2 - 3$ (Figure 1).

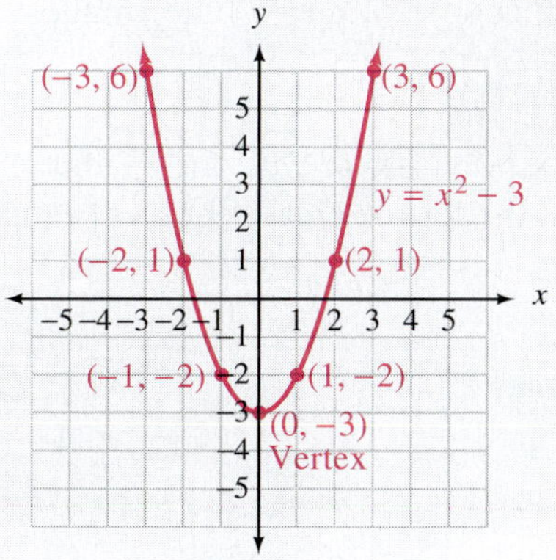

FIGURE 1

This graph is an example of a *parabola.* All equations of the form $y = ax^2 + bx + c$, $a \neq 0$ have parabolas for graphs.

Although it is always possible to graph parabolas by making a table of values of x and y that satisfy the equation, there are other methods that are faster and, in some cases, more accurate.

The important points associated with the graph of a parabola are the highest (or lowest) point on the graph and the x-intercepts. The y-intercepts can also be useful.

Intercepts for Parabolas

The graph of the equation $y = ax^2 + bx + c$ crosses the y-axis at $y = c$ since substituting $x = 0$ into $y = ax^2 + bx + c$ yields $y = c$.

Since the graph crosses the x-axis when $y = 0$, the x-intercepts are those values of x that are solutions to the quadratic equation $0 = ax^2 + bx + c$.

The Vertex of a Parabola

The highest or lowest point on a parabola is called the *vertex.* The vertex for the graph of $y = ax^2 + bx + c$ will occur when

$$x = \frac{-b}{2a}$$

To see this, we must transform the right side of $y = ax^2 + bx + c$ into an expression that contains x in just one of its terms. This is accomplished by completing the square on the first two terms. Here is what it looks like:

$$y = ax^2 + bx + c$$

$$y = a\left(x^2 + \frac{b}{a}x\right) + c$$

$$y = a\left[x^2 + \frac{b}{a}x + \left(\frac{b}{2a}\right)^2\right] + c - a\left(\frac{b}{2a}\right)^2$$

$$y = a\left(x + \frac{b}{2a}\right)^2 + \frac{4ac - b^2}{4a}$$

NOTE

What we are doing here is attempting to explain why the vertex of a parabola always has an x-coordinate of $\frac{-b}{2a}$. But the explanation may not be easy to understand the first time you see it. It may be helpful to look over the examples in this section and the notes that accompany these examples and then come back and read over this discussion again.

It may not look like it, but this last line indicates that the vertex of the graph of $y = ax^2 + bx + c$ has an x-coordinate of $\frac{-b}{2a}$. Since a, b, and c are constants, the only quantity that is varying in the last expression is the x in $\left(x + \frac{b}{2a}\right)^2$. Since the quantity $\left(x + \frac{b}{2a}\right)^2$ is the square of $x + \frac{b}{2a}$, the smallest it will ever be is 0, and that will happen when $x = \frac{-b}{2a}$.

We can use the vertex point along with the x- and y-intercepts to sketch the graph of any equation of the form $y = ax^2 + bx + c$. Here is a summary of the preceding information.

Graphing Parabolas I

The graph of $y = ax^2 + bx + c$, $a \neq 0$, will have

1. A y-intercept at $y = c$
2. x-intercepts (if they exist) at

$$x = \frac{-b \pm \sqrt{b^2 - 4ac}}{2a}$$

3. A vertex when $x = \frac{-b}{2a}$

PRACTICE PROBLEMS

1. Graph $y = x^2 - 2x - 3$.

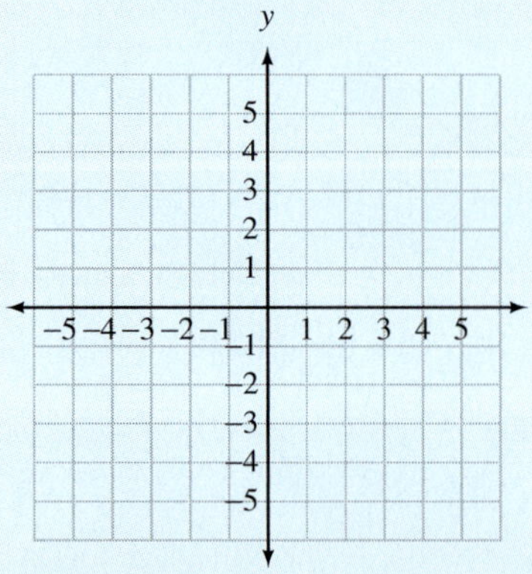

Answers

1–4. See Solutions to Selected Practice Problems.

EXAMPLE 1

Sketch the graph of $y = x^2 - 6x + 5$.

Solution To find the x-intercepts, we let $y = 0$ and solve for x:

$$0 = x^2 - 6x + 5$$

$$0 = (x - 5)(x - 1)$$

$$x = 5 \quad \text{or} \quad x = 1$$

To find the coordinates of the vertex, we first find

$$x = \frac{-b}{2a} = \frac{-(-6)}{2(1)} = 3$$

The x-coordinate of the vertex is 3. To find the y-coordinate, we substitute 3 for x in our original equation:

$$y = 3^2 - 6(3) + 5 = 9 - 18 + 5 = -4$$

The graph crosses the x-axis at 1 and 5 and has its vertex at $(3, -4)$. Plotting these points and connecting them with a smooth curve, we have the graph shown in Figure 2. The graph is a parabola that opens up, so we say the graph is *concave up.* The vertex is the lowest point on the graph. (Note that the graph crosses the y-axis at 5, which is the value of y we obtain when we let $x = 0$.)

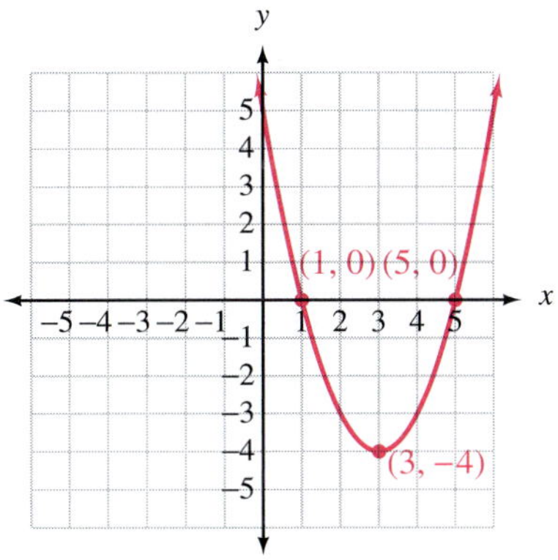

FIGURE 2

Finding the Vertex by Completing the Square

Another way to locate the vertex of the parabola in Example 1 is by completing the square on the first two terms on the right side of the equation $y = x^2 - 6x + 5$. In this case, we would do so by adding 9 to and subtracting 9 from the right side of the equation. This amounts to adding 0 to the equation, so we know we haven't changed its solutions. This is what it looks like:

$$y = (x^2 - 6x \qquad) + 5$$

$$y = (x^2 - 6x + \mathbf{9}) + 5 - \mathbf{9}$$

$$y = (x - 3)^2 - 4$$

You may have to look at this last equation a while to see this, but when $x = 3$, then $y = (x - 3)^2 - 4 = 0^2 - 4 = -4$ is the smallest y will ever be. And that is why the vertex is at $(3, -4)$. As a matter of fact, this is the same kind of reasoning we used when we derived the formula $x = \frac{-b}{2a}$ for the x-coordinate of the vertex.

EXAMPLE 2 Graph $y = -x^2 - 2x + 3$.

Solution To find the x-intercepts, we let $y = 0$:

$$0 = -x^2 - 2x + 3$$

$$0 = x^2 + 2x - 3 \quad \textbf{Multiply each side by } -1$$

$$0 = (x + 3)(x - 1)$$

$$x = -3 \quad \text{or} \quad x = 1$$

The x-coordinate of the vertex is given by

$$x = \frac{-b}{2a} = \frac{-(-2)}{2(-1)} = \frac{2}{-2} = -1$$

2. Graph $y = -x^2 + 2x + 8$.

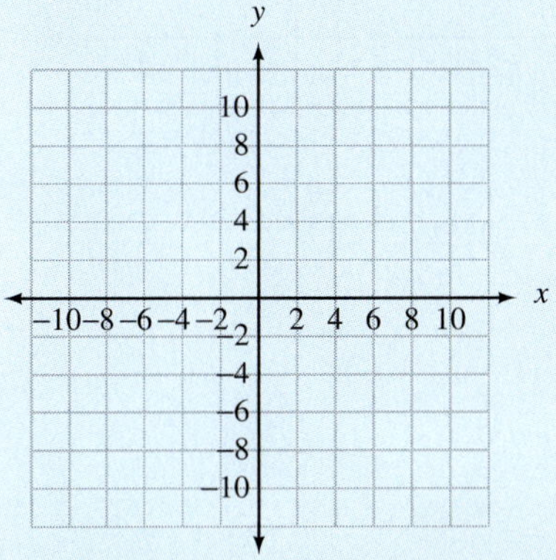

To find the y-coordinate of the vertex, we substitute -1 for x in our original equation to get

$$y = -(-1)^2 - 2(-1) + 3 = -1 + 2 + 3 = 4$$

Our parabola has x-intercepts at -3 and 1 and a vertex at $(-1, 4)$. Figure 3 shows the graph. We say the graph is *concave down* since it opens downward.

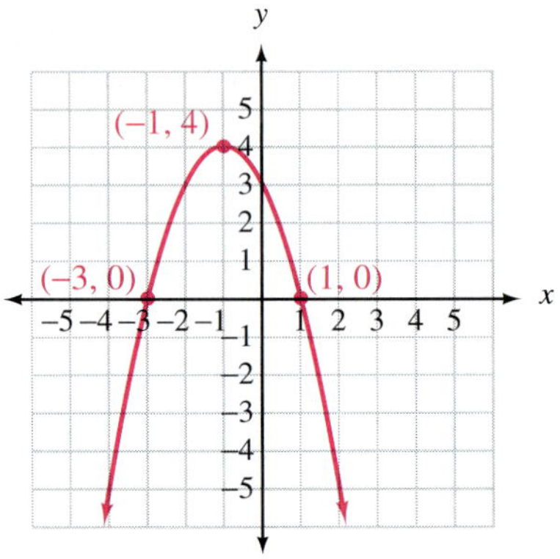

FIGURE 3

Again, we could have obtained the coordinates of the vertex by completing the square on the first two terms on the right side of our equation. To do so, we must first factor -1 from the first two terms. (Remember, the leading coefficients must be 1 to complete the square.) When we complete the square, we add 1 inside the parentheses, which actually decreases the right side of the equation by -1 since everything in the parentheses is multiplied by -1. To make up for it, we add 1 outside the parentheses.

$$y = -1(x^2 + 2x) + 3$$

$$y = -1(x^2 + 2x + \mathbf{1}) + 3 + \mathbf{1}$$

$$y = -1(x + 1)^2 + 4$$

The last line tells us that the *largest* value of y will be 4, and that will occur when $x = -1$.

3. Graph $y = 2x^2 - 4x + 1$.

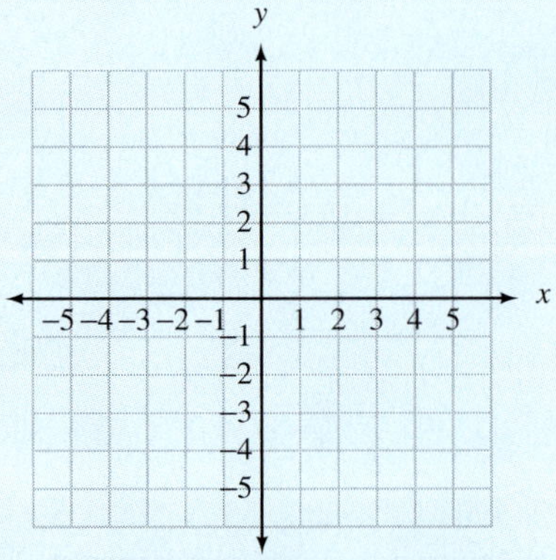

EXAMPLE 3

Graph $y = 3x^2 - 6x + 1$.

Solution To find the x-intercepts, we let $y = 0$ and solve for x:

$$0 = 3x^2 - 6x + 1$$

Since the right side of this equation does not factor, we can look at the discriminant to see what kind of solutions are possible. The discriminant for this equation is

$$b^2 - 4ac = 36 - 4(3)(1) = 24$$

Since the discriminant is a positive number but not a perfect square, the equation will have irrational solutions. This means that the x-intercepts are irrational numbers and will have to be approximated with decimals using the quadratic formula. Rather than use the quadratic formula, we will find some other points on the graph, but first let's find the vertex.

Here are both methods of finding the vertex:

Using the formula that gives us the x-coordinate of the vertex, we have:

$$x = \frac{-b}{2a} = \frac{-(-6)}{2(3)} = 1$$

Substituting 1 for x in the equation gives us the y-coordinate of the vertex:

$$y = 3 \cdot 1^2 - 6 \cdot 1 + 1 = -2$$

To complete the square on the right side of the equation, we factor 3 from the first two terms, add 1 inside the parentheses, and add -3 outside the parentheses (this amounts to adding 0 to the right side):

$$y = 3(x^2 - 2x \qquad) + 1$$

$$y = 3(x^2 - 2x + \mathbf{1}) + 1 - \mathbf{3}$$

$$y = 3(x - 1)^2 - 2$$

In either case, the vertex is $(1, -2)$.

If we can find two points, one on each side of the vertex, we can sketch the graph. Let's let $x = 0$ and $x = 2$ since each of these numbers is the same distance from $x = 1$ and $x = 0$ will give us the y-intercept.

When $x = 0$

$$\begin{aligned} y &= 3(0)^2 - 6(0) + 1 \\ &= 0 - 0 + 1 \\ &= 1 \end{aligned}$$

When $x = 2$

$$\begin{aligned} y &= 3(2)^2 - 6(2) + 1 \\ &= 12 - 12 + 1 \\ &= 1 \end{aligned}$$

The two points just found are $(0, 1)$ and $(2, 1)$. Plotting these two points along with the vertex $(1, -2)$, we have the graph shown in Figure 4.

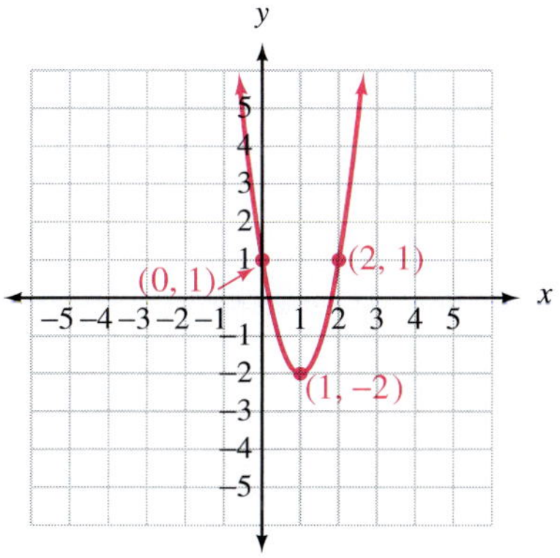

FIGURE 4

EXAMPLE 4

Graph $y = -2x^2 + 6x - 5$.

Solution Letting $y = 0$, we have

$$0 = -2x^2 + 6x - 5$$

Again, the right side of this equation does not factor. The discriminant is $b^2 - 4ac = 36 - 4(-2)(-5) = -4$, which indicates that the solutions are complex numbers. This means that our original equation does not have x-intercepts. The graph does not cross the x-axis.

Let's find the vertex.

4. Graph $y = -x^2 + 4x - 5$.

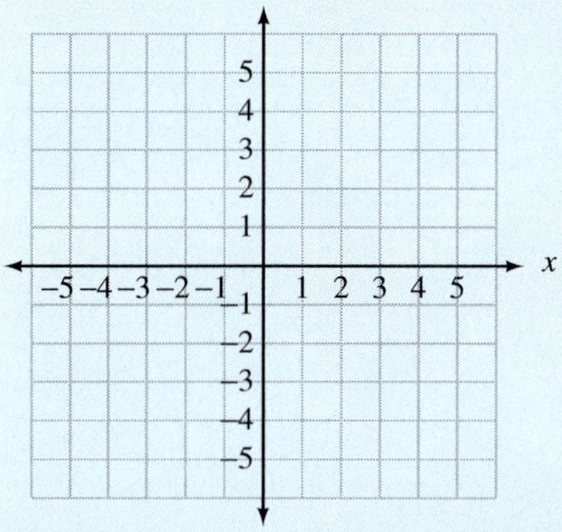

Using our formula for the x-coordinate of the vertex, we have

$$x = \frac{-b}{2a} = \frac{-6}{2(-2)} = \frac{6}{4} = \frac{3}{2}$$

To find the y-coordinate, we let $x = \frac{3}{2}$:

$$\begin{aligned} y &= -2\left(\frac{3}{2}\right)^2 + 6\left(\frac{3}{2}\right) - 5 \\ &= \frac{-18}{4} + \frac{18}{2} - 5 \\ &= \frac{-18 + 36 - 20}{4} \\ &= -\frac{1}{2} \end{aligned}$$

Finding the vertex by completing the square is a more complicated matter. To make the coefficient of x^2 a 1, we must factor -2 from the first two terms. To complete the square inside the parentheses, we add $\frac{9}{4}$. Since each term inside the parentheses is multiplied by -2, we add $\frac{9}{2}$ outside the parentheses so that the net result is the same as adding 0 to the right side:

$$y = -2(x^2 - 3x \qquad) - 5$$

$$y = -2\left(x^2 - 3x + \mathbf{\frac{9}{4}}\right) - 5 + \mathbf{\frac{9}{2}}$$

$$y = -2\left(x - \frac{3}{2}\right)^2 - \frac{1}{2}$$

The vertex is $(\frac{3}{2}, -\frac{1}{2})$. Since this is the only point we have so far, we must find two others, Let's let $x = 3$ and $x = 0$ since each point is the same distance from $x = \frac{3}{2}$ and on either side:

When $x = 3$

$$\begin{aligned} y &= -2(3)^2 + 6(3) - 5 \\ &= -18 + 18 - 5 \\ &= -5 \end{aligned}$$

When $x = 0$

$$\begin{aligned} y &= -2(0)^2 + 6(0) - 5 \\ &= 0 + 0 - 5 \\ &= -5 \end{aligned}$$

The two additional points on the graph are $(3, -5)$ and $(0, -5)$. Figure 5 shows the graph.

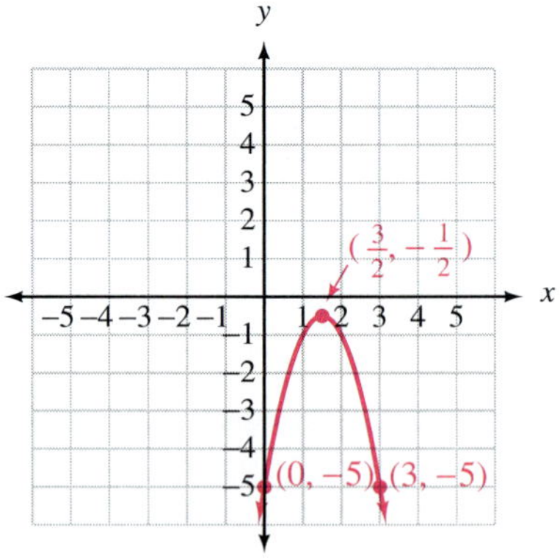

FIGURE 5

The graph is concave down. The vertex is the highest point on the graph. ■

By looking at the equations and graphs in Examples 1 through 4, we can conclude that the graph of $y = ax^2 + bx + c$ will be concave up when a is positive and concave down when a is negative. Taking this even further, if $a > 0$, then the vertex is the lowest point on the graph, and if $a < 0$, the vertex is the highest point on the graph. Finally, if we complete the square on x in the equation $y = ax^2 + bx + c$, $a \neq 0$, we can rewrite the equation of our parabola as $y = a(x - h)^2 + k$. When the equation is in this form, the vertex is at the point (h, k). Here is a summary:

Graphing Parabolas II

The graph of

$$y = a(x - h)^2 + k, a \neq 0$$

will be a parabola with a vertex at (h, k). The vertex will be the highest point on the graph when $a < 0$, and it will be the lowest point on the graph when $a > 0$.

EXAMPLE 5

A company selling copies of an accounting program for home computers finds that it will make a weekly profit of P dollars from selling x copies of the program, according to the equation

$$P(x) = -0.1x^2 + 27x - 500$$

How many copies of the program should it sell to make the largest possible profit, and what is the largest possible profit?

Solution Since the coefficient of x^2 is negative, we know the graph of this parabola will be concave down, meaning that the vertex is the highest point of the curve. We find the vertex by first finding its x-coordinate:

$$x = \frac{-b}{2a} = \frac{-27}{2(-0.1)} = \frac{27}{0.2} = 135$$

This represents the number of programs the company needs to sell each week to make a maximum profit. To find the maximum profit, we substitute 135 for x in the original equation. (A calculator is helpful for these kinds of calculations.)

$$\begin{aligned} P(135) &= -0.1(135)^2 + 27(135) - 500 \\ &= -0.1(18{,}225) + 3{,}645 - 500 \\ &= -1{,}822.5 + 3{,}645 - 500 \\ &= 1{,}322.5 \end{aligned}$$

The maximum weekly profit is \$1,322.50 and is obtained by selling 135 programs a week.

5. Find the largest value of y if
$y = -0.01x^2 + 12x - 400$

EXAMPLE 6

An art supply store finds that they can sell x sketch pads each week at p dollars each, according to the equation $x = 900 - 300p$. Graph the revenue equation $R = xp$. Then use the graph to find the price p that will bring in the maximum revenue. Finally, find the maximum revenue.

Solution As it stands, the revenue equation contains three variables. Since we are asked to find the value of p that gives us the maximum value of R, we rewrite the equation using just the variables R and p. Since $x = 900 - 300p$, we have

$$R = xp = (900 - 300p)p$$

The graph of this equation is shown in Figure 6. The graph appears in the first quadrant only since R and p are both positive quantities.

6. Repeat Example 6 if the number of sketch pads they can sell each week is related to the price by the equation
$x = 800 - 200p$

Answer

5. $y = 3{,}200$

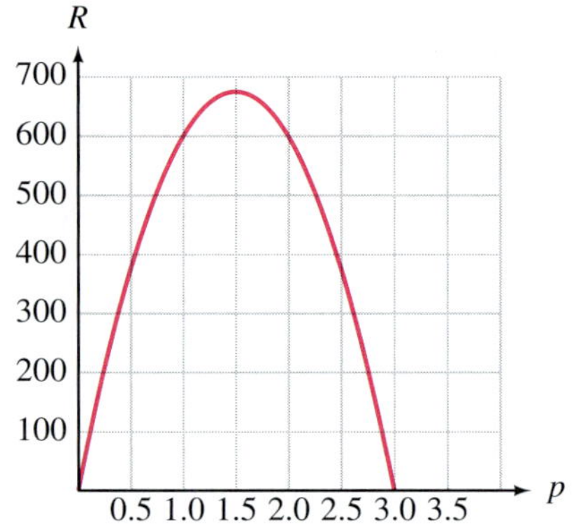

FIGURE 6

From the graph we see that the maximum value of R occurs when $p = \$1.50$. We can calculate the maximum value of R from the equation:

When $p = 1.5$

The equation $R = (900 - 300p)p$

becomes

$$R = (900 - 300 \cdot 1.5)1.5$$
$$= (900 - 450)1.5$$
$$= 450 \cdot 1.5$$
$$= 675$$

The maximum revenue is \$675. It is obtained by setting the price of each sketch pad at $p = \$1.50$.

USING TECHNOLOGY

Graphing Calculators

If you have been using a graphing calculator for some of the material in this course, you are well aware that your calculator can draw all the graphs in this section very easily. It is important, however, that you be able to recognize and sketch the graph of any parabola by hand. It is a skill that all successful intermediate algebra students should possess, even if they are proficient in the use of a graphing calculator. My suggestion is that you work the problems in this section and problem set without your calculator. Then use your calculator to check your results.

Finding the Equation from the Graph

EXAMPLE 7 At the 1997 Washington County Fair in Oregon, David Smith, Jr., The Bullet, was shot from a cannon. As a human cannonball, he reached a height of 70 feet before landing in a net 160 feet from the cannon. Sketch the graph of his path, and then find the equation of the graph.

Solution We assume that the path taken by the human cannonball is a parabola. If the origin of the coordinate system is at the opening of the cannon, then the net that catches him will be at 160 on the x-axis. Figure 7 shows the graph:

7. How would the equation in Example 7 look if the maximum height were 80 feet? (Assume the maximum distance is still 160 feet.)

Answer

6. See Solutions to Selected Practice Problems.

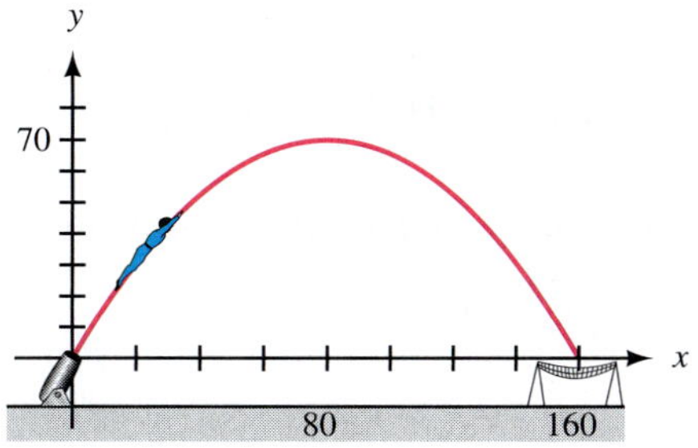

FIGURE 7

Since the curve is a parabola, we know the equation will have the form

$$y = a(x - h)^2 + k$$

Since the vertex of the parabola is at (80, 70), we can fill in two of the three constants in our equation, giving us

$$y = a(x - 80)^2 + 70$$

To find a, we note that the landing point will be (160, 0). Substituting the coordinates of this point into the equation, we solve for a:

$$0 = a(160 - 80)^2 + 70$$

$$0 = a(80)^2 + 70$$

$$0 = 6400a + 70$$

$$a = -\frac{70}{6400} = -\frac{7}{640}$$

The equation that describes the path of the human cannonball is

$$y = -\tfrac{7}{640}(x - 80)^2 + 70 \text{ for } 0 \le x \le 160$$

USING TECHNOLOGY

Graphing Calculators

Graph the equation found in Example 7 by graphing it on a graphing calculator using the window shown below. (We will use this graph later in the book to find the angle between the cannon and the horizontal.)

Window: X from 0 to 180, increment 20

Y from 0 to 80, increment 10

On the TI-83, an increment of 20 for X means Xscl=20.

Answer

7. $y = -\frac{1}{80}(x - 80)^2 + 80$

Linking Objectives and Examples

Next to each **objective**, list the examples that are best described by that objective.

A	
B	
C	

Getting Ready for Class

After reading through the preceding section, respond in your own words and in complete sentences.

1. What is a parabola?
2. What part of the equation of a parabola determines whether the graph is concave up or concave down?
3. Suppose $f(x) = ax^2 + bx + c$ is the equation of a parabola. Explain how $f(4) = 1$ relates to the graph of the parabola.
4. A line can be graphed with two points. How many points are necessary to get a reasonable sketch of a parabola? Explain.

Problem Set 8.5

INTERMEDIATE Algebra $f(x)$ Now™

For each of the following equations, give the x-intercepts and the coordinates of the vertex, and sketch the graph.

1. $y = x^2 + 2x - 3$

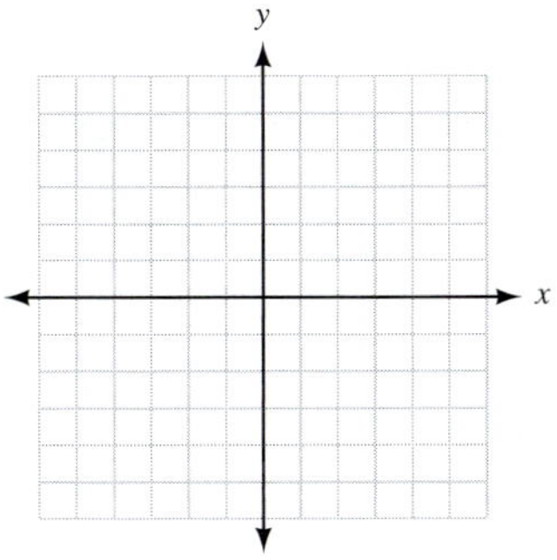

2. $y = x^2 - 2x - 3$

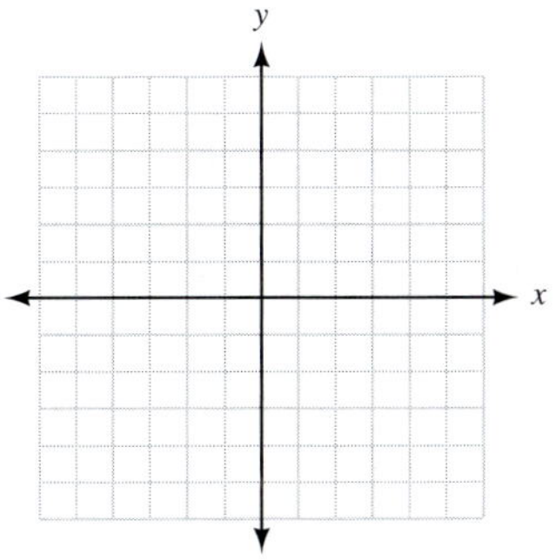

3. $y = -x^2 - 4x + 5$

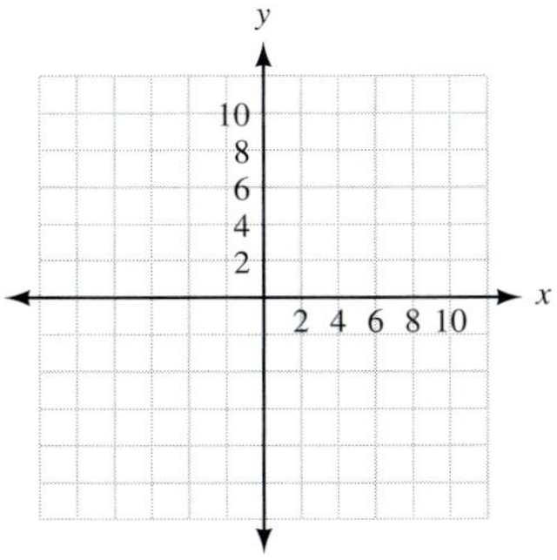

4. $y = x^2 + 4x - 5$

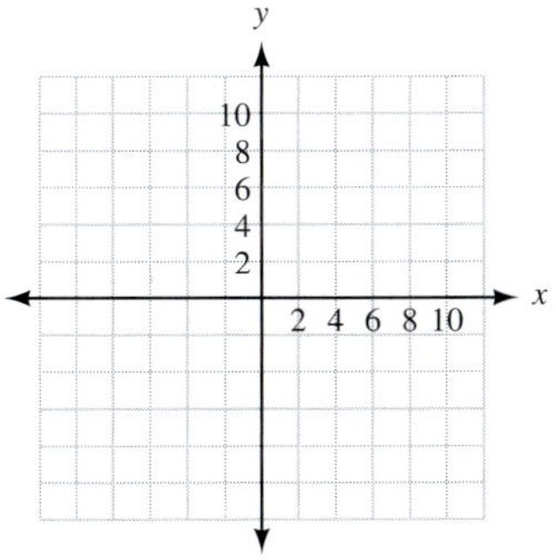

5. $y = x^2 - 1$

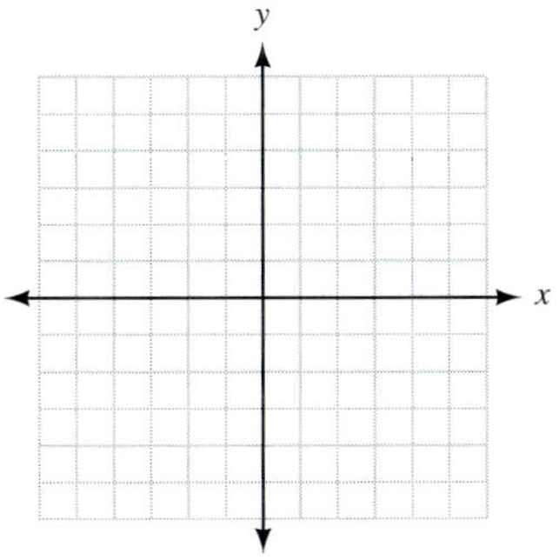

6. $y = x^2 - 4$

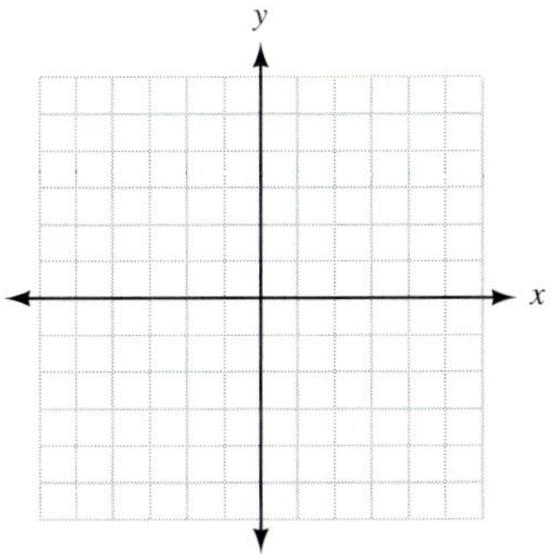

7. $y = -x^2 + 9$

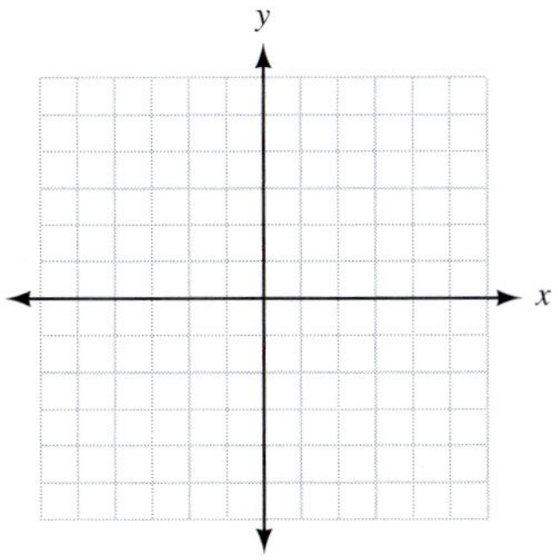

8. $y = -x^2 + 1$

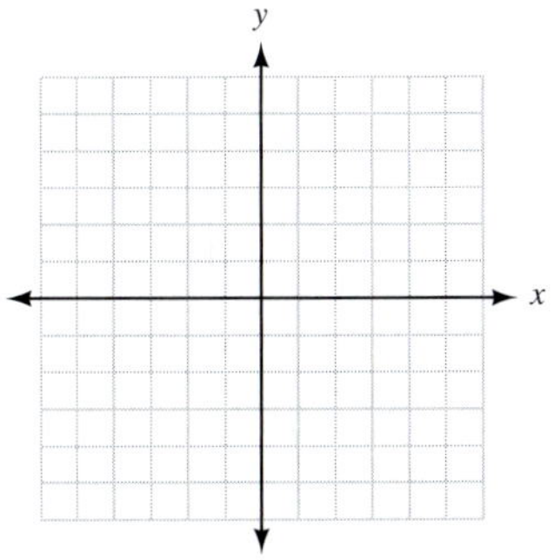

9. $y = 2x^2 - 4x - 6$

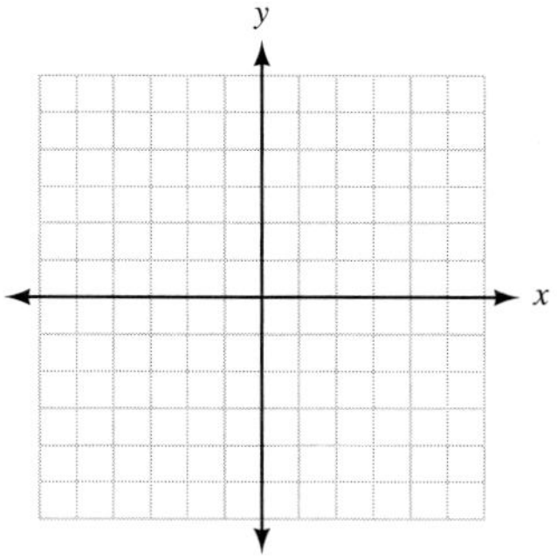

10. $y = 2x^2 + 4x - 6$

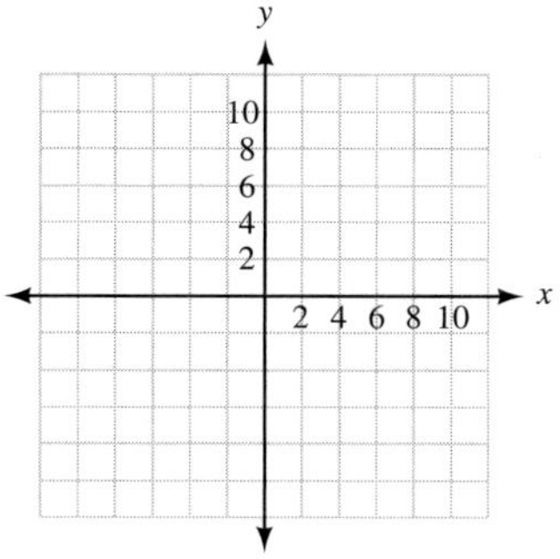

11. $y = x^2 - 2x - 4$

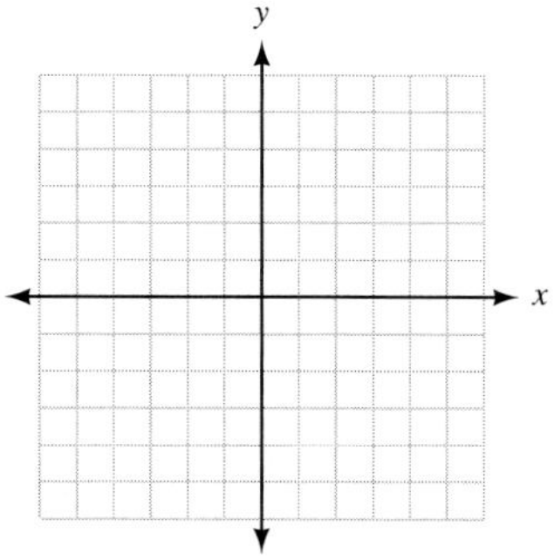

12. $y = x^2 - 2x - 2$

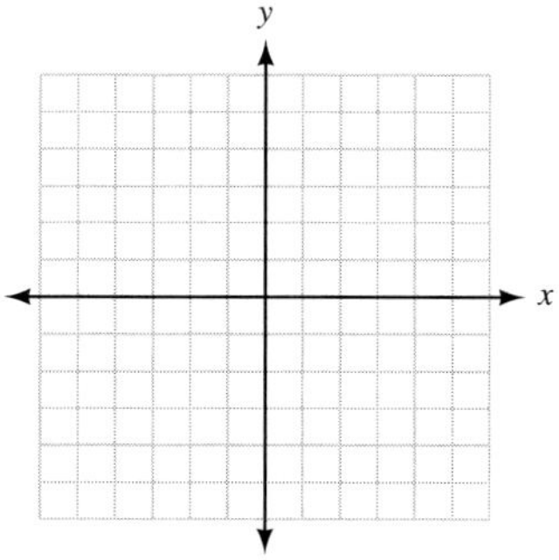

Find the vertex and any two convenient points to sketch the graphs of the following.

13. $y = x^2 - 4x - 4$

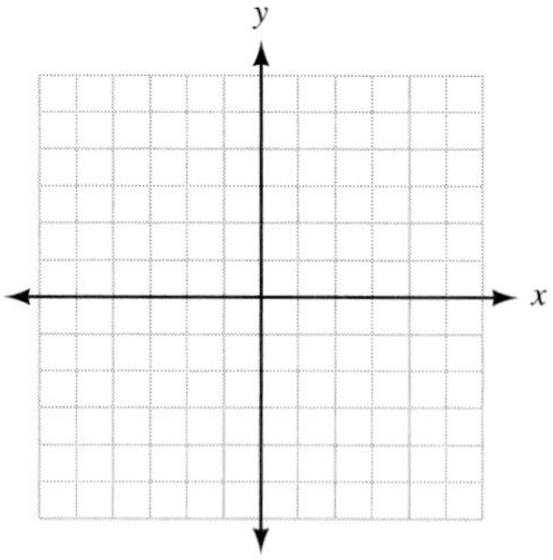

14. $y = x^2 - 2x + 3$

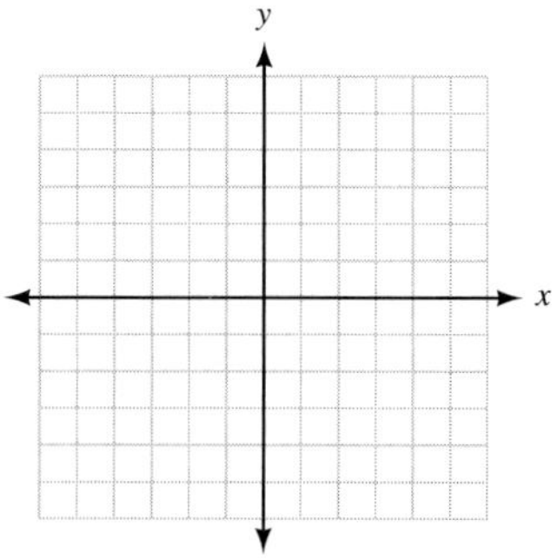

15. $y = -x^2 + 2x - 5$

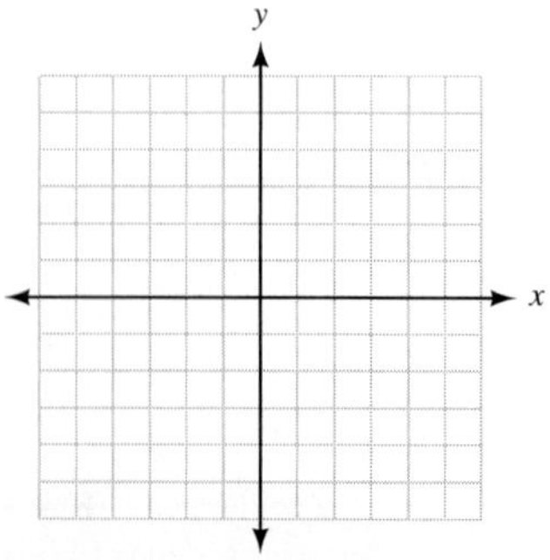

16. $y = -x^2 + 4x - 2$

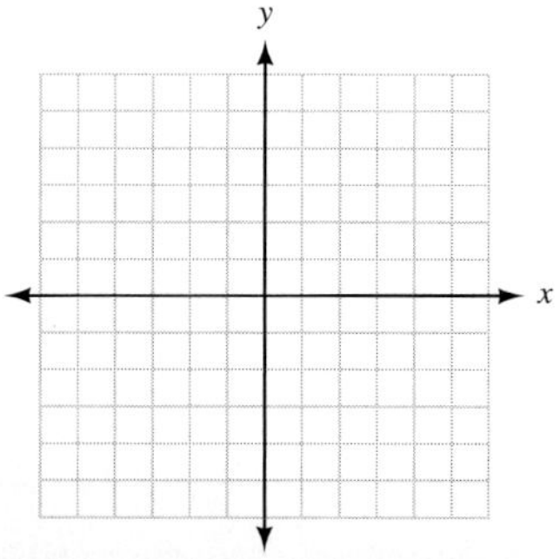

17. $y = x^2 + 1$

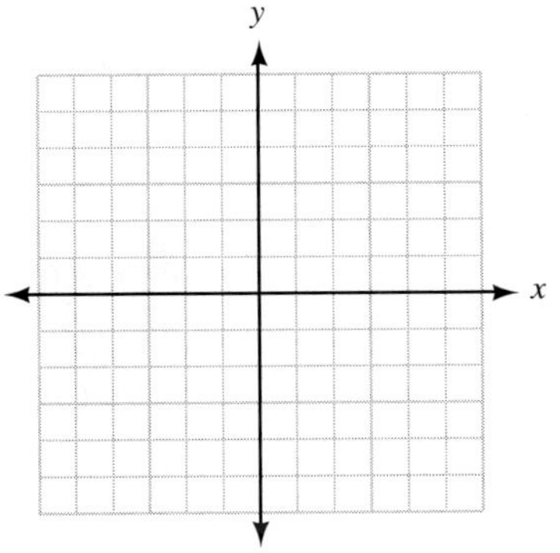

18. $y = x^2 + 4$

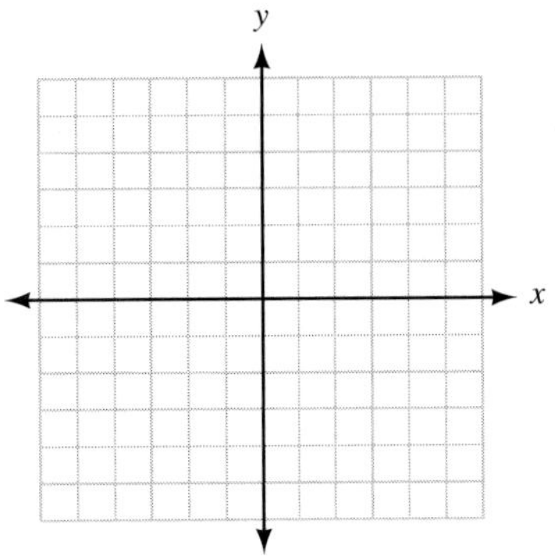

19. $y = -x^2 - 3$

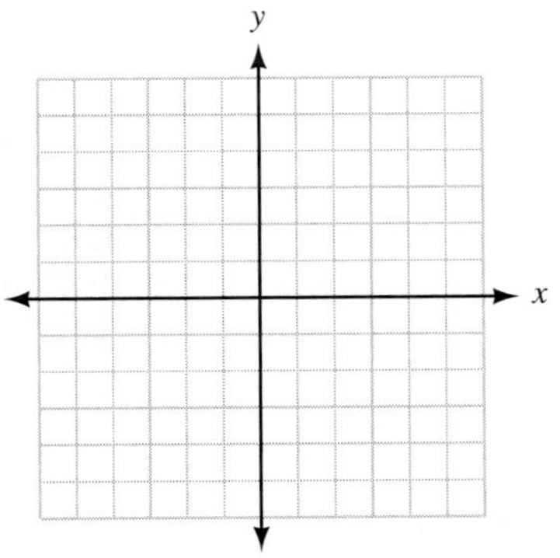

20. $y = -x^2 - 2$

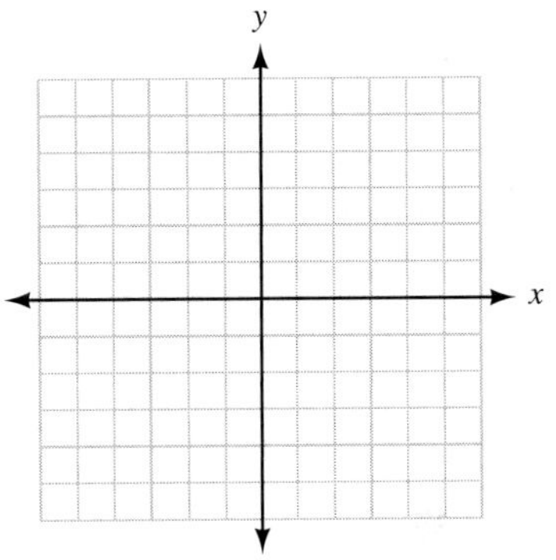

For each of the following equations, find the coordinates of the vertex, and indicate whether the vertex is the highest point on the graph or the lowest point on the graph. (Do not graph.)

21. $y = x^2 - 6x + 5$

22. $y = -x^2 + 6x - 5$

23. $y = -x^2 + 2x + 8$

24. $y = x^2 - 2x - 8$

25. $y = 12 + 4x - x^2$

26. $y = -12 - 4x + x^2$

27. $y = -x^2 - 8x$

28. $y = x^2 + 8x$

Applying the Concepts

29. Maximum Profit A company finds that it can make a profit of P dollars each month by selling x patterns, according to the formula $P(x) = -0.002x^2 + 3.5x - 800$. How many patterns must it sell each month to have a maximum profit? What is the maximum profit?

30. Maximum Profit A company selling picture frames finds that it can make a profit of P dollars each month by selling x frames, according to the formula $P(x) = -0.002x^2 + 5.5x - 1{,}200$. How many frames must it sell each month to have a maximum profit? What is the maximum profit?

Kathleen Olson

31. Maximum Height Chaudra is tossing a softball into the air with an underhand motion. The distance of the ball above her hand at any time is given by the function

$$h(t) = 32t - 16t^2 \qquad \text{for } 0 \le t \le 2$$

where $h(t)$ is the height of the ball (in feet) and t is the time (in seconds). Find the times at which the ball is in her hand and the maximum height of the ball.

32. Maximum Area Justin wants to fence 3 sides of a rectangular exercise yard for his dog. The fourth side of the exercise yard will be a side of the house. He has 80 feet of fencing available. Find the dimensions of the exercise yard that will enclose the maximum area.

33. Maximum Revenue A company that manufactures typewriter ribbons knows that the number of ribbons x it can sell each week is related to the price p of each ribbon by the equation $x = 1{,}200 - 100p$. Graph the revenue equation $R = xp$. Then use the graph to find the price p that will bring in the maximum revenue. Finally, find the maximum revenue.

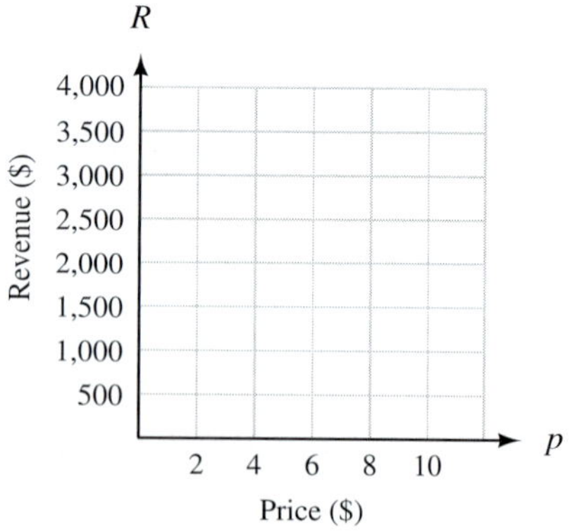

34. Maximum Revenue A company that manufactures diskettes for home computers finds that it can sell x diskettes each day at p dollars per diskette, according to the equation $x = 800 - 100p$. Graph the revenue equation $R = xp$. Then use the graph to find the price p that will bring in the maximum revenue. Finally, find the maximum revenue.

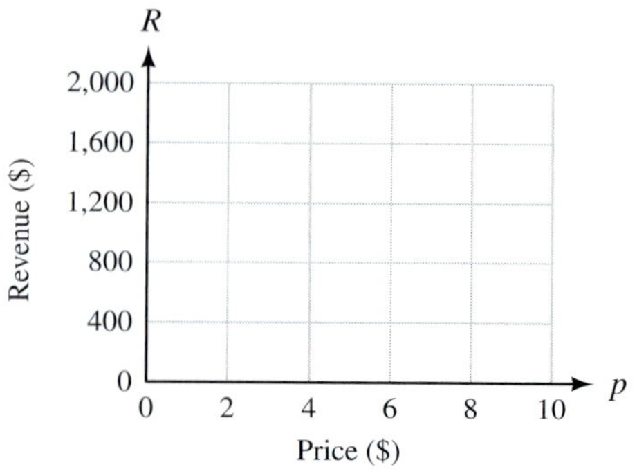

35. Maximum Revenue The relationship between the number of calculators x a company sells each day and the price p of each calculator is given by the equation $x = 1{,}700 - 100p$. Graph the revenue equation $R = xp$, and use the graph to find the price p that will bring in the maximum revenue. Then find the maximum revenue.

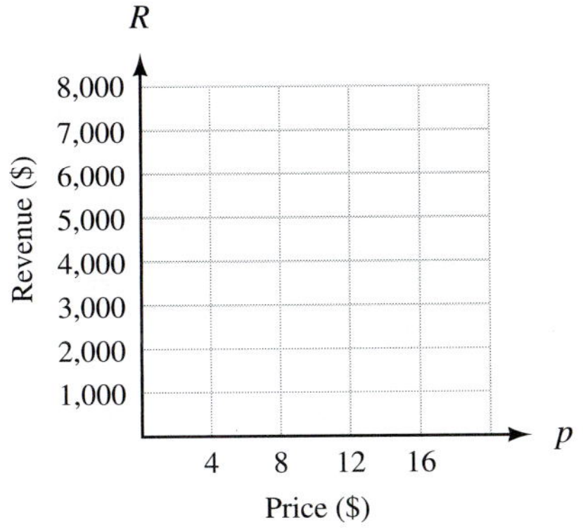

36. Maximum Revenue The relationship between the number x of pencil sharpeners a company sells each week and the price p of each sharpener is given by the equation $x = 1{,}800 - 100p$. Graph the revenue equation $R = xp$, and use the graph to find the price p that will bring in the maximum revenue. Then find the maximum revenue.

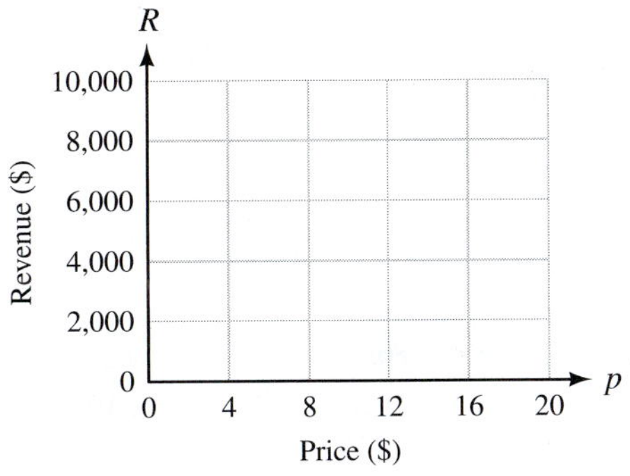

37. Human Cannonball A human cannonball is shot from a cannon at the county fair. He reaches a height of 60 feet before landing in a net 180 feet from the cannon. Sketch the graph of his path, and then find the equation of the graph.

38. Human Cannonball Suppose the equation gives the height $h(x)$ of the human cannonball after he has traveled x feet horizontally. What is $h(30)$, and what does it represent?

$$h(x) = -\frac{7}{640}(x - 80)^2 + 70$$

The Image Bank/Getty Images

Getting Ready for the Next Section

Solve.

39. $x^2 - 2x - 8 = 0$

40. $x^2 - x - 12 = 0$

41. $6x^2 - x = 2$

42. $3x^2 - 5x = 2$

43. $x^2 - 6x + 9 = 0$

44. $x^2 + 8x + 16 = 0$

Maintaining Your Skills

Perform the indicated operations.

45. $(3 - 5i) - (2 - 4i)$

46. $2i(5 - 6i)$

47. $(3 + 2i)(7 - 3i)$

48. $(4 + 5i)^2$

49. $\dfrac{i}{3 + i}$

50. $\dfrac{2 + 3i}{2 - 3i}$

Extending the Concepts

Finding the Equation from the Graph For each problem below, the graph is a parabola. In each case, find an equation in the form $y = a(x - h)^2 + k$ that describes the graph.

51.

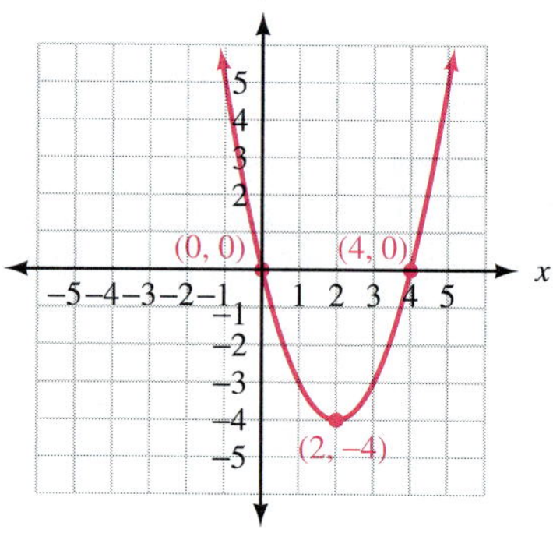

52.

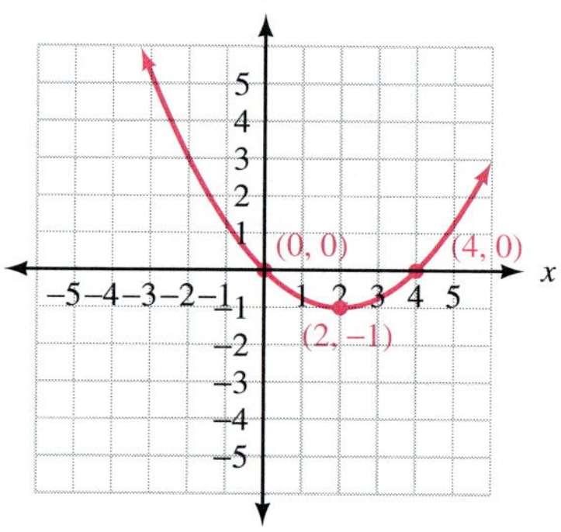

Improving Your Quantitative Literacy

53. Interpreting Graphs The graph below shows the different paths taken by the human cannonball when his velocity out of the cannon is 50 miles/hour and his cannon is inclined at varying angles.

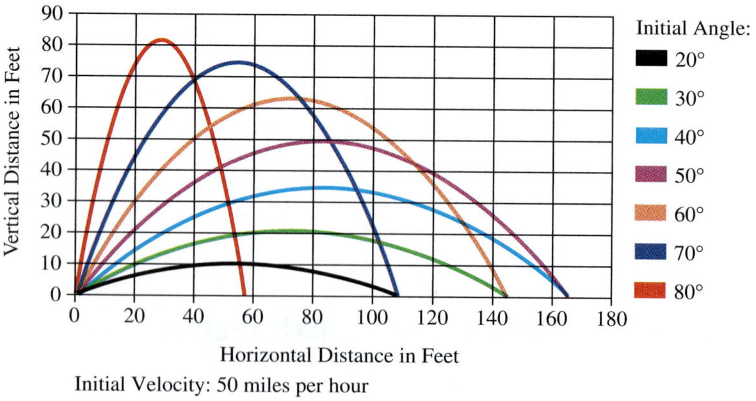

a. If his landing net is placed 108 feet from the cannon, at what angle should the cannon be inclined so that he lands in the net?

b. Approximately where do you think he would land if the cannon was inclined at 45°?

c. If the cannon was inclined at 45°, approximately what height do you think he would attain?

d. Do you think there is another angle for which he would travel the same distance he travels at 80°? Give an estimate of that angle.

e. The fact that every landing point can come from two different paths makes us think that the equations that give us the landing points must be what type of equations?

Quadratic Inequalities

8.6

OBJECTIVE

A	Solve quadratic inequalities and graph the solution set.

Quadratic inequalities in one variable are inequalities of the form

$$ax^2 + bx + c < 0 \qquad ax^2 + bx + c > 0$$

$$ax^2 + bx + c \le 0 \qquad ax^2 + bx + c \ge 0$$

where a, b, and c are constants, with $a \neq 0$. The technique we will use to solve inequalities of this type involves graphing. Suppose, for example, we wish to find the solution set for the inequality $x^2 - x - 6 > 0$. We begin by factoring the left side to obtain

$$(x - 3)(x + 2) > 0$$

We have two real numbers $x - 3$ and $x + 2$ whose product $(x - 3)(x + 2)$ is greater than zero; that is, their product is positive. The only way the product can be positive is either if both factors, $(x - 3)$ and $(x + 2)$, are positive or if they are both negative. To help visualize where $x - 3$ is positive and where it is negative, we draw a real number line and label it accordingly:

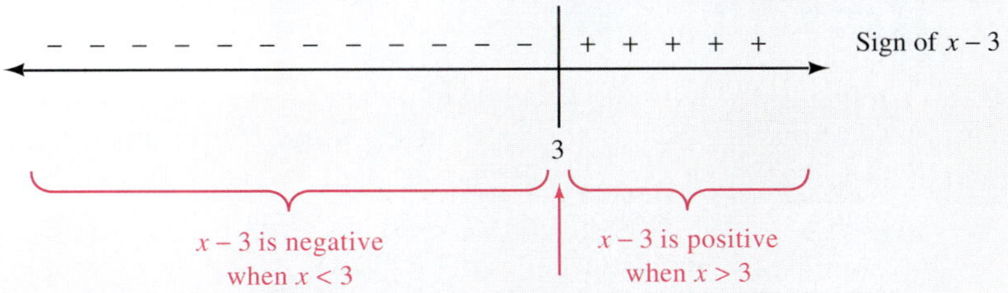

Here is a similar diagram showing where the factor $x + 2$ is positive and where it is negative:

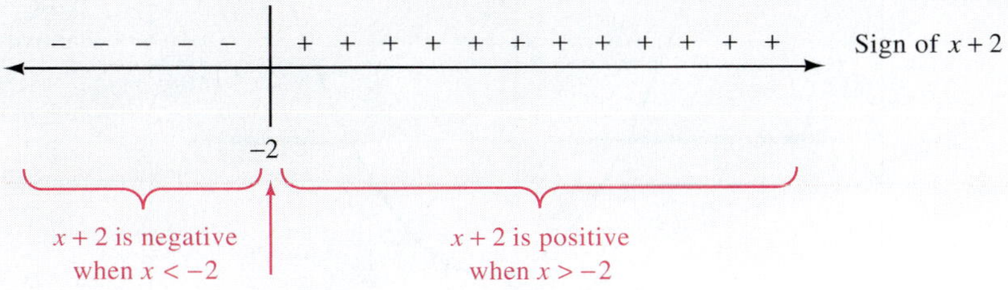

Drawing the two number lines together and eliminating the unnecessary numbers, we have

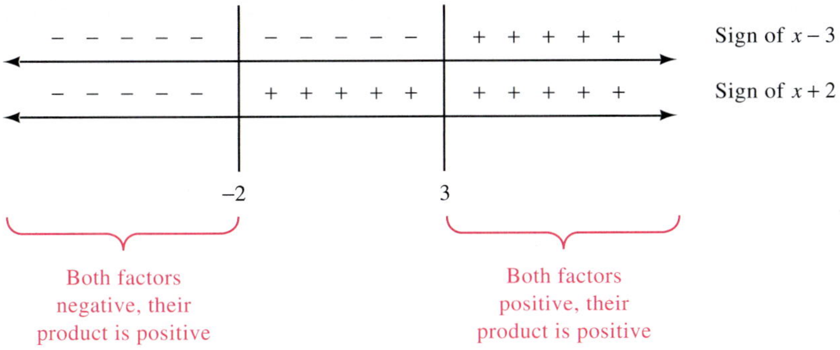

We can see from the preceding diagram that the graph of the solution to $x^2 - x - 6 > 0$ is

$$x < -2 \quad \text{or} \quad x > 3$$

USING TECHNOLOGY

Graphical Solutions to Quadratic Inequalities

We can solve the preceding problem by using a graphing calculator to visualize where the product $(x - 3)(x + 2)$ is positive. First, we graph the function $y = (x - 3)(x + 2)$ as shown in Figure 1

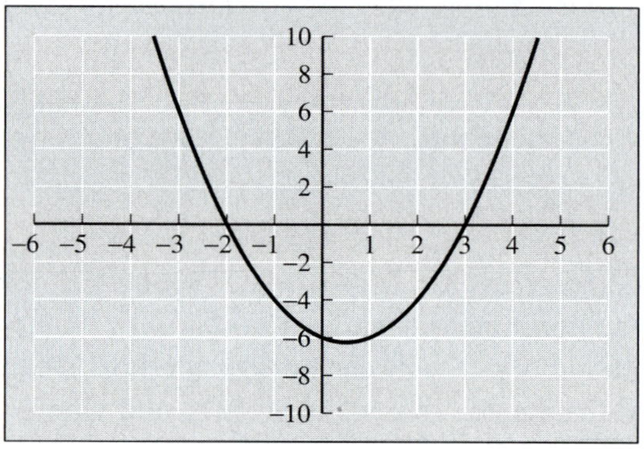

FIGURE 1

Next, we observe where the graph is above the x-axis. As you can see, the graph is above the x-axis to the right of 3 and to the left of −2, as shown in Figure 2.

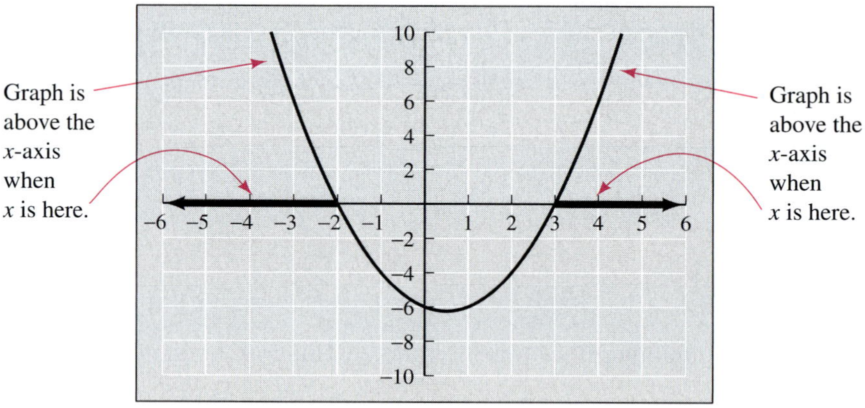

FIGURE 2

When the graph is above the x-axis, we have points whose y-coordinates are positive. Since these y-coordinates are the same as the expression $(x - 3)(x + 2)$, the values of x for which the graph of $y = (x - 3)(x + 2)$ is above the x-axis are the values of x for which the inequality $(x - 3)(x + 2) > 0$ is true. Our solution set is therefore

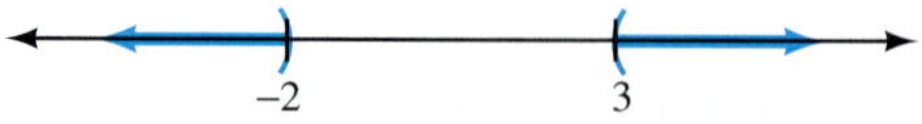

$x < -2$ or $x > 3$

EXAMPLE 1

Solve for x: $x^2 - 2x - 8 \le 0$.

Algebraic Solution We begin by factoring:

$$x^2 - 2x - 8 \le 0$$
$$(x - 4)(x + 2) \le 0$$

The product $(x - 4)(x + 2)$ is negative or zero. The factors must have opposite signs. We draw a diagram showing where each factor is positive and where each factor is negative:

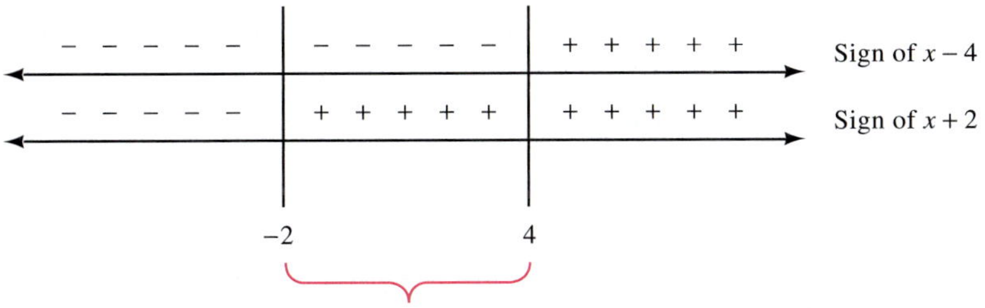

From the diagram we have the graph of the solution set:

$$-2 \le x \le 4$$

Graphic Solution To solve this inequality with a graphing calculator, we graph the function $y = (x - 4)(x + 2)$ and observe where the graph is below the x-axis. These points have negative y-coordinates, which means that the product $(x - 4)(x + 2)$ is negative for these points. Figure 3 shows the graph of $y = (x - 4)(x + 2)$, along with the region on the x-axis where the graph contains points with negative y-coordinates.

As you can see, the graph is below the x-axis when x is between -2 and 4. Since our original inequality includes the possibility that $(x - 4)(x + 2)$ is 0, we include the endpoints, -2 and 4, with our solution set.

$$-2 \le x \le 4$$

EXAMPLE 2

Solve for x: $6x^2 - x \ge 2$.

Algebraic Solution

$$6x^2 - x \ge 2$$
$$6x^2 - x - 2 \ge 0 \leftarrow \textbf{Standard form}$$
$$(3x - 2)(2x + 1) \ge 0$$

The product is positive, so the factors must agree in sign. Here is the diagram showing where that occurs:

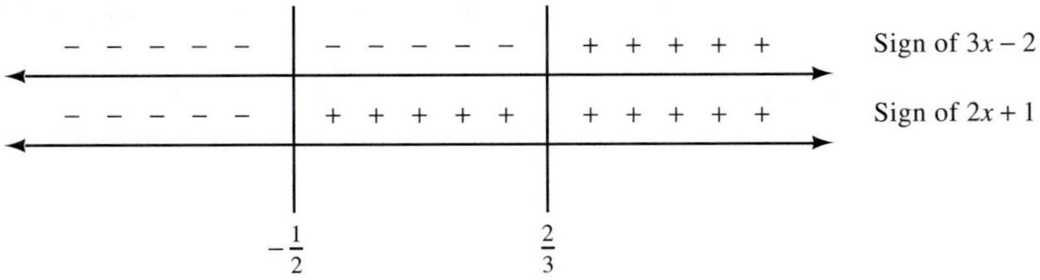

Since the factors agree in sign below $-\frac{1}{2}$ and above $\frac{2}{3}$, the graph of the solution set is

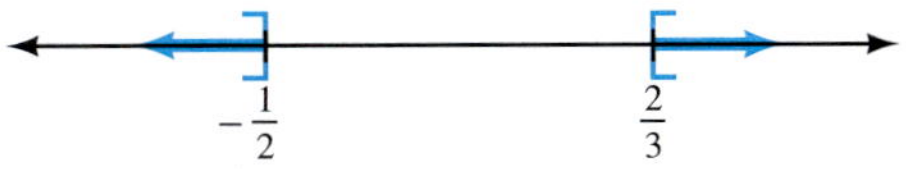

$$x \le -\frac{1}{2} \quad \text{or} \quad x \ge \frac{2}{3}$$

PRACTICE PROBLEMS

1. Solve for x:
$$x^2 + 2x - 15 \le 0$$

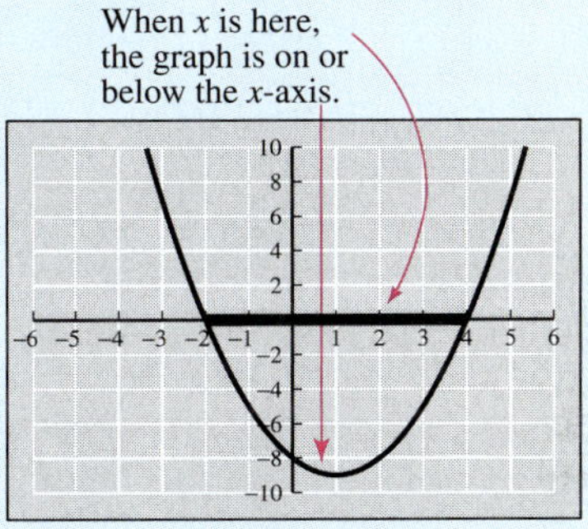

FIGURE 3

2. Solve for x:
$$2x^2 + 3x > 2$$

Answer

1. See Solutions to Selected Practice Problems.

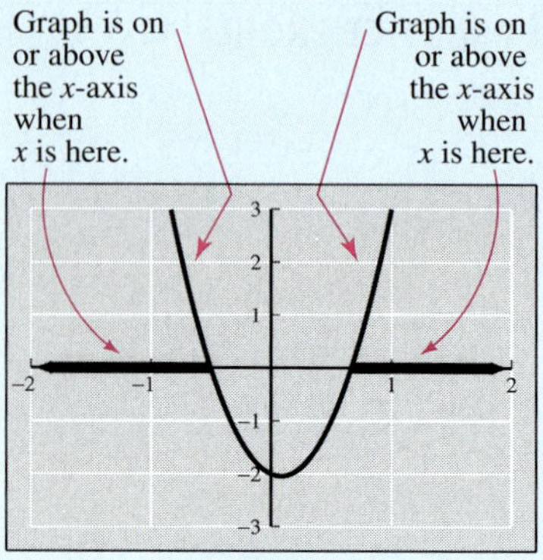

FIGURE 4

Graphical Solution To solve this inequality with a graphing calculator, we graph the function $y = (3x - 2)(2x + 1)$ and observe where the graph is above the x-axis. These are the points that have positive y-coordinates, which means that the product $(3x - 2)(2x + 1)$ is positive for these points. Figure 4 shows the graph of $y = (3x - 2)(2x + 1)$, along with the regions on the x-axis where the graph is on or above the x-axis.

To find the points where the graph crosses the x-axis, we need to use either the Trace and Zoom features to zoom in on each point, or the calculator function that finds the intercepts automatically (on the TI-82/83 this is the root/zero function under the CALC key). Whichever method we use, we will obtain the following result:

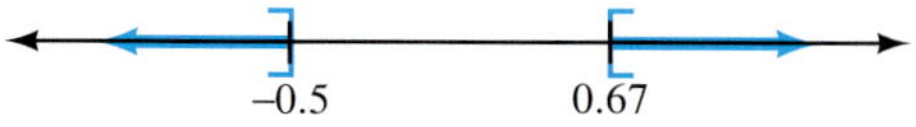

$$x \le -0.5 \quad \text{or} \quad x \ge 0.67$$

3. Solve $x^2 + 10x + 25 < 0$.

EXAMPLE 3

Solve $x^2 - 6x + 9 \ge 0$.

Algebraic Solution

$$x^2 - 6x + 9 \ge 0$$

$$(x - 3)^2 \ge 0$$

This is a special case in which both factors are the same. Since $(x - 3)^2$ is always positive or zero, the solution set is all real numbers; that is, any real number that is used in place of x in the original inequality will produce a true statement.

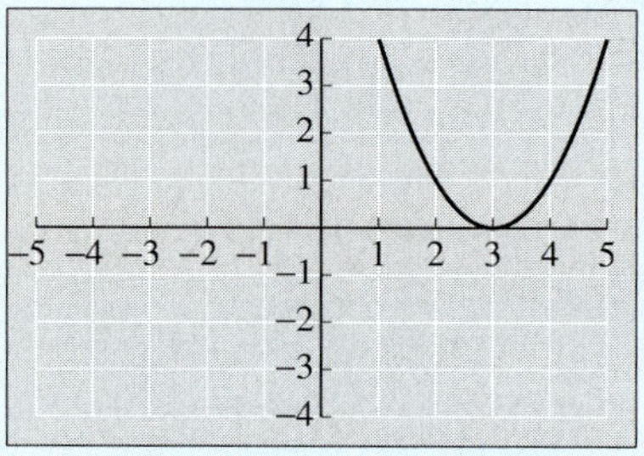

FIGURE 5

Graphical Solution The graph of $y = (x - 3)^2$ is shown in Figure 5. Notice that it touches the x-axis at 3 and is above the x-axis everywhere else. This means that every point on the graph has a y-coordinate greater than or equal to 0, no matter what the value of x. The conclusion that we draw from the graph is that the inequality $(x - 3)^2 \ge 0$ is true for all values of x.

Linking Objectives and Examples

Next to each **objective**, list the examples that are best described by that objective.

A	

Getting Ready for Class

After reading through the preceding section, respond in your own words and in complete sentences.

1. What is the first step in solving a quadratic inequality?
2. How do you show that the endpoint of a line segment is not part of the graph of a quadratic inequality?
3. How would you use the graph of $y = ax^2 + bx + c$ to help you find the graph of $ax^2 + bx + c < 0$?
4. Can a quadratic inequality have exactly one solution? Give an example.

Answers

2. See Solutions to Selected Practice Problems.
3. No solution

Problem Set 8.6 INTERMEDIATE Algebra f(x) Now™

Solve each of the following inequalities and graph the solution set.

1. $x^2 + x - 6 > 0$

2. $x^2 + x - 6 < 0$

3. $x^2 - x - 12 \leq 0$

4. $x^2 - x - 12 \geq 0$

5. $x^2 + 5x \geq -6$

6. $x^2 - 5x > 6$

7. $6x^2 < 5x - 1$

8. $4x^2 \geq -5x + 6$

9. $x^2 - 9 < 0$

10. $x^2 - 16 \geq 0$

11. $4x^2 - 9 \geq 0$

12. $9x^2 - 4 < 0$

13. $2x^2 - x - 3 < 0$

14. $3x^2 + x - 10 \geq 0$

15. $x^2 - 4x + 4 \geq 0$

16. $x^2 - 4x + 4 < 0$

17. $x^2 - 10x + 25 < 0$

18. $x^2 - 10x + 25 > 0$

19. $(x - 2)(x - 3)(x - 4) > 0$

20. $(x - 2)(x - 3)(x - 4) < 0$

21. $(x + 1)(x + 2)(x + 3) \leq 0$

22. $(x + 1)(x + 2)(x + 3) \geq 0$

23. The graph of $y = x^2 - 4$ is shown in Figure 6. Use the graph to write the solution set for each of the following:

a. $x^2 - 4 < 0$

b. $x^2 - 4 > 0$

c. $x^2 - 4 = 0$

24. The graph of $y = 4 - x^2$ is shown in Figure 7. Use the graph to write the solution set for each of the following:

a. $4 - x^2 < 0$

b. $4 - x^2 > 0$

c. $4 - x^2 = 0$

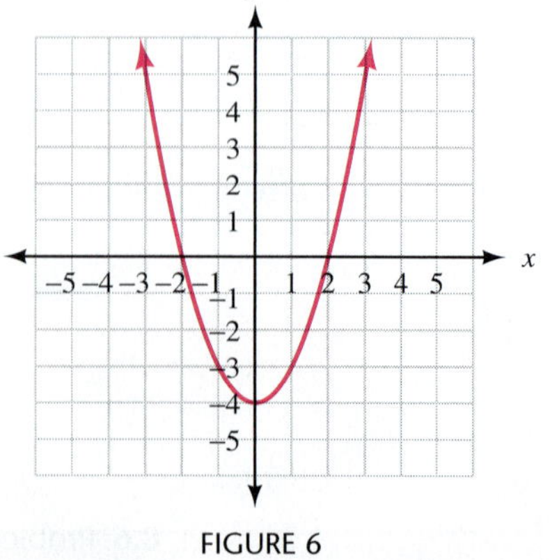

FIGURE 6

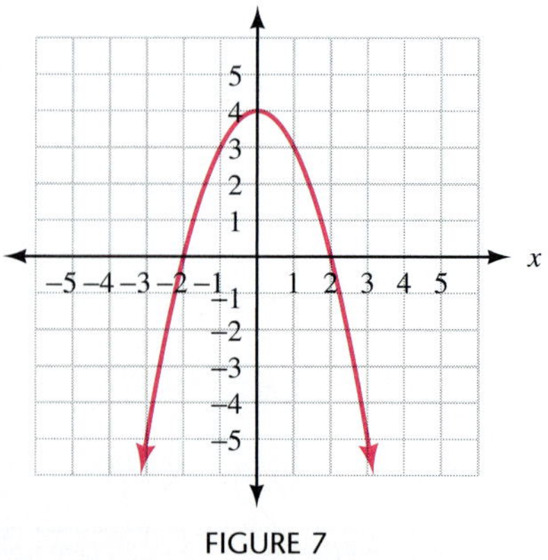

FIGURE 7

25. The graph of $y = x^2 - 3x - 10$ is shown in Figure 8. Use the graph to write the solution set for each of the following:

a. $x^2 - 3x - 10 < 0$

b. $x^2 - 3x - 10 > 0$

c. $x^2 - 3x - 10 = 0$

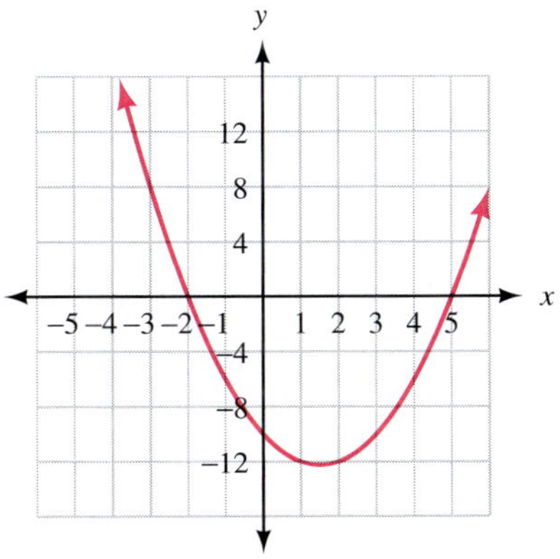

FIGURE 8

26. The graph of $y = x^2 + x - 12$ is shown in Figure 9. Use the graph to write the solution set for each of the following:

a. $x^2 + x - 12 < 0$

b. $x^2 - x - 12 > 0$

c. $x^2 + x - 12 = 0$

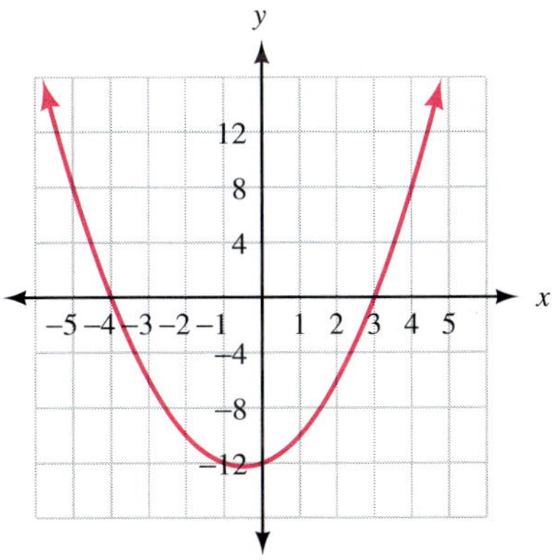

FIGURE 9

27. The graph of $y = x^3 - 3x^2 - x + 3$ is shown in Figure 10. Use the graph to write the solution set for each of the following:

a. $x^3 - 3x^2 - x + 3 < 0$

b. $x^3 - 3x^2 - x + 3 > 0$

c. $x^3 - 3x^2 - x + 3 = 0$

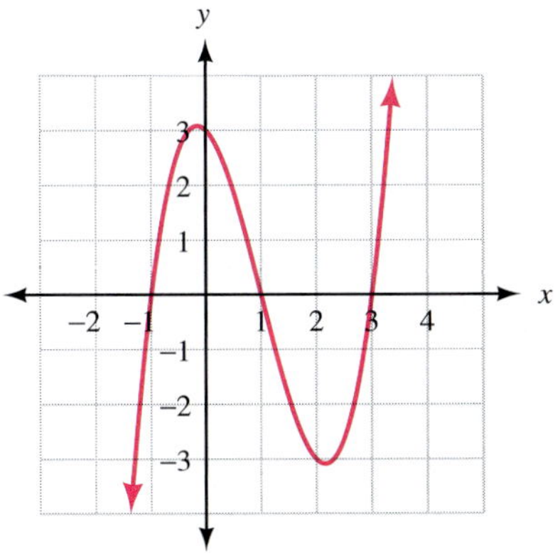

FIGURE 10

28. The graph of $y = x^3 + 4x^2 - 4x - 16$ is shown in Figure 11. Use the graph to write the solution set for each of the following:

a. $x^3 + 4x^2 - 4x - 16 < 0$

b. $x^3 + 4x^2 - 4x - 16 > 0$

c. $x^3 + 4x^2 - 4x - 16 = 0$

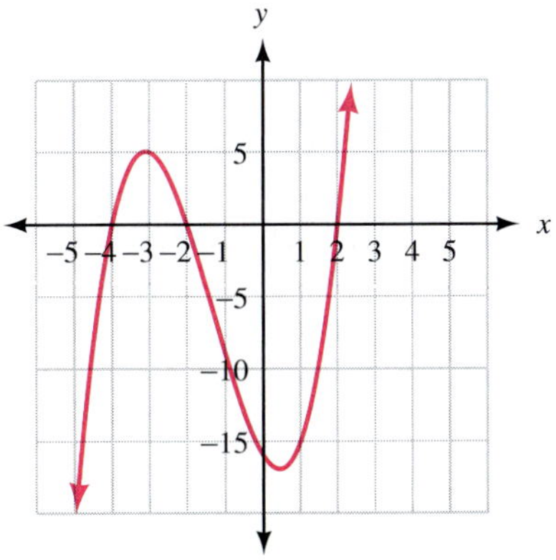

FIGURE 11

Applying the Concepts

29. Dimensions of a Rectangle The length of a rectangle is 3 inches more than twice the width. If the area is to be at least 44 square inches, what are the possibilities for the width?

30. Dimensions of a Rectangle The length of a rectangle is 5 inches less than 3 times the width. If the area is to be less than 12 square inches, what are the possibilities for the width?

31. Revenue A manufacturer of portable radios knows that the weekly revenue produced by selling x radios is given by the equation $R = 1{,}300p - 100p^2$, where p is the price of each radio. What price should she charge for each radio if she wants her weekly revenue to be at least \$4,000?

32. Revenue A manufacturer of small calculators knows that the weekly revenue produced by selling x calculators is given by the equation $R = 1{,}700p - 100p^2$, where p is the price of each calculator. What price should be charged for each calculator if the revenue is to be at least \$7,000 each week?

Maintaining Your Skills

Use a calculator to evaluate. Give answers to 4 decimal places.

33. $\dfrac{50{,}000}{32{,}000}$

34. $\dfrac{2.4362}{1.9758} - 1$

35. $\dfrac{1}{2}\left(\dfrac{4.5926}{1.3876} - 2\right)$

36. $1 + \dfrac{0.06}{12}$

Solve each equation.

37. $\sqrt{3t - 1} = 2$

38. $\sqrt{4t + 5} + 7 = 3$

39. $\sqrt{x + 3} = x - 3$

40. $\sqrt{x + 3} = \sqrt{x} - 3$

Graph each equation.

41. $y = \sqrt[3]{x} - 1$

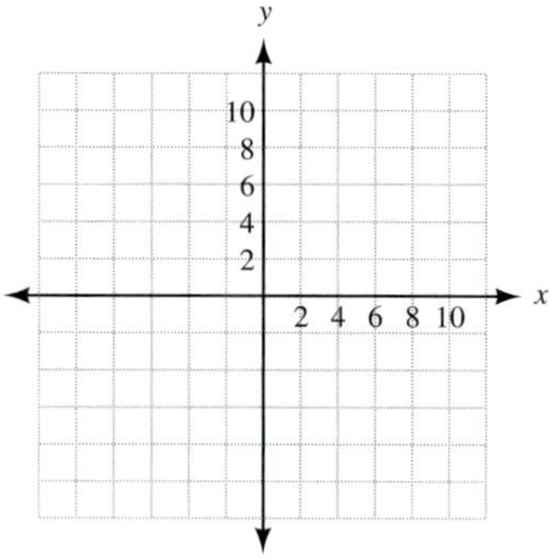

42. $y = \sqrt[3]{x - 1}$

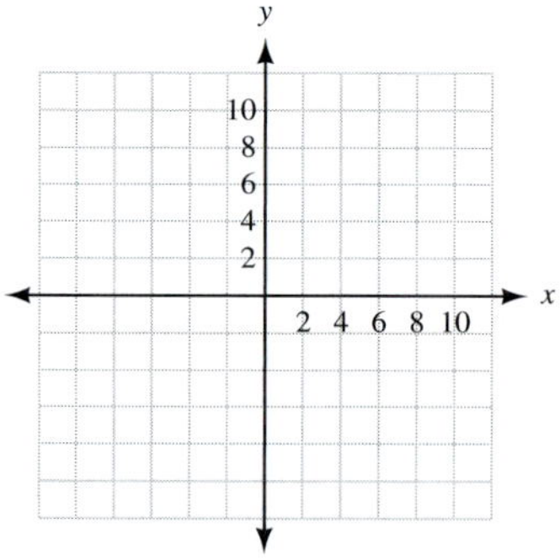

Extending the Concepts

Graph the solution set for each inequality.

43. $x^2 - 2x - 1 < 0$

44. $x^2 - 6x + 7 < 0$

45. $x^2 - 8x + 13 > 0$

46. $x^2 - 10x + 18 > 0$

Chapter 8 Summary

Theorem 1 [8.1]

If $a^2 = b$, where b is a real number, then

$$a = \sqrt{b} \quad \text{or} \quad a = -\sqrt{b}$$

which can be written as $a = \pm\sqrt{b}$.

EXAMPLES

1. If $(x - 3)^2 = 25$
   then $x - 3 = \pm 5$
   $x = 3 \pm 5$
   $x = 8$ or $x = -2$

To Solve a Qaudratic Equation by Completing the Square [8.1]

Step 1: Write the equation in the form $ax^2 + bx = c$.

Step 2: If $a \neq 1$, divide through by the constant a so the coefficient of x^2 is 1.

Step 3: Complete the square on the left side by adding the square of $\frac{1}{2}$ the coefficient of x to both sides.

Step 4: Write the left side of the equation as the square of a binomial. Simplify the right side if possible.

Step 5: Apply Theorem 1, and solve as usual.

2. Solve $x^2 - 6x - 6 = 0$

$$x^2 - 6x = 6$$
$$x^2 - 6x + \mathbf{9} = 6 + \mathbf{9}$$
$$(x - 3)^2 = 15$$
$$x - 3 = \pm\sqrt{15}$$
$$x = 3 \pm \sqrt{15}$$

The Quadratic Theorem [8.2]

For any quadratic equation in the form $ax^2 + bx + c = 0$, $a \neq 0$, the two solutions are

$$x = \frac{-b \pm \sqrt{b^2 - 4ac}}{2a}$$

This last expression is known as the *quadratic formula.*

3. If $2x^2 + 3x - 4 = 0$, then

$$x = \frac{-3 \pm \sqrt{9 - 4(2)(-4)}}{2(2)}$$
$$= \frac{-3 \pm \sqrt{41}}{4}$$

The Discriminant [8.3]

The expression $b^2 - 4ac$ that appears under the radical sign in the quadratic formula is known as the *discriminant.*

We can classify the solutions to $ax^2 + bx + c = 0$:

THE SOLUTIONS ARE	WHEN THE DISCRIMINANT IS
Two complex numbers containing *i*	Negative
One rational number	Zero
Two rational numbers	A positive perfect square
Two irrational numbers	A positive number, but not a perfect square

4. The discriminant for
   $x^2 + 6x + 9 = 0$
   is $D = 36 - 4(1)(9) = 0$, which means the equation has one rational solution.

Equations Quadratic in Form [8.4]

There are a variety of equations whose form is quadratic. We solve most of them by making a substitution so the equation becomes quadratic, and then solving the equation by factoring or the quadratic formula. For example,

The equation	*is quadratic in*
$(2x - 3)^2 + 5(2x - 3) - 6 = 0$	$2x - 3$
$4x^4 - 7x^2 - 2 = 0$	x^2
$2x - 7\sqrt{x} + 3 = 0$	$\sqrt{x}$

5. The equation $x^4 - x^2 - 12 = 0$ is quadratic in x^2. Letting $y = x^2$ we have

$$y^2 - y - 12 = 0$$
$$(y - 4)(y + 3) = 0$$
$$y = 4 \quad \text{or} \quad y = -3$$

Resubstituting x^2 for y, we have

$$x^2 = 4 \quad \text{or} \quad x^2 = -3$$
$$x = \pm 2 \quad \text{or} \quad x = \pm i\sqrt{3}$$

Graphing Parabolas [8.5]

The graph of any equation of the form

$$y = ax^2 + bx + c \qquad a \neq 0$$

is a *parabola.* The graph is *concave up* if $a > 0$ and *concave down* if $a < 0$. The highest or lowest point on the graph is called the *vertex* and always has an x-coordinate of $x = \frac{-b}{2a}$.

6. The graph of $y = x^2 - 4$ will be a parabola. It will cross the x-axis at 2 and -2, and the vertex will be $(0, -4)$.

Quadratic Inequalities [8.6]

We solve quadratic inequalities by manipulating the inequality to get 0 on the right side and then factoring the left side. We then make a diagram that indicates where the factors are positive and where they are negative. From this sign diagram and the original inequality we graph the appropriate solution set.

7. Solve $x^2 - 2x - 8 > 0$. We factor and draw the sign diagram:

$$(x - 4)(x + 2) > 0$$

− − − −	− − − −	+ + + +	$(x - 4)$
− − − −	+ + + +	+ + + +	$(x + 2)$
	-2	4	

The solution is $x < -2$ or $x > 4$.

Chapter 8 Review/Test

The problems below form a comprehensive review of the material in this chapter. They can be used to study for exams. If you would like to take a practice test on this chapter, you can use the odd-numbered problems. Give yourself an hour and work as many of the odd-numbered problems as possible. When you are finished, or when an hour has passed, check your answers with the answers in the back of the book. You can use the even-numbered problems for a second practice test.

Solve each equation. [8.1]

1. $(2t - 5)^2 = 25$

2. $(3t - 2)^2 = 4$

3. $(3y - 4)^2 = -49$

4. $(2x + 6)^2 = 12$

Solve by completing the square. [8.1]

5. $2x^2 + 6x - 20 = 0$

6. $3x^2 + 15x = -18$

7. $a^2 + 9 = 6a$

8. $a^2 + 4 = 4a$

9. $2y^2 + 6y = -3$

10. $3y^2 + 3 = 9y$

Solve each equation. [8.2]

11. $\frac{1}{6}x^2 + \frac{1}{2}x - \frac{5}{3} = 0$

12. $8x^2 - 18x = 0$

13. $4t^2 - 8t + 19 = 0$

14. $100x^2 - 200x = 100$

15. $0.06a^2 + 0.05a = 0.04$

16. $9 - 6x = -x^2$

17. $(2x + 1)(x - 5) - (x + 3)(x - 2) = -17$

18. $2y^3 + 2y = 10y^2$

19. $5x^2 = -2x + 3$

20. $x^3 - 27 = 0$

21. $3 - \dfrac{2}{x} + \dfrac{1}{x^2} = 0$

22. $\dfrac{1}{x - 3} + \dfrac{1}{x + 2} = 1$

23. Profit The total cost (in dollars) for a company to produce x items per week is $C = 7x + 400$. The revenue for selling all x items is $R = 34x - 0.1x^2$. How many items must it produce and sell each week for its weekly profit to be $1,300? [8.2]

24. Profit The total cost (in dollars) for a company to produce x items per week is $C = 70x + 300$. The revenue for selling all x items is $R = 110x - 0.5x^2$. How many items must it produce and sell each week for its weekly profit to be $300? [8.2]

Use the discriminant to find the number and kind of solutions for each equation. [8.3]

25. $2x^2 - 8x = -8$

26. $4x^2 - 8x = -4$

27. $2x^2 + x - 3 = 0$

28. $5x^2 + 11x = 12$

29. $x^2 - x = 1$

30. $x^2 - 5x = -5$

31. $3x^2 + 5x = -4$

32. $4x^2 - 3x = -6$

Determine k so that each equation has exactly one real solution. [8.3]

33. $25x^2 - kx + 4 = 0$

34. $4x^2 + kx + 25 = 0$

35. $kx^2 + 12x + 9 = 0$

36. $kx^2 - 16x + 16 = 0$

37. $9x^2 + 30x + k = 0$

38. $4x^2 + 28x + k = 0$

For each of the following problems, find an equation that has the given solutions. [8.3]

39. $x = 3, x = 5$

40. $x = -2, x = 4$

41. $y = \frac{1}{2}, y = -4$

42. $t = 3, t = -3, t = 5$

Find all solutions. [8.4]

43. $(x - 2)^2 - 4(x - 2) - 60 = 0$

44. $6(2y + 1)^2 - (2y + 1) - 2 = 0$

45. $x^4 - x^2 = 12$

46. $x - \sqrt{x} - 2 = 0$

47. $2x - 11\sqrt{x} = -12$

48. $\sqrt{x + 5} = \sqrt{x} + 1$

49. $\sqrt{y + 21} + \sqrt{y} = 7$

50. $\sqrt{y + 9} - \sqrt{y - 6} = 3$

51. Projectile Motion An object is tossed into the air with an upward velocity of 10 feet per second from the top of a building h feet high. The time it takes for the object to hit the ground below is given by the formula $16t^2 - 10t - h = 0$. Solve this formula for t. [8.4]

52. Projectile Motion An object is tossed into the air with an upward velocity of v feet per second from the top of a 10-foot wall. The time it takes for the object to hit the ground below is given by the formula $16t^2 - vt - 10 = 0$. Solve this formula for t. [8.4]

Solve each inequality and graph the solution set. [8.6]

53. $x^2 - x - 2 < 0$

54. $3x^2 - 14x + 8 \le 0$

55. $2x^2 + 5x - 12 \ge 0$

56. $(x + 2)(x - 3)(x + 4) > 0$

Find the x-intercepts, if they exist, and the vertex for each parabola. Then use them to sketch the graph. [8.5]

57. $y = x^2 - 6x + 8$

58. $y = x^2 - 4$

Chapter 8 Projects

Quadratic Functions

MAXIMUM VOLUME OF A BOX

Group Project

Number of People 5

Time Needed 30 minutes

Equipment Graphing calculator and five pieces of graph paper

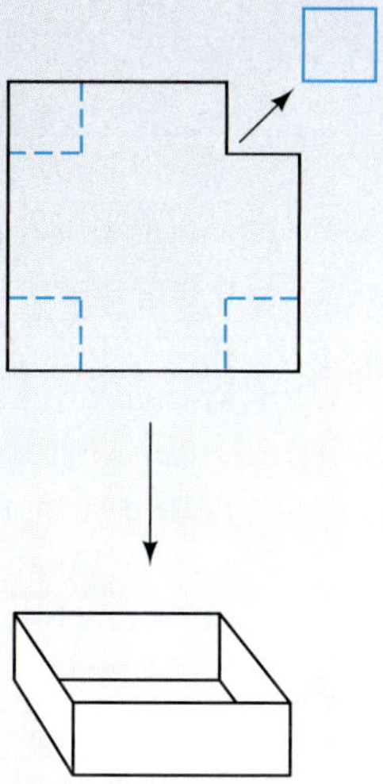

Background For many people, having a concrete model to work with allows them to visualize situations that they would have difficulty with if they had only a written description to work with. The purpose of this project is to rework a problem we have worked previously but this time with a concrete model.

Procedure You are going to make boxes of varying dimensions from rectangles that are 11 centimeters wide and 17 centimeters long.

1. Cut a rectangle from your graph paper that is 11 squares for the width and 17 squares for the length. Pretend that each small square is 1 centimeter by 1 centimeter. Do this with five pieces of paper.
2. One person tears off one square from each corner of their paper, then folds up the sides to form a box. Write down the length, width, and height of this box. Then calculate its volume.
3. The next person tears a square that is two units on a side from each corner of their paper, then folds up the sides to form a box. Write down the length, width, and height of this box. Then calculate its volume.
4. The next person follows the same procedure, tearing a still larger square from each corner of their paper. This continues until the squares that are to be torn off are larger than the original piece of paper.
5. Enter the data from each box you have created into the table above. Then graph all the points (x, V) from the table.
6. Using what you have learned from filling in the table, write a formula for the volume of the box that is created when a square of side x is cut from each corner of the original piece of paper.
7. Graph the equation you found in Problem 6 on a graphing calculator. Use the result to connect the points you plotted in the graph.
8. Use the graphing calculator to find the value of x that will give the maximum volume of the box. Your answer should be accurate to the nearest hundredth.

TABLE 1

VOLUME OF A BOX

SIDE OF SQUARE x(cm)	LENGTH OF BOX L	WIDTH OF BOX W	HEIGHT OF BOX H	VOLUME OF BOX V
1				
2				
3				
4				
5				

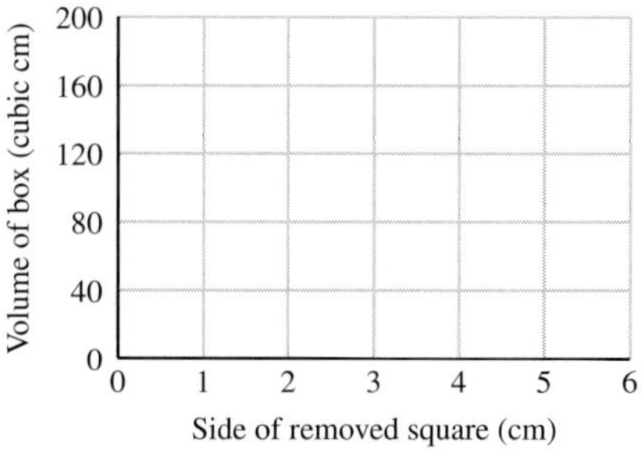

ARLIE O. PETTERS

Research Project

It can seem at times as if all the mathematicians of note lived 100 or more years ago. However, that is not the case. There are mathematicians doing research today, who are discovering new mathematical ideas and extending what is known about mathematics. One of the current group of research mathematicians is Arlie O. Petters. Dr. Petters earned his Ph.D. from the Massachusetts Institute of Technology in 1991 and currently is working in the mathematics department at Duke University. Use the Internet to find out more about Dr. Petters' education, awards, and research interests. Then use your results to give a profile, in essay form, of a present-day working mathematician.

Arlie O. Petters, Massachusetts Institute of Technology

APPENDIX A

Resources

By gathering resources early in the term, before you need help, the information about these resources will be available to you when they are needed.

Instructor

Knowing the contact information for your instructor is very important. You may already have this information from the course syllabus. It is a good idea to write it down again.

Name ______________________ Office Location ______________________
Available Hours: M ________ T ________ W ________ TH ________ F ________
Phone Number ____________ ext. ________ E-mail Address ______________ @ ________

Tutoring Center

Many schools offer tutoring, free of charge to their students. If this is the case at your school, find out when and where tutoring is offered.

Tutoring Location ______________________ Phone Number ____________ ext. ________
Available Hours: M ________ T ________ W ________ TH ________ F ________

Computer Lab

Many schools offer a computer lab where students can use the online resources and software available with their textbook. Other students using the same software and websites as you can be very helpful. Find out where the computer lab at your school is located.

Computer Lab Location ______________________ Phone Number ____________ ext. ________
Available Hours: M ________ T ________ W ________ TH ________ F ________

Videotapes

A complete set of video tapes is available to your school. These tapes feature the author of your textbook presenting full-length, 15- to 20-minute lessons from every section of your textbook. If you miss class, or find yourself behind, these tapes will prove very useful.

Videotape Location ______________________ Phone Number ____________ ext. ________
Available Hours: M ________ T ________ W ________ TH ________ F ________

Classmates

Form a study group and meet on a regular basis. When you meet try to speak to each other using proper mathematical language. That is, use the words that you see in the definition and property boxes in your textbook.

Name ______________________ Phone ____________ E-mail ______________
Name ______________________ Phone ____________ E-mail ______________
Name ______________________ Phone ____________ E-mail ______________

Synthetic Division

Synthetic division is a short form of long division with polynomials. We will consider synthetic division only for those cases in which the divisor is of the form $x + k$, where k is a constant.

Let's begin by looking over an example of long division with polynomials.

$$\begin{array}{r} 3x^2 - 2x + 4 \\ x + 3\overline{)3x^3 + 7x^2 - 2x - 4} \\ \underline{3x^3 + 9x^2 } \\ -2x^2 - 2x \\ \underline{-2x^2 - 6x} \\ 4x - 4 \\ \underline{4x + 12} \\ -16 \end{array}$$

We can rewrite the problem without showing the variable since the variable is written in descending powers and similar terms are in alignment. It looks like this:

$$\begin{array}{r} 3 \quad -2 \quad +4 \\ 1 + 3\overline{)3 \quad 7 \quad -2 \quad -4} \\ \underline{(3) \quad +9} \\ -2 \quad (-2) \\ \underline{(-2) \quad -6} \\ 4 \quad (-4) \\ \underline{(4) \quad 12} \\ -16 \end{array}$$

We have used parentheses to enclose the numbers that are repetitions of the numbers above them. We can compress the problem by eliminating all repetitions except the first one:

$$\begin{array}{r} 3 \quad -2 \quad 4 \\ 1 + 3\overline{)3 \quad 7 \quad -2 \quad -4} \\ \underline{9 \quad -6 \quad 12} \\ 3 \quad -2 \quad 4 \quad -16 \end{array}$$

The top line is the same as the first three terms of the bottom line, so we eliminate the top line. Also, the 1 that was the coefficient of x in the original problem can be eliminated since we will consider only division problems where the divisor is of the form $x + k$. The following is the most compact form of the original division problem:

$$\begin{array}{r} +3\overline{)3 \quad 7 \quad -2 \quad -4} \\ \underline{9 \quad -6 \quad 12} \\ 3 \quad -2 \quad 4 \quad -16 \end{array}$$

If we check over the problem, we find that the first term in the bottom row is exactly the same as the first term in the top row—and it always will be in problems of this type. Also, the last three terms in the bottom row come from multiplication by +3 and then subtraction. We can get an equivalent result by multiplying by −3 and adding. The problem would then look like this:

$$\begin{array}{r|rrrr} -3 & 3 & 7 & -2 & -4 \\ & \downarrow & -9 & 6 & -12 \\ \hline & 3 & -2 & 4 & \fbox{-16} \end{array}$$

We have used the brackets ⌋⌊ to separate the divisor and the remainder. This last expression is synthetic division. It is an easy process to remember. Simply change the sign of the constant term in the divisor, then bring down the first term of the dividend. The process is then just a series of multiplications and additions, as indicated in the following diagram by the arrows:

$$\begin{array}{r|rrrr} -3 & 3 & 7 & -2 & -4 \\ & \downarrow & -9 & 6 & -12 \\ \hline & 3 & -2 & 4 & \boxed{-16} \end{array}$$

The last term of the bottom row is always the remainder.

Here are some additional examples of synthetic division with polynomials.

PRACTICE PROBLEMS

1. Divide $x^3 + 2x^2 - 8x + 1$ by $x + 2$.

2. Divide $\dfrac{2x^3 - 5x^2 + 3}{x - 3}$.

3. Divide $\dfrac{x^3 + 8}{x + 2}$.

Answers

1. $x^2 - 8 + \dfrac{17}{x+2}$

2. $2x^2 + x + 3 + \dfrac{12}{x-3}$

3. $x^2 - 2x + 4$

EXAMPLE 1

Divide $x^4 - 2x^3 + 4x^2 - 6x + 2$ by $x - 2$.

Solution We change the sign of the constant term in the divisor to get +2 and then complete the procedure:

$$\begin{array}{r|rrrrr} +2 & 1 & -2 & 4 & -6 & 2 \\ & \downarrow & 2 & 0 & 8 & 4 \\ \hline & 1 & 0 & 4 & 2 & \boxed{6} \end{array}$$

From the last line we have the answer:

$$1x^3 + 0x^2 + 4x + 2 + \frac{6}{x - 2}$$

EXAMPLE 2

Divide $\dfrac{3x^3 - 4x + 5}{x + 4}$.

Solution Since we cannot skip any powers of the variable in the polynomial $3x^3 - 4x + 5$, we rewrite it as $3x^3 + 0x^2 - 4x + 5$ and proceed as we did in Example 1:

$$\begin{array}{r|rrrr} -4 & 3 & 0 & -4 & 5 \\ & \downarrow & -12 & 48 & -176 \\ \hline & 3 & -12 & 44 & \boxed{-171} \end{array}$$

From the synthetic division, we have

$$\frac{3x^3 - 4x + 5}{x + 4} = 3x^2 - 12x + 44 - \frac{171}{x + 4}$$

EXAMPLE 3

Divide $\dfrac{x^3 - 1}{x - 1}$.

Solution Writing the numerator as $x^3 + 0x^2 + 0x - 1$ and using synthetic division, we have

$$\begin{array}{r|rrrr} +1 & 1 & 0 & 0 & -1 \\ & \downarrow & 1 & 1 & 1 \\ \hline & 1 & 1 & 1 & \boxed{0} \end{array}$$

which indicates

$$\frac{x^3 - 1}{x - 1} = x^2 + x + 1$$

Problem Set B

INTERMEDIATE Algebra f(x) Now™

Use synthetic division to find the following quotients.

1. $\dfrac{x^2 - 5x + 6}{x + 2}$

2. $\dfrac{x^2 + 8x - 12}{x - 3}$

3. $\dfrac{3x^2 - 4x + 1}{x - 1}$

4. $\dfrac{4x^2 - 2x - 6}{x + 1}$

5. $\dfrac{x^3 + 2x^2 + 3x + 4}{x - 2}$

6. $\dfrac{x^3 - 2x^2 - 3x - 4}{x - 2}$

7. $\dfrac{3x^3 - x^2 + 2x + 5}{x - 3}$

8. $\dfrac{2x^3 - 5x^2 + x + 2}{x - 2}$

9. $\dfrac{2x^3 + x - 3}{x - 1}$

10. $\dfrac{3x^3 - 2x + 1}{x - 5}$

11. $\dfrac{x^4 + 2x^2 + 1}{x + 4}$

12. $\dfrac{x^4 - 3x^2 + 1}{x - 4}$

13. $\dfrac{x^5 - 2x^4 + x^3 - 3x^2 - x + 1}{x - 2}$

14. $\dfrac{2x^5 - 3x^4 + x^3 - x^2 + 2x + 1}{x + 2}$

15. $\dfrac{x^2 + x + 1}{x - 1}$

16. $\dfrac{x^2 + x + 1}{x + 1}$

17. $\dfrac{x^4 - 1}{x + 1}$

18. $\dfrac{x^4 + 1}{x - 1}$

19. $\dfrac{x^3 - 1}{x - 1}$

20. $\dfrac{x^3 - 1}{x + 1}$

APPENDIX C

Conditional Statements

Consider the two statements below:

Statement 1: If I study, then I will get good grades.
Statement 2: If x is a negative number, then $-x$ is a positive number.

In Statement 1, if we let A represent the phrase "I study" and B represent the phrase "I will get good grades," then Statement 1 has the form

If A, then B

Likewise, in Statement 2, if A is the phrase "x is a negative number" and B is "$-x$ is a positive number," then Statement 2 has the form

If A, then B

Each statement has the same form: If A, then B. We call this the "if/then" form, and any statement that has this form is called a *conditional statement.* For every conditional statement, the first phrase, A, is called the *hypothesis,* and the second phrase, B, is called the *conclusion.* All conditional statements can be written in the form

If *hypothesis,* then *conclusion*

Notation

A shorthand way to write an "if/then" statement is with the implies symbol

$$A \Rightarrow B$$

This statement is read "A implies B." It is equivalent to saying "If A, then B."

EXAMPLE 1

Identify the hypothesis and conclusion in each statement.

a. If a and b are positive numbers, then $-a(-b) = ab$.
b. If it is raining, then the streets are wet.
c. $C = 90° \Rightarrow c^2 = a^2 + b^2$.

Solution

a. *Hypothesis:* a and b are positive numbers
Conclusion: $-a(-b) = ab$
b. *Hypothesis:* It is raining
Conclusion: The streets are wet
c. *Hypothesis:* $C = 90°$
Conclusion: $c^2 = a^2 + b^2$

For each conditional statement $A \Rightarrow B$, we can find three related statements that may or may not be true depending on whether or not the original conditional statement is true.

DEFINITION

For every conditional statement $A \Rightarrow B$, there exist the following associated statements:

The converse:	$B \Rightarrow A$	If B, then A
The inverse:	not $A \Rightarrow$ not B	If not A, then not B
The contrapositive:	not $B \Rightarrow$ not A	If not B, then not

PRACTICE PROBLEMS

1. Identify the hypothesis and conclusion in each conditional statement.
   a. If a and b are negative numbers, then $a + b < 0$.
   b. If it is a bird, then it can fly.
   c. $a = -5 \Rightarrow a^2 = 25$.

Answers

1. **a.** Hypothesis: a and b are negative numbers
   Conclusion: $a + b < 0$
   b. Hypothesis: it is a bird
   Conclusion: it can fly
   c. Hypothesis: $a = -5$
   Conclusion: $a^2 = 25$

Visit our web site at http://www.brookscole.com

2. For the statement below, write the converse, the inverse, and the contrapositive.

   If it is a bird, then it can fly.

EXAMPLE 2 For the statement below, write the converse, the inverse, and the contrapositive.

If it is raining, then the streets are wet.

Solution It is sometimes easier to work a problem like this if we write out the phrases, A, B, not A, and not B.

Let A = it is raining; then not A = it is not raining.

Let B = the streets are wet; then not B = the streets are not wet.

Here are the three associated statements:

The converse: (If B, then A.) If the streets are wet, then it is raining.
The inverse: (If not A, then not B.) If it is not raining, then the streets are not wet.
The contrapositive: (If not B, then not A.) If the streets are not wet, then it is not raining.

True or False?

Next, we want to answer this question: "If a conditional statement is true, which, if any, of the associated statements are true also?" Consider the statement below.

If it is a square, then it has four sides.

We know from our experience with squares that this is a true statement. Now, is the converse necessarily true? Here is the converse:

If it has four sides, then it is a square.

Obviously the converse is *not* true because there are many four-sided figures that are not squares—rectangles, parallelograms, and trapezoids, to mention a few. So, the converse of a true conditional statement is not necessarily true itself.

Next, we consider the inverse of our original statement.

If it is not a square, then it does not have four sides.

Again, the inverse is not true since there are many nonsquare figures that do have four sides. For example, a 3-inch by 5-inch rectangle fits that description.

Finally, we consider the contrapositive:

If it does not have four sides, then it is not a square.

As you can see, the contrapositive is true; that is, if something doesn't have four sides, it can't possibly be a square.

The preceding discussion leads us to the following theorem.

Theorem

If a conditional statement is true, then so is its contrapositive; that is, the two statements

If A, then B and If not B, then not A

are equivalent. One can't be true without the other being true also; that is, they are either both true or both false.

The theorem doesn't mention the inverse and the converse because they are true or false independent of the original statement; that is, knowing that a

Answer

2. See Solutions to Selected Practice Problems.

conditional statement is true tells us that the contrapositive is also true—but the truth of the inverse and the converse does not follow from the truth of the original statement.

The next two examples are intended to clarify the preceding discussion and our theorem. As you read through them, be careful not to let your intuition, experience, or opinion get in the way.

EXAMPLE 3 If the statement "If you are guilty, then you will be convicted" is true, give another statement that also must be true.

Solution From our theorem and the discussion preceding it, we know that the contrapositive of a true conditional statement is also true. Here is the contrapositive of our original statement:

If you are not convicted, then you are not guilty

Remember, we are not asking for your opinion; we are simply asking for another conditional statement that must be true if the original statement is true. The answer is *always* the contrapositive. Now, you may be wondering about the converse:

If you are convicted, then you must be guilty

It may be that the converse is actually true. But if it is, it is not because of the original conditional statement; that is, the truth of the converse *does not follow* from the truth of the original statement. ■

3. If the statement "If you are asleep, then your eyes are closed" is true, give another statement that also must be true.

EXAMPLE 4 If the following statement is true, what other conditional statement also must be true?

If $a = b$, then $a^2 = b^2$

Solution Again, every true conditional statement has a true contrapositive. Therefore, the statement below is also true:

If $a^2 \neq b^2$, then $a \neq b$

In this case, we know from experience that the original statement is true; that is, if two numbers are equal, then so are their squares. We also know from experience that the contrapositive is true; if the squares of two numbers are not equal, then the numbers themselves can't be equal. Do you think the inverse and converse are true also? Here is the converse:

If $a^2 = b^2$, then $a = b$

The converse is not true. If a is -3 and b is 3, then a^2 and b^2 are equal, but a and b are not. This same kind of reasoning will show that the inverse is not necessarily true. This example then gives further evidence that our theorem is true: A true conditional statement has a true contrapositive. No conclusion can be drawn about the inverse or the converse. ■

4. If the statement below is true, what other conditional statement also must be true?

If $a = 3$, then $a^2 = 9$

Everyday Language

In everyday life, we don't always use the "if/then" form exactly as we have illustrated it here. Many times we use shortened, reversed, or otherwise altered forms of "if/then" statements. For instance, each of the following statements is a variation of the "if/then" form, and each carries the same meaning.

Answers

3–4. See Solutions to Selected Practice Problems.

If it is raining, then the streets are wet.

If it is raining, the streets are wet.

When it rains, the streets get wet.

The streets are wet if it is raining.

The streets are wet because it is raining.

Rain will make the streets wet.

5. Write the following statement in "if/then" form.

History repeats itself.

EXAMPLE 5 Write the following statement in "if/then" form.

Romeo loves Juliet

Solution We must be careful that we do not change the meaning of the statement when we write it in "if/then" form. Here is an "if/then" form that has the same meaning as the original statement:

If he is Romeo, then he loves Juliet

We can see that it would be incorrect to rewrite the original statement as

If she is Juliet, then she loves Romeo

because the original statement is Romeo loves Juliet, not Juliet loves Romeo. It would also be incorrect to rewrite our statement as either

If he is not Romeo, then he does not love Juliet

or

If he loves Juliet, then he is Romeo

because people other than Romeo may also love Juliet. (The preceding statements are actually the inverse and converse, respectively, of the original statement.) Finally, another statement that has the same meaning as our original statement is

If he does not love Juliet, then he is not Romeo

because this is the contrapositive of our original statement, and we know that the contrapositive is always true when the original statement is true. ■

Answer

5. See Solutions to Selected Practice Problems.

Problem Set C

INTERMEDIATE Algebra $f(x)$ Now™

For each conditional statement below, state the hypothesis and the conclusion.

1. If you argue for your limitations, then they are yours.

2. If you think you can, then you can.

3. If x is an even number, then x is divisible by 2.

4. If x is an odd number, then x is not divisible by 2.

5. If a triangle is equilateral, then all of its angles are equal.

6. If a triangle is isosceles, then two of its angles are equal.

7. If $x + 5 = -2$, then $x = -7$.

8. If $x - 5 = -2$, then $x = 3$

For each of the following conditional statements, give the converse, the inverse, and the contrapositive.

9. If $a = 8$, then $a^2 = 64$.

10. If $x = y$, then $x^2 = y^2$.

11. If $\frac{a}{b} = 1$, then $a = b$.

12. If $a + b = 0$, then $a = -b$.

13. If it is a square, then it is a rectangle.

14. If you live in a glass house, then you shouldn't throw stones.

15. If better is possible, then good is not enough.

16. If a and b are positive, then ab is positive.

For each statement below, write an equivalent statement in "if/then" form.

17. $E \Rightarrow F$

18. $a^3 = b^3 \Rightarrow a = b$

19. Misery loves company.

20. Rollerblading is not a crime.

21. The squeaky wheel gets the grease.

22. The girl who can't dance says the band can't play.

In each of the following problems a conditional statement is given. If the conditional statement is true, which of the three statements that follow it also must be true?

23. If you heard it first, then you heard it on Eyewitness News.

a. If you heard it on Eyewitness News, then you heard it first.

b. If you didn't hear it first, then you didn't hear it on Eyewitness News.

c. If you didn't hear it on Eyewitness News, then you didn't hear it first.

24. If it is raining, then the streets are wet.

a. If the streets are wet, then it is raining.

b. If the streets are not wet, then it is not raining.

c. If it is not raining, then the streets are not wet.

25. If $C = 90°$, then $c^2 = a^2 + b^2$.

a. If $c^2 \neq a^2 + b^2$, then $C \neq 90°$.

b. If $c^2 = a^2 + b^2$, then $C = 90°$.

c. If $C \neq 90°$, then $c^2 \neq a^2 + b^2$.

26. If you graduate from college, then you will get a good job.

a. If you get a good job, then you graduated from college.

b. If you do not get a good job, then you did not graduate from college.

c. If you do not graduate from college, then you will not get a good job.

27. If you get a B average, then your car insurance will cost less.
  a. If your car insurance costs less, then you got a B average.
  b. If you do not get a B average, then your car insurance will not cost less.
  c. If your car insurance does not cost less, then you did not get a B average.

28. If you go out without a sweater, then you will get sick.
  a. If you do not get sick, then you went out with a sweater.
  b. If you go out with a sweater, then you will not get sick.
  c. If you get sick, then you went out without a sweater.

29. If a and b are negative, then $a + b$ is negative.
  a. If a and b are not both negative, then $a + b$ is not negative.
  b. If $a + b$ is not negative, then a and b are not both negative.
  c. If $a + b$ is negative, then a and b are negative.

30. If it is a square, then all its sides have the same length.
  a. If all its sides are not the same length, then it is not a square.
  b. If all its sides have the same length, then it is a square.
  c. If it is not a square, then its sides are not all the same length

APPENDIX D

Algebra and Composition with Functions

A company produces and sells copies of an accounting program for home computers. The price they charge for the program is related to the number of copies sold by the demand function

$$p(x) = 35 - 0.1x$$

We find the revenue for this business by multiplying the number of items sold by the price per item. When we do so, we are forming a new function by combining two existing functions; that is, if $n(x) = x$ is the number of items sold and $p(x) = 35 - 0.1x$ is the price per item, then revenue is

$$R(x) = n(x) \cdot p(x) = x(35 - 0.1x) = 35x - 0.1x^2$$

In this case, the revenue function is the product of two functions. When we combine functions in this manner, we are applying our rules for algebra to functions.

To carry this situation further, we know the profit function is the difference between two functions. If the cost function for producing x copies of the accounting program is $C(x) = 8x + 500$, then the profit function is

$$P(x) = R(x) - C(x) = (35x - 0.1x^2) - (8x + 500) = -500 + 27x - 0.1x^2$$

The relationship between these last three functions is shown visually in Figure 1.

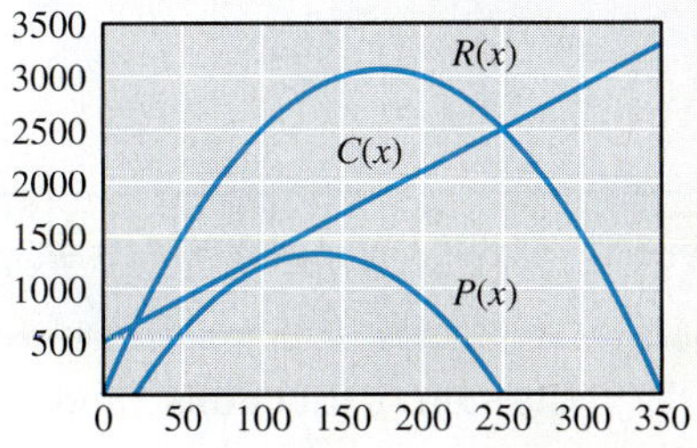

FIGURE 1

Again, when we combine functions in the manner shown, we are applying our rules for algebra to functions. To begin this section, we take a formal look at addition, subtraction, multiplication, and division with functions.

If we are given two functions f and g with a common domain, we can define four other functions as follows.

DEFINITION

$(f + g)(x) = f(x) + g(x)$ The function $f + g$ is the sum of the functions f and g.

$(f - g)(x) = f(x) - g(x)$ The function $f - g$ is the difference of the functions f and g.

$(fg)(x) = f(x)g(x)$ The function fg is the product of the functions f and g.

$\left(\frac{f}{g}\right)(x) = \frac{f(x)}{g(x)}$ The function f/g is the quotient of the functions f and g, where $g(x) \neq 0$.

PRACTICE PROBLEMS

1. If $f(x) = x^2 - 4$ and $g(x) = x + 2$, find formulas for
   a. $f + g$
   b. $f - g$
   c. fg
   d. $\frac{f}{g}$

2. Let $f(x) = 3x + 2$, $g(x) = 3x^2 - 10x - 8$, and $h(x) = x - 4$. Find
   a. $f + g$
   b. fh
   c. fg

Answers

1. **a.** $x^2 + x - 2$ **b.** $x^2 - x - 6$ **c.** $x^3 + 2x^2 - 4x - 8$ **d.** $x - 2$

EXAMPLE 1 If $f(x) = 4x^2 + 3x + 2$ and $g(x) = 2x^2 - 5x - 6$, write the formula for the functions $f + g$, $f - g$, fg, and f/g.

Solution The function $f + g$ is defined by

$$\begin{aligned}(f + g)(x) &= f(x) + g(x)\\ &= (4x^2 + 3x + 2) + (2x^2 - 5x - 6)\\ &= 6x^2 - 2x - 4\end{aligned}$$

The function $f - g$ is defined by

$$\begin{aligned}(f - g)(x) &= f(x) - g(x)\\ &= (4x^2 + 3x + 2) - (2x^2 - 5x - 6)\\ &= 4x^2 + 3x + 2 - 2x^2 + 5x + 6\\ &= 2x^2 + 8x + 8\end{aligned}$$

The function fg is defined by

$$\begin{aligned}(fg)(x) &= f(x)g(x)\\ &= (4x^2 + 3x + 2)(2x^2 - 5x - 6)\\ &= 8x^4 - 20x^3 - 24x^2 + 6x^3 - 15x^2 - 18x + 4x^2 - 10x - 12\\ &= 8x^4 - 14x^3 - 35x^2 - 28x - 12\end{aligned}$$

The function f/g is defined by

$$\begin{aligned}\left(\frac{f}{g}\right)(x) &= \frac{f(x)}{g(x)}\\ &= \frac{4x^2 + 3x + 2}{2x^2 - 5x - 6}\end{aligned}$$

■

EXAMPLE 2 Let $f(x) = 4x - 3$, $g(x) = 4x^2 - 7x + 3$, and $h(x) = x - 1$. Find $f + g$, fh, fg, and g/f.

Solution The function $f + g$, the sum of functions f and g, is defined by

$$\begin{aligned}(f + g)(x) &= f(x) + g(x)\\ &= (4x - 3) + (4x^2 - 7x + 3)\\ &= 4x^2 - 3x\end{aligned}$$

The function fh, the product of functions f and h, is defined by

$$\begin{aligned}(fh)(x) &= f(x)h(x)\\ &= (4x - 3)(x - 1)\\ &= 4x^2 - 7x + 3\\ &= g(x)\end{aligned}$$

The product of the functions f and g, fg, is given by

$$\begin{aligned}(fg)(x) &= f(x)g(x)\\ &= (4x - 3)(4x^2 - 7x + 3)\\ &= 16x^3 - 28x^2 + 12x - 12x^2 + 21x - 9\\ &= 16x^3 - 40x^2 + 33x - 9\end{aligned}$$

The quotient of the functions g and f, g/f, is defined as

$$\left(\frac{g}{f}\right)(x) = \frac{g(x)}{f(x)}$$

$$= \frac{4x^2 - 7x + 3}{4x - 3}$$

Factoring the numerator, we can reduce to lowest terms:

$$\left(\frac{g}{f}\right)(x) = \frac{(4x - 3)(x - 1)}{4x - 3}$$

$$= x - 1$$

$$= h(x)$$

d. $\frac{g}{f}$

EXAMPLE 3 If f, g, and h are the same functions defined in Example 2, evaluate $(f + g)(2)$, $(fh)(-1)$, $(fg)(0)$, and $(g/f)(5)$.

Solution We use the formulas for $f + g$, fh, fg, and g/f found in Example 2:

$$(f + g)(2) = 4(2)^2 - 3(2)$$

$$= 16 - 6$$

$$= 10$$

$$(fh)(-1) = 4(-1)^2 - 7(-1) + 3$$

$$= 4 + 7 + 3$$

$$= 14$$

$$(fg)(0) = 16(0)^3 - 40(0)^2 + 33(0) - 9$$

$$= 0 - 0 + 0 - 9$$

$$= -9$$

$$\frac{g}{f}(5) = 5 - 1$$

$$= 4$$

3. Use the functions from Practice Problem 2 to find

a. $(f + g)(3)$

b. $(fh)(-1)$

c. $(fg)(0)$

d. $\left(\frac{g}{f}\right)(10)$

Composition of Functions

In addition to the four operations used to combine functions shown so far in this section, there is a fifth way to combine two functions to obtain a new function. It is called *composition of functions.* To illustrate the concept, recall the definition of training heart rate: Training heart rate, in beats per minute, is resting heart rate plus 60% of the difference between maximum heart rate and resting heart rate. If your resting heart rate is 70 beats per minute, then your training heart rate is a function of your maximum heart rate M

$$T(M) = 70 + 0.6(M - 70) = 70 + 0.6M - 42 = 28 + 0.6M$$

But your maximum heart rate is found by subtracting your age in years from 220. So, if x represents your age in years, then your maximum heart rate is

$$M(x) = 220 - x$$

Therefore, if your resting heart rate is 70 beats per minute and your age in years is x, then your training heart rate can be written as a function of x.

$$T(x) = 28 + 0.6(220 - x)$$

Answers

2. **a.** $3x^2 - 7x - 6$ **b.** $g(x)$ **c.** $9x^3 - 24x^2 - 44x - 16$ **d.** $h(x)$

3. **a.** 0 **b.** 5 **c.** -16 **d.** 6

This last line is the composition of functions T and M. We input x into function M, which outputs $M(x)$. Then, we input $M(x)$ into function T, which outputs $T(M(x))$, which is the training heart rate as a function of age x. Here is a diagram of the situation, which is called a function map:

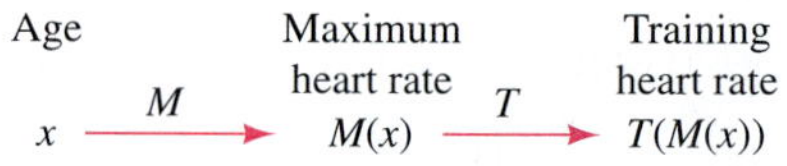

FIGURE 2

Now let's generalize the preceding ideas into a formal development of composition of functions. To find the composition of two functions f and g, we first require that the range of g have numbers in common with the domain of f. Then the composition of f with g, $f \circ g$, is defined this way:

$$(f \circ g)(x) = f(g(x))$$

To understand this new function, we begin with a number x, and we operate on it with g, giving us $g(x)$. Then we take $g(x)$ and operate on it with f, giving us $f(g(x))$. The only numbers we can use for the domain of the composition of f with g are numbers x in the domain of g, for which $g(x)$ is in the domain of f. The diagrams in Figure 3 illustrate the composition of f with g.

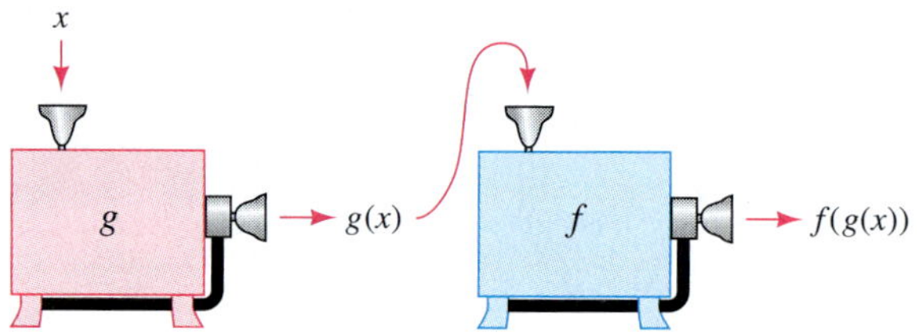

Function machines

FIGURE 3

Composition of functions is not commutative. The composition of f with g, $f \circ g$, may therefore be different from the composition of g with f, $g \circ f$.

$$(g \circ f)(x) = g(f(x))$$

Again, the only numbers we can use for the domain of the composition of g with f are numbers in the domain of f, for which $f(x)$ is in the domain of g. The diagrams in Figure 4 illustrate the composition of g with f.

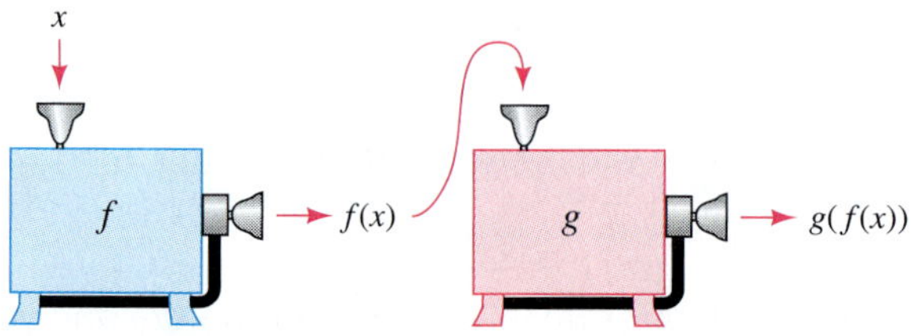

Function machines

FIGURE 4

EXAMPLE 4 If $f(x) = x + 5$ and $g(x) = x^2 - 2x$, find $(f \circ g)(x)$ and $(g \circ f)(x)$.

Solution The composition of f with g is

$$\begin{aligned}(f \circ g)(x) &= f(g(x)) \\ &= f(x^2 - 2x) \\ &= (x^2 - 2x) + 5 \\ &= x^2 - 2x + 5\end{aligned}$$

The composition of g with f is

$$\begin{aligned}(g \circ f)(x) &= g(f(x)) \\ &= g(x + 5) \\ &= (x + 5)^2 - 2(x + 5) \\ &= (x^2 + 10x + 25) - 2(x + 5) \\ &= x^2 + 8x + 15\end{aligned}$$

4. If $f(x) = x - 4$ and $g(x) = x^2 + 3x$, find
   a. $(f \circ g)(x)$
   b. $(g \circ f)(x)$

Answer

4 **a.** $x^2 + 3x - 4$ **b.** $x^2 - 5x + 4$

Problem Set D

INTERMEDIATE Algebra $f(x)$ Now™

Let $f(x) = 4x - 3$ and $g(x) = 2x + 5$. Write a formula for each of the following functions.

1. $f + g$
2. $f - g$
3. $g - f$
4. $g + f$
5. fg
6. f/g
7. g/f
8. ff

If the functions f, g, and h are defined by $f(x) = 3x - 5$, $g(x) = x - 2$ and $h(x) = 3x^2 - 11x + 10$, write a formula for each of the following functions.

9. $g + f$
10. $f + h$
11. $g + h$
12. $f - g$
13. $g - f$
14. $h - g$
15. fg
16. gf
17. fh
18. gh
19. h/f
20. h/g
21. f/h
22. g/h
23. $f + g + h$
24. $h - g + f$
25. $h + fg$
26. $h - fg$

Let $f(x) = 2x + 1$, $g(x) = 4x + 2$, and $h(x) = 4x^2 + 4x + 1$, and find the following.

27. $(f + g)(2)$

28. $(f - g)(-1)$

29. $(fg)(3)$

30. $(f/g)(-3)$

31. $(h/g)(1)$

32. $(hg)(1)$

33. $(fh)(0)$

34. $(h - g)(-4)$

35. $(f + g + h)(2)$

36. $(h - f + g)(0)$

37. $(h + fg)(3)$

38. $(h - fg)(5)$

39. Let $f(x) = x^2$ and $g(x) = x + 4$, and find

a. $(f \circ g)(5)$

b. $(g \circ f)(5)$

c. $(f \circ g)(x)$

d. $(g \circ f)(x)$

40. Let $f(x) = 3 - x$ and $g(x) = x^3 - 1$, and find

a. $(f \circ g)(0)$

b. $(g \circ f)(0)$

c. $(f \circ g)(x)$

d. $(g \circ f)(x)$

41. Let $f(x) = x^2 + 3x$ and $g(x) = 4x - 1$, and find

a. $(f \circ g)(0)$

b. $(g \circ f)(0)$

c. $(f \circ g)(x)$

d. $(g \circ f)(x)$

42. Let $f(x) = (x - 2)^2$ and $g(x) = x + 1$, and find the following

a. $(f \circ g)(-1)$

b. $(g \circ f)(-1)$

c. $(f \circ g)(x)$

d. $(g \circ f)(x)$

For each of the following pairs of functions f and g, show that $(f \circ g)(x) = (g \circ f)(x) = x$.

43. $f(x) = 5x - 4$ and $g(x) = \frac{x + 4}{5}$

44. $f(x) = \frac{x}{6} - 2$ and $g(x) = 6x + 12$

Applying the Concepts

45. Profit, Revenue, and Cost A company manufactures and sells prerecorded videotapes. Here are the equations they use in connection with their business.

Number of tapes sold each day: $n(x) = x$

Selling price for each tape: $p(x) = 11.5 - 0.05x$

Daily fixed costs: $f(x) = 200$

Daily variable costs: $v(x) = 2x$

Find the following functions.

a. Revenue $= R(x) =$ the product of the number of tapes sold each day and the selling price of each tape.

b. Cost $= C(x) =$ the sum of the fixed costs and the variable costs.

c. Profit $= P(x) =$ the difference between revenue and cost.

d. Average cost $= \overline{C}(x) =$ the quotient of cost and the number of tapes sold each day.

46. Profit, Revenue, and Cost A company manufactures and sells diskettes for home computers. Here are the equations they use in connection with their business.

Number of diskettes sold each day: $n(x) = x$

Selling price for each diskette: $p(x) = 3 - \frac{1}{300}x$

Daily fixed costs: $f(x) = 200$

Daily variable costs: $v(x) = 2x$

Find the following functions.

a. Revenue $= R(x) =$ the product of the number of diskettes sold each day and the selling price of each diskette.

b. Cost $= C(x) =$ the sum of the fixed costs and the variable costs.

c. Profit $= P(x) =$ the difference between revenue and cost.

d. Average cost $= C(x) =$ the quotient of cost and the number of diskettes sold each day.

47. Training Heart Rate Find the training heart rate function, $T(M)$, for a person with a resting heart rate of 62 beats per minute, then find the following.

a. Find the maximum heart rate function, $M(x)$, for a person x years of age.

b. What is the maximum heart rate for a 24-year-old person?

c. What is the training heart rate for a 24-year-old person with a resting heart rate of 62 beats per minute?

d. What is the training heart rate for a 36-year-old person with a resting heart rate of 62 beats per minute?

e. What is the training heart rate for a 48-year-old person with a resting heart rate of 62 beats per minute?

48. Training Heart Rate Find the training heart rate function, $T(M)$, for a person with a resting heart rate of 72 beats per minute, then finding the following to the nearest whole number.

a. Find the maximum heart rate function, $M(x)$, for a person x years of age.

b. What is the maximum heart rate for a 20-year-old person?

c. What is the training heart rate for a 20-year-old person with a resting heart rate of 72 beats per minute?

d. What is the training heart rate for a 30-year-old person with a resting heart rate of 72 beats per minute?

e. What is the training heart rate for a 40-year-old person with a resting heart rate of 72 beats per minute?

Solutions to Selected Practice Problems

Solutions to most of the practice problems are shown here. Before you look here to see where you have made a mistake, you should try the problem you are working on twice. If you do not get the correct answer the second time you work the problem, then the solution here should show you where you went wrong.

Chapter 1

Section 1.1

2. $4^2 = 4 \cdot 4 = 16$

3. $2^4 = 2 \cdot 2 \cdot 2 \cdot 2 = 16$

4. $3^3 = 3 \cdot 3 \cdot 3 = 27$

5. $6 + 2(3 + 4) = 6 + 2(7)$
$= 6 + 14$
$= 20$

6. $5 \cdot 3^2 - 2 \cdot 4^2 = 5 \cdot 9 - 2 \cdot 16$
$= 45 - 32$
$= 13$

7. $30 - (2 \cdot 3^2 - 8) = 30 - (2 \cdot 9 - 8)$
$= 30 - (18 - 8)$
$= 30 - 10$
$= 20$

8. $60 + 20 \div 2 - 40 = 60 + 10 - 40$
$= 70 - 40$
$= 30$

9. $3 + 5[2 + (7 \cdot 2 - 10)] = 3 + 5[2 + (14 - 10)]$
$= 3 + 5(2 + 4)$
$= 3 + 5(6)$
$= 3 + 30$
$= 33$

10. When $a = 0$, $(0 + 4)^2 = 4^2 = 16$
When $a = 1$, $(1 + 4)^2 = 5^2 = 25$
When $a = 2$, $(2 + 4)^2 = 6^2 = 36$
When $a = 3$, $(3 + 4)^2 = 7^2 = 49$
When $a = 4$, $(4 + 4)^2 = 8^2 = 64$

11. When $x = 4$ and $y = 1$, then
$2x + 7y - 5 = 2 \cdot 4 + 7 \cdot 1 - 5$
$= 8 + 7 - 5$
$= 15 - 5$
$= 10$

12. True. The dots mean the set continues indefinitely in the same manner.

13. True, since every member of the finite set {2, 4, 8} is a member of the infinite set {2, 4, 6, . . .}.

14. The union of A and B is the set of all elements that are either in A or in B. Therefore, $A \cup B = \{1, 2, 3\} \cup \{0, 3, 7\} = \{0, 1, 2, 3, 7\}$.

15. The intersection of A and B is the set of elements in both A and B. Therefore, $A \cap B = \{1, 2, 3\} \cap \{0, 3, 7\} = \{3\}$.

Section 1.2

1.
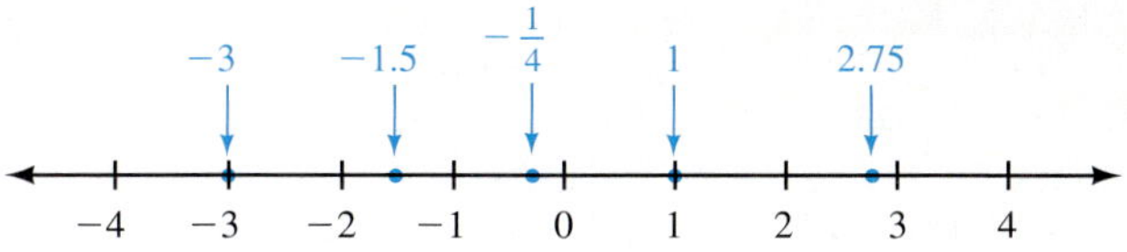

2.

3.

4.
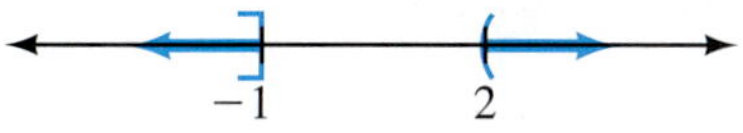

5.

6.

7.

8. If t is more than 10 but less than 20, then $t > 10$ and $t < 20$, or equivalently, $10 < t < 20$.

9. If W ranges from a low of 125 to a high of 131, then W is greater than or equal to 125 and less than or equal to 131, that is, $W \geq 125$ and $W \leq 131$, or equivalently, $125 \leq W \leq 131$.

11. $420 = 42 \cdot 10$
$= 6 \cdot 7 \cdot 2 \cdot 5$
$= 2 \cdot 3 \cdot 7 \cdot 2 \cdot 5$
$= 2^2 \cdot 3 \cdot 5 \cdot 7$

12. $\dfrac{231}{770} = \dfrac{3 \cdot 7 \cdot 11}{2 \cdot 5 \cdot 7 \cdot 11} = \dfrac{3}{2 \cdot 5} = \dfrac{3}{10}$

Section 1.3

2. $\frac{3}{7} \cdot \frac{2}{5} = \frac{3 \cdot 2}{7 \cdot 5} = \frac{6}{35}$ **3.** $9 \cdot \frac{1}{4} = \frac{9}{1} \cdot \frac{1}{4} = \frac{9}{4}$ **4.** $\left(\frac{3}{4}\right)^3 = \frac{3}{4} \cdot \frac{3}{4} \cdot \frac{3}{4} = \frac{27}{64}$ **5.** The reciprocal of 5 is $\frac{1}{5}$ because $5 \cdot \frac{1}{5} = \frac{5}{1} \cdot \frac{1}{5} = \frac{5}{5} = 1$.

6. The reciprocal of $\frac{1}{4}$ is 4 because $\frac{1}{4} \cdot 4 = \frac{1}{4} \cdot \frac{4}{1} = \frac{4}{4} = 1$. **7.** The reciprocal of $\frac{3}{4}$ is $\frac{4}{3}$ because $\frac{3}{4} \cdot \frac{4}{3} = \frac{12}{12} = 1$.

8. The reciprocal of x is $\frac{1}{x}$ because $x \cdot \frac{1}{x} = \frac{x}{1} \cdot \frac{1}{x} = \frac{x}{x} = 1$. **17.** $5 + (7 + y) = (5 + 7) + y = 12 + y$ **18.** $3(2x) = (3 \cdot 2)x = 6x$

19. $\frac{1}{3}(3a) = \left(\frac{1}{3} \cdot 3\right)a = a$ **20.** $5\left(\frac{1}{5}x\right) = \left(5 \cdot \frac{1}{5}\right)x = 1x = x$ **21.** $9\left(\frac{2}{3}x\right) = \left(9 \cdot \frac{2}{3}\right)x = 6x$ **22.** $7(6x + 8) = 7(6x) + 7(8) = 42x + 56$ **23.** $9(7x + 11y) = 9(7x) + 9(11y) = 63x + 99y$

24. $\frac{1}{3}(3x + 6) = \frac{1}{3}(3x) + \frac{1}{3}(6) = x + 2$ **25.** $4(5y + 2) + 8 = 4(5y) + 4(2) + 8 = 20y + 8 + 8 = 20y + 16$ **26.** $a\left(2 - \frac{1}{a}\right) = a \cdot 2 - a \cdot \frac{1}{a} = 2a - 1$

27. $5\left(\frac{1}{5}x + 8\right) = 5 \cdot \frac{1}{5}x + 5 \cdot 8 = x + 40$ **28.** $12\left(\frac{2}{3}x + \frac{3}{4}y\right) = 12 \cdot \frac{2}{3}x + 12 \cdot \frac{3}{4}y = 8x + 9y$ **29.** $4x + 9x = (4 + 9)x = 13x$ **30.** $y + 5y = (1 + 5)y = 6y$

31. $\frac{3}{14} + \frac{7}{30} = \frac{3}{2 \cdot 7} + \frac{7}{2 \cdot 3 \cdot 5} = \frac{\mathbf{3 \cdot 5}}{\mathbf{3 \cdot 5}} \cdot \frac{3}{2 \cdot 7} + \frac{7}{2 \cdot 3 \cdot 5} \cdot \frac{\mathbf{7}}{\mathbf{7}} = \frac{45}{210} + \frac{49}{210} = \frac{94}{210} = \frac{47}{105}$

32. $10x + 3 + 7x + 12 = (10x + 7x) + (3 + 12) = (10 + 7)x + (3 + 12) = 17x + 15$

33. $9 + 5(4y + 8) + 10y = 9 + 20y + 40 + 10y = (20y + 10y) + (9 + 40) = 30y + 49$

Section 1.4

2. $7 - 4 = 7 + (-4) = 3$ **3.** $-6 - 3 = -6 + (-3) = -9$ **4.** $8 - (-2) = 8 + 2 = 10$ **5.** $-4 - (-6) = -4 + 6 = 2$

6. $-3 - (-5) = -3 + 5 = 2$ **7.** $(-7 - 4) + (-8) = -11 + (-8) = -19$ **13.** $(-6 - 4)(8 - 12) = (-10)(-4) = 40$

14. $3 - 7(9 - 5) - 2 = 3 - 7(4) - 2 = 3 - 28 - 2 = -25 - 2 = -27$

15. $3(5 - 8)^3 - 2(-1 - 1)^2 = 3(-3)^3 - 2(-2)^2 = 3(-27) - 2(4) = -89$

16. $2\left(\frac{x}{2}\right) = 2\left(\frac{1}{2}x\right) = \left(2 \cdot \frac{1}{2}\right)x = x$ **17.** $2\left(\frac{x}{2} - 3\right) = 2 \cdot \frac{x}{2} - 2 \cdot 3 = x - 6$ **18.** $\frac{-6 - 6}{-5 - 3} = \frac{-12}{-8} = \frac{3}{2}$ **19.** $\frac{5(-6) + 3(-2)}{4(-3) + 3} = \frac{-30 + (-6)}{-12 + 3} = \frac{-36}{-9} = 4$ **20.** $\frac{3^3 - 4^3}{3^2 + 4^2} = \frac{27 - 64}{9 + 16} = \frac{-37}{25} = -\frac{37}{25}$

21. $2(5y - 1) - y = 10y - 2 - y = 9y - 2$ **22.** $6 - 2(5x + 1) + 4x = 6 - 10x - 2 + 4x = -6x + 4$ **23.** $4(3a + 1) - (7a - 6) = 12a + 4 - 7a + 6 = 5a + 10$

24. $\frac{3}{5} \div \frac{6}{7} = \frac{3}{5} \cdot \frac{7}{6} = \frac{21}{30} = \frac{7}{10}$ **25.** $12 \div \frac{3}{4} = \frac{12}{1} \cdot \frac{4}{3} = \frac{48}{3} = 16$ **26.** $-\frac{5}{6} \div 10 = -\frac{5}{6} \cdot \frac{1}{10} = -\frac{5}{60} = -\frac{1}{12}$

27. a. $7[4(-1) - 6] = 7(-4 - 6) = 7(-10) = -70$
b. $28(-1) - 42 = -28 - 42 = -70$
c. $28(-1) - 6 = -28 - 6 = -34$

Section 1.5

1. a. If the sequence is formed by multiplying by 3 each time, the next term will be $3(135) = 405$.
b. Turning the triangle a quarter turn to the right each time, the next triangle will point to the right.
c. The sequence of cubes. The next term will be $5^3 = 125$.

2. a. Add 10 each time.
b. Add $\frac{1}{2}$ each time.
c. Add 5 each time.

3. a. Multiply by 2 each time.
b. Multiply by -3 each time.
c. Multiply by $\frac{1}{2}$ each time.

4. Add 34 and 55, the result is 89.

Chapter 2

Section 2.1

1. $4(3) - 2 \stackrel{?}{=} 10$

$12 - 2 = 10$

$10 = 10$

2. $3(5) + 1 \stackrel{?}{=} 16$

$15 + 1 = 16$

$16 = 16$

$4(5) - 6 \stackrel{?}{=} 14$

$20 - 6 = 14$

$14 = 14$

3.

$$\frac{2}{3}x + 4 = -8$$
$$\frac{2}{3}x + 4 - \mathbf{4} = -8 - \mathbf{4}$$
$$\frac{2}{3}x = -12$$
$$\frac{\mathbf{3}}{\mathbf{2}} \cdot \frac{2}{3}x = \frac{\mathbf{3}}{\mathbf{2}}(-12)$$
$$x = -18$$

4. $3a - 3 = -5a + 9$

$$3a + \mathbf{5a} - 3 = -5a + \mathbf{5a} + 9$$
$$8a - 3 = 9$$
$$8a - 3 + \mathbf{3} = 9 + \mathbf{3}$$
$$8a = 12$$
$$\frac{\mathbf{1}}{\mathbf{8}} \cdot 8a = \frac{\mathbf{1}}{\mathbf{8}} \cdot 12$$
$$a = \frac{12}{8} = \frac{3}{2}$$

5. a. $-x = \frac{2}{3}$

$$-1(-x) = -1\left(\frac{2}{3}\right)$$
$$x = -\frac{2}{3}$$

b. $-y = -4$

$$-1(-y) = -1(-4)$$
$$y = 4$$

6.

$$\frac{3}{5}x + \frac{1}{3} = -\frac{5}{6}$$
$$\frac{3}{5}x + \frac{1}{3} + \left(-\frac{\mathbf{1}}{\mathbf{3}}\right) = -\frac{5}{6} + \left(-\frac{\mathbf{1}}{\mathbf{3}}\right)$$
$$\frac{3}{5}x = -\frac{7}{6}$$
$$\frac{\mathbf{5}}{\mathbf{3}} \cdot \frac{3}{5}x = \frac{\mathbf{5}}{\mathbf{3}}\left(-\frac{7}{6}\right)$$
$$x = -\frac{35}{18}$$

Or eliminating the fractions in the beginning we can solve it this way:

$$30 \cdot \frac{3}{5}x + 30 \cdot \frac{1}{3} = 30\left(-\frac{5}{6}\right)$$
$$18x + 10 = -25$$
$$18x = -35$$
$$x = -\frac{35}{18}$$

7. a. Working with the decimals.

$$0.08x + 0.10(8{,}000 - x) = 680$$
$$0.08x + 800 - 0.10x = 680$$
$$-0.02x + 800 = 680$$
$$-0.02x + 800 + (\mathbf{-800}) = 680 + (\mathbf{-800})$$
$$-0.02x = -120$$
$$x = \frac{-120}{-0.02} = 6{,}000$$

b. Eliminating the decimals by multiplying each side by 100.

$$100(0.08x) + 100(0.10)(8{,}000 - x) = 100(680)$$
$$8x + 10(8{,}000 - x) = 68{,}000$$
$$8x + 80{,}000 - 10x = 68{,}000$$
$$-2x + 80{,}000 = 68{,}000$$
$$-2x = -12{,}000$$
$$x = \frac{-12{,}000}{-2}$$
$$= 6{,}000$$

8.

$$2(5y - 3) + 4 = 9y - 6$$
$$10y - 6 + 4 = 9y - 6$$
$$10y - 2 = 9y - 6$$
$$10y + (\mathbf{-9y}) - 2 = 9y + (\mathbf{-9y}) - 6$$
$$y - 2 = -6$$
$$y - 2 + \mathbf{2} = -6 + \mathbf{2}$$
$$y = -4$$

9. $6 - 2(5x - 1) + 4x = 20$

$$6 - 10x + 2 + 4x = 20$$
$$-6x + 8 = 20$$
$$-6x + 8 + (\mathbf{-8}) = 20 + (\mathbf{-8})$$
$$-6x = 12$$
$$-\frac{\mathbf{1}}{\mathbf{6}}(-6x) = -\frac{\mathbf{1}}{\mathbf{6}}(12)$$
$$x = -2$$

10.

$$3(5x + 1) = 10 + 15x$$
$$15x + 3 = 10 + 15x$$
$$15x + (\mathbf{-15x}) + 3 = 10 + 15x + (\mathbf{-15x})$$
$$3 = 10$$

A false statement indicates there is no solution to our original equation.

11.

$$-4 + 8x = 2(4x - 2)$$
$$-4 + 8x = 8x - 4$$
$$-4 + 8x + (\mathbf{-8x}) = 8x + (\mathbf{-8x}) - 4$$
$$-4 = -4$$

A true statement indicates all real numbers are solutions to our equation.

Section 2.2

1. When $x = -3$, we have
$$\begin{aligned} 2(-3) - 3y &= 6 \\ -6 - 3y &= 6 \\ -3y &= 12 \\ y &= -4 \end{aligned}$$

2. When $x = 375$, we have
$$\begin{aligned} 375 &= 900 - 300p \\ -525 &= -300p \\ p &= \frac{-525}{-300} = \$1.75 \end{aligned}$$

3. $$\begin{aligned} 486.7 &= (3.14)(6^2) + 2(3.14)(6)h \\ 486.7 &= 113.04 + 37.68h \\ 373.66 &= 37.68h \end{aligned}$$
$h = 9.9$ centimeters to the nearest tenth

4. $$\begin{aligned} P &= 2w + 2l \\ P - 2w &= 2l \\ \frac{P - 2w}{2} &= l \end{aligned}$$
or $\frac{P}{2} - w = l$

5. $$\begin{aligned} ax + 5 &= cx + 3 \\ ax - cx &= 3 - 5 \\ (a - c)x &= -2 \\ x &= \frac{-2}{a - c} \end{aligned}$$

6. $$\begin{aligned} \frac{y-2}{x-0} &= \frac{4}{3} \\ \frac{y-2}{x} &= \frac{4}{3} \\ x \cdot \frac{y-2}{x} &= \frac{4}{3} \cdot x \\ y - 2 &= \frac{4}{3}x \\ y &= \frac{4}{3}x + 2 \end{aligned}$$

7. $$\begin{aligned} \frac{y+8}{x-7} &= 6 \\ (x-7) \cdot \frac{y+8}{x-7} &= 6 \cdot (x-7) \\ y + 8 &= 6x - 42 \\ y &= 6x - 50 \end{aligned}$$

8. $$\begin{aligned} x &= 0.25(74) \\ x &= 18.5 \end{aligned}$$

9. $$\begin{aligned} x \cdot 84 &= 21 \\ x &= \frac{21}{84} \\ x &= 0.25 \\ x &= 25\% \end{aligned}$$

10. $$\begin{aligned} 35 &= 0.40x \\ \frac{35}{0.40} &= x \\ 87.5 &= x \end{aligned}$$

11. $$\begin{aligned} a_1 &= 3(1) - 2 = 1 \\ a_2 &= 3(2) - 2 = 4 \\ a_3 &= 3(3) - 2 = 7 \\ a_4 &= 3(4) - 2 = 10 \end{aligned}$$

12. $$\begin{aligned} a_1 &= \frac{1}{1+1} = \frac{1}{2} \\ a_2 &= \frac{2}{2+1} = \frac{2}{3} \\ a_3 &= \frac{3}{3+1} = \frac{3}{4} \\ a_4 &= \frac{4}{4+1} = \frac{4}{5} \end{aligned}$$

Section 2.3

1. $$\begin{aligned} 2x + 2(4x - 10) &= 12.5 \\ 2x + 8x - 20 &= 12.5 \\ 10x - 20 &= 12.5 \\ 10x &= 32.5 \\ x &= 3.25 \end{aligned}$$
The length is $x = 3.25$ feet; the width is $4x - 10 = 4(3.25) - 10 = 3$ feet.

2. $$\begin{aligned} x + 0.0725x &= 23{,}466.30 \\ 1.0725x &= 23{,}466.30 \\ x &= \frac{23{,}466.30}{1.0725} \\ &= 21{,}880 \end{aligned}$$
The price of the car is \$21,880.00.

3. $$\begin{aligned} x + 5x &= 180 \\ 6x &= 180 \\ x &= 30 \end{aligned}$$
The angles are 30° and 150°.

4.

	DOLLARS AT 8%	DOLLARS AT 10%	TOTAL
Number	x	$8{,}000 - x$	8,000
Interest	$0.08x$	$0.10(8{,}000 - x)$	680

$$\begin{aligned} 0.08x + 0.10(8{,}000 - x) &= 680 \\ 8x + 10(8{,}000 - x) &= 68{,}000 \\ 8x + 80{,}000 - 10x &= 68{,}000 \\ -2x + 80{,}000 &= 68{,}000 \\ -2x &= -12{,}000 \\ x &= 6{,}000 \end{aligned}$$
\$6,000 at 8% and \$2,000 at 10%

5.

WIDTH	LENGTH	AREA (IN²)
1	$\ell = 7 - 1 = 6$	6
2	$\ell = 7 - 2 = 5$	10
3	$\ell = 7 - 3 = 4$	12
4	$\ell = 7 - 4 = 3$	12
5	$\ell = 7 - 5 = 2$	10
6	$\ell = 7 - 6 = 1$	6

Section 2.4

1. $$\begin{aligned} 4x - 2 &> 3x + 4 \\ 4x + (\mathbf{-3x}) - 2 &> 3x + (\mathbf{-3x}) + 4 \\ x - 2 &> 4 \\ x - 2 + \mathbf{2} &> 4 + \mathbf{2} \\ x &> 6 \end{aligned}$$

Interval notation $(6, \infty)$

2. $$\begin{aligned} -3y - 2 &< 7 \\ -3y &< 9 \\ -\frac{\mathbf{1}}{\mathbf{3}}(-3y) &> -\frac{\mathbf{1}}{\mathbf{3}}(9) \\ y &> -3 \end{aligned}$$

Interval notation $(-3, \infty)$

3. $$\begin{aligned} 3(2x + 5) &\le -3x \\ 6x + 15 &\le -3x \\ 9x + 15 &\le 0 \\ 9x &\le -15 \\ \frac{\mathbf{1}}{\mathbf{9}}(9x) &\le \frac{\mathbf{1}}{\mathbf{9}}(-15) \\ x &\le -\frac{5}{3} \end{aligned}$$

Interval notation $(-\infty, -\frac{5}{3}]$

4. $$\begin{aligned} -7 &\le 2x + 1 \le 7 \\ -8 &\le 2x \le 6 \\ -4 &\le x \le 3 \end{aligned}$$

Interval notation $[-4, 3]$

5. $$\begin{aligned} 3t - 6 &\le -3 \quad \text{or} \quad 3t - 6 \ge 3 \\ 3t &\le 3 \quad \text{or} \quad 3t \ge 9 \\ t &\le 1 \quad \text{or} \quad t \ge 3 \end{aligned}$$

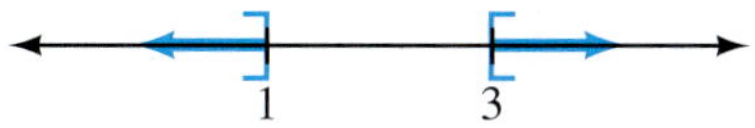

Interval notation $(-\infty, 1] \cup [3, \infty)$

6. $$\begin{aligned} 1{,}300 - 100p &\ge 800 \\ -100p &\ge -500 \\ p &\le 5 \end{aligned}$$

The price is \$5 or less.

7. $$\begin{aligned} -18 \le \quad & F \quad \le 22 \\ -18 \le \quad & \frac{9}{5}C + 32 \le 22 \\ -50 \le \quad & \frac{9}{5}C \quad \le -10 \\ \frac{5}{9}(-50) \le \quad & \frac{5}{9}\left(\frac{9}{5}C\right) \le \frac{5}{9}(-10) \\ -27.8^\circ \le \quad & C \quad \le -5.6^\circ \end{aligned}$$ to the nearest tenth

Section 2.5

1. If $|x| = 3$
then $x = 3$ or $x = -3$

2. If $|3x - 6| = 9$
$$\begin{aligned} \text{then } 3x - 6 &= 9 \quad \text{or} \quad 3x - 6 = -9 \\ 3x &= 15 \quad \text{or} \quad 3x = -3 \\ x &= 5 \quad \text{or} \quad x = -1 \end{aligned}$$

3. If $|4x - 3| + 2 = 3$
then $|4x - 3| = 1$
$$\begin{aligned} 4x - 3 &= 1 \quad \text{or} \quad 4x - 3 = -1 \\ 4x &= 4 \quad \text{or} \quad 4x = 2 \\ x &= 1 \quad \text{or} \quad x = \frac{1}{2} \end{aligned}$$

4. $|7a - 1| = -2$ has no solution since the left side is positive or zero and the right side is negative.

5. $|x + 3| = |x + 8|$
$$\begin{aligned} x + 3 &= x + 8 \quad \text{or} \quad x + 3 = -(x + 8) \\ 3 &= 8 \qquad\qquad\quad x + 3 = -x - 8 \\ &\qquad\qquad\qquad\quad 2x + 3 = -8 \\ &\qquad\qquad\qquad\quad 2x = -11 \\ &\qquad\qquad\qquad\quad x = -\frac{11}{2} \end{aligned}$$
No solution here

Section 2.6

1. If $|2x + 1| \le 7$
$$\begin{aligned} \text{then } -7 &\le 2x + 1 \le 7 \\ -8 &\le 2x \le 6 \\ -4 &\le x \le 3 \end{aligned}$$

2. If $|4a - 3| < 5$
$$\begin{aligned} \text{then } -5 &< 4a - 3 < 5 \\ -2 &< 4a < 8 \\ -\frac{1}{2} &< a < 2 \end{aligned}$$

3. If $|x + 2| > 7$
$$\begin{aligned} \text{then } x + 2 &< -7 \quad \text{or} \quad x + 2 > 7 \\ x &< -9 \quad \text{or} \quad x > 5 \end{aligned}$$

4. If $|3t - 6| \ge 3$
$$\begin{aligned} \text{then } 3t - 6 &\le -3 \quad \text{or} \quad 3t - 6 \ge 3 \\ 3t &\le 3 \quad \text{or} \quad 3t \ge 9 \\ t &\le 1 \quad \text{or} \quad t \ge 3 \end{aligned}$$

5. If $|2x + 5| - 2 < 9$
$$\begin{aligned} \text{then } |2x + 5| &< 11 \\ -11 &< 2x + 5 < 11 \\ -16 &< 2x < 6 \\ -8 &< x < 3 \end{aligned}$$

6. If $|5 - 2t| > 3$
$$\begin{aligned} \text{then } 5 - 2t &< -3 \quad \text{or} \quad 5 - 2t > 3 \\ -2t &< -8 \quad \text{or} \quad -2t > -2 \end{aligned}$$
$$-\frac{1}{2}(-2t) > -\frac{1}{2}(-8) \text{ or } -\frac{1}{2}(-2t) < -\frac{1}{2}(-2)$$
$$t > 4 \quad \text{or} \quad t < 1$$

7. $|8y + 3| \le -5$ has no solution since the left side cannot be negative and any number less than -5 must be negative.

8. $|2x - 9| \ge -7$ All real numbers satisfy this inequality since the left side is positive or zero, either of which is always greater than -7.

Chapter 3

Section 3.1

1.
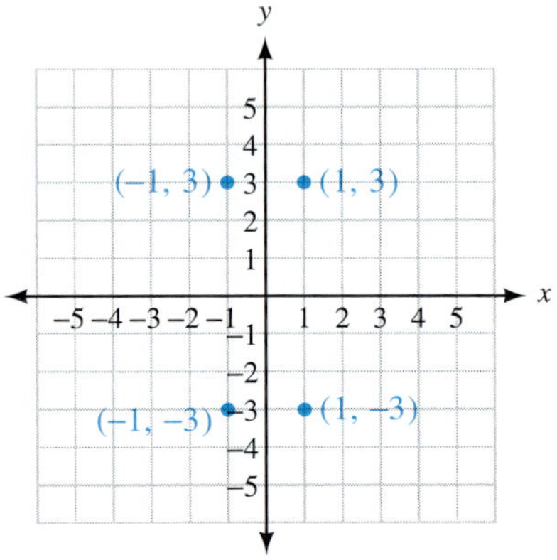

2.
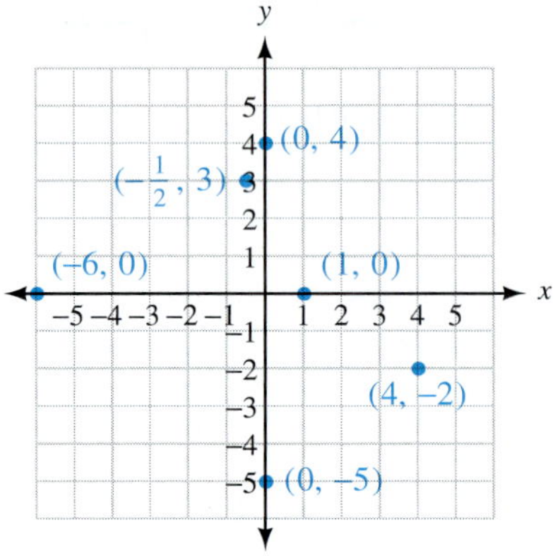

3.
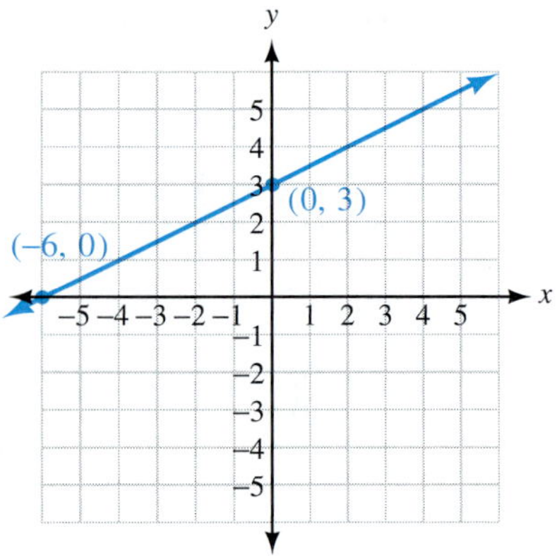

4.
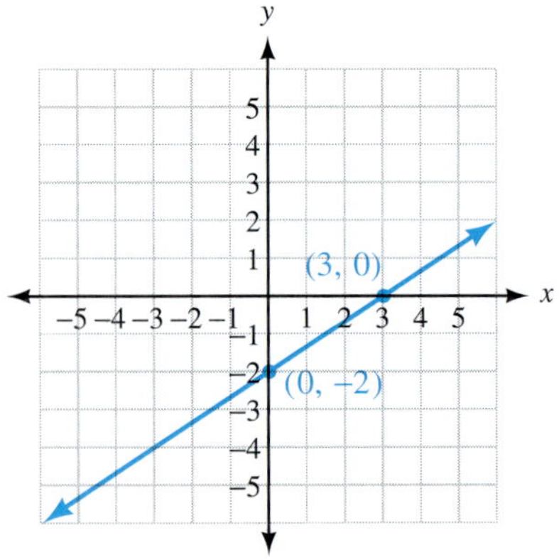

5.
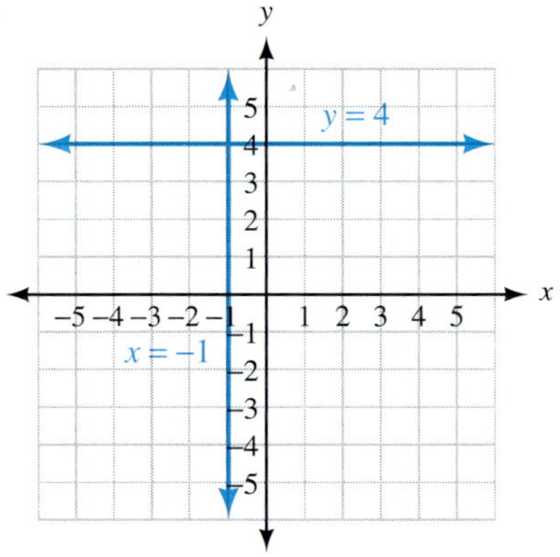

Section 3.2

1. When $x = 0, y = -5$, so the point $(0, -5)$ is on the graph. When $x = 2, y = 1$, so the point $(2, 1)$ is a second point on the graph. The ratio of rise to run going from $(0, -5)$ to $(2, 1)$ is $\frac{6}{2} = 3$. The slope of $y = 3x - 5$ is 3.

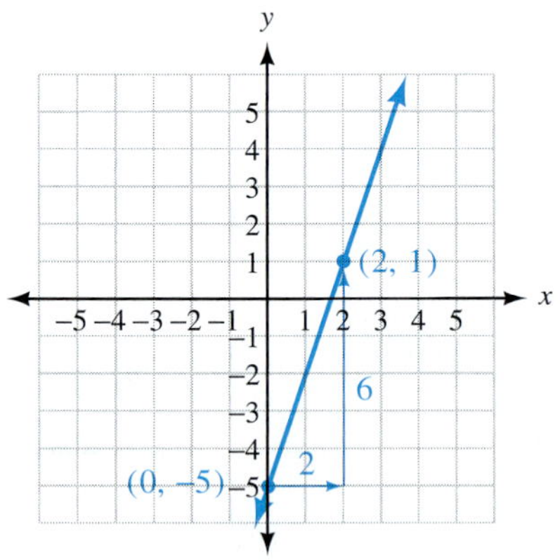

2. $m = \dfrac{4 - (-2)}{3 - 1} = \dfrac{6}{2} = 3$

3. $m = \dfrac{-3 - (-3)}{2 - (-1)} = \dfrac{0}{3} = 0$

4. $m = \dfrac{2.93 - 1.25}{2005 - 1985} = \dfrac{1.68}{20} = 0.084$ dollars/year = 8.4 cents/year

Section 3.3

1.
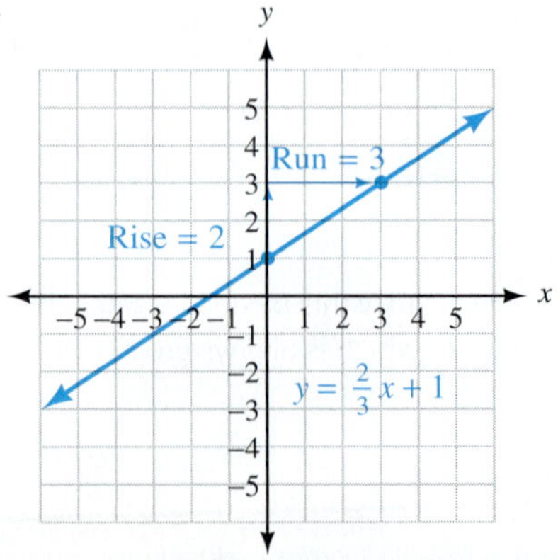

2. $4x - 5y = 7$

$$-5y = -4x + 7$$

$$y = \frac{4}{5}x - \frac{7}{5}$$

Slope $= \frac{4}{5}$, y-intercept $= -\frac{7}{5}$

3. $-3x + 2y = -6$

$$2y = 3x - 6$$

$$y = \frac{3}{2}x - 3$$

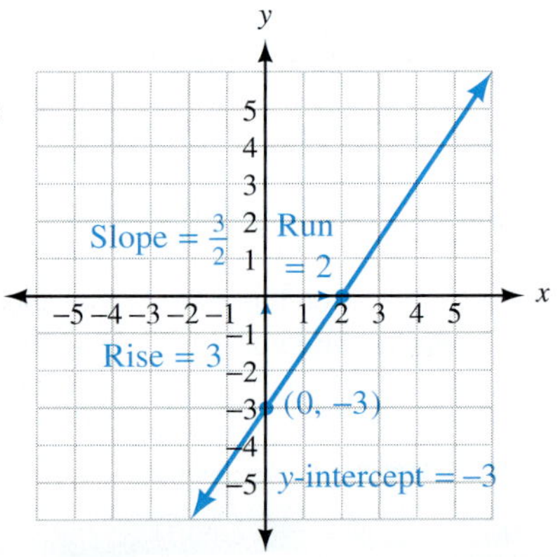

4. $m = 3,\ (x_1, y_1) = (-1, 2)$

$$y - y_1 = m(x - x_1)$$
$$y - 2 = 3(x + 1)$$
$$y - 2 = 3x + 3$$
$$y = 3x + 5$$

5. $m = \dfrac{5 - (-3)}{2 - 6} = \dfrac{8}{-4} = -2$

let $(x_1, y_1) = (2, 5)$

then $y - 5 = -2(x - 2)$

$$y - 5 = -2x + 4$$
$$y = -2x + 9$$

6. $3x - y = 2$

$$-y = -3x + 2$$
$$y = 3x - 2$$

The slope of a line perpendicular to this line is $-\frac{1}{3}$.

Using $m = -\frac{1}{3}$ and $(x_1, y_1) = (3, 2)$ we have

$$y - 2 = -\frac{1}{3}(x - 3)$$
$$y - 2 = -\frac{1}{3}x + 1$$
$$y = -\frac{1}{3}x + 3$$
$$3y = -x + 9$$
$$x + 3y = 9$$

Section 3.4

1.

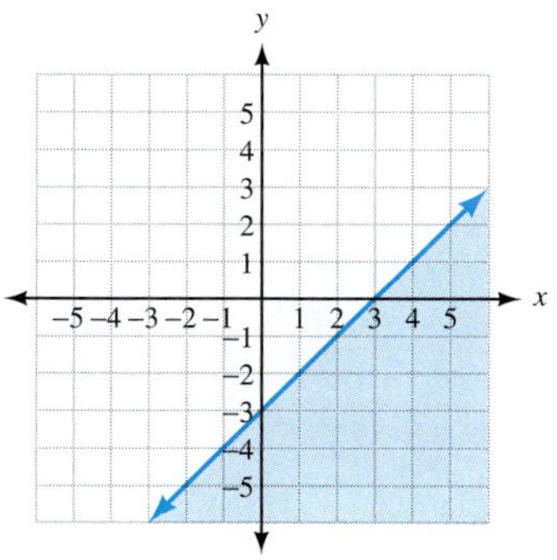

2.

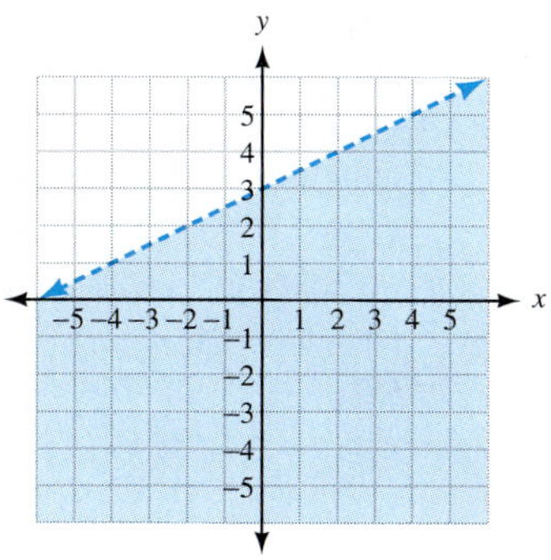

3.

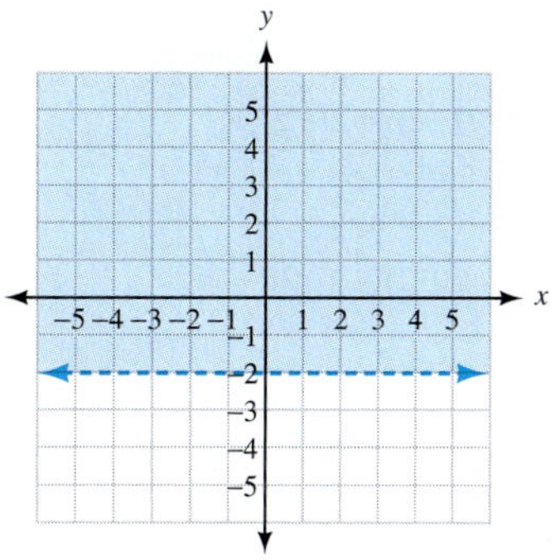

4.

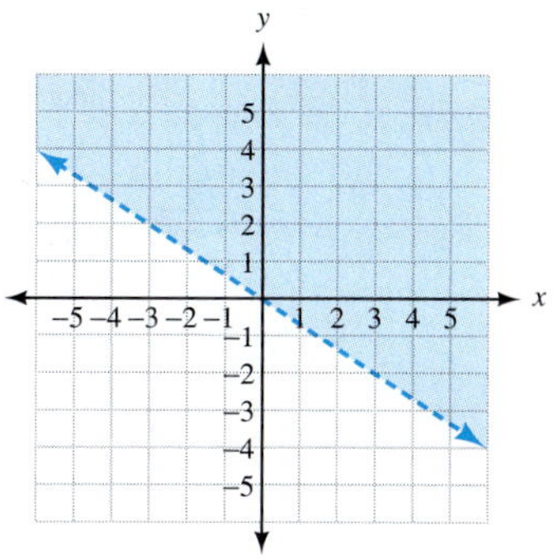

Section 3.5

1.

x	y
0	0
10	80
20	160
30	240
40	320

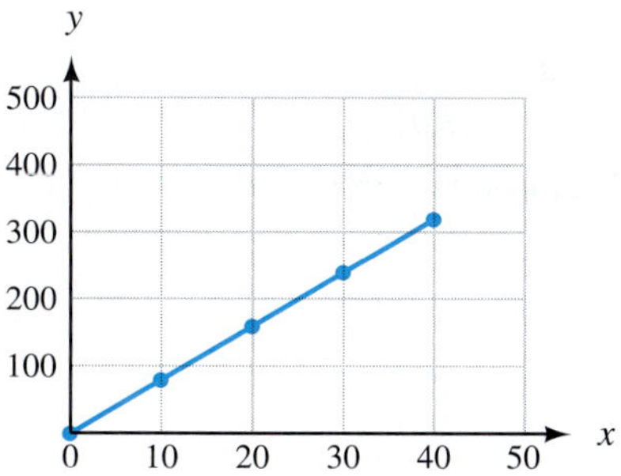

4.

t	h
0	0
$\frac{1}{2}$	20
1	32
$\frac{3}{2}$	36
2	32
$\frac{5}{2}$	20
3	0

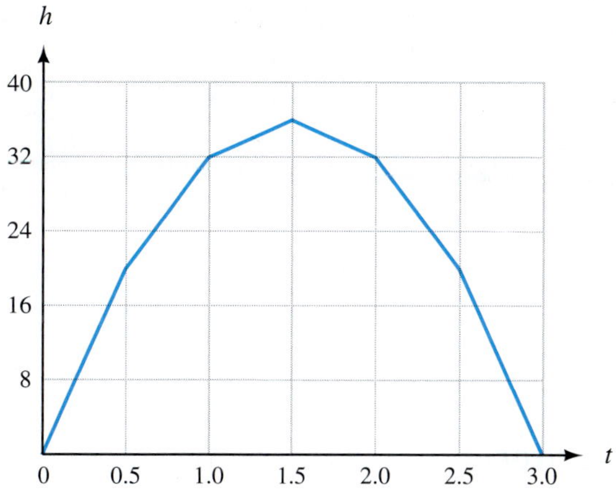

5.

x	y
5	−3
0	−2
−3	−1
−4	0
−3	1
0	2
5	3

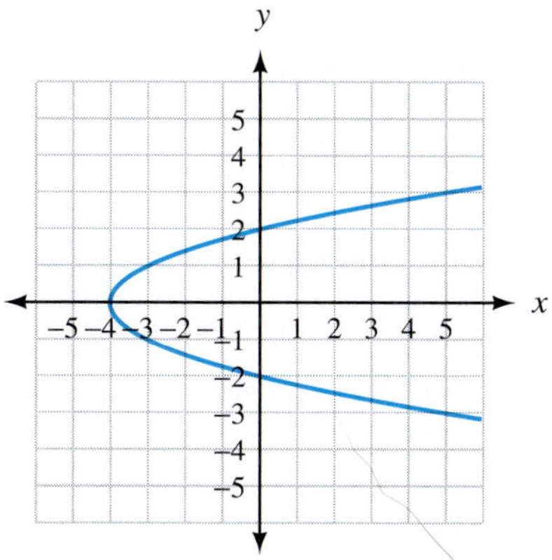

6.

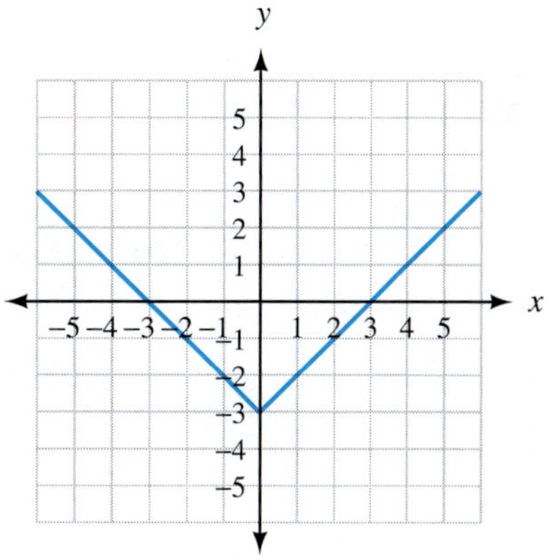

Section 3.6

1. a. $f(0) = 8(0)$
$= 0$

b. $f(5) = 8(5)$
$= 40$

c. $f(10.5) = 8(10.5)$
$= 84$

2. a. $s(8) = \frac{88}{8}$
$= 11$

b. $s(11) = \frac{88}{11}$
$= 8$

3. a. $C(5) = 80\left(\frac{1}{2}\right)^{5/5}$
$= 80\left(\frac{1}{2}\right)^{1}$
$= 40$

b. $C(10) = 80\left(\frac{1}{2}\right)^{10/5}$
$= 80\left(\frac{1}{2}\right)^{2}$
$= 80\left(\frac{1}{4}\right)$
$= 20$

4. a. $C(5) = 2\pi(5)$
$= 10\pi$
≈ 31.4 inches

b. $A(5) = \pi(5)^2$
$= 25\pi$
≈ 78.5 inches^2

5. a. $f(0) = 4(0)^2 - 3$
$= -3$

b. $f(3) = 4(3)^2 - 3$
$= 4(9) - 3$
$= 36 - 3$
$= 33$

c. $f(-2) = 4(-2)^2 - 3$
$= 4(4) - 3$
$= 16 - 3$
$= 13$

6. a. $f(5) = 2(5) + 1$
$= 10 + 1$
$= 11$

b. $g(5) = (5)^2 - 3$
$= 25 - 3$
$= 22$

c. $f(-2) = 2(-2) + 1$
$= -4 + 1$
$= -3$

d. $g(-2) = (-2)^2 - 3$
$= 4 - 3$
$= 1$

e. $f(a) = 2(a) + 1$
$= 2a + 1$

f. $g(a) = (a)^2 - 3$
$= a^2 - 3$

8. a. $f[g(2)] = f[4(2) + 1]$
$= f[8 + 1]$
$= f(9)$
$= 3(9)^2$
$= 3(81)$
$= 243$

b. $g[f(2)] = g[3(2)^2]$
$= g[3(4)]$
$= g(12)$
$= 4(12) + 1$
$= 48 + 1$
$= 49$

Section 3.7

1. $y = Kx$
$24 = K \cdot 8$
$\Rightarrow K = 3$
$y = 3x$
$y = 3 \cdot 2$
$= 6$

2. $d(2.5) = 16(2.5)^2$
$= 16(6.25)$
$= 100$ feet

3. $V = \frac{2{,}400}{P}$
$150 = \frac{2{,}400}{P}$
$150P = 2{,}400$
$P = \frac{2{,}400}{150}$
$= 16$ pounds per square inch

4. $y = Kxz^2$
$81 = K(2)\,9^2$
$81 = 162K$
$\Rightarrow K = \frac{1}{2}$
$y = \frac{1}{2}xz^2$
$= \frac{1}{2}(4)(4^2)$
$= 32$

5. $R = \frac{0.0005l}{d^2}$
$= \frac{0.0005(300)}{(0.25)^2}$
$= 2.4$ ohms

Chapter 4

Section 4.1

1.
$$\begin{array}{rl} 2x - y = 7 & \xrightarrow{\text{4 times each side}} \\ 3x + 4y = -6 & \xrightarrow[\text{no change}]{} \end{array} \quad \begin{array}{rcr} 8x - 4y &=& 28 \\ 3x + 4y &=& -6 \\ \hline 11x &=& 22 \\ x &=& 2 \end{array}$$

Substituting $x = 2$ into $2x - y = 7$ gives us $y = -3$. The solution is $(2, -3)$.

2.
$$\begin{array}{rl} 3x - 2y = -8 & \xrightarrow{\text{2 times each side}} \\ -2x + 3y = 7 & \xrightarrow[\text{3 times each side}]{} \end{array} \quad \begin{array}{rcr} 6x - 4y &=& -16 \\ -6x + 9y &=& 21 \\ \hline 5y &=& 5 \\ y &=& 1 \end{array}$$

Substituting $y = 1$ into $3x - 2y = -8$ gives us $x = -2$. The solution is $(-2, 1)$.

3.
$$\begin{array}{rl} 3x - 5y = 2 & \xrightarrow{\text{2 times each side}} \\ 2x + 4y = 1 & \xrightarrow[\text{-3 times each side}]{} \end{array} \quad \begin{array}{rcr} 6x - 10y &=& 4 \\ -6x - 12y &=& -3 \\ \hline -22y &=& 1 \\ y &=& -\frac{1}{22} \end{array}$$

$$\begin{array}{rl} 3x - 5y = 2 & \xrightarrow{\text{4 times each side}} \\ 2x + 4y = 1 & \xrightarrow[\text{5 times each side}]{} \end{array} \quad \begin{array}{rcr} 12x - 20y &=& 8 \\ 10x + 20y &=& 5 \\ \hline 22x &=& 13 \\ x &=& \frac{13}{22} \end{array}$$

Solution: $\left(\frac{13}{22}, -\frac{1}{22}\right)$

4. $$\begin{aligned} 2x + 7y &= 3 \xrightarrow{-2 \text{ times each side}} & -4x - 14y &= -6 \\ 4x + 14y &= 1 \xrightarrow[\text{no change}]{} & 4x + 14y &= 1 \\ & & \hline 0 &= -5 \end{aligned}$$

We have eliminated both variables and are left with a false statement, indicating that the lines are parallel. There is no solution to the system.

5. $$\begin{aligned} 2x + 7y &= 3 \xrightarrow{-2 \text{ times each side}} & -4x - 14y &= -6 \\ 4x + 14y &= 6 \xrightarrow[\text{no change}]{} & 4x + 14y &= 6 \\ & & \hline 0 &= 0 \end{aligned}$$

We have eliminated both variables and are left with a true statement. The lines coincide. Any ordered pair that satisfies one of the equations will satisfy the other.

6. $$\begin{aligned} \frac{1}{3}x + \frac{1}{2}y &= 4 \xrightarrow{\text{times } 6} & 2x + 3y &= 24 \\ \frac{2}{3}x - \frac{1}{4}y &= 3 \xrightarrow[\text{times } 12]{} & 8x - 3y &= 36 \\ & & \hline 10x \quad &= 60 \\ & & x &= 6 \end{aligned}$$

Substituting $x = 6$ into any equation with both variables gives $y = 4$. The solution is (6, 4).

7. $4x - 2y = -2$

$y = x + 3$

Substituting $x + 3$ for y in the first equation gives us

$$\begin{aligned} 4x - 2(x + 3) &= -2 \\ 4x - 2x - 6 &= -2 \\ 2x - 6 &= -2 \\ 2x &= 4 \\ x &= 2 \end{aligned}$$

When $x = 2, y = 2 + 3 = 5$.

The solution is (2, 5).

8. $5x - 3y = -4$

$x + 2y = 7$

Solving the second equation for x gives $x = -2y + 7$. Substituting this for x in the first equation we have

$$\begin{aligned} 5(-2y + 7) - 3y &= -4 \\ -10y + 35 - 3y &= -4 \\ -13y + 35 &= -4 \\ -13y &= -39 \\ y &= 3 \end{aligned}$$

Putting $y = 3$ into either of the first two equations gives $x = 1$. The solution is (1, 3).

Section 4.2

1. $$\begin{aligned} x + 2y + z &= 2 \quad (1) \\ x + y - z &= 6 \quad (2) \\ x - y + 2z &= -7 \quad (3) \end{aligned}$$

Adding equations (1) and (2) yields

$2x + 3y = 8 \quad (4)$

Adding twice (2) to (3) yields

$3x + y = 5 \quad (5)$

We solve the system made up of equations (4) and (5) as follows.

$$\begin{aligned} 2x + 3y &= 8 \xrightarrow{\text{no change}} & 2x + 3y &= 8 \\ 3x + y &= 5 \xrightarrow[\text{times } -3]{} & -9x - 3y &= -15 \\ & & \hline -7x \quad &= -7 \\ & & x &= 1 \end{aligned}$$

Substituting $x = 1$ into equation (4) or (5) gives us $y = 2$. Substituting $x = 1$ and $y = 2$ into equation (1), (2), or (3) gives us $z = -3$. The solution is (1, 2, −3).

2. $$\begin{aligned} 3x - 2y + z &= 2 \quad (1) \\ 3x + y + 3z &= 7 \quad (2) \\ x + 4y - z &= 4 \quad (3) \end{aligned}$$

Adding (1) and (3) yields

$4x + 2y = 6 \quad (4)$

Adding three times (3) to (2) yields

$6x + 13y = 19 \quad (5)$

We solve the system made up of (4) and (5) as follows:

$$\begin{aligned} 4x + 2y &= 6 \xrightarrow{\text{times } 6} & 24x + 12y &= 36 \\ 6x + 13y &= 19 \xrightarrow[\text{times } -4]{} & -24x - 52y &= -76 \\ & & \hline -40y &= -40 \\ & & y &= 1 \end{aligned}$$

Putting $y = 1$ into equation (4) or (5) gives us $x = 1$. When x is 1 and y is 1 in equation (1), (2), or (3), then $z = 1$. The solution is (1, 1, 1).

3. $$\begin{aligned} 3x + 5y - 2z &= 1 \quad (1) \\ 6x + 10y - 4z &= 2 \quad (2) \\ x - 8y + z &= 4 \quad (3) \end{aligned}$$

Equation (2) is twice equation (1), indicating that the planes they represent coincide with one another. There is either no solution to the system or an infinite number of solutions.

4. $$\begin{aligned} 3x - y + 2z &= 4 \quad (1) \\ 6x - 2y + 4z &= 2 \quad (2) \\ 5x - 3y + 7z &= 5 \quad (3) \end{aligned}$$

If we add −2 times equation (1) to equation (2), we are left with the false statement $0 = -6$. This means that the planes represented by these equations are parallel. There is therefore no solution to the system.

5. $$\begin{aligned} x + 2y &= 0 \quad (1) \\ 3y + z &= -3 \quad (2) \\ 2x - z &= 5 \quad (3) \end{aligned}$$

Adding equations (2) and (3) gives us equation (4) which is

$2x + 3y = 2 \quad (4)$

Equations (4) and (1) form a system in two variables.

$$\begin{aligned} x + 2y &= 0 \xrightarrow{\text{times } -2} & -2x - 4y &= 0 \\ 2x + 3y &= 2 \xrightarrow[\text{no change}]{} & 2x + 3y &= 2 \\ & & \hline -y &= 2 \\ & & y &= -2 \end{aligned}$$

When $y = -2$ in equation (1), x becomes 4. When we substitute −2 for y in equation (2), z is 3. The solution is (4, −2, 3).

Section 4.3

1. a. $\begin{vmatrix} 2 & 1 \\ 4 & 3 \end{vmatrix} = 6 - 4 = 2$

b. $\begin{vmatrix} 4 & -2 \\ 0 & 3 \end{vmatrix} = 12 - 0 = 12$

2. $\begin{vmatrix} -3 & x \\ 2 & x \end{vmatrix} = 20$

$$-3x - 2x = 20$$
$$-5x = 20$$
$$x = -4$$

3. (products: $-5 \quad -8 \quad 0$ above; $2 \quad 0 \quad 6$ below)

$$\begin{vmatrix} 2 & 0 & -1 \\ 3 & 1 & 2 \\ 5 & -2 & 1 \end{vmatrix} \begin{matrix} 2 & 0 \\ 3 & 1 \\ 5 & -2 \end{matrix} = 2 + 0 + 6 - (-5 - 8 + 0) = 8 - (-13)$$
$$= 21$$

4. $\begin{vmatrix} 2 & 0 & -1 \\ 3 & 1 & 2 \\ 5 & -2 & 1 \end{vmatrix} = 2\begin{vmatrix} 1 & 2 \\ -2 & 1 \end{vmatrix} - 0\begin{vmatrix} 3 & 2 \\ 5 & 1 \end{vmatrix} - 1\begin{vmatrix} 3 & 1 \\ 5 & -2 \end{vmatrix}$

$$= 2(1 + 4) - 0(3 - 10) - 1(-6 - 5)$$
$$= 2(5) - 0 - 1(-11)$$
$$= 10 + 11$$
$$= 21$$

5. $\begin{vmatrix} 0 & 4 & -2 \\ 3 & 1 & 1 \\ 1 & -2 & 0 \end{vmatrix} = -4\begin{vmatrix} 3 & 1 \\ 1 & 0 \end{vmatrix} + 1\begin{vmatrix} 0 & -2 \\ 1 & 0 \end{vmatrix} + 2\begin{vmatrix} 0 & -2 \\ 3 & 1 \end{vmatrix}$

$$= -4(0 - 1) + 1(0 + 2) + 2(0 + 6)$$
$$= -4(-1) + 1(2) + 2(6)$$
$$= 4 + 2 + 12$$
$$= 18$$

Section 4.4

1. $3x - 5y = 2$
$2x + 4y = 1$

$$D = \begin{vmatrix} 3 & -5 \\ 2 & 4 \end{vmatrix} = 22$$

$$D_x = \begin{vmatrix} 2 & -5 \\ 1 & 4 \end{vmatrix} = 13$$

$$D_y = \begin{vmatrix} 3 & 2 \\ 2 & 1 \end{vmatrix} = -1$$

$$x = \frac{D_x}{D} = \frac{13}{22}$$

$$y = \frac{D_y}{D} = -\frac{1}{22}$$

The solution is $\left(\frac{13}{22}, -\frac{1}{22}\right)$.

2. $x + 2y + z = 2$
$x + y - z = 6$
$x - y + z = -4$

$$D = \begin{vmatrix} 1 & 2 & 1 \\ 1 & 1 & -1 \\ 1 & -1 & 1 \end{vmatrix} = -6$$

$$D_x = \begin{vmatrix} 2 & 2 & 1 \\ 6 & 1 & -1 \\ -4 & -1 & 1 \end{vmatrix} = -6$$

$$D_y = \begin{vmatrix} 1 & 2 & 1 \\ 1 & 6 & -1 \\ 1 & -4 & 1 \end{vmatrix} = -12$$

$$D_z = \begin{vmatrix} 1 & 2 & 2 \\ 1 & 1 & 6 \\ 1 & -1 & -4 \end{vmatrix} = 18$$

$$x = \frac{D_x}{D} = \frac{-6}{-6} = 1$$

$$y = \frac{D_y}{D} = \frac{-12}{-6} = 2$$

$$z = \frac{D_z}{D} = \frac{18}{-6} = -3$$

The solution is $(1, 2, -3)$.

3. $x + y = 3$
$2x - z = 3$
$y + 2z = 9$

$$D = \begin{vmatrix} 1 & 1 & 0 \\ 2 & 0 & -1 \\ 0 & 1 & 2 \end{vmatrix} = -3$$

$$D_x = \begin{vmatrix} 3 & 1 & 0 \\ 3 & 0 & -1 \\ 9 & 1 & 2 \end{vmatrix} = -12$$

$$D_y = \begin{vmatrix} 1 & 3 & 0 \\ 2 & 3 & -1 \\ 0 & 9 & 2 \end{vmatrix} = 3$$

$$D_z = \begin{vmatrix} 1 & 1 & 3 \\ 2 & 0 & 3 \\ 0 & 1 & 9 \end{vmatrix} = -15$$

$$x = \frac{D_x}{D} = \frac{-12}{-3} = 4$$

$$y = \frac{D_y}{D} = \frac{3}{-3} = -1$$

$$z = \frac{D_z}{D} = \frac{-15}{-3} = 5$$

The solution is $(4, -1, 5)$.

Section 4.5

1. Let x and y represent the two numbers. Then,
$x + y = 14$
$y = 2x - 1$
The two numbers are 5 and 9.

2. Let x = the number of adult tickets sold and y = the number of children's tickets sold.
$x + y = 750$
$2x + 1y = 1{,}090$
The number of adult tickets is 340 and the number of children's tickets is 410.

3. Let x = the amount invested at 6% and y = the amount invested at 7%.
$x + y = 12{,}000$
$0.06x + 0.07y = 790$
To solve this system, multiply the top equation by -6 and the bottom equation by 100 to get
$-6x - 6y = -72{,}000$
$6x + 7y = 79{,}000$
The amount invested at 6% is \$5,000 and the amount invested at 7% is \$7,000.

4. Let x = the number of gallons of 30% solution and y = the number of gallons of 70% solution.

$$x + y = 16$$
$$0.30x + 0.70y = 0.60(16) = 9.6$$

Multiply the top equation by -3 and the bottom equation by 10 to get

$$-3x - 3y = -48$$
$$3x + 7y = 96$$

which yields 4 gallons of 30% solution and 12 gallons of 70% solution.

5. Let x = the speed of the boat in still water and y = the speed of the current.

	d	*r*	*t*
Upstream	18	$x - y$	3
Downstream	20	$x + y$	2

$$18 = 3(x - y)$$
$$20 = 2(x + y)$$

or
$$3x - 3y = 18$$
$$2x + 2y = 20$$

The speed of the boat in still water is 8 miles per hour, whereas the speed of the current is 2 miles per hour.

6. Let x = the number of nickels,
y = the number of dimes,
and z = the number of quarters.

$$x + y + z = 15$$
$$5x + 10y + 25z = 110$$
$$x = 4y - 1$$

Rewriting the third equation, the system can be written as

$$x + y + z = 15$$
$$5x + 10y + 25z = 110$$
$$x - 4y = -1$$

There are 11 nickels, 3 dimes, and 1 quarter.

7. The new data are summarized in Table 1.

TABLE 1

CORRESPONDING TEMPERATURES

IN DEGREES FAHRENHEIT	IN DEGREES CELSIUS
95	35
167	75

If we assume the relationship is linear, then the formula that relates the two temperature scales can be written in slope-intercept form as

$$C = mF + b$$

Substituting the values from the table into this equation gives us two equations in two unknowns.

$$35 = 95m + b \xrightarrow{\text{Times } -1} -35 = -95m - b$$
$$75 = 167m + b \xrightarrow[\text{No change}]{} 75 = 167m + b$$
$$40 = 72m$$
$$m = \frac{40}{72} = \frac{5}{9}$$

To find the value of b, we substitute $m = \frac{5}{9}$ into $35 = 95m + b$ and solve for b.

$$35 = 95\left(\frac{5}{9}\right) + b$$

$$35 = \frac{475}{9} + b$$

$$b = 35 - \frac{475}{9} = \frac{315}{9} - \frac{475}{9} = -\frac{160}{9}$$

The equation that gives C in terms of F is

$$C = \frac{5}{9}F - \frac{160}{9}$$

Section 4.6

1.

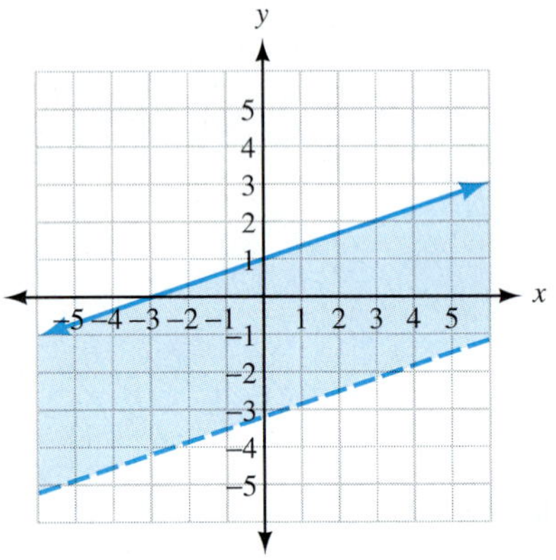

2.

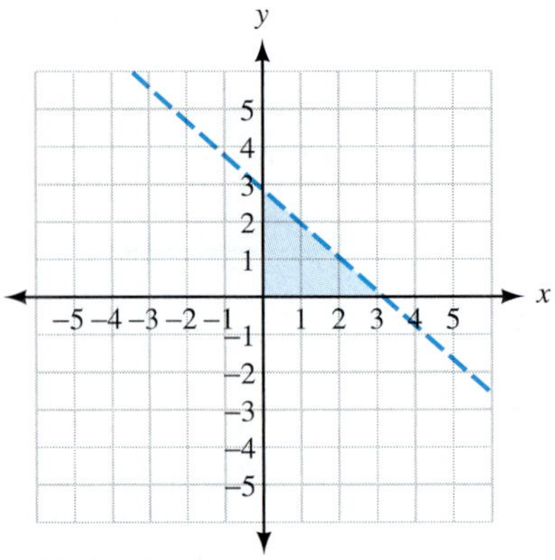

3.

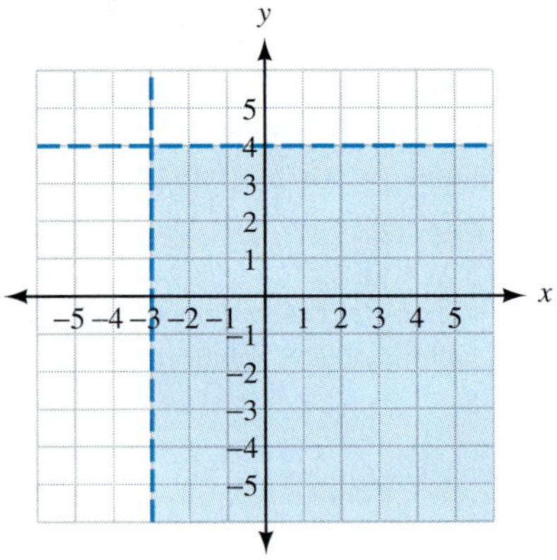

4.

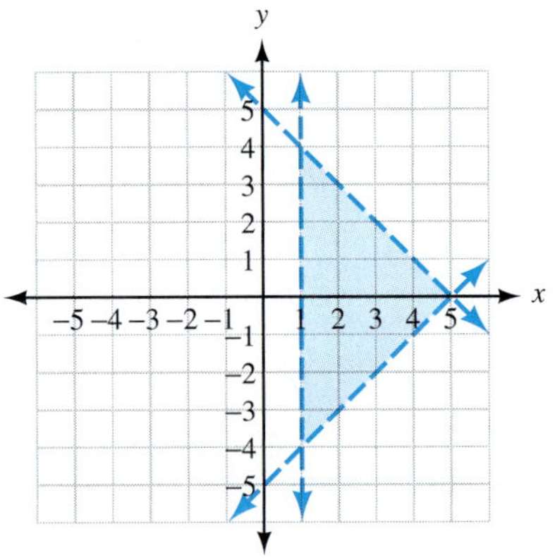

5. The horizontal line would move down to $y = 300$ from $y = 500$.

Chapter 5

Section 5.1

1. $x^5 \cdot x^4 = (x \cdot x \cdot x \cdot x \cdot x)(x \cdot x \cdot x \cdot x) = x^9$

2. $(4^5)^2 = 4^5 \cdot 4^5 = 4^{10}$

3. $(2x)^3 = (2x)(2x)(2x)$
$= 8x^3$

4. $(-2x^3)(4x^5) = (-2 \cdot 4)(x^3 \cdot x^5)$
$= -8x^8$

5. $(-3y^3)^3(2y^6) = (-27y^9)(2y^6)$
$= -54y^{15}$

6. $(a^2)^3(a^3b^4)^2(b^5)^2 = a^6 \cdot a^6 \cdot b^8 \cdot b^{10}$
$= a^{12}b^{18}$

7. $3^{-2} = \dfrac{1}{3^2} = \dfrac{1}{9}$

8. $(-5)^{-3} = \dfrac{1}{(-5)^3} = -\dfrac{1}{125}$

9. $\left(\dfrac{2}{3}\right)^{-2} = \dfrac{1}{\left(\dfrac{2}{3}\right)^2} = \dfrac{1}{\dfrac{4}{9}} = \dfrac{9}{4}$

10. $\dfrac{3^7}{3^4} = 3^{7-4} = 3^3 = 27$

11. $\dfrac{x^3}{x^{10}} = x^{3-10} = x^{-7} = \dfrac{1}{x^7}$

12. $\dfrac{a^5}{a^{-7}} = a^{5-(-7)} = a^{12}$

13. $\dfrac{m^{-3}}{m^{-5}} = m^{-3-(-5)} = m^{-3+5} = m^2$

16. $\dfrac{(x^2)^{-4}(x^3)^6}{(x^{-3})^5} = \dfrac{x^{-8}x^{18}}{x^{-15}} = \dfrac{x^{10}}{x^{-15}} = x^{10-(-15)} = x^{25}$

17. $\dfrac{18a^7b^{-4}}{36a^2b^{-8}} = \dfrac{18}{36} \cdot \dfrac{a^7}{a^2} \cdot \dfrac{b^{-4}}{b^{-8}} = \dfrac{1}{2} \cdot a^5 \cdot b^4 = \dfrac{a^5b^4}{2}$

18. $\dfrac{(3x^{-2}y^7)^2}{(x^5y^{-2})^{-4}} = \dfrac{9x^{-4}y^{14}}{x^{-20}y^8} = 9x^{16}y^6$

21. $(3 \times 10^7)(2 \times 10^{-4}) = (3 \cdot 2) \times (10^7)(10^{-4})$
$= 6 \times 10^3$

22. $\dfrac{3.9 \times 10^6}{1.3 \times 10^{-4}} = \dfrac{3.9}{1.3} \times \dfrac{10^6}{10^{-4}}$
$= 3 \times 10^{10}$

23. $\dfrac{(2.4 \times 10^6)(1.8 \times 10^{-4})}{1.2 \times 10^{-3}} = \dfrac{(2.4)(1.8)}{1.2} \times \dfrac{10^6 \cdot 10^{-4}}{10^{-3}}$
$= 3.6 \times 10^5$

Section 5.2

6. $(3x^2 + 2x - 5) + (2x^2 - 7x + 3) = (3x^2 + 2x^2) + (2x - 7x) + (-5 + 3)$
$= (3 + 2)x^2 + (2 - 7)x + (-5 + 3)$
$= 5x^2 - 5x - 2$

7.
$$\begin{array}{r} x^3 + 7x^2 + 3x + 2 \\ \underline{-3x^3 - 2x^2 + 3x - 1} \\ -2x^3 + 5x^2 + 6x + 1 \end{array}$$

8. $(4x^2 - 2x + 7) - (7x^2 - 3x + 1) = 4x^2 - 2x + 7 - 7x^2 + 3x - 1$
$= -3x^2 + x + 6$

9. $(7x - 4) - (3x + 5) = 7x - 4 - 3x - 5$
$= 4x - 9$

10. $2x - 4[6 - (5x + 3)]$
$= 2x - 4(6 - 5x - 3)$
$= 2x - 4(-5x + 3)$
$= 2x + 20x - 12$
$= 22x - 12$

11. $(9x - 4) - [(2x + 5) - (x + 3)]$
$= 9x - 4 - (2x + 5 - x - 3)$
$= 9x - 4 - (x + 2)$
$= 9x - 4 - x - 2$
$= 8x - 6$

12. $2(-2)^3 - 3(-2)^2 + 4(-2) - 8$
$= 2(-8) - 3(4) + 4(-2) - 8$
$= -16 - 12 - 8 - 8$
$= -44$

13. $P = -500 + 27x - 0.1x^2$
$= -500 + 27(170) - 0.1(170)^2$
$= -500 + 27(170) - 0.1(28{,}900)$
$= -500 + 4{,}590 - 2{,}890$
$= \$1{,}200$

Section 5.3

1. $5x^2(3x^2 - 4x + 2) = 5x^2(3x^2) + 5x^2(-4x) + 5x^2(2)$
$= 15x^4 - 20x^3 + 10x^2$

2. $(4x - 1)(x + 3) = (4x - 1)x + (4x - 1)3$
$= 4x^2 - x + 12x - 3$
$= 4x^2 + 11x - 3$

3.
$$\begin{array}{r} x^2 - 5x + 6 \\ 3x + 2 \\ \hline 3x^3 - 15x^2 + 18x \\ 2x^2 - 10x + 12 \\ \hline 3x^3 - 13x^2 + 8x + 12 \end{array}$$

4. $(2a - 3b)(5a - b) = 10a^2 - 2ab - 15ab + 3b^2$
F O I L
$= 10a^2 - 17ab + 3b^2$

5. $(6 - 3t)(2 + 5t) = 12 + 30t - 6t - 15t^2$
F O I L
$= 12 + 24t - 15t^2$

6. $\left(3x + \frac{1}{4}\right)\left(4x - \frac{1}{3}\right) = 12x^2 - x + x - \frac{1}{12}$
F O I L
$= 12x^2 - \frac{1}{12}$

7. $(a^4 - 2)(a^4 + 5) = a^8 + 5a^4 - 2a^4 - 10$
F O I L
$= a^8 + 3a^4 - 10$

8. $(7x - 2)(3y + 8) = 21xy + 56x - 6y - 16$
F O I L

9. $(3x - 2)^2 = (3x - 2)(3x - 2)$
$= 9x^2 - 6x - 6x + 4$
$= 9x^2 - 12x + 4$

11. $(4t + 5)^2 = (4t)^2 + 2(4t)(5) + 5^2$
$= 16t^2 + 40t + 25$

12. $(5x - 3y)^2 = (5x)^2 - 2(5x)(3y) + (3y)^2$
$= 25x^2 - 30xy + 9y^2$

13. $(6 - a^4)^2 = 6^2 - 2(6)(a^4) + (a^4)^2$
$= 36 - 12a^4 + a^8$

14. $(4x - 3)(4x + 3) = 16x^2 + 12x - 12x - 9$
$= 16x^2 - 9$

19. $R(p) = xp = (250 - 25p)p$
$= 250p - 25p^2$

To find $R(x)$, we solve the equation $x = 250 - 25p$ for p.
$250 - 25p = x$
$-25p = x - 250$
$p = \frac{x - 250}{-25}$
$= 10 - 0.04x$

We find $R(x)$ by substituting $10 - 0.04x$ for p in the formula $R(x) = xp$;
$R(x) = xp = x(10 - 0.04x)$
$= 10x - 0.04x^2$

20. $P(x) = R(x) - C(x)$
$= 10x - 0.04x^2 - (210 + 1.5x)$
$= -0.04x^2 + 8.5x - 210$
$P(130) = -0.04(130)^2 + 8.5(130) - 210$
$= -0.04(16{,}900) + 8.5(130) - 210$
$= -676 + 1{,}105 - 210$
$= \$219$

Section 5.4

1. $15a^7 - 25a^5 + 30a^3$
$= 5a^3(3a^4) - 5a^3(5a^2) + 5a^3(6)$
$= 5a^3(3a^4 - 5a^2 + 6)$

2. $12x^4y^5 - 9x^3y^4 - 15x^5y^3$
$= 3x^3y^3(4xy^2) - 3x^3y^3(3y) - 3x^3y^3(5x^2)$
$= 3x^3y^3(4xy^2 - 3y - 5x^2)$

3. $4(a + b)^4 - 6(a + b)^3 + 16(a + b)^2$
$= 2(a + b)^2 \cdot 2(a + b)^2 - 2(a + b)^2 \cdot 3(a + b) + 2(a + b)^2 \cdot 8$
$= 2(a + b)^2[2(a + b)^2 - 3(a + b) + 8]$

4. $R = 80.5x - 0.35x^2$
$= x(80.5 - 0.35x)$
$\Rightarrow p = 80.5 - 0.35x$

5. $ab^3 + b^3 + 6a + 6 = b^3(a + 1) + 6(a + 1)$
$= (a + 1)(b^3 + 6)$

6. $15 - 3y^2 - 5x^2 + x^2y^2 = 3(5 - y^2) - x^2(5 - y^2)$
$= (5 - y^2)(3 - x^2)$

7. $x^3 + 5x^2 + 3x + 15 = x^2(x + 5) + 3(x + 5)$
$= (x + 5)(x^2 + 3)$

Section 5.5

1. The leading coefficient is 1. We need two numbers whose sum is -1 and whose product is -12. The numbers are -4 and $+3$.

$$x^2 - x - 12 = (x - 4)(x + 3)$$

2. We need two numbers whose product is $-15y^2$ and whose sum is $+2y$. The numbers are $+5y$ and $-3y$.

$$x^2 + 2xy - 15y^2 = (x + 5y)(x - 3y)$$

3. Since there is no pair of integers whose product is 1 and whose sum is 1, the trinomial $x^2 + x + 1$ is not factorable. We say it is a prime polynomial.

4. $5x^2 + 25x + 30 = 5(x^2 + 5x + 6)$
$= 5(x + 3)(x + 2)$

5. We list all possible factors along with their products as follows:

POSSIBLE FACTORS	FIRST TERM	MIDDLE TERM	LAST TERM
$(3x + 2)(x - 1)$	$3x^2$	$-x$	-2
$(3x - 2)(x + 1)$	$3x^2$	$+x$	-2
$(3x + 1)(x - 2)$	$3x^2$	$-5x$	-2
$(3x - 1)(x + 2)$	$3x^2$	$+5x$	-2

The first line has the correct middle term:

$$3x^2 - x - 2 = (3x + 2)(x - 1)$$

6. $15x^4 + x^2 - 2 = (5x^2 + 2)(3x^2 - 1)$

7. $3x^2(x - 2) - 7x(x - 2) + 2(x - 2)$
$= (x - 2)(3x^2 - 7x + 2)$
$= (x - 2)(3x - 1)(x - 2)$
$= (x - 2)^2(3x - 1)$

8. $ac = -24$
$b = -5$
Two numbers whose product is -24 and whose sum is -5 are -8 and 3.
$8x^2 - 5x - 3 = 8x^2 - 8x + 3x - 3$
$= 8x(x - 1) + 3(x - 1)$
$= (x - 1)(8x + 3)$

9. $ac = 24$
$b = 11$
Two numbers whose product is 24 and whose sum is 11 are 8 and 3.
$4x^2 + 11x + 6 = 4x^2 + 8x + 3x + 6$
$= 4x(x + 2) + 3(x + 2)$
$= (x + 2)(4x + 3)$

10. $ac = -60$
$b = -4$
Two numbers whose product is -60 and whose sum is -4 are -10 and 6.
$4x^2 - 4x - 15 = 4x^2 - 10x + 6x - 15$
$= 2x(2x - 5) + 3(2x - 5)$
$= (2x - 5)(2x + 3)$

Section 5.6

5. $(y + 2)^2 + 8(y + 2) + 16 = [(y + 2) + 4]^2$
$= (y + 6)^2$

6. $27x^2 - 36x + 12 = 3(9x^2 - 12x + 4)$
$= 3(3x - 2)^2$

7. $x^2 - 16 = x^2 - 4^2 = (x + 4)(x - 4)$

8. $64 - t^2 = 8^2 - t^2 = (8 + t)(8 - t)$

9. $25x^2 - 36y^2 = (5x)^2 - (6y)^2 = (5x + 6y)(5x - 6y)$

10. $9x^6 - 1 = (3x^3)^2 - 1^2 = (3x^3 + 1)(3x^3 - 1)$

11. $x^2 - \dfrac{25}{64} = x^2 - \left(\dfrac{5}{8}\right)^2 = \left(x + \dfrac{5}{8}\right)\left(x - \dfrac{5}{8}\right)$

12. $x^4 - 81 = (x^2)^2 - 9^2$
$= (x^2 - 9)(x^2 + 9)$
$= (x - 3)(x + 3)(x^2 + 9)$

13. $(x - 4)^2 - 9 = [(x - 4) - 3][(x - 4) + 3]$
$= (x - 7)(x - 1)$

14. $x^2 - 6x + 9 - y^2 = (x - 3)^2 - y^2$
$= (x - 3 + y)(x - 3 - y)$

15. $x^3 + 5x^2 - 4x - 20 = x^2(x + 5) - 4(x + 5)$
$= (x + 5)(x^2 - 4)$
$= (x + 5)(x - 2)(x + 2)$

16. We use the distributive property to expand the product:

$$\begin{aligned}(x - 3)(x^2 + 3x + 9) &= x^3 + 3x^2 + 9x \\ &\qquad -3x^2 - 9x - 27 \\ &= x^3 \qquad\qquad\quad - 27\end{aligned}$$

17. The first term is the cube of 3 and the second term is the cube of x. Therefore,

$$\begin{aligned}27 + x^3 = 3^3 + x^3 &= (3 + x)[3^2 - 3(x) + x^2] \\ &= (3 + x)(9 - 3x + x^2)\end{aligned}$$

18. $8x^3 + y^3 = (2x)^3 + y^3$
$= (2x + y)[(2x)^2 - (2x)(y) + y^2]$
$= (2x + y)(4x^2 - 2xy + y^2)$

19. $a^3 - \frac{1}{27} = a^3 - \left(\frac{1}{3}\right)^3 = \left(a - \frac{1}{3}\right)\left[a^2 + a\left(\frac{1}{3}\right) + \left(\frac{1}{3}\right)^2\right]$
$= \left(a - \frac{1}{3}\right)\left(a^2 + \frac{1}{3}a + \frac{1}{9}\right)$

20. $x^6 - 1 = (x^3)^2 - 1^2 = (x^3 + 1)(x^3 - 1)$
$= (x + 1)(x^2 - x + 1)(x - 1)(x^2 + x + 1)$

Section 5.7

1. $3x^8 - 27x^6 = 3x^6(x^2 - 9)$
$= 3x^6(x + 3)(x - 3)$

2. $4x^4 + 40x^3 + 100x^2 = 4x^2(x^2 + 10x + 25)$
$= 4x^2(x + 5)^2$

3. $y^4 + 36y^2 = y^2(y^2 + 36)$

4. $6x^2 - x - 15 = (3x - 5)(2x + 3)$

5. $3x^5 - 81x^2 = 3x^2(x^3 - 27)$
$= 3x^2(x - 3)(x^2 + 3x + 9)$

6. $3a^2b^3 + 6a^2b^2 - 3a^2b = 3a^2b(b^2 + 2b - 1)$

7. $x^2 - 10x + 25 - b^2 = (x - 5)^2 - b^2$
$= (x - 5 + b)(x - 5 - b)$

Section 5.8

1. $x^2 - x - 6 = 0$
$(x - 3)(x + 2) = 0$
$x - 3 = 0$ or $x + 2 = 0$
$x = 3$ or $x = -2$

2. $6 \cdot \frac{1}{2}x^3 = 6 \cdot \frac{5}{6}x^2 + 6 \cdot \frac{1}{3}x$
$3x^3 = 5x^2 + 2x$
$3x^3 - 5x^2 - 2x = 0$
$x(3x^2 - 5x - 2) = 0$
$x(3x + 1)(x - 2) = 0$
$x = 0$ or $3x + 1 = 0$ or $x - 2 = 0$
$x = 0$ or $x = -\frac{1}{3}$ or $x = 2$

3. $100x^2 = 500x$
$100x^2 - 500x = 0$
$100x(x - 5) = 0$
$100x = 0$ or $x - 5 = 0$
$x = 0$ or $x = 5$

4. $(x + 1)(x + 2) = 12$
$x^2 + 3x + 2 = 12$
$x^2 + 3x - 10 = 0$
$(x + 5)(x - 2) = 0$
$x + 5 = 0$ or $x - 2 = 0$
$x = -5$ or $x = 2$

5. $x^3 + 5x^2 - 4x - 20 = 0$
$x^2(x + 5) - 4(x + 5) = 0$
$(x + 5)(x^2 - 4) = 0$
$(x + 5)(x + 2)(x - 2) = 0$
$x + 5 = 0$ or $x + 2 = 0$ or $x - 2 = 0$
$x = -5$ or $x = -2$ or $x = 2$

6. Let x = one integer and $x + 3$ = the other.
$x^2 + (x + 3)^2 = 29$
$x^2 + x^2 + 6x + 9 = 29$
$2x^2 + 6x + 9 = 29$
$2x^2 + 6x - 20 = 0$
$x^2 + 3x - 10 = 0$
$(x + 5)(x - 2) = 0$
$x + 5 = 0$ or $x - 2 = 0$
$x = -5$ or $x = 2$
$x + 3 = -5 + 3$ $\quad x + 3 = 2 + 3$
$= -2$ $\quad = 5$
The two integers are −5 and −2 or 2 and 5.

7.

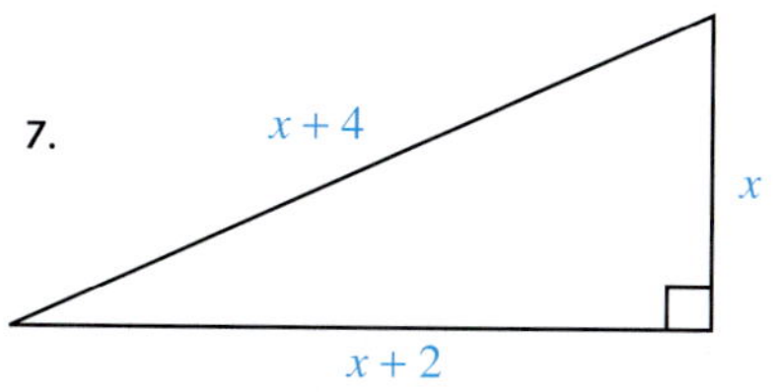

$(x + 4)^2 = (x + 2)^2 + x^2$
$x^2 + 8x + 16 = x^2 + 4x + 4 + x^2$
$x^2 + 8x + 16 = 2x^2 + 4x + 4$
$0 = x^2 - 4x - 12$
$0 = (x - 6)(x + 2)$
$x - 6 = 0$ or $x + 2 = 0$
$x = 6$ or $x = -2$
Since the length of a side cannot be negative, the shortest side is 6. The other two sides are 6 + 2 = 8 and 6 + 4 = 10.

8. If $v = 64$ and $h = 64$
then $h = vt - 16t^2$
becomes $64 = 64t - 16t^2$
$16t^2 - 64t + 64 = 0$
$t^2 - 4t + 4 = 0$
$(t - 2)^2 = 0$
$t = 2$ seconds

9. $R = (1{,}300 - 100p)p$
If $R = 3{,}600$, then
$3{,}600 = (1{,}300 - 100p)p$
$3{,}600 = 1{,}300p - 100p^2$
$100p^2 - 1{,}300p + 3{,}600 = 0$
$p^2 - 13p + 36 = 0$
$(p - 4)(p - 9) = 0$
$p - 4 = 0$ or $p - 9 = 0$
$p = 4$ or $p = 9$
The price can be set at either \$4 or \$9.

Chapter 6

Section 6.1

1. $\dfrac{x^2-9}{x+3} = \dfrac{\cancel{(x+3)}(x-3)}{\cancel{x+3}}$

$= x - 3$

2. $\dfrac{y^2-y-6}{y^2-4} = \dfrac{(y-3)\cancel{(y+2)}}{(y-2)\cancel{(y+2)}}$

$= \dfrac{y-3}{y-2}$

3. $\dfrac{3a^3+3}{6a^2-6a+6} = \dfrac{3(a^3+1)}{6(a^2-a+1)}$

$= \dfrac{3(a+1)\cancel{(a^2-a+1)}}{6\cancel{(a^2-a+1)}}$

$= \dfrac{a+1}{2}$

4. $\dfrac{x^2+4x+ax+4a}{x^2+ax+4x+4a} = \dfrac{x(x+4)+a(x+4)}{x(x+a)+4(x+a)}$

$= \dfrac{\cancel{(x+4)}\cancel{(x+a)}}{\cancel{(x+a)}\cancel{(x+4)}}$

$= 1$

5. When $a = 7$ and $b = 4$

the expression $\dfrac{a-b}{b-a}$ becomes

$\dfrac{7-4}{4-7} = \dfrac{3}{-3} = -1$

6. $\dfrac{7-x}{x^2-49} = \dfrac{-1\cancel{(x-7)}}{(x+7)\cancel{(x-7)}} = \dfrac{-1}{x+7}$

7. $C = 165(3.14) = 518$ feet to the nearest foot

$r = \dfrac{d}{t} = \dfrac{518}{40} = 13.0$ feet per second

9. a. $f(0) = \dfrac{-0+6}{-0-3}$

$= \dfrac{6}{-3}$

$= -2$

b. $f(-6) = \dfrac{-6+6}{-6-3}$

$= \dfrac{0}{-9}$

$= 0$

c. $f(6) = \dfrac{6+6}{6-3}$

$= \dfrac{12}{3}$

$= 4$

d. $f(3) = \dfrac{3+6}{3-3}$

$= \dfrac{9}{0}$

Undefined

10.

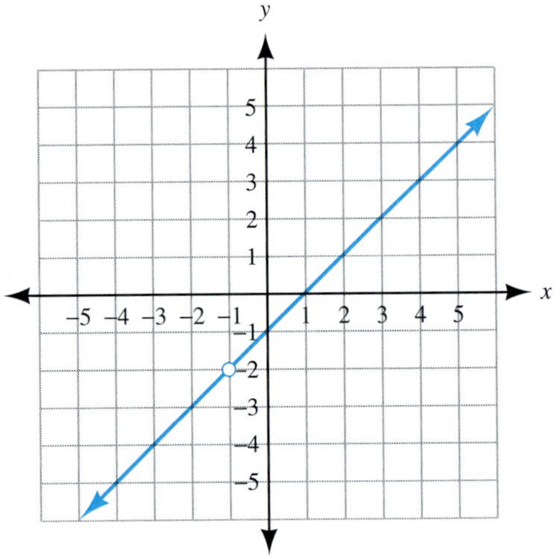

This graph does not contain the point $(-1, -2)$, whereas the graph of $y = x - 1$ does.

Section 6.2

1. $\dfrac{12x^4-18x^3+24x^2}{6x} = \dfrac{12x^4}{6x} - \dfrac{18x^3}{6x} + \dfrac{24x^2}{6x}$

$= 2x^3 - 3x^2 + 4x$

2. $\dfrac{27x^4y^7-81x^5y^3}{-9x^3y^2} = \dfrac{27x^4y^7}{-9x^3y^2} - \dfrac{81x^5y^3}{-9x^3y^2}$

$= -3xy^5 + 9x^2y$

3. $\dfrac{12a^5+8a^4+16a^3+4a^2}{8a^4} = \dfrac{12a^5}{8a^4} + \dfrac{8a^4}{8a^4} + \dfrac{16a^3}{8a^4} + \dfrac{4a^2}{8a^4}$

$= \dfrac{3a}{2} + 1 + \dfrac{2}{a} + \dfrac{1}{2a^2}$

4. $\dfrac{2x^2-5xy+3y^2}{x-y} = \dfrac{(2x-3y)\cancel{(x-y)}}{\cancel{x-y}}$

$= 2x - 3y$

5. $\dfrac{f(x)-f(a)}{x-a} = \dfrac{(2x+3)-(2a+3)}{x-a}$

$= \dfrac{2x-2a}{x-a}$

$= \dfrac{2(x-a)}{x-a}$

$= 2$

6. $\dfrac{f(x)-f(a)}{x-a} = \dfrac{(x^2-1)-(a^2-1)}{x-a}$

$= \dfrac{x^2-a^2}{x-a}$

$= \dfrac{(x+a)(x-a)}{x-a}$

$= x + a$

7.
$$\begin{array}{r} 215 \\ 35\overline{)7{,}546} \\ \underline{7\,0\downarrow\,\,} \\ 54 \\ \underline{35\downarrow} \\ 196 \\ \underline{175} \\ 21 \end{array}$$

Answer: $215 + \frac{3}{5}$

8.
$$\begin{array}{r} 3x + 1 \\ x - 3\overline{)\,3x^2 - 8x - 1} \\ - \quad + \qquad \\ \underline{\cancel{+}\,3x^2 \cancel{-}\,9x \qquad} \\ + \ x - 1 \\ - \quad + \\ \underline{\cancel{+}\ x \cancel{-}\,3} \\ 2 \end{array}$$

Answer: $3x + 1 + \dfrac{2}{x - 3}$

9.
$$\begin{array}{r} 3x^2 + 6x + 15 \\ x - 2\overline{)\,3x^3 + 0x^2 + 3x + 1} \\ - \quad + \qquad\qquad\qquad \\ \underline{\cancel{+}\,3x^3 \cancel{-}\,6x^2 \qquad\qquad\quad} \\ + 6x^2 + \ 3x \qquad \\ - \quad + \qquad\qquad \\ \underline{\cancel{+}\,6x^2 \cancel{-}\,12x \qquad} \\ + 15x + 1 \\ - \quad + \\ \underline{\cancel{+}\,15x \cancel{-}\,30} \\ 31 \end{array}$$

Answer: $3x^2 + 6x + 15 + \dfrac{31}{x - 2}$

10.
$$\begin{array}{r} 2x - 3y \\ x - y\overline{)\,2x^2 - 5xy + 3y^2} \\ - \quad + \qquad\quad \\ \underline{\cancel{+}\,2x^2 \cancel{-}\,2xy \qquad\quad} \\ - 3xy + 3y^2 \\ + \quad - \\ \underline{\cancel{-}\,3xy \cancel{+}\,3y^2} \\ 0 \end{array}$$

Answer: $2x - 3y$

Section 6.3

1. $\dfrac{3}{4} \cdot \dfrac{12}{27} = \dfrac{3 \cdot 12}{4 \cdot 27}$

$= \dfrac{\cancel{3} \cdot \cancel{2} \cdot \cancel{2} \cdot \cancel{3}}{\cancel{2} \cdot \cancel{2} \cdot \cancel{3} \cdot \cancel{3} \cdot 3}$

$= \dfrac{1}{3}$

2. $\dfrac{6x^4}{4y^9} \cdot \dfrac{12y^5}{3x^2} = \dfrac{\overset{2}{\cancel{6}} \cdot \overset{3}{\cancel{12}}x^4y^5}{\cancel{3} \cdot \cancel{4}x^2y^9}$

$= \dfrac{6x^2}{y^4}$

3. $\dfrac{x + 5}{x^2 - 25} \cdot \dfrac{x - 5}{x^2 - 10x + 25} = \dfrac{\cancel{(x + 5)}\,\cancel{(x - 5)}}{\cancel{(x + 5)}\,\cancel{(x - 5)}\,(x - 5)^2}$

$= \dfrac{1}{(x - 5)^2}$

4. $\dfrac{3y^2 - 3y}{3y - 12} \cdot \dfrac{y^2 - 2y - 8}{y^2 + 3y + 2} = \dfrac{\cancel{3}y(y - 1)\cancel{(y - 4)}\,\cancel{(y + 2)}}{\cancel{3}\cancel{(y - 4)}\,(y + 1)\cancel{(y + 2)}}$

$= \dfrac{y(y - 1)}{y + 1}$

5. $\dfrac{5}{9} \div \dfrac{10}{27} = \dfrac{5}{9} \cdot \dfrac{27}{10}$

$= \dfrac{\cancel{5} \cdot \cancel{3} \cdot \cancel{3} \cdot 3}{\cancel{3} \cdot \cancel{3} \cdot 2 \cdot \cancel{5}}$

$= \dfrac{3}{2}$

6. $\dfrac{9x^4}{4y^3} \div \dfrac{3x^2}{8y^5} = \dfrac{9x^4}{4y^3} \cdot \dfrac{8y^5}{3x^2}$

$= \dfrac{\overset{3}{\cancel{9}} \cdot \overset{2}{\cancel{8}}x^4y^5}{\cancel{3} \cdot \cancel{4}x^2y^3}$

$= 6x^2y^2$

7. $\dfrac{xy^2 - y^3}{x^2 - y^2} \div \dfrac{x^3 + y^3}{x^2 + 2xy + y^2} = \dfrac{xy^2 - y^3}{x^2 - y^2} \cdot \dfrac{x^2 + 2xy + y^2}{x^3 + y^3}$

$= \dfrac{y^2\cancel{(x - y)}\,\cancel{(x + y)}\,\cancel{(x + y)}}{\cancel{(x + y)}\,\cancel{(x - y)}\,\cancel{(x + y)}\,(x^2 - xy + y^2)}$

$= \dfrac{y^2}{x^2 - xy + y^2}$

8. $\dfrac{a^2 + 3a - 4}{a - 4} \cdot \dfrac{a + 3}{a^2 - 4a + 3} \div \dfrac{a + 1}{a^2 - 2a - 3}$

$= \dfrac{(a^2 + 3a - 4)(a + 3)(a^2 - 2a - 3)}{(a - 4)(a^2 - 4a + 3)(a + 1)}$

$= \dfrac{(a + 4)\cancel{(a - 1)}\,(a + 3)\cancel{(a - 3)}\,\cancel{(a + 1)}}{(a - 4)\cancel{(a - 3)}\,\cancel{(a - 1)}\,\cancel{(a + 1)}}$

$= \dfrac{(a + 4)(a + 3)}{a - 4}$

9. $\dfrac{xa + xb - ya - yb}{xa + 2x + ya + 2y} \cdot \dfrac{xa + 2x + ya + 2y}{xa + xb + ya + yb}$

$= \dfrac{x(a + b) - y(a + b)}{x(a + 2) + y(a + 2)} \cdot \dfrac{x(a + 2) + y(a + 2)}{x(a + b) + y(a + b)}$

$= \dfrac{\cancel{(a + b)}\,(x - y)\cancel{(a + 2)}\,\cancel{(x + y)}}{\cancel{(a + 2)}\,(x + y)\cancel{(a + b)}\,\cancel{(x + y)}}$

$= \dfrac{x - y}{x + y}$

10. $(5x^2 - 45) \cdot \dfrac{3}{5x - 15} = \dfrac{5x^2 - 45}{1} \cdot \dfrac{3}{5x - 15}$

$= \dfrac{\cancel{5}(x + 3)\cancel{(x - 3)}\,3}{\cancel{5}\cancel{(x - 3)}}$

$= 3(x + 3)$

Section 6.4

1. $\frac{3}{8} + \frac{1}{8} = \frac{3+1}{8}$

$= \frac{4}{8}$

$= \frac{1}{2}$

2. $\frac{x}{x^2-9} + \frac{3}{x^2-9} = \frac{x+3}{x^2-9}$

$= \frac{\cancel{x+3}}{\cancel{(x+3)}(x-3)}$

$= \frac{1}{x-3}$

3. $\frac{2x-7}{x-2} - \frac{x-5}{x-2} = \frac{2x-7-(x-5)}{x-2}$

$= \frac{2x-7-x+5}{x-2}$

$= \frac{x-2}{x-2}$

$= 1$

4. $\frac{3}{10} + \frac{11}{42} = \frac{3}{2\cdot 5} + \frac{11}{2\cdot 3\cdot 7}$

$= \frac{3}{2\cdot 5}\cdot\frac{\mathbf{3\cdot 7}}{\mathbf{3\cdot 7}} + \frac{11}{2\cdot 3\cdot 7}\cdot\frac{\mathbf{5}}{\mathbf{5}}$

$= \frac{63}{2\cdot 3\cdot 5\cdot 7} + \frac{55}{2\cdot 3\cdot 5\cdot 7}$

$= \frac{118}{2\cdot 3\cdot 5\cdot 7}$

$= \frac{\cancel{2}\cdot 59}{\cancel{2}\cdot 3\cdot 5\cdot 7}$

$= \frac{59}{105}$

5. $\frac{-3}{x^2-2x-8} + \frac{4}{x^2-16} = \frac{-3}{(x-4)(x+2)} + \frac{4}{(x-4)(x+4)}$

$= \frac{-3}{(x-4)(x+2)}\cdot\frac{\mathbf{x+4}}{\mathbf{x+4}} + \frac{4}{(x-4)(x+4)}\cdot\frac{\mathbf{x+2}}{\mathbf{x+2}}$

$= \frac{-3x-12+4x+8}{(x-4)(x+4)(x+2)}$

$= \frac{\cancel{x-4}}{\cancel{(x-4)}(x+4)(x+2)}$

$= \frac{1}{(x+4)(x+2)}$

6. $\frac{x-4}{2x-6} + \frac{3}{x^2-9} = \frac{x-4}{2(x-3)} + \frac{3}{(x+3)(x-3)}$

$= \frac{x-4}{2(x-3)}\cdot\frac{\mathbf{x+3}}{\mathbf{x+3}} + \frac{3}{(x+3)(x-3)}\cdot\frac{\mathbf{2}}{\mathbf{2}}$

$= \frac{(x-4)(x+3)+3\cdot 2}{2(x+3)(x-3)}$

$= \frac{x^2-x-12+6}{2(x+3)(x-3)}$

$= \frac{x^2-x-6}{2(x+3)(x-3)}$

$= \frac{\cancel{(x-3)}(x+2)}{2(x+3)\cancel{(x-3)}}$

$= \frac{x+2}{2(x+3)}$

7. $\frac{2x-4}{x^2+5x+4} - \frac{x-4}{x^2+6x+8} = \frac{2x-4}{(x+4)(x+1)} - \frac{x-4}{(x+4)(x+2)}$

$= \frac{2x-4}{(x+4)(x+1)}\cdot\frac{\mathbf{x+2}}{\mathbf{x+2}} - \frac{x-4}{(x+4)(x+2)}\cdot\frac{\mathbf{x+1}}{\mathbf{x+1}}$

$= \frac{(2x-4)(x+2)-(x-4)(x+1)}{(x+4)(x+1)(x+2)}$

$= \frac{(2x^2-8)-(x^2-3x-4)}{(x+4)(x+1)(x+2)}$

$= \frac{x^2+3x-4}{(x+4)(x+1)(x+2)}$

$= \frac{\cancel{(x+4)}(x-1)}{\cancel{(x+4)}(x+1)(x+2)}$

$= \frac{x-1}{(x+1)(x+2)}$

8. $\dfrac{x^2}{x-4}+\dfrac{x+12}{4-x}=\dfrac{x^2}{x-4}+\dfrac{x+12}{4-x}\cdot\dfrac{-1}{-1}$

$=\dfrac{x^2}{x-4}+\dfrac{-x-12}{x-4}$

$=\dfrac{x^2-x-12}{x-4}$

$=\dfrac{\cancel{(x-4)}(x+3)}{\cancel{x-4}}$

$=x+3$

9. $2+\dfrac{25}{5x-1}=\dfrac{2}{1}+\dfrac{25}{5x-1}$

$=\dfrac{2}{1}\cdot\dfrac{\mathbf{5x-1}}{\mathbf{5x-1}}+\dfrac{25}{5x-1}$

$=\dfrac{10x-2+25}{5x-1}$

$=\dfrac{10x+23}{5x-1}$

10. $\dfrac{1}{x}+\dfrac{1}{3x}=\dfrac{3}{3x}+\dfrac{1}{3x}=\dfrac{4}{3x}$

Section 6.5

1. $\dfrac{\frac{2}{3}}{\frac{5}{6}}=\dfrac{2}{3}\cdot\dfrac{6}{5}=\dfrac{12}{15}=\dfrac{4}{5}$

2. $\dfrac{\frac{1}{x}-\frac{1}{3}}{\frac{1}{x}+\frac{1}{3}}=\dfrac{\left(\frac{1}{x}-\frac{1}{3}\right)\mathbf{3x}}{\left(\frac{1}{x}+\frac{1}{3}\right)\mathbf{3x}}$

$=\dfrac{\frac{1}{x}(3x)-\frac{1}{3}(3x)}{\frac{1}{x}(3x)+\frac{1}{3}(3x)}$

$=\dfrac{3-x}{3+x}$

3. $\dfrac{\frac{x+5}{x^2-16}}{\frac{x^2-25}{x-4}}=\dfrac{x+5}{x^2-16}\cdot\dfrac{x-4}{x^2-25}$

$=\dfrac{(x+5)(x-4)}{(x+4)(x-4)(x+5)(x-5)}$

$=\dfrac{1}{(x+4)(x-5)}$

4. $\dfrac{1-\frac{9}{x^2}}{1-\frac{1}{x}-\frac{6}{x^2}}=\dfrac{\mathbf{x^2}\cdot\left(1-\frac{9}{x^2}\right)}{\mathbf{x^2}\cdot\left(1-\frac{1}{x}-\frac{6}{x^2}\right)}$

$=\dfrac{x^2\cdot 1-x^2\cdot\frac{9}{x^2}}{x^2\cdot 1-x^2\cdot\frac{1}{x}-x^2\cdot\frac{6}{x^2}}$

$=\dfrac{x^2-9}{x^2-x-6}$

$=\dfrac{(x+3)\cancel{(x-3)}}{\cancel{(x-3)}(x+2)}$

$=\dfrac{x+3}{x+2}$

5. $2+\dfrac{5}{x-\frac{1}{5}}=2+\dfrac{5}{x-\frac{1}{5}}\cdot\dfrac{\mathbf{5}}{\mathbf{5}}$

$=\dfrac{2}{1}+\dfrac{25}{5x-1}$

$=\dfrac{2}{1}\cdot\dfrac{\mathbf{5x-1}}{\mathbf{5x-1}}+\dfrac{25}{5x-1}$

$=\dfrac{10x-2+25}{5x-1}$

$=\dfrac{10x+23}{5x-1}$

Section 6.6

1. $\dfrac{x}{3}+1=\dfrac{1}{2}$ LCD = 6

$\mathbf{6}\left(\dfrac{x}{3}+1\right)=\mathbf{6}\cdot\dfrac{1}{2}$

$6\cdot\dfrac{x}{3}+6\cdot 1=6\cdot\dfrac{1}{2}$

$2x+6=3$

$2x=-3$

$x=-\dfrac{3}{2}$

2. $\dfrac{2}{a+5}=\dfrac{1}{3}$ LCD = $3(a+5)$

$\mathbf{3(a+5)}\cdot\dfrac{2}{a+5}=\mathbf{3(a+5)}\cdot\dfrac{1}{3}$

$6=a+5$

$1=a$

3. $$\frac{x}{x+1} - \frac{1}{2} = \frac{-1}{x+1} \qquad \text{LCD} = 2(x+1)$$

$$\mathbf{2(x+1)}\left[\frac{x}{x+1} - \frac{1}{2}\right] = \mathbf{2(x+1)} \cdot \frac{-1}{x+1}$$

$$2(x+1) \cdot \frac{x}{x+1} - 2(x+1) \cdot \frac{1}{2} = 2(x+1) \cdot \frac{-1}{x+1}$$

$$2x - (x+1) = 2(-1)$$
$$2x - x - 1 = -2$$
$$x - 1 = -2$$
$$x = -1$$

The only possible solution is $x = -1$, but when $x = -1$, the original equation has two undefined terms. There is no solution to the equation.

4. $$\frac{x}{x^2-9} - \frac{1}{x+3} = \frac{1}{4x-12}$$

$$\frac{x}{(x+3)(x-3)} - \frac{1}{x+3} = \frac{1}{4(x-3)} \qquad \text{LCD} = 4(x+3)(x-3)$$

$$\mathbf{4(x+3)(x-3)} \cdot \frac{x}{(x+3)(x-3)} - \mathbf{4(x+3)(x-3)} \cdot \frac{1}{x+3} = \mathbf{4(x+3)(x-3)} \cdot \frac{1}{4(x-3)}$$

$$4x - 4(x-3) = x + 3$$
$$4x - 4x + 12 = x + 3$$
$$12 = x + 3$$
$$9 = x$$

5. $$1 - \frac{2}{x} = \frac{8}{x^2} \qquad \text{LCD} = x^2$$

$$\mathbf{x^2}\left(1 - \frac{2}{x}\right) = \mathbf{x^2} \cdot \frac{8}{x^2}$$

$$x^2 \cdot 1 - x^2 \cdot \frac{2}{x} = x^2 \cdot \frac{8}{x^2}$$

$$x^2 - 2x = 8$$
$$x^2 - 2x - 8 = 0$$
$$(x-4)(x+2) = 0$$
$$x - 4 = 0 \quad \text{or} \quad x + 2 = 0$$
$$x = 4 \quad \text{or} \quad x = -2$$

6. $$\frac{y+1}{3(y+4)} = \frac{8}{(y+4)(y-4)}$$

The LCD is $3(y+4)(y-4)$. Multiplying each side by the LCD gives us

$$(y+1)(y-4) = 8 \cdot 3$$
$$y^2 - 3y - 4 = 24$$
$$y^2 - 3y - 28 = 0$$
$$(y-7)(y+4) = 0$$
$$y = 7 \quad \text{or} \quad y = -4$$

The only solution is 7 because the original equation is undefined when y is -4.

7. $$x = \frac{y+2}{y-1}$$

$$x(y-1) = y + 2$$
$$xy - x = y + 2$$
$$xy - y = x + 2$$
$$y(x-1) = x + 2$$
$$y = \frac{x+2}{x-1}$$

8. $$\frac{1}{a} = \frac{1}{x} + \frac{1}{b}$$

$$axb \cdot \frac{1}{a} = axb \cdot \frac{1}{x} + axb \cdot \frac{1}{b}$$

$$xb = ab + ax$$
$$xb - ax = ab$$
$$x(b-a) = ab$$
$$x = \frac{ab}{b-a}$$

Section 6.7

1. Let x = one of the numbers and $3x$ = the other number.

$$\frac{1}{x} + \frac{1}{3x} = \frac{4}{3} \qquad \text{LCD} = 3x$$

$$\mathbf{3x} \cdot \frac{1}{x} + \mathbf{3x} \cdot \frac{1}{3x} = \mathbf{3x} \cdot \frac{4}{3}$$

$$3 + 1 = 4x$$
$$4 = 4x$$
$$1 = x$$

The two numbers are 1 and 3.

2.

	d	*r*	*t*
Upstream	1	$15 - x$	$\frac{1}{15-x}$
Downstream	2	$15 + x$	$\frac{2}{15+x}$

$$\frac{1}{15-x} = \frac{2}{15+x} \qquad \text{LCD} = (15-x)(15+x)$$

$$1(15+x) = 2(15-x)$$
$$15 + x = 30 - 2x$$
$$3x = 15$$
$$x = 5 \text{ miles per hour}$$

The speed of the current is 5 miles per hour.

3.

	d	r	t
Upstream	8	$x-2$	$\frac{8}{x-2}$
Downstream	8	$x+2$	$\frac{8}{x+2}$

$$\frac{8}{x-2}+\frac{8}{x+2}=3 \qquad \text{LCD}=(x+2)(x-2)$$

$$\mathbf{(x+2)(x-2)}\cdot\frac{8}{x-2}+\mathbf{(x+2)(x-2)}\cdot\frac{8}{x+2}=\mathbf{(x+2)(x-2)}\cdot 3$$

$$(x+2)\cdot 8+(x-2)\cdot 8=(x+2)(x-2)\cdot 3$$
$$8x+16+8x-16=3x^2-12$$
$$16x=3x^2-12$$
$$0=3x^2-16x-12$$
$$0=(3x+2)(x-6)$$
$$3x+2=0 \quad\text{or}\quad x-6=0$$
$$x=-\frac{2}{3} \quad\text{or}\quad x=6$$

The speed of the boat in still water is 6 miles per hour. (The $-\frac{2}{3}$ cannot be a solution because it is negative.)

4. Let x = the length of time it takes to fill the sink with both the drain and faucet open.

$$\frac{1}{3}-\frac{1}{4}=\frac{1}{x} \qquad \text{LCD}=12x$$
$$4x-3x=12$$
$$x=12 \text{ minutes}$$

5. $13.0 \text{ feet per second} = \frac{13.0 \text{ feet}}{1 \text{ second}}\cdot\frac{1 \text{ mile}}{5{,}280 \text{ feet}}\cdot\frac{60 \text{ second}}{1 \text{ minute}}\cdot\frac{60 \text{ minutes}}{1 \text{ hour}}$

$= \frac{13.0\cdot 60\cdot 60 \text{ miles}}{5{,}280 \text{ hours}}$

$= 8.9$ miles per hour

6. $1{,}100 \text{ feet per minute} = \frac{1{,}100 \text{ feet}}{1 \text{ minute}}\cdot\frac{1 \text{ mile}}{5{,}280 \text{ feet}}\cdot\frac{60 \text{ minutes}}{1 \text{ hour}}$

$= \frac{1{,}100\cdot 60 \text{ miles}}{5{,}280 \text{ hours}}$

$= 12.5$ miles per hour

Chapter 7

Section 7.1

1. The positive square root of 36 is 6 because 6 is the positive number with the property $6^2 = 36$. The negative square root of 36 is -6 since -6 is the negative number whose square is 36. The square roots of 36 are 6 and -6.
2. $\sqrt[3]{-64} = -4$ because $(-4)^3 = (-4)(-4)(-4) = -64$.
3. $\sqrt{-25}$ is not a real number since there is no real number whose square is -25.
4. $-\sqrt{4} = -2$; this is the negative square root of 4.
5. $\sqrt[5]{-1} = -1$ because $(-1)^5 = (-1)(-1)(-1)(-1)(-1) = -1$.
6. $\sqrt[4]{-16}$ is not a real number since there is no real number we can raise to the fourth power and obtain -16.

10. $9^{1/2} = \sqrt{9} = 3$ 11. $27^{1/3} = \sqrt[3]{27} = 3$ 12. $-49^{1/2} = -\sqrt{49} = -7$ 13. $(-49)^{1/2} = \sqrt{-49}$ which is not a real number.

14. $\left(\frac{16}{25}\right)^{1/2} = \sqrt{\frac{16}{25}} = \frac{4}{5}$

15. $\sqrt[3]{8x^3y^9} = (8x^3y^9)^{1/3}$
$= 8^{1/3}(x^3)^{1/3}(y^9)^{1/3}$
$= 2xy^3$

16. $\sqrt[4]{81a^4b^8} = (81a^4b^8)^{1/4}$
$= 81^{1/4}(a^4)^{1/4}(b^8)^{1/4}$
$= 3ab^2$

17. $9^{3/2} = (9^{1/2})^3 = 3^3 = 27$

18. $16^{3/4} = (16^{1/4})^3 = 2^3 = 8$ 19. $8^{-2/3} = (8^{1/3})^{-2} = 2^{-2} = \frac{1}{4}$ 20. $\left(\frac{16}{81}\right)^{-3/4} = \left(\frac{81}{16}\right)^{3/4} = \left[\left(\frac{81}{16}\right)^{1/4}\right]^3 = \left(\frac{3}{2}\right)^3 = \frac{27}{8}$

21. $x^{1/2}\cdot x^{1/4} = x^{1/2+1/4} = x^{3/4}$ 22. $(y^{3/5})^{5/6} = y^{(3/5)(5/6)} = y^{1/2}$ 23. $\frac{z^{3/4}}{z^{2/3}} = z^{3/4-2/3} = z^{1/12}$

24. $\frac{(x^{1/3}y^{-3})^6}{x^4y^{10}} = \frac{x^2y^{-18}}{x^4y^{10}}$
$= x^{-2}y^{-28}$
$= \frac{1}{x^2y^{28}}$

Section 7.2

1. $x^{1/3}(x^{2/3} - x^{1/3}) = x^{1/3} \cdot x^{2/3} - x^{1/3} \cdot x^{1/3}$
$= x^1 - x^{2/3}$
$= x - x^{2/3}$

2. $(x^{3/5} + 2)(x^{3/5} - 7) = x^{3/5}x^{3/5} - 7x^{3/5} + 2x^{3/5} - 14$
$= x^{6/5} - 5x^{3/5} - 14$

3. $(5a^{1/2} - 4b^{1/2})(3a^{1/2} - b^{1/2}) = 15a - 5a^{1/2}b^{1/2} - 12a^{1/2}b^{1/2} + 4b$
$= 15a - 17a^{1/2}b^{1/2} + 4b$

4. $(t^{1/3} + 3)^2 = (t^{1/3})^2 + 6t^{1/3} + 9$
$= t^{2/3} + 6t^{1/3} + 9$

5. $(x^{5/2} - 3^{1/2})(x^{5/2} + 3^{1/2}) = (x^{5/2})^2 - (3^{1/2})^2$
$= x^5 - 3$

6.
$$\begin{array}{r} a^{2/3} - a^{1/3}b^{1/3} + b^{2/3} \\ a^{1/3} + b^{1/3} \\ \hline a \quad - a^{2/3}b^{1/3} + a^{1/3}b^{2/3} \\ + a^{2/3}b^{1/3} - a^{1/3}b^{2/3} + b \\ \hline a \qquad\qquad\qquad\qquad + b \end{array}$$

7. $\dfrac{36x^{3/4}y^{1/4} - 18x^{5/4}y^{1/4}}{6x^{1/4}y^{1/4}} = \dfrac{36x^{3/4}y^{1/4}}{6x^{1/4}y^{1/4}} - \dfrac{18x^{5/4}y^{1/4}}{6x^{1/4}y^{1/4}}$
$= 6x^{1/2} - 3x$

8. $8(x-3)^{3/2} - 6(x-3)^{1/2} = 2(x-3)^{1/2}[4(x-3) - 3]$
$= 2(x-3)^{1/2}(4x - 15)$

9. $x^{2/3} - 4x^{1/3} - 21$
$= (x^{1/3})^2 - 4x^{1/3} - 21$
$= (x^{1/3} - 7)(x^{1/3} + 3)$

10. $6x^{2/3} + 19x^{1/3} + 10$
$= 6(x^{1/3})^2 + 19x^{1/3} + 10$
$= (3x^{1/3} + 2)(2x^{1/3} + 5)$

11. $(x^2 - 3)^{1/2} - \dfrac{x^2}{(x^2-3)^{1/2}} = \dfrac{(x^2-3)^{1/2}}{1} \cdot \dfrac{\mathbf{(x^2-3)^{1/2}}}{\mathbf{(x^2-3)^{1/2}}} - \dfrac{x^2}{(x^2-3)^{1/2}}$
$= \dfrac{x^2 - 3 - x^2}{(x^2-3)^{1/2}} = \dfrac{-3}{(x^2-3)^{1/2}}$

12. $r = \left(\dfrac{800}{600}\right)^{1/3} - 1$
$= 1.101 - 1$
$= 0.101$ or 10.1%

Section 7.3

1. $\sqrt{18} = \sqrt{9 \cdot 2}$
$= \sqrt{9}\sqrt{2}$
$= 3\sqrt{2}$

2. $\sqrt{50x^2y^3} = \sqrt{25x^2y^2 \cdot 2y}$
$= \sqrt{25x^2y^2}\sqrt{2y}$
$= 5xy\sqrt{2y}$

3. $\sqrt[3]{54a^4b^3} = \sqrt[3]{27a^3b^3 \cdot 2a}$
$= \sqrt[3]{27a^3b^3}\sqrt[3]{2a}$
$= 3ab\sqrt[3]{2a}$

4. $\sqrt{75x^5y^8} = \sqrt{25x^4y^8 \cdot 3x}$
$= \sqrt{25x^4y^8}\sqrt{3x}$
$= 5x^2y^4\sqrt{3x}$

5. $\sqrt[4]{48a^8b^5c^4} = \sqrt[4]{16a^8b^4c^4 \cdot 3b}$
$= \sqrt[4]{16a^8b^4c^4}\sqrt[4]{3b}$
$= 2a^2bc\sqrt[4]{3b}$

6. $\sqrt{\dfrac{5}{9}} = \dfrac{\sqrt{5}}{\sqrt{9}}$
$= \dfrac{\sqrt{5}}{3}$

7. $\sqrt{\dfrac{2}{3}} = \dfrac{\sqrt{2}}{\sqrt{3}}$
$= \dfrac{\sqrt{2}}{\sqrt{3}} \cdot \dfrac{\mathbf{\sqrt{3}}}{\mathbf{\sqrt{3}}}$
$= \dfrac{\sqrt{6}}{3}$

8. $\dfrac{5}{\sqrt{2}} = \dfrac{5}{\sqrt{2}} \cdot \dfrac{\mathbf{\sqrt{2}}}{\mathbf{\sqrt{2}}}$
$= \dfrac{5\sqrt{2}}{2}$

9. $\dfrac{3\sqrt{5x}}{\sqrt{2y}} = \dfrac{3\sqrt{5x}}{\sqrt{2y}} \cdot \dfrac{\mathbf{\sqrt{2y}}}{\mathbf{\sqrt{2y}}}$
$= \dfrac{3\sqrt{10xy}}{2y}$

10. $\dfrac{5}{\sqrt[3]{9}} = \dfrac{5}{\sqrt[3]{3^2}}$
$= \dfrac{5}{\sqrt[3]{3^2}} \cdot \dfrac{\mathbf{\sqrt[3]{3}}}{\mathbf{\sqrt[3]{3}}}$
$= \dfrac{5\sqrt[3]{3}}{\sqrt[3]{3^3}}$
$= \dfrac{5\sqrt[3]{3}}{3}$

11. $\sqrt{\dfrac{48x^3y^4}{7z}} = \dfrac{\sqrt{48x^3y^4}}{\sqrt{7z}}$
$= \dfrac{\sqrt{16x^2y^4}\sqrt{3x}}{\sqrt{7z}}$
$= \dfrac{4xy^2\sqrt{3x}}{\sqrt{7z}}$
$= \dfrac{4xy^2\sqrt{3x}}{\sqrt{7z}} \cdot \dfrac{\mathbf{\sqrt{7z}}}{\mathbf{\sqrt{7z}}}$
$= \dfrac{4xy^2\sqrt{21xz}}{7z}$

12. $\sqrt{16x^2} = 4|x|$

13. $\sqrt{25x^3} = 5|x|\sqrt{x}$

14. $\sqrt{x^2 + 10x + 25}$
$= \sqrt{(x+5)^2}$
$= |x + 5|$

15. $\sqrt{2x^3 + 7x^2}$
$= \sqrt{x^2(2x+7)}$
$= |x|\sqrt{2x+7}$

16. $\sqrt[3]{(-3)^3}$
$= \sqrt[3]{-27}$
$= -3$

17. $\sqrt[3]{(-1)^3}$
$= \sqrt[3]{-1}$
$= -1$

Section 7.4

1. $3\sqrt{5} - 2\sqrt{5} + 4\sqrt{5} = (3 - 2 + 4)\sqrt{5}$
$= 5\sqrt{5}$

2. $4\sqrt{50} + 3\sqrt{8} = 4\sqrt{25 \cdot 2} + 3\sqrt{4 \cdot 2}$
$= 4\sqrt{25}\sqrt{2} + 3\sqrt{4}\sqrt{2}$
$= 4 \cdot 5\sqrt{2} + 3 \cdot 2\sqrt{2}$
$= 20\sqrt{2} + 6\sqrt{2}$
$= 26\sqrt{2}$

3. $4\sqrt{18x^2y} - 3x\sqrt{50y} = 4\sqrt{9x^2 \cdot 2y} - 3x\sqrt{25 \cdot 2y}$
$= 4\sqrt{9x^2}\,\sqrt{2y} - 3x\sqrt{25}\,\sqrt{2y}$
$= 4 \cdot 3x\sqrt{2y} - 3x \cdot 5\sqrt{2y}$
$= 12x\sqrt{2y} - 15x\sqrt{2y}$
$= -3x\sqrt{2y}$

4. $2\sqrt[3]{27a^2b^4} + 3b\sqrt[3]{125a^2b} = 2\sqrt[3]{27b^3 \cdot a^2b} + 3b\sqrt[3]{125 \cdot a^2b}$
$= 2\sqrt[3]{27b^3}\,\sqrt[3]{a^2b} + 3b\,\sqrt[3]{125}\,\sqrt[3]{a^2b}$
$= 2 \cdot 3b\sqrt[3]{a^2b} + 3b \cdot 5\sqrt[3]{a^2b}$
$= 6b\sqrt[3]{a^2b} + 15b\sqrt[3]{a^2b}$
$= 21b\sqrt[3]{a^2b}$

5. $\frac{\sqrt{5}}{3} + \frac{1}{\sqrt{5}} = \frac{\sqrt{5}}{3} + \frac{1}{\sqrt{5}} \cdot \frac{\mathbf{\sqrt{5}}}{\mathbf{\sqrt{5}}}$
$= \frac{\sqrt{5}}{3} + \frac{\sqrt{5}}{5}$
$= \left(\frac{1}{3} + \frac{1}{5}\right)\sqrt{5}$
$= \left(\frac{5}{15} + \frac{3}{15}\right)\sqrt{5}$
$= \frac{8}{15}\sqrt{5}$
$= \frac{8\sqrt{5}}{15}$

6. First we construct a golden rectangle from a square of side 6.

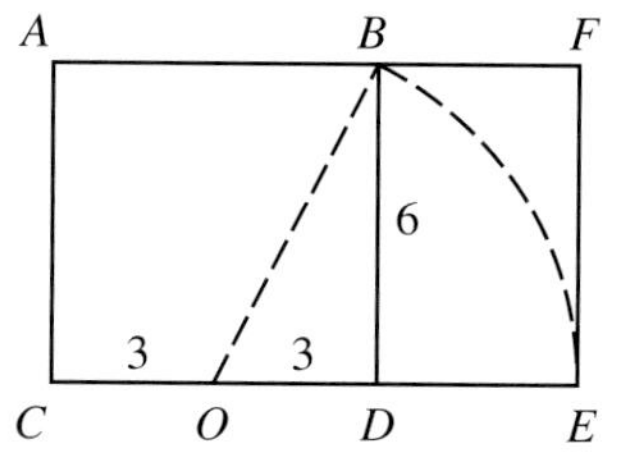

The length of the diagonal OB is found from the Pythagorean theorem.

$$OB = \sqrt{3^2 + 6^2} = \sqrt{9 + 36} = \sqrt{45} = 3\sqrt{5}$$

The ratio of the length to the width for the rectangle is the golden ratio.

$$\text{Golden ratio} = \frac{CE}{EF} = \frac{3 + 3\sqrt{5}}{6}$$
$$= \frac{\cancel{3}(1 + \sqrt{5})}{\cancel{3} \cdot 2}$$
$$= \frac{1 + \sqrt{5}}{2}$$

Section 7.5

1. We can rearrange the order and grouping of the numbers by applying the commutative and associative properties of multiplication.
$(7\sqrt{3})(5\sqrt{11}) = (7 \cdot 5)\,\sqrt{3} \cdot \sqrt{11})$
$= 35\sqrt{33}$

2. $\sqrt{2}(3\sqrt{5} - 4\sqrt{2}) = \sqrt{2} \cdot 3\sqrt{5} - \sqrt{2} \cdot 4\sqrt{2}$
$= 3\sqrt{10} - 4\sqrt{4}$
$= 3\sqrt{10} - 4 \cdot 2$
$= 3\sqrt{10} - 8$

3. $(\sqrt{2} + \sqrt{7})(\sqrt{2} - 3\sqrt{7}) = \sqrt{2}\sqrt{2} - \sqrt{2} \cdot 3\sqrt{7} + \sqrt{7}\sqrt{2} - \sqrt{7} \cdot 3\sqrt{7}$
$= 2 - 3\sqrt{14} + \sqrt{14} - 21$
$= -19 - 2\sqrt{14}$

4. $(\sqrt{x} + 5)^2 = (\sqrt{x} + 5)(\sqrt{x} + 5)$
$= \sqrt{x}\sqrt{x} + 5\sqrt{x} + 5\sqrt{x} + 25$
$= x + 10\sqrt{x} + 25$

5. $(5\sqrt{a} - 3\sqrt{b})^2 = (5\sqrt{a})^2 - 2 \cdot 5\sqrt{a} \cdot 3\sqrt{b} + (3\sqrt{b})^2$
$= 25a - 30\sqrt{ab} + 9b$

6. $(\sqrt{x+3} - 1)^2 = (\sqrt{x+3} - 1)(\sqrt{x+3} - 1)$
$= x + 3 - 2\sqrt{x+3} + 1$
$= x + 4 - 2\sqrt{x+3}$

7. $(\sqrt{5} + \sqrt{3})(\sqrt{5} - \sqrt{3}) = (\sqrt{5})^2 - (\sqrt{3})^2$
$= 5 - 3$
$= 2$

8. $\frac{3}{\sqrt{7} - \sqrt{3}} = \frac{3}{\sqrt{7} - \sqrt{3}} \cdot \frac{\mathbf{\sqrt{7} + \sqrt{3}}}{\mathbf{\sqrt{7} + \sqrt{3}}}$
$= \frac{3\sqrt{7} + 3\sqrt{3}}{7 - 3}$
$= \frac{3\sqrt{7} + 3\sqrt{3}}{4}$

9. $\frac{\sqrt{10} - 3}{\sqrt{10} + 3} = \frac{\sqrt{10} - 3}{\sqrt{10} + 3} \cdot \frac{\mathbf{\sqrt{10} - 3}}{\mathbf{\sqrt{10} - 3}}$
$= \frac{10 - 6\sqrt{10} + 9}{10 - 9}$
$= \frac{19 - 6\sqrt{10}}{1}$
$= 19 - 6\sqrt{10}$

10. First we construct a golden rectangle from a square of side 4.

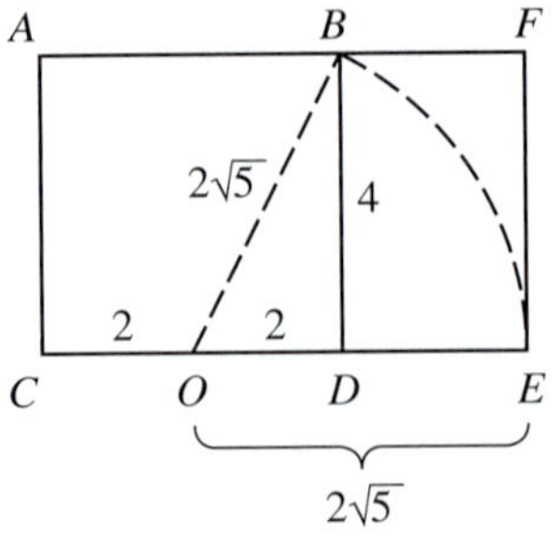

Next, we find expressions for the length and width of the smaller rectangle.

$$\text{Length} = EF = 4$$
$$\text{Width} = DE = 2\sqrt{5} - 2$$

Next, we find the ratio of length to width.

$$\text{Ratio of length to width} = \frac{EF}{DE} = \frac{4}{2\sqrt{5}-2} = \frac{2 \cdot 2}{2(\sqrt{5}-1)} = \frac{2}{\sqrt{5}-1}$$

From this point on the rest of the problem looks like the solution to Example 10 in the text.

Section 7.6

1.
$$\begin{aligned} \sqrt{2x+4} &= 4 \\ (\sqrt{2x+4})^2 &= 4^2 \\ 2x+4 &= 16 \\ 2x &= 12 \\ x &= 6 \end{aligned}$$

2. $\sqrt{7x-3} = -5$ has no solution since the left side is a positive number or 0 and the right side is a negative number.

3.
$$\begin{aligned} \sqrt{4x+5} + 2 &= 7 \\ \sqrt{4x+5} &= 5 \\ (\sqrt{4x+5})^2 &= 5^2 \\ 4x+5 &= 25 \\ 4x &= 20 \\ x &= 5 \end{aligned}$$

4.
$$\begin{aligned} t-6 &= \sqrt{t-4} \\ (t-6)^2 &= (\sqrt{t-4})^2 \\ t^2 - 12t + 36 &= t - 4 \\ t^2 - 13t + 40 &= 0 \\ (t-5)(t-8) &= 0 \end{aligned}$$
$$t - 5 = 0 \quad \text{or} \quad t - 8 = 0$$
$$t = 5 \quad \text{or} \quad t = 8$$

Only 8 checks in the original equation.

5.
$$\begin{aligned} \sqrt{x-9} &= \sqrt{x} - 3 \\ (\sqrt{x-9})^2 &= (\sqrt{x}-3)^2 \\ x - 9 &= x - 6\sqrt{x} + 9 \\ -9 &= -6\sqrt{x} + 9 \\ -18 &= -6\sqrt{x} \\ 3 &= \sqrt{x} \\ 3^2 &= (\sqrt{x})^2 \\ 9 &= x \end{aligned}$$

6.
$$\begin{aligned} \sqrt{x+4} &= 2 - \sqrt{3x} \\ (\sqrt{x+4})^2 &= (2-\sqrt{3x})^2 \\ x+4 &= 4 - 4\sqrt{3x} + 3x \\ -2x &= -4\sqrt{3x} \\ x &= 2\sqrt{3x} \\ x^2 &= (2\sqrt{3x})^2 \\ x^2 &= 4 \cdot 3x \\ x^2 - 12x &= 0 \\ x(x-12) &= 0 \end{aligned}$$
$$x = 0 \quad \text{or} \quad x - 12 = 0$$
$$x = 12$$

Checking each solution in the original equation shows that $x = 12$ is extraneous. The only solution is $x = 0$.

7.
$$\begin{aligned} \sqrt{x+2} &= \sqrt{x+3} - 1 \\ (\sqrt{x+2})^2 &= (\sqrt{x+3}-1)^2 \\ x+2 &= x+3 - 2\sqrt{x+3} + 1 \\ 2 &= 3 - 2\sqrt{x+3} + 1 \\ 2 &= 4 - 2\sqrt{x+3} \\ -2 &= -2\sqrt{x+3} \\ 1 &= \sqrt{x+3} \\ 1^2 &= (\sqrt{x+3})^2 \\ 1 &= x+3 \\ -2 &= x \end{aligned}$$

8.
$$\begin{aligned} \sqrt[3]{3x-7} &= 2 \\ (\sqrt[3]{3x-7})^3 &= 2^3 \\ 3x - 7 &= 8 \\ 3x &= 15 \\ x &= 5 \end{aligned}$$

9. $y = \sqrt{x} + 3$

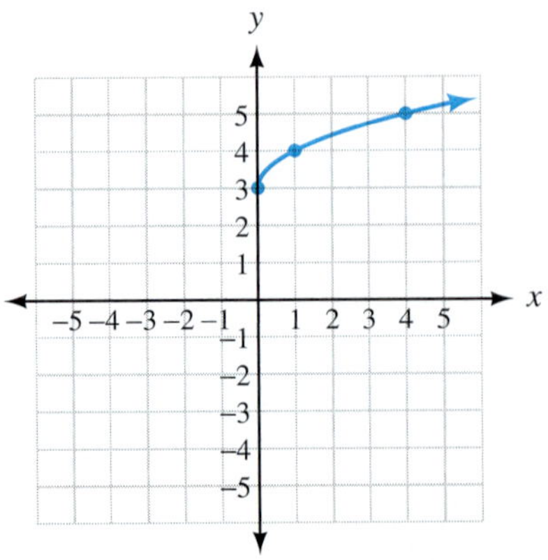

$y = \sqrt{x+3}$

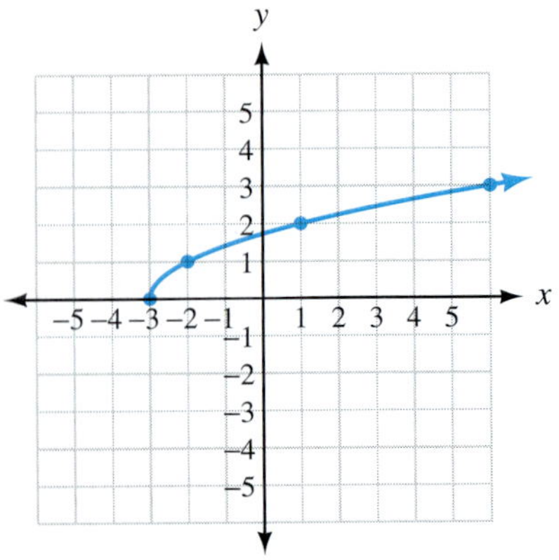

Section 7.7

1. $\sqrt{-36} = i\sqrt{36} = i \cdot 6 = 6i$ **2.** $-\sqrt{-64} = -i\sqrt{64} = -i \cdot 8 = -8i$ **3.** $\sqrt{-18} = i\sqrt{18} = i \cdot 3\sqrt{2} = 3i\sqrt{2}$ **4.** $-\sqrt{-19} = -i\sqrt{19}$

5. $i^{20} = (i^2)^{10} = (-1)^{10} = 1$ **6.** $i^{23} = (i^2)^{11} \cdot i = (-1)^{11} \cdot i = -i$ **7.** $i^{50} = (i^2)^{25} = (-1)^{25} = -1$

8. $4x + 7i = 8 - 14yi$

$\Rightarrow 4x = 8$ and $7 = -14y$

$x = 2$ $\quad y = -\dfrac{1}{2}$

9. $(2x - 1) + 9i = 5 + (4y + 1)i$

$\Rightarrow 2x - 1 = 5$ and $9 = 4y + 1$

$2x = 6 \quad 8 = 4y$

$x = 3 \quad 2 = y$

10. $(2 + 6i) + (3 - 4i) = (2 + 3) + (6i - 4i)$

$= 5 + 2i$

11. $(6 + 5i) - (4 + 3i) = 6 + 5i - 4 - 3i$

$= 2 + 2i$

12. $(7 - i) - (8 - 2i) = 7 - i - 8 + 2i$

$= -1 + i$

13. $(2 + 3i)(1 - 4i) = 2 \cdot 1 - 2 \cdot 4i + 3i \cdot 1 - 3i \cdot 4i$

$= 2 - 8i + 3i - 12i^2$

$= 2 - 5i - 12(-1)$

$= 2 - 5i + 12$

$= 14 - 5i$

14. $-3i(2 + 3i) = -3i \cdot 2 - 3i \cdot 3i$

$= -6i - 9i^2$

$= -6i - 9(-1)$

$= -6i + 9$

$= 9 - 6i$

15. $(2 + 4i)^2 = 2^2 + 2 \cdot 2 \cdot 4i + (4i)^2$

$= 4 + 16i - 16$

$= -12 + 16i$

16. $(3 - 5i)(3 + 5i) = 3^2 - (5i)^2$

$= 9 - 25i^2$

$= 9 + 25$

$= 34$

17. $\dfrac{3 + 2i}{2 - 5i} = \dfrac{3 + 2i}{2 - 5i} \cdot \dfrac{\mathbf{2 + 5i}}{\mathbf{2 + 5i}}$

$= \dfrac{6 + 15i + 4i + 10i^2}{4 - 25i^2}$

$= \dfrac{6 + 19i - 10}{4 + 25}$

$= \dfrac{-4 + 19i}{29}$

$= -\dfrac{4}{29} + \dfrac{19}{29}i$

18. $\dfrac{3 + 2i}{i} = \dfrac{3 + 2i}{i} \cdot \dfrac{\mathbf{-i}}{\mathbf{-i}}$

$= \dfrac{-3i - 2i^2}{-i^2}$

$= \dfrac{-3i + 2}{1}$

$= 2 - 3i$

Chapter 8

Section 8.1

1. $(3x + 2)^2 = 16$

$3x + 2 = \pm 4$

$3x = -2 \pm 4$

$x = \dfrac{-2 \pm 4}{3}$

$x = \dfrac{-2 + 4}{3}$ or $x = \dfrac{-2 - 4}{3}$

$x = \dfrac{2}{3}$ or $x = -2$

2. $(4x - 3)^2 = -50$

$4x - 3 = \pm\sqrt{-50}$

$4x - 3 = \pm 5i\sqrt{2}$

$4x = 3 \pm 5i\sqrt{2}$

$x = \dfrac{3 \pm 5i\sqrt{2}}{4}$

3. $x^2 + 10x + 25 = 20$

$(x + 5)^2 = 20$

$x + 5 = \pm\sqrt{20} = \pm 2\sqrt{5}$

$x = -5 \pm 2\sqrt{5}$

4. $x^2 + 3x - 4 = 0$

$x^2 + 3x = 4$

$x^2 + 3x + \mathbf{\dfrac{9}{4}} = 4 + \mathbf{\dfrac{9}{4}}$

$\left(x + \dfrac{3}{2}\right)^2 = \dfrac{25}{4}$

$x + \dfrac{3}{2} = \pm\dfrac{5}{2}$

$x = -\dfrac{3}{2} \pm \dfrac{5}{2}$

$x = -\dfrac{3}{2} + \dfrac{5}{2}$ or $x = -\dfrac{3}{2} - \dfrac{5}{2}$

$x = 1$ or $x = -4$

5. $5x^2 - 3x + 2 = 0$

$5x^2 - 3x = -2$

$x^2 - \dfrac{3}{5}x = -\dfrac{2}{5}$

$x^2 - \dfrac{3}{5}x + \mathbf{\dfrac{9}{100}} = -\dfrac{2}{5} + \mathbf{\dfrac{9}{100}}$

$\left(x - \dfrac{3}{10}\right)^2 = -\dfrac{31}{100}$

$x - \dfrac{3}{10} = \pm\sqrt{-\dfrac{31}{100}}$

$x - \dfrac{3}{10} = \pm\dfrac{i\sqrt{31}}{10}$

$x = \dfrac{3}{10} \pm \dfrac{i\sqrt{31}}{10}$

$x = \dfrac{3 \pm i\sqrt{31}}{10}$

6. Using Figure 3 in the text as a guide, we have $AC = 2$, so $AB = 4$. If $BC = x$, then by the Pythagorean theorem we have

$x^2 + 2^2 = 4^2$

$x^2 + 4 = 16$

$x^2 = 12$

$x = 2\sqrt{3}$

The sides are 2, $2\sqrt{3}$, and 4 inches.

7. $4{,}330^2 = x^2 + 960^2$

$18{,}748{,}900 = x^2 + 921{,}600$

$x^2 = 18{,}748{,}900 - 921{,}600$

$x^2 = 17{,}827{,}300$

$x = \sqrt{17{,}827{,}300}$

$x = 4{,}222$ feet to the nearest foot

Section 8.2

1. $6x^2 + 7x + 2 = 0$

$a = 6, b = 7, c = 2$

$$x = \frac{-7 \pm \sqrt{7^2 - 4(6)(2)}}{2(6)}$$

$$= \frac{-7 \pm \sqrt{49 - 48}}{12}$$

$$= \frac{-7 \pm \sqrt{1}}{12}$$

$$= \frac{-7 \pm 1}{12}$$

$$x = \frac{-7 + 1}{12} \quad \text{or} \quad x = \frac{-7 - 1}{12}$$

$$x = -\frac{1}{2} \quad \text{or} \quad x = -\frac{2}{3}$$

2. $\frac{x^2}{2} + x = \frac{1}{3}$ LCD = 6

$$\mathbf{6} \cdot \frac{x^2}{2} + \mathbf{6} \cdot x = \mathbf{6} \cdot \frac{1}{3}$$

$$3x^2 + 6x = 2$$

$$3x^2 + 6x - 2 = 0$$

$a = 3, b = 6, c = -2$

$$x = \frac{-6 \pm \sqrt{6^2 - 4(3)(-2)}}{2(3)}$$

$$= \frac{-6 \pm \sqrt{36 + 24}}{6}$$

$$= \frac{-6 \pm \sqrt{60}}{6}$$

$$= \frac{-6 \pm 2\sqrt{15}}{6}$$

$$= \frac{\cancel{2}(-3 \pm \sqrt{15})}{\cancel{2} \cdot 3}$$

$$= \frac{-3 \pm \sqrt{15}}{3}$$

3. $\frac{1}{x + 4} - \frac{1}{x} = \frac{1}{2}$ LCD = $2x(x + 4)$

$$\mathbf{2x(x+4)}\left(\frac{1}{x+4} - \frac{1}{x}\right) = \mathbf{2x(x+4)} \cdot \frac{1}{2}$$

$$\mathbf{2x(x+4)} \cdot \frac{1}{x+4} - \mathbf{2x(x+4)} \cdot \frac{1}{x} = \mathbf{2x(x+4)} \cdot \frac{1}{2}$$

$$2x - 2(x + 4) = x(x + 4)$$

$$2x - 2x - 8 = x^2 + 4x$$

$$-8 = x^2 + 4x$$

$$0 = x^2 + 4x + 8$$

$a = 1, b = 4, c = 8$

$$x = \frac{-4 \pm \sqrt{4^2 - 4(1)(8)}}{2(1)}$$

$$= \frac{-4 \pm \sqrt{16 - 32}}{2}$$

$$= \frac{-4 \pm \sqrt{-16}}{2}$$

$$= \frac{-4 \pm 4i}{2}$$

$$= \frac{2(-2 \pm 2i)}{2}$$

$$= -2 \pm 2i$$

4. $8t^3 - 27 = 0$

$$(2t - 3)(4t^2 + 6t + 9) = 0$$

$$2t - 3 = 0 \quad \text{or} \quad 4t^2 + 6t + 9 = 0$$

$$2t = 3$$

$$t = \frac{3}{2}$$

$$t = \frac{-6 \pm \sqrt{36 - 4(4)(9)}}{2(4)}$$

$$= \frac{-6 \pm \sqrt{-108}}{8}$$

$$= \frac{-6 \pm 6i\sqrt{3}}{8}$$

$$= \frac{-3 \pm 3i\sqrt{3}}{4}$$

5. $12 = 32t - 16t^2$

$16t^2 - 32t + 12 = 0$

$4t^2 - 8t + 3 = 0$ Divide by 4.

$$t = \frac{8 \pm \sqrt{64 - 4(4)(3)}}{2(4)}$$

$$= \frac{8 \pm \sqrt{16}}{8}$$

$$= \frac{8 \pm 4}{8}$$

$$t = \frac{8 + 4}{8} = \frac{12}{8} = \frac{3}{2} \quad \text{or} \quad t = \frac{8 - 4}{8} = \frac{4}{8} = \frac{1}{2}$$

Note: Since the solutions are rational numbers, we could have solved the equation by factoring.

6. $P = \$1{,}320$

$P = -500 + 27x - 0.1x^2$

$1{,}320 = -500 + 27x - 0.1x^2$

$0.1x^2 - 27x + 1{,}820 = 0$

$$x = \frac{27 \pm \sqrt{(-27)^2 - 4(0.1)(1{,}820)}}{(2)(0.1)}$$

$$= \frac{27 \pm \sqrt{729 - 728}}{0.2}$$

$$= \frac{27 \pm \sqrt{1}}{0.2}$$

$$= \frac{27 + 1}{0.2} \quad \text{or} \quad = \frac{27 - 1}{0.2}$$

$$= \frac{28}{0.2} \quad \text{or} \quad = \frac{26}{0.2}$$

$$x = 140 \quad \text{or} \quad x = 130$$

Section 8.3

1. $x^2 - 3x - 28 = 0$

$b^2 - 4ac = (-3)^2 - 4(1)(-28)$
$= 9 + 112$
$= 121$

which is a positive number that is a perfect square. Our equation, therefore, has two rational solutions.

2. $x^2 - 6x + 9 = 0$

$b^2 - 4ac = (-6)^2 - 4(1)(9)$
$= 36 - 36$
$= 0$

which means our equation has exactly one rational solution.

3. $3x^2 - 2x + 4 = 0$

$b^2 - 4ac = (-2)^2 - 4(3)(4)$
$= 4 - 48$
$= -44$

which is a negative number, implying that our equation has two complex solutions.

4. $x^2 + 1 = 4x$

$x^2 - 4x + 1 = 0$

$b^2 - 4ac = (-4)^2 - 4(1)(1)$
$= 16 - 4$
$= 12$

which is a positive number but not a perfect square. Our equation, therefore, has two irrational solutions.

5. $9x^2 + kx = -4$

$9x^2 + kx + 4 = 0$

$b^2 - 4ac = k^2 - 4(9)(4) = 0$

$k^2 - 144 = 0$

$k^2 = 144$

$k = \pm 12$

6. $t = -2, t = 2, t = 3$

$(t + 2)(t - 2)(t - 3) = 0$

$(t^2 - 4)(t - 3) = 0$

$t^3 - 3t^2 - 4t + 12 = 0$

7. $x = -\dfrac{3}{4}$ $\quad x = \dfrac{1}{5}$

$4x = -3$ $\quad 5x = 1$

$4x + 3 = 0$ $\quad 5x - 1 = 0$

$(4x + 3)(5x - 1) = 0$

$20x^2 + 11x - 3 = 0$

Section 8.4

1. $(x - 2)^2 - 3(x - 2) - 10 = 0$

let $y = x - 2$

$y^2 - 3y - 10 = 0$

$(y - 5)(y + 2) = 0$

$y - 5 = 0$ or $y + 2 = 0$

$y = 5$ or $y = -2$

now replace y with $x - 2$

$x - 2 = 5$ or $x - 2 = -2$

$x = 7$ or $x = 0$

2. $6x^4 - 13x^2 = 5$

$6x^4 - 13x^2 - 5 = 0$

$(3x^2 + 1)(2x^2 - 5) = 0$

$3x^2 + 1 = 0$ or $2x^2 - 5 = 0$

$3x^2 = -1$ or $2x^2 = 5$

$x^2 = -\dfrac{1}{3}$ or $x^2 = \dfrac{5}{2}$

$x = \pm\dfrac{i\sqrt{3}}{3}$ or $x = \pm\dfrac{\sqrt{10}}{2}$

3. $x - \sqrt{x} - 12 = 0$

let $y = \sqrt{x}$

$y^2 - y - 12 = 0$

$(y - 4)(y + 3) = 0$

$y - 4 = 0$ or $y + 3 = 0$

$y = 4$ or $y = -3$

now replace y with $\sqrt{x}$

$\sqrt{x} = 4$ or $\sqrt{x} = -3$

$x = 16$ or $x = 9$

Checking each solution in the original equation will show that 9 is extraneous. The only solution is $x = 16$.

4. $16t^2 - 10t - h = 0$

$$t = \frac{10 \pm \sqrt{100 - 4(16)(-h)}}{2(16)}$$

$$= \frac{10 \pm \sqrt{100 + 64h}}{32}$$

$$= \frac{10 \pm 2\sqrt{25 + 16h}}{32}$$

$$= \frac{5 \pm \sqrt{25 + 16h}}{16}$$

5. $AB = 2, BC = x$

If $\dfrac{AB}{BC} = \dfrac{BC}{AC}$

then $\dfrac{2}{x} = \dfrac{x}{x + 2}$

$2(x + 2) = x^2$

$2x + 4 = x^2$

$0 = x^2 - 2x - 4$

$$x = \frac{2 \pm \sqrt{4 - 4(1)(-4)}}{2}$$

$$= \frac{2 \pm \sqrt{20}}{2}$$

$$= \frac{2 \pm 2\sqrt{5}}{2}$$

$$= \frac{2(1 \pm \sqrt{5})}{2}$$

$= 1 \pm \sqrt{5}$

Since x must be positive, we have $x = 1 + \sqrt{5}$, which is twice the golden ratio.

Section 8.5

1. To find the x-intercepts we let $y = 0$

$$x^2 - 2x - 3 = 0$$
$$(x - 3)(x + 1) = 0$$
$$x = 3, x = -1$$

The x-coordinate of the vertex is

$$x = \frac{-b}{2a} = \frac{-(-2)}{2(1)} = 1$$

The y-coordinate of the vertex is

$$y = 1^2 - 2(1) - 3 = -4$$

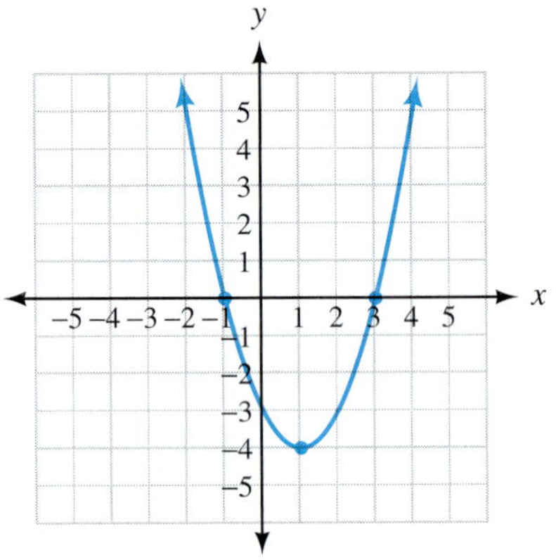

2. x-intercepts:

$$-x^2 + 2x + 8 = 0$$
$$x^2 - 2x - 8 = 0$$
$$(x - 4)(x + 2) = 0$$
$$x = 4, x = -2$$

Vertex:

$$x = \frac{-b}{2a} = \frac{-2}{2(-1)} = 1$$
$$y = -(1)^2 + 2(1) + 8 = 9$$

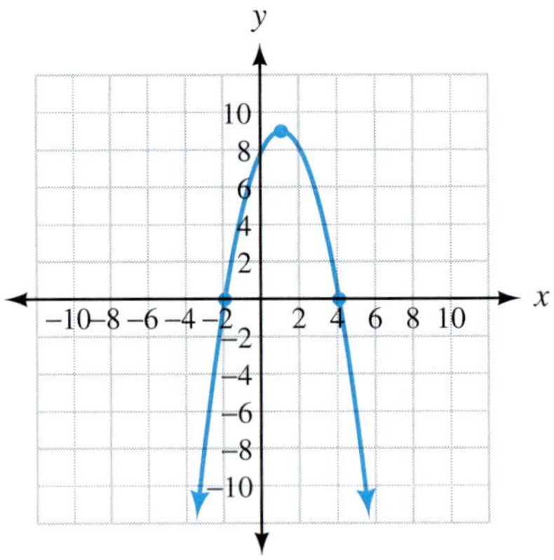

3. The x-intercepts are found by using the quadratic formula on $2x^2 - 4x + 1 = 0$. They are irrational numbers, approximately 0.3 and 1.7. Two other points on the graph are (0, 1) and (2, 1).

Vertex:

$$x = \frac{-b}{2a} = \frac{-(-4)}{2(2)} = 1$$
$$y = 2(1)^2 - 4(1) + 1 = -1$$

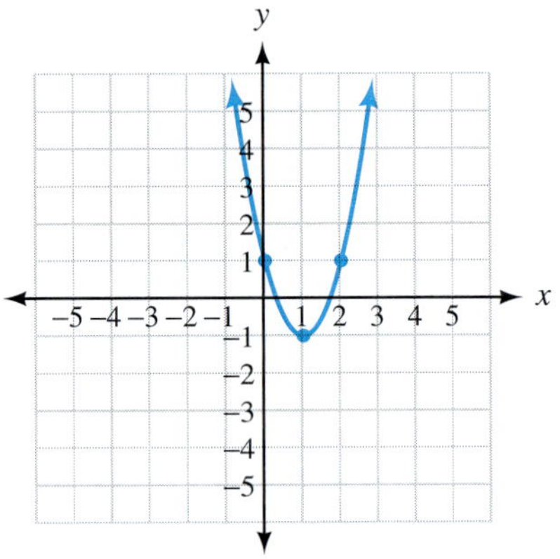

4. The graph does not cross the x-axis because the equation $-x^2 + 4x - 5 = 0$ does not have real solutions. Two points on the graph are (0, −5) and (4, −5).

Vertex:

$$x = \frac{-b}{2a} = \frac{-4}{2(-1)} = 2$$
$$y = -2^2 + 4(2) - 5 = -1$$

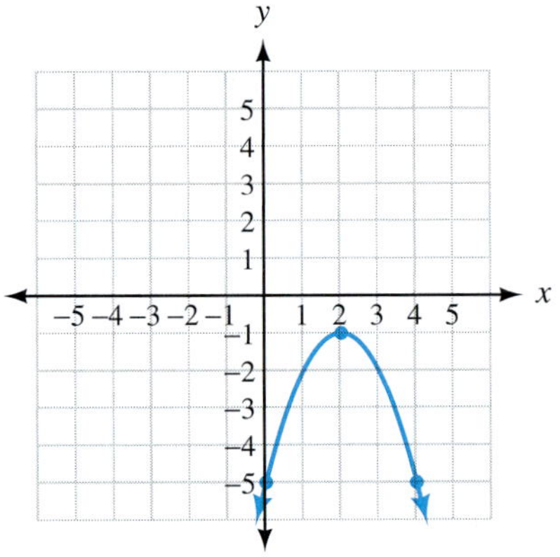

5. $y = -0.01x^2 + 12x - 400$

$$x = \frac{-b}{2a} = \frac{-12}{2(-0.01)} = \frac{-12}{-0.02} = 600$$
$$y = -0.01(600)^2 + 12(600) - 400$$
$$= -0.01(360{,}000) + 7{,}200 - 400$$
$$= -3{,}600 + 7{,}200 - 400$$
$$= 3{,}200$$

6.

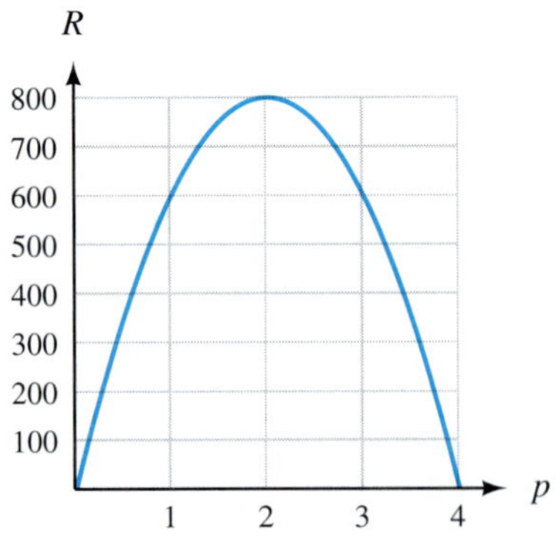

From the graph we see that the maximum value of R occurs when $p = \$2.00$. We can calculate the maximum value of R from the equation:

When $p = 2$

the equation $R = (800 - 200p)p$

becomes $R = (800 - 200 \cdot 2)2$

$= (800 - 400)2$

$= 400 \cdot 2$

$= 800$

The maximum revenue is \$800. It is obtained by setting the price of each sketch pad at $p = \$2.00$.

7. $y = a(x - h)^2 + k$
Substitute 80 for both h and k.
$y = a(x - 80)^2 + 80$
Substitute 0 for both x and y.
$0 = a(0 - 80)^2 + 80$
Solve for a to get $a = -\frac{1}{80}$.
Equation is $y = -\frac{1}{80}(x - 80)^2 + 80$.

Section 8.6

1. $x^2 + 2x - 15 \le 0$
$(x + 5)(x - 3) \le 0$

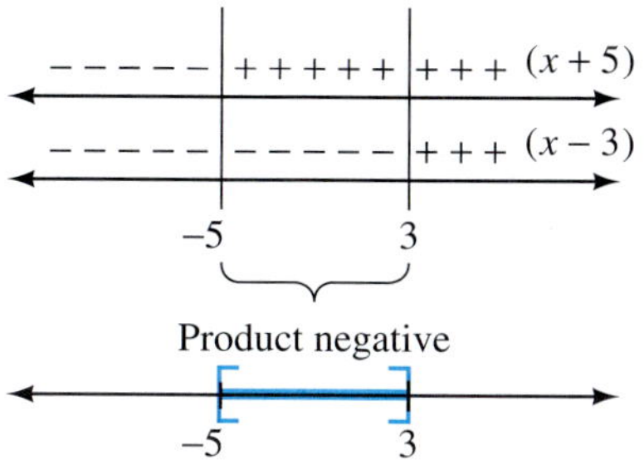

2. $2x^2 + 3x > 2$
$2x^2 + 3x - 2 > 0$
$(2x - 1)(x + 2) > 0$

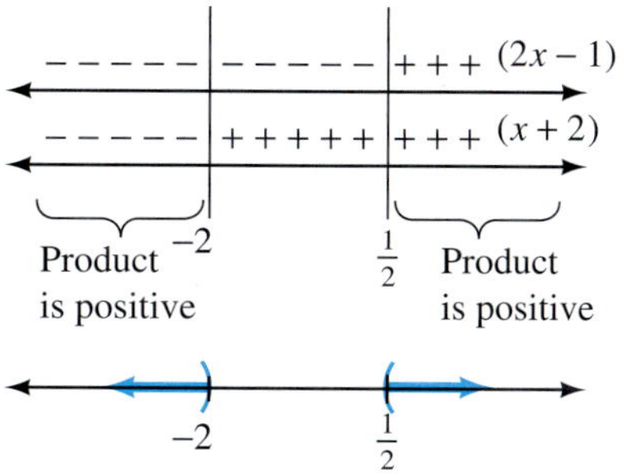

3. $x^2 + 10x + 25 < 0$
$(x + 5)^2 < 0$
No value of x will satisfy this inequality because the left side always will be positive or 0, but never less than 0.

Chapter 9

Section 9.1

1. a. $f(0) = 4^0 = 1$
b. $f(1) = 4^1 = 4$
c. $f(2) = 4^2 = 16$
d. $f(3) = 4^3 = 64$
e. $f(-1) = 4^{-1} = \frac{1}{4}$
f. $f(-2) = 4^{-2=} = \frac{1}{4^2} = \frac{1}{16}$

2. $A(12) = 1{,}200 \cdot 2^{-(12)/8}$
≈ 424.3 micrograms
$A(24) = 1{,}200 \cdot 2^{-(24)/8}$
$= 150$ micrograms

3.

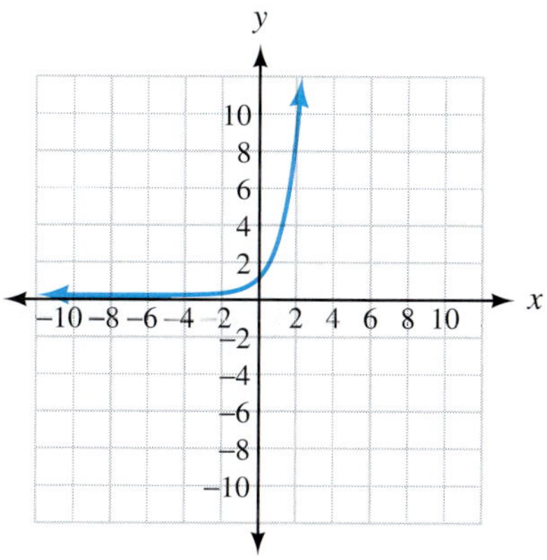

4.

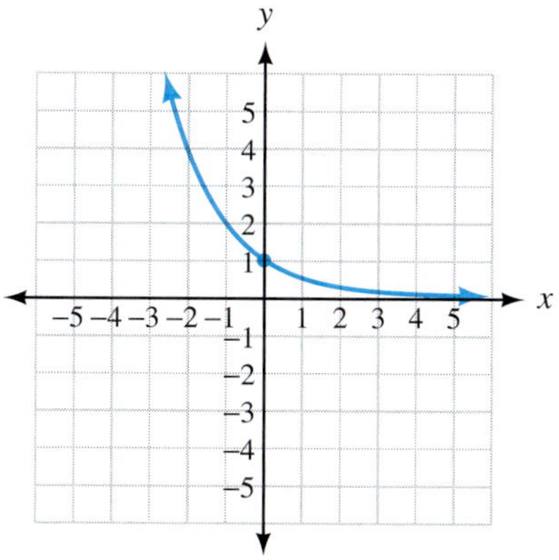

5. a. $A(5) = 600\left(1 + \frac{0.06}{12}\right)^{12 \cdot 5}$
$= 600\left(1 + \frac{0.06}{12}\right)^{60}$
≈ 809.31

b.

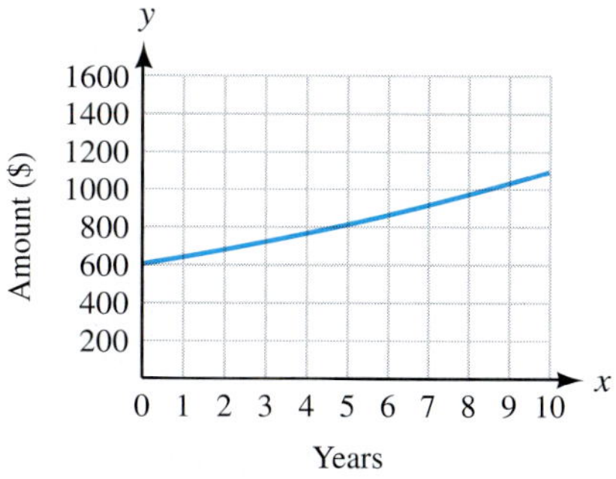

From the graph, $x = 8.5$ when $y = 1{,}000$.

6. $A(5) = 600e^{0.06 \cdot 5}$
$= 600e^{0.3}$
≈ 809.92

Section 9.2

1. The inverse of $y = 4x + 1$ is $x = 4y + 1$. Solving for y we have

$$4y + 1 = x$$
$$4y = x - 1$$
$$y = \frac{x-1}{4}$$

2. The equation of the inverse is $x = y^2 + 1$ or $y = \pm\sqrt{x-1}$.

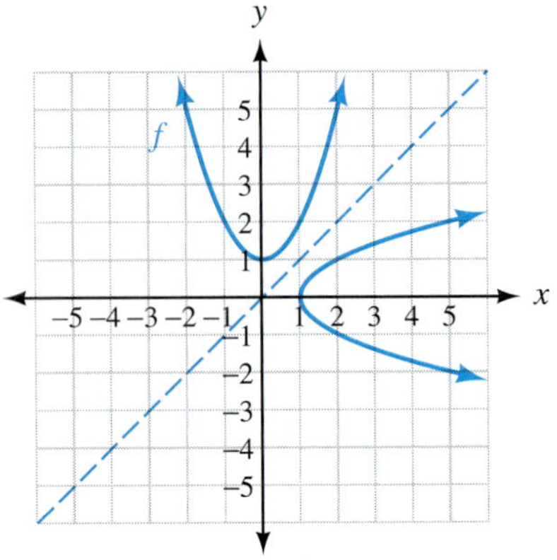

3. $x = \frac{y-3}{y+1}$ is the inverse of g.

$$x(y + 1) = y - 3$$
$$xy + x = y - 3$$
$$xy - y = -x - 3$$
$$y(x - 1) = -x - 3$$
$$y = \frac{-x-3}{x-1}$$

Multiply numerator and denominator by -1 to get $y = \frac{3+x}{1-x}$.

4.

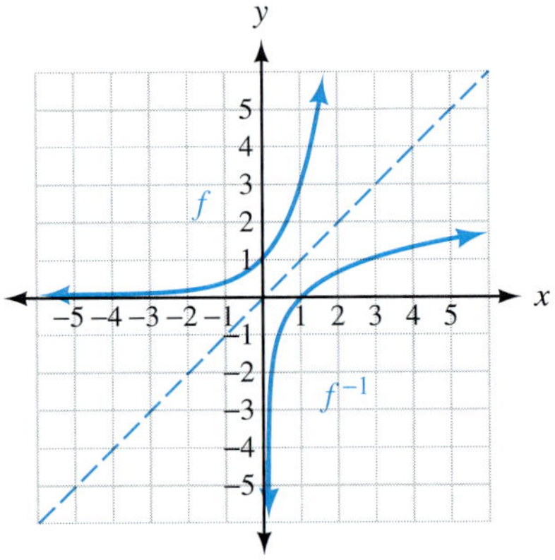

Section 9.3

1. $\log_2 x = 3$ is equivalent to $x = 2^3 = 8$

2. $\log_x 5 = 2$ is equivalent to $x^2 = 5$

$$x = \sqrt{5} \text{ only}$$

since the base in a logarithmic statement cannot be negative.

3. $\log_9 27 = x$ is equivalent

to $9^x = 27$

$$(3^2)^x = 3^3$$
$$3^{2x} = 3^3$$

so $2x = 3$

$$x = \frac{3}{2}$$

4. First graph $y = 3^x$ and then reflect it about the line $y = x$ to obtain the graph of $y = \log_3 x$.

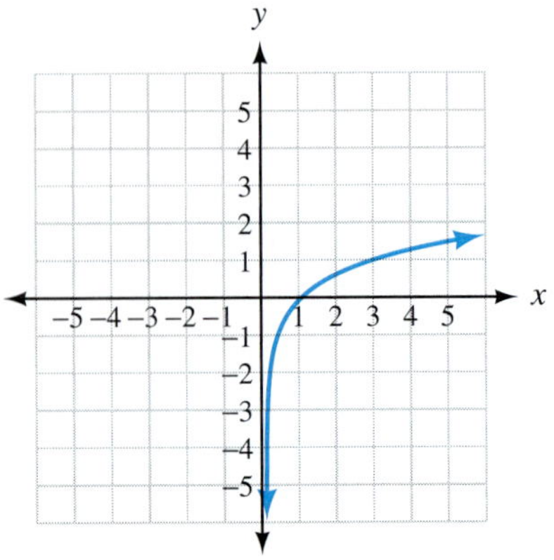

5. $\log_3 27 = \log_3 3^3$
$= 3$

6. $\log_{10} 1{,}000 = \log_{10} 10^3$
$= 3$

7. $\log_6 6 = \log_6 6^1$
$= 1$

8. $\log_3 1 = \log_3 3^0$
$= 0$

9. $\log_2 (\log_8 8) = \log_2 1$
$= 0$

10. $\log_{10} T = 6$
$\Rightarrow T = 10^6 = 1{,}000{,}000$

Section 9.4

1. $\log_3 \frac{5a}{b} = \log_3 5a - \log_3 b$
$= \log_3 5 + \log_3 a - \log_3 b$

2. $\log_{10} \frac{x^2}{\sqrt[3]{y}} = \log_{10} \frac{x^2}{y^{1/3}}$
$= \log_{10} x^2 - \log_{10} y^{1/3}$
$= 2 \log_{10} x - \frac{1}{3} \log_{10} y$

3. $3 \log_4 x + \log_4 y - 2 \log_4 z$
$= \log_4 x^3 + \log_4 y - \log_4 z^2$
$= \log_4 x^3 y - \log_4 z^2$
$= \log_4 \frac{x^3 y}{z^2}$

4. $\log_2 (x + 3) + \log_2 x = 2$
$\log_2 (x + 3)x = 2$
$(x + 3)x = 2^2$
$x^2 + 3x = 4$
$x^2 + 3x - 4 = 0$
$(x - 1)(x + 4) = 0$
$x = 1$ or $x = -4$
Since $x = -4$ would make both terms of our original equation undefined, the only solution is $x = 1$.

Section 9.5

1. When using a scientific calculator, we simply enter the number 27,600 and press the key labeled [log].
$\log 27{,}600 \approx 4.4409$

4. When using a scientific calculator, we enter the number 3.9786 and press the key labeled [10^x].
$\log x = 3.9786$
$\Rightarrow x = 10^{3.9786}$
$\approx 9{,}519$

5. $\log x = -1.5901$
$\Rightarrow x = 10^{-1.5901}$
≈ 0.0257

6. $\log T = 5.5$
$\Rightarrow T = 10^{5.5} \approx 3.16 \times 10^5$

7. $4.1 = -\log[H^+]$
$\Rightarrow \log[H^+] = -4.1$
$\Rightarrow [H^+] = 10^{-4.1}$
$\approx 7.9 \times 10^{-5}$

8. $pH = -\log[1.8 \times 10^{-5}]$
$\approx -(-4.7)$
$= 4.7$

9. a. $\ln e^2 = 2 \ln e = 2(1) = 2$
b. $\ln e^4 = 4 \ln e = 4(1) = 4$
c. $\ln e^{-2} = -2 \ln e = -2(1) = -2$
d. $\ln e^x = x \ln e = x(1) = x$

10. $\ln Pe^{rt} = \ln P + \ln e^{rt}$
$= \ln P + rt \ln e$
$= \ln P + rt$

11. a. $\ln 35 = \ln 5 \cdot 7 = \ln 5 + \ln 7 = 1.6094 + 1.9459 = 3.5553$
b. $\ln 0.2 = \ln \frac{2}{10} = \ln \frac{1}{5} = \ln 1 - \ln 5 = 0 - 1.6094 = -1.6094$
c. $\ln 49 = \ln 7^2 = 2 \ln 7 = 2(1.9459) = 3.8918$

Section 9.6

1. $12^{x+2} = 20$
$\log 12^{x+2} = \log 20$
$(x + 2) \log 12 = \log 20$
$x + 2 = \frac{\log 20}{\log 12}$
$x = \frac{\log 20}{\log 12} - 2$
$\approx \frac{1.3010}{1.0792} - 2$
≈ -0.7945 or -0.7944

2. $10{,}000 = 5{,}000(1 + 0.11)^t$
$10{,}000 = 5{,}000(1.11)^t$
$2 = (1.11)^t$
$\log 2 = t \log 1.11$
$t = \frac{\log 2}{\log 1.11}$
$\approx \frac{0.3010}{0.0453}$
≈ 6.64 years

3. $\log_6 14 = \frac{\log 14}{\log 6}$
$\approx \frac{1.1461}{0.7782}$
≈ 1.4728 or 1.4729, depending on whether you round your intermediate answers.

4. $75{,}000 = 32{,}000x^{0.05t}$
$2.34 \approx e^{0.05t}$
$\ln 2.34 \approx 0.05t \ln e$
$\ln 2.34 \approx 0.05t$
$t \approx \frac{\ln 2.34}{0.05} \approx 17.04$

Chapter 10

Section 10.1

1. $d = \sqrt{(-4-2)^2 + (1-5)^2}$
$= \sqrt{36+16}$
$= \sqrt{52}$
$= 2\sqrt{13}$

2. $\sqrt{(x-3)^2 + (2+1)^2} = \sqrt{10}$
$(x-3)^2 + 9 = 10$
$(x-3)^2 = 1$
$x - 3 = \pm 1$
$x = 3 \pm 1$
$x = 3 + 1$ or $x = 3 - 1$
$= 4$ $\quad = 2$
Both solutions check.

3. $(x-4)^2 + [y-(-3)]^2 = 2^2$
$(x-4)^2 + (y+3)^2 = 4$

4. Center = (0, 0), radius = 5
$(x-0)^2 + (y-0)^2 = 5^2$
$x^2 + y^2 = 25$

5.

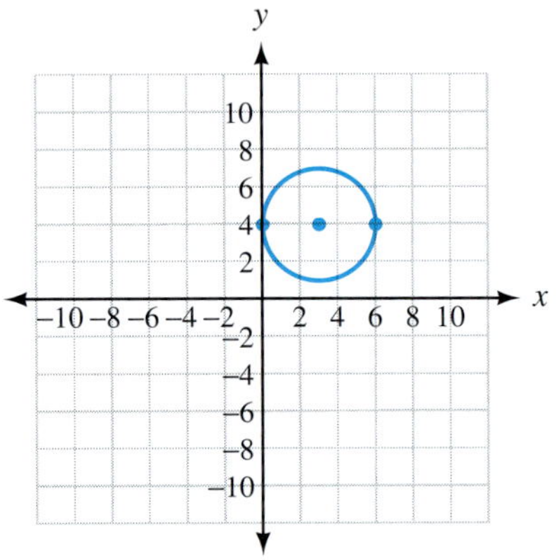

6.

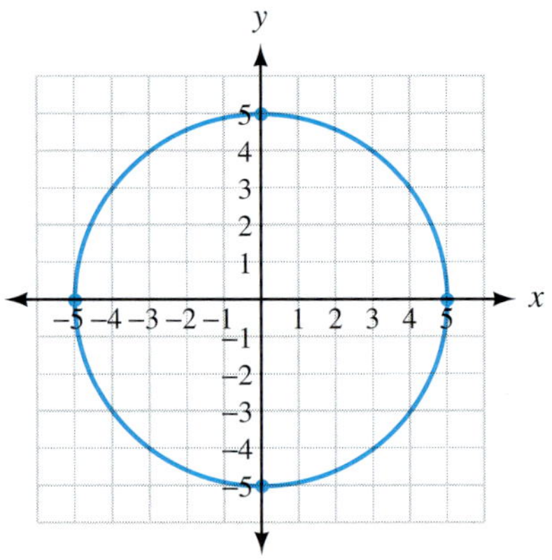

7. $x^2 + y^2 - 6x + 4y - 3 = 0$
$x^2 - 6x \quad + y^2 + 4y = 3$
$x^2 - 6x + \mathbf{9} + y^2 + 4y + \mathbf{4} = 3 + \mathbf{9} + \mathbf{4}$
$(x-3)^2 + (y+2)^2 = 16$

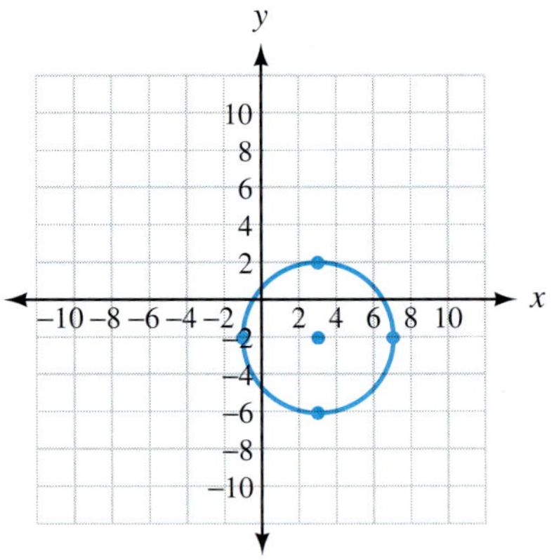

Section 10.2

1.

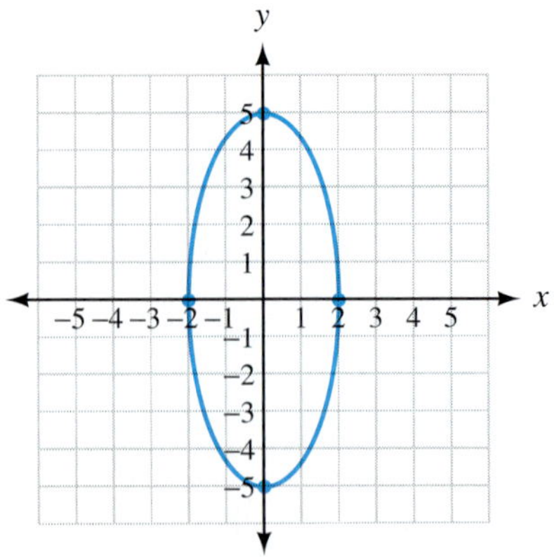

2.

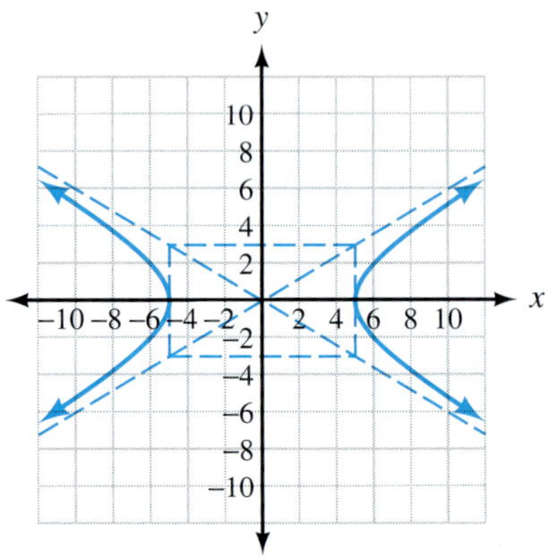

Section 10.3

1.

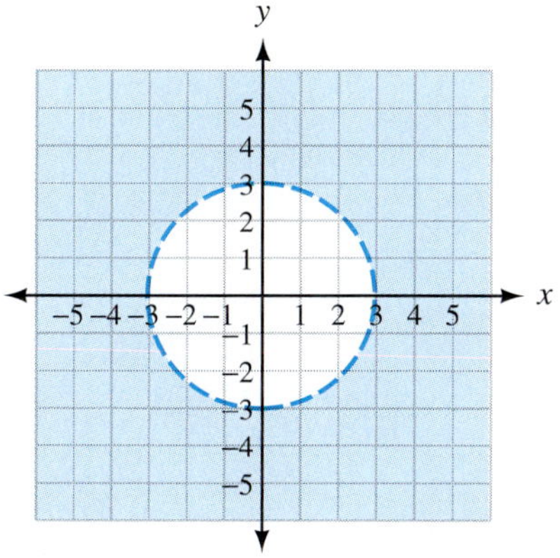

2.

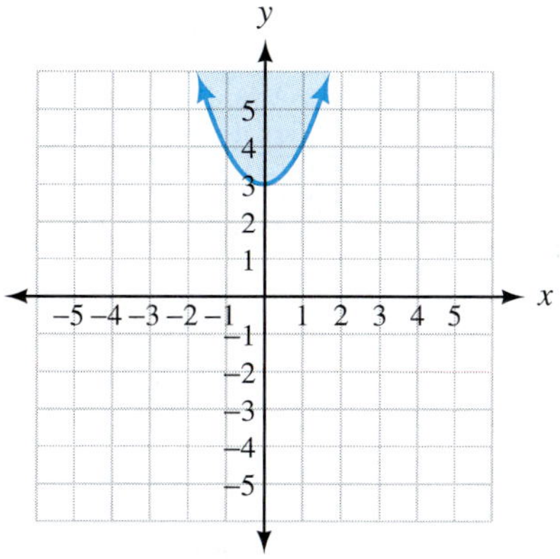

3.

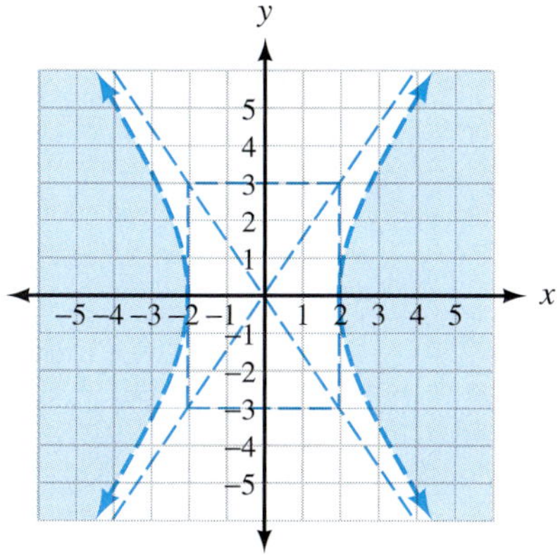

4. $x^2 + y^2 = 9$

$x - y = 3$

Solving the bottom equation for x gives $x = y + 3$. Substituting $y + 3$ for x in the top equation we have

$$(y + 3)^2 + y^2 = 9$$
$$y^2 + 6y + 9 + y^2 = 9$$
$$2y^2 + 6y = 0$$
$$2y(y + 3) = 0$$

$y = 0$ or $y = -3$

Substituting these values of y into $x = y + 3$ gives $x = 3$ and $x = 0$, respectively. The two solutions are (3, 0) and (0, −3).

5. $16x^2 - 4y^2 = 64 \xrightarrow{\text{no change}} 16x^2 - 4y^2 = 64$

$x^2 + y^2 = 4 \xrightarrow[\text{times 4}]{} 4x^2 + 4y^2 = 16$

$$20x^2 = 80$$
$$x^2 = 4$$
$$x = \pm 2$$

Using the second equation in the original system, we have

When $x = 2$ $\quad 2^2 + y^2 = 4 \Rightarrow y = 0$

When $x = -2$ $\quad (-2)^2 + y^2 = 4 \Rightarrow y = 0$

The two solutions are (2, 0) and (−2, 0).

6. $x^2 + y^2 = 4$

$y = x^2 - 4$

Substituting $x^2 - 4$ from the second equation for y in the first equation yields

$$x^2 + (x^2 - 4)^2 = 4$$
$$x^2 + x^4 - 8x^2 + 16 = 4$$
$$x^4 - 7x^2 + 16 = 4$$
$$x^4 - 7x^2 + 12 = 0$$
$$(x^2 - 4)(x^2 - 3) = 0$$

$x^2 = 4$ or $x^2 = 3$

$x = \pm 2$ or $x = \pm\sqrt{3}$

Substituting these four values of x into the second equation in our system gives us four solutions: (2, 0), (−2, 0), $(\sqrt{3}, -1)$, $(-\sqrt{3}, -1)$.

7. Let x and y represent the two numbers.

$x^2 + y^2 = 58$

$y = x^2 - 2$

Substituting $x^2 - 2$ into the first equation in place of y gives us

$$x^2 + (x^2 - 2)^2 = 58$$
$$x^2 + x^4 - 4x^2 + 4 = 58$$
$$x^4 - 3x^2 - 54 = 0$$
$$(x^2 - 9)(x^2 + 6) = 0$$

$x^2 = 9$ or $x^2 = -6$

$x = \pm 3$ or $x = \pm i\sqrt{6}$

When $x = \pm 3, y = 7$

When $x = \pm i\sqrt{6}, y = -8$

but since we have no way of comparing the size of a complex number like $i\sqrt{6}$ with a real number like −8, we cannot check the complex solutions. The only solutions are 3 and 7 and −3 and 7.

8. $x^2 + y^2 \geq 9$

$\frac{x^2}{4} + \frac{y^2}{25} \leq 1$

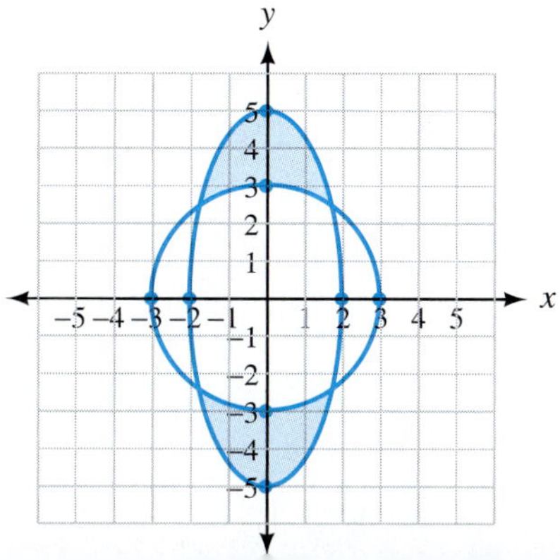

Appendix B

1.

$\underline{-2}\rfloor$	1	2	−8	1
	↓	−2	0	16
	1	0	−8	$\boxed{17}$

Answer: $x^2 - 8 + \dfrac{17}{x+2}$

2.

$\underline{3}\rfloor$	2	−5	0	3
	↓	6	3	9
	2	1	3	$\boxed{12}$

Answer: $2x^2 + x + 3 + \dfrac{12}{x-3}$

3.

$\underline{-2}\rfloor$	1	0	0	8
	↓	−2	4	−8
	1	−2	4	$\boxed{0}$

Answer: $x^2 - 2x + 4$

Appendix C

2. The converse: (If B, then A.) If it can fly, then it is a bird.
The inverse: (If not A, then not B.) If it is not a bird, then it cannot fly.
The contrapositive: (If not B, then not A.) If it cannot fly, then it is not a bird.

3. The contrapositive also must be true: If your eyes are opened, then you are not asleep.

4. The contrapositive also must be true: If $a^2 \neq 9$, then $a \neq 3$.

5. If it is history, then it repeats itself.

Appendix D

1. a.
$$\begin{aligned}(f+g)(x) &= f(x) + g(x)\\ &= (x^2-4) + (x+2)\\ &= x^2 + x - 2\end{aligned}$$

b.
$$\begin{aligned}(f-g)(x) &= f(x) - g(x)\\ &= (x^2-4) - (x+2)\\ &= x^2 - x - 6\end{aligned}$$

c.
$$\begin{aligned}(fg)(x) &= f(x)\,g(x)\\ &= (x^2-4)(x+2)\\ &= x^3 + 2x^2 - 4x - 8\end{aligned}$$

d.
$$\begin{aligned}\left(\frac{f}{g}\right)(x) &= \frac{f(x)}{g(x)}\\ &= \frac{x^2-4}{x+2}\\ &= \frac{(x+2)(x-2)}{x+2}\\ &= x - 2\end{aligned}$$

2. a.
$$\begin{aligned}(f+g)(x) &= f(x) + g(x)\\ &= 3x + 2 + 3x^2 - 10x - 8\\ &= 3x^2 - 7x - 6\end{aligned}$$

b.
$$\begin{aligned}(fh)(x) &= f(x)h(x)\\ &= (3x+2)(x-4)\\ &= 3x^2 - 10x - 8\\ &= g(x)\end{aligned}$$

c.
$$\begin{aligned}(fg)(x) &= f(x)g(x)\\ &= (3x+2)(3x^2-10x-8)\\ &= 9x^3 - 24x^2 - 44x - 16\end{aligned}$$

d.
$$\begin{aligned}\left(\frac{g}{f}\right)(x) &= \frac{g(x)}{f(x)}\\ &= \frac{3x^2-10x-8}{3x+2}\\ &= \frac{(3x+2)(x-4)}{3x+2}\\ &= x - 4\\ &= h(x)\end{aligned}$$

3. a.
$$\begin{aligned}(f+g)(3) &= 3(3)^2 - 7(3) - 6\\ &= 27 - 21 - 6\\ &= 0\end{aligned}$$

b.
$$\begin{aligned}(fh)(-1) &= 3(-1)^2 - 10(-1) - 8\\ &= 3 + 10 - 8\\ &= 5\end{aligned}$$

c.
$$\begin{aligned}(fg)(0) &= 9(0)^3 - 24(0)^2 - 44(0) - 16\\ &= -16\end{aligned}$$

d.
$$\begin{aligned}\left(\frac{g}{f}\right)(10) &= 10 - 4\\ &= 6\end{aligned}$$

4. a.
$$\begin{aligned}(f\circ g)(x) &= f(g(x))\\ &= f(x^2+3x)\\ &= (x^2+3x) - 4\\ &= x^2 + 3x - 4\end{aligned}$$

b.
$$\begin{aligned}(g\circ f)(x) &= g(f(x))\\ &= g(x-4)\\ &= (x-4)^2 + 3(x-4)\\ &= x^2 - 8x + 16 + 3x - 12\\ &= x^2 - 5x + 4\end{aligned}$$

Answers to Odd-Numbered Problems and Chapter Reviews

Chapter 1

Getting Ready for Chapter 1

1. 3.8 **2.** 0.5 **3.** 1.8 **4.** 3 **5.** 4 **6.** 44 **7.** 10 **8.** 21 **9.** 210 **10.** 231 **11.** 630 **12.** 13,500 **13.** 3.6 **14.** 0.35 **15.** 215.6 **16.** 0.2052 **17.** 785 **18.** 87 **19.** 320 **20.** 3

Problem Set 1.1

1. $x + 5 = 2$ **3.** $6 - x = y$ **5.** $2t < y$ **7.** $x + y < x - y$ **9.** $3(x - 5) > y$ **11.** 36 **13.** 100 **15.** 8 **17.** 16 **19.** 10,000 **21.** 121 **23. a.** 19 **b.** 27 **c.** 27 **25. a.** 16 **b.** 12 **c.** 18 **27. a.** 33 **b.** 33 **29. a.** 144 **b.** 74 **c.** 144 **31. a.** 23 **b.** 41 **c.** 65 **33. a.** 39 **b.** 7 **c.** 5 **35. a.** 48 **b.** 24 **37. a.** 41 **b.** 95 **39.** 5,431 **41.** 138 **43.** 78 **45.** 152 **47.** 11 **49.** 87 **51.** 4 **53.** 22 **55.** 16 **57.** 625 **59.** 1 **61.** 0.6 **63.** 8,700 **65.** 5.046 **67.** 5 **69.** 128 **71.** 625 **73.** 31 **75.** 1,000 **77.** 750 **79.** 0.5 **81.** 1.6 **83.** 320 **85.** 1.7917 **87.** 2.0793 **89.** 785 **91.** 185.12 **93.** 12.3106 **95.** −0.1587 **97.** 196 **99.** 650 **101. a.** 7 **b.** 10 **c.** 25 **d.** 32 **103. a.** 121 **b.** 121 **c.** 101 **d.** 81 **105. a.** 1 **b.** 2,400 **c.** 0.52 **107.** {0, 1, 2, 3, 4, 5, 6} **109.** {2, 4} **111.** {1, 3, 5} **113.** {0, 1, 2, 3, 4, 5, 6} **115.** {0, 2} **117.** {0, 6} **119.** {0, 1, 2, 3, 4, 5, 6, 7} **121.** {1, 2, 4, 5} **123. a.** 0.03 seconds **b.** 0.37 seconds

Problem Set 1.2

1.

3.
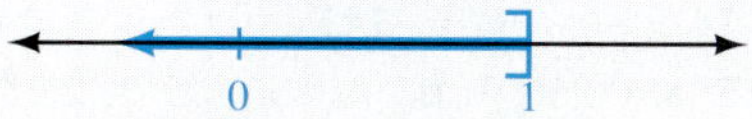

5.
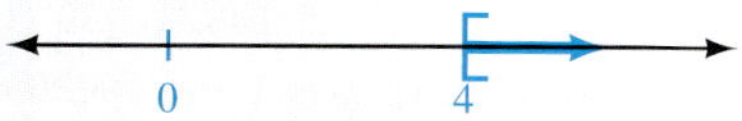

7.
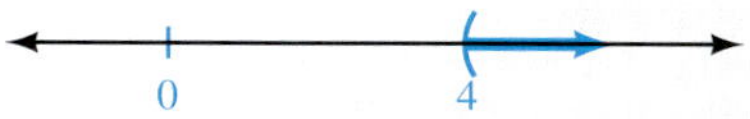

9.

11.

13.

15.

17.

19.

21. ∅ **23.**

25.

27.

29.

31.
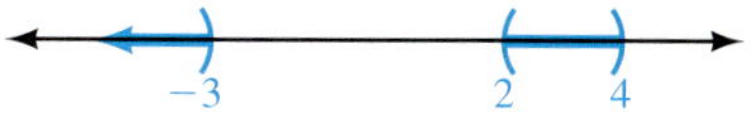

33.

35.

37. $x \ge 5$ **39.** $x \le -3$ **41.** $x \le 4$ **43.** $-4 < x < 4$ **45.** $-4 \le x \le 4$ **47.** 1, 2 **49.** $-6, -5.2, 0, 1, 2, 2.3, \frac{9}{2}$ **51.** $-\sqrt{7}, -\pi, \sqrt{17}$ **53.** 0, 1, 2 **55.** $2^2 \cdot 3 \cdot 5$ **57.** $2 \cdot 7 \cdot 19$ **59.** $3 \cdot 37$ **61.** $3^2 \cdot 41$ **63.** $\frac{3}{7}$ **65.** $\frac{11}{21}$ **67.** $\frac{3}{5}$ **69.** $\frac{5}{6}$ **71.** $\frac{5}{9}$ **73.** $\frac{3}{4}$ **75.** 40 **77.** $\frac{5}{11}$ **79.** −3 and 7 **81.** $3 < x < 13$ **83.** $x < 3$ or $x > 13$

85. a. $10 = 3 + 7$ Other answers are possible. **b.** $16 = 5 + 11$ Other answers are possible. **c.** $24 = 5 + 19$ Other answers are possible. **d.** $36 = 17 + 19$ Other answers are possible.

87. $50 \le F \le 270$ **89. a.** 98% **b.** 72% **c.** 54% **91.** 120 **93.** $6! = 6 \cdot 5 \cdot 4 \cdot 3 \cdot 2 \cdot 1 = 6 \cdot 5!$

Problem Set 1.3

1. $4, -4, \frac{1}{4}$ **3.** $-\frac{1}{2}, \frac{1}{2}, -2$ **5.** $5, -5, \frac{1}{5}$ **7.** $\frac{3}{8}, -\frac{3}{8}, \frac{8}{3}$ **9.** $-\frac{1}{6}, \frac{1}{6}, -6$ **11.** $3, -3, \frac{1}{3}$ **13.** −1, 1 **15.** 0 **17.** 2 **19.** $\frac{3}{4}$ **21.** π **23.** −4 **25.** −2 **27.** $-\frac{3}{4}$ **29.** $\frac{21}{40}$ **31.** 2 **33.** $\frac{8}{27}$ **35.** $\frac{1}{10{,}000}$ **37.** $\frac{72}{385}$ **39.** 1 **41.** $6 + x$ **43.** $a + 8$ **45.** $15y$ **47.** x **49.** a **51.** x **53.** $3x + 18$ **55.** $12x + 8$ **57.** $15a + 10b$ **59.** $\frac{4}{3}x + 2$ **61.** $2 + y$ **63.** $40t + 8$ **65.** $9x + 3y - 6z$ **67.** $3x + 7y$ **69.** $6x + 7y$ **71.** $3x + 1$ **73.** $2x - 1$ **75.** $x + 2$ **77.** $a - 3$ **79.** $x + 24$ **81.** $3x - 2y$

83. $3x + 8y$ **85.** $8x + 5y$ **87.** $15x + 10$ **89.** $8y + 32$ **91.** $15t + 9$ **93.** $28x + 11$ **95.** $\frac{7}{15}$ **97.** $\frac{29}{35}$ **99.** $\frac{35}{144}$ **101.** $\frac{949}{1,260}$ **103.** $\frac{47}{105}$ **105.** 15 **107.** $14a + 7$ **109.** $6y + 6$ **111.** $12x + 2$ **113.** $8y + 11$ **115.** $24a + 15$ **117.** $11x + 20$ **119.** Commutative property of addition **121.** Commutative property of multiplication **123.** Additive inverse **125.** Commutative property of addition **127.** Associative and commutative properties of multiplication **129.** Commutative and associative properties of addition **131.** Distributive property **133.** $7(x + 2) = 7x + 14$; $7x + 7(2) = 7x + 14$ **135.** $x(y + 4) = xy + 4x$; $xy + 4x$ **137.** $12 + \frac{1}{4}(12) = 15$ **139. a.** Answers will vary. **b.** 2001 **c.** Negative

Problem Set 1.4

1. 4 **3.** -4 **5.** -10 **7.** -4 **9.** $\frac{19}{12}$ **11.** $-\frac{32}{105}$ **13.** -8 **15.** -12 **17.** $-7x$ **19.** 13 **21.** -14 **23.** $6a$ **25.** -15 **27.** 15 **29.** -24 **31.** $-10x$ **33.** x **35.** y **37.** $-8x + 6$ **39.** $-3a + 4$ **41.** -14 **43.** 18 **45.** 16 **47.** -19 **49.** 50 **51.** 20 **53.** -2 **55.** 1 **57.** -30 **59.** 18 **61.** 277 **63.** -73 **65.** 39 **67.** 11 **69.** -5 **71.** 11 **73.** 7 **75.** -44 **77.** 2 **79.** $\frac{4}{3}$ **81.** 0 **83.** Undefined **85.** $-\frac{2}{3}$ **87.** $x - 5$ **89.** $x - 7$ **91.** $-6x + 9y$ **93.** $3x - 18$ **95.** $3a + 6$ **97.** $3x + 60$ **99.** $11y$ **101.** $14x + 12$ **103.** $7m - 15$ **105.** $-2x + 9$ **107.** $7y + 10$ **109.** $-20x + 5$ **111.** $-11x + 10$ **113.** -2 **115.** Undefined **117.** 0 **119.** $-\frac{2}{3}$ **121.** 32 **123.** 64 **125.** $-\frac{1}{18}$ **127.** $\frac{5}{3}$ **129.** 11 **131.** 12

133.

a	b	SUM $a + b$	DIFFERENCE $a - b$	PRODUCT ab	QUOTIENT $\frac{a}{b}$
3	12	15	-9	36	$\frac{1}{4}$
-3	12	9	-15	-36	$-\frac{1}{4}$
3	-12	-9	15	-36	$-\frac{1}{4}$
-3	-12	-15	9	36	$\frac{1}{4}$

135.

x	$3(5x - 2)$	$15x - 6$	$15x - 2$
-2	-36	-36	-32
-1	-21	-21	-17
0	-6	-6	-2
1	9	9	13
2	24	24	28

137. a. 1 **b.** $\frac{3}{2}$ **c.** -1 **d.** 135 **139.** 1.5283 **141.** -0.0794 **143.** -0.0714 **145.** 3.4 **147.** 1.6 **149.** 1,200 **151.** 190 **153.** Santa Fe 10:10 P.M., Detroit 3:30 A.M. **155. a.** 00:00:10 **b.** 00:02:20 **c.** 00:22:53

Problem Set 1.5

1. 5 **3.** 10 **5.** 25 **7.** 29 **9.** 23 **11.** △ **13.** ⊙ **15.** 17, 21 **17.** $-2, -3$ **19.** $-4, -7$ **21.** $-\frac{1}{2}, -\frac{3}{4}$ **23.** $\frac{5}{2}, 3$ **25.** 27 **27.** -270 **29.** $\frac{1}{8}$ **31.** $\frac{5}{2}$ **33.** -625 **35.** $-\frac{1}{125}$ **37. a.** 12 **b.** 16 **39.** 144 **41.** 2, 3, 5, among others **43.** 2, 8, 34

45.

TWO NUMBERS a and b	THEIR PRODUCT ab	THEIR SUM $a + b$
1, -24	-24	-23
-1, 24	-24	23
2, -12	-24	-10
-2, 12	-24	10
3, -8	-24	-5
-3, 8	-24	5
4, -6	-24	-2
-4, 6	-24	2

47. 41, 37.5, 34, 30.5, 27, 23.5; yes **49.** 41, 45.5, 50, 54.5, 59, 63.5; yes
51. The patient on Antidepressant 1 misses his morning dose; less than half of the antidepressant will remain in the body. The patient on Antidepressant 2 still has most of the medication in his body even after missing a dose because of the relatively long (5-day) half-life.

53.

HOURS SINCE DISCONTINUING	CONCENTRATION (ng/mL)
0	60
4	30
8	15
12	7.5
16	3.75

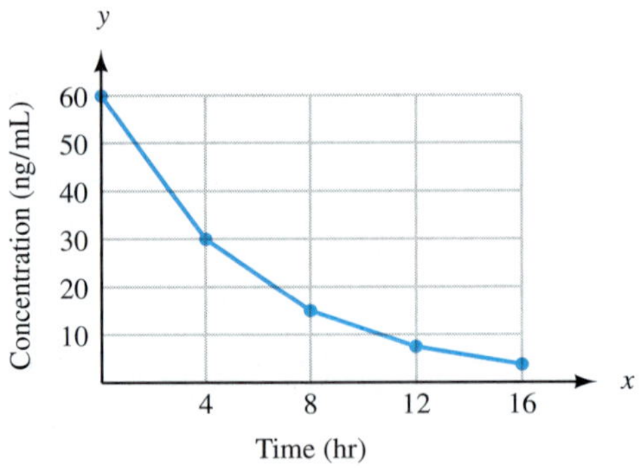

55.

ELEVATION (ft)	BOILING POINT (°F)
−2,000	**215.6**
−1,000	**213.8**
0	**212**
1,000	**210.2**
2,000	**208.4**
3,000	**206.6**

57. **a.** The patient taking his medication
b. The patient stops taking his medication
c. 50 ng/mL
d. 4, 8, and 12 hours

59.

TEMPERATURE (FAHRENHEIT)	SHELF-LIFE (DAYS)
32°	24
40°	10
50°	2
60°	1
70°	$\frac{1}{2}$

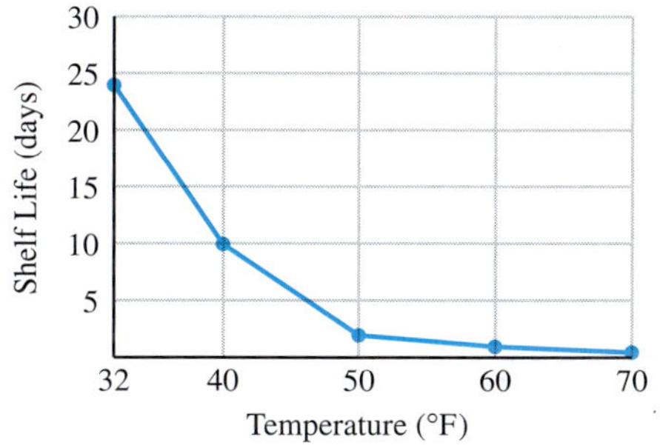

Chapter 1 Review/Test

1. $x + 2$ **2.** $x - 2$ **3.** $\frac{x}{2}$ **4.** $2x$ **5.** $2(x + y)$ **6.** $2x + y$ **7.** 27 **8.** 125 **9.** 64 **10.** 1 **11.** 32 **12.** 81 **13.** 17 **14.** 4 **15.** 13 **16.** 7 **17.** 9 **18.** 32 **19.** 30 **20.** 43 **21.** {1, 2, 3, 4, 5, 6} **22.** {1, 3} **23.** {5} **24.** {6}
25.

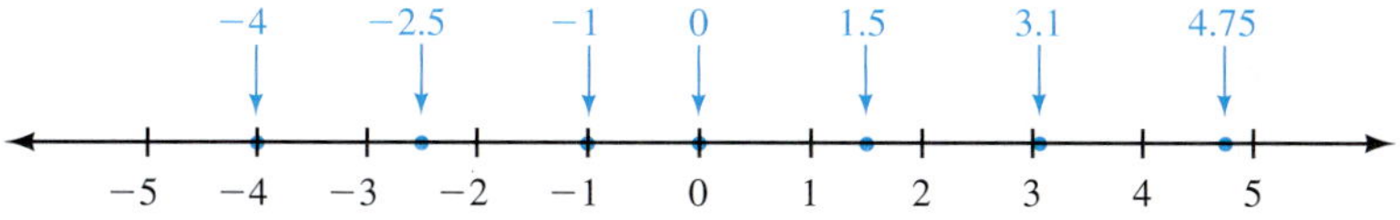

26.

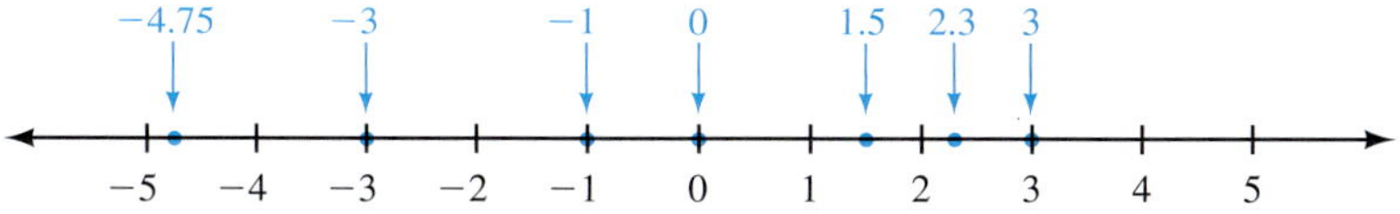

27. $-2, \frac{1}{2}$ **28.** $\frac{2}{5}, -\frac{5}{2}$ **29.** 3 **30.** −5

31. 4 **32.** 6 **33.** −7, 0, 5 **34.** $-7, -4.2, 0, \frac{3}{4}, 5$ **35.** $-\sqrt{3}, \pi$ **36.** $2^2 \cdot 3^2 \cdot 11^2$ **37.** $\frac{11}{13}$ **38.** 1 **39.** $\frac{27}{64}$ **40.** 2

41.

42.

43.

44.

45. $x \geq 4$ **46.** $x \leq 5$ **47.** $0 < x < 8$

48. $0 \leq x \leq 8$ **49.** $2y$ **50.** $3x$ **51.** $8x + 5$ **52.** $y + 2$ **53.** a **54.** c **55.** a **56.** b, d **57.** a, c **58.** f **59.** e **60.** g **61.** 2 **62.** −2 **63.** 1 **64.** 17 **65.** 3 **66.** −6 **67.** 2 **68.** 66 **69.** $-\frac{5}{6}$ **70.** $-\frac{5}{6}$ **71.** −42 **72.** 30 **73.** $21x$ **74.** $-6x$ **75.** $-6x + 10$ **76.** $-6x + 21$ **77.** $-x + 3$ **78.** $-15x + 3$ **79.** $-\frac{5}{6}$ **80.** −36 **81.** $\frac{1}{10}$ **82.** $-\frac{2}{7}$ **83.** −13 **84.** −17 **85.** 0 **86.** −1 **87.** − 36 **88.** − 34 **89.** 16 **90.** −24 **91.** 2 **92.** 39 **93.** 0 **94.** $-2y + 9$ **95.** $-18x - 14$ **96.** $5a - 22$ **97.** −1, arithmetic **98.** 8 **99.** −1, arithmetic **100.** $\frac{1}{16}$, geometric

Chapter 2

Getting Ready for Chapter 2

1. −4 **2.** −9 **3.** −12 **4.** $-\frac{7}{8}$ **5.** $-\frac{7}{8}$ **6.** $-\frac{3}{8}$ **7.** x **8.** x **9.** $16x$ **10.** $16x + 12$ **11.** $5x$ **12.** $0.01x + 500$

13. $7y - 3$ **14.** $-7x + 14$ **15.** **a.** 3 **b.** 5 **c.** 9 **16.** **a.** $\frac{1}{2}$ **b.** $\frac{1}{3}$ **c.** $\frac{1}{4}$ **17.** $2x - 3$ **18.** $4x - 10$ **19.** $x + y = 180$ **20.** $a + b = 90$

Problem Set 2.1

1. 8 **3.** 5 **5.** 2 **7.** −7 **9.** $-\frac{9}{2}$ **11.** −4 **13.** $-\frac{4}{3}$ **15.** −4 **17.** −2 **19.** $\frac{3}{4}$ **21.** 12 **23.** −10 **25.** 7 **27.** −7 **29.** 3 **31.** $\frac{4}{5}$ **33.** −4 **35.** 1 **37.** −3 **39.** 4 **41.** 0 **43.** 17 **45.** 2 **47.** 6 **49.** $-\frac{4}{3}$ **51.** $-\frac{3}{2}$ **53.** $\frac{5}{3}$ **55.** 12 **57.** 6,000 **59.** 5,000 **61.** $\frac{3}{2}$ **63.** 1 **65.** No solution **67.** All real numbers are solutions. **69.** No solution **71.** 6

73. $-\frac{1}{2}$ **75.** $-\frac{2}{3}$ **77.** $-\frac{7}{640}$ **79.** 2 **81.** 20 **83.** 0.7 **85.** 2,400 **87.** 24 **89.** 6 **91.** 3 **93.** $\frac{8}{15}$ **95.** $-\frac{5}{11}$

97. a. $\$6.60 = \$0.40n + \$1.80$ **b.** 12 miles **99. a.** $3{,}937{,}000 = 1125A$ **b.** 3,500 square miles **101.** 0.5 **103.** 62.5 **105.** 0 **107.** 1.25 **109.** 13 **111. a.** 3 **b.** 5 **c.** 9 **113.** Commutative **115.** Associative **117.** Commutative and associative **119.** Multiplicative identity **121.** Commutative **123.** Additive identity

Problem Set 2.2

1. -3 **3.** 0 **5.** $\frac{3}{2}$ **7.** 4 **9.** $\frac{8}{5}$ **11.** 2 **13.** $-\frac{7}{640}$ **15.** 675 **17. a.** 0 **b.** 0 **19. a.** 23 **b.** 23 **21.** $-\frac{1}{2}$ **23.** 3 **25.** 2,400 **27.** $\ell = \frac{A}{w}$ **29.** $t = \frac{I}{pr}$ **31.** $T = \frac{PV}{nR}$ **33.** $x = \frac{y-b}{m}$ **35.** $F = \frac{9}{5}C + 32$ **37.** $v = \frac{h - 16t^2}{t}$ **39.** $d = \frac{A-a}{n-1}$ **41.** $y = -\frac{2}{3}x + 2$ **43.** $y = \frac{3}{5}x + 3$ **45.** $y = \frac{1}{3}x + 2$ **47.** $x = \frac{5}{a-b}$ **49.** $P = \frac{A}{1+rt}$ **51.** $y = -\frac{1}{4}x + 2$ **53.** $y = \frac{3}{5}x - 3$ **55.** $y = \frac{1}{2}x + \frac{3}{2}$ **57.** $y = -2x - 5$ **59. a.** $y = 4x - 1$ **b.** $y = -\frac{1}{2}x$ **c.** $y = -3$ **61.** 20.52 **63.** 25% **65.** 925

67. 4, 7, 10, 13, 16 **69.** 4, 7, 12, 19, 28 **71.** $\frac{1}{4}, \frac{2}{5}, \frac{1}{2}, \frac{4}{7}, \frac{5}{8}$ **73.** $\frac{1}{1}, \frac{1}{4}, \frac{1}{9}, \frac{1}{16}, \frac{1}{25} \to 1, \frac{1}{4}, \frac{1}{9}, \frac{1}{16}, \frac{1}{25}$ **75.** 2, 4, 8, 16, 32 **77.** $2, \frac{3}{2}, \frac{4}{3}, \frac{5}{4}, \frac{6}{5}$ **79.** 1991 **81. a.** 1.8 tons **b.** 1.2 tons **83.** \$5.00 **85.** \$10.00 **87.** 2 centimeters **89.** 2 inches **91.** $2x - 3$ **93.** $x + y = 180$ **95.** 30 **97.** 8.5 **99.** 6,000 **101.** 20 **103.** 31 **105.** 8 **107.** 1 **109.** Shar = 128.4 beats per min, Sara = 140.4 beats per min

Problem Set 2.3

1. 10 feet by 20 feet **3.** 7 feet **5.** 5 inches **7.** 4 meters **9.** \$92.00 **11.** \$200.00 **13.** \$3,260.66 **15.** \$99.6 million **17.** 150 calories **19.** 20°, 160° **21. a.** 20.4°, 69.6° **b.** 38.4°, 141.6° **23.** 27°, 72°, 81° **25.** 34°, 44°, 102° **27.** 43°, 43°, 94° **29.** \$6,000 at 8%, \$3,000 at 9% **31.** \$5,000 at 12%, \$10,000 at 10% **33.** \$4,000 at 8%, \$2,000 at 9% **35.** Shell: \$700, Other: \$2,150 **37.** 30 fathers, 45 sons **39.** \$54 **41.** 44 minutes

43.

t	0	$\frac{1}{4}$	1	$\frac{7}{4}$	2
h	0	7	1	7	0

45.

HOT COFFEE SALES	
YEAR	SALES (BILLIONS OF DOLLARS)
2005	7
2006	7.5
2007	8
2008	8.6
2009	9.2

47.

AGE (YEARS)	MAXIMUM HEART RATE (BEATS PER MINUTE)
18	202
19	201
20	200
21	199
22	198
23	197

49.

RESTING HEART RATE (BEATS PER MINUTE)	TRAINING HEART RATE (BEATS PER MINUTE)
60	144
62	145
64	146
68	147
70	148
72	149

51.

w	l	A
2	22	44
4	20	80
6	18	108
8	16	128
10	14	140
12	12	144

53.

55.

57. -5 **59.** 6

61.

63.

65.

67.

Problem Set 2.4

1. $x \le \frac{3}{2}$

3. $x > 4$

5. $x \ge -5$

7. $x < 4$

9. $x \ge -6$

11. $x \ge 4$

13. $x < -3$ 
15. $m \geq -1$ 

17. $x \geq -3$ 
19. $y \leq \frac{7}{2}$ 

21. $x < 6$ 
23. $y \geq -52$ 
25. $(-\infty, -2]$ **27.** $[1, \infty)$

29. $(-\infty, 3)$ **31.** $(-\infty, -1]$ **33.** $[-17, \infty)$ **35.** $x > 435$ **37.** $[3, 7]$ 

39. $(-4, 2)$ 
41. $[4, 6]$ 

43. $(-4, 2)$ 
45. $(-3, 3)$ 

47. $(-\infty, -7] \cup [-3, \infty)$ 
49. $(-\infty, -1] \cup [\frac{3}{5}, \infty)$ 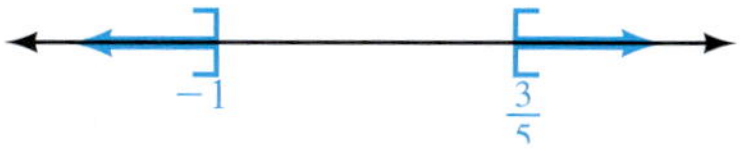

51. $(-\infty, -10) \cup (6, \infty)$ 
53. $p \leq 2$; set the price at \$2.00 or less per pad

55. $p > 1.25$; charge more than \$1.25 per pad **57.** During the years 1990 through 1994 **59.** Less than 55 minutes
61. 35° to 45° Celsius; $35° \leq C \leq 45°$ **63.** −25° to −10° Celsius; $-25° \leq C \leq -10°$ **65.** −3 **67.** 15 **69.** No solution **71.** −1
73. 3 **75.** −3 **77.** The distance between x and 0 on the number line **79.** $x < \frac{c-b}{a}$ **81.** $\frac{-c-b}{a} < x < \frac{c-b}{a}$
83. **a.** 339 **b.** $x \geq$ \$125,000 **c.** The actual percent is likely to be between 15% and 23%.

Problem Set 2.5

1. −4, 4 **3.** −2, 2 **5.** ∅ **7.** −1, 1 **9.** ∅ **11.** −6, 6 **13.** −3, 7 **15.** $\frac{17}{3}, \frac{7}{3}$ **17.** 2, 4 **19.** $-\frac{5}{2}, \frac{5}{6}$ **21.** −1, 5
23. ∅ **25.** −4, 20 **27.** −4, 8 **29.** $-\frac{10}{3}, \frac{2}{3}$ **31.** ∅ **33.** $-1, \frac{3}{2}$ **35.** 5, 25
37. −30, 26 **39.** −12, 28 **41.** $-5, \frac{3}{5}$
43. $1, \frac{1}{9}$ **45.** $-\frac{1}{2}$ **47.** 0 **49.** $-\frac{1}{2}$ **51.** $-\frac{1}{6}, -\frac{7}{4}$ **53.** All real numbers **55.** All real numbers **57.** 1987 and 1995
59. $x < 4$ **61.** $-\frac{11}{3} \leq a$ **63.** $t \leq -\frac{3}{2}$

65. 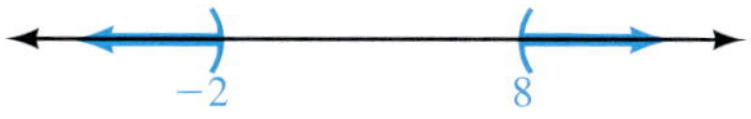
67. 

69. $\frac{19}{15}$ **71.** 40 **73.** −2 **75.** 0 **77.** $-\frac{2}{3}$

Problem Set 2.6

1. $-3 < x < 3$ 
3. $x \leq -2$ or $x \geq 2$ 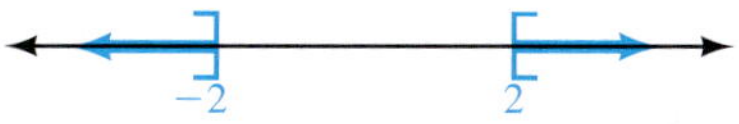

5. $-3 < x < 3$ 
7. $t < -7$ or $t > 7$ 
9. ∅

11. All real numbers **13.** $-4 < x < 10$ 

15. $a \leq -9$ or $a \geq -1$ 
17. ∅ **19.** $-1 < x < 5$ 

21. $y \leq -5$ or $y \geq -1$ 
23. $k \leq -5$ or $k \geq 2$ 

25. $-1 < x < 7$ 
27. $a \leq -2$ or $a \geq 1$ 

29. $-6 < x < \frac{8}{3}$ 
31. $x < 2$ or $x > 8$ 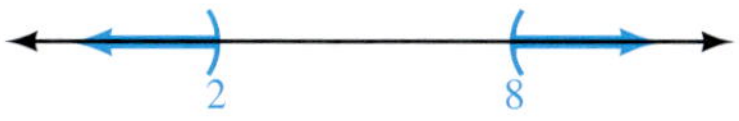

33. $x \leq -3$ or $x \geq 12$ 
35. $x < 2$ or $x > 6$ 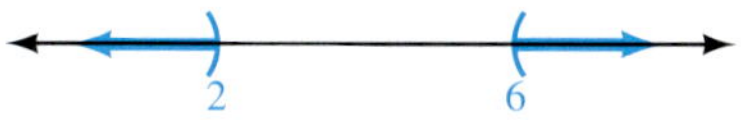

37. $0.99 < x < 1.01$ **39.** $x \leq -\frac{3}{5}$ or $x \geq -\frac{2}{5}$ **41.** $-\frac{1}{6} \leq x \leq \frac{3}{2}$ **43.** $-0.05 < x < 0.25$ **45.** $|x| \leq 4$ **47.** $|x - 5| \leq 1$
49. $|x - 65| \leq 10$ **51.** −6 **53.** 66 **55.** −13 **57.** 11 **59.** −4 **61.** 13 **63.** 16 **65.** −1

Chapter 2 Review/Test

1. 10 **2.** 2 **3.** 2 **4.** −3 **5.** −3 **6.** 0 **7.** $\frac{2}{3}$ **8.** $\frac{10}{13}$ **9.** $\frac{5}{2}$ **10.** 1 **11.** $-\frac{5}{11}$ **12.** $-\frac{2}{9}$ **13.** $h = 17$ **14.** $t = 20$ **15.** $p = \frac{I}{rt}$ **16.** $x = \frac{y - b}{m}$ **17.** $y = \frac{4}{3}x - 4$ **18.** $v = \frac{d - 16t^2}{t}$ **19.** 3, 6, 9, 12; increasing **20.** 4, 5, 6, 7; increasing **21.** −2, 4, −8, 16 **22.** 1, $\frac{1}{2}$, $\frac{1}{3}$, $\frac{1}{4}$; decreasing **23.** 4 feet by 12 feet **24.** 3 meters, 4 meters, 5 meters **25.** **a.** 3,780 bricks **b.** 625 feet **26.** \$24,875.24 **27.** $(-\infty, \frac{1}{2})$ **28.** $(-\infty, 8]$ **29.** $(-\infty, 12]$ **30.** $(-1, \infty)$ **31.** [2, 6] **32.** [0, 1] **33.** $(-\infty, -\frac{3}{2}] \cup [3, \infty)$ **34.** $(-\infty, 1) \cup [4, \infty)$ **35.** −2, 2 **36.** −4, 4 **37.** 2, 4 **38.** −1, 4 **39.** $-\frac{3}{2}$, 3 **40.** 5, 9 **41.** −1, 1 **42.** 0

43.

44.
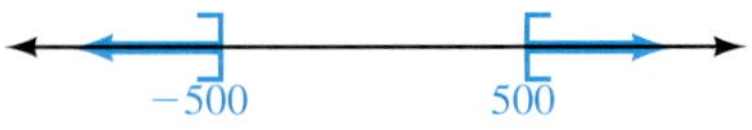

45. ∅

46.

47. ∅ **48.** ∅ **49.** ∅ **50.** All real numbers except 0 **51.** ∅ **52.** All real numbers

Chapter 3

Getting Ready for Chapter 3

1. 324 **2.** 0.0005 **3.** 3 **4.** −9 **5.** 3 **6.** 0 **7.** 113 **8.** 7 **9.** 3 **10.** 4 **11.** $-\frac{3}{50}$ **12.** $-\frac{3}{5}$ **13.** $y = \frac{2}{3}x - \frac{5}{3}$ **14.** $y = -\frac{1}{2}x + \frac{7}{2}$ **15.** (5, 0), (4, 3) **16.** (0, 0), (2, 0)

17.

t	0	1	2	3	4
h	0	48	64	48	0

18.

t	4	6	8	10
h	15	10	7.5	6

19. 135 **20.** $\frac{1}{2}$

Problem Set 3.1

1.
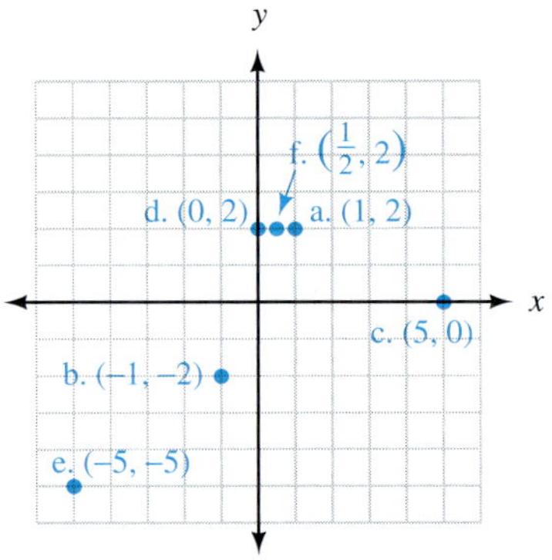

3. A (4, 1) B (−4, 3) C (−2, −5) D (2, −2) E (0, 5) F (−4, 0) G (1, 0)

5. x-intercept = 3, y-intercept = −2
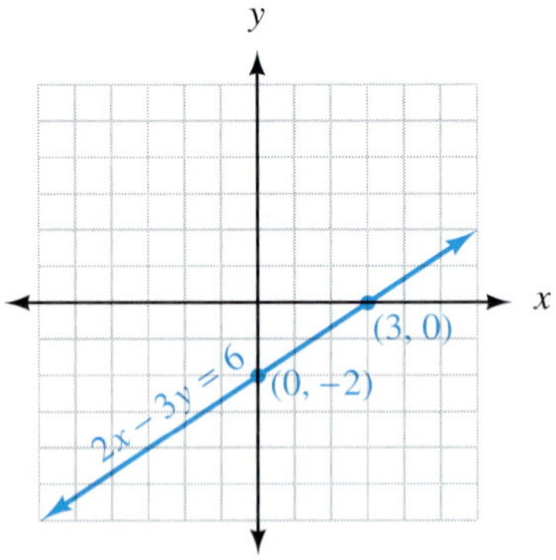

7. x-intercept = 5, y-intercept = −4
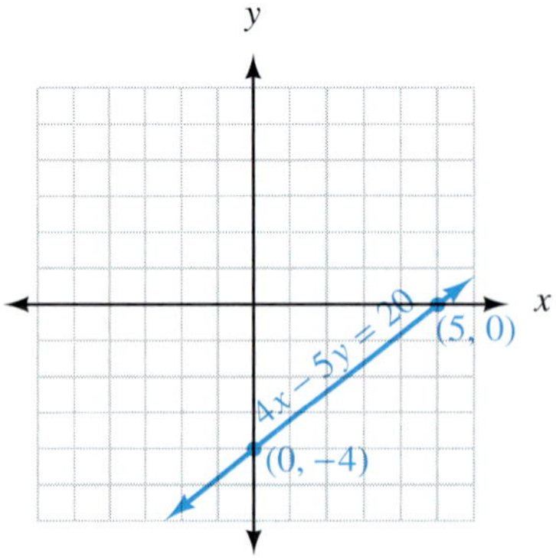

9.
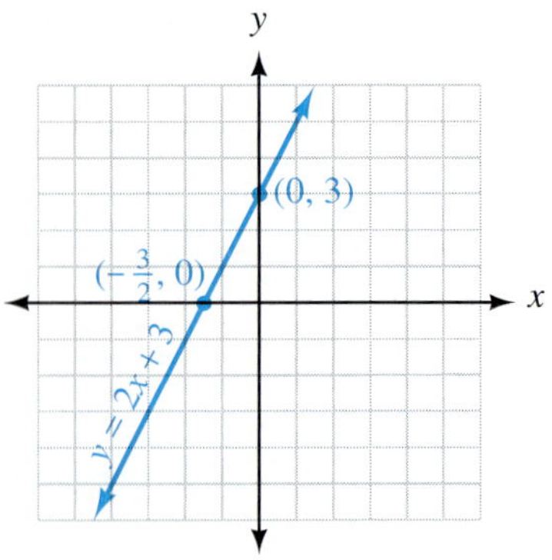

11. Table b

13.

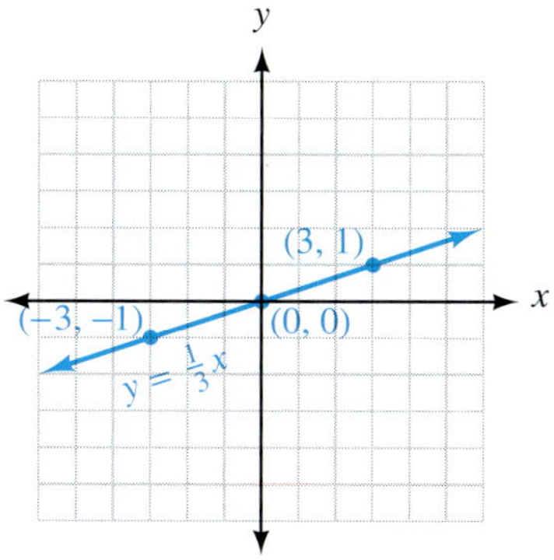

15.

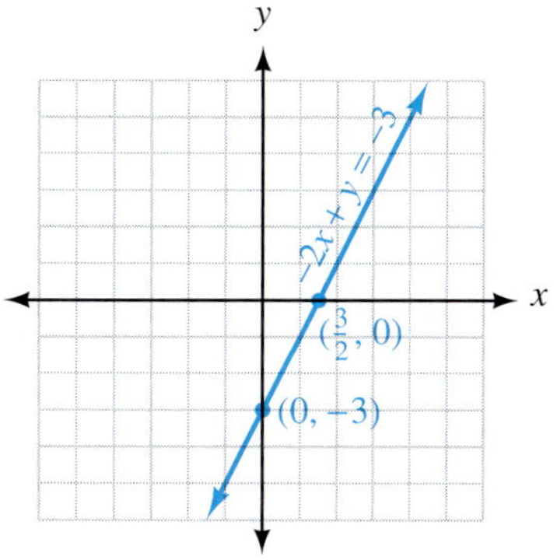

17.

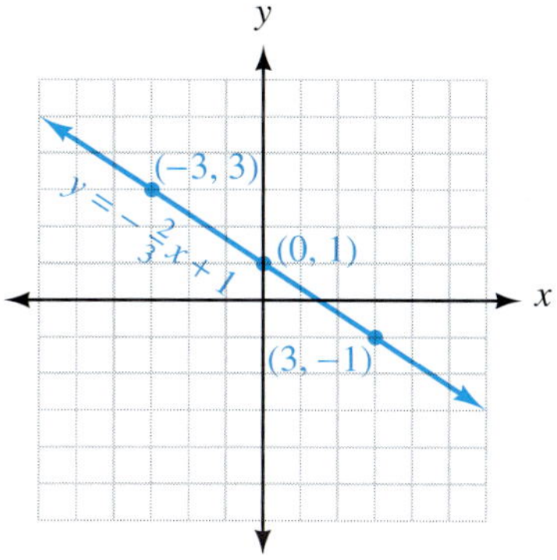

19.

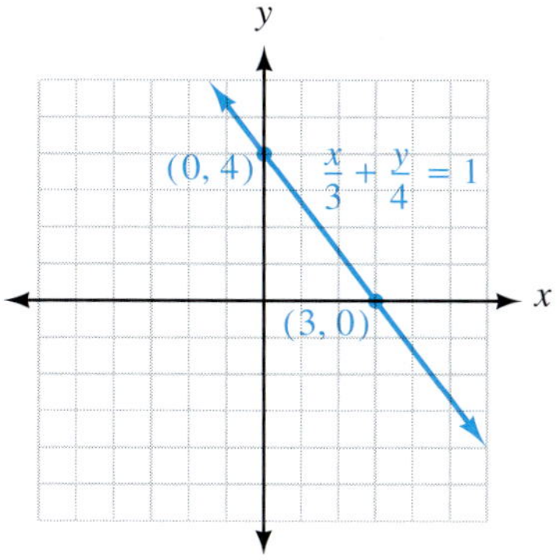

21. Equation b.

23. a.

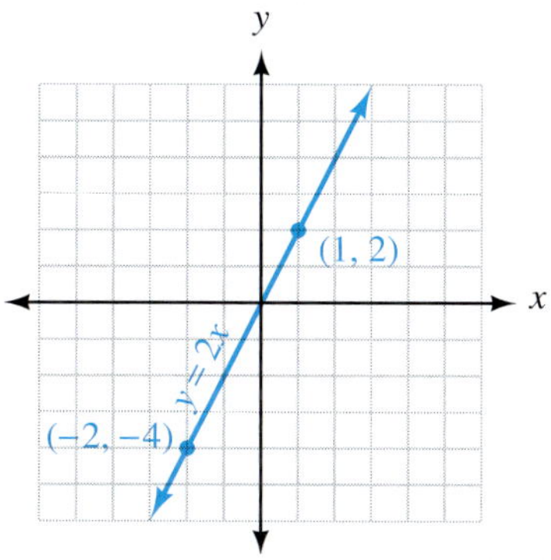

b.

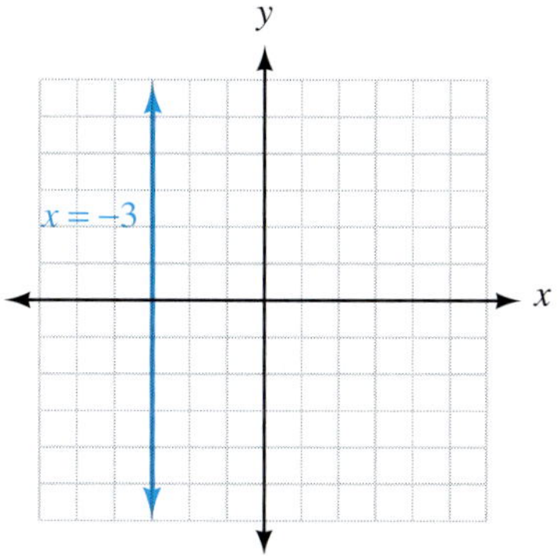

c.

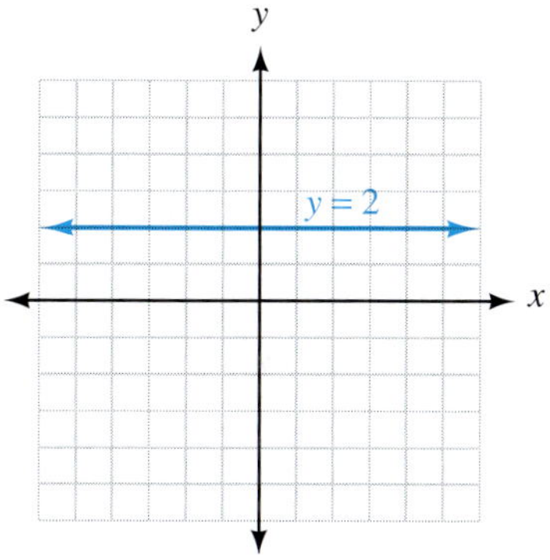

25. a.

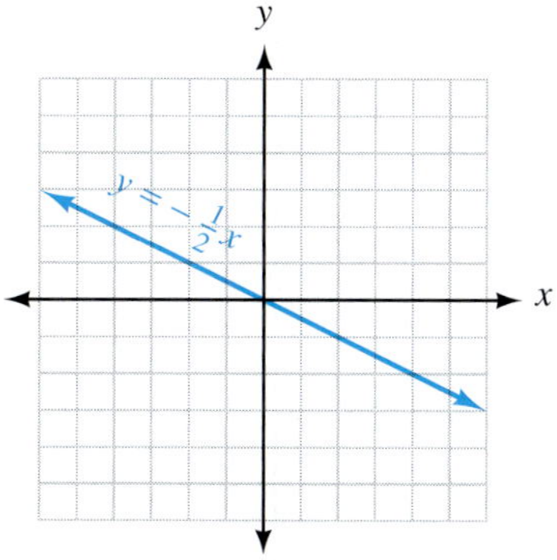

b.

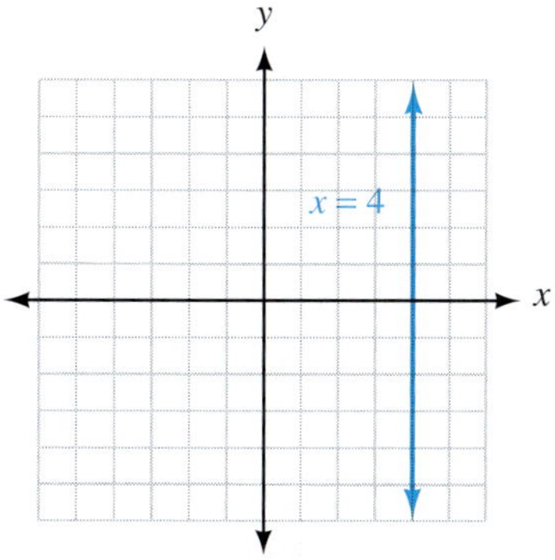

c.

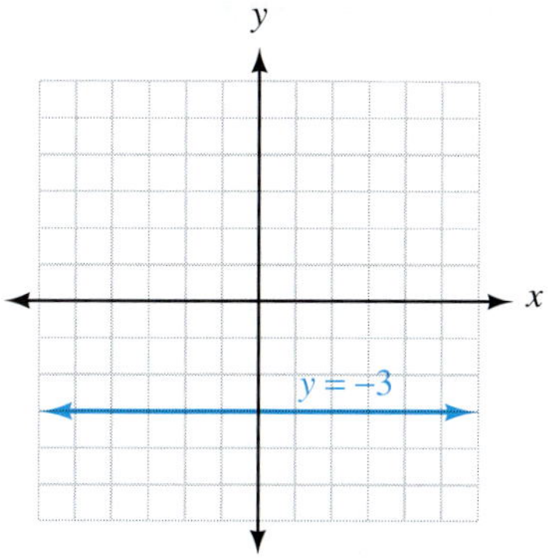

27.

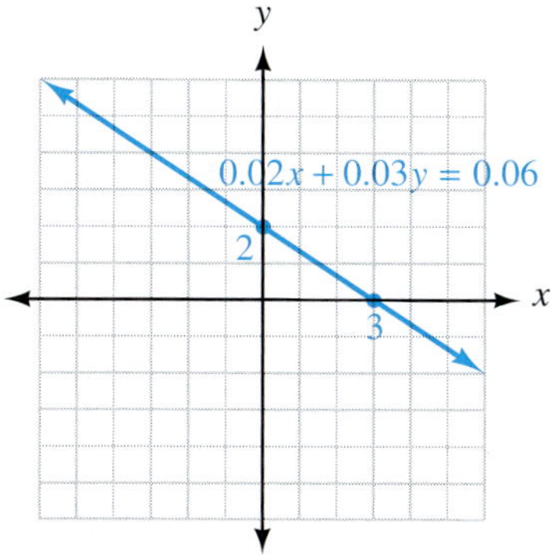

29.

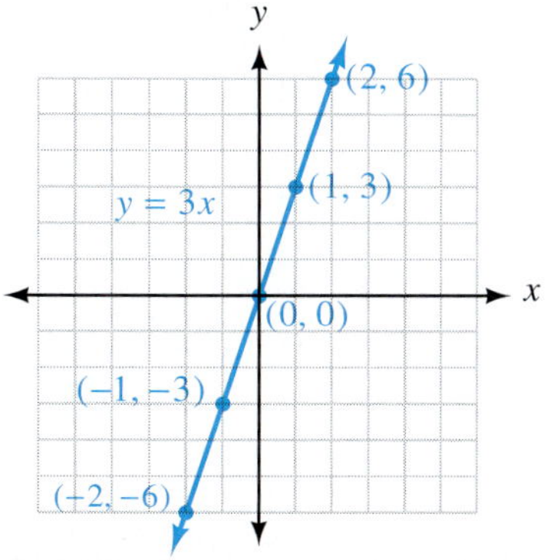

31.

TIME (HR)	DISTANCE (FT)
0	0
4	6
8	4
12	10
16	8
20	14
24	12

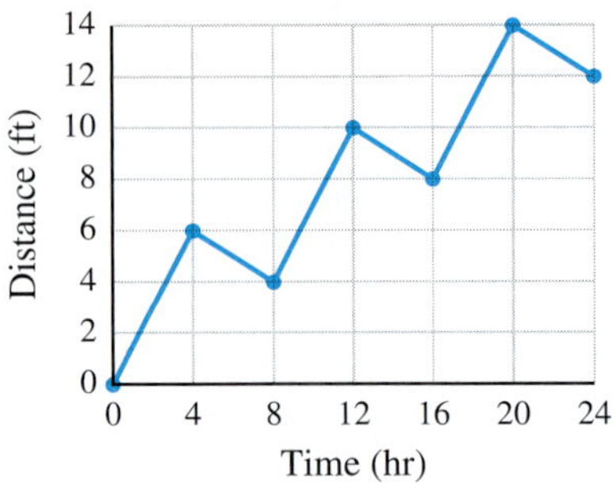

33.

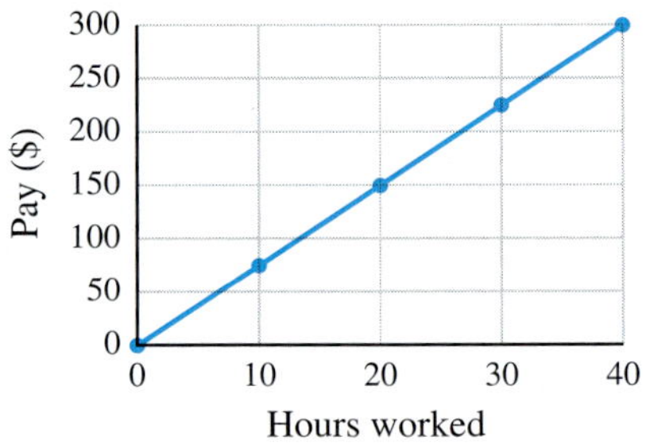

35.

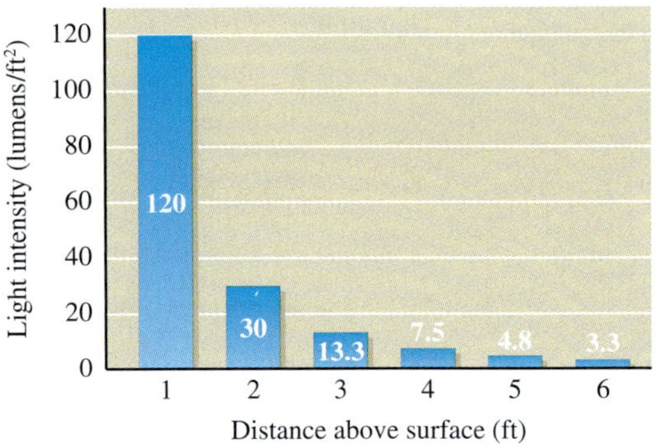

37. **a.** About 35% **b.** About 70% **c.** 2000 and 2001 **39.** $-\frac{6}{100}$ **41.** 1 **43.** $-\frac{4}{3}$ **45.** Undefined **47.** **a.** $\frac{3}{2}$ **b.** $-\frac{3}{2}$

49. 2 **51.** −2 **53.** 1,200 **55.** 2 **57.** $\frac{8}{15}$

Problem Set 3.2

1. $\frac{3}{2}$ **3.** No slope **5.** $\frac{2}{3}$

7.

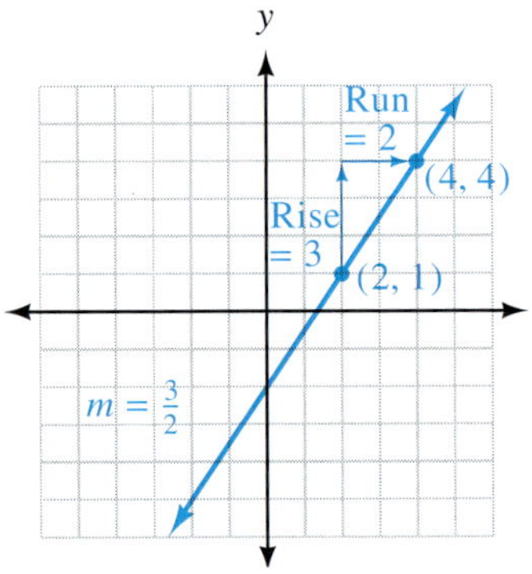

9.

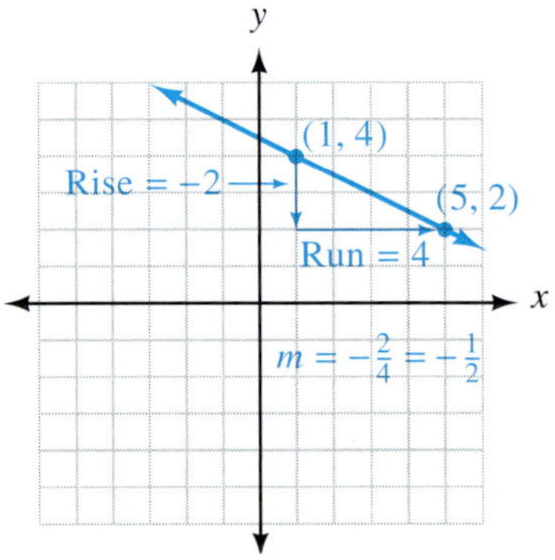

11.

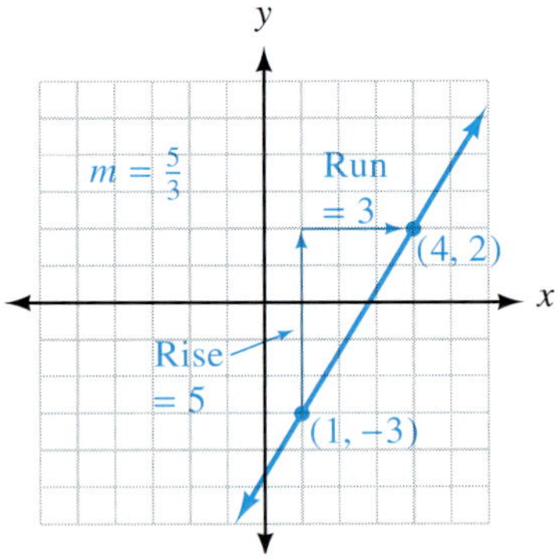

13.

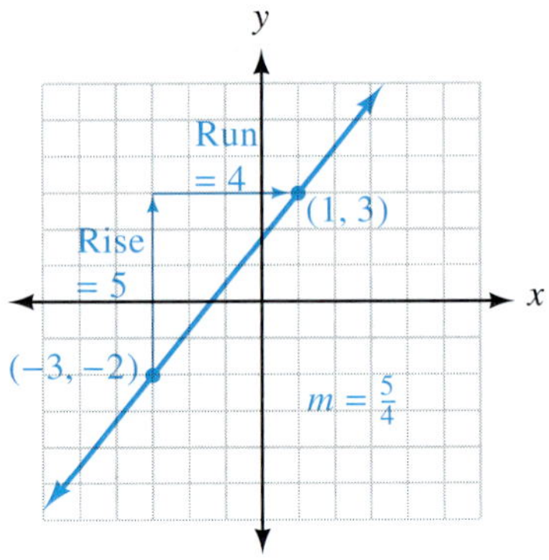

15.

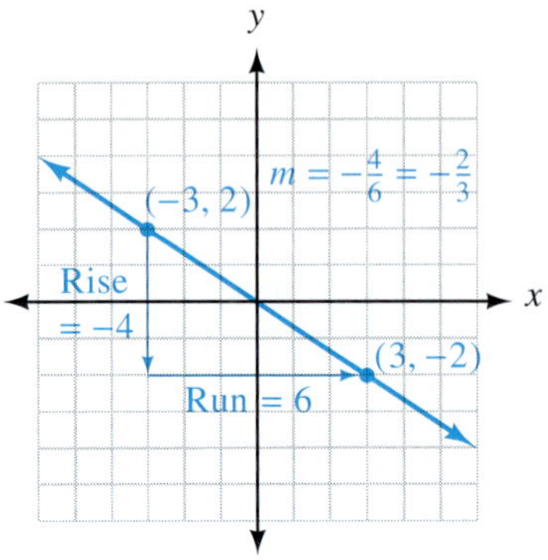

17.

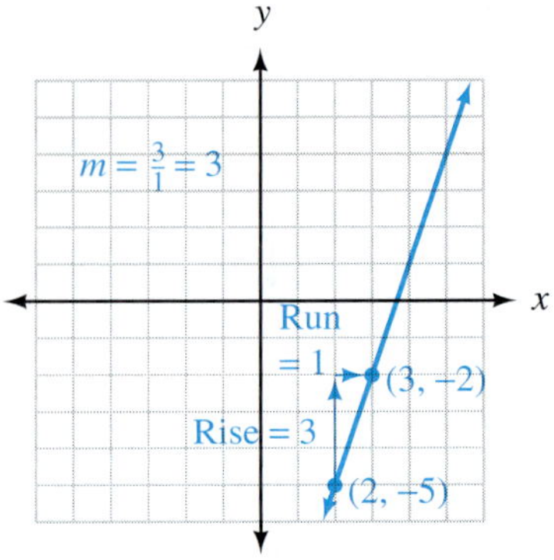

19.

x	y
0	2
3	0
Slope = $-\frac{2}{3}$	

21.

x	y
0	−5
3	−3
Slope = $\frac{2}{3}$	

23.

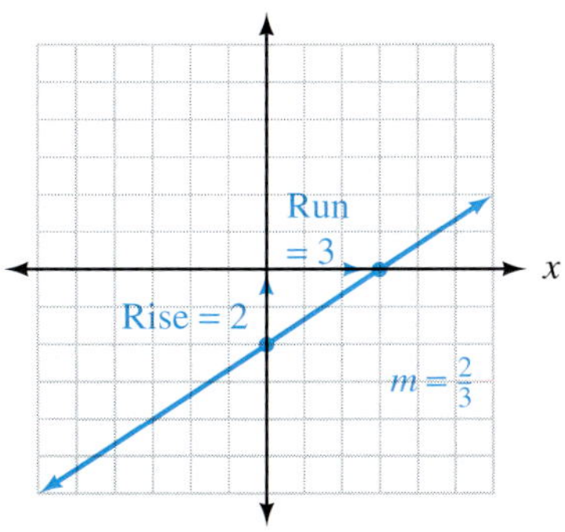

25.

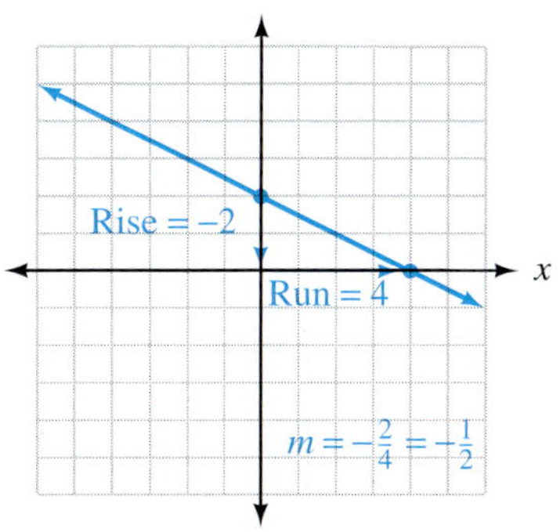

27. **a.** yes **b.** no **29.** $\frac{1}{5}$ **31.** 0 **33.** 8 **35.** **a.** 15,800 feet **b.** $-\frac{7}{100}$

37. 6.5 percent/year. Over the 10 years from 1995 to 2005, the percent of adults that access the Internet has increased at an average rate of 6.5% each year.

39. **a.** −4,768.6 cases/year. New cases of rubella decreased at an average of 4,768.6 cases per year from 1969 to 1979.
b. −999.1 cases/year. New cases of rubella decreased at an average of 999.1 cases per year from 1979 to 1989.
c. The slope is 0 meaning that the number of new cases of German Measles is constant at 9 cases/year.

41. −1 **43.** −2 **45.** $y = mx + b$ **47.** $y = -2x - 5$ **49.** 5 **51.** 0 **53.** $y = -\frac{3}{2}x + 6$ **55.** $t = \frac{A - P}{Pr}$

57. **a.** 36 km/hr with no wind, 24 km/hr against the wind, 48 km/hr with the wind
b. The speed against the wind is 12 km/hr slower. **c.** The speed with the wind is 12 km/hr faster. **d.** 12 km/hr

Problem Set 3.3

1. $y = 2x + 3$ **3.** $y = x - 5$ **5.** $y = \frac{1}{2}x + \frac{3}{2}$ **7.** $y = 4$

9. Slope = 3, y-intercept = −2, perpendicular slope = $-\frac{1}{3}$

11. Slope = $\frac{2}{3}$, y-intercept = −4, perpendicular slope = $-\frac{3}{2}$

13. Slope = $-\frac{4}{5}$, y-intercept = 4, perpendicular slope = $\frac{5}{4}$

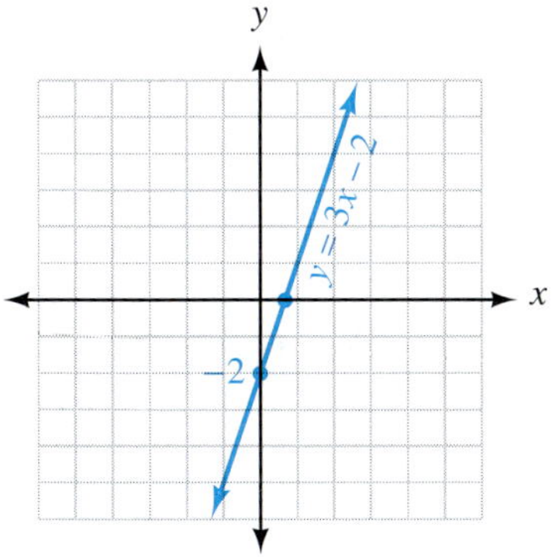

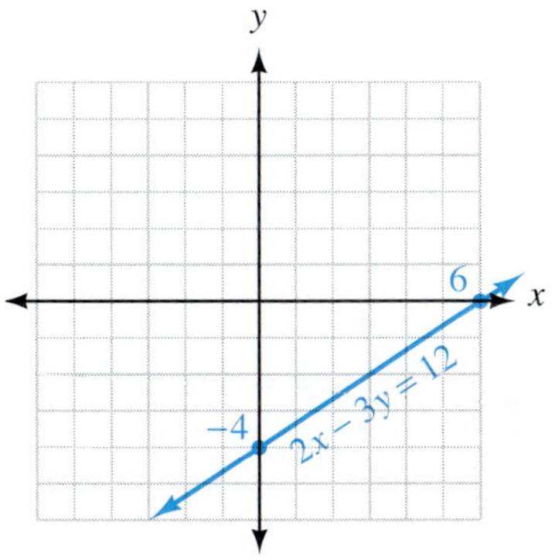

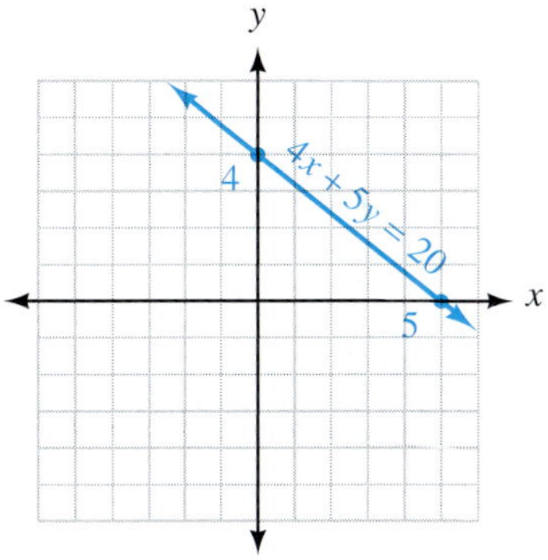

15. Slope = $\frac{1}{2}$, y-intercept = −4, $y = \frac{1}{2}x - 4$ **17.** Slope = $-\frac{2}{3}$, y-intercept = 3, $y = -\frac{2}{3}x + 3$ **19.** $y = 2x - 1$ **21.** $y = -\frac{1}{2}x - 1$

23. $y = -3x + 1$ **25.** $x - y = 2$ **27.** $2x - y = 3$ **29.** $6x - 5y = 3$ **31.** (0, −4), (2, 0); $y = 2x - 4$ **33.** (−2, 0) (0, 4); $y = 2x + 4$

35. Slope = 0, y-intercept = −2

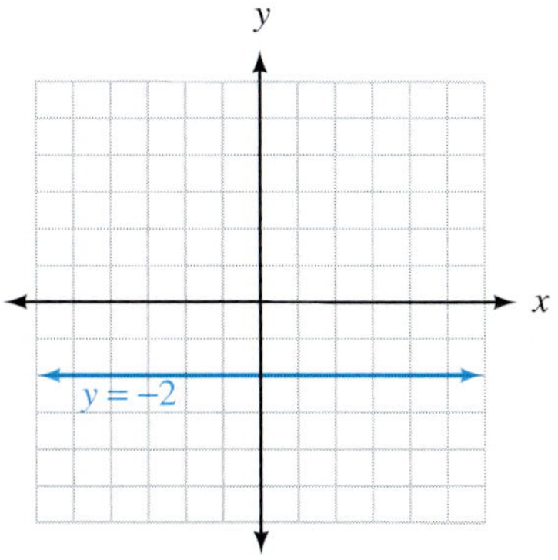

37. $y = 3x + 7$ **39.** $y = -\frac{5}{2}x - 13$ **41.** $y = \frac{1}{4}x + \frac{1}{4}$ **43.** $y = -\frac{2}{3}x + 2$ **45.** **b.** 86° F **47.** **a.** \$190,000 **b.** \$19 **c.** \$6.50

49. **a.** $y = \frac{69}{2}x - 68{,}994$ **b.** 213 books **51.** (0, 0), (4, 0) **53.** (0, 0), (2, 0) **55.** 23 inches, 5 inches **57.** \$46.50

59. **a.** $y = 6.5x - 12{,}958.5$ **b.** In 2006 80.5% of adults will be online. **c.** When we try to estimate the number of adults online in 2010 we get a number larger than 100%, which is impossible.

Problem Set 3.4

1.

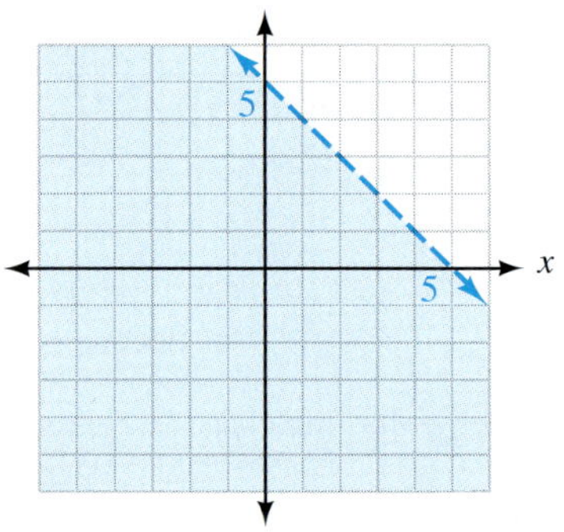

3.

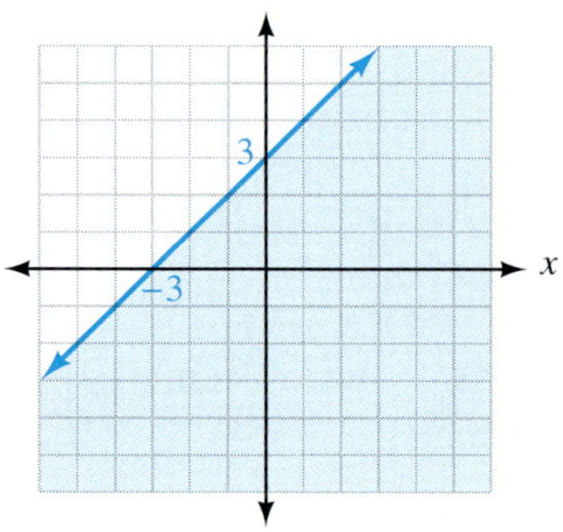

5.

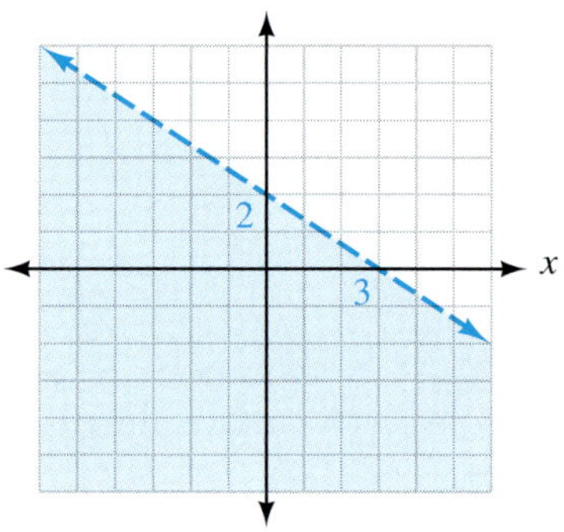

7.

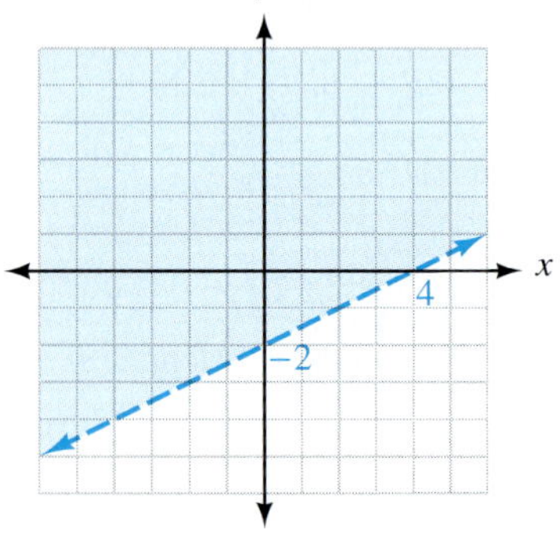

9.

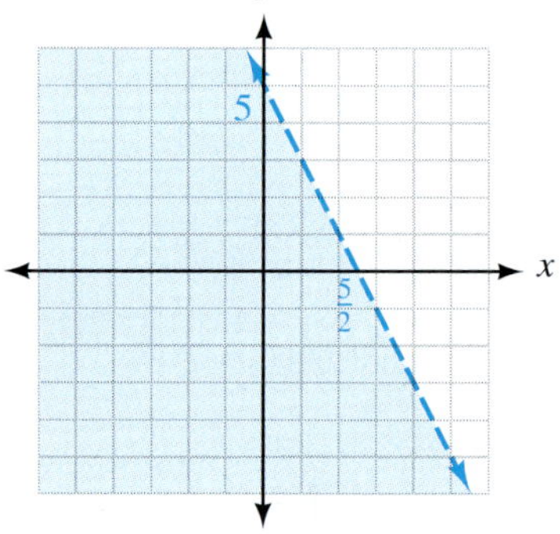

11.

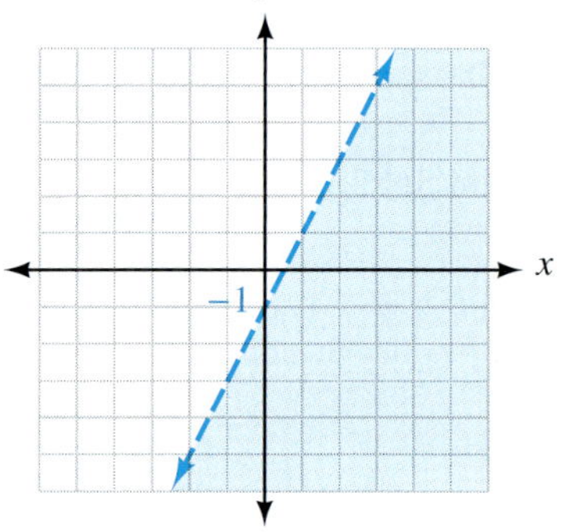

13. $x + y > 4$ **15.** $-x + 2y \le 4$

17.

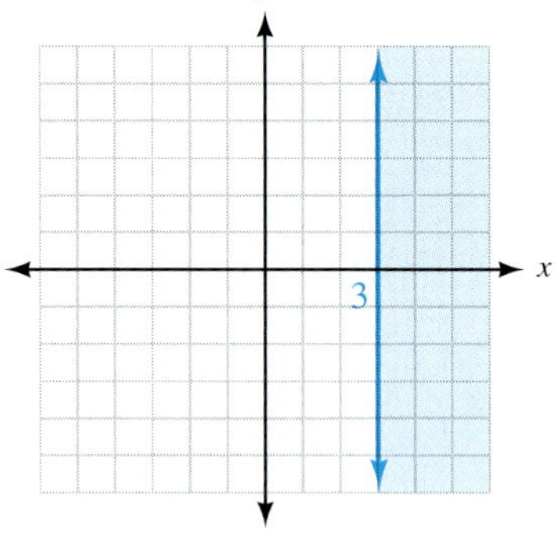

19.

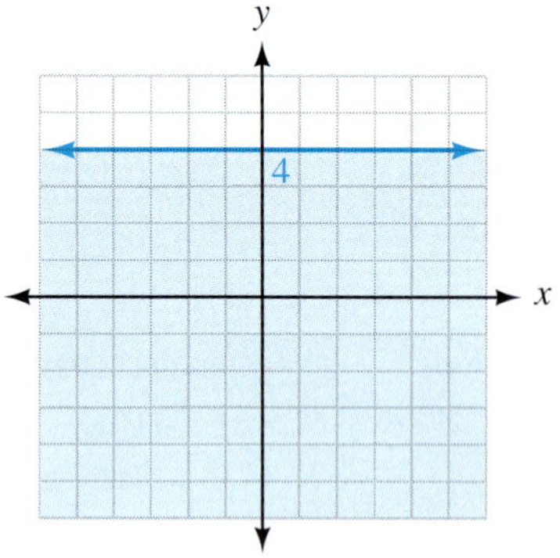

21.

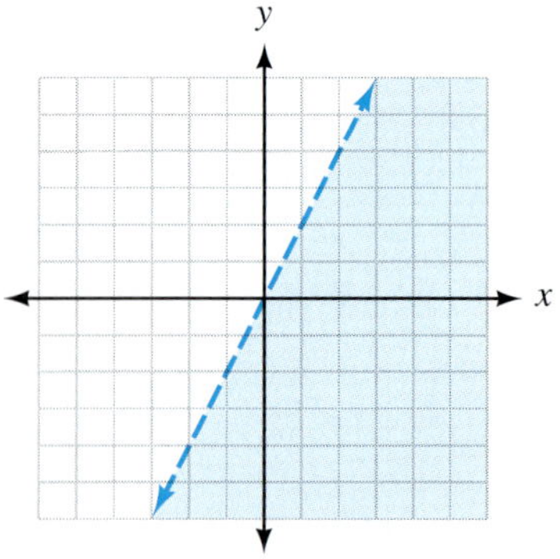

23.

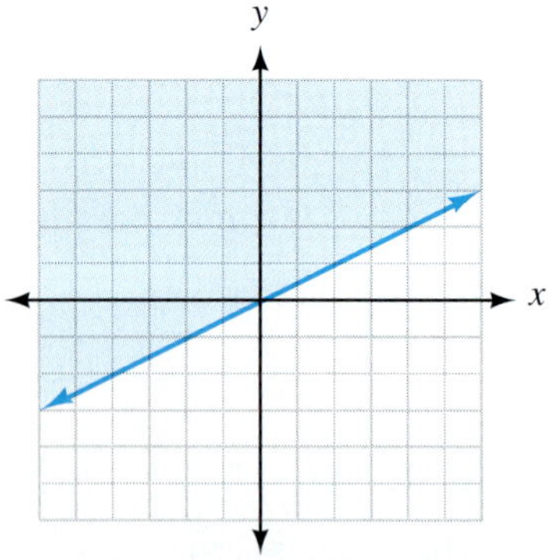

25.

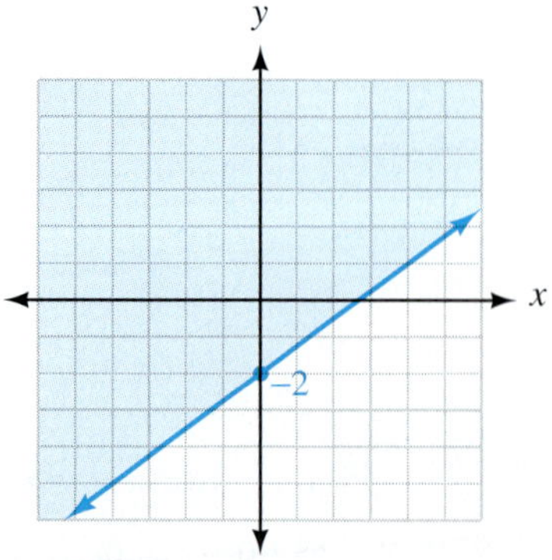

27.

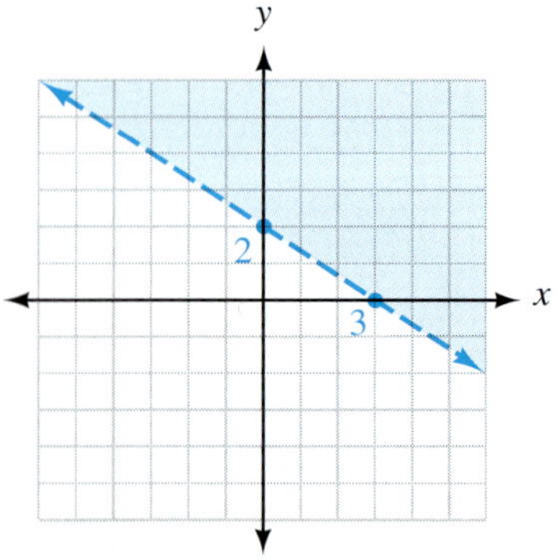

29.

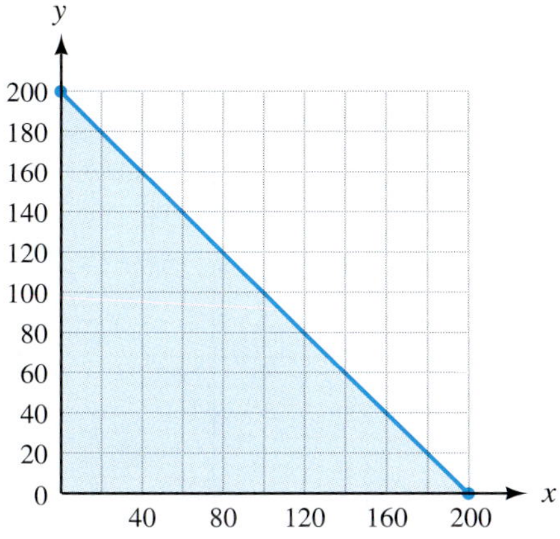

31.

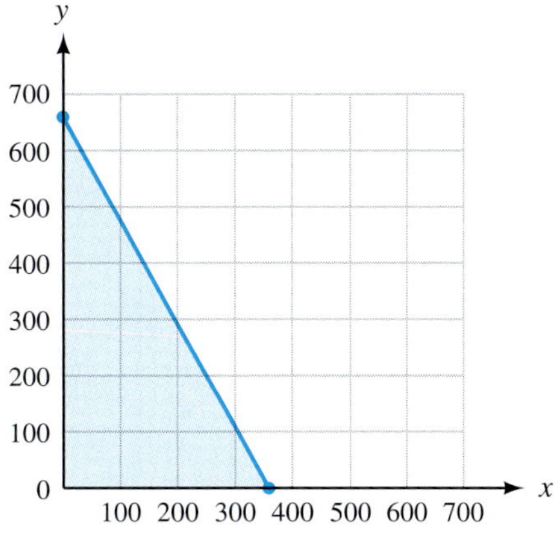

33.

x	y
0	0
10	75
20	150

35.

x	y
0	0
1	1
1	−1

37. $y \le 7$ **39.** $t > -2$ **41.** $-1 < t < 2$

43.

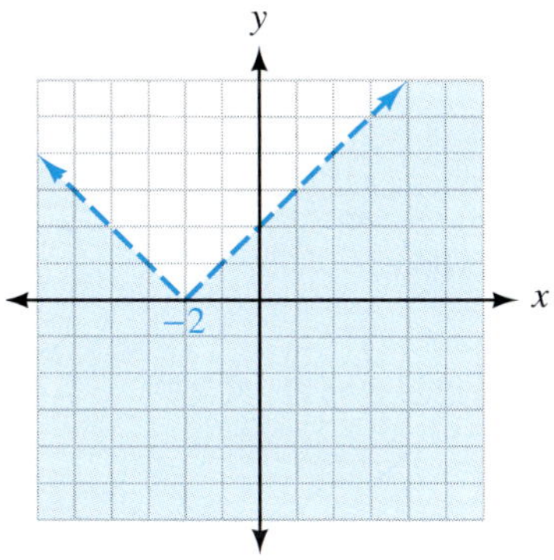

45.

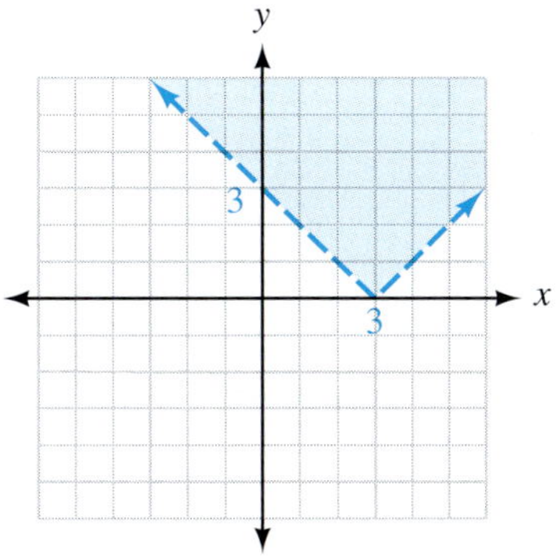

Problem Set 3.5

1. Domain = {1, 2, 4}; Range = {3, 5, 1}; a function **3.** Domain = {−1, 1, 2}; range = {3, −5}; a function

5. Domain = {7, 3}; Range = {−1, 4}; not a function **7.** Yes **9.** No **11.** No **13.** Yes **15.** Yes

17. Domain = All real numbers; Range = $\{y \mid y \ge -1\}$; a function

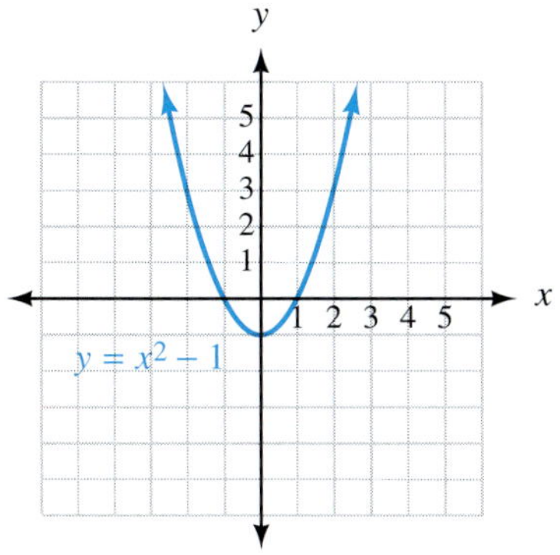

19. Domain = All real numbers; Range = $\{y \mid y \ge 4\}$; a function

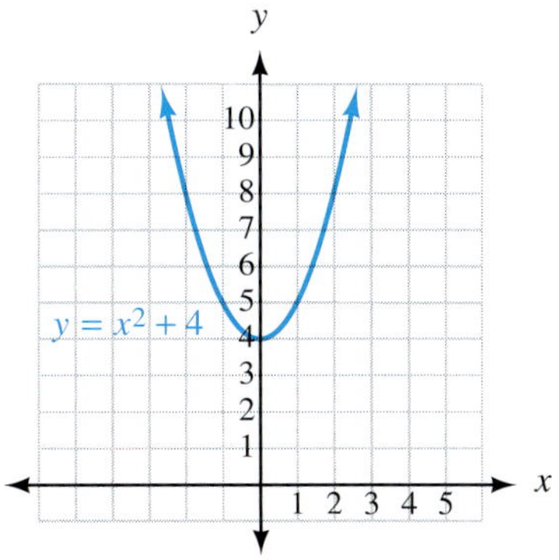

21. Domain = $\{x \mid x \ge -1\}$; Range = All real numbers; not a function

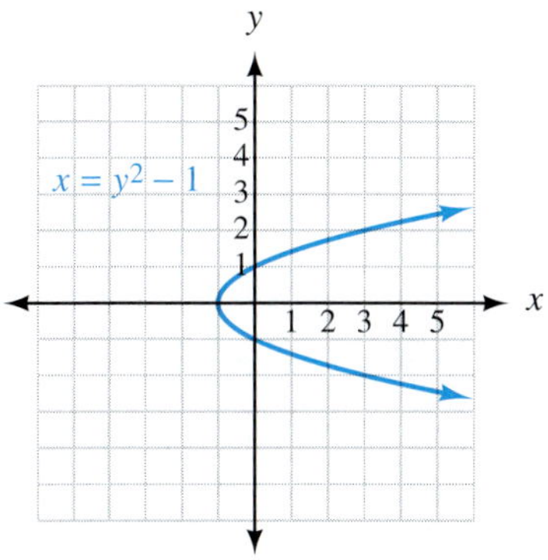

23. Domain = $\{x \mid x \geq 4\}$; Range = All real numbers; not a function

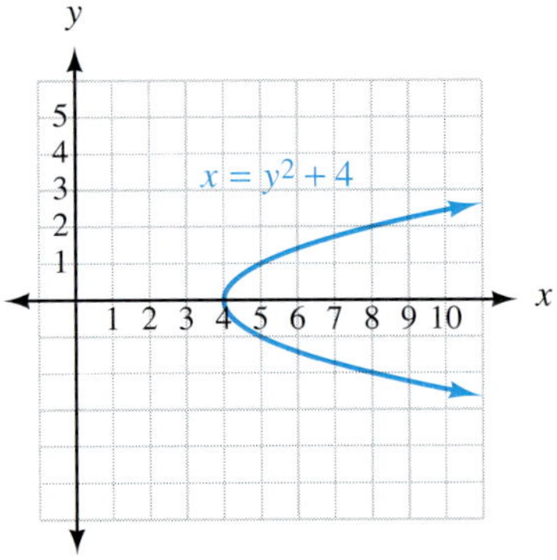

25. Domain = All real numbers; Range = $\{y \mid y \geq 0\}$; a function

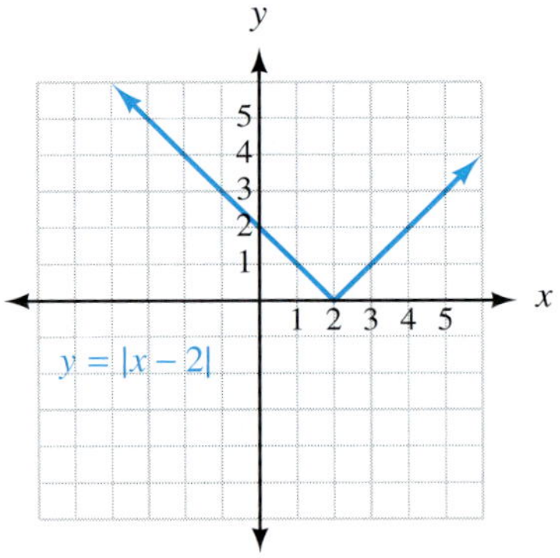

27. Domain = All real numbers; Range = $\{y \mid y \geq -2\}$; a function

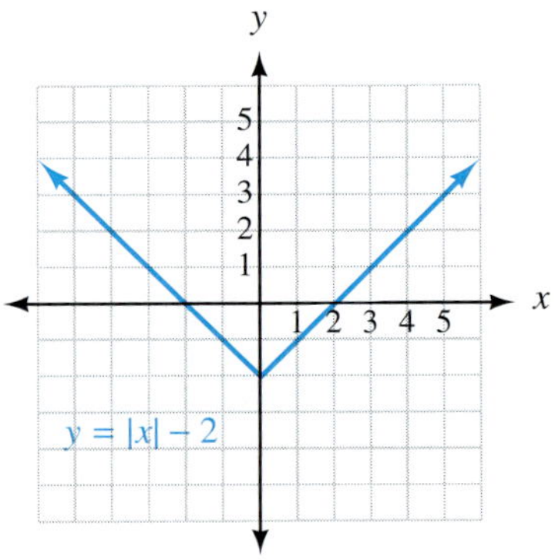

29. a. $y = 8.5x$ for $10 \leq x \leq 40$

b.

HOURS WORKED	FUNCTION RULE	GROSS PAY ($)
x	$y = 8.5x$	y
10	$y = 8.5(10) = 85$	85
20	$y = 8.5(20) = 170$	170
30	$y = 8.5(30) = 255$	255
40	$y = 8.5(40) = 340$	340

c.

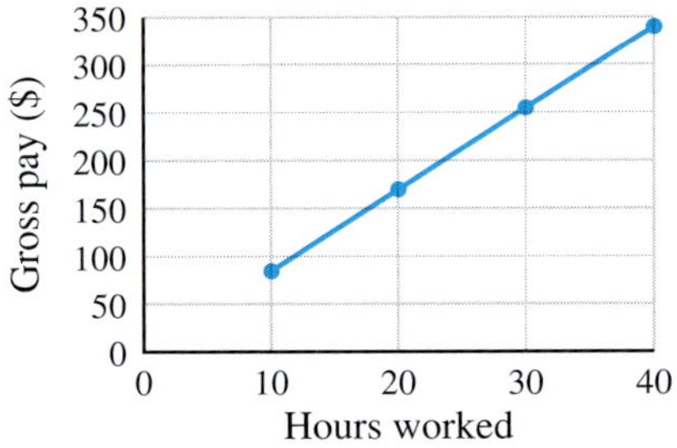

d. Domain = $\{x \mid 10 \leq x \leq 40\}$; Range = $\{y \mid 85 \leq y \leq 340\}$ **e.** Minimum = \$85; Maximum = \$340

31. a.

TIME (SEC)	FUNCTION RULE	DISTANCE (FT)
t	$h = 16t - 16t^2$	h
0	$h = 16(0) - 16(0)^2$	0
0.1	$h = 16(0.1) - 16(0.1)^2$	1.44
0.2	$h = 16(0.2) - 16(0.2)^2$	2.56
0.3	$h = 16(0.3) - 16(0.3)^2$	3.36
0.4	$h = 16(0.4) - 16(0.4)^2$	3.84
0.5	$h = 16(0.5) - 16(0.5)^2$	4
0.6	$h = 16(0.6) - 16(0.6)^2$	3.84
0.7	$h = 16(0.7) - 16(0.7)^2$	3.36
0.8	$h = 16(0.8) - 16(0.8)^2$	2.56
0.9	$h = 16(0.9) - 16(0.9)^2$	1.44
1	$h = 16(1) - 16(1)2$	0

b. Domain = $\{t \mid 0 \leq t \leq 1\}$; Range = $\{h \mid 0 \leq h \leq 4\}$

c.

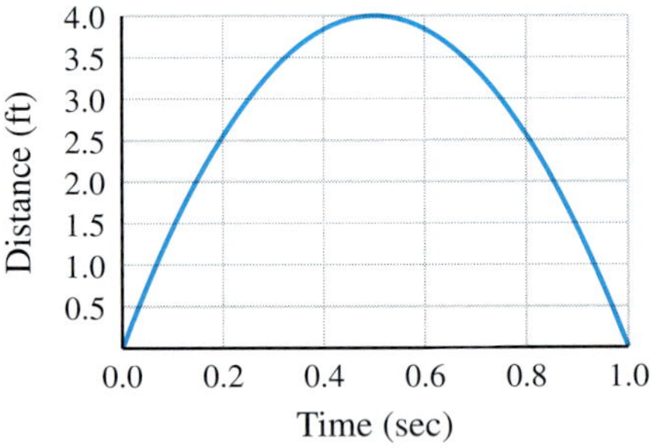

33. a.

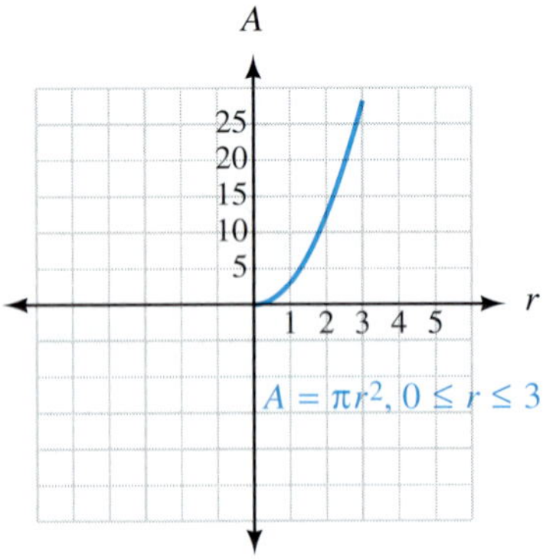

b. Domain = $\{r \mid 0 \leq r \leq 3\}$ Range = $\{A \mid 0 \leq A \leq 9\pi\}$

35. a. Yes **b.** Domain = $\{t \mid 0 \leq t \leq 6\}$; Range = $\{h \mid 0 \leq h \leq 60\}$ **c.** $t = 3$ **d.** $h = 60$ **e.** $t = 6$ **37.** 150 **39.** 113 **41.** -9 **43. a.** 6 **b.** 7.5 **45. a.** 27 **b.** 6 **47.** 10 **49.** -14 **51.** 1 **53.** -3 **55.** $-\frac{6}{5}$ **57.** $-\frac{7}{640}$ **59. a.** Fig 11 **b.** Fig 12 **c.** Fig 10 **d.** Fig 9

Problem Set 3.6

1. -1 **3.** -11 **5.** 2 **7.** 4 **9.** 35 **11.** -13 **13.** 1 **15.** -9 **17.** 8 **19.** 19 **21.** 16 **23.** 0 **25.** $3a^2 - 4a + 1$

27. 4 **29.** 0 **31.** 2 **33.** −8 **35.** −1 **37.** $2a^2 - 8$ **39.** $2b^2 - 8$ **41.** 0 **43.** −2 **45.** −3

47.

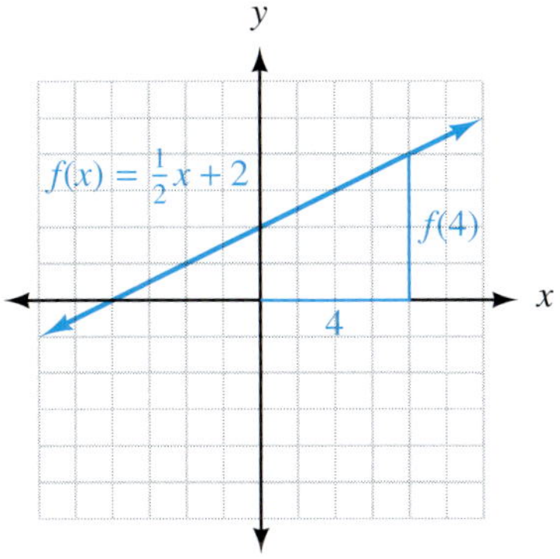

49. $x = 4$ **51.**

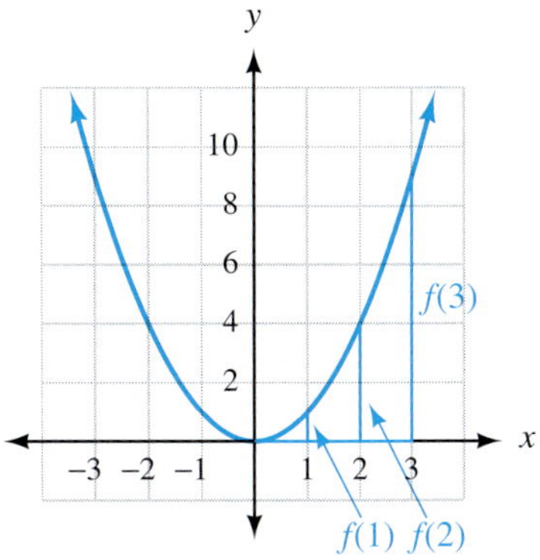

53. $V(3) = 300$, the painting is worth \$300 in 3 years; $V(6) = 600$, the painting is worth \$600 in 6 years.

55. $P(x) = 2x + 2(2x + 3) = 6x + 6$, where $x > 0$

57. **a.** \$5,625 **b.** \$1,500 **c.** $\{t \mid 0 \le t \le 5\}$ **d.**
e. $\{V(t) \mid 1{,}500 \le V(t) \le 18{,}000\}$ **f.** 2.42 years

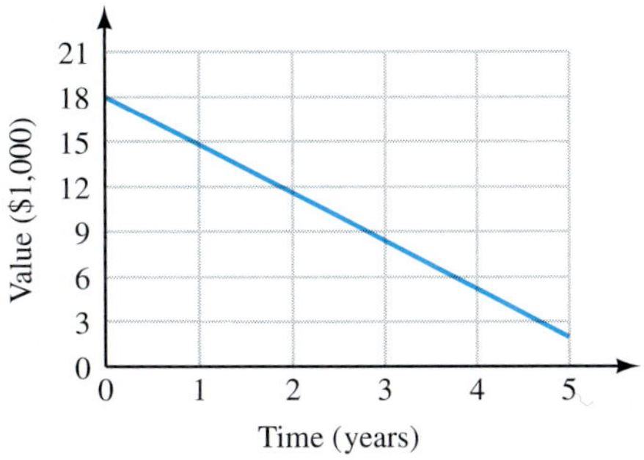

59. 196 **61.** 4 **63.** 1.6 **65.** 3 **67.** 2,400 **69.** $-\frac{2}{3}, 4$ **71.** −2, 1 **73.** ∅ **75.** **a.** 2 **b.** 0 **c.** 1 **d.** 4

77. **a.**

Weight (ounces)	0.6	1.0	1.1	2.5	3.0	4.8	5.0	5.3
Cost (cents)	39	39	63	87	87	135	135	159

b. More than 2 ounces, but not more than 3 ounces; $2 < x \le 3$

c. domain: $\{x \mid 0 < x \le 6\}$ **d.** range: $\{c \mid c = 39, 63, 87, 111, 135, 159\}$

Problem Set 3.7

1. 30 **3.** 5 **5.** −6 **7.** $\frac{1}{2}$ **9.** 40 **11.** 225 **13.** $\frac{81}{5}$ **15.** 40.5 **17.** 64 **19.** 8 **21.** $\frac{50}{3}$ pounds

23. **a.** $T = 4P$ **c.** 70 pounds per square inch **25.** 12 pounds per square inch **27.** **a.** $f = \frac{80}{d}$ **c.** An f-stop of 8

b.

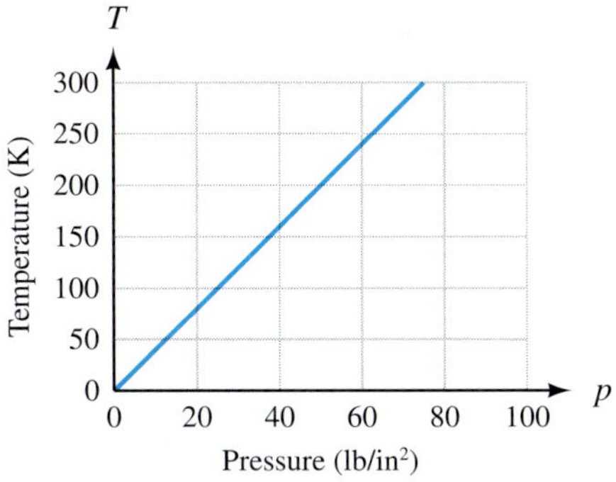

b.

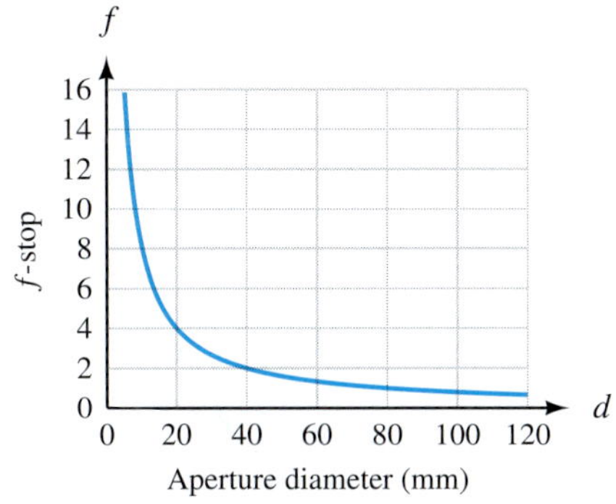

29. $\frac{1{,}504}{15}$ square inches **31.** 1.5 ohms **33.** 1.28 meters **35.** $F = G\dfrac{m_1m_2}{d^2}$

37. 12 **39.** 28 **41.** $-\frac{7}{4}$ **43.** $w = \frac{P - 21}{2}$ **45.** $t \ge -6$ **47.** $x < 6$ **49.** 6, 2 **51.** ∅

53. $x \le -9$ or $x \ge -1$

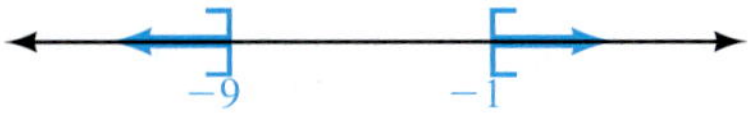

55. All real numbers **57.** **a.** Square of speed **b.** $d = 0.0675s^2$ **c.** About 204 feet from the cannon **d.** About 7.5 feet further away

Chapter 3 Review/Test

1.

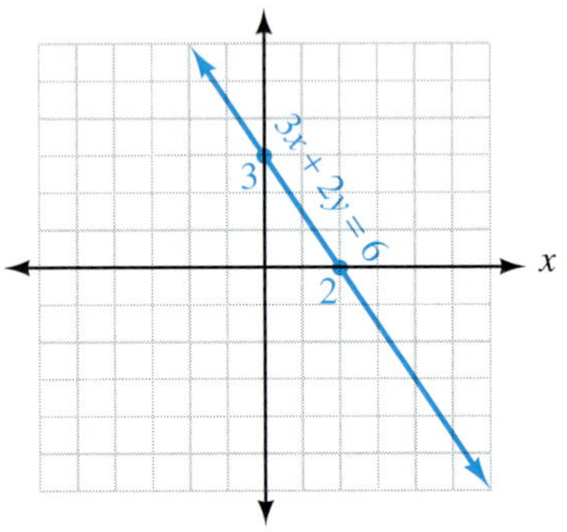

2.

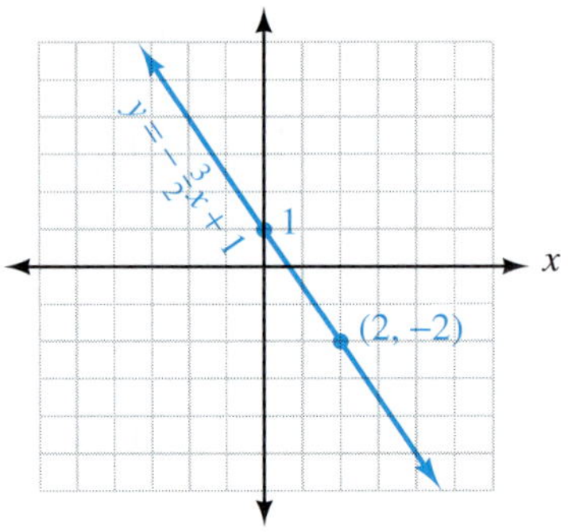

3.

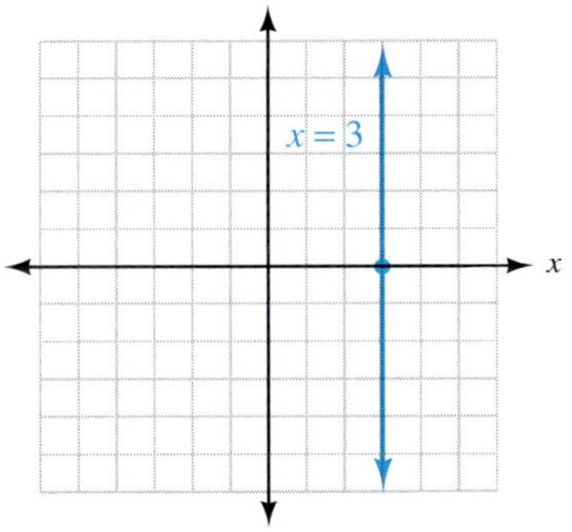

4. −2 **5.** 0 **6.** 3 **7.** 5 **8.** −5 **9.** 6 **10.** $y = 3x + 5$ **11.** $y = -2x$ **12.** $m = 3, b = -6$ **13.** $m = \frac{2}{3}, b = -3$
14. $y = 2x$ **15.** $y = -\frac{1}{3}x$ **16.** $y = 2x + 1$ **17.** $y = 7$ **18.** $y = -\frac{3}{2}x - \frac{17}{2}$ **19.** $y = 2x - 7$ **20.** $y = \frac{1}{3}x - \frac{2}{3}$

21.

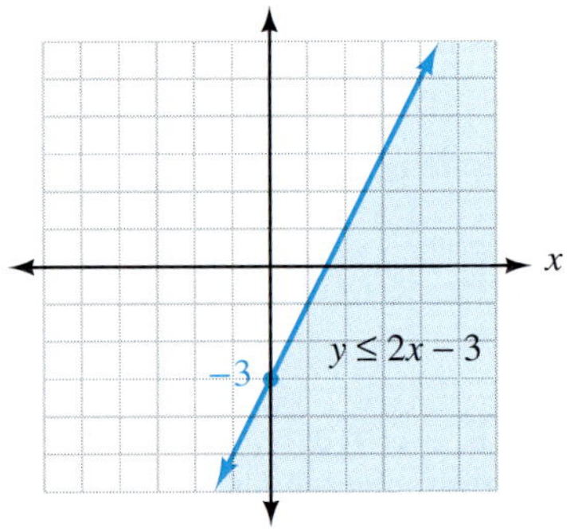

22.

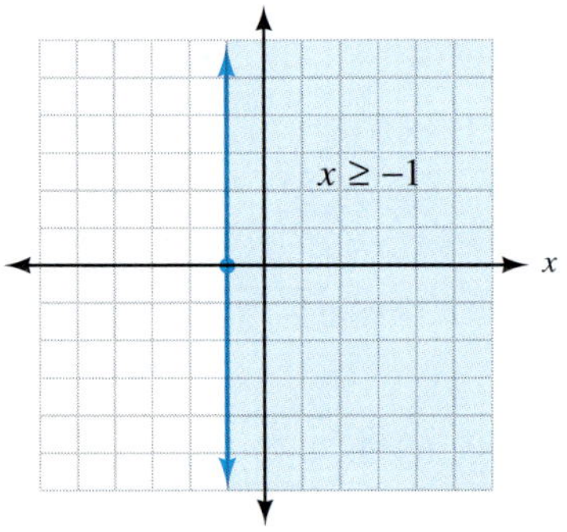

23. Domain = {2, 3, 4}; Range = {4, 3, 2}; a function

24. Domain = {6, −4, −2}; Range = {3, 0}; a function **25.** 0 **26.** 1 **27.** 1 **28.** $3a + 2$ **29.** 1 **30.** 31 **31.** 24 **32.** 6
33. 4 **34.** 25 **35.** 84 pounds **36.** 16 footcandles

Chapter 4

Getting Ready for Chapter 4

1. 540 **2.** 750 **3.** −2 **4.** −6 **5.** 4 **6.** 29 **7.** −4 **8.** 18 **9.** $-6x + 9y$ **10.** $12x + 15y$ **11.** $3x - 2y$
12. $3x + 8y$ **13.** $3x + 7y$ **14.** $6x + 7y$ **15.** 2 **16.** 3 **17.** −1 **18.** $\frac{8}{15}$ **19.** 4 **20.** $x > 435$

Problem Set 4.1

1. (4, 3)

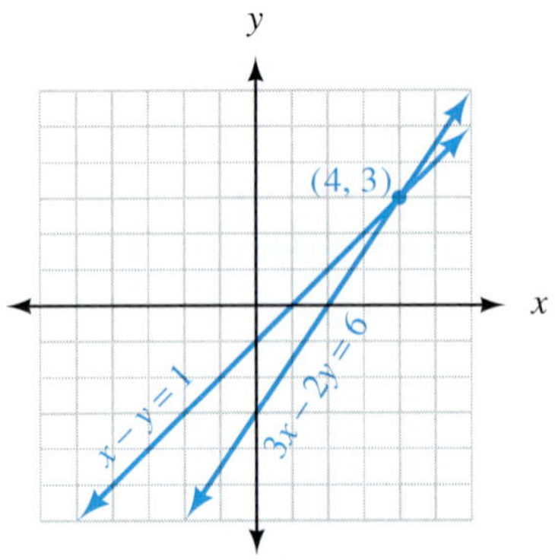

3. (−5, −6)

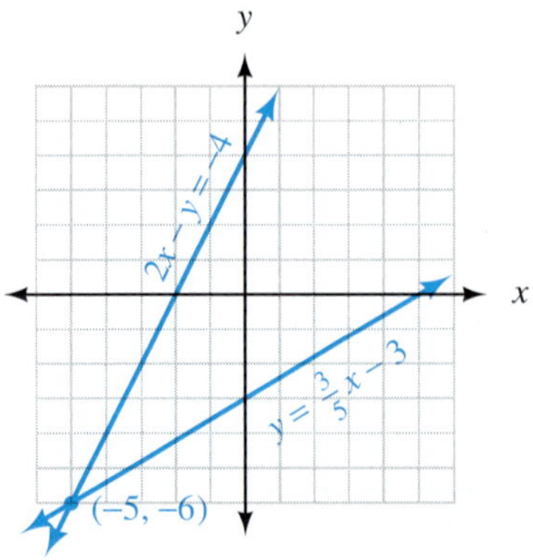

5. (4, 2)

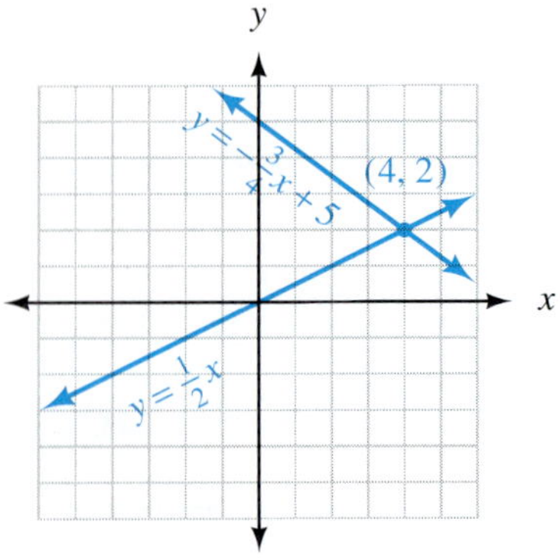

7. Lines are parallel; there is no solution.

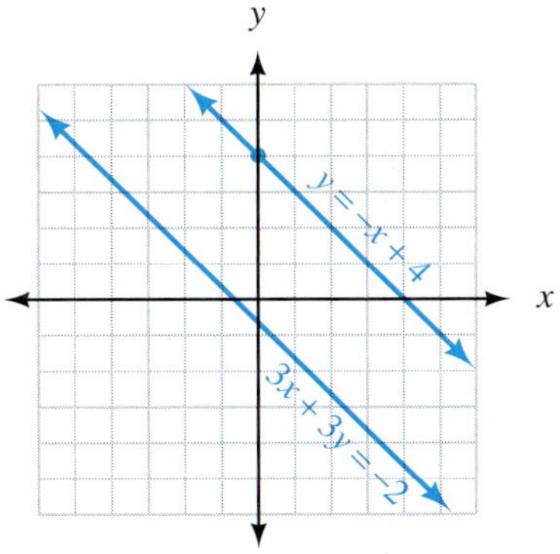

9. Lines coincide; any solution to one of the equations is a solution to the other. **11.** (2, 3) **13.** (1, 1)

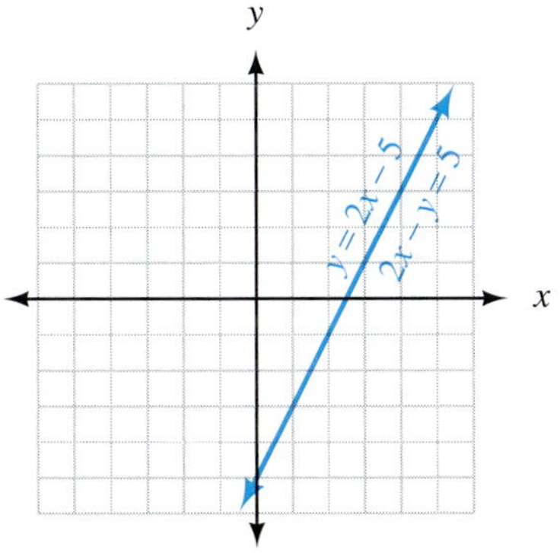

15. Lines coincide: $\{(x, y) \mid 3x - 2y = 6\}$ **17.** $(1, -\frac{1}{2})$ **19.** $(\frac{1}{2}, -3)$ **21.** $(-\frac{8}{3}, 5)$ **23.** (2, 2) **25.** Parallel lines; ∅ **27.** (12, 30) **29.** (10, 24) **31.** (3, −3) **33.** $(\frac{4}{3}, -2)$ **35.** $y = 5,\ z = 2$ **37.** (2, 4) **39.** Lines coincide: $\{(x, y) \mid 2x - y = 5\}$ **41.** Lines coincide: $\{(x, y) \mid x = \frac{3}{2}y\}$ **43.** $(-\frac{15}{43}, -\frac{27}{43})$ **45.** $(\frac{60}{43}, \frac{46}{43})$ **47.** $(\frac{9}{41}, -\frac{11}{41})$ **49.** (6,000, 4,000) **51.** 2 **53.** (4, 0) **55.** −10 **57.** $3y + 2z$ **59.** 1 **61.** 3 **63.** $10x - 2z$ **65.** $9x + 3y - 6z$ **67.** 3 **69.** $m = \frac{2}{3}, b = -2$ **71.** $y = \frac{2}{3}x + 6$ **73.** $y = \frac{2}{3}x - 2$ **75.** $a = 3, b = -2$ **77.** **a.** 3 **b.** between 1988 and 1991 **c.** The point of intersection is the year in which the number of alcohol and nonalcohol related traffic fatalities were equal and each accounted for 50% of fatal car crashes.

Problem Set 4.2

1. (1, 2, 1) **3.** (2, 1, 3) **5.** (2, 0, 1) **7.** $(\frac{1}{2}, \frac{2}{3}, -\frac{1}{2})$ **9.** No solution, inconsistent system **11.** (4, −3, −5) **13.** No unique solution **15.** (4, −5, −3) **17.** No unique solution **19.** $(\frac{1}{2}, 1, 2)$ **21.** $(\frac{1}{2}, \frac{1}{3}, \frac{1}{4})$ **23.** (1, 3, 1) **25.** (−1, 2, −2) **27.** 4 amp, 3 amp, 1 amp **29.** −2 **31.** 11 **33.** −14 **35.** −4 **37.** 147 **39.** 50 **41.** 100 **43.** (1, 2, 3, 4)

Problem Set 4.3

1. 3 **3.** 5 **5.** −1 **7.** 0 **9.** 2 **11.** −3 **13.** −2 **15.** −3 **17.** 3 **19.** 0 **21.** 3 **23.** 8 **25.** 6 **27.** −228 **29.** $\begin{vmatrix} y & x \\ m & 1 \end{vmatrix} = y - mx = b; y = mx + b$ **31.** **a.** $y = 0.3x + 3.4$ **b.** $y = 0.3(2) + 3.4$; \$4 billion **33.** −10 **35.** 3 **37.** $-\frac{5}{11}$ **39.** 13 **41.** 22 **43.** 13 **45.** domain = {1, 3, 4}; range = {2, 4}; is a function **47.** domain = {1, 2, 3}; range = {1, 2, 3}; is a function **49.** is a function **51.** not a function **53.** 4 **55.** 4

Problem Set 4.4

1. (3, 1) **3.** Lines are parallel; ∅ **5.** $(-\frac{15}{43}, -\frac{27}{43})$ **7.** $(\frac{60}{43}, \frac{46}{43})$ **9.** (3, −1, 2) **11.** $(\frac{1}{2}, \frac{5}{2}, 1)$ **13.** No unique solution **15.** $(-\frac{10}{91}, -\frac{9}{13}, \frac{107}{91})$ **17.** $(\frac{71}{13}, -\frac{12}{13}, \frac{24}{13})$ **19.** (3, 1, 2) **21.** $x = 50$ items **23.** $3x + 2$ **25.** $-\frac{160}{9}$ **27.** 320 **29.** $2x + 5y$ **31.** 6 **33.** $y = 5, z = 2$ **35.** 3 **37.** 0 **39.** 1 **41.** 3 **43.** $x = -\frac{1}{a - b} = \frac{1}{b - a}, y = \frac{1}{a - b}$ **45.** $x = \frac{1}{a^2 + ab + b^2}, y = \frac{a + b}{a^2 + ab + b^2}$ **47.** $x + 2y = 1$, $3x + 4y = 0$

Problem Set 4.5

1. $y = 2x + 3, x + y = 18$; the two numbers are 5 and 13. **3.** 10, 16 **5.** 1, 3, 4

7. Let x = the number of adult tickets and y = the number of children's tickets.

$x + y = 925$ 225 adult and 700 children's tickets

$2x + y = 1{,}150$

9. Let x = the amount invested at 6% and y = the amount invested at 7%.

$x + y = 20{,}000$ He has $12,000 at 6% and $8,000 at 7%.

$0.06x + 0.07y = 1{,}280$

11. $4,000 at 6%, $8,000 at 7.5% **13.** $200 at 6%, $1,400 at 8%, $600 at 9% **15.** 3 gallons of 50%, 6 gallons of 20% **17.** 5 gallons of 20%, 10 gallons of 14% **19.** 12.5 lbs of 40%, 37.5 lbs of 60% **21.** boat; 9 mph; current: 3 mph **23.** airplane: 270 mph; wind: 30 mph **25.** 12 nickels, 8 dimes **27.** 3 nickels, 3 dimes, 3 quarters **29.** 110 nickels **31.** $x = -200p + 700$; 100 items **33.** $h = -16t^2 + 64t + 80$ **35.** No **37.** (4, 0) **39.** $x > 435$

41.

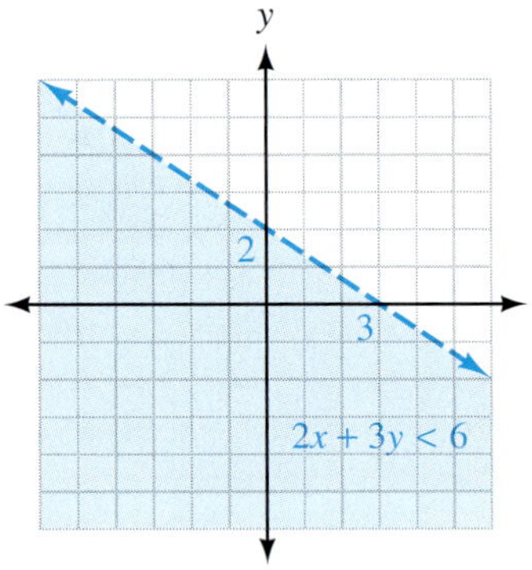

43.

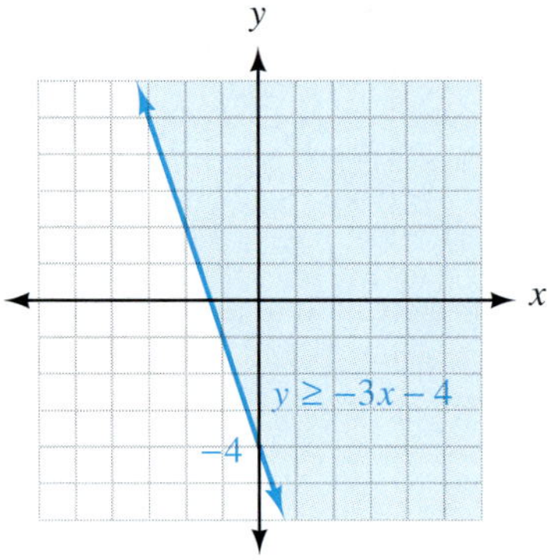

45.

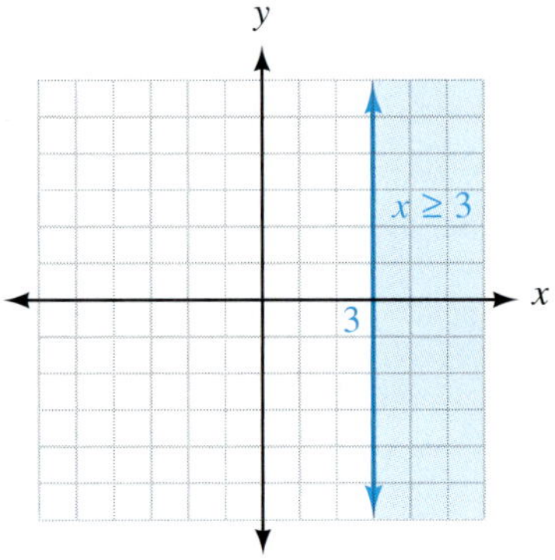

47.

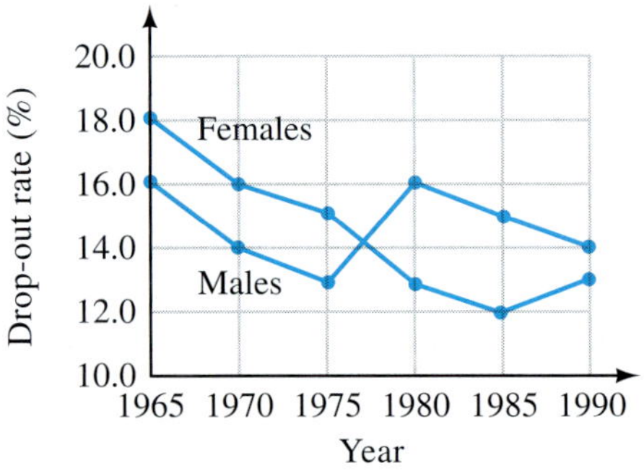

49. a. $M = 0.6x - 1172$ **b.** $F = -0.4x + 805$ **c.** year 1977

Problem Set 4.6

1.

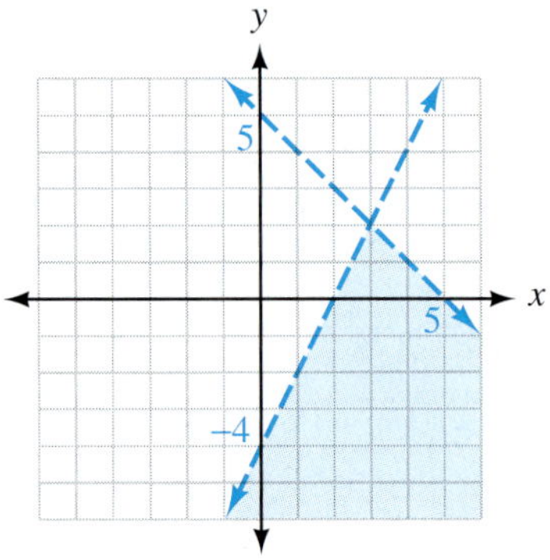

3.

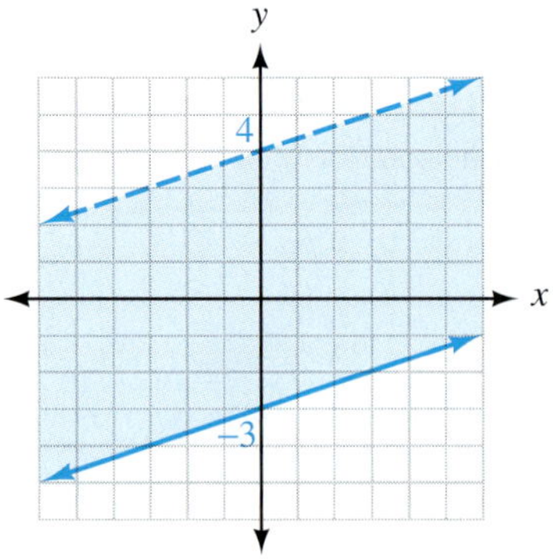

5.

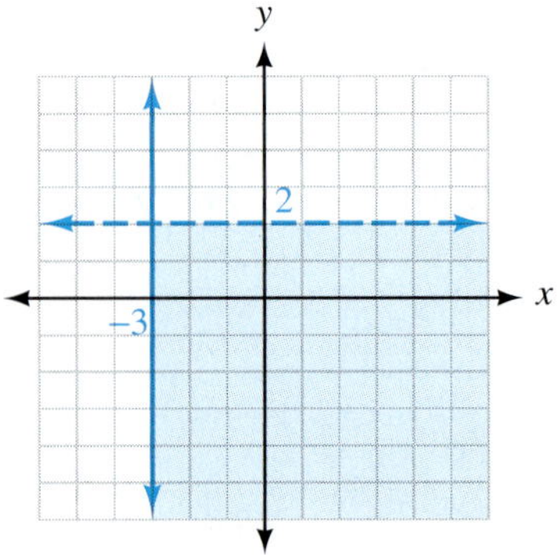

7.

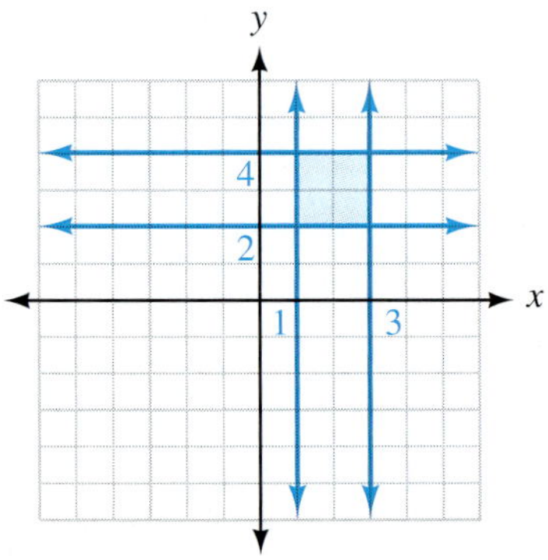

9.

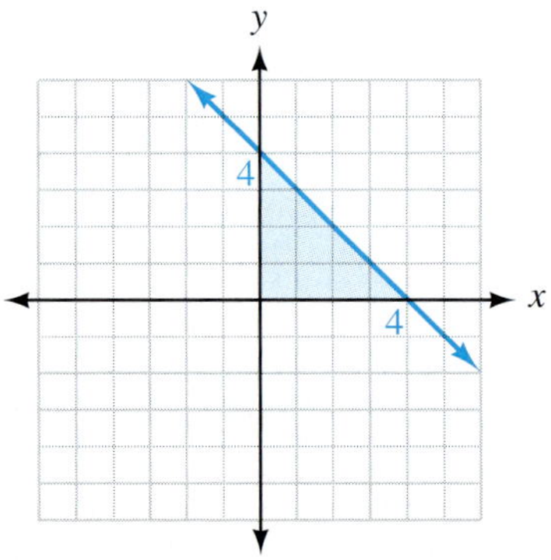

11.

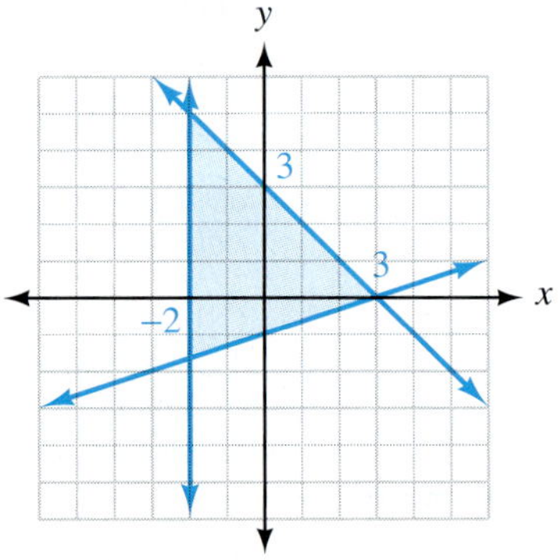

13.

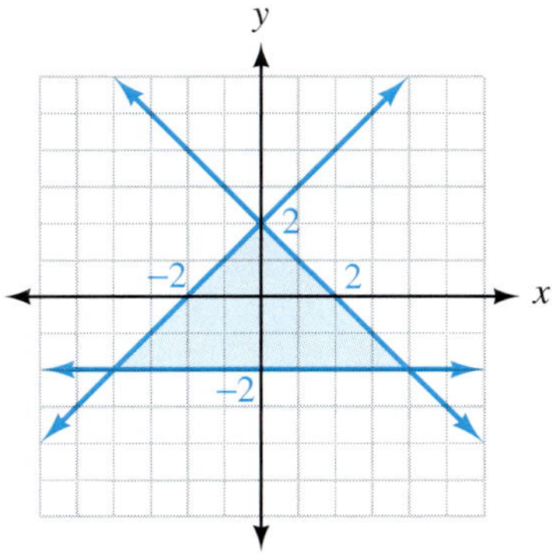

15.

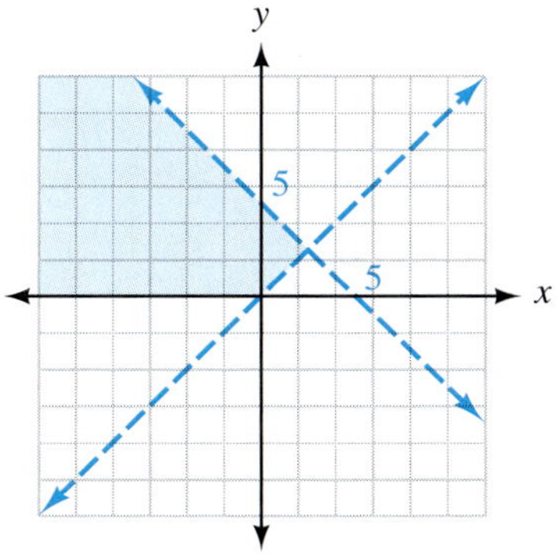

17.

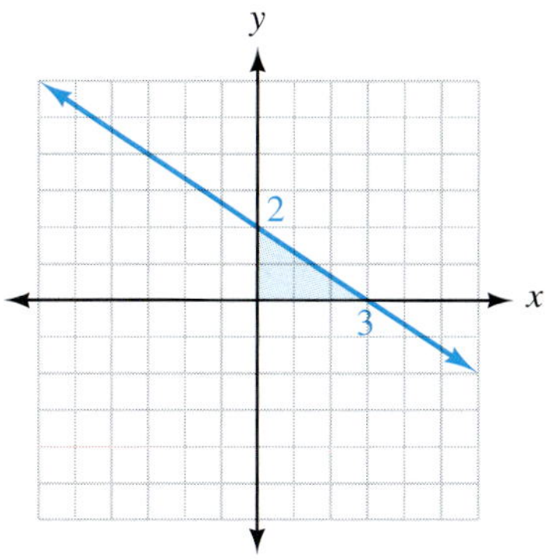

19. $x + y \le 4$
$-x + y < 4$

21. $x + y \ge 4$
$-x + y < 4$

23. **a.** $0.55x + 0.65y \le 40$
$x \ge 2y$
$x > 15$
$y \ge 0$

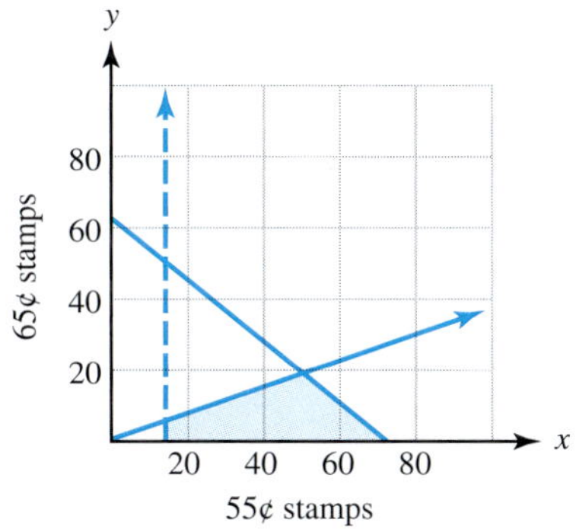

b. 10 65-cent stamps

25. x-intercept = 3, y-intercept = 6, slope = −2

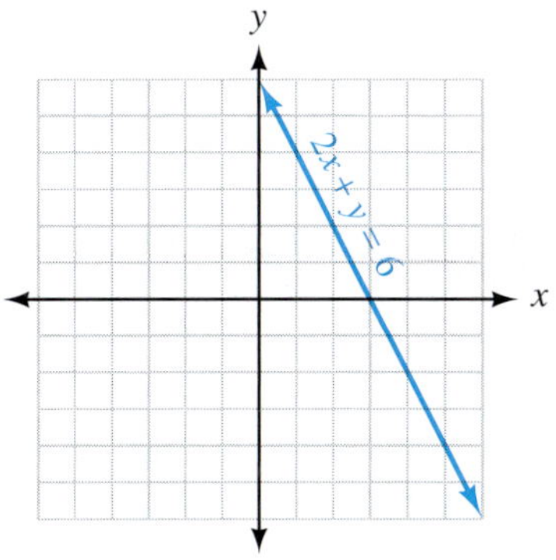

27. x-intercept = −2, no y-intercept, no slope

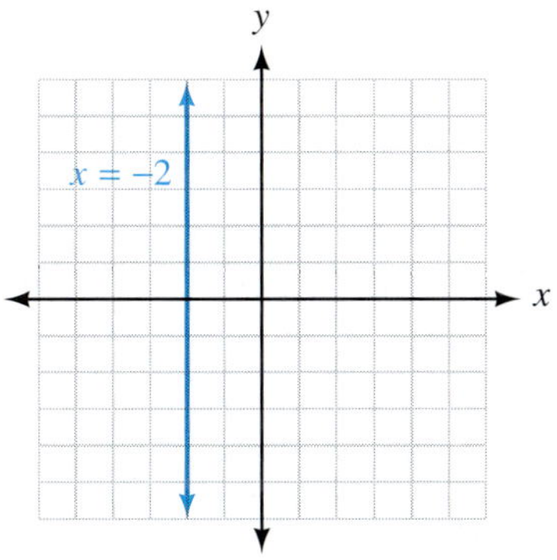

29. $y = -\frac{3}{7}x + \frac{5}{7}$ **31.** $x = 4$ **33.** Domain = All real numbers; Range = $\{y \mid y \ge -9\}$; a funcion **35.** −4 **37.** 4 **39.** $\frac{81}{4}$

Chapter 4 Review/Test

1. (6, −2) **2.** (2, −4) **3.** No solution (parallel lines) **4.** (1, −1) **5.** (0, 1) **6.** No unique solution (lines coincide) **7.** (3, 1) **8.** $(\frac{3}{2}, 0)$ **9.** (3, 5) **10.** (4, 8) **11.** (3, 12) **12.** (3, −2) **13.** $(\frac{3}{2}, \frac{1}{2})$ **14.** (4, 1) **15.** (6, −2) **16.** (7, −4) **17.** (−5, 3) **18.** No solution (parallel lines) **19.** (3, −1, 4) **20.** $(2, \frac{1}{2}, -3)$ **21.** $(-1, \frac{1}{2}, \frac{3}{2})$ **22.** $(2, \frac{1}{3}, \frac{2}{3})$ **23.** No unique solution (dependent system) **24.** No unique solution (dependent system) **25.** (2, −1, 4) **26.** No unique solution (dependent system) **27.** 23 **28.** −3 **29.** −3 **30.** −16 **31.** 36 **32.** −2 **33.** $\frac{4}{7}$ **34.** −4 **35.** $(\frac{7}{29}, -\frac{19}{29})$ **36.** $(\frac{34}{41}, -\frac{18}{41})$ **37.** No unique solution (lines coincide) **38.** $(\frac{17}{27}, -\frac{11}{81})$ **39.** $(-\frac{50}{33}, -\frac{23}{33})$ **40.** $(\frac{11}{14}, -\frac{1}{14}, \frac{11}{14})$ **41.** $(\frac{113}{46}, \frac{59}{23}, -\frac{7}{23})$ **42.** No solution (inconsistent) **43.** 47 adult, 80 children **44.** 12 dimes, 8 quarters **45.** \$5,000 at 12%, \$7,000 at 15% **46.** Boat: 12 mph; current: 2 mph

47.

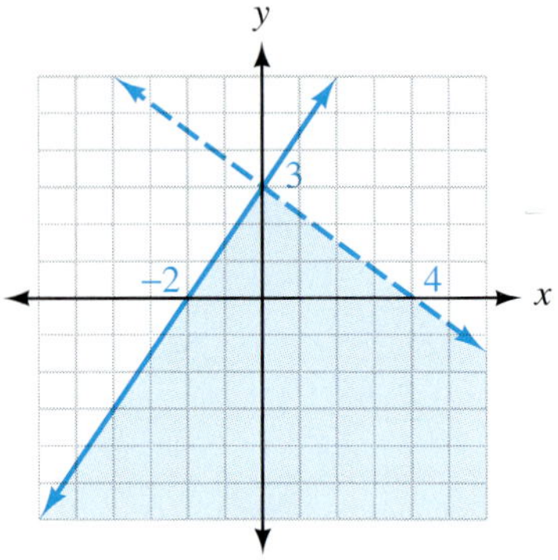

48.

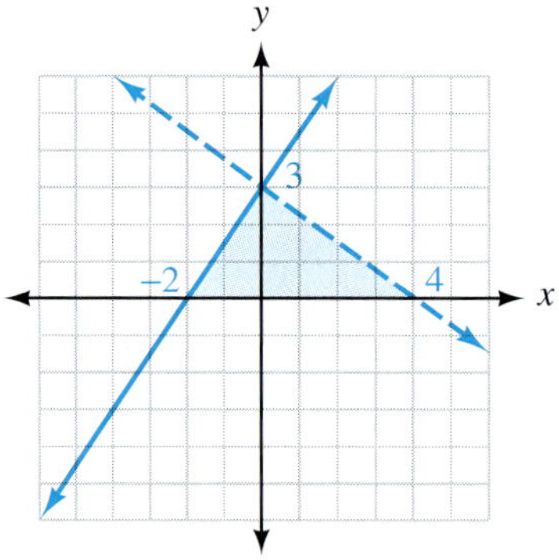

49.

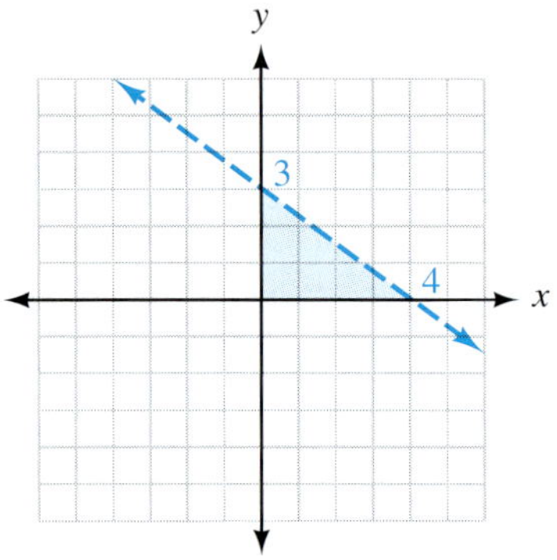

50.

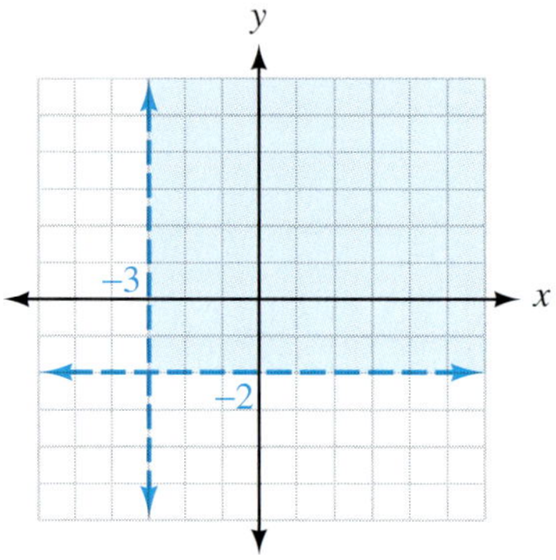

Chapter 5

Getting Ready for Chapter 5

1. 3 **2.** 2 **3.** −16 **4.** −7 **5.** −32 **6.** −6 **7.** $\frac{1}{2}$ **8.** $\frac{1}{2}$ **9.** 81 **10.** 8 **11.** 1,000 **12.** 10,000 **13.** −8 **14.** −27 **15.** $\frac{1}{25}$ **16.** $\frac{1}{9}$ **17.** $-\frac{1}{8}$ **18.** $-\frac{1}{125}$ **19.** $\frac{9}{4}$ **20.** $\frac{16}{9}$ **21.** 4,520 **22.** 39,100 **23.** 3.76 **24.** 2.74

Problem Set 5.1

1. 16 **3.** −16 **5.** −0.027 **7.** 32 **9.** $\frac{1}{8}$ **11.** $\frac{25}{36}$ **13.** x^9 **15.** 64 **17.** $-\frac{8}{27}x^6$ **19.** $-6a^6$ **21.** $\frac{1}{9}$ **23.** $-\frac{1}{32}$ **25.** $\frac{16}{9}$ **27.** 17 **29.** x^3 **31.** $\frac{a^6}{b^{15}}$ **33.** $\frac{8}{125y^{18}}$ **35.** $\frac{1}{5}$ **37.** $\frac{24a^{12}c^6}{b^3}$ **39.** $\frac{8x^{22}}{81y^{23}}$ **41.** $\frac{1}{x^{10}}$ **43.** a^{10} **45.** $\frac{1}{t^6}$ **47.** x^{12} **49.** x^{18} **51.** $\frac{1}{x^{22}}$ **53.** $\frac{a^3b^7}{4}$ **55.** $\frac{y^{38}}{x^{16}}$ **57.** $\frac{16y^{16}}{x^8}$ **59.** x^4y^6 **61.** 3.78×10^5 **63.** 4.9×10^3 **65.** 3.7×10^{-4} **67.** 4.95×10^{-3} **69.** 5,340 **71.** 7,800,000 **73.** 0.00344 **75.** 0.49 **77.** 8×10^4 **79.** 2×10^9 **81.** 2.5×10^{-6} **83.** $80x^3y^6$ **85.** $72x^3y^3$ **87.** $15xy$ **89.** $12x^7y^6$ **91.** $54a^6b^2c^4$ **93.** $2x^3$ **95.** $4x$ **97.** $-4xy^4$ **99.** $-2x^2y^2$ **101.** 50 **103.** 7.9×10^{-5} **105.** 6.3×10^8 **107. a.** 7.3×10^4 **b.** 1.94×10^5 **c.** 1.9×10^6 **109.** 1.003×10^{19} miles **111. a.** 4.22×10^{11} **b.** \$7,033 per household **113.** $5x$ **115.** $8x^2$ **117.** $2x^3$ **119.** −5 **121.** $-2x + 3$ **123.** $9x + 6$ **125.** 1,200 **127.** (1, 2) **129.** $(\frac{5}{2}, -1)$ **131.** (0, 3) **133.** (3, 4) **135.** $\frac{1}{x^3}$ **137.** y^3 **139.** x^5

141.

NUMBER OF BYTES		
UNIT	EXPONENTIAL FORM	SCIENTIFIC NOTATION
Kilobyte	$2^{10} = 1,024$	1.024×10^3
Megabyte	$2^{20} \approx 1,048,000$	1.048×10^6
Gigabyte	$2^{30} \approx 1,074,000,000$	1.074×10^9
Terabyte	$2^{40} \approx 1,099,500,000,000$	1.0995×10^{12}

Problem Set 5.2

1. Trinomial, 2, 5 **3.** Binomial, 1, 3 **5.** Trinomial, 2, 8 **7.** Polynomial, 3, 4 **9.** Monomial, 0, $-\frac{3}{4}$ **11.** Trinomial, 3, 6 **13.** $7x + 1$ **15.** $2x^2 + 7x - 15$ **17.** $12a^2 - 7ab - 10b^2$ **19.** $x^2 - 13x + 3$ **21.** $\frac{1}{4}x^2 - \frac{7}{12}x - \frac{1}{4}$ **23.** $-y^3 - y^2 - 4y + 7$ **25.** $2x^3 + x^2 - 3x - 17$

27. $\frac{1}{14}x^2 + \frac{1}{7}xy + \frac{5}{7}y^2$ **29.** $-3a^3 + 6a^2b - 5ab^2$ **31.** $-3x$ **33.** $3x^2 - 12xy$ **35.** $17x^5 - 12$ **37.** $14a^2 - 2ab + 8b^2$ **39.** $2 - x$
41. $10x - 5$ **43.** $9x - 35$ **45.** $9y - 4x$ **47.** $9a + 2$ **49.** -2 **51.** 208 **53.** -15 **55. a.** $3x - 3a$ **b.** $2x - 2a$
57. 240 feet for both **59.** $P(x) = -300 + 40x - 0.5x^2$; $P(60) = \$300$ **61.** $P(x) = -800 + 3.5x - 0.002x^2$; $P(1{,}000) = \$700$
63. $2x^2 + 7x - 15$ **65.** $6x^3 - 11x^2y + 11xy^2 - 12y^3$ **67.** $-12x^4$ **69.** $20x^5$ **71.** a^6 **73.** 650 **75.** $x - 5$ **77.** $x - 7$
79. $5x - 1$ **81.** $x - 1$ **83.** $3x + 8y$ **85. a.** 5 **b.** -4 **c.** -3 **d.** 3 **e.** 4 **f.** 0 **g.** -4 **h.** 4

Problem Set 5.3

1. $12x^3 - 10x^2 + 8x$ **3.** $-3a^5 + 18a^4 - 21a^2$ **5.** $2a^5b - 2a^3b^2 + 2a^2b^4$ **7.** $x^2 - 2x - 15$ **9.** $6x^4 - 19x^2 + 15$
11. $x^3 + 9x^2 + 23x + 15$ **13.** $a^3 - b^3$ **15.** $8x^3 + y^3$ **17.** $2a^3 - a^2b - ab^2 - 3b^3$ **19.** $x^2 + x - 6$ **21.** $6a^2 + 13a + 6$
23. $20 - 2t - 6t^2$ **25.** $x^6 - 2x^3 - 15$ **27.** $20x^2 - 9xy - 18y^2$ **29.** $18t^2 - \frac{2}{9}$ **31.** $25x^2 + 20xy + 4y^2$ **33.** $25 - 30t^3 + 9t^6$
35. $4a^2 - 9b^2$ **37.** $9r^4 - 49s^2$ **39.** $\frac{1}{9}x^2 - \frac{4}{25}$ **41.** $x^3 - 6x^2 + 12x - 8$ **43.** $x^3 - \frac{3}{2}x^2 + \frac{3}{4}x - \frac{1}{8}$ **45.** $3x^3 - 18x^2 + 33x - 18$
47. $a^2b^2 + b^2 + 8a^2 + 8$ **49.** $3x^2 + 12x + 14$ **51.** $24x$ **53.** $x^2 + 4x - 5$ **55.** $4a^2 - 30a + 56$ **57.** $32a^2 + 20a - 18$
59. $(x + y)^2 + (x + y) - 20 = x^2 + 2xy + y^2 + x + y - 20$ **61.** $2^4 - 3^4 = -65$; $(2 - 3)^4 = 1$; $(2^2 + 3^2)(2 + 3)(2 - 3) = -65$
63. $R(p) = 900p - 300p^2$; $R(x) = 3x - \frac{x^2}{300}$; \$672 **65.** $R(p) = 350p - 10p^2$; $R(x) = 35x - \frac{x^2}{10}$; \$1,852.50
67. $P(x) = -500 + 30x - \frac{x^2}{10}$; $P(60) = \$940$ **69.** $A = 100 + 400r + 600r^2 + 400r^3 + 100r^4$ **71.** $8a^2$ **73.** -48 **75.** $2a^3b$
77. $-3b^2$ **79.** $-y^4$ **81.** (1, 2, 3) **83.** (1, 3, 1)

Problem Set 5.4

1. $5x^2(2x - 3)$ **3.** $9y^3(y^3 + 2)$ **5.** $3ab(3a - 2b)$ **7.** $7xy^2(3y^2 + x)$ **9.** $3(a^2 - 7a + 10)$ **11.** $4x(x^2 - 4x - 5)$
13. $10x^2y^2(x^2 + 2xy - 3y^2)$ **15.** $xy(-x + y - xy)$ **17.** $2xy^2z(2x^2 - 4xz + 3z^2)$ **19.** $5abc(4abc - 6b + 5ac)$ **21.** $(a - 2b)(5x - 3y)$
23. $3(x + y)^2(x^2 - 2y^2)$ **25.** $(x + 5)(2x^2 + 7x + 6)$ **27.** $(x + 1)(3y + 2a)$ **29.** $(xy + 1)(x + 3)$ **31.** $(x - 2)(3y^2 + 4)$
33. $(x - a)(x - b)$ **35.** $(b + 5)(a - 1)$ **37.** $(b^2 + 1)(a^4 - 5)$ **39.** $(x + 3)(x^2 - 4)$ **41.** $(x + 2)(x^2 - 25)$ **43.** $(2x + 3)(x^2 - 4)$
45. $(x + 3)(4x^2 - 9)$ **47.** 6 **49.** $P(1 + r) + P(1 + r)r = (1 + r)(P + Pr) = (1 + r)P(1 + r) = P(1 + r)^2$ **51.** $p = 11.5 - 0.05x$; \$5.25
53. \$28.50 **55.** $3x^2(x^2 - 3xy - 6y^2)$ **57.** $(x - 3)(2x^2 - 4x - 3)$ **59.** $3x^2 + 5x - 2$ **61.** $3x^2 - 5x + 2$ **63.** $x^2 + 5x + 6$
65. $6y^2 + y - 35$ **67.** $20 - 19a + 3a^2$ **69.**

TWO NUMBERS a AND b	THEIR PRODUCT ab	THEIR SUM $a + b$
1, −24	−24	−23
−1, 24	−24	23
2, −12	−24	−10
−2, 12	−24	10
3, −8	−24	−5
−3, 8	−24	5
4, −6	−24	−2
−4, 6	−24	2

71. 36 **73.** 0

Problem Set 5.5

1. $(x + 3)(x + 4)$ **3.** $(x + 3)(x - 4)$ **5.** $(y + 3)(y - 2)$ **7.** $(2 - x)(8 + x)$ **9.** $(2 + x)(6 + x)$ **11.** $3(a - 2)(a - 5)$
13. $4x(x - 5)(x + 1)$ **15.** $(x + 2y)(x + y)$ **17.** $(a + 6b)(a - 3b)$ **19.** $(x - 8a)(x + 6a)$ **21.** $(x - 6b)^2$ **23.** $3(x - 3y)(x + y)$
25. $2a^3(a^2 + 2ab + 2b^2)$ **27.** $10x^2y^2(x + 3y)(x - y)$ **29.** $(2x - 3)(x + 5)$ **31.** $(2x - 5)(x + 3)$ **33.** $(2x - 3)(x - 5)$
35. $(2x - 5)(x - 3)$ **37.** Prime **39.** $(2 + 3a)(1 + 2a)$ **41.** $15(4y + 3)(y - 1)$ **43.** $x^2(3x - 2)(2x + 1)$ **45.** $10r(2r - 3)^2$
47. $(4x + y)(x - 3y)$ **49.** $(2x - 3a)(5x + 6a)$ **51.** $(3a + 4b)(6a - 7b)$ **53.** $200(1 + 2t)(3 - 2t)$ **55.** $y^2(3y - 2)(3y + 5)$
57. $2a^2(3 + 2a)(4 - 3a)$ **59.** $2x^2y^2(4x + 3y)(x - y)$ **61.** $100(3x^2 + 1)(x^2 + 3)$ **63.** $(5a^2 + 3)(4a^2 + 5)$ **65.** $3(3 + 4r^2)(1 - r^2)$
67. $(x + 5)(2x + 3)(x + 2)$ **69.** $(2x + 3)(x + 5)(x + 2)$ **71.** $9x^2 - 25y^2$ **73.** $a + 250$
75. $y = 2(2x - 1)(x + 5)$; $y = 0$ when $x = \frac{1}{2}$ or $x = -5$, $y = 42$ when $x = 2$ **77.** $\frac{5}{8}$ **79.** x^3 **81.** $4x^2$ **83.** $\frac{1}{2}$ **85.** x^2 **87.** $3x$
89. $2y$ **91.** $4x^2 - 9$ **93.** $4x^2 - 12x + 9$ **95.** $8x^3 - 27$ **97.** $(2x^3 + 5y^2)(4x^3 + 3y^2)$ **99.** $(3x - 5)(x + 100)$
101. $(\frac{1}{4}x + 1)(\frac{1}{2}x + 2)$ **103.** $(2x + 0.5)(x + 0.5)$

Problem Set 5.6

1. $(x - 3)^2$ **3.** $(a - 6)^2$ **5.** $(5 - t)^2$ **7.** $(\frac{1}{3}x + 3)^2$ **9.** $(2y^2 - 3)^2$ **11.** $(4a + 5b)^2$ **13.** $(\frac{1}{5} + \frac{1}{4}t^2)^2$ **15.** $4(2x - 3)^2$

17. $3a(5a+1)^2$ **19.** $(x+2+3)^2=(x+5)^2$ **21.** $(x+3)(x-3)$ **23.** $(7x+8y)(7x-8y)$ **25.** $(2a+\frac{1}{2})(2a-\frac{1}{2})$ **27.** $(x+\frac{3}{5})(x-\frac{3}{5})$
29. $(3x+4y)(3x-4y)$ **31.** $10(5+t)(5-t)$ **33.** $(x^2+9)(x+3)(x-3)$ **35.** $(3x^3+1)(3x^3-1)$ **37.** $(4a^2+9)(2a+3)(2a-3)$
39. $(\frac{1}{9}+\frac{y^2}{4})(\frac{1}{3}+\frac{y}{2})(\frac{1}{3}-\frac{y}{2})$ **41.** $(x-y)(x+y)(x^2+xy+y^2)(x^2-xy+y^2)$ **43.** $2a(a-2)(a+2)(a^2+2a+4)(a^2-2a+4)$
45. $(x+1)(x-5)$ **47.** $y(y+8)$ **49.** $(x-5+y)(x-5-y)$ **51.** $(a+4+b)(a+4-b)$ **53.** $(x+y+a)(x+y-a)$
55. $(x+3)(x+2)(x-2)$ **57.** $(x+2)(x+5)(x-5)$ **59.** $(2x+3)(x+2)(x-2)$ **61.** $(x+3)(2x+3)(2x-3)$ **63.** $(x-y)(x^2+xy+y^2)$
65. $(a+2)(a^2-2a+4)$ **67.** $(3+x)(9-3x+x^2)$ **69.** $(y-1)(y^2+y+1)$ **71.** $10(r-5)(r^2+5r+25)$ **73.** $(4+3a)(16-12a+9a^2)$
75. $(2x-3y)(4x^2+6xy+9y^2)$ **77.** $(t+\frac{1}{3})(t^2-\frac{1}{3}t+\frac{1}{9})$ **79.** $(3x-\frac{1}{3})(9x^2+x+\frac{1}{9})$ **81.** $(4a+5b)(16a^2-20ab+25b^2)$
83. 30 and -30 **85.** $y(y^2+25)$ **87.** $2ab^3(b^2+4b+1)$ **89.** $(2x-3)(2x+a)$ **91.** $(x+2)(x-2)$ **93.** $(x-3)^2$
95. $(3a-4)(2a-1)$ **97.** $(x+2)(x^2-2x+4)$ **99.** $(-\frac{15}{43}, -\frac{27}{43})$ **101.** $(1, 3, 1)$

Problem Set 5.7

1. $(x+9)(x-9)$ **3.** $(x-3)(x+5)$ **5.** $(x+2)(x+3)^2$ **7.** $(x^2+2)(y^2+1)$ **9.** $2ab(a^2+3a+1)$ **11.** Does not factor: prime
13. $3(2a+5)(2a-5)$ **15.** $(3x-2y)^2$ **17.** $(5-t)^2$ **19.** $4x(x^2+4y^2)$ **21.** $2y(y+5)^2$ **23.** $a^4(a+2b)(a^2-2ab+4b^2)$
25. $(t+3+x)(t+3-x)$ **27.** $(x+5)(x+3)(x-3)$ **29.** $5(a+b)^2$ **31.** Does not factor: prime **33.** $3(x+2y)(x+3y)$
35. $(3a+\frac{1}{3})^2$ **37.** $(x-3)(x-7)^2$ **39.** $(x+8)(x-8)$ **41.** $(2-5x)(4+3x)$ **43.** $a^5(7a+3)(7a-3)$ **45.** $(r+\frac{1}{5})(r-\frac{1}{5})$
47. Does not factor: prime **49.** $100(x-3)(x+2)$ **51.** $a(5a+3)(5a+1)$ **53.** $(3x^2+1)(x^2-5)$ **55.** $3a^2b(2a-1)(4a^2+2a+1)$
57. $(4-r)(16+4r+r^2)$ **59.** $5x^2(2x+3)(2x-3)$ **61.** $100(2t+3)(2t-3)$ **63.** $2x^3(4x-5)(2x-3)$
65. $(y+1)(y-1)(y^2-y+1)(y^2+y+1)$ **67.** $2(5+a)(5-a)$ **69.** $3x^2y^2(2x+3y)^2$ **71.** $(x-2+y)(x-2-y)$
73. $2x^2+2x+1$ **75.** t^2-4t+3 **77.** $(x-6)(x+4)$ **79.** $x(2x+1)(x-3)$ **81.** $(x+2)(x-3)(x+3)$ **83.** 6 **85.** $-\frac{1}{2}$
87. 60 geese; 48 ducks **89.** 150 oranges; 144 apples

Problem Set 5.8

1. $6, -1$ **3.** $0, 2, 3$ **5.** $\frac{1}{3}, -4$ **7.** $\frac{2}{3}, \frac{3}{2}$ **9.** $5, -5$ **11.** $0, -3, 7$ **13.** $-4, \frac{5}{2}$ **15.** $0, \frac{4}{3}$ **17.** $-\frac{1}{5}, \frac{1}{3}$ **19.** $0, -\frac{4}{3}, \frac{4}{3}$
21. $-10, 0$ **23.** $-5, 1$ **25.** $1, 2$ **27.** $-2, 3$ **29.** $-2, \frac{1}{4}$ **31.** $-3, -2, 2$ **33.** $-2, -5, 5$ **35.** $-\frac{3}{2}, -2, 2$ **37.** $-3, -\frac{3}{2}, \frac{3}{2}$
39. $-2, \frac{5}{3}$ **41.** $-1, 9$ **43.** $0, -3$ **45.** $4, 2$ **47.** $-4, -2$ **49.** $-5, -3$ or $3, 5$ **51.** $-5, -4$ or $4, 5$ **53.** 24 feet
55. 6, 8, 10 **57.** 2 feet, 8 feet **59.** 18 inches, 4 inches **61.** 0 and 2 seconds **63.** 1 and 2 seconds
65. 0 and $\frac{3}{2}$ seconds **67.** 2 and 3 seconds **69.** \$4 or \$8 **71.** \$7 or \$10 **73.** $(1, 2)$ **75.** $(15, 12)$ **77.** $(3, -2, 1)$
79. \$4,000 at 5%, \$8,000 at 6%

81.

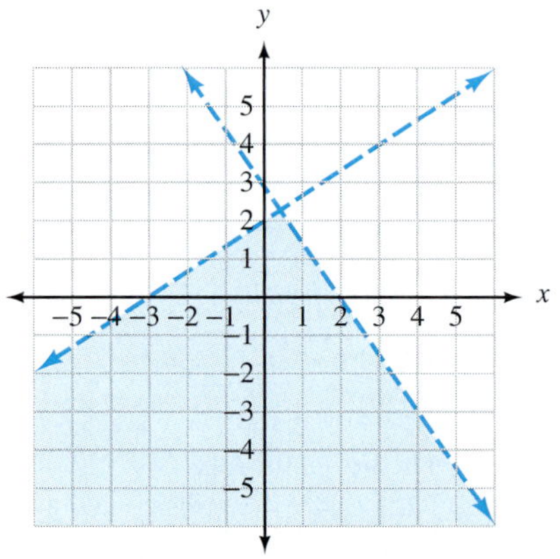

83.

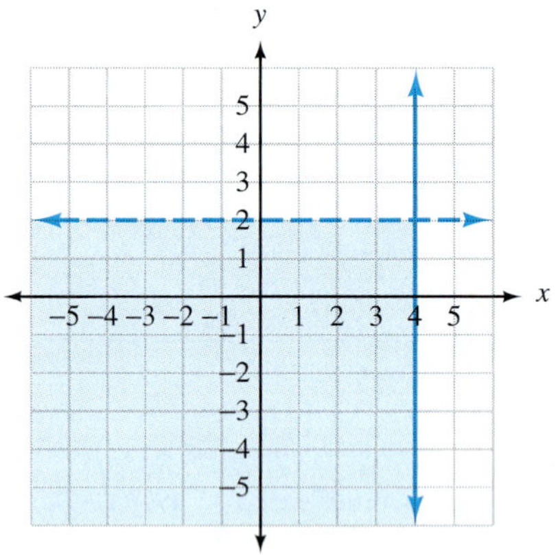

Chapter 5 Review/Test

1. x^{10} **2.** $25x^6$ **3.** $-32x^{18}y^8$ **4.** $\frac{1}{8}$ **5.** $\frac{9}{4}$ **6.** $\frac{1}{2}$ **7.** 3.45×10^7 **8.** 3.57×10^{-3} **9.** 44,500 **10.** 0.000445 **11.** $\frac{1}{a^9}$
12. $\frac{x^{12}}{4}$ **13.** x^2 **14.** 8×10^{-2} **15.** 4×10^{-10} **16.** $2x^2-5x+7$ **17.** $2x^3-2x^2-2x-4$ **18.** x^2-2x-3 **19.** $30x+12$
20. 15 **21.** $12x^3-6x^2+3x$ **22.** $2a^4b^3+4a^3b^4+2a^2b^5$ **23.** $18-9y+y^2$ **24.** $6x^4+5x^2-4$ **25.** $2t^3-4t^2-6t$ **26.** x^3+27
27. $8x^3-27$ **28.** a^4-4a^2+4 **29.** $9x^2+30x+25$ **30.** $16x^2-24xy+9y^2$ **31.** $x^2-\frac{1}{9}$ **32.** $4a^2-b^2$ **33.** x^3-3x^2+3x-1
34. $x^{2m}-4$ **35.** $3xy(2x^3-3y^3+6x^2y^2)$ **36.** $4(x+y)^2(x^2-2y^2)$ **37.** $(4x^2+5)(2-y)$ **38.** $(1-y)(1+y)(x^3+8b^2)$
39. $(x-2)(x-3)$ **40.** $2x(x+5)(x-3)$ **41.** $(5a-4b)(4a-5b)$ **42.** $x^2(3x+2)(2x-5)$ **43.** $3y(4x+5)(2x-3)$
44. $(x^2+4)(x+2)(x-2)$ **45.** $3(a^2+3)^2$ **46.** $(a-2)(a^2+2a+4)$ **47.** $5x(x+3y)^2$ **48.** $3ab(a-3b)(a+3b)$
49. $(x-5+y)(x-5-y)$ **50.** $(6-5a)(6+5a)$ **51.** $(x+3)(x-3)(x+4)$ **52.** $-3, -2$ **53.** $-\frac{1}{2}, \frac{4}{5}$ **54.** $-\frac{5}{3}, \frac{5}{3}$ **55.** $0, -2$
56. $-3, 6$ **57.** $-4, 3, -3$ **58.** $-10, -8$ or $8, 10$ **59.** $-5, -4$ or $4, 5$ **60.** 3, 4, 5 **61.** 6, 8, 10

Chapter 6

Getting Ready for Chapter 6

1. $\frac{3}{4}$ **2.** $\frac{2}{5}$ **3.** 40 **4.** 40 **5.** 785 **6.** 518.1 **7.** $x - 5$ **8.** $b - a$ **9.** $8x^2 - 16x$
10. -2 **11.** 0 **12.** Undefined **13.** $-4xy^4$ **14. a.** 16.7 **b.** 10 **c.** 7.1 **15.** $(x + a)(x - a)$ **16.** $(x - 3)(x - 3)$ **17.** $x^2(x - y)$
18. $y(x - 1)$ **19.** $2(y - 1)(y + 1)$ **20.** $(2y - 3)(4y^2 + 6y + 9)$

Problem Set 6.1

1. $g(0) = -3, g(-3) = 0, g(3) = 3, g(-1) = -1$, $g(1)$ is undefined **3.** $h(0) = -3, h(-3) = 3, h(3) = 0, h(-1)$ is undefined, $h(1) = -1$
5. $\{x \mid x \neq 1\}$ **7.** $\{t \mid t \neq 4, t \neq -4\}$ **9.** $\frac{x-4}{6}$ **11.** $\frac{4x-3y}{x(x+y)}$ **13.** $(a^2 + 9)(a + 3)$ **15.** $\frac{a-6}{a+6}$ **17.** $\frac{2y+3}{y+1}$ **19.** $\frac{a^2-ab+b^2}{a-b}$
21. $\frac{2(x-1)}{x}$ **23.** $\frac{2x+3y}{2x+y}$ **25.** $\frac{x+3}{y-4}$ **27.** $\frac{x+b}{x-2b}$ **29.** $x + 2$ **31.** $\frac{2x^2-5}{3x-2}$ **33.** -1 **35.** $-(y + 6)$
37. $\frac{-(3a+1)}{3a-1}$ or $-\frac{3a+1}{3a-1}$ **39.** 3 **41.** $x + a$ **43.** 2; 2 **45.** Undefined; 4 **47.** 1; 1 **49.** 3; 3

51.

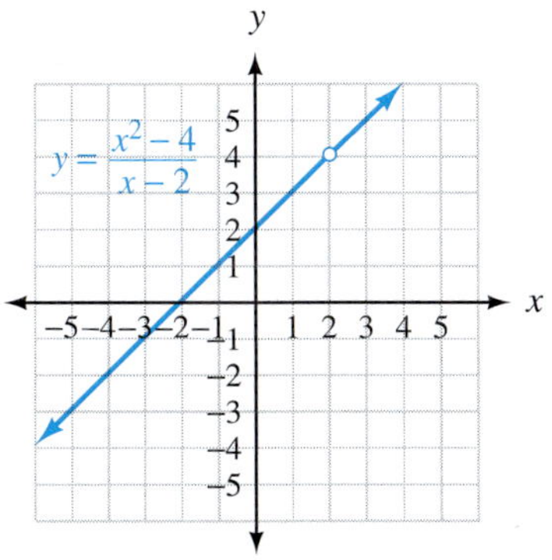

The graph of $y = x + 2$ includes the point (2, 4) while the other graph does not.

53.

WEEKS x	WEIGHT (POUNDS) $W(x)$
0	200
1	194
4	184
12	173
24	168

55. 6.8 feet per second
57. a. Domain = $\{t \mid 20 \leq t \leq 50\}$

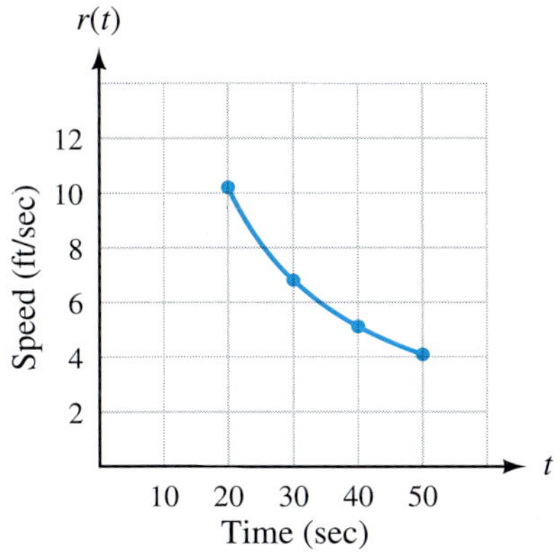

59. a. Domain = $\{d \mid 1 \leq d \leq 6\}$

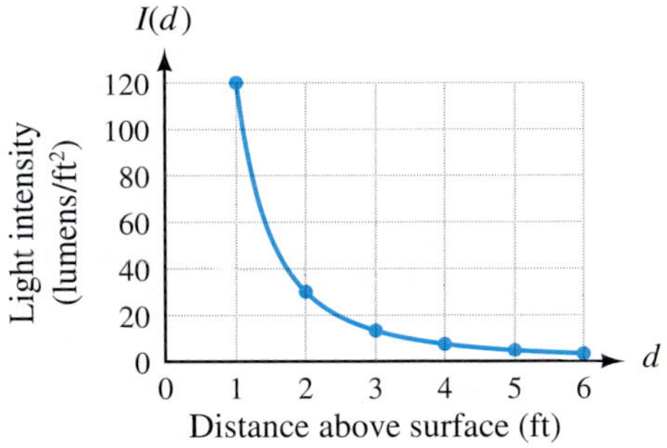

61. 3,768 inches per minute; 2,826 inches per minute **63.** $2x^3$ **65.** $-2x^2y^2$ **67.** 185.12 **69.** $4x^3 - 8x^2$ **71.** $4x^3 - 6x - 20$
73. $-3x + 9$ **75.** $(x + a)(x - a)$ **77.** $(x + y)(x - 7y)$ **79.** $3x^2 - 7x + 4$ **81.** 9 **83.** $8x^2$
85. a. 2 **b.** -4 **c.** Undefined **d.** 2 **e.** 1 **f.** -6 **g.** 4 **h.** -3

Problem Set 6.2

1. $2x^2 - 4x + 3$ **3.** $-2x^2 - 3x + 4$ **5.** $2y^2 + \frac{5}{2} - \frac{3}{2y^2}$ **7.** $-\frac{5}{2}x + 4 + \frac{3}{x}$ **9.** $4ab^3 + 6a^2b$ **11.** $-xy + 2y^2 + 3xy^2$ **13.** $x + 2$
15. $a - 3$ **17.** $5x + 6y$ **19.** $x^2 + xy + y^2$ **21.** $(y^2 + 4)(y + 2)$ **23.** $(x + 2)(x + 5)$ **25.** $x - 7 + \frac{7}{x+2}$ **27.** $2x^2 - 5x + 1 + \frac{4}{x+1}$
29. $y^2 - 3y - 13$ **31.** $3y^2 + 6y + 8 + \frac{37}{2y-4}$ **33.** $a^3 + 2a^2 + 4a + 6 + \frac{17}{a-2}$ **35.** $y^3 + 2y^2 + 4y + 8$ **37.** 4 **39.** 5 **41.** $x + a$
43. $x + a$ **47. a.**

x	1	5	10	15	20
$C(x)$	2.15	2.75	3.50	4.25	5.00

b. $\overline{C}(x) = \frac{2}{x} + 0.15$

c.

x	1	5	10	15	20
$\overline{C}(x)$	2.15	0.55	0.35	0.28	0.25

d. It decreases.

49. $\frac{2}{3}$ **51.** $20x^2y^2$ **53.** $72x^4y^5$ **55.** $(x+2)(x-2)$ **57.** $x^2(x-y)$ **59.** $2(y-1)(y+1)$ **61.** $\frac{21}{10}$ **63.** $\frac{11}{8}$ **65.** $\frac{1}{18}$ **67.** 32 **69.** 17

71. $\frac{x^{16}}{y^{22}}$ **73.** $4x^3 - x^2 + 3$ **75.** $0.5x^2 - 0.4x + 0.3$ **77.** $\frac{3}{2}x - \frac{5}{2} + \frac{1}{2x+4}$ **79.** $\frac{2}{3}x + \frac{1}{3} + \frac{2}{3x-1}$

Problem Set 6.3

1. $\frac{1}{6}$ **3.** $\frac{9}{4}$ **5.** $\frac{1}{2}$ **7.** $\frac{15y}{x^2}$ **9.** $\frac{b}{a}$ **11.** $\frac{2y^5}{z^3}$ **13.** $\frac{x+3}{x+2}$ **15.** $y+1$ **17.** $\frac{3(x+4)}{x-2}$ **19.** 1 **21.** $\frac{(a-2)(a+2)}{a-5}$

23. $\frac{9t^2-6t+4}{4t^2-2t+1}$ **25.** $\frac{x+3}{x+4}$ **27.** $\frac{5a-b}{9a^2+15ab+25b^2}$ **29.** 2 **31.** $\frac{x(x-1)}{x^2+1}$ **33.** $\frac{(a+4b)(a-3b)}{(a-4b)(a+5b)}$ **35.** $\frac{2y-1}{2y-3}$

37. $\frac{(y-2)(y+1)}{(y+2)(y-1)}$ **39.** $\frac{x-1}{x+1}$ **41.** $\frac{x-2}{x+3}$ **43.** $3x$ **45.** $2(x+5)$ **47.** $x-2$ **49.** $-(y-4)$ **51.** $(a-5)(a+1)$

53.

NUMBER OF COPIES	PRICE PER COPY ($)
1	20.33
10	9.33
20	6.40
50	4.00
100	3.05

55. $305.00 **57.** $x^2 - 6x + 9$

59. $\frac{2}{3}$ **61.** $\frac{47}{105}$ **63.** $x-7$ **65.** $(x+1)(x-1)$ **67.** $2(x+5)$ **69.** $(a-b)(a^2+ab+b^2)$

71. $10x^5 + 8x^3 - 6x^2$ **73.** $12a^2 + 11a - 5$ **75.** $12xy - 6x + 28y - 14$ **77.** $9 - 6t^2 + t^4$ **79.** $3x^3 + 18x^2 + 33x + 18$

Problem Set 6.4

1. $\frac{5}{4}$ **3.** $\frac{1}{3}$ **5.** $\frac{41}{24}$ **7.** $\frac{19}{144}$ **9.** $\frac{31}{24}$ **11.** 1 **13.** -1 **15.** $\frac{1}{x+y}$ **17.** 1 **19.** $\frac{a^2+2a-3}{a^3}$ **21.** 1 **23.** $\frac{4-3t}{2t^2}$

25. $\frac{1}{2}$ **27.** $\frac{1}{5}$ **29.** $\frac{x+3}{2(x+1)}$ **31.** $\frac{a-b}{a^2+ab+b^2}$ **33.** $\frac{2y-3}{4y^2+6y+9}$ **35.** $\frac{2(2x-3)}{(x-3)(x-2)}$ **37.** $\frac{1}{2t-7}$ **39.** $\frac{4}{(a-3)(a+1)}$

41. $\frac{-4x^2}{(2x+1)(2x-1)(4x^2+2x+1)}$ **43.** $\frac{2}{(2x+3)(4x+3)}$ **45.** $\frac{a}{(a+4)(a+5)}$ **47.** $\frac{x+1}{(x-2)(x+3)}$ **49.** $\frac{x-1}{(x+1)(x+2)}$

51. $\frac{1}{(x+2)(x+1)}$ **53.** $\frac{1}{(x+2)(x+3)}$ **55.** $\frac{4x+5}{2x+1}$ **57.** $\frac{22-5t}{4-t}$ **59.** $\frac{2x^2+3x-4}{2x+3}$ **61.** $\frac{2x-3}{2x}$ **63.** $\frac{1}{2}$ **65.** $x + \frac{4}{x} = \frac{x^2+4}{x}$

67. $\frac{51}{10} = 5.1$ **69.** **a.** $T = 120$ months **b.** The two objects will never meet. **71.** $\frac{6}{5}$ **73.** $x+2$ **75.** $3-x$

77. $(x+2)(x-2)$ **79.** 5.4×10^4 **81.** 3.4×10^{-4} **83.** 6,440 **85.** 0.00644 **87.** 1.2×10^4 **89.** $\frac{x-1}{x+3}$

Problem Set 6.5

1. $\frac{9}{8}$ **3.** $\frac{2}{15}$ **5.** $\frac{119}{20}$ **7.** $\frac{1}{x+1}$ **9.** $\frac{a+1}{a-1}$ **11.** $\frac{y-x}{y+x}$ **13.** $\frac{1}{(x+5)(x-2)}$ **15.** $\frac{1}{a^2-a+1}$ **17.** $\frac{x+3}{x+2}$ **19.** $\frac{a+3}{a-2}$

21. $\frac{x+4}{x+2}$ **23.** $\frac{a-1}{a+1}$ **25.** $-\frac{x}{3}$ **27.** $\frac{y^2+1}{2y}$ **29.** $\frac{-x^2+x-1}{x-1}$ **31.** $\frac{5}{3}$ **33.** **a.** $\frac{4}{ax}$ **b.** $-\frac{1}{(x+1)(a+1)}$ **c.** $-\frac{a+x}{a^2x^2}$

35. $(a^{-1}+b^{-1})^{-1} = \left(\frac{1}{a}+\frac{1}{b}\right)^{-1} = \left(\frac{a+b}{ab}\right)^{-1} = \frac{ab}{a+b}$ **37.** **a.** As v approaches 0, the denominator approaches 1. **b.** $v = \frac{fs}{h} - s$

39. $xy - 2y$ **41.** $3x - 18$ **43.** ab **45.** $(y+5)(y-5)$ **47.** $x(a+b)$ **49.** 2 **51.** -15 **53.** 5 **55.** 1 **57.** $-3, 4$

59. 2, 3

Problem Set 6.6

1. $-\frac{35}{3}$ **3.** $-\frac{18}{5}$ **5.** $\frac{36}{11}$ **7.** 2 **9.** 5 **11.** 2 **13.** $-3, 4$ **15.** $1, -\frac{4}{3}$ **17.** Possible solution -1, which does not check; ∅

19. 5 **21.** $-\frac{1}{2}, \frac{5}{3}$ **23.** $\frac{2}{3}$ **25.** 18 **27.** Possible solution 4, which does not check; ∅

29. Possible solutions 3 and -4; only -4 checks; -4 **31.** -6 **33.** -5 **35.** $\frac{53}{17}$ **37.** Possible solutions 1 and 2; only 2 checks; 2

39. Possible solution 3 , which does not check; ∅ **41.** $\frac{22}{3}$ **43.** 2 **45.** 1, 5 **47.** $x = \frac{ab}{a-b}$ **49.** $R = \frac{R_1R_2}{R_1+R_2}$ **51.** $y = \frac{x-3}{x-1}$

53. $y = \frac{1-x}{3x-2}$ **57.**

TIME t(sec)	SPEED OF KAYAK RELATIVE TO THE WATER v(m/sec)	CURRENT OF THE RIVER c(m/sec)
240	4	1
300	4	2
514	4	3
338	3	1
540	3	2
N/A	3	3

59. 2,358 **61.** 12.3 **63.** 3 **65.** 9, −1 **67.** 60 **69.** 5 **71.** 8 meters, 13 meters **73.** −6, −5 or 5, 6 **75.** 3, 4, 5
77. $-\frac{2}{3}, -\frac{5}{2}, \frac{5}{2}$ **79.** −2, 2, 3

Problem Set 6.7

As you can see, in addition to the answers to the problems we have included some of the equations used to solve the problems. Remember, you should attempt the problems on your own before looking here to check your answers or equations.

1. $\frac{1}{x} + \frac{1}{3x} = \frac{20}{3}$; $\frac{1}{5}$ and $\frac{3}{5}$ **3.** $x + \frac{1}{x} = \frac{10}{3}$; 3 or $\frac{1}{3}$ **5.** $\frac{1}{x} + \frac{1}{x+1} = \frac{7}{12}$; 3, 4 **7.** $\frac{7+x}{9+x} = \frac{5}{6}$; 3
9. Let x = speed of current; $\frac{1.5}{5-x} = \frac{3}{5+x}$; $\frac{5}{3}$ miles per hour **11.** $\frac{8}{x+2} + \frac{8}{x-2} = 3$; 6 miles per hour
13. Train A: 75 miles per hour, Train B: 60 miles per hour; let x = speed of B; $\frac{150}{x+15} = \frac{120}{x}$
15. 540 miles per hour; let x = speed of 747; $\frac{810}{270} - 1\frac{1}{2} = \frac{810}{x}$ **17.** 54 miles per hour; let x = usual speed; $\frac{270}{x+6} + \frac{1}{2} = \frac{270}{x}$
19. Let x = time to fill the tank with both open; $\frac{1}{8} - \frac{1}{16} = \frac{1}{x}$; 16 hours **21.** $\frac{15}{2}$ minutes; $\frac{1}{3} - \frac{1}{5} = \frac{1}{x}$
23. Let x = time to fill with hot water; $\frac{1}{x} + \frac{1}{3.5} = \frac{1}{2.1}$; $5\frac{1}{4}$ minutes **25.** 51.1 acres **27.** 5.9 miles per hour
29. 20.7 miles per hour **31.** 4.6 miles per hour **33.** 3.6 miles per hour **35.** $\frac{1}{3}\left[\left(x + \frac{2}{3}x\right) + \frac{1}{3}\left(x + \frac{2}{3}x\right)\right] = 10, x = \frac{27}{2}$ **37.** $\frac{2}{3a}$
39. $(x-3)(x+2)$ **41.** 1 **43.** $\frac{3-x}{3+x}$ **45.** Possible solution 3, which does not check; ∅

Chapter 6 Review/Test

1. $\frac{25x^2}{7y^3}$ **2.** $\frac{a(a-b)}{4}$ **3.** $\frac{x-5}{x+5}$ **4.** $\frac{a+1}{a-1}$ **5.** $3x + 2 + \frac{4}{x}$ **6.** $-9b + 5a - 7a^2b^2$ **7.** $x^{3n} - x^{2n}$ **8.** $x + 2$ **9.** $5x + 6y$
10. $y^3 + 2y^2 + 4y + 8$ **11.** $4x + 1 - \frac{2}{2x-7}$ **12.** $y^2 - 3y - 13$ **13.** $\frac{9}{5}$ **14.** $\frac{3x}{4y^2}$ **15.** $\frac{x-1}{x^2+1}$ **16.** 1 **17.** $\frac{x+2}{x-2}$
18. $(2x-3)(x+3)$ **19.** $\frac{31}{30}$ **20.** −1 **21.** $\frac{x^2+x+1}{x^3}$ **22.** $\frac{1}{(y+4)(y+3)}$ **23.** $\frac{x-1}{2(x+1)(x+2)}$ **24.** $\frac{15x-2}{5x-2}$ **25.** 5
26. $\frac{1}{a^2-a+1}$ **27.** $\frac{x^2+x+1}{x^2+1}$ **28.** $\frac{x+3}{x+2}$ **29.** 6 **30.** 1 **31.** −6 **32.** Possible solution −3, which does not check; ∅
33. $\frac{22}{3}$ **34.** Possible solutions 4 and −5; only 4 checks **35.** Car: 30 miles per hour; truck: 20 miles per hour
36. 7.5 miles per hour **37.** 742 miles per hour

Chapter 7

Getting Ready for Chapter 7

1. 64 **2.** 64 **3.** −64 **4.** −64 **5.** −8 **6.** −8 **7.** 4 **8.** 2 **9.** $\frac{7}{6}$ **10.** $\frac{1}{3}$ **11.** −22 **12.** 4 **13.** $\frac{1}{2}$
14. $\frac{17}{4} = 4\frac{1}{4}$ **15.** x^5 **16.** y^6 **17.** $243r^{40}s^{100}$ **18.** x **19.** $\frac{1}{27}$ **20.** $\frac{16}{81}$

Problem Set 7.1

1. 12 **3.** Not a real number **5.** −7 **7.** −3 **9.** 2 **11.** Not a real number **13.** 0.2 **15.** 0.2 **17.** $6a^4$ **19.** $3a^4$
21. $2x^2y$ **23.** $2a^3b^5$ **25.** 6 **27.** −3 **29.** 2 **31.** −2 **33.** 2 **35.** $\frac{9}{5}$ **37.** 9 **39.** 125 **41.** $\frac{1}{3}$ **43.** $\frac{1}{27}$ **45.** $\frac{6}{5}$
47. $\frac{8}{27}$ **49.** 7 **51.** $\frac{3}{4}$ **53.** $x^{4/5}$ **55.** a **57.** $\frac{1}{x^{2/5}}$ **59.** $x^{1/6}$ **61.** $x^{9/25}y^{1/2}z^{1/5}$ **63.** $\frac{b^{7/4}}{a^{1/8}}$ **65.** $y^{3/10}$ **67.** $\frac{1}{a^2b^4}$ **69.** $\frac{s^{1/2}}{r^{20}}$
71. $10b^3$ **73.** 25 mph **75.** **a.** 424 picometers **b.** 600 picometers **c.** 6×10^{-10} meters **77.** **a.** B **b.** A **c.** C **d.** (0, 0) and (1, 1)
79. $x^6 - x^5$ **81.** $12a^2 - 11ab + 2b^2$ **83.** $x^6 - 4$ **85.** $3x - 4x^3y$ **87.** $(x-5)(x+2)$ **89.** $(3x-2)(2x+5)$ **91.** x^2 **93.** t
95. $x^{1/3}$ **97.** $x^6 - x^3$ **99.** $x^2 + 2x - 15$ **101.** $x^4 - 10x^2 + 25$ **103.** $x^3 - 27$ **105.** $(9^{1/2} + 4^{1/2})^2 = (3+2)^2 = 5^2 = 25 \neq 9 + 4$
107. $\sqrt{\sqrt{a}} = (a^{1/2})^{1/2} = a^{1/4} = \sqrt[4]{a}$

Problem Set 7.2

1. $x + x^2$ **3.** $a^2 - a$ **5.** $6x^3 - 8x^2 + 10x$ **7.** $12x^2 - 36y^2$ **9.** $x^{4/3} - 2x^{2/3} - 8$ **11.** $a - 10a^{1/2} + 21$ **13.** $20y^{2/3} - 7y^{1/3} - 6$
15. $10x^{4/3} + 21x^{2/3}y^{1/2} + 9y$ **17.** $t + 10t^{1/2} + 25$ **19.** $x^3 + 8x^{3/2} + 16$ **21.** $a - 2a^{1/2}b^{1/2} + b$ **23.** $4x - 12x^{1/2}y^{1/2} + 9y$ **25.** $a - 3$
27. $x^3 - y^3$ **29.** $t - 8$ **31.** $4x^3 - 3$ **33.** $x + y$ **35.** $a - 8$ **37.** $8x + 1$ **39.** $t - 1$ **41.** $2x^{1/2} + 3$ **43.** $3x^{1/3} - 4y^{1/3}$
45. $3a - 2b$ **47.** $3(x-2)^{1/2}(4x-11)$ **49.** $5(x-3)^{7/5}(x-6)$ **51.** $3(x+1)^{1/2}(3x^2+3x+2)$ **53.** $(x^{1/3}-2)(x^{1/3}-3)$
55. $(a^{1/5}-4)(a^{1/5}+2)$ **57.** $(2y^{1/3}+1)(y^{1/3}-3)$ **59.** $(3t^{1/5}+5)(3t^{1/5}-5)$ **61.** $(2x^{1/7}+5)^2$ **63.** $\frac{3+x}{x^{1/2}}$ **65.** $\frac{x+5}{x^{1/3}}$ **67.** $\frac{x^3+3x^2+1}{(x^3+1)^{1/2}}$
69. $\frac{-4}{(x^2+4)^{1/2}}$ **71.** 15.8% **73.** 5 **75.** 6 **77.** $4x^2y$ **79.** $5y$ **81.** 3 **83.** 2 **85.** $2ab$ **87.** 25 **89.** $48x^4y^2$ **91.** $4x^6y^6$

93. $\frac{1}{x^2+9}$ **95.** $3x - 4x^3y$ **97.** $5x - 4$ **99.** $x^2 + 5x + 25$

Problem Set 7.3

1. $2\sqrt{2}$ **3.** $7\sqrt{2}$ **5.** $12\sqrt{2}$ **7.** $4\sqrt{5}$ **9.** $4\sqrt{3}$ **11.** $15\sqrt{3}$ **13.** $3\sqrt[3]{2}$ **15.** $4\sqrt[3]{2}$ **17.** $6\sqrt[3]{2}$ **19.** $2\sqrt[5]{2}$ **21.** $3x\sqrt{2x}$ **23.** $2y\sqrt[4]{2y^3}$ **25.** $2xy^2\sqrt[3]{5xy}$ **27.** $4abc^2\sqrt{3b}$ **29.** $2bc\sqrt[3]{6a^2c}$ **31.** $2xy^2\sqrt[5]{2x^3y^2}$ **33.** $3xy^2z\sqrt[5]{x^2}$ **35.** $2\sqrt{3}$ **37.** $\sqrt{-20}$, which is not a real number **39.** $\frac{\sqrt{11}}{2}$ **41.** $\frac{2\sqrt{3}}{3}$ **43.** $\frac{5\sqrt{6}}{6}$ **45.** $\frac{\sqrt{2}}{2}$ **47.** $\frac{\sqrt{5}}{5}$ **49.** $2\sqrt[3]{4}$ **51.** $\frac{2\sqrt[3]{3}}{3}$ **53.** $\frac{\sqrt[4]{24x^2}}{2x}$ **55.** $\frac{\sqrt[4]{8y^3}}{y}$ **57.** $\frac{\sqrt[3]{36xy^2}}{3y}$ **59.** $\frac{\sqrt[3]{6xy^2}}{3y}$ **61.** $\frac{\sqrt[4]{2x}}{2x}$ **63.** $\frac{3x\sqrt{15xy}}{5y}$ **65.** $\frac{5xy\sqrt{6xz}}{2z}$ **67.** $\frac{2ab\sqrt[3]{6ac^2}}{3c}$ **69.** $\frac{2xy^2\sqrt[3]{3z^2}}{3z}$ **71.** $5|x|$ **73.** $3|xy|\sqrt{3x}$ **75.** $|x-5|$ **77.** $|2x+3|$ **79.** $2|a(a+2)|$ **81.** $2|x|\sqrt{x-2}$ **83.** $\sqrt{9+16} = \sqrt{25} = 5$; $\sqrt{9} + \sqrt{16} = 3 + 4 = 7$ **85.** $5\sqrt{13}$ feet **87. a.** 13 feet **b.** $2\sqrt{14} \approx 7.5$ feet **91.** $7x$ **93.** $27xy^2$ **95.** $\frac{5}{6}x$ **97.** $3\sqrt{2}$ **99.** $5y\sqrt{3xy}$ **101.** $2a\sqrt[3]{ab^2}$ **103.** $\frac{y^3}{x^2}$ **105.** 1 **107.** $\frac{4x^2-6x+9}{9x^2-3x+1}$ **109.** $12\sqrt[3]{5}$ **111.** $6\sqrt[3]{49}$ **113.** $\frac{\sqrt[10]{a^7}}{a}$ **115.** $\frac{\sqrt[20]{a^9}}{a}$ **117.**

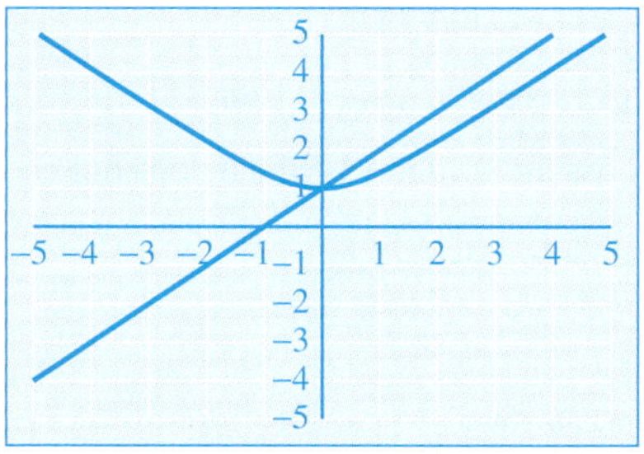

119. About $\frac{3}{4}$ of a unit **121.** $x = 0$

Problem Set 7.4

1. $7\sqrt{5}$ **3.** $-x\sqrt{7}$ **5.** $\sqrt[3]{10}$ **7.** $9\sqrt[5]{6}$ **9.** 0 **11.** $\sqrt{5}$ **13.** $-32\sqrt{2}$ **15.** $-3x\sqrt{2}$ **17.** $-2\sqrt[3]{2}$ **19.** $8x\sqrt[3]{xy^2}$ **21.** $3a^2b\sqrt{3ab}$ **23.** $11ab\sqrt[3]{3a^2b}$ **25.** $10xy\sqrt[4]{3y}$ **27.** $\sqrt{2}$ **29.** $\frac{8\sqrt{5}}{15}$ **31.** $\frac{(x-1)\sqrt{x}}{x}$ **33.** $\frac{3\sqrt{2}}{2}$ **35.** $\frac{5\sqrt{6}}{6}$ **37.** $\frac{8\sqrt[3]{25}}{5}$ **39.** $\sqrt{12} \approx 3.464$; $2\sqrt{3} = 2(1.732) \approx 3.464$ **41.** $\sqrt{8} + \sqrt{18} \approx 2.828 + 4.243 = 7.071$; $\sqrt{50} \approx 7.071$; $\sqrt{26} \approx 5.099$ **47.** $\frac{\sqrt{3}}{2}$ **49.** 6 **51.** $4x^2 + 3xy - y^2$ **53.** $x^2 + 6x + 9$ **55.** $x^2 - 4$ **57.** $6\sqrt{2}$ **59.** 6 **61.** $9x$ **63.** $\frac{\sqrt{6}}{2}$ **65.** 1 **67.** $\frac{13-3t}{3-t}$ **69.** $\frac{6}{4x+3}$ **71.** $\frac{x-y}{x^2+xy+y^2}$

Problem Set 7.5

1. $3\sqrt{2}$ **3.** $10\sqrt{21}$ **5.** 720 **7.** 54 **9.** $\sqrt{6} - 9$ **11.** $24 + 6\sqrt[3]{4}$ **13.** $7 + 2\sqrt{6}$ **15.** $x + 2\sqrt{x} - 15$ **17.** $34 + 20\sqrt{3}$ **19.** $19 + 8\sqrt{3}$ **21.** $x - 6\sqrt{x} + 9$ **23.** $4a - 12\sqrt{ab} + 9b$ **25.** $x + 4\sqrt{x-4}$ **27.** $x - 6\sqrt{x-5} + 4$ **29.** 1 **31.** $a - 49$ **33.** $25 - x$ **35.** $x - 8$ **37.** $10 + 6\sqrt{3}$ **39.** $\frac{\sqrt{3}+1}{2}$ **41.** $\frac{5-\sqrt{5}}{4}$ **43.** $\frac{x+3\sqrt{x}}{x-9}$ **45.** $\frac{10+3\sqrt{5}}{11}$ **47.** $\frac{3\sqrt{x}+3\sqrt{y}}{x-y}$ **49.** $2 + \sqrt{3}$ **51.** $\frac{11-4\sqrt{7}}{3}$ **53.** $\frac{a+2\sqrt{ab}+b}{a-b}$ **55.** $\frac{x+4\sqrt{x}+4}{x-4}$ **57.** $\frac{5-\sqrt{21}}{4}$ **59.** $\frac{\sqrt{x}-3x+2}{1-x}$ **63.** $10\sqrt{3}$ **65.** $x + 6\sqrt{x} + 9$ **67.** 75 **69.** $\frac{5\sqrt{2}}{4}$ seconds; $\frac{5}{2}$ seconds **75.** $t^2 + 10t + 25$ **77.** x **79.** 7 **81.** $-4, -3$ **83.** $-6, -3$ **85.** $-5, -2$ **87.** Yes **89.** No **91.** $-\frac{1}{8}$ **93.** $\frac{y-2}{y+2}$ **95.** $\frac{2x+1}{2x-1}$

Problem Set 7.6

1. 4 **3.** ∅ **5.** 5 **7.** ∅ **9.** $\frac{39}{2}$ **11.** ∅ **13.** 5 **15.** 3 **17.** $-\frac{32}{3}$ **19.** 3, 4 **21.** $-1, -2$ **23.** -1 **25.** ∅ **27.** 7 **29.** 0, 3 **31.** -4 **33.** 8 **35.** 0 **37.** 9 **39.** 0 **41.** 8 **43.** Possible solution 9, which does not check; ∅ **45.** Possible solutions 0 and 32; only 0 checks; 0 **47.** Possible solutions -2 and 6; only 6 checks; 6

49.
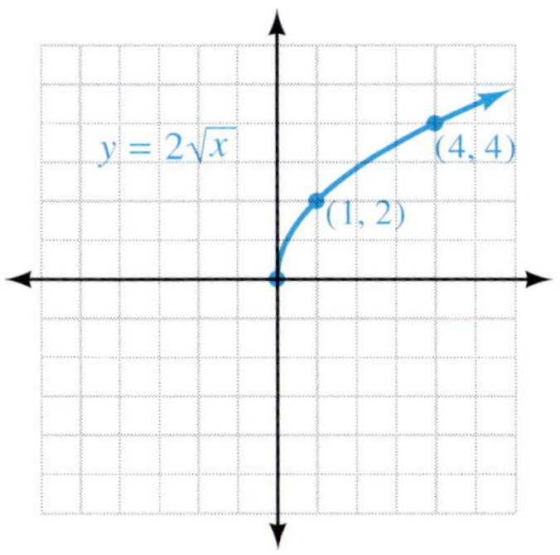

51.
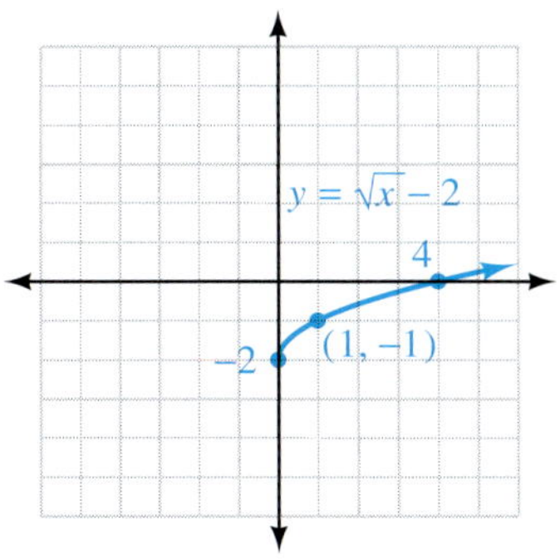

53.
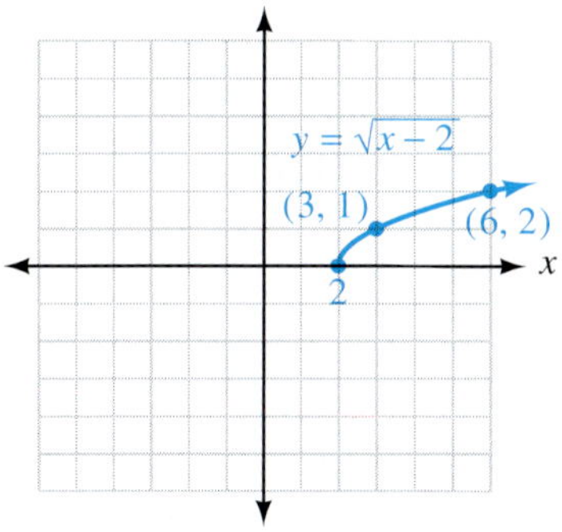

55.
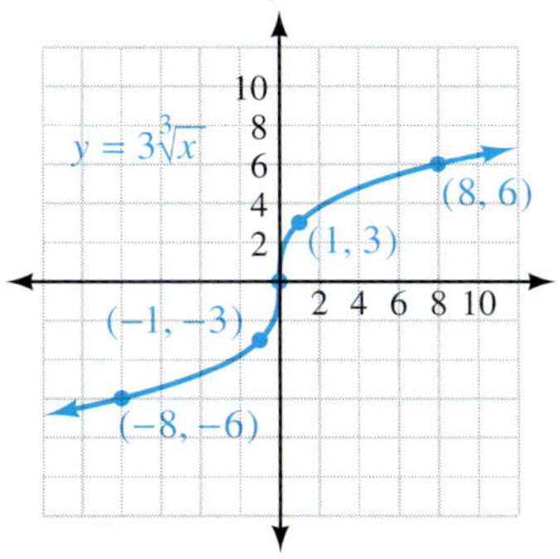

57.
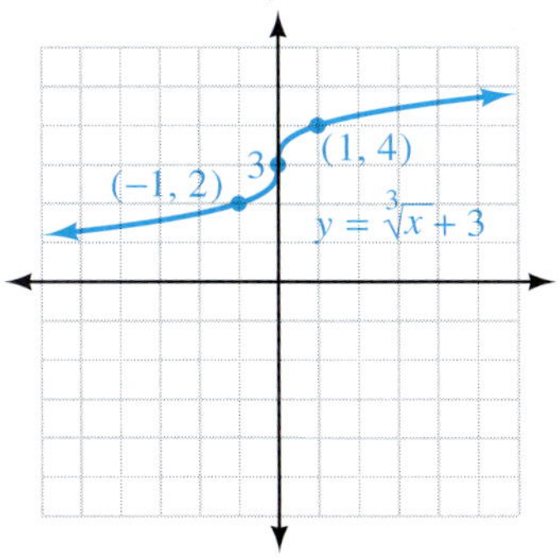

59.
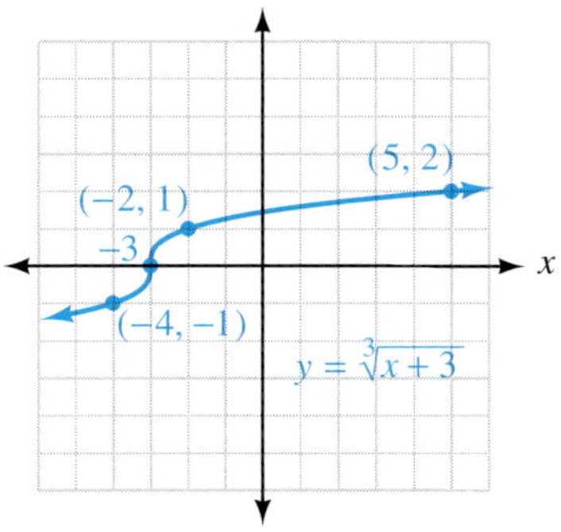

61. $h = 100 - 16t^2$ **63.** $\frac{392}{121} \approx 3.24$ feet **65.** 5 meters **67.** 2,500 meters **69.** 5 **71.** $2\sqrt{3}$ **73.** -1 **75.** 1 **77.** 4 **79.** 2 **81.** $10 - 2x$ **83.** $2 - 3x$ **85.** $6 + 7x - 20x^2$ **87.** $8x - 12x^2$ **89.** $4 + 12x + 9x^2$ **91.** $4 - 9x^2$ **93.** $\sqrt{6} - 2$ **95.** $x + 10\sqrt{x} + 25$ **97.** $\dfrac{x - 3\sqrt{x}}{x - 9}$

Problem Set 7.7

1. $6i$ **3.** $-5i$ **5.** $6i\sqrt{2}$ **7.** $-2i\sqrt{3}$ **9.** 1 **11.** -1 **13.** $-i$ **15.** $x = 3, y = -1$ **17.** $x = -2, y = -\frac{1}{2}$ **19.** $x = -8, y = -5$ **21.** $x = 7, y = \frac{1}{2}$ **23.** $x = \frac{3}{7}, y = \frac{2}{5}$ **25.** $5 + 9i$ **27.** $5 - i$ **29.** $2 - 4i$ **31.** $1 - 6i$ **33.** $2 + 2i$ **35.** $-1 - 7i$ **37.** $6 + 8i$ **39.** $2 - 24i$ **41.** $-15 + 12i$ **43.** $18 + 24i$ **45.** $10 + 11i$ **47.** $21 + 23i$ **49.** $-2 + 2i$ **51.** $2 - 11i$ **53.** $-21 + 20i$ **55.** $-2i$ **57.** $-7 - 24i$ **59.** 5 **61.** 40 **63.** 13 **65.** 164 **67.** $-3 - 2i$ **69.** $-2 + 5i$ **71.** $\frac{8}{13} + \frac{12}{13}i$ **73.** $-\frac{18}{13} - \frac{12}{13}i$ **75.** $-\frac{5}{13} + \frac{12}{13}i$ **77.** $\frac{13}{15} - \frac{2}{5}i$ **79.** $R = -11 - 7i$ ohms **81.** $-\frac{3}{2}$ **83.** $-3, \frac{1}{2}$ **85.** $\frac{5}{4}$ or $\frac{4}{5}$

Chapter 7 Review/Test

1. 7 **2.** -3 **3.** 2 **4.** 27 **5.** $2x^3y^2$ **6.** $\frac{1}{16}$ **7.** x^2 **8.** a^2b^4 **9.** $a^{7/20}$ **10.** $a^{5/12}b^{8/3}$ **11.** $12x + 11x^{1/2}y^{1/2} - 15y$ **12.** $a^{2/3} - 10a^{1/3} + 25$ **13.** $4x^{1/2} + 2x^{5/6}$ **14.** $2(x-3)^{1/4}(4x-13)$ **15.** $\dfrac{x+5}{x^{1/4}}$ **16.** $2\sqrt{3}$ **17.** $5\sqrt{2}$ **18.** $2\sqrt[3]{2}$ **19.** $3x\sqrt{2}$ **20.** $4ab^2c\sqrt{5a}$ **21.** $2abc\sqrt[4]{2bc^2}$ **22.** $\dfrac{3\sqrt{2}}{2}$ **23.** $3\sqrt[3]{4}$ **24.** $\dfrac{4x\sqrt{21xy}}{7y}$ **25.** $\dfrac{2y\sqrt[3]{45x^2z^2}}{3z}$ **26.** $-2x\sqrt{6}$ **27.** $3\sqrt{3}$ **28.** $\dfrac{8\sqrt{5}}{5}$ **29.** $7\sqrt{2}$ **30.** $11a^2b\sqrt{3ab}$ **31.** $-4xy\sqrt[3]{xz^2}$ **32.** $\sqrt{6} - 4$ **33.** $x - 5\sqrt{x} + 6$ **34.** $3\sqrt{5} + 6$ **35.** $6 + \sqrt{35}$ **36.** $\dfrac{63 + 12\sqrt{7}}{47}$ **37.** 0 **38.** 3 **39.** 5 **40.** $\varnothing$

41.

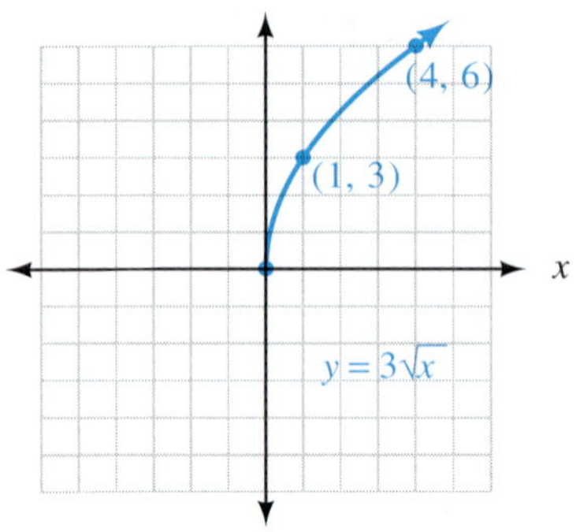

42.

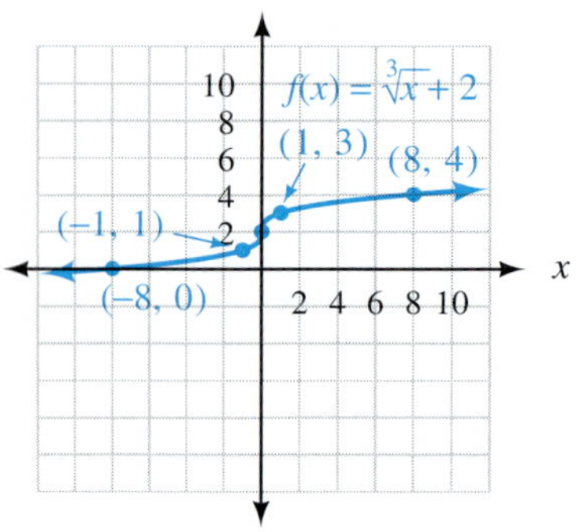

43. 1 **44.** $-i$ **45.** $x = -\frac{3}{2}, y = -\frac{1}{2}$ **46.** $x = -2, y = -4$ **47.** $9 + 3i$ **48.** $-7 + 4i$ **49.** $-6 + 12i$ **50.** $5 + 14i$
51. $12 + 16i$ **52.** 25 **53.** $1 - 3i$ **54.** $-\frac{6}{5} + \frac{3}{5}i$ **55.** 22.5 feet **56.** 65 yards

Chapter 8

Getting Ready for Chapter 8

1. 4 **2.** −1 **3.** $\frac{33}{4} = 8\frac{1}{4}$ **4.** $-\frac{5}{9}$ **5.** ±5 **6.** ±3 **7.** $\pm 2\sqrt{3}$ **8.** $\pm 3\sqrt{2}$ **9.** 9 **10.** 4 **11.** $\frac{25}{4}$ **12.** $\frac{49}{4}$

13. $\frac{16}{9}$ **14.** $\frac{1}{25}$ **15.** $(x - 3)^2$ **16.** $(x - 4)^2$ **17.** $(x + 6)^2$ **18.** $(5t + 1)(25t^2 - 5t + 1)$ **19.** $\frac{b^2 - 4ac}{4a^2}$ **20.** $9a^2 - 6a - 3$

Problem Set 8.1

1. ±5 **3.** $\pm\frac{\sqrt{3}}{2}$ **5.** $\pm 2i\sqrt{3}$ **7.** $\pm\frac{3\sqrt{5}}{2}$ **9.** −2, 3 **11.** $\frac{-3 \pm 3i}{2}$ **13.** $-4 \pm 3i\sqrt{3}$ **15.** $\frac{3 \pm 2i}{2}$ **17.** ±1
19. $x^2 + 12x + 36 = (x + 6)^2$ **21.** $x^2 - 4x + 4 = (x - 2)^2$ **23.** $a^2 - 10a + 25 = (a - 5)^2$ **25.** $x^2 + 5x + \frac{25}{4} = (x + \frac{5}{2})^2$

27. $y^2 - 7y + \frac{49}{4} = (y - \frac{7}{2})^2$ **29.** −6, 2 **31.** −3, −9 **33.** $1 \pm 2i$ **35.** $4 \pm \sqrt{15}$ **37.** $\frac{5 \pm \sqrt{37}}{2}$ **39.** $1 \pm \sqrt{5}$ **41.** $\frac{4 \pm \sqrt{13}}{3}$

43. $\frac{3 \pm i\sqrt{71}}{8}$ **45.** $\frac{\sqrt{3}}{2}$ inch, 1 inch **47.** $\sqrt{2}$ inches **49.** 781 feet **51.** 7.3% to the nearest tenth **53.** $20\sqrt{2} \approx 28$ feet tenth

55. 13 **57.** 7 **59.** $\frac{1}{2}$ **61.** 13 **63.** $(3t - 2)(9t^2 + 6t + 4)$ **65.** $3\sqrt{5}$ **67.** $3y^2\sqrt{3y}$ **69.** $3x^2y\sqrt[3]{2y^2}$ **71.** $\frac{3\sqrt{2}}{2}$ **73.** $\sqrt[3]{2}$

75. a. 10 **b.** 12

Problem Set 8.2

1. a. $\frac{-2 \pm \sqrt{10}}{3}$ **b.** $\frac{2 \pm \sqrt{10}}{3}$ **c.** $\frac{-2 \pm i\sqrt{2}}{3}$ **d.** $\frac{-2 \pm \sqrt{10}}{2}$ **e.** $\frac{2 \pm i\sqrt{2}}{2}$ **3. a.** $1 \pm i$ **b.** $1 \pm 2i$ **c.** $-1 \pm i$ **5.** 1, 2

7. $\frac{2 \pm i\sqrt{14}}{3}$ **9.** 0, 5 **11.** $0, -\frac{4}{3}$ **13.** $\frac{3 \pm \sqrt{5}}{4}$ **15.** $-3 \pm \sqrt{17}$ **17.** $\frac{-1 \pm i\sqrt{5}}{2}$ **19.** 1 **21.** $\frac{1 \pm i\sqrt{47}}{6}$ **23.** $4 \pm \sqrt{2}$

25. $\frac{1}{2}, 1$ **27.** $-\frac{1}{2}, 3$ **29.** $\frac{-1 \pm i\sqrt{7}}{2}$ **31.** $1 \pm \sqrt{2}$ **33.** $\frac{-3 \pm \sqrt{5}}{2}$ **35.** 3, −5 **37.** $2, -1 \pm i\sqrt{3}$ **39.** $-\frac{3}{2}, \frac{3 \pm 3i\sqrt{3}}{4}$

41. $\frac{1}{5}, \frac{-1 \pm i\sqrt{3}}{10}$ **43.** $0, \frac{-1 \pm i\sqrt{5}}{2}$ **45.** $0, 1 \pm i$ **47.** $0, \frac{-1 \pm i\sqrt{2}}{3}$ **49.** 2 seconds **51.** $40 \pm 20 = 20$ or 60 items

53. 0.49 centimeter (8.86 cm is impossible)
55. a. $\ell + w = 10$, $\ell w = 15$
b. 8.16 yards, 1.84 yards
c. Two answers are possible because either dimension (long or short) may be considered the length.

57. 169 **59.** 0 **61.** ±12 **63.** $x^2 - x - 6$ **65.** $x^3 - 4x^2 - 3x + 18$ **67.** $4y + 1 + \frac{-2}{2y - 7}$ **69.** $x^2 + 7x + 12$ **71.** 5

73. $\frac{27}{125}$ **75.** $\frac{1}{4}$ **77.** $21x^3y$ **79.** $-2\sqrt{3}, \sqrt{3}$ **81.** $\frac{-1 \pm \sqrt{3}}{\sqrt{2}} = \frac{-\sqrt{2} \pm \sqrt{6}}{2}$ **83.** $-2i, i$

Problem Set 8.3

1. $D = 16$, two rational **3.** $D = 0$, one rational **5.** $D = 5$, two irrational **7.** $D = 17$, two irrational **9.** $D = 36$, two rational
11. $D = 116$, two irrational **13.** ±10 **15.** ±12 **17.** 9 **19.** −16 **21.** $\pm 2\sqrt{6}$ **23.** $x^2 - 7x + 10 = 0$ **25.** $t^2 - 3t - 18 = 0$

27. $y^3 - 4y^2 - 4y + 16 = 0$ **29.** $2x^2 - 7x + 3 = 0$ **31.** $4t^2 - 9t - 9 = 0$ **33.** $6x^3 - 5x^2 - 54x + 45 = 0$ **35.** $10a^2 - a - 3 = 0$ **37.** $9x^3 - 9x^2 - 4x + 4 = 0$ **39.** $x^4 - 13x^2 + 36 = 0$ **41.** $x^2 + 4x - 5$ **43.** $4a^2 - 30a + 56$ **45.** $32a^2 + 20a - 18$ **47.** $\pm\frac{1}{2}$ **49.** no solution **51.** 1 **53.** $-2, 4$ **55.** $-2, \frac{1}{4}$ **57.** $a^{11/2} - a^{9/2}$ **59.** $x^3 - 6x^{3/2} + 9$ **61.** $6x^{1/2} - 5x$ **63.** $5(x-3)^{1/2}(2x-9)$ **65.** $(2x^{1/3} - 3)(x^{1/3} - 4)$ **67.** $-1, 5$ **69.** $1 + \sqrt{2} \approx 2.41; 1 - \sqrt{2} \approx -0.41$ **71.** $-1, \frac{1}{2}, 1$ **73.** $\frac{1}{2}, \sqrt{2} \approx 1.41, -\sqrt{2} \approx -1.41$

Problem Set 8.4

1. 1, 2 **3.** $-8, -\frac{5}{2}$ **5.** $\pm 3, \pm i\sqrt{3}$ **7.** $\pm 2i, \pm i\sqrt{5}$ **9.** $\frac{7}{2}, 4$ **11.** $-\frac{9}{8}, \frac{1}{2}$ **13.** $\pm\frac{\sqrt{30}}{6}, \pm i$ **15.** $\pm\frac{\sqrt{21}}{3}, \pm\frac{i\sqrt{21}}{3}$ **17.** 4, 25 **19.** Possible solutions 25 and 9; only 25 checks; 25 **21.** Possible solutions $\frac{25}{9}$ and $\frac{49}{4}$; only $\frac{25}{9}$ checks; $\frac{25}{9}$ **23.** 27, 38 **25.** 4, 12 **27.** $t = \frac{v \pm \sqrt{v^2 + 64h}}{32}$ **29.** $x = \frac{-4 \pm 2\sqrt{4-k}}{k}$ **31.** $x = -y$ **33.** $t = \frac{1 \pm \sqrt{1+h}}{4}$

35. a.

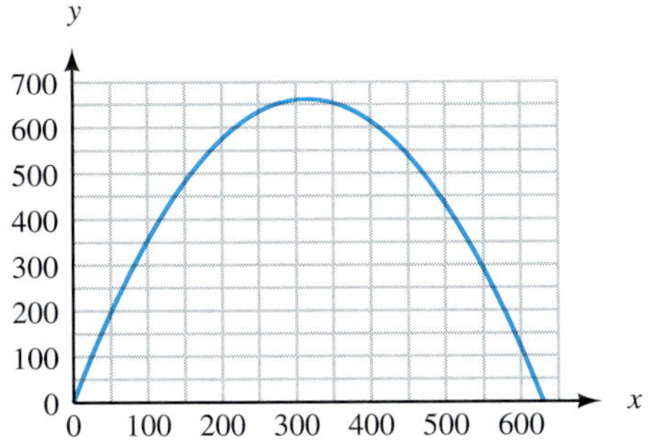

b. 630 feet

39. -2 **41.** 1,322.5 **43.** $-\frac{7}{640}$ **45.** 1, 5 **47.** $-3, 1$ **49.** $\frac{3}{2} \pm \frac{1}{2}i$ **51.** 9; 3 **53.** 1; 1 **55.** $3\sqrt{7}$ **57.** $5\sqrt{2}$ **59.** $39x^2y\sqrt{5x}$ **61.** $-11 + 6\sqrt{5}$ **63.** $x + 4\sqrt{x} + 4$ **65.** $\frac{7 + 2\sqrt{7}}{3}$ **67.** x-intercepts $= -2, 0, 2$; y-intercept $= 0$ **69.** x-intercepts $= -3, -\frac{1}{3}, 3$; y-intercept $= -9$ **71.** $\frac{1}{2}$ and -1

Problem Set 8.5

1. x-intercepts $= -3, 1$;
Vertex $= (-1, -4)$

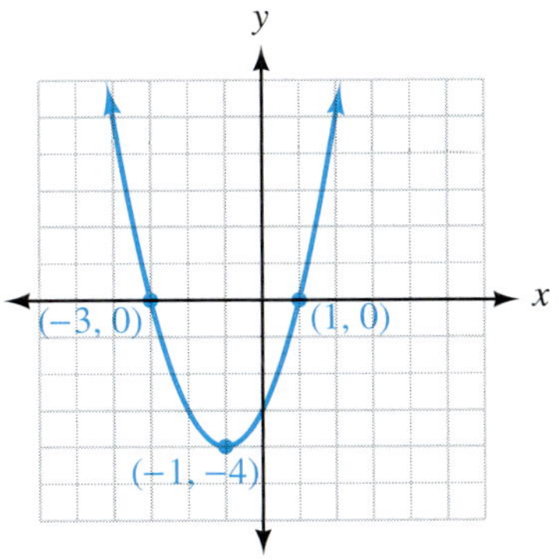

3. x-intercepts $= -5, 1$;
Vertex $= (-2, 9)$

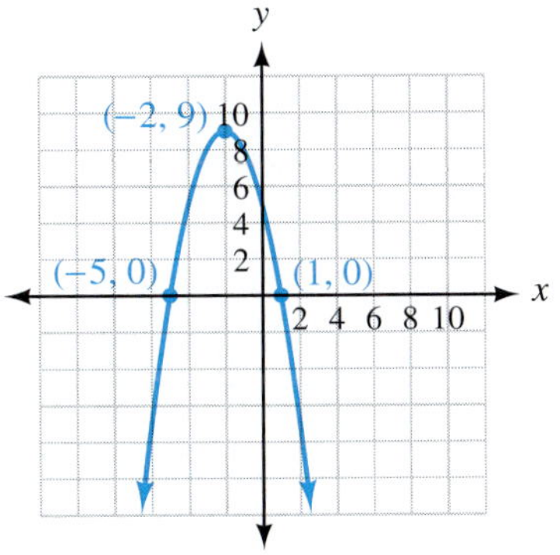

5. x-intercepts $= -1, 1$;
Vertex $= (0, -1)$

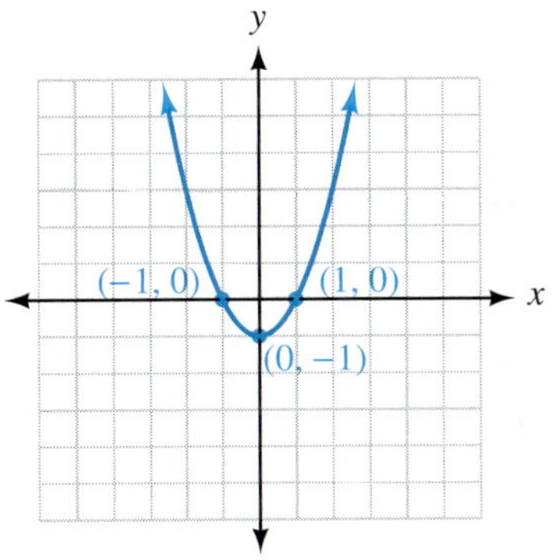

7. x-intercepts $= 3, -3$;
Vertex $= (0, 9)$

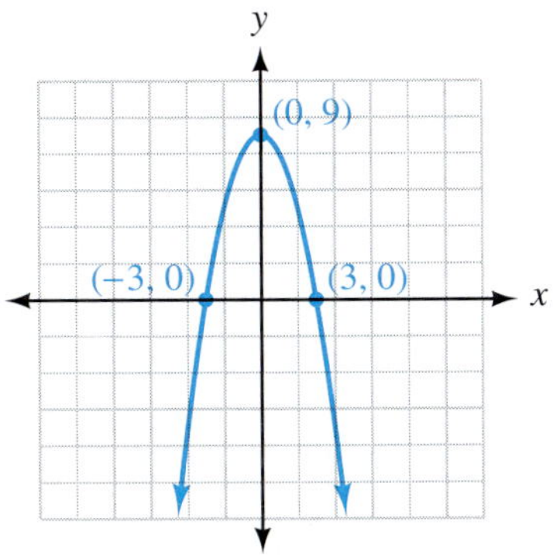

9. x-intercepts $= -1, 3$;
Vertex $= (1, -8)$

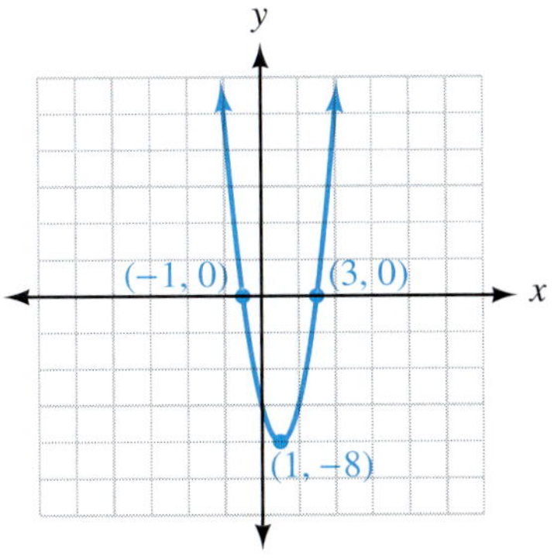

11. x-intercepts $= 1 + \sqrt{5}, 1 - \sqrt{5}$;
Vertex $= (1, -5)$

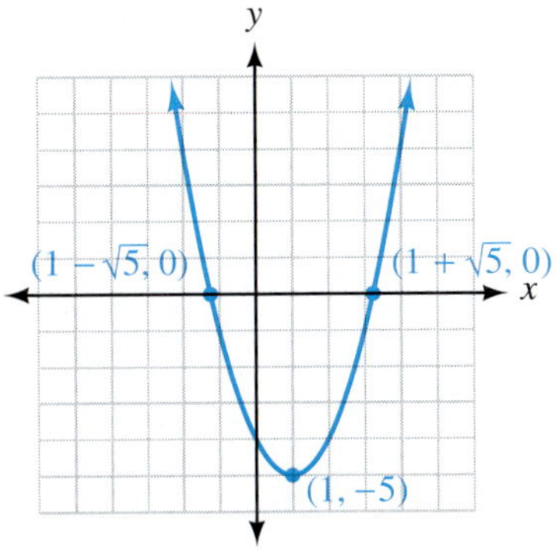

13. Vertex = (2, −8)

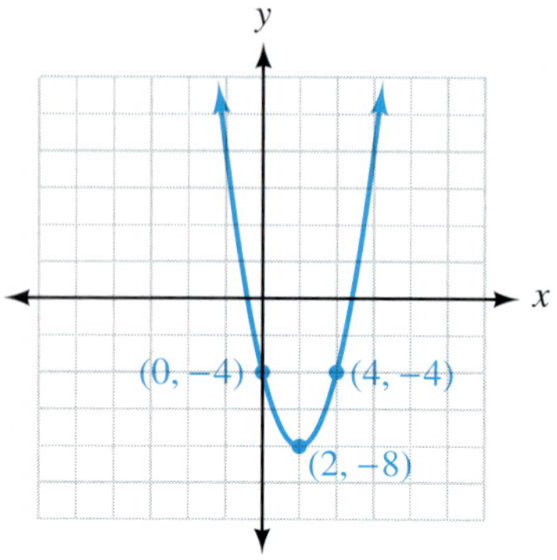

15. Vertex = (1, −4)

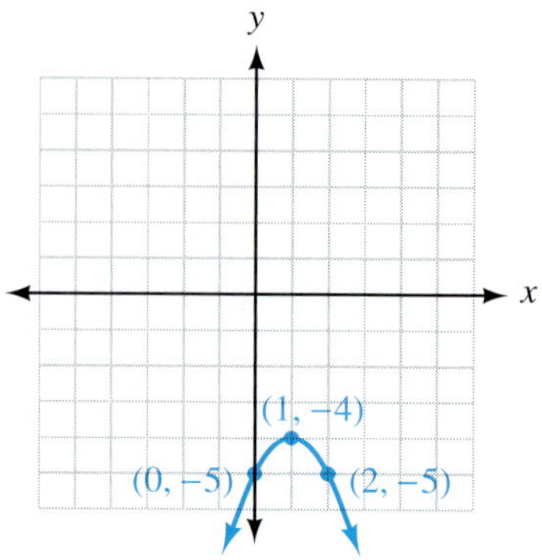

17. Vertex = (0, 1)

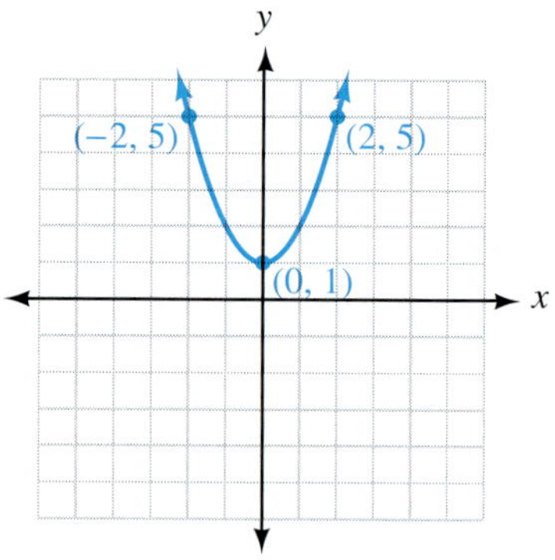

19. Vertex = (0, −3)

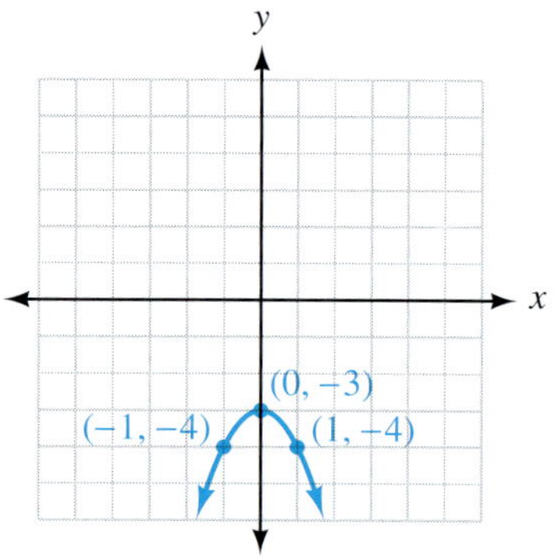

21. (3, −4) lowest **23.** (1, 9) highest

25. (2, 16) highest **27.** (−4, 16) highest **29.** 875 patterns; maximum profit \$731.25
31. The ball is in her hand when $h(t) = 0$, which means $t = 0$ or $t = 2$ seconds. Maximum height is $h(1) = 16$ feet.
33. Maximum $R = \$3,600$ when $p = \$6.00$ **35.** Maximum $R = \$7,225$ when $p = \$8.50$

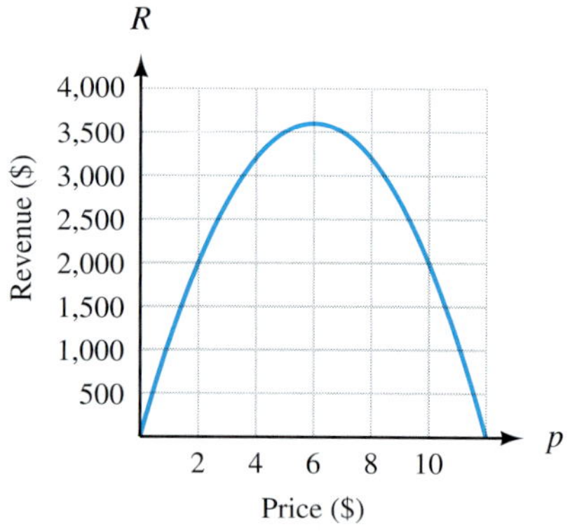

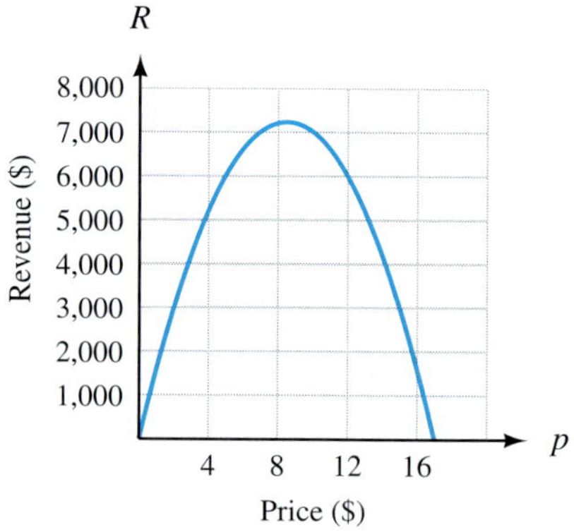

37. $y = -\frac{1}{135}(x - 90)^2 + 60$

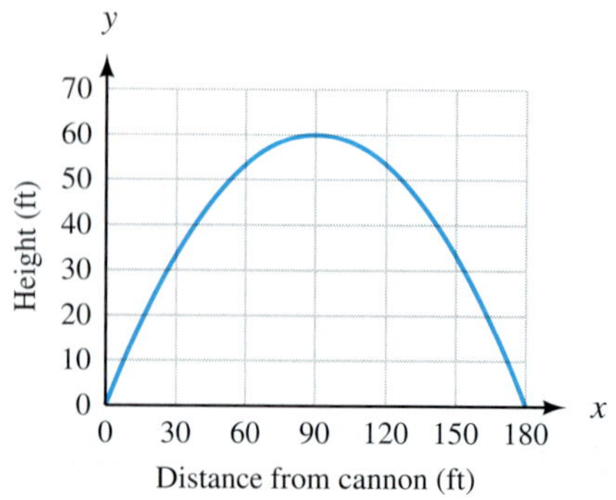

39. −2, 4 **41.** $-\frac{1}{2}, \frac{2}{3}$ **43.** 3 **45.** $1 - i$ **47.** $27 + 5i$ **49.** $\frac{1}{10} + \frac{3}{10}i$ **51.** $y = (x - 2)^2 - 4$
53. **a.** 20° or 70° **b.** At approximately 180 feet **c.** Approximately 42 or 43 feet **d.** Approximately 10° **e.** Quadratic

Problem Set 8.6

1. 

3. 

5. 

7. 

9. 

11. 

13. 

15. All real numbers **17.** No solution; ∅ **19.** 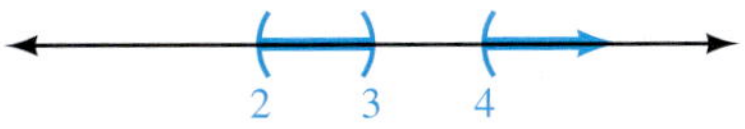

21. 

23. a. $-2 < x < 2$ **b.** $x < -2$ or $x > 2$ **c.** $x = -2$ or $x = 2$

25. a. $-2 < x < 5$ **b.** $x < -2$ or $x > 5$ **c.** $x = -2$ or $x = 5$
27. a. $x < -1$ or $1 < x < 3$ **b.** $-1 < x < 1$ or $x > 3$ **c.** $x = -1$ or $x = 1$ or $x = 3$ **29.** $x \geq 4$; the width is at least 4 inches
31. $5 \leq p \leq 8$; charge at least \$5 but no more than \$8 for each radio **33.** 1.5625 **35.** 0.6549 **37.** $\frac{5}{3}$
39. Possible solutions 1 and 6; only 6 checks; 6

41. 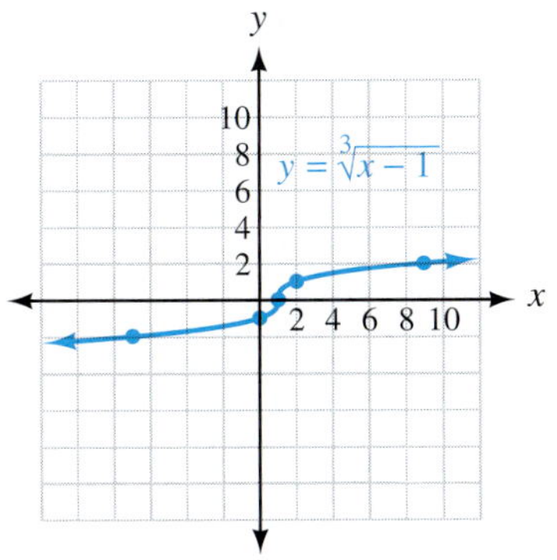

43. 

45. 

Chapter 8 Review/Test

1. 0, 5 **2.** 0, $\frac{4}{3}$ **3.** $\frac{4 \pm 7i}{3}$ **4.** $-3 \pm \sqrt{3}$ **5.** −5, 2 **6.** −3, −2 **7.** 3 **8.** 2 **9.** $\frac{-3 \pm \sqrt{3}}{2}$ **10.** $\frac{3 \pm \sqrt{5}}{2}$ **11.** −5, 2

12. 0, $\frac{9}{4}$ **13.** $\frac{2 \pm i\sqrt{15}}{2}$ **14.** $1 \pm \sqrt{2}$ **15.** $-\frac{4}{3}, \frac{1}{2}$ **16.** 3 **17.** $5 \pm \sqrt{7}$ **18.** 0, $\frac{5 \pm \sqrt{21}}{2}$ **19.** $-1, \frac{3}{5}$ **20.** 3, $\frac{-3 \pm 3i\sqrt{3}}{2}$

21. $\frac{1 \pm i\sqrt{2}}{3}$ **22.** $\frac{3 \pm \sqrt{29}}{2}$ **23.** 100 or 170 items **24.** 20 or 60 items **25.** $D = 0$; 1 rational **26.** $D = 0$; 1 rational

27. $D = 25$; 2 rational **28.** $D = 361$; 2 rational **29.** $D = 5$; 2 irrational **30.** $D = 5$; 2 irrational **31.** $D = -23$; 2 complex
32. $D = -87$; 2 complex **33.** ±20 **34.** ±20 **35.** 4 **36.** 4 **37.** 25 **38.** 49 **39.** $x^2 - 8x + 15 = 0$ **40.** $x^2 - 2x - 8 = 0$
41. $2y^2 + 7y - 4 = 0$ **42.** $t^3 - 5t^2 - 9t + 45 = 0$ **43.** −4, 12 **44.** $-\frac{3}{4}, -\frac{1}{6}$ **45.** $\pm 2, \pm i\sqrt{3}$

46. Possible solutions 4 and 1, only 4 checks; 4 **47.** $\frac{9}{4}$, 16 **48.** 4 **49.** 4 **50.** 7 **51.** $t = \frac{5 \pm \sqrt{25 + 16h}}{16}$

52. $t = \frac{v \pm \sqrt{v^2 + 640}}{32}$ **53.** 

54. 

55. 

56. 

57.

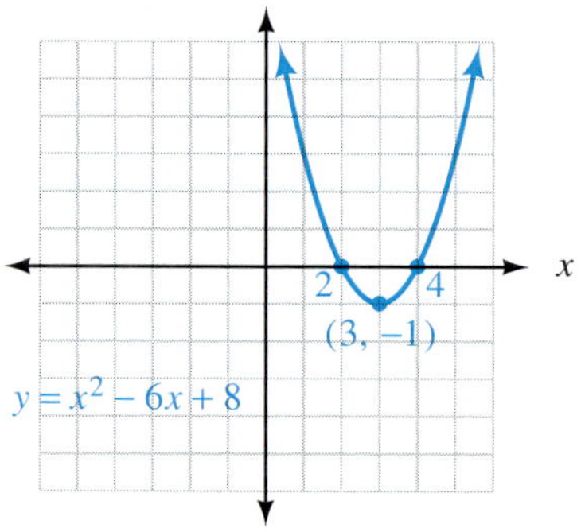

58.

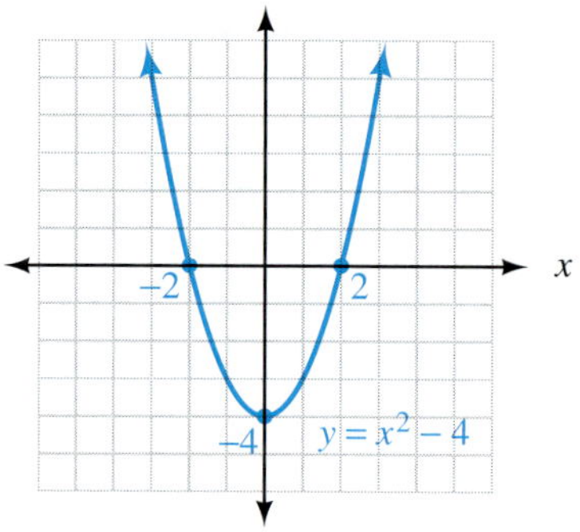

Chapter 9

Getting Ready for Chapter 9

1. 1 **2.** 4 **3.** $\frac{1}{4}$ **4.** 9 **5.** 1 **6.** $\frac{1}{9}$ **7.** 9 **8.** 0.4204 **9.** 0.1768 **10.** 504.5378 **11.** 265.1650 **12.** 742.9737 **13.** 905.6808 **14.** 1.0075 **15.** 1.02 **16.** −4, 1 **17.** $\frac{101}{24}$ **18.** $y = \frac{x+3}{2}$ **19.** $y = \pm\sqrt{x+2}$ **20.** $-\frac{6}{5}, -2$

Problem Set 9.1

1. 1 **3.** 2 **5.** $\frac{1}{27}$ **7.** 13

9.

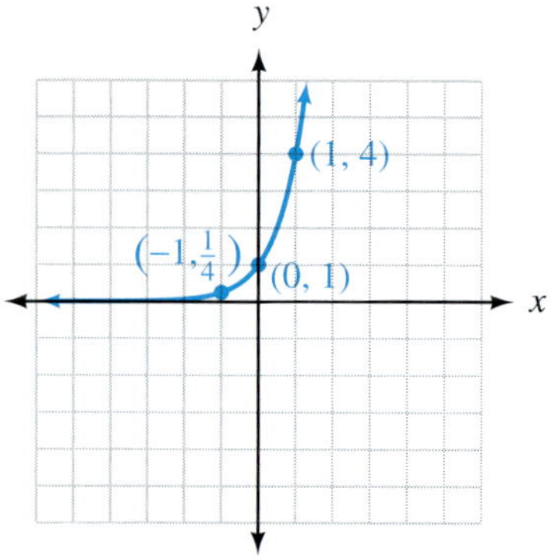

11.

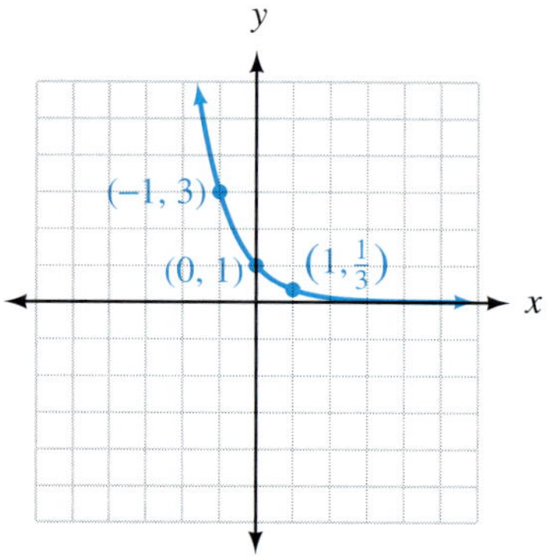

13.

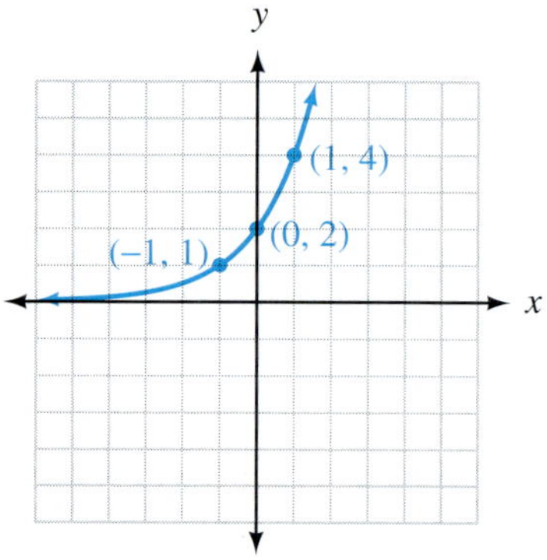

15.

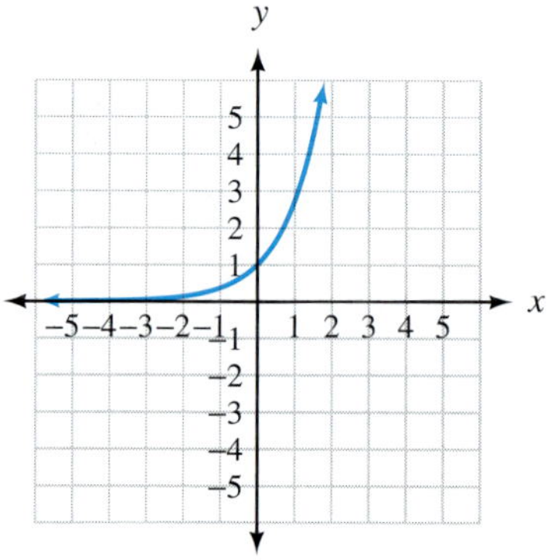

17.

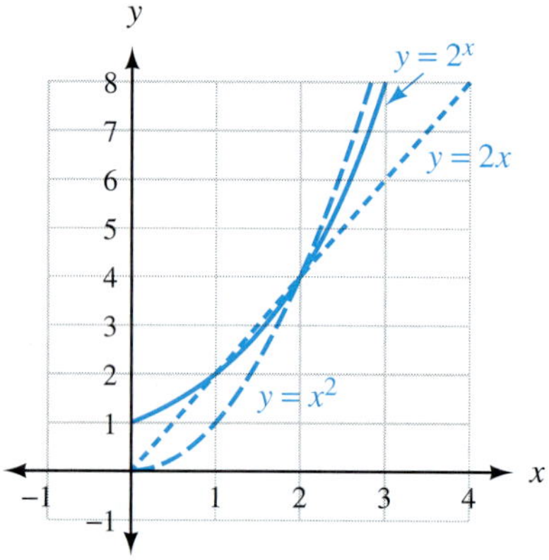

19.

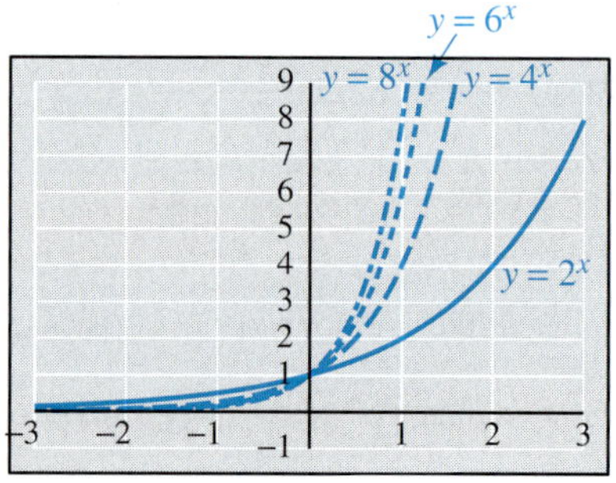

21. $h = 6 \cdot (\frac{2}{3})^n$; 5th bounce: $6(\frac{2}{3})^5 \approx 0.79$ feet **23.** After 8 days, 700 micrograms; After 11 days, $1{,}400 \cdot 2^{-11/8} \approx 539.8$ micrograms

25. **a.** $A(t) = 1{,}200\left(1 + \frac{.06}{4}\right)^{4t}$ **b.** \$1,932.39 **c.** About 11.64 years **d.** \$1,939.29

27. **a.** The function underestimated the expenditures by \$69 billion.
b. \$3,015 billion
\$3,347 billion
\$3,715 billion

29.

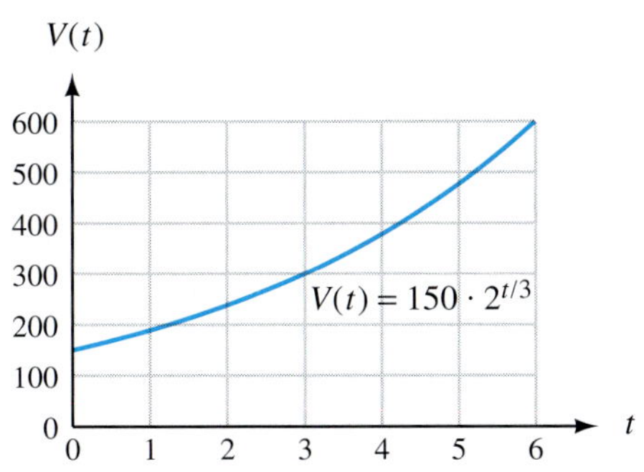

31. After 6 years

33. a. \$129,138.48 **b.** $\{t \mid 0 \le t \le 6\}$

c.

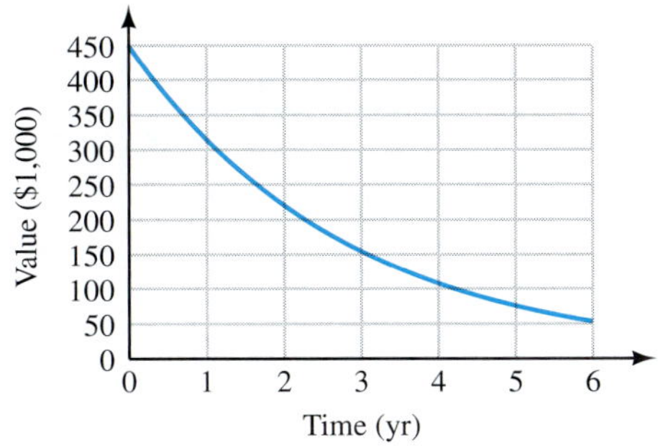

d. $\{V(t) \mid 52{,}942.05 \le V(t) \le 450{,}000\}$
e. After approximately 4 years 8 months

35. $y = \dfrac{x+3}{2}$ **37.** $y = \pm\sqrt{x+2}$ **39.** $y = \dfrac{2x-4}{x-1}$ **41.** $y = x^2 + 3$

43. $D = \{1, 3, 4\}$, $R = \{2, 4, 1\}$, a function **45.** $\{x \mid x \ge -\frac{1}{3}\}$ **47.** -18 **49.** 2

51.

t	$s(t)$
0	0
1	83.3
2	138.9
3	175.9
4	200.6
5	217.1
6	228.1

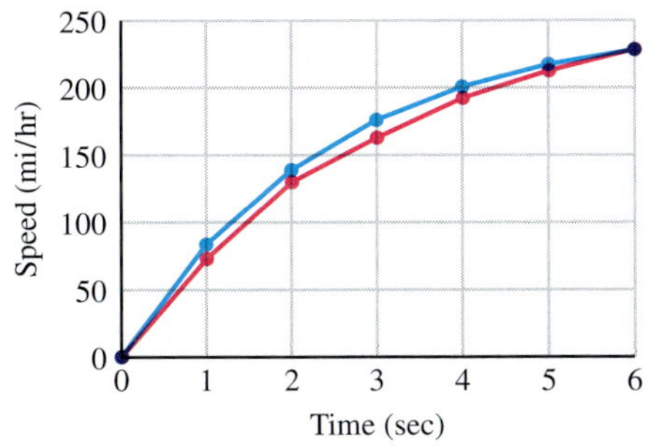

Problem Set 9.2

1. $f^{-1}(x) = \dfrac{x+1}{3}$ **3.** $f^{-1}(x) = \sqrt[3]{x}$ **5.** $f^{-1}(x) = \dfrac{x-3}{x-1}$ **7.** $f^{-1}(x) = 4x + 3$ **9.** $f^{-1}(x) = 2(x+3) = 2x + 6$

11. $f^{-1}(x) = \dfrac{1 - x}{3x - 2}$

13.

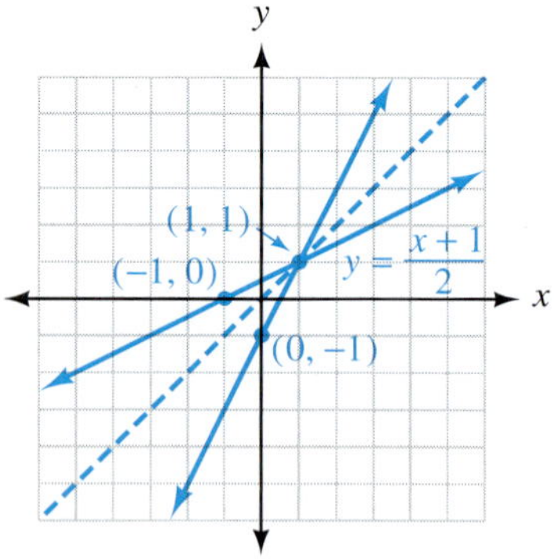

15.

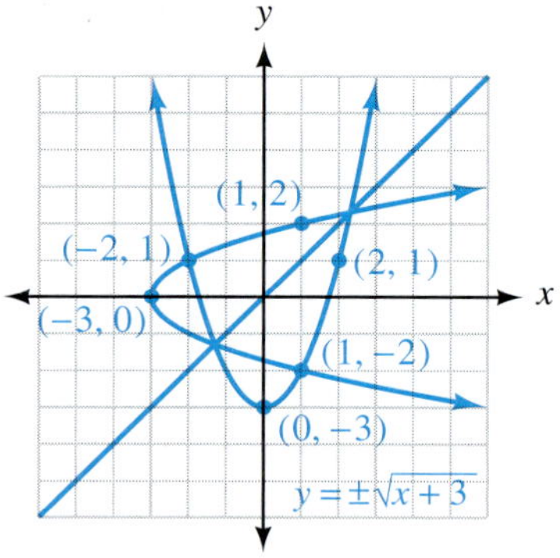

17.

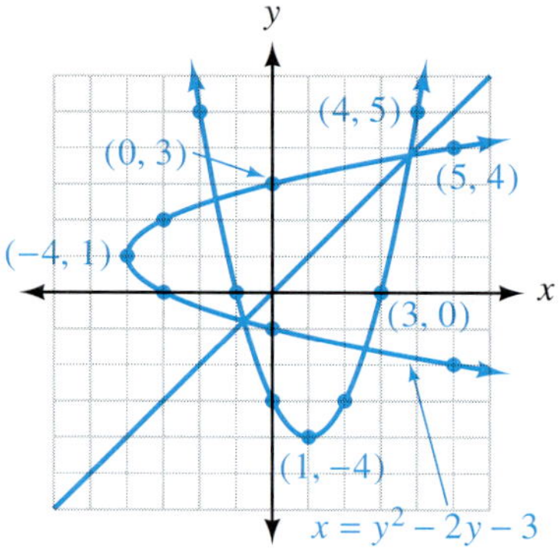

19.

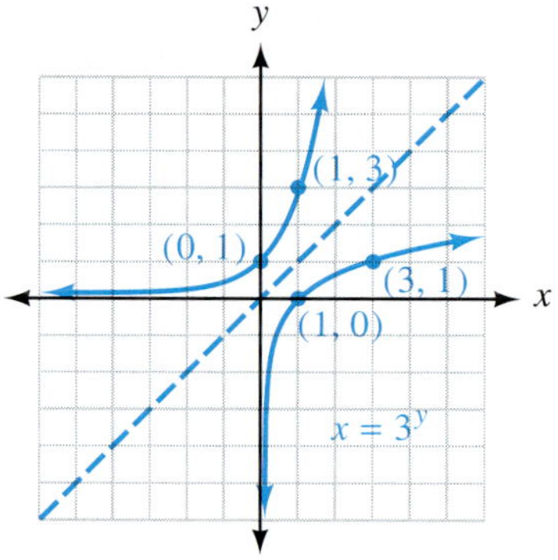

21.

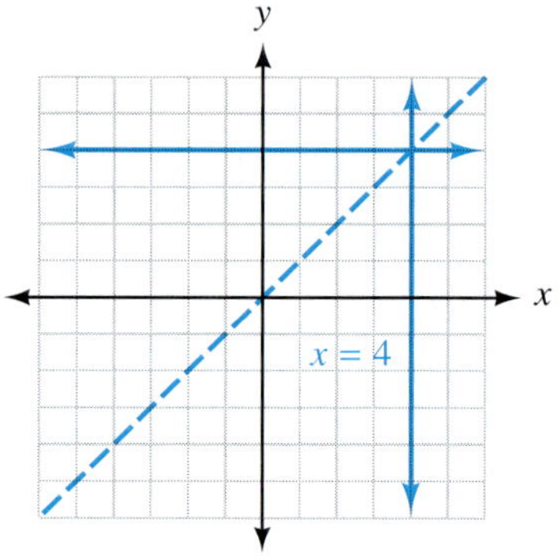

23.

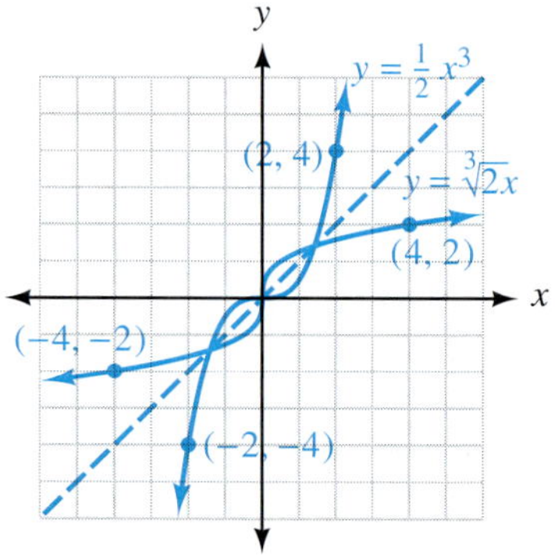

25.

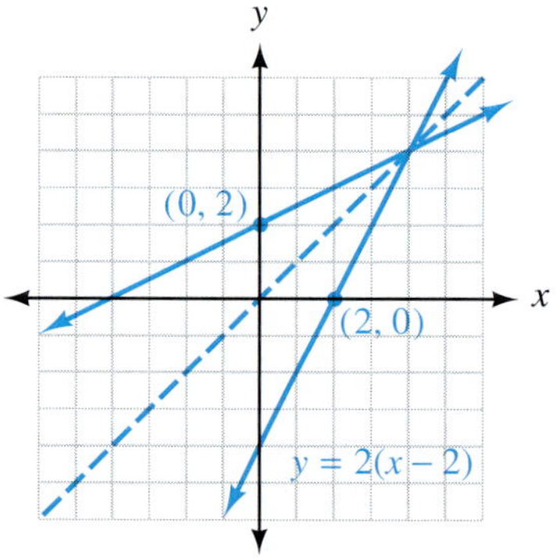

27.

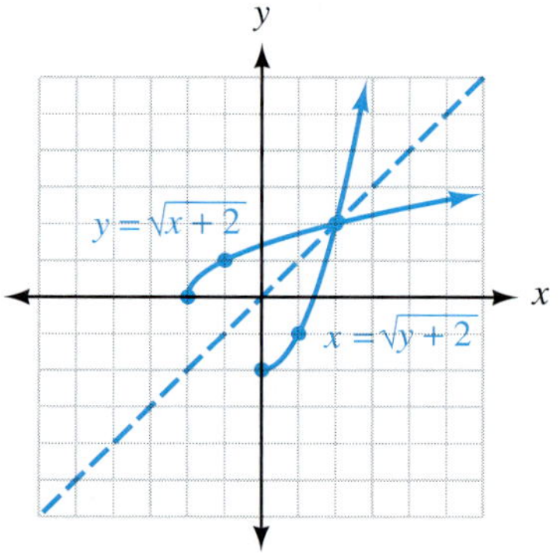

29. a. Yes **b.** No **c.** Yes **31. a.** 4 **b.** $\frac{4}{3}$ **c.** 2 **d.** 2 **33.** $f^{-1}(x) = \frac{1}{x}$
35. a. −3 **b.** −6 **c.** 2 **d.** 3 **e.** −2 **f.** 3 **g.** They are inverses of each other. **37.** $\frac{1}{9}$ **39.** $\frac{2}{3}$ **41.** $\sqrt[3]{4}$ **43.** 3 **45.** 4
47. 4 **49.** 1 **51.** −2, 3 **53.** 25 **55.** $5 \pm \sqrt{17}$ **57.** $1 \pm i$ **59.** $f^{-1}(x) = \dfrac{x - 5}{3}$ **61.** $f^{-1}(x) = \sqrt[3]{x - 1}$ **63.** $f^{-1}(x) = \dfrac{2x - 4}{x - 1}$
65. a. 1 **b.** 2 **c.** 5 **d.** 0 **e.** 1 **f.** 2 **g.** 2 **h.** 5

Problem Set 9.3

1. $\log_2 16 = 4$ **3.** $\log_5 125 = 3$ **5.** $\log_{10} 0.01 = -2$ **7.** $\log_2 \frac{1}{32} = -5$ **9.** $\log_{1/2} 8 = -3$ **11.** $\log_3 27 = 3$ **13.** $10^2 = 100$
15. $2^6 = 64$ **17.** $8^0 = 1$ **19.** $10^{-3} = 0.001$ **21.** $6^2 = 36$ **23.** $5^{-2} = \frac{1}{25}$ **25.** 9 **27.** $\frac{1}{125}$ **29.** 4 **31.** $\frac{1}{3}$ **33.** 2 **35.** $\sqrt[3]{5}$

37.

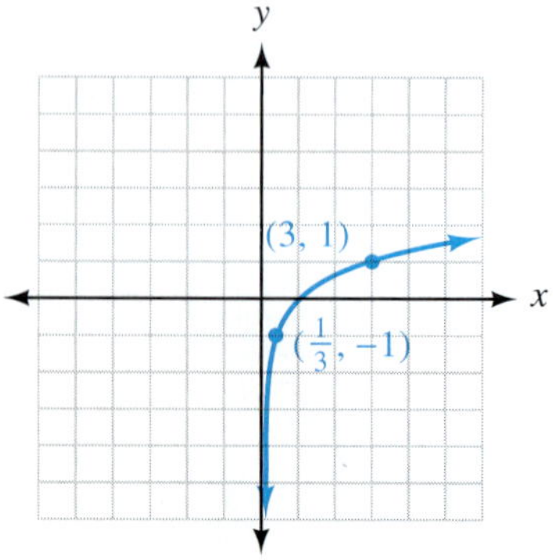

39.

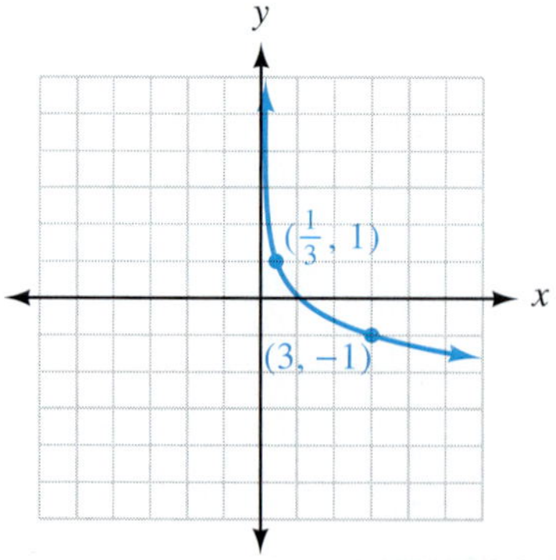

41.

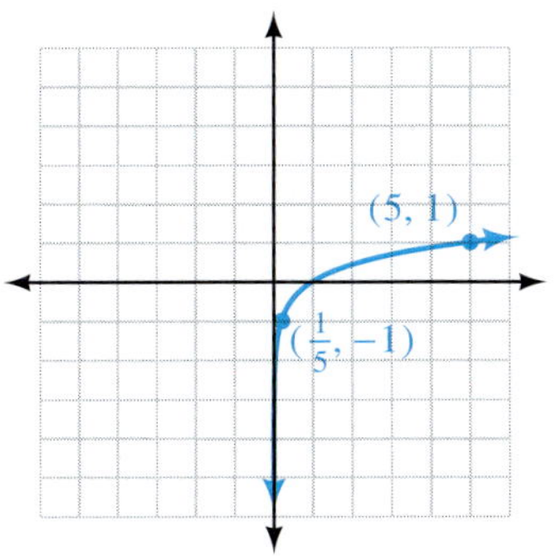

43.

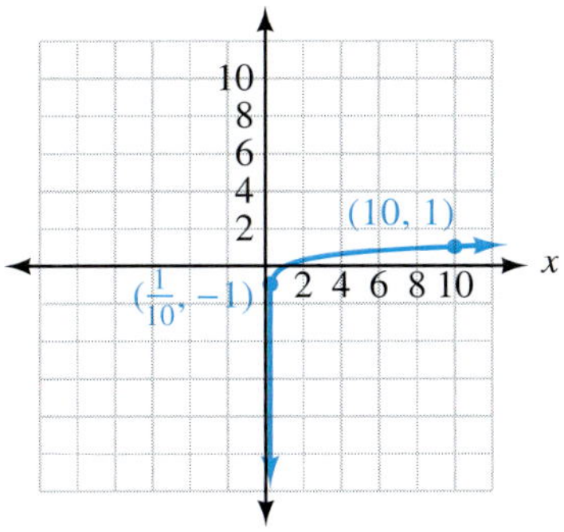

45. 4 **47.** $\frac{3}{2}$ **49.** 3 **51.** 1 **53.** 0

55. 0 **57.** $\frac{1}{2}$ **59.** 7 **61.** 10^{-6} **63.** 2 **65.** 10^8 times as large **67.** 4 **69.** $-4, 2$ **71.** $-\frac{11}{8}$ **73.** $2^3 = (x + 2)(x)$

75. $3^4 = \dfrac{x-2}{x+1}$ **77.** 25; 5 **79.** 1; 1 **81.** $\pm 3i$ **83.** $\pm 2i$ **85.** $\dfrac{-2 \pm \sqrt{10}}{2}$ **87.** $\dfrac{2 \pm i\sqrt{3}}{2}$ **89.** 5, $\dfrac{-5 \pm 5i\sqrt{3}}{2}$

91. 0, 5, −1 **93.** $\dfrac{3 \pm \sqrt{29}}{2}$ **95. a.**

x	$f(x)$
−1	$\frac{1}{8}$
0	1
1	8
2	64

b.

x	$f^{-1}(x)$
$\frac{1}{8}$	−1
1	0
8	1
64	2

c. $f(x) = 8^x$ **d.** $f^{-1}(x) = \log_8 x$

Problem Set 9.4

1. $\log_3 4 + \log_3 x$ **3.** $\log_6 5 - \log_6 x$ **5.** $5 \log_2 y$ **7.** $\frac{1}{3}\log_9 z$ **9.** $2\log_6 x + 4\log_6 y$ **11.** $\frac{1}{2}\log_5 x + 4\log_5 y$
13. $\log_b x + \log_b y - \log_b z$ **15.** $\log_{10} 4 - \log_{10} x - \log_{10} y$ **17.** $2\log_{10} x + \log_{10} y - \frac{1}{2}\log_{10} z$ **19.** $3\log_{10} x + \frac{1}{2}\log_{10} y - 4\log_{10} z$
21. $\frac{2}{3}\log_b x + \frac{1}{3}\log_b y - \frac{4}{3}\log_b z$ **23.** $\log_b xz$ **25.** $\log_3 \dfrac{x^2}{y^3}$ **27.** $\log_{10} \sqrt{x}\sqrt[3]{y}$ **29.** $\log_2 \dfrac{x^3\sqrt{y}}{z}$ **31.** $\log_2 \dfrac{\sqrt{x}}{y^3z^4}$
33. $\log_{10} \dfrac{x^{3/2}}{y^{3/4}z^{4/5}}$ **35.** $\frac{2}{3}$ **37.** 18 **39.** Possible solutions −1 and 3; only 3 checks; 3 **41.** 3
43. Possible solutions −2 and 4; only 4 checks; 4 **45.** Possible solutions −1 and 4; only 4 checks; 4
47. Possible solutions $-\frac{5}{2}$ and $\frac{5}{3}$; only $\frac{5}{3}$ checks; $\frac{5}{3}$ **49.** $D = 10(\log_{10} I - \log_{10} I_0)$ **51.** 1 **53.** 1 **55.** 4 **57.** 2.5×10^{-6} **59.** 51
61. $\frac{x^2}{3} + \frac{y^2}{36}$ **63.** $\frac{x^2}{4} + \frac{y^2}{25}$ **65.** $D = -7$; two complex **67.** $x^2 - 2x - 15 = 0$ **69.** $3y^2 - 11y + 6 = 0$

Problem Set 9.5

1. 2.5775 **3.** 1.5775 **5.** 3.5775 **7.** −1.4225 **9.** 4.5775 **11.** 2.7782 **13.** 3.3032 **15.** −2.0128 **17.** −1.5031
19. −0.3990 **21.** 759 **23.** 0.00759 **25.** 1,430 **27.** 0.00000447 **29.** 0.0000000918 **31.** 10^{10} **33.** 10^{-10} **35.** 10^{20}
37. $\frac{1}{100}$ **39.** 1,000 **41.** 1 **43.** 5 **45.** x **47.** $\ln 10 + 3t$ **49.** $\ln A - 2t$ **51.** 2.7080 **53.** −1.0986 **55.** 2.1972
57. 2.7724 **59. a.** 1.1886 **b.** 0.2218 **61. a.** 0.6667 **b.** −0.3010 **63.** 3.19 **65.** 1.78×10^{-5} **67.** 3.16×10^5
69. 2.00×10^8 **71.** 10 times larger **73.**

LOCATION	DATE	MAGNITUDE, M	SHOCK WAVE, T
Moresby Island	January 23	4.0	1×10^4
Vancouver Island	April 30	5.3	1.99×10^5
Quebec City	June 29	3.2	1.58×10^3
Mould Bay	November 13	5.2	1.58×10^5
St. Lawrence	December 14	3.7	5.01×10^3

75. 12.9% **77.** 5.3% **79.** e **81.** 0.7 **83.** 3.125 **85.** 1.2575 **87.** $t \cdot \log(1.05)$ **89.** $0.05t$ **91.** 0, −3 **93.** −4, −2
95. $\pm 2, \pm i\sqrt{2}$ **97.** 1, $\frac{9}{4}$

Problem Set 9.6

1. 1.4650 **3.** 0.6826 **5.** −1.5440 **7.** −0.6477 **9.** −0.3333 **11.** 2.0000 **13.** −0.1845 **15.** 0.1845 **17.** 1.6168
19. 2.1131 **21.** 1.3333 **23.** 0.7500 **25.** 1.3917 **27.** 0.7186 **29.** 2.6356 **31.** 4.1632 **33.** 5.8435 **35.** −1.0642
37. 2.3026 **39.** 10.7144 **41.** 11.7 years **43.** 9.25 years **45.** 8.75 years **47.** 18.58 years **49.** 11.55 years
51. 18.3 years **53.** lowest point: (−2, −23) **55.** highest point: $(\frac{3}{2}, 9)$ **57.** 2 seconds, 64 feet
59. 13.9 years later or toward the end of 2007 **61.** 27.5 years **63.** $t = \dfrac{1}{r}\ln\dfrac{A}{P}$ **65.** $t = \dfrac{1}{k}\dfrac{\log P - \log A}{\log 2}$ **67.** $t = \dfrac{\log A - \log P}{\log(1 - r)}$

Chapter 9 Review/Test

1. 16 **2.** $\frac{1}{2}$ **3.** $\frac{1}{9}$ **4.** -5 **5.** $\frac{5}{6}$ **6.** 7 **7.**

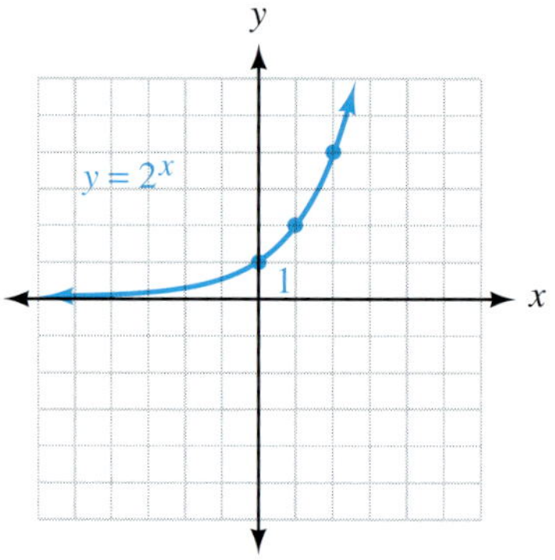

8.

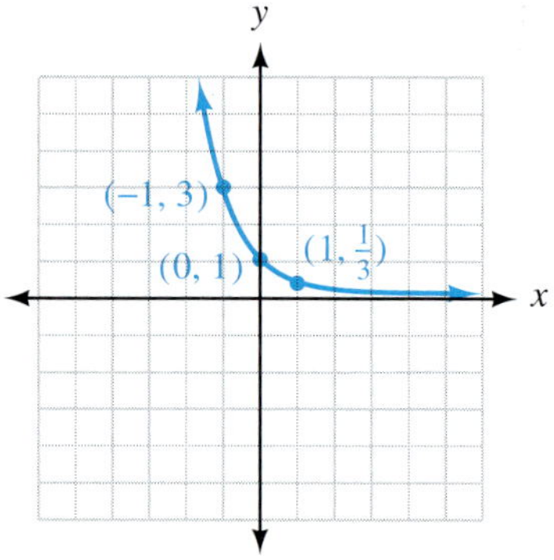

9.

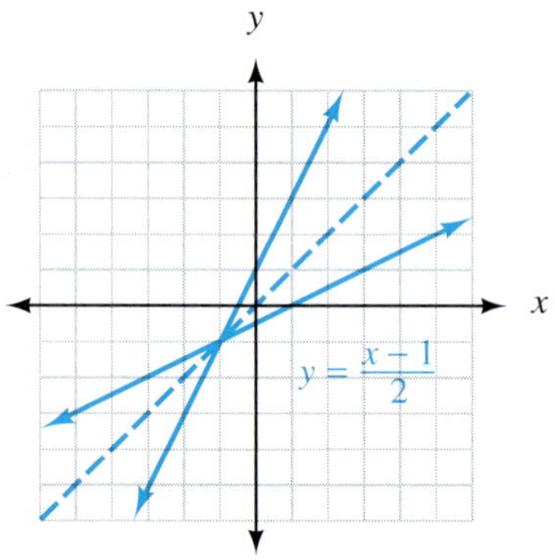

10.

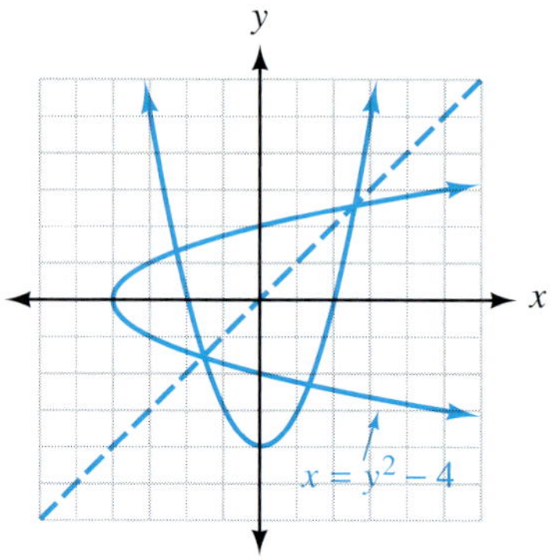

11. $f^{-1}(x) = \frac{x-3}{2}$

12. $y = \pm\sqrt{x+1}$ **13.** $f^{-1}(x) = 2x - 4$ **14.** $y = \pm\sqrt{\frac{4-x}{2}}$ **15.** $\log_3 81 = 4$ **16.** $\log_7 49 = 2$ **17.** $\log_{10} 0.01 = -2$
18. $\log_2 \frac{1}{8} = -3$ **19.** $2^3 = 8$ **20.** $3^2 = 9$ **21.** $4^{1/2} = 2$ **22.** $4^1 = 4$ **23.** 25 **24.** $\frac{3}{4}$ **25.** 10
26.

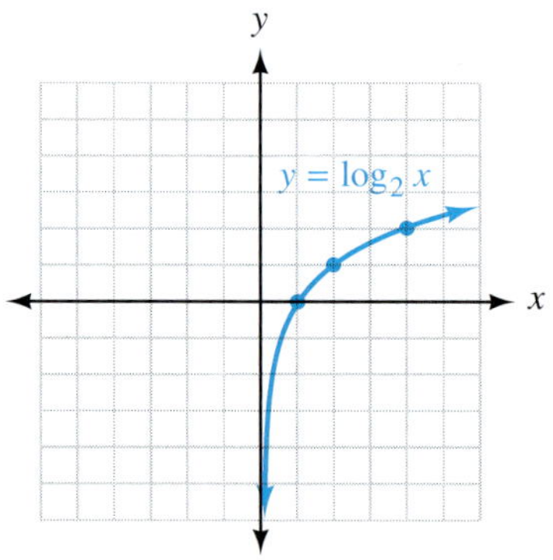

27.

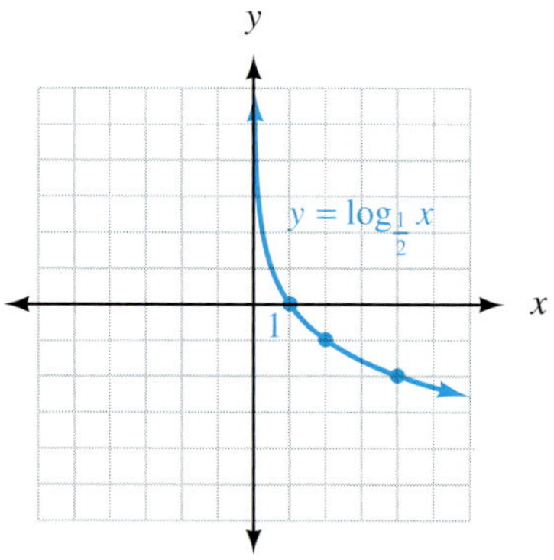

28. 2 **29.** $\frac{2}{3}$ **30.** 0 **31.** $\log_2 5 + \log_2 x$

32. $\log_{10} 2 + \log_{10} x - \log_{10} y$ **33.** $\frac{1}{2}\log_a x + 3\log_a y - \log_a z$ **34.** $2\log_{10} x - 3\log_{10} y - 4\log_{10} z$ **35.** $\log_2 xy$ **36.** $\log_3 \frac{x}{4}$
37. $\log_a \frac{25}{3}$ **38.** $\log_2 \frac{x^3y^2}{z^4}$ **39.** 2 **40.** 6 **41.** Possible solutions -1 and 3; only 3 checks; 3 **42.** 3
43. Possible solutions -2 and 3; only 3 checks; 3 **44.** Possible solutions -1 and 4; only 4 checks; 4 **45.** 2.5391 **46.** -0.1469
47. 9,230 **48.** 0.0251 **49.** 1 **50.** 0 **51.** 2 **52.** -4 **53.** 2.1 **54.** 5.1 **55.** 2.0×10^{-3} **56.** 3.2×10^{-8} **57.** $\frac{3}{2}$
58. $x = \frac{1}{3}\left(\frac{\log 5}{\log 4} - 2\right) = -0.28$ **59.** 0.75 **60.** 2.43 **61.** About 4.67 years **62.** About 9.25 years

Chapter 10

Getting Ready for Chapter 10

1. $\sqrt{29}$ **2.** $2\sqrt{5}$ **3.** $\sqrt{13}$ **4.** $\sqrt{26}$ **5.** ± 4 **6.** $\pm 5i$ **7.** 4, 2 **8.** -4, 2 **9.** 4, 2 **10.** -4, 2 **11.** 9; 3
12. 4; 2 **13.** 1; 1 **14.** $\frac{1}{4}$; $\frac{1}{2}$ **15.** $(5y + 6)(y + 2)$ **16.** $(3x - 4)(x + 7)$ **17.** $\frac{x^2}{3} + \frac{y^2}{36}$ **18.** $\frac{x^2}{16} + \frac{y^2}{4}$ **19.** $y = -\frac{1}{3}x + 1$
20. $y = -2x + 3$

Problem Set 10.1

1. 5 **3.** $\sqrt{106}$ **5.** $\sqrt{61}$ **7.** $\sqrt{130}$ **9.** 3 or -1 **11.** 3 **13.** $(x-2)^2+(y-3)^2=16$ **15.** $(x-3)^2+(y+2)^2=9$
17. $(x+5)^2+(y+1)^2=5$ **19.** $x^2+(y+5)^2=1$ **21.** $x^2+y^2=4$

23. Center = (0, 0), Radius = 2

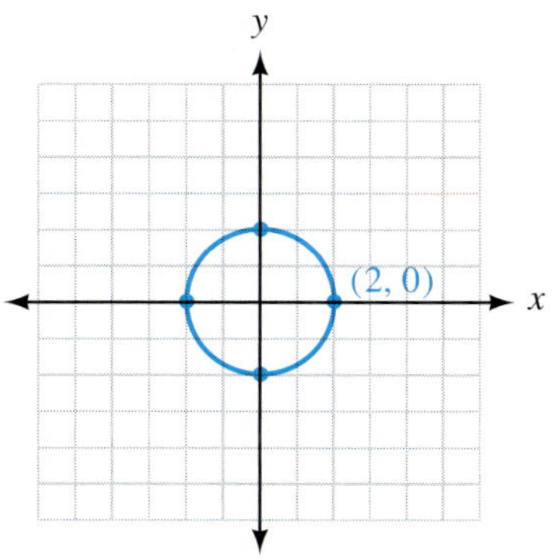

25. Center = (1, 3), Radius = 5

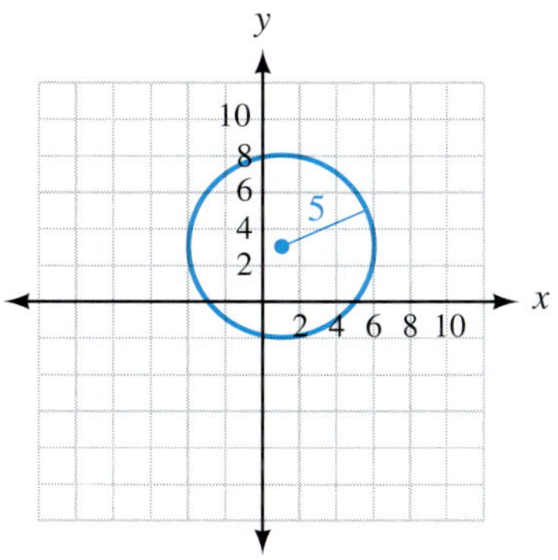

27. Center = (−2, 4), Radius = $2\sqrt{2}$

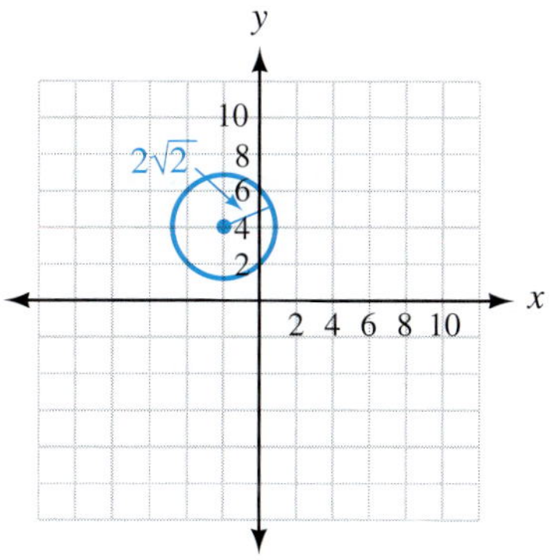

29. Center = (−1, −1), Radius = 1

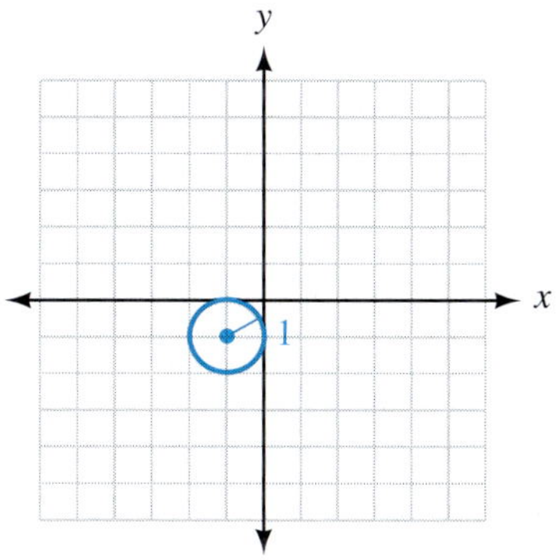

31. Center = (0, 3), Radius = 4

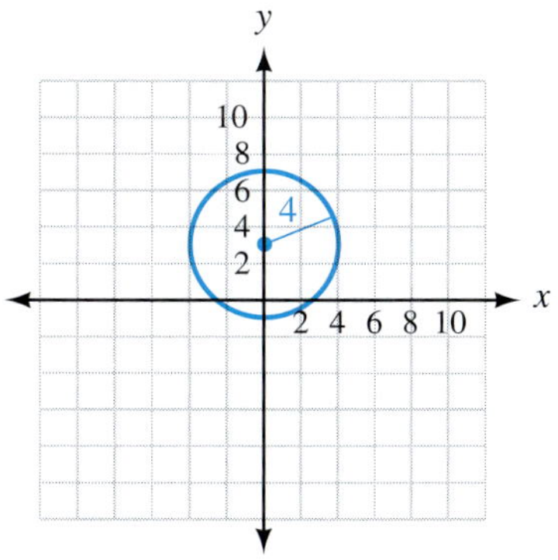

33. Center = (2, 3), Radius = 3

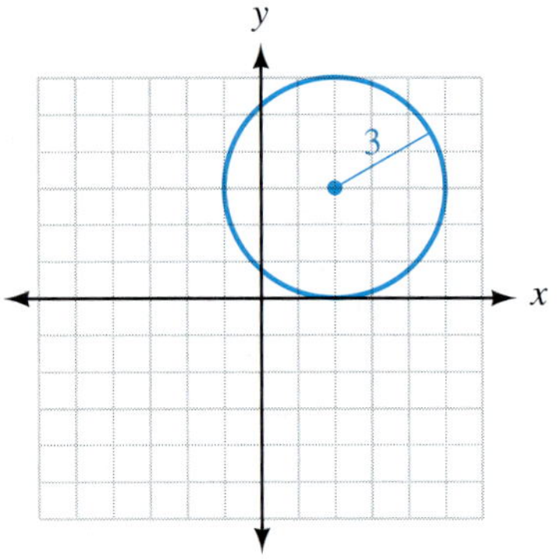

35. Center = $(1, -\frac{1}{2})$, Radius = 2

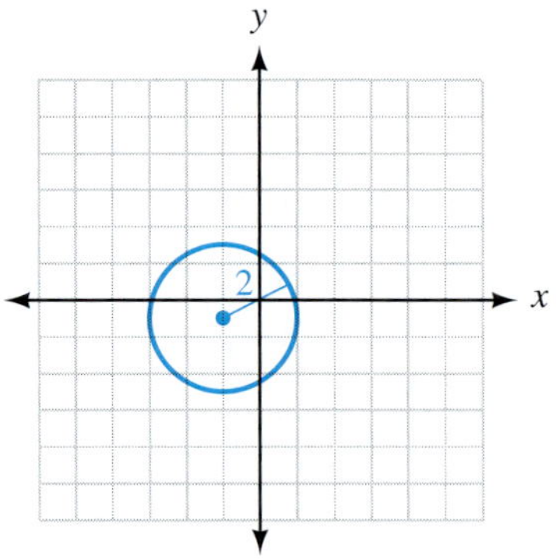

37. $(x-3)^2 + (y-4)^2 = 25$

39. A: $(x-\frac{1}{2})^2 + (y-1)^2 = \frac{1}{4}$
B: $(x-1)^2 + (y-1)^2 = 1$
C: $(x-2)^2 + (y-1)^2 = 4$

41. A: $(x+8)^2 + y^2 = 64$
B: $x^2 + y^2 = 64$
C: $(x-8)^2 + y^2 = 64$

43. $(x-500)^2 + (y-132)^2 = 120^2$ **45.** $y = \pm 3$ **47.** $y = \pm 2i$, no real solution **49.** $x = \pm 3i$, no real solution **51.** $\frac{x^2}{9} + \frac{y^2}{4}$
53. x-intercept 4, y-intercept -3 **55.** ± 2.4 **57. a.** $\frac{1}{9}$ **b.** $\frac{1}{3}$ **c.** 1 **d.** 3 **e.** 9 **59.** $f^{-1}(x) = \dfrac{x-3}{2}$ **61.** $y = \pm\sqrt{x+4}$
63. $(x-2)^2 + (y-3)^2 = 4$ **65.** $(x-2)^2 + (y-3)^2 = 4$ **67.** 5 **69.** 5
71. $y = \sqrt{9-x^2}$ corresponds to the top half; $y = -\sqrt{9-x^2}$ to the bottom half.

Problem Set 10.2

1.

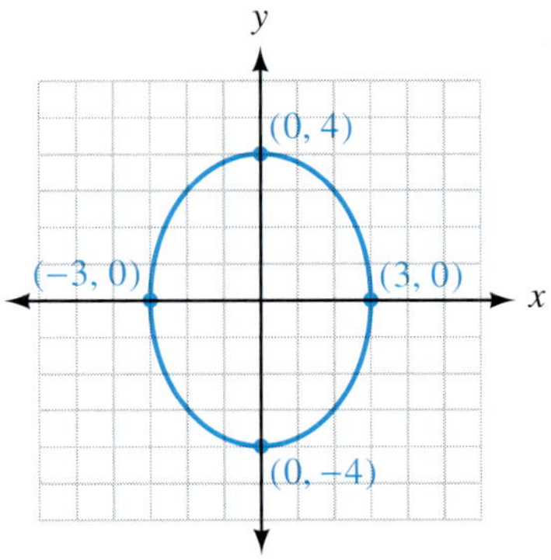

3.

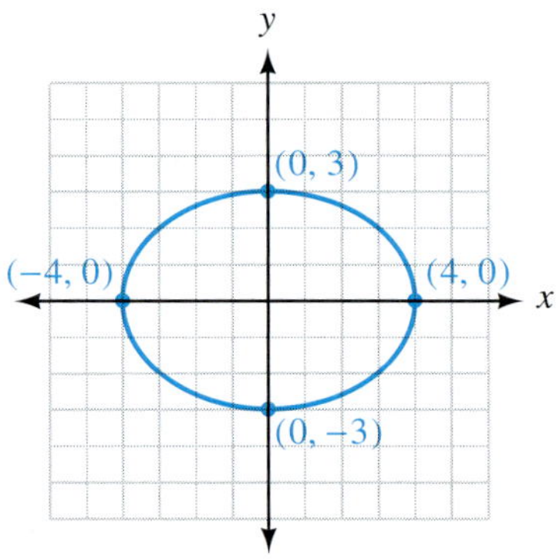

5.

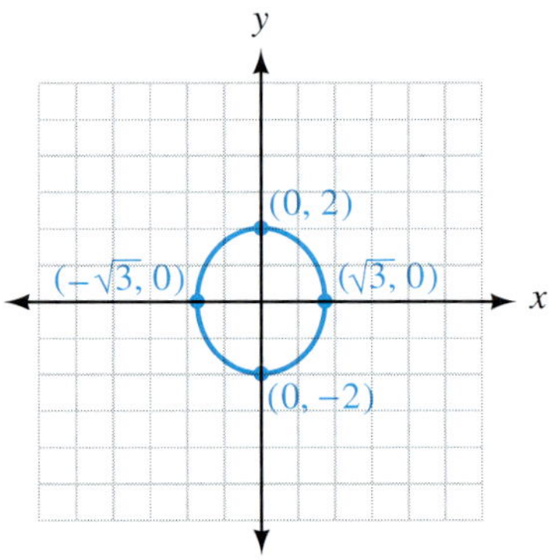

7.

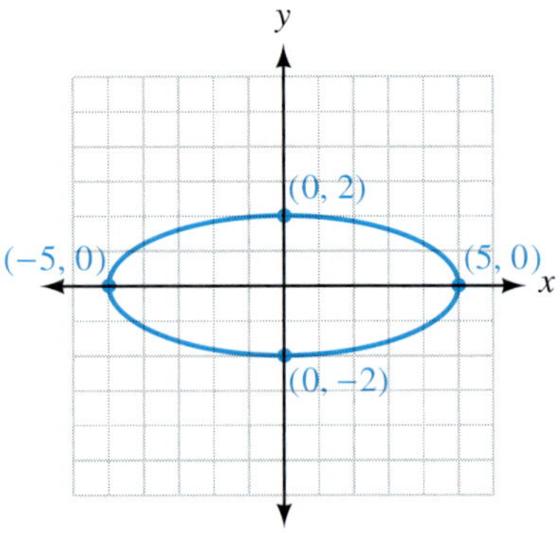

9.

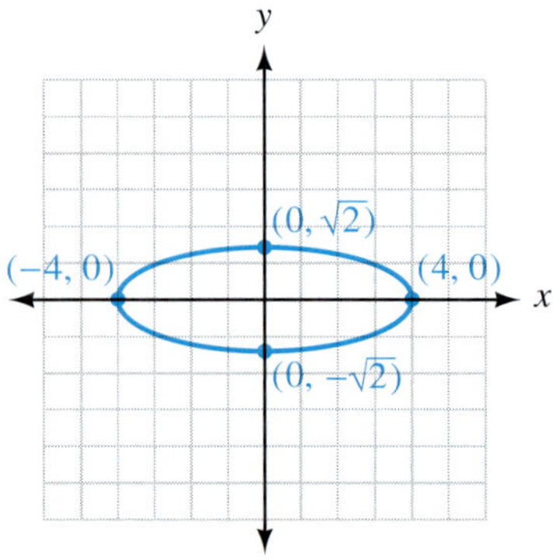

11.

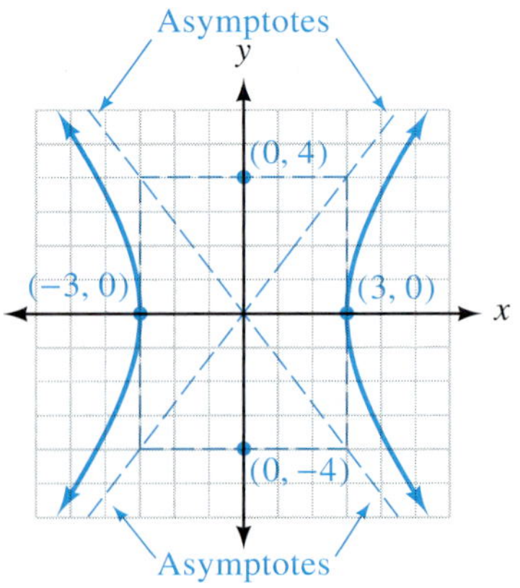

13.

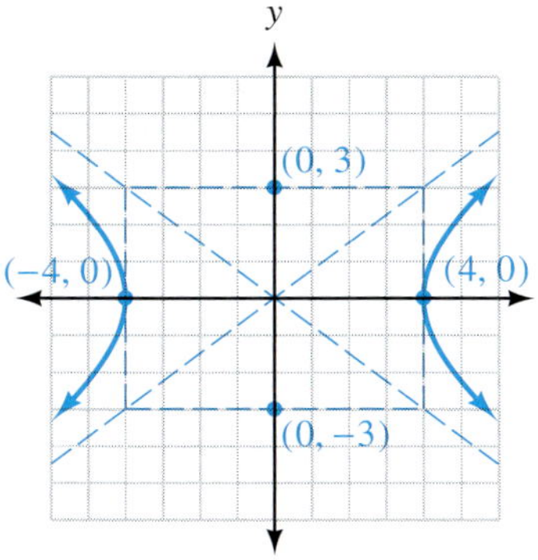

15.

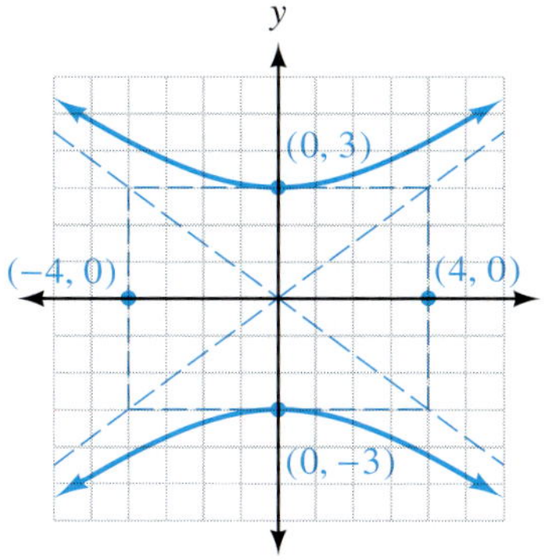

17.

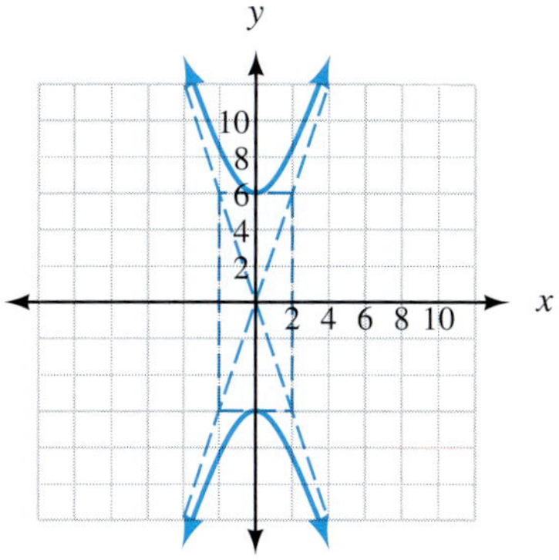

19.

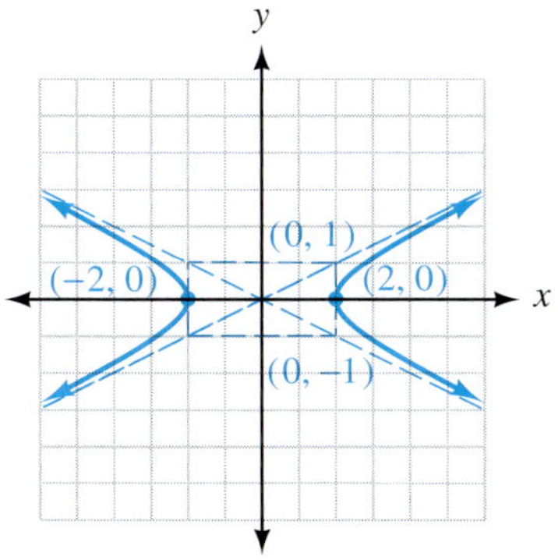

21.

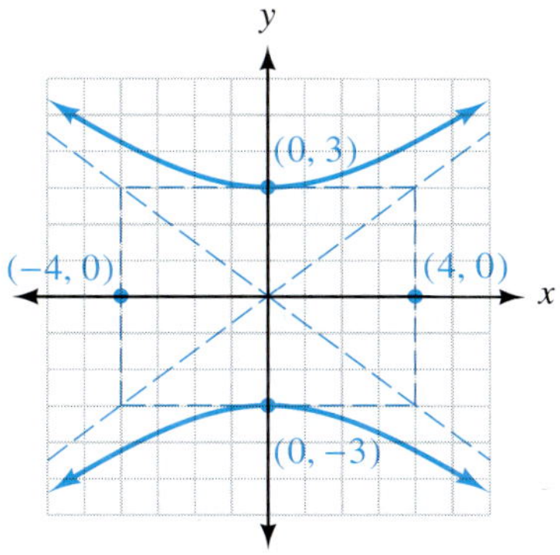

23. x-intercepts = ±3, y-intercepts = ±2 **25.** x-intercepts = ±0.2, no y-intercepts **27.** x-intercepts = $\pm\frac{3}{5}$, y-intercepts = $\pm\frac{2}{5}$
29. ±1.8 **31.** ±3.2 **33.** $y = \frac{3}{4}x, y = -\frac{3}{4}x$ **35.** $\frac{x^2}{16} + \frac{y^2}{4} = 1$
37. The equation is $\frac{x^2}{20^2} + \frac{y^2}{10^2} = 1$. A 6-foot man could walk upright under the arch anywhere between 16 feet to the left and 16 feet the right of the center. **39.** About 5.3 feet wide **41.** (0, 0) **43.** $4y^2 + 16y + 16$ **45.** $x = 2y + 4$ **47.** $-x^2 + 6$
49. $(5y + 6)(y + 2)$ **51.** ±2 **53.** ±2 **55.** $\log_{10} 100 = 2$ **57.** $3^4 = 81$ **59.** 27 **61.** 5 **63.** 1 **65.** 2.7604
67. −1.2396 **69.** 445 **71.** 4.45×10^{-8}

Problem Set 10.3

1.

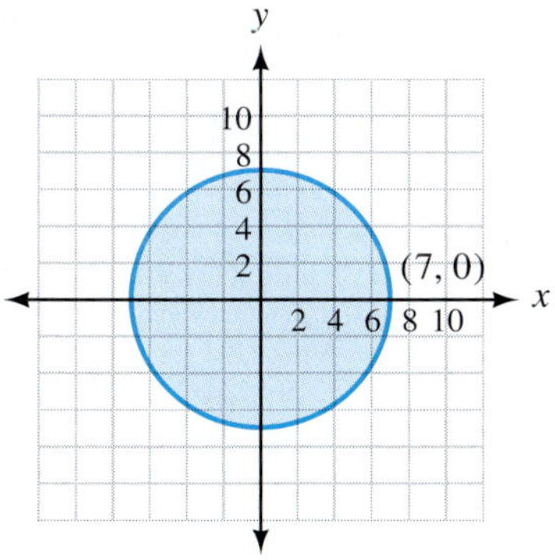

3.

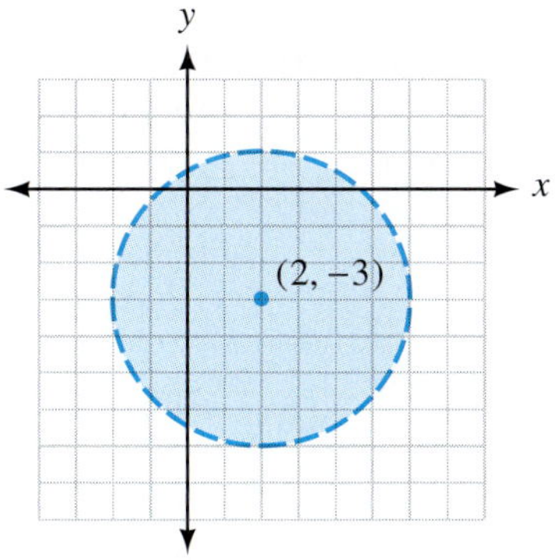

5.

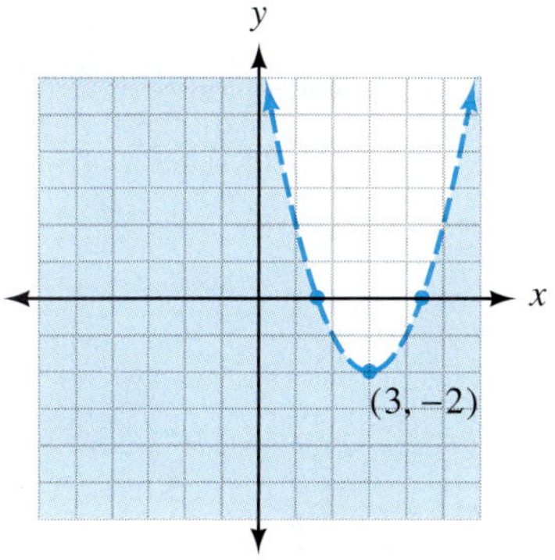

7.

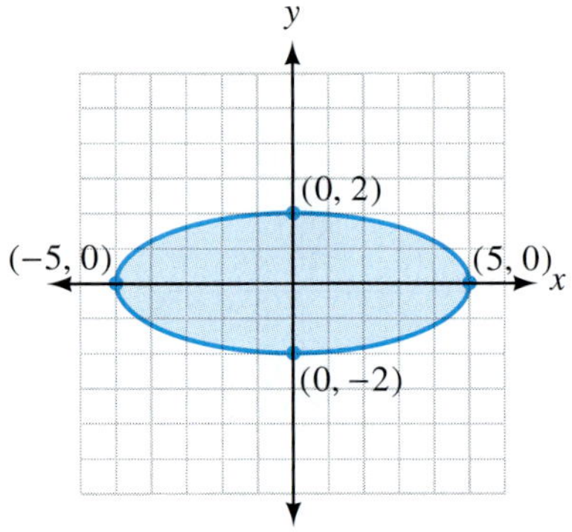

9. (0, 3), $(\frac{12}{5}, -\frac{9}{5})$ **11.** (0, 4), $(\frac{16}{5}, \frac{12}{5})$ **13.** (5, 0), (−5, 0) **15.** (0, −3), $(\sqrt{5}, 2)$, $(-\sqrt{5}, 2)$ **17.** (0, −4), $(\sqrt{7}, 3)$, $(-\sqrt{7}, 3)$
19. (−4, 11), $(\frac{5}{2}, \frac{5}{4})$ **21.** (−4, 5), (1, 0) **23.** (2, −3), (5, 0) **25.** (3, 0), (−3, 0) **27.** (4, 0), (0, −4)
29. a. $(-4, 4\sqrt{3})$ and $(-4, -4\sqrt{3})$ **b.** $(4, 4\sqrt{3})$ and $(4, -4\sqrt{3})$

31.

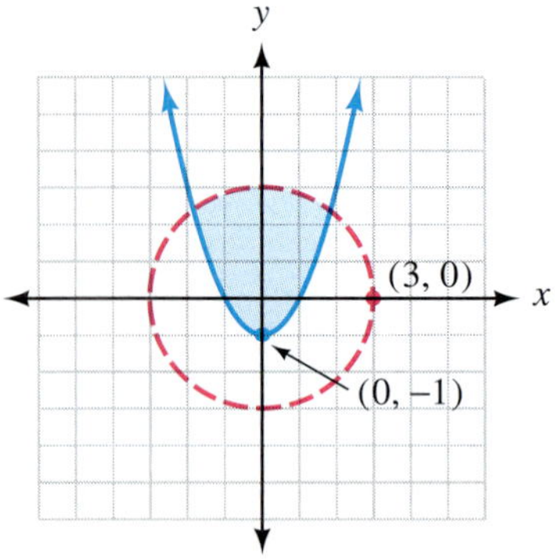

33.

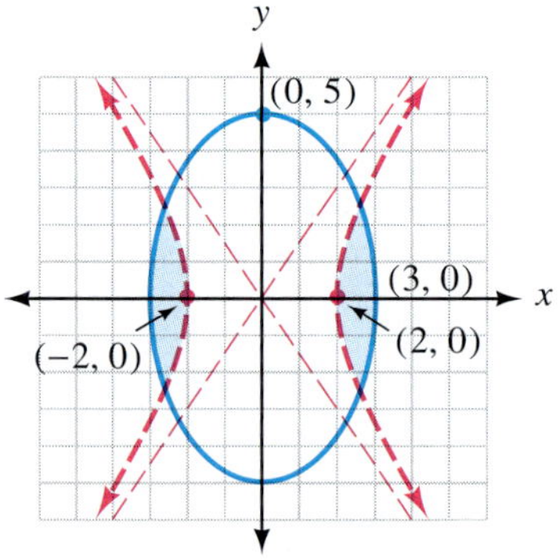

35. No intersection

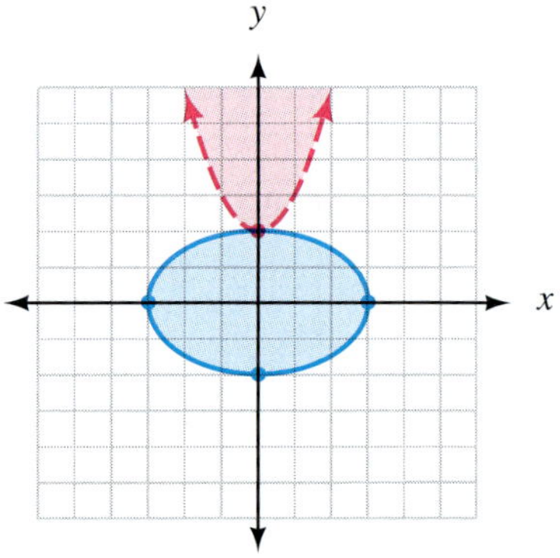

37. $x^2 + y^2 < 16$
$y > 4 - \frac{1}{4}x^2$
39. 8, 5 or −8, −5 or 8, −5 or −8, 5 **41.** 6, 3 or 13, −4 **43.** $3 \log_2 x + \log_2 y$ **45.** $\log_{10} \frac{x}{y^2}$ **47.** 80 **49.** $x = \frac{\log_7}{\log_5} \approx 1.21$
51. Approximately 7 years **53.** 2.16 **55.** 6.36

Chapter 10 Review/Test

1. $\sqrt{10}$ **2.** $\sqrt{13}$ **3.** 5 **4.** 9 **5.** −2, 6 **6.** −12, 4 **7.** $(x - 3)^2 + (y - 1)^2 = 4$ **8.** $(x - 3)^2 + (y + 1)^2 = 16$
9. $(x + 5)^2 + y^2 = 9$ **10.** $(x + 3)^2 + (y - 4)^2 = 18$ **11.** $x^2 + y^2 = 25$ **12.** $x^2 + y^2 = 9$ **13.** $(x + 2)^2 + (y - 3)^2 = 25$
14. $(x + 6)^2 + (y - 8)^2 = 100$

15. (0, 0); $r = 2$

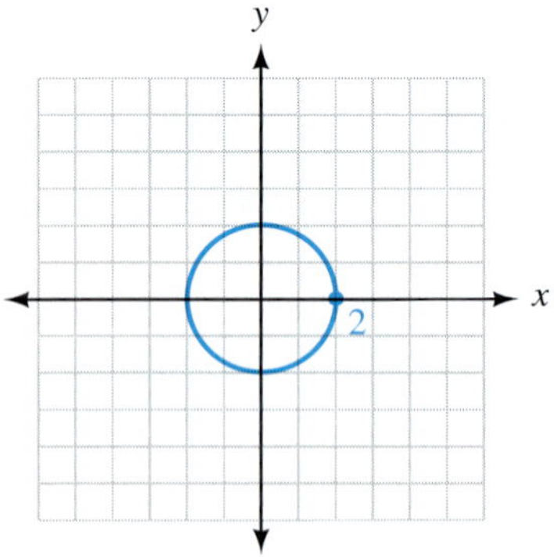

16. (3, −1); $r = 4$

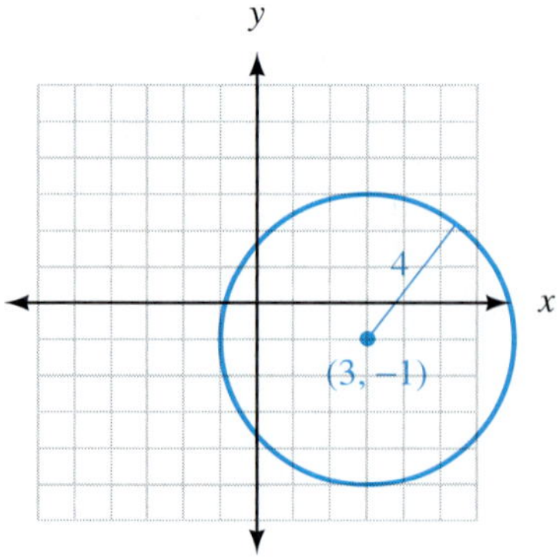

17. (3, −2); $r = 3$

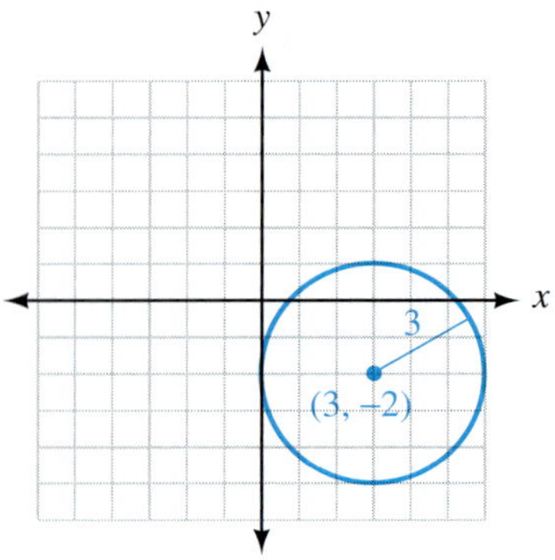

18. $(-2, 1)$; $r = 3$

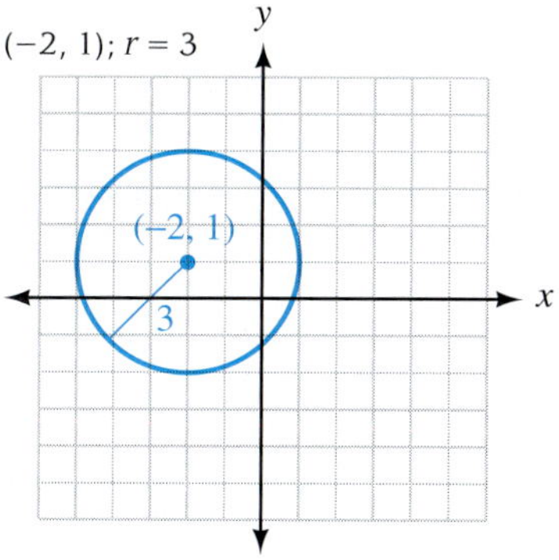

19.

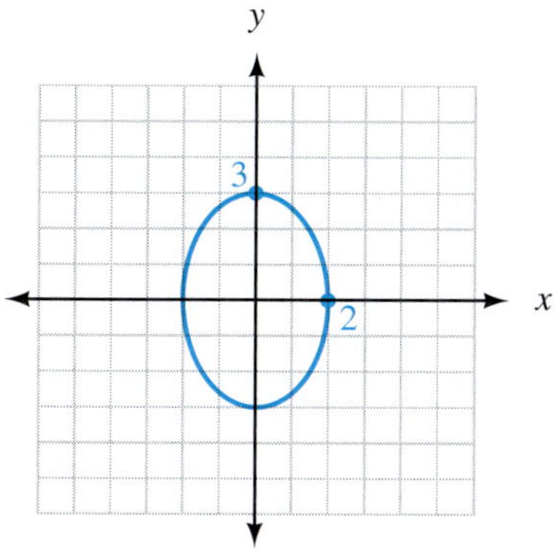

20.

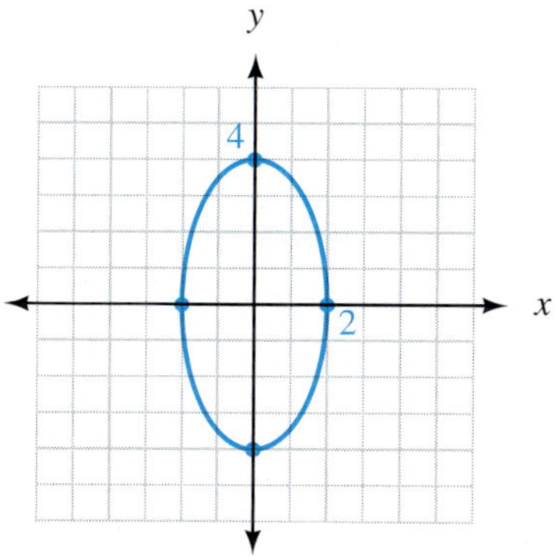

21.

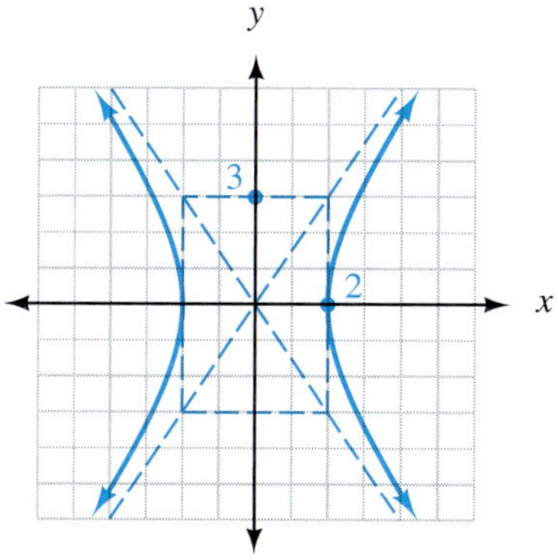

22.

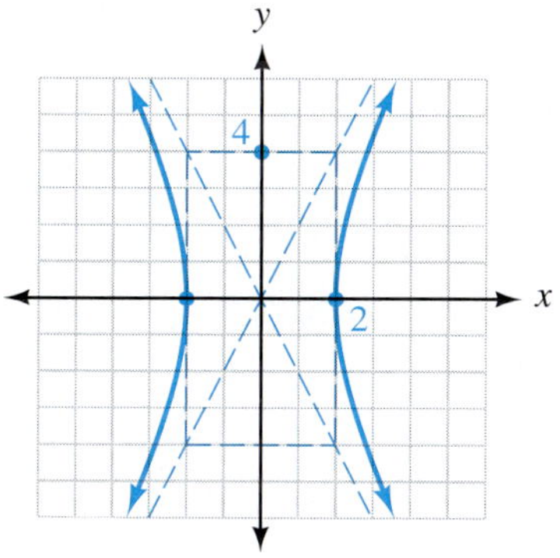

23.

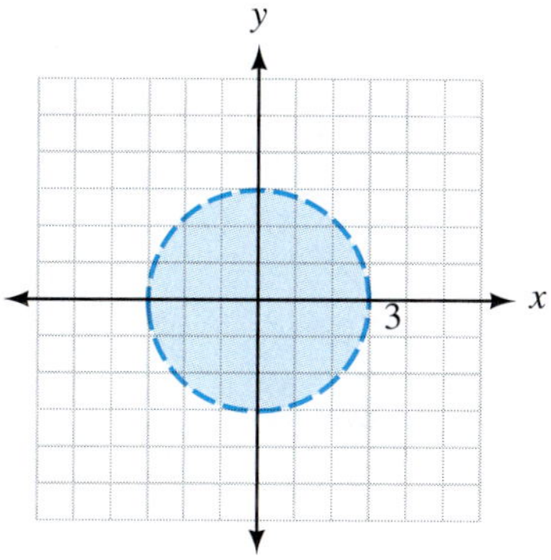

24.

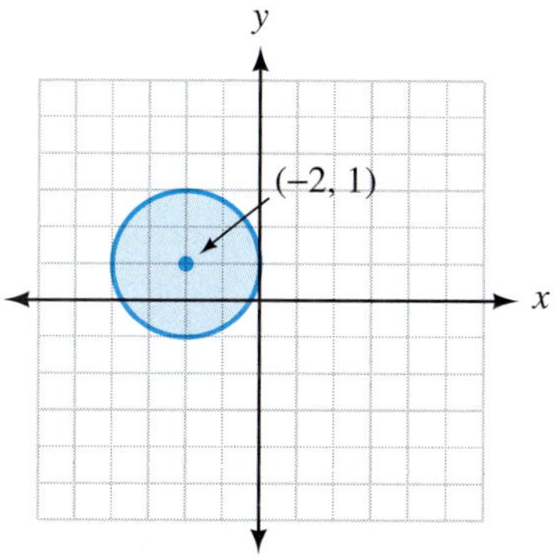

25.

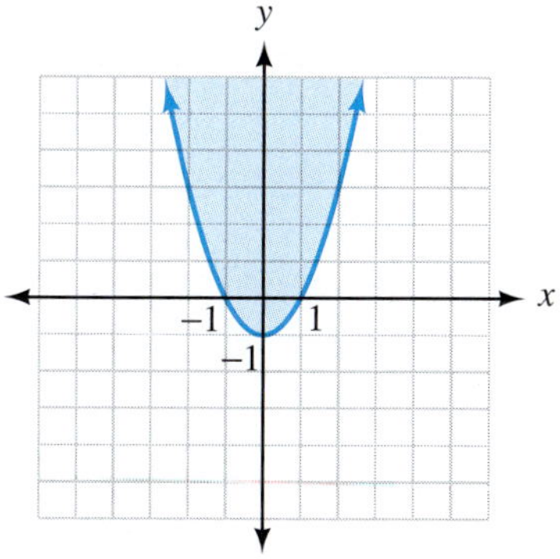

26.

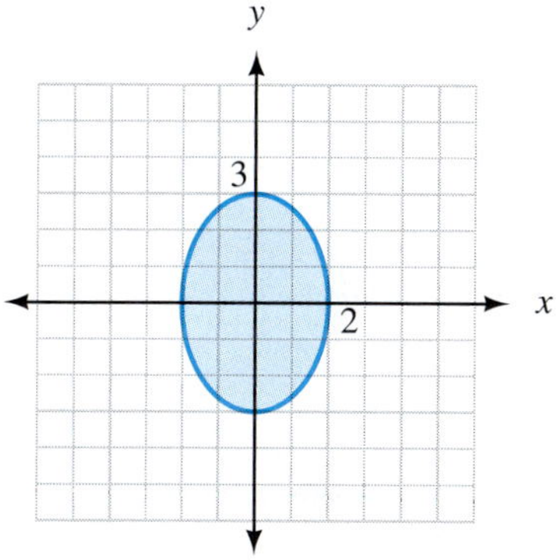

27.

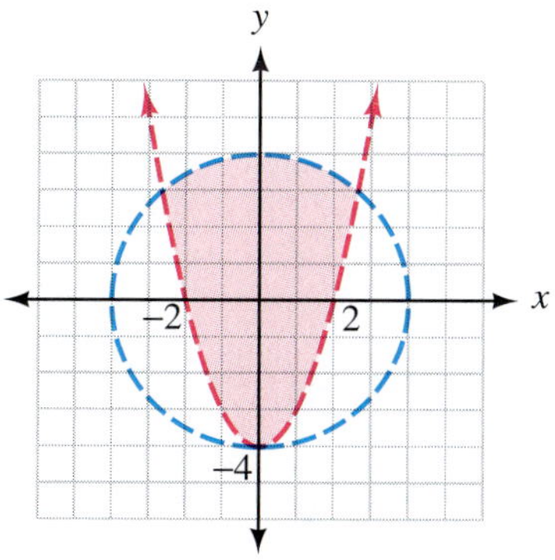

28.

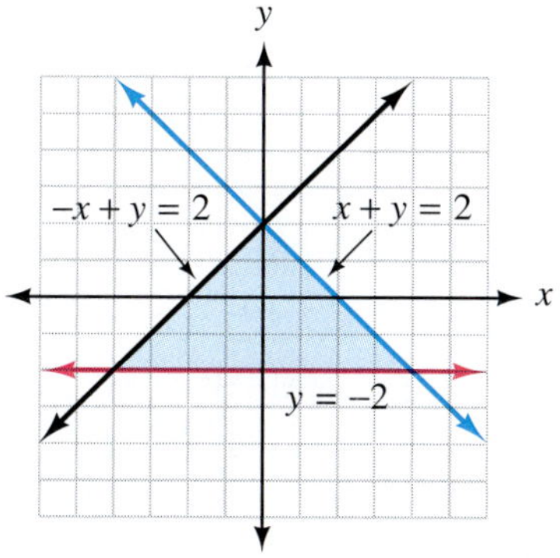

29. $(0, 4)$, $\left(\frac{16}{5}, -\frac{12}{5}\right)$ **30.** $(0, -2)$, $(\sqrt{3}, 1)$, $(-\sqrt{3}, 1)$ **31.** $(-2, 0)$, $(2, 0)$ **32.** $\left(-\sqrt{7}, -\frac{\sqrt{6}}{2}\right)$, $\left(-\sqrt{7}, \frac{\sqrt{6}}{2}\right)$, $\left(\sqrt{7}, -\frac{\sqrt{6}}{2}\right)$, $\left(\sqrt{7}, \frac{\sqrt{6}}{2}\right)$

Appendix B

1. $x - 7 + \frac{20}{x+2}$ **3.** $3x - 1$ **5.** $x^2 + 4x + 11 + \frac{26}{x-2}$ **7.** $3x^2 + 8x + 26 + \frac{83}{x-3}$ **9.** $2x^2 + 2x + 3$

11. $x^3 - 4x^2 + 18x - 72 + \dfrac{289}{x+4}$ **13.** $x^4 + x^2 - x - 3 - \dfrac{5}{x-2}$ **15.** $x + 2 + \dfrac{3}{x-1}$ **17.** $x^3 - x^2 + x - 1$ **19.** $x^2 + x + 1$

Appendix C

1. Hypothesis: You argue for your limitations.
Conclusion: They are yours.
3. Hypothesis: x is an even number.
Conclusion: x is divisible by 2.
5. Hypothesis: A triangle is equilateral.
Conclusion: All of its angles are equal.
7. Hypothesis: $x + 5 = -2$
Conclusion: $x = -7$
9. Converse: If $a^2 = 64$, then $a = 8$
Inverse: If $a \neq 8$, then $a^2 \neq 64$
Contrapositive: If $a^2 \neq 64$, then $a \neq 8$
11. Converse: If $a = b$, then $\frac{a}{b} = 1$
Inverse: If $\frac{a}{b} \neq 1$, then $a \neq b$
Contrapositive: If $a \neq b$, then $\frac{a}{b} \neq 1$
13. Converse: If it is a rectangle, then it is a square.
Inverse: If it is not a square, then it is not a rectangle.
Contrapositive: If it is not a rectangle, then it is not a square.
15. Converse: If good is not enough, then better is possible.
Inverse: If better is not possible, then good is enough.
Contrapositive: If good is enough, then better is not possible.
17. If E, then F **19.** If it is misery, then it loves company. **21.** If it is the squeaky wheel, then it gets the grease. **23.** (c)
25. (a) **27.** (c) **29.** (b)

Appendix D

1. $6x + 2$ **3.** $-2x + 8$ **5.** $8x^2 + 14x - 15$ **7.** $\dfrac{2x+5}{4x-3}$ **9.** $4x - 7$ **11.** $3x^2 - 10x + 8$ **13.** $-2x + 3$

15. $3x^2 - 11x + 10$ **17.** $9x^3 - 48x^2 + 85x - 50$ **19.** $\dfrac{3x^2 - 11x + 10}{3x - 5} = \dfrac{(3x-5)(x-2)}{3x-5} = x - 2$

21. $\dfrac{3x-5}{3x^2 - 11x + 10} = \dfrac{3x-5}{(3x-5)(x-2)} = \dfrac{1}{x-2}$ **23.** $3x^2 - 7x + 3$ **25.** $6x^2 - 22x + 20$ **27.** 15 **29.** 98 **31.** $\frac{3}{2}$ **33.** 1
35. 40 **37.** 147 **39. a.** 81 **b.** 29 **c.** $(x + 4)^2$ **d.** $x^2 + 4$ **41. a.** -2 **b.** -1 **c.** $16x^2 + 4x - 2$ **d.** $4x^2 + 12x - 1$

45. a. $R(x) = 11.5x - 0.05x^2$ **b.** $C(x) = 2x + 200$ **c.** $P(x) = -0.05x^2 + 9.5x - 200$ **d.** $\overline{C}(x) = 2 + \dfrac{200}{x}$

47. a. $M(x) = 220 - x$ **b.** 196 **c.** 142 **d.** 135 **e.** 128

Credits

This page constitutes an extension of the copyright page. We have made every effort to trace the ownership of all copyrighted material and to secure permission from copyright holders. In the event of any question arising as to the use of any material, we will be pleased to make the necessary corrections in future printings. Thanks are due to the following authors, publishers, and agents for permission to use the material indicated.

Chapter 1. 1: David Young-Wolff/PhotoEdit **3:** Thomson-Brooks/Cole **38:** The British Museum/The Bridgeman Art Library **52:** Royalty-Free/Corbis **56:** David Young-Wolff/PhotoEdit **61:** Paul S. Bowen **62:** David Young-Wolff/PhotoEdit **71:** Bettmann/Corbis

Chapter 2. 73: Yellow Dog Productions/Getty Images **87:** Jan Halaska/Index Stock Imagery/PictureQuest **100:** Tom & Dee Ann McCarthy/Corbis **125:** Courtesy of Amtrak **148:** Smithsonian Institution Libraries, Photo No. 92-338

Chapter 3. 149: David Young-Wolff/PhotoEdit **189:** Thomson-Brooks/Cole **226, top:** PhotoDisc/Getty Images **226, bottom:** Michael Grecco/The Image Bank/Getty Images **240:** The Image Bank/Getty Images **248, left, right:** David Eugene Smith Collection/Rare Book and Manuscript Library/Columbia University

Chapter 4. 249: Spencer Grant/PhotoEdit **309:** Peter Jones/Reuters New Media/Corbis

Chapter 5. 323: Jack Kurtz/The Image Works **334, bottom left:** Stockbyte/SuperStock **334, center and bottom right:** NASA **336:** Edge Tech Corporation

Chapter 6. 403: Michael A. Keller/zefa/Corbis **480:** The British Museum/The Bridgeman Art Library **488, left:** Hulton Archive/Getty Images **488, right:** Selznick Films/The Kobal Collection

Chapter 7. 489: Charles Rotkin/Corbis **564:** Culver Pictures

Chapter 8. 565: The Image Bank/Getty Images **588, left:** Pixtal/SuperStock **588, right:** Kathleen Olson **589:** Library of Congress **609:** Royalty-Free/Corbis **624:** Kathleen Olson **625:** The Image Bank/Getty Images **640:** Arlie O. Petters, Massachusetts Institute of Technology

Chapter 9. 641: Royalty-Free/Corbis **653, left:** Sandy Felsenthal/Corbis **652, top:** "Coca-Cola" is a registered trademark of The Coca-Cola Company. Used with the express permission of The Coca-Cola Company. **652, bottom:** Christian Pierre/SuperStock **653, right:** Lester Lefkowitz/Getty Images **667:** Reuters/Jose Miguel Gomez/Getty Images **712:** 20th Century Fox/The Kobal Collection

Chapter 10. 713: Tom Stewart/Corbis **716, left:** Adam Woolfitt/Corbis **716, right:** Christopher Cormack/Corbis **727, 729:** NASA **735:** National Park Service **754:** Bettmann/Corbis

Index